Advances in Electrical Engineering and Computational Science

Lecture Notes in Electrical Engineering

Volume 39

For further volumes:
http://www.springer.com/series/7818

Sio-Iong Ao · Len Gelman
Editors

Advances in Electrical Engineering and Computational Science

Springer

Editors
Dr. Sio-Iong Ao
Harvard School of Engineering
and Applied Sciences
Harvard University
60 Oxford Street
Cambridge
MA 02138, USA
siao@seas.harvard.edu

Dr. Len Gelman
Cranfield University
School of Engineering
Dept. of Process and Systems
Engineering
Cranfield, Beds
United Kingdom MK43 0AL

ISBN 978-90-481-2310-0 e-ISBN 978-90-481-2311-7

DOI 10.1007/978-90-481-2311-7

Library of Congress Control Number: 2009926266

Printed on acid-free paper

9 8 7 6 5 4 3 2 1

springer.com

Preface

A large international conference on Advances in Electrical Engineering and Computational Science was held in London, UK, July 2–4, 2008, under the World Congress on Engineering (WCE 2008). The WCE 2008 was organized by the International Association of Engineers (IAENG); the Congress details are available at: http://www.iaeng.org/WCE2008. IAENG is a non-profit international association for engineers and computer scientists, which was founded originally in 1968. The World Congress on Engineering serves as good platforms for the engineering community to meet with each other and exchange ideas. The conferences have also struck a balance between theoretical and application development. The conference committees have been formed with over 200 members who are mainly research center heads, faculty deans, department heads, professors, and research scientists from over 30 countries. The conferences are truly international meetings with a high level of participation from many countries. The response to the Congress has been excellent. There have been more than 1,000 manuscript submissions for the WCE 2008. All submitted papers have gone through the peer review process, and the overall acceptance rate is 57%.

This volume contains 61 revised and extended research articles written by prominent researchers participating in the conference. Topics covered include Control Engineering, Network Management, Wireless Networks, Biotechnology, Signal Processing, Computational Intelligence, Computational Statistics, Internet Computing, High Performance Computing, and industrial applications. The book offers the state of the art of tremendous advances in electrical engineering and computational science and also serves as an excellent reference work for researchers and graduate students working on electrical engineering and computational science.

Dr. Sio-Iong Ao
Prof. Len Gelman

Contents

1 **Acoustic Behavior of Squirrel Cage Induction Motors** 1
C. Grabner
1.1 Introduction 1
1.2 Acoustic Noise Physics 2
1.2.1 Sound Power 2
1.2.2 Sound Pressure Level in Decibel 2
1.2.3 Sound Pressure Level and Human Hearing 3
1.2.4 Human Hearing Mechanism 3
1.2.5 Loudness Level in Phon 3
1.2.6 Weighing the Sound Pressure Level 5
1.3 Mutation of Stator and Rotor Parts 5
1.4 Direct Comparison of Sound Pressure Levels for Different Motor Designs 6
1.4.1 Evaluation of Design Impacts to the Equivalent Sound Pressure Level 7
1.4.2 Fast Fourier Sound Pressure Spectrum for Each Motor Design at Rated-Load 8
1.5 Conclusion 10
References 11

2 **Model Reduction of Weakly Nonlinear Systems** 13
Marissa Condon and Georgi G. Grahovski
2.1 Introduction 13
2.2 Perturbative Approximation of Nonlinear Systems 14
2.3 KRYLOV-Based Model Reduction 16
2.4 Scale Invariance Property 17
2.5 KRYLOV Reduction of Perturbative Representation 19
2.6 Illustrative Example 19
2.7 Conclusions 21
References 22

3 Investigation of a Multizone Drift Doping Based Lateral Bipolar Transistor on Buried Oxide Thick Step ... 23
Sajad A. Loan, S. Qureshi, and S. Sundar Kumar Iyer
3.1 Introduction ... 23
3.2 Ideal, Conventional and Proposed Devices ... 24
3.3 Simulation Results and Discussion ... 25
3.4 Conclusion ... 32
References ... 32

4 Development of a Battery Charging Management Unit for Renewable Power Generation ... 33
Yong Yin, Jihong Wang, Xing Luo, and Shen Guo
4.1 Introduction ... 33
4.2 Energy Conversion Modes ... 35
4.3 Lead Acid Battery ... 35
4.4 Battery Charging and Discharging Strategies ... 37
4.4.1 Membership Function Database ... 38
4.4.2 Fuzzification ... 40
4.4.3 Fuzzy Rule Evaluation ... 41
4.4.4 Defuzzification and Simulation ... 42
4.5 System Implementation ... 43
4.6 Conclusions and Future Works ... 45
References ... 46

5 Optical CDMA Transceiver Architecture: Polarization Modulation with Dual-Balanced Detection ... 47
Mohammad Massoud Karbassian and Hooshang Ghafouri-Shiraz
5.1 Introduction ... 47
5.2 POLSK–OCDMA Transmitter ... 48
5.3 Analysis of POLSK–OCDMA Receiver ... 50
5.4 Discussion of Results ... 55
5.5 Conclusion ... 56
References ... 57

6 Template Based: A Novel STG Based Logic Synthesis for Asynchronous Control Circuits ... 59
Sufian Sudeng and Arthit Thongtak
6.1 Introduction ... 59
6.2 Petri-Net and Signal Transition Graph ... 61
6.3 Asynchronous DMA Controller Specification ... 62
6.4 Synthesis Using State Based Method ... 64
6.5 Synthesis Using Structural Encoding Method ... 66
6.6 Synthesis Using State Based Method ... 67
6.7 Result ... 72
6.8 Concluding Remarks ... 73
References ... 74

7 A Comparison of Induction Motor Speed Estimation Using Conventional MRAS and an AI-Based MRAS Parallel System 75
Chao Yang and Professor John W. Finch
7.1 Introduction 75
7.2 Speed Estimation Using Conventional Model Reference Adaptive System 76
7.3 Artificial Intelligence-Based Model Reference Adaptive System .. 77
7.4 MRAS Based Two-Layer ANN Speed Estimator with Dynamic Reference Model 78
7.5 Simulation Results and Discussion 81
7.6 Conclusion 83
References 84

8 A New Fuzzy-Based Additive Noise Removal Filter 87
M. Wilscy and Madhu S. Nair
8.1 Introduction 87
8.2 Conventional Noise Removal Techniques 89
8.3 Proposed Fuzzy Noise Removal Filter 90
8.3.1 Fuzzy Sub-filter I 90
8.3.2 Fuzzy Sub-filter II 92
8.4 Results and Discussion 93
8.5 Conclusion 96
References 96

9 Enhancement of Weather Degraded Color Images and Video Sequences Using Wavelet Fusion 99
Jisha John and M. Wilscy
9.1 Introduction 99
9.2 Atmospheric Scattering Models 101
9.3 Contrast Correction 102
9.3.1 Normalized Brightness and Airlight 102
9.3.2 Correcting Contrast 102
9.4 Intensity Based Enhancement 104
9.5 Wavelet Fusion 104
9.6 Video Enhancement 105
9.7 Performance Analysis 105
9.8 Results and Discussion 106
9.9 Conclusion 107
References 108

10 A GA-Assisted Brain Fiber Tracking Algorithm for DT-MRI Data .. 111
L.M. San-José-Revuelta
10.1 Introduction 111
10.2 Brief Tracking Algorithm Description 113
10.2.1 Basic Concepts 113

10.2.2 Anisotropy and Local Probability 114
10.2.3 Eigenvectors and Direction Considerations 114
10.2.4 Path Probabilities 115
10.2.5 Final Criterion and Pool of "Future Seeds" 115
10.2.6 Parameters to Be Estimated 115
10.3 Proposed GA for Parameter Estimation 116
10.3.1 Estimation Procedure 116
10.3.2 GA Description 116
10.3.3 Genetic Operators 117
10.3.4 Elitism and Entropy Dependent Operators 117
10.4 Numerical Results 118
10.4.1 Synthetic Images 118
10.4.2 Real Images 120
10.4.3 Final Remarks 121
10.5 Conclusions 121
References 122

11 A Bridge-Ship Collision Avoidance System Based on FLIR Image Sequences 123
Jun Liu, Hong Wei, Xi-Yue Huang, Nai-Shuai He, and Ke Li
11.1 Introduction 123
11.2 The FLIR Video Surveillance System 125
11.3 The Detection Algorithm for Moving Ships 126
11.3.1 Extracting the ROS 126
11.3.2 Calculating the Multi-scale Fractal Feature 126
11.3.3 Segmenting the Fractal Feature Image 129
11.3.4 Detecting Moving Ships 129
11.4 Experimental Results and Discussion 130
11.5 Conclusion 130
References 133

12 A Gray Level Feature Detector and Its Hardware Architecture 135
Neeta Nain, Rajesh Kumar, and Bhavitavya Bhadviya
12.1 Introduction 135
12.2 Feature Point Detection Model 136
12.3 Hardware Implementation 139
12.3.1 Architecture 1: A Complete Hardwired Approach 139
12.3.2 Architecture 2: A Hardware Software Co-design 141
12.4 Synthesis Results and Analysis 142
12.5 Conclusions 144
References 144

13 SVD and DWT-SVD Domain Robust Watermarking using Differential Evolution Algorithm 147
Veysel Aslantas
13.1 Introduction 147
13.2 SVD Domain and DWT-SVD Domain Watermarking Techniques 149
13.3 Optimal Watermarkings using DE 151
13.3.1 Basic Concepts of DE 151
13.3.2 Optimisation of Scaling Factors 152
13.4 Results 154
13.5 Conclusions 155
References 158

14 Design and Performance Evaluation of a Prototype Large Ring PET Scanner 161
M. Monjur Ahasan and David J Parker
14.1 Introduction 161
14.2 The Macropet Design 162
14.3 Performance Evaluation 164
14.3.1 Spatial Resolution 164
14.3.2 Sensitivity 165
14.3.3 Scatter Fraction 166
14.3.4 Count Rate Performance 168
14.3.5 Imaging Studies 169
14.4 Conclusions 171
References 171

15 Robust Wavelet-Based Video Watermarking Using Edge Detection . . 173
Tamás Polyák and Gábor Fehér
15.1 Introduction 173
15.2 The Watermark Embedding Process 174
15.3 The Watermark Detection Process 177
15.4 Experimental Results 177
15.4.1 Quality Results 178
15.4.2 Robustness of the Algorithm 179
15.5 Conclusion 181
References 182

16 High Speed Soft Computing Based Circuit for Edges Detection in Images 183
Nashaat M. Hussein and Angel Barriga
16.1 Introduction 183
16.2 Edge Detection Algorithm 184
16.2.1 The Filter Stage 184
16.2.2 The Threshold Stage 185
16.2.3 The Edge Detection Stage 187

16.3 Hardware Implementation 188
16.3.1 The Threshold Generation Circuit 189
16.3.2 Design of the Edge Detection System 190
16.4 Conclusion 193
References 193

17 A Comparison Between 3D OSEM and FBP Image Reconstruction Algorithms in SPECT 195
Khalid S. Alzimami, Salem A. Sassi, and Nicholas M. Spyrou
17.1 Introduction 195
17.2 Background 196
17.2.1 Analytical Reconstruction Algorithms: Filtered Backprojection 197
17.2.2 Iterative Reconstruction Algorithms 198
17.2.3 Physical Factors Affecting Quantitative and Qualitative Accuracy 199
17.3 A Comparison of 3D OSEM with FBP Image Reconstruction Algorithms 201
17.3.1 Methods 201
17.3.2 Results and Discussions 201
17.4 Summary 204
References 205

18 Performance Improvement of Wireless MAC Using Non-Cooperative Games 207
Debarshi Kumar Sanyal, Matangini Chattopadhyay, and Samiran Chattopadhyay
18.1 Introduction 207
18.2 Background 209
18.3 Game-Theoretic Model of Medium Access Control 209
18.4 Distributed Mechanisms to Achieve Nash Equilibrium 212
18.5 Performance Evaluation 213
18.5.1 CAP 214
18.5.2 AT 214
18.5.3 CO 215
18.5.4 SW 215
18.6 Discussion 217
18.7 Conclusion 217
References 218

19 Performance Evaluation of Mobile Ad Hoc Networking Protocols 219
Nadia Qasim, Fatin Said, and Hamid Aghvami
19.1 Introduction 219
19.2 Mobile Ad Hoc Network's Routing Protocols 221
19.3 Simulation Setup 222
19.4 Simulation Environment 223
19.5 Simulation Results 224

19.5.1 Throughput 225
19.5.2 End to End Delays 225
19.5.3 Media Access Delay 227
19.5.4 Packet Delivery Ratio 227
19.6 Conclusion 228
References 229

20 IEEE 802.11E Block Acknowledgement Policies 231
O. Cabral, A. Segarra, and F. J. Velez
20.1 Introduction 231
20.2 IEEE 802.11e and Block Acknowledgement Description 232
20.2.1 IEEE 802.11e User Priorities and Access Categories 232
20.2.2 Block Acknowledgement 232
20.3 State of the Art 234
20.4 System, Scenario and Assumptions 235
20.5 Results 237
20.5.1 Block Acknowledgement with Standalone Services 237
20.5.2 Block Acknowledgement with Mixtures of Applications 237
20.6 Conclusions 241
References 241

21 Routing in a Custom-Made IEEE 802.11E Simulator 243
João Miguel Ferro and Fernando J. Velez
21.1 Introduction 243
21.2 Previous Work 244
21.3 Overview 245
21.4 Results 246
21.5 Conclusions 252
References 253

22 Web-Based Management of Distributed Services 255
George Oikonomou and Theodore Apostolopoulos
22.1 Introduction 255
22.2 Related Work 256
22.3 WebDMF: A Web-Based Management Framework for Distributed Services 257
22.3.1 Management and Service Nodes 258
22.3.2 Management Representative 258
22.3.3 Domains 260
22.3.4 CIM Schemas and Operations 260
22.4 Implementation and Performance Evaluation 262
22.4.1 Implementation Details 262
22.4.2 Performance Evaluation 263
22.5 Conclusions 264
References 265

23 Image Index Based Digital Watermarking Technique for Ownership Claim and Buyer Fingerprinting ... 267
Sarabjeet S. Bedi, Rabia Bano, and Shekhar Verma
23.1 Introduction ... 267
23.2 Image Key and Buyer Fingerprint ... 269
23.2.1 Image Key ... 269
23.2.2 Buyer Fingerprint ... 269
23.3 Proposed Technique ... 270
23.3.1 Basic Idea ... 270
23.3.2 Generation of Watermark ... 270
23.3.3 Insertion of Watermark ... 270
23.3.4 Extraction of Watermark ... 271
23.4 Results and Discussion ... 272
23.5 Conclusion ... 276
References ... 276

24 Reverse Engineering: EDOWA Worm Analysis and Classification ... 277
Madihah Mohd Saudi, Emran Mohd Tamil, Andrea J Cullen, Mike E Woodward, and Mohd Yamani Idna Idris
24.1 Introduction ... 277
24.2 Method of Testing ... 278
24.2.1 Worm Analysis Process ... 279
24.2.2 Loading Specimen ... 279
24.3 EDOWA Classification ... 281
24.3.1 Infection ... 282
24.3.2 Activation ... 283
24.3.3 Payload ... 284
24.3.4 Operating Algorithms ... 285
24.3.5 Propagation ... 287
24.4 Conclusion ... 287
References ... 288

25 Reconfigurable Hardware Implementation of a GPS-Based Vehicle Tracking System ... 289
Adnan Yaqzan, Issam Damaj, and Rached Zantout
25.1 Introduction ... 289
25.2 Upgrading the *Aram* Locator *GPS* System ... 290
25.3 The *FPGA*-Based *Aram* System ... 291
25.3.1 The Intersystem Process 1 ... 293
25.3.2 The Intersystem Process 2 ... 294
25.3.3 The Memory Block ... 294
25.3.4 Communication Protocols: I2C and UART ... 296
25.4 Performance Analysis and Evaluation ... 296
25.5 Conclusion ... 298
References ... 298

26 Unknown Malicious Identification ... 301
Ying-xu. Lai and Zeng-hui. Liu
26.1 Introduction ... 301
26.2 Related Work ... 302
26.3 Naïve Bayesian and Application ... 303
26.3.1 String Extraction and Elimination ... 303
26.3.2 Naïve Bayes ... 304
26.4 Increment Naïve Bayes ... 305
26.4.1 Naïve Bayes ... 305
26.4.2 Multi-naïve Bayes ... 305
26.4.3 Half Increment Bayes (HIB) ... 306
26.4.4 Complexity ... 308
26.5 Experimental Results ... 309
26.5.1 Experimental Design ... 309
26.5.2 Performance Analysis ... 309
26.6 Conclusion ... 311
References ... 311

27 Understanding Programming Language Semantics for the Real World ... 313
Trong Wu
27.1 Introduction ... 313
27.2 Extension of the Applications to a Real World ... 315
27.3 Modularization and Communication in the Chaotic World ... 316
27.4 Parallel or Concurrency Is the Nature of the World ... 319
27.5 Accuracy and Efficiency Are Required by This Sophicificate World ... 321
27.6 Exception Handling Is the Safe Guard in the Dangerous World ... 324
27.7 Conclusion ... 326
References ... 327

28 Analysis of Overhead Control Mechanisms in Mobile AD HOC Networks ... 329
S. Gowrishankar, T.G. Basavaraju, and SubirKumarSarkar
28.1 Introduction ... 329
28.2 Theoretical Analysis of Overhead in Hierarchical Routing Scheme ... 330
28.3 Theoretical Analysis and Overhead Minimizing Techniques for AD HOC Networks Using Clustering Mechanisms ... 332
28.4 Minimizing Overhead in AD HOC Networks by Header Compression ... 336
28.5 Minimizing Overhead for AD HOC Networks Connected to Internet ... 337
References ... 339

29 Context Aware In-Vehicle Infotainment Interfaces 343
H. Sharma and A.K. Ramani
29.1 In-Vehicle Infotainment 344
29.1.1 Infotainment System Interaction 344
29.2 Context Aware Interaction 345
29.2.1 Context Aware Design 345
29.2.2 Context Management 346
29.3 Interaction Architecture 346
29.3.1 Layered Model 346
29.3.2 Component Model 347
29.3.3 Framework Core 350
29.3.4 Distribution and Logical Mobility 351
29.3.5 Application Model 351
29.4 Application Development 351
29.4.1 Application Model 352
29.4.2 Evaluation 352
29.5 Summary 353
References 354

30 Facial Expression Analysis Using PCA 355
V. Praseeda Lekshmi, M. Sasikumar, Divya S. Vidyadharan, and S. Naveen
30.1 Introduction 355
30.2 Background and Related Work 356
30.3 Method 357
30.3.1 Training stage 357
30.3.2 Face Recognition and Expression Classification 358
30.4 Results 361
30.5 Conclusion 363
References 363

31 The Design of a USART IP Core 365
A. H. El-Mousa, N. Anssari, A. Al-Suyyagh, and H. Al-Zubi
31.1 Introduction 365
31.2 USART Theory of Operation 366
31.2.1 Modes of Operation 367
31.3 Specifications of the Developed USART 368
31.4 USART System Design Methodology 368
31.4.1 Transmitter Module 370
31.4.2 Receiver Module 372
31.5 Testing and Verification Procedures 373
References 376

32 Multilayer Perceptron Training Optimization for High Speed Impacts Classification 377
Angel Garcia-Crespo, Belen Ruiz-Mezcua, Israel Gonzalez-Carrasco, and Jose Luis Lopez-Cuadrado
32.1 Introduction 377
32.2 Ballistic Impact 378
32.3 Use of ANN in Impacts Situations 380
32.3.1 Impact Scenario Parameters 381
32.3.2 Network Structure 382
32.3.3 Input Data 383
32.4 Solution Proposed 383
32.4.1 Randomness Elimination 384
32.4.2 Determination of the Influence of the Independent Variables 385
32.5 Evaluation 385
32.6 Conclusions 387
References 388

33 Developing Emotion-Based Pet Robots 389
W.-P. Lee, J.-W. Kuo, and P.-C. Lai
33.1 Introduction 389
33.2 Building Emotion-Based Pet Robots 390
33.2.1 A User-Oriented Framework 390
33.2.2 Control Architecture 391
33.2.3 User–Robot Interaction 392
33.2.4 Emotion Model 394
33.3 Experiments and Results 395
33.3.1 Implementation 395
33.3.2 Building Emotion-Based Behavior Selector 396
33.4 Conclusions and Future Work 398
References 400

34 Designing Short Term Trading Systems with Artificial Neural Networks 401
Bruce Vanstone, Gavin Finnie, and Tobias Hahn
34.1 Introduction 401
34.2 Review of Literature 402
34.3 Methodology 403
34.4 Results 405
34.5 Conclusions 408
References 409

35 Reorganising Artificial Neural Network Topologies 411
Thomas D. Jorgensen, Barry Haynes, and Charlotte Norlund
35.1 Introduction 411
35.2 Background 412

35.3 Neuroscientific Foundations . . . 413
35.4 Reorganising Neural Networks . . . 413
35.4.1 Neural Complexity . . . 414
35.4.2 Using the Complexity Measure . . . 415
35.4.3 The Simulated Track and Robot . . . 415
35.5 Experiments and Results . . . 416
35.5.1 The Simulation Environment . . . 417
35.5.2 The Fitness Function . . . 417
35.5.3 The Benchmark and the Reorganised Networks . . . 417
35.5.4 Evaluation and Comparison of the Proposed Methods . . . 419
35.6 Conclusion . . . 420
References . . . 421

36 Design and Implementation of an E-Learning Model by Considering Learner's Personality and Emotions . . . 423
S. Fatahi, M. Kazemifard, and N. Ghasem-Aghaee
36.1 Introduction . . . 423
36.2 Previous Works . . . 424
36.3 Psychological Principles . . . 424
36.4 Proposed Model . . . 425
36.5 Virtual Classmate Model . . . 427
36.6 Simulation of Proposed Model . . . 430
36.7 Implementation . . . 430
36.8 Results . . . 432
36.9 Conclusion and Future Works . . . 433
References . . . 433

37 A Self Progressing Fuzzy Rule-Based System for Optimizing and Predicting Machining Process . . . 435
Asif Iqbal and Naeem U. Dar
37.1 Introduction . . . 435
37.2 System Configuration . . . 437
37.3 The Self-Development Mode . . . 438
37.3.1 Data Acquisition . . . 438
37.3.2 Self-Development of Fuzzy Sets . . . 439
37.3.3 Self-Development of Prediction Rule-Base . . . 439
37.3.4 Self-Development of Optimization Rule-Base . . . 442
37.4 Optimization and Prediction of Machining Process . . . 442
37.5 Conclusion . . . 446
References . . . 446

38 Selection of Ambient Light for Laser Digitizing of Quasi-Lambertian Surfaces . . . 447
D. Blanco, P. Fernández, E. Cuesta, C. M. Suárez, and N. Beltrán
38.1 Introduction . . . 447
38.2 Objectives . . . 449

38.3 Configuration of the Tests 449
38.4 Experimental Procedure 451
38.5 Analysis Criteria 452
38.6 Results Discussion 453
38.7 Conclusions 457
References 457

39 Ray-Tracing Techniques Applied to the Accessibility Analysis for the Automatic Contact and Non Contact Inspection 459
B. J. Álvarez, P. Fernández, J.C. Rico, and G.Valiño
39.1 Introduction 459
39.2 Analysis Methodology 460
39.3 Analysis Considering Inspection Devices as Infinite Half-Lines 460
39.4 Interference Analysis Considering Real Dimensions of the Inspection Device 462
39.5 Clustering 464
39.6 Application Results 465
39.6.1 Inspection Process by Means of a Touch-Trigger Probe 465
39.6.2 Scanning Process by Means of a Laser Stripe System 466
39.7 Conclusions 468
References 469

40 Detecting Session Boundaries to Personalize Search Using a Conceptual User Context 471
Mariam Daoud, Mohand Boughanem, and Lynda Tamine-Lechani
40.1 Introduction 471
40.2 Related Works 472
40.3 Detecting Session Boundaries to Personalize Search 473
40.3.1 A Conceptual User Context Modelling for Personalizing Search 473
40.3.2 How to Detect Session Boundaries 475
40.4 Experimental Evaluation 476
40.4.1 Experimental Data Sets 476
40.4.2 Experimental Design and Results 477
40.5 Conclusion and Outlook 481
References 481

41 Mining Weather Information in Dengue Outbreak: Predicting Future Cases Based on Wavelet, SVM and GA 483
Yan Wu, Gary Lee, Xiuju Fu, Harold Soh, and Terence Hung
41.1 Introduction 484
41.2 Model Construction 485
41.2.1 Data Representation 485
41.2.2 Wavelet Decomposition 486
41.2.3 Genetic Algorithm 487
41.2.4 Support Vector Machines 487

41.2.5 Regression Learning . . . 488
41.2.6 Model Assessment Criteria . . . 488
41.3 Results and Discussions . . . 488
41.4 Conclusion . . . 492
References . . . 493

42 PC_Tree: Prime-Based and Compressed Tree for Maximal Frequent Patterns Mining . . . 495
Mohammad Nadimi-Shahraki, Norwati Mustapha, Md Nasir B Sulaiman, and Ali B Mamat
42.1 Problem of Maximal Frequent Patterns Mining . . . 495
42.2 Related Work . . . 496
42.3 Proposed Method . . . 497
42.3.1 Database Encoding Technique . . . 497
42.3.2 PC_Tree Construction . . . 498
42.4 Experimental Results . . . 501
42.5 Conclusion and Future Works . . . 503
References . . . 504

43 Towards a New Generation of Conversational Agents Based on Sentence Similarity . . . 505
Karen O'Shea, Dr. Zuhair Bandar, and Dr. Keeley Crockett
43.1 Introduction . . . 505
43.2 Sentence Similarity Measure . . . 507
43.3 Traditional CA Scripting . . . 509
43.4 CA Scripting Using Sentence Similarity . . . 510
43.5 Experimental Methodology . . . 511
43.5.1 Domain . . . 511
43.5.2 Experiments . . . 511
43.6 Results and Discussion . . . 512
43.7 Conclusions and Further Work . . . 513
References . . . 514

44 Direction-of-Change Financial Time Series Forecasting Using Neural Networks: A Bayesian Approach . . . 515
Andrew A. Skabar
44.1 Introduction . . . 515
44.2 MLPs for Financial Prediction . . . 516
44.3 Bayesian Methods for MLPs . . . 517
44.4 Empirical Results . . . 519
44.4.1 Input Pre-processing . . . 519
44.4.2 Hypothesis Testing . . . 520
44.4.3 Setting the Priors . . . 521
44.4.4 MCMC Sampling . . . 523
44.5 Conclusions . . . 523
References . . . 524

45 A Dynamic Modeling of Stock Prices and Optimal Decision Making Using MVP Theory ... 525
Ramin Rajabioun and Ashkan Rahimi-Kian
45.1 Introduction ... 526
45.2 Modeling and Prediction ... 527
45.3 Virtual Stock Market ... 529
45.4 The Market Simulation Results ... 532
45.5 Conclusion ... 534
References ... 536

46 A Regularized Unconstrained Optimization in the Bond Portfolio Valuation and Hedging ... 539
Yury Gryazin and Michael Landrigan
46.1 Introduction ... 539
46.2 Option-Adjusted Spread Analysis ... 541
46.3 The Spread Variance Minimization Approach ... 542
46.4 Numerical Method ... 543
46.5 Numerical Results ... 544
46.6 Conclusions ... 548
References ... 549

47 Approximation of Pareto Set in Multi Objective Portfolio Optimization ... 551
I. Radziukyniene and A. Zilinskas
47.1 Introduction ... 551
47.2 Multi-Objective Portfolio Optimization Problem ... 552
47.3 Multi-Objective Optimization by the Method of Adjustable Weights ... 553
47.4 Evolutionary Methods for Multi-Objective Optimization ... 556
47.5 Description of Experimental Investigation ... 558
47.6 Discussion on Experimental Results ... 559
47.7 Conclusions ... 561
References ... 561

48 The Scaling Approach for Credit Limit Management and Its Application ... 563
M.Y. Konovalikhhin, D.O. Sergienko and O.V. Balaeva
48.1 Introduction ... 563
48.2 Analysis ... 565
48.2.1 Model Description ... 565
48.2.2 B3 Coefficient Calculation ... 566
48.2.3 B2 Coefficient Calculation ... 569
48.2.4 B1 Coefficient Calculation ... 570
48.3 Analysis Results ... 571
48.4 Conclusions ... 573
References ... 573

49 Expected Tail Loss Efficient Frontiers for CDOS of Bespoke Portfolios Under One-Factor Copula Marginal Distributions 575
Diresh Jewan, Renkuan Guo, and Gareth Witten
49.1 Introduction 575
49.2 Bespoke CDO Mechanics 576
49.3 Heavy-Tail Modelling and Copulas 577
49.3.1 Stable Paretian Distributions 577
49.3.2 Copulas 578
49.4 Credit Risk Measures 578
49.4.1 Convex and Coherent Risk Measures 579
49.4.2 Expected-Tail-Loss 579
49.4.3 Portfolio Optimisation Under Expected-Tail-Loss 580
49.5 Copula Marginal ETL Efficient Frontier for the CDO Collateral .. 581
49.6 Concluding Remarks 585
References 585

50 Data Mining for Retail Inventory Management 587
Pradip Kumar Bala
50.1 Introduction 587
50.2 Literature Review 589
50.3 Purchase Dependencies in Retail Sale in the Context of Inventory Management 590
50.4 Model Development 591
50.5 Illustration with an Example 592
50.6 Case Discussion 594
50.7 Conclusions 596
References 597

51 Economic Process Capability Index for Product Design and Process Planning 599
Angus Jeang
51.1 Introduction 599
51.2 Process Capability Indices (PCI) 601
51.3 Proposed Process Capability Index 603
51.3.1 Single Quality Characteristic 603
51.3.2 Multiple Quality Characteristics 605
51.4 Summary 608
References 608

52 Comparing Different Approaches for Design of Experiments (DoE) .. 611
Martín Tanco, Elisabeth Viles, and Lourdes Pozueta
52.1 Introduction 611
52.2 Approaches to Design of Experiments 612
52.2.1 The Classical Approach 612
52.2.2 The Taguchi Approach 614
52.2.3 The Shainin Approach 615

52.3 Limitations of the Different Approaches 616
52.3.1 The Classical Approach 616
52.3.2 Taguchi 617
52.3.3 Shainin 618
52.4 Conclusions and Recommendations 618
References 619

53 Prevention of Workpice Form Deviations in CNC Turning Based on Tolerance Specifications 623
Gonzalo Valiño Riestra, David Blanco Fernandez, Braulio Jose Álvarez, Sabino Mateos Diaz, and Natalia Beltrán Delgado
53.1 Introduction 624
53.2 Form Errors in Turning 624
53.3 Calculation of Radial Deviations 625
53.4 Relationship Between Deviations and Tolerances 627
53.4.1 Length Tolerance 627
53.4.2 Diametrical Tolerance 629
53.4.3 Geometrical Tolerance of Flatness 629
53.4.4 Geometrical Tolerance of Cylindricity 630
53.4.5 Geometrical Tolerance of Profile of a Surface 630
53.4.6 Geometrical Tolerance of Parallelism 631
53.4.7 Geometrical Tolerance of Perpendicularity 631
53.4.8 Geometrical Tolerance of Coaxiality 631
53.4.9 Geometrical Tolerance of Circular Runout 631
53.4.10 Geometrical Tolerance of Total Runout 632
53.5 Deviations as Optimization Conditions 632
53.6 Conclusions 632
References 633

54 Protein–Protein Interaction Prediction Using Homology and Inter-domain Linker Region Information 635
Nazar Zaki
54.1 Introduction 635
54.1.1 The Importance of Protein–Protein Interaction 635
54.1.2 Current Methods to Predict PPI 636
54.1.3 Computational Approaches to Predict PPI 636
54.2 Method 637
54.2.1 Similarity Measures Between Protein Sequences 637
54.2.2 Identify and Eliminate Inter-domain Linker Regions 639
54.2.3 Detecting Domain Matches and Associated Structural Relationships in Proteins 641
54.3 Experimental Work 641
54.4 Results and Discussion 642
54.5 Conclusion 644
References 644

55 Cellular Computational Model of Biology 647
Alex H. Leo and Gang Sun
55.1 Introduction 647
55.2 List of Variables 650
55.3 Determining mRNA Polymerases, mR(n + 1) 650
55.4 Determining β-Galactosidases, $G(n+1)$ 651
55.5 Determining Lactoses with in and out the Cell, Lext(n + 1) and L(n + 1) 653
55.6 Determining ATP Molecules, A(n) 654
55.7 The System Structure 656
55.8 In Comparison to Traditional Biological Experiment 657
55.9 Conclusion 657
55.10 Discussion 657
References 658

56 Patient Monitoring: Wearable Device for Patient Monitoring 659
Robert G. Lupu, Andrei Stan, and Florina Ungureanu
56.1 Introduction 659
56.2 Wearable Monitoring Device 661
56.3 Functionality 662
56.4 Firmware Issue 665
56.5 Conclusion 668
References 668

57 Face Recognition and Expression Classification 669
Praseeda Lekshmi.V, Dr.M.Sasikumar, and Divya S Vidyadharan
57.1 Introduction 669
57.2 Background and Related Work 670
57.3 Discrete Cosine Transform 670
57.4 Radial Basis Function 671
57.5 Method 671
57.5.1 Facial Region Identification 672
57.5.2 Face Recognition 674
57.5.3 Facial Expression Analysis 676
57.6 Results 677
57.7 Conclusion 678
References 679

58 Spiking Neurons and Synaptic Stimuli: Neural Response Comparison Using Coincidence-Factor 681
Mayur Sarangdhar and Chandrasekhar Kambhampati
58.1 Introduction 681
58.2 Neuronal Model and Synapse 682
58.2.1 The Neuron Model 682

58.2.2 The Synaptic Current ... 683
58.2.3 The Total External Current ... 684
58.3 Comparison of Two Spike Trains ... 684
58.3.1 Responses of the Neuron ... 684
58.3.2 Comparison of Responses ... 685
58.3.3 Coincidence-Factor ... 686
58.3.4 Two-Dimensional Analysis ... 688
58.3.5 Binary Clustering ... 688
58.4 Conclusions ... 690
References ... 690

59 Overcoming Neuro-Muscular Arm Impairment by Means of Passive Devices ... 693
Federico Casolo, Simone Cinquemani, and Matteo Cocetta
59.1 Introduction ... 693
59.2 Background ... 694
59.3 Experimental Test Apparatus Set Up ... 695
59.4 Experimental Tests ... 697
59.5 Device Evolution ... 699
59.5.1 Simulations ... 701
59.6 Conclusions ... 702
References ... 704

60 EEG Classification of Mild and Severe Alzheimer's Disease Using Parallel Factor Analysis Method ... 705
Charles-Francois Vincent Latchoumane, Francois-Benois Vialatte, Jaeseung Jeong, and Andrzej Cichocki
60.1 Introduction ... 705
60.1.1 Diagnosis of Alzheimer's Disease and EEGs ... 705
60.1.2 Multi-way Array Decomposition ... 706
60.2 Subjects and EEG Recordings ... 707
60.3 Method ... 707
60.3.1 Three-Way Tensor for Analysis ... 707
60.3.2 Classification in Divide and Conquer Scheme ... 708
60.3.3 The PARAllel FACtor Analysis (PARAFAC) ... 709
60.3.4 Filter Estimation and Feature Extraction ... 710
60.3.5 From Filter to Feature ... 711
60.4 Results ... 711
60.5 Discussion ... 713
References ... 714

61 Feature Selection of Gene Expression Data Based on Fuzzy Attribute Clustering ... 717
Elham Chitsaz, Mohammad Taheri, and Seraj D. Katebi
61.1 Introduction ... 717
61.2 Related Work ... 718

61.3 The Proposed Fuzzy Approach 720
61.4 Experimental Results 722
61.4.1 Stability 722
61.4.2 Classification Accuracy 724
61.5 Conclusion 725
References 726

Chapter 1
Acoustic Behavior of Squirrel Cage Induction Motors

C. Grabner

Abstract The evaluation of distinct influences to the acoustic noise emission of industrial standard squirrel cage motors is fairly a very sensitive and complex topic. Main interest is thereby given to the influence of the complete mechanical induction motor design to the almost manifold acoustic response under sinusoidal power supply. This includes the investigation of unexpected effects even caused by passing critical speed values which are corresponding to mechanical resonance frequencies. To take remedial measures, totally-closed stator topologies and rotor skewing opportunities, have been deeply investigated in a comparatively manner. The performed comparatively measurement of the speed dependent equivalent continuous A-weighted surface sound pressure level gives a good overview about the principal design influences.

Keywords Acoustic Behavior · Motor Design · Induction Motor · Sound Pressure Level

1.1 Introduction

As the motor design process has to consider multi-physical effects, such as, e.g. electrical, thermal and mechanical aspects, a very deep and complex analysis has to take place in order to find the best technical and commercial choice for the product. Thus, extended investigations to the acoustic behavior of the complete speed variable induction drive system, as it is schematically depicted in Fig. 1.1, have been performed in the lab to obtain reproducible results with high accuracy.

C. Grabner
Research and Development, Siemens AG, Frauenauracherstrasse 80, D-91056 Erlangen,
E-mail: Grabner.Christian@SIEMENS.COM

S.I. Ao and L. Gelman (Eds.), *Advances in Electrical Engineering and Computational Science*, Lecture Notes in Electrical Engineering.

Fig. 1.1 A 750 W 4-pole squirrel cage induction motor in FS80

1.2 Acoustic Noise Physics

Sound is the mechanical vibration of a gaseous medium through which the energy is transferred away from the source by progressive sound waves. Whenever an object vibrates, a small portion of the energy involved is lost to the surrounding area as unwanted sound – the acoustic noise [1, 2].

1.2.1 Sound Power

Any source of noise is commonly characterized by the emitted sound power, measured in watts. This fundamental physical property of the source itself is often known as absolute parameter for the acoustic output.

The sound power can be found conceptually by adding the product of the areas times the acoustic intensities for the areas on any hypothetical surface that contains the source. Since equipment for measuring of acoustic intensity is generally not available, sound power inevitably has to be determined indirectly by taking the spatial average of the sound pressure squared measured on equal areas on the surrounding surface.

1.2.2 Sound Pressure Level in Decibel

As the acoustic intensity, the power passing through a unit area in space, is proportional in the far field to the square of the sound pressure, a convenient scale for the measurement can be defined as sound pressure level

$$L_p(t) = 10\log\left(\frac{p(t)}{p_0}\right)^2 = 20\log\left(\frac{p(t)}{p_0}\right), \tag{1.1}$$

in decibel, with p_0 as the reference sound pressure of 20 μ Pa. The measured sound pressure p (t) in (1) depends on many insecure factors, such as e.g. orientation and distance of receiver, temperature and velocity gradients inside the involved medium.

The analysis of (1) in the frequency domain is fortunately done with the discrete fast Fourier-analysis. Therefore, the time-periodical signal (1) is sampled as $L_{p,n}$ and further processed at a distinct number of N samples as

$$\hat{L}_{p,\upsilon} = \sum_{n=0}^{N-1} L_{p,n} e^{-j(2\pi n/N)\upsilon}, \quad \upsilon = 0, 1, 2 \ldots N-1 \tag{1.2}$$

in order to obtain the Fourier coefficients $\hat{L}_{p,\upsilon}$ of the interested frequency-domain.

1.2.3 Sound Pressure Level and Human Hearing

Unfortunately, no simple relationship exists between the measured physical sound pressure level and the human perception of the same sound. The treatment of acoustic noise and its effects is therefore a complicated problem, which must take a wide variety of parameters into account to achieve good correlations between measurement and the resultant human reaction.

1.2.4 Human Hearing Mechanism

The quietest sound at 1,000 Hz which can be heard by the average person is found in Fig. 1.2 to be about 20 µ Pa and this value has been standardized as the nominal hearing threshold for the purpose of sound level measuring. At the other end of the scale the threshold of pain occurs at a sound pressure of approximately 100 Pa.

An inspection of Fig. 1.2 shows that a pure tone having a sound pressure level of, e.g. 20 dB and a frequency of 1,000 Hz is plainly audible, whereas one having the same sound pressure level but a frequency of 100 Hz is well below the threshold of audibility and cannot be heard at all.

1.2.5 Loudness Level in Phon

The loudness of a pure tone of constant sound pressure level, perhaps the simplest acoustic signal of all, varies with its frequency, even though the sound pressure may be the same in every case.

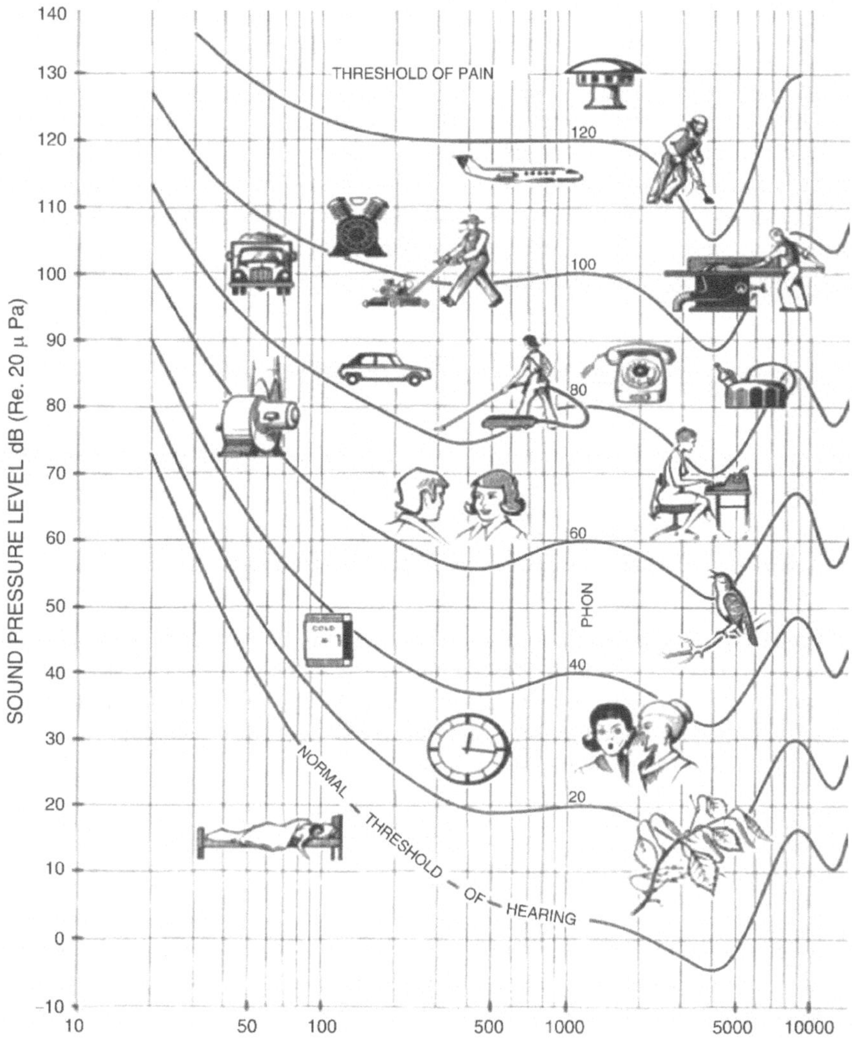

Fig. 1.2 Typical sound pressure levels in decibel and equal loudness levels in phon for different noise sources

Although our hearing mechanism is not well-adapted for making quantity measurements of the relative loudness of different sounds, there is a fair agreement between observers, when two pure tones of different frequencies appear to be equally load. It is therefore possible to establish in Fig. 1.2 contour plots of equal loudness in phone. These subjective felled loudness curves are obtained by alternately sounding a reference tone of 1,000 Hz and a second tone of some other frequency. The intensity level of the second tone is then adjusted to the value that makes the two tones appear equally load.

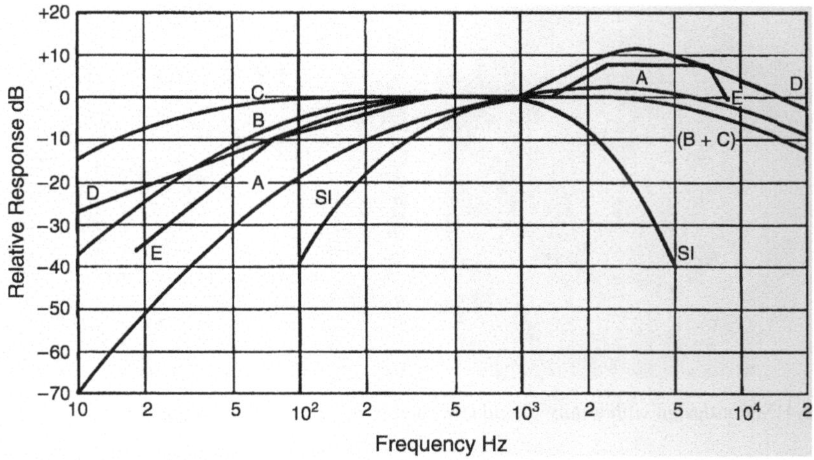

Fig. 1.3 Standardized weighting curves for sound level measurement

A pure ton having a frequency of 100 Hz and a sound pressure level of about 35 dB sounds as load as a pure 1,000 Hz tone whose intensity level is 20 dB, and hence the loudness level of the 100 Hz tone is by definition 20 phon. The 1,000 Hz frequency is thus the reference for all loudness measurements, and all contours of equal loudness expressed in phon have the same numerical value as the sound pressure level in decibel at 1,000 Hz.

1.2.6 Weighing the Sound Pressure Level

Due to the human ears assessment of loudness, the defined purely physical sound pressure terminus has to be modified by an implicit weighting process in a way which corresponds to the more complex response of the human being. Therefore, among several possibilities, distinguished by B, C, D, E or SI-weighting in Fig. 1.3, the A-weighted level has be found to be the most suitable for modifying the frequency response to follow approximately the equal loudness of 20 phon in Fig. 1.2.

1.3 Mutation of Stator and Rotor Parts

The stator part of a squirrel cage induction motor is typically manufactured so far with semi-closed stator slots [3]. However, continual progress in the industrial automation process allows the utilization of a novel stator construction with totally-closed stator slots as shown in Fig. 1.4. The squirrel cage rotor could alternatively be built up with a skewed or even non-skewed rotor as depicted in Fig. 1.5. Based on the proposed different stator and rotor components, four motor designs, as listed in

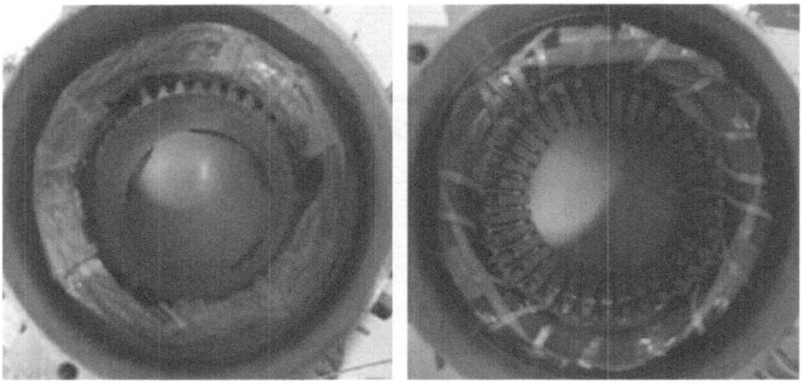

Fig. 1.4 Stator design with totally-closed (*left*) and semi-closed (*right*) stator slots

Fig. 1.5 Skewed (*left*) and un-skewed (*right*) rotor

Table 1.1 Test results with regard to their suitability for acoustic noise minimization

Combination	Stator slot design	Rotor design
Motor ①	Totally-closed	Skewed
Motor ②	Totally-closed	Un-skewed
Motor ③	Semi-closed	Un-skewed
Motor ④	Semi-closed	Skewed

Table 1.1, have been consecutively tested with regard to their suitability for acoustic noise minimization.

1.4 Direct Comparison of Sound Pressure Levels for Different Motor Designs

All investigated motor designs are operated at the rated electrical voltage level of 400 V. They have a rated power of approximately 750 W at the nominal speed of about 1,395 rpm. As all listed samples in Table 1.1, are assisted with the same motor fan, a direct comparison of the acoustic tests results is feasible.

The noise measurement has been carried out for each motor design over a wide speed range beginning from the lower value of 500 rpm up to rated-speed of 1,395 rpm at constant mechanical rated-torque of 5 Nm. Within increasing higher speed ranges, the drive works in the field weakening range and so the load has been continuously reduced.

After reaching the physically quasi-steady operational state for each adjusted speed value, the measured data set has been stored. Thus, purely dynamic acoustic noise in fact of a transient real-time run-up is herewith not considered.

1.4.1 Evaluation of Design Impacts to the Equivalent Sound Pressure Level

The depicted sound pressure level courses in Figs. 1.6 and 1.7, which are representing the skewed rotor designs ① and ④, show a very smooth and continuous track behavior, which fortunately avoids extensively noise peaks over the total speed range. Thus, the utilization of the standard skewing technology avoids the generation of undesired significant noise peak over the complete speed range.

As obviously from Figs. 1.6 and 1.7, the course of the proposed novel motor combination ① has advantageously the lowest sound pressure level in comparison to the usually design ④ at several speed values.

The introduced novel motor topology of the standard squirrel cage induction motor with totally-closed slot tips is therefore suitable to reduce undesired noise emission not only at the nominal speed of 1,395 rpm. The emitted noise of the proposed motor design ① is reduced by the amount of 10 dB at no-load in Fig. 1.6 and about 8 dB at rated-load in Fig. 1.7 in comparison to the state of the art design ④.

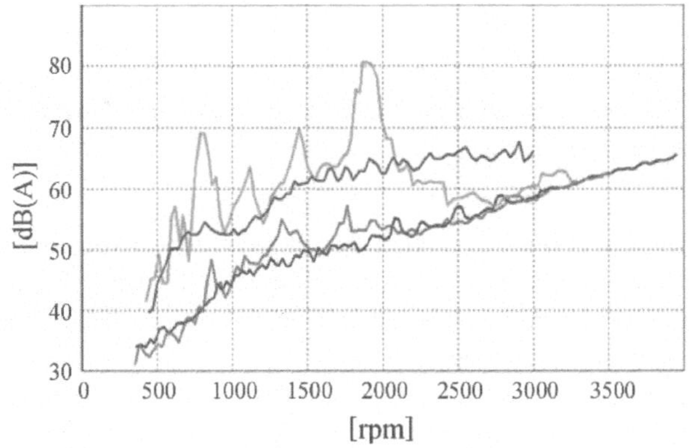

Fig. 1.6 Measured sound pressure level versus speed for the motor ① (blue), motor ② (red), motor ③ (green) and motor ④ (violet) at sinusoidal power supply and no-load

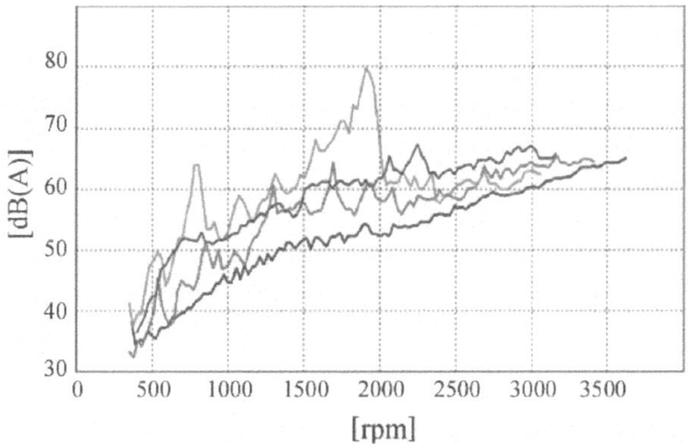

Fig. 1.7 Measured sound pressure level versus speed for the motor ① (blue), motor ② (red), motor ③ (green) and motor ④ (violet) at sinusoidal power supply and rated-load

Contrarily, in case of both test objects ② and ③, a very discontinuous and even speed sensitive sound level behaviors characterized by some very distinct noise peaks is found from Figs. 1.6 and 1.7.

Especially the varying operational state from no-load to rated-load causes extended magnetic saturation effects within the totally-closed stator tooth tip regions. This fact has, in dependency on the motor design ② or ③ unexpected inverse impacts to the emitted noise levels. There are some interesting secondary aspects occurring at higher speed ranges. Considering the test object ③, the maximal occurring peak of 81 dB in Fig. 1.6 at 1,842 rpm is slightly reduced at the same speed value to 79 dB in Fig. 1.7. A completely contrarily observation could be for motor ②. The noise peak of 55 dB at 1,670 rpm at no-load is significantly shifted to much higher levels with arising additional peaks of 62 dB at the same speed in case of the rated-load condition in Fig. 1.7.

1.4.2 Fast Fourier Sound Pressure Spectrum for Each Motor Design at Rated-Load

The description of the time-depended sound pressure in the frequency domain allows a deeper inside into the fundamental aspects of noise generation. The harmonic sound pressure frequency components of the motors ① to ④ are depicted in Figs. 1.8–1.11 over the extended speed range.

The utilization of a skewed rotor type in motor ① obviously results in Fig. 1.8 in green shaded lower range spectral magnitudes. Contrarily, the test results of the skewed design ④ given in Fig. 1.11 show a high 58 dB peak at the frequency of 1,900 Hz.

A distinct noise peak of electromagnetic origin of the un-skewed design ② is visible in Fig. 1.9 at the speed 1,670 rpm with the frequency of 940 Hz.

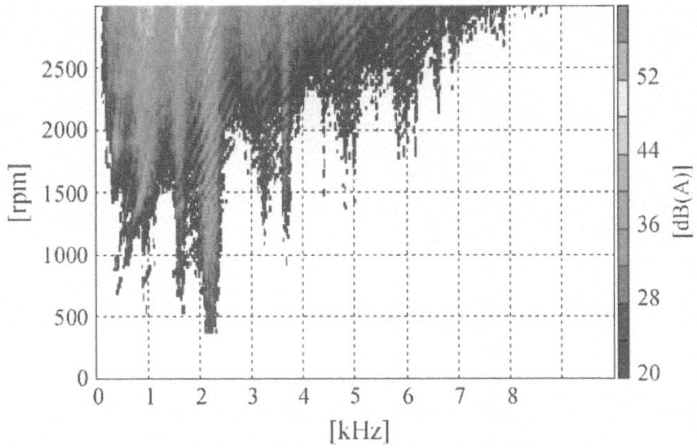

Fig. 1.8 FFT of the sound pressure level for motor ① at rated-load depicted for the speed range 500–3,000 rpm

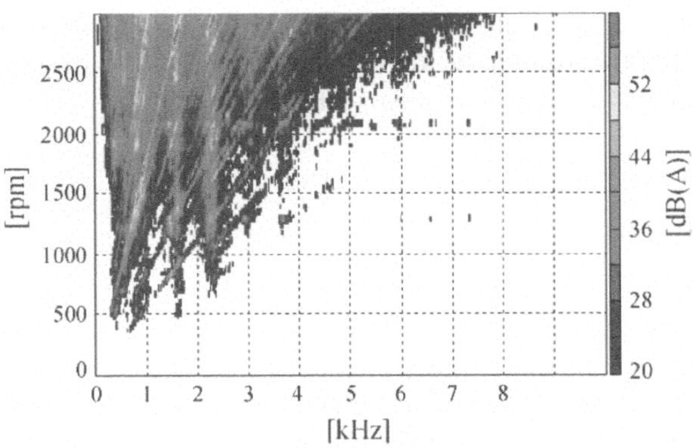

Fig. 1.9 FFT of the sound pressure level for motor ② at rated-load depicted for the speed range 500–3,000 rpm

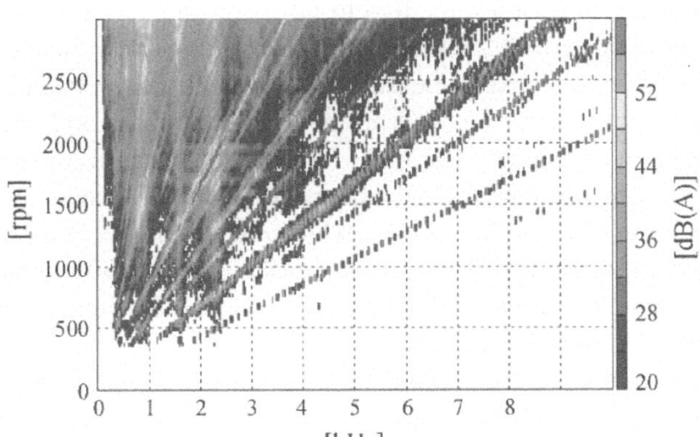

Fig. 1.10 FFT of the sound pressure level for motor ③ at rated-load depicted for the speed range 500–3,000 rpm

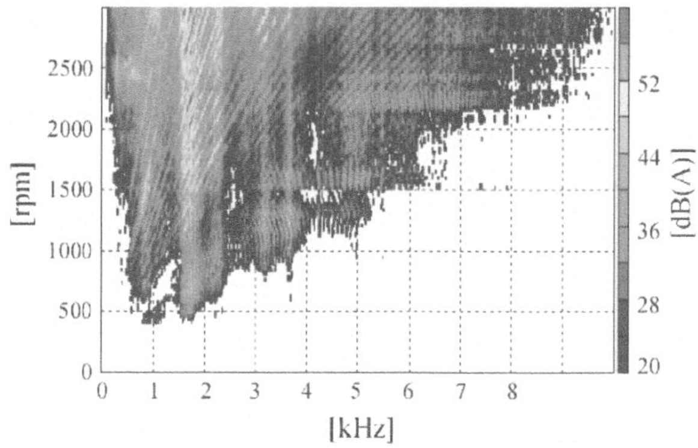

Fig. 1.11 FFT of the sound pressure level for motor ④ at rated-load depicted for the speed range 500–3,000 rpm

Some very distinct spectral components due to stator slot harmonics of motor ③ at the speed of 1,900 rpm could be identified in Fig. 1.10 with magnitudes of 59 dB at the surrounding of the single frequency of 2,300 Hz.

1.5 Conclusion

The evaluation of distinct influences to the acoustic noise emission of industrial standard squirrel cage motors is fairly a very sensitive and complex topic.

Main interest is thereby given to the influence of the complete mechanical induction motor design to the almost manifold acoustic response under sinusoidal power supply. This includes the investigation of unexpected effects even caused by passing critical speed values which are corresponding to mechanical resonance frequencies. To take remedial measures, totally-closed stator topologies and rotor skewing opportunities, have been deeply investigated in a comparatively manner.

The performed comparatively measurement of the speed dependent equivalent continuous A-weighted surface sound pressure level gives a good overview about the principal design influences. The main focus at standard drives is thereby only given to the occurring noise peaks at the interested nominal speed.

The modification of induction motors, which are commonly assisted by skewed rotors, by applying novel totally-closed stator slot designs, is the most favorable possibility for reducing the emitted acoustic noise level. Fortunately, the magnetic utilization could almost be kept equal.

If the rotor is alternatively manufactured in an un-skewed kind, the acoustic noise emission would reach the same level as with utilized semi-closed slots in combination with a skewed rotor.

References

1. L.E. Kinsler and A.R. Frey, *Fundamentals of acoustics*, Wiley:, New York/London, 1962.
2. M. Bruneau, *Fundamentals of acoustics*, ISTE, London, 2006.
3. H. Jordan, *Der geräuscharme Elektromotor*, Verlag W. Girardet, Essen, 1950.

Chapter 2
Model Reduction of Weakly Nonlinear Systems

Marissa Condon and Georgi G. Grahovski

Abstract In general, model reduction techniques fall into two categories – moment –matching and Krylov techniques and balancing techniques. The present contribution is concerned with the former. The present contribution proposes the use of a perturbative representation as an alternative to the bilinear representation [4]. While for weakly nonlinear systems, either approximation is satisfactory, it will be seen that the perturbative method has several advantages over the bilinear representation. In this contribution, an improved reduction method is proposed. Illustrative examples are chosen, and the errors obtained from the different reduction strategies will be compared.

Keywords Model Reduction · Weakly Nonlinear Systems · Perturbative Approximation · Nonlinear Circuit

2.1 Introduction

As the level of detail and intricacy of dynamical system models continues to grow, the essential behaviour of such systems can be lost in a cloud of complexity. Furthermore, numerical simulations of such systems can consume extensive memory resources and take inordinate amounts of time. Consequently, to cope with the growing complexity and dimensionality of systems, model reduction has become a vital aspect of modern system simulation. Model reduction techniques for linear systems are well studied (e.g. [1, 2] and references therein). However, the study of nonlinear systems is much more complicated and the development of model reduction methods for large-scale nonlinear systems represents a formidable challenge.

M. Condon (✉)
School of Electronic Engineering, Dublin City University, Glasnevin, Dublin 9, Ireland,
E-mail: Marissa.Condon@dcu.ie

S.I. Ao and L. Gelman (Eds.), *Advances in Electrical Engineering and Computational Science*, Lecture Notes in Electrical Engineering.

In general, model reduction techniques fall into two categories – moment – matching and Krylov techniques and balancing techniques. The present contribution is concerned with the former. The advantage of Krylov-based methods is that matrix–vector multiplications are all that are involved in the formation of the projection matrices that are used to project the system onto a reduced system. Also sparse linear solvers and iterative approaches can be used to improve the computational efficiency [3].

To be amenable to the application of Krylov methods, the nonlinear function describing the system must be approximated in a suitable manner. A widely used approach is the utilisation of the bilinear representation. However, the present contribution proposes the use of a perturbative representation as an alternative to the bilinear representation [4]. While for weakly nonlinear systems, either approximation is satisfactory, it will be seen that the perturbative method has several advantages over the bilinear representation. The use of the perturbative representation in reduction techniques based on Krylov methods has been addressed in [4]. In this contribution, an improved reduction method is proposed. Illustrative examples are chosen, which are typical of examples employed for comparing model reduction approaches [4]. The errors obtained from the different reduction strategies will be compared.

2.2 Perturbative Approximation of Nonlinear Systems

Let the nonlinear dynamical system under consideration be of the form:

$$\begin{aligned} \dot{x}(t) &= f(x(t)) + Bu(t) \\ y(t) &= Cx(t), \end{aligned} \tag{2.1}$$

where $f : \mathbb{R}^n \to \mathbb{R}^n$ is a non-linear function, $x \in \mathbb{R}^n$ are the state-space variables. The initial condition is $x(0) = x_0$ and $u(t), y(t) \in \mathbb{R}$. $B, C \in \mathbb{R}^n$ are constant vectors (C is a vector-row and B is a vector-column). It is assumed that $x = 0$ is a stable equilibrium point of the system (2.1) and $x_0 = 0$. Under this assumption, $f(x)$ can be expanded in a generalised Taylor's series about $x = 0$:

$$f(x) = A_1 x^{(1)} + A_2 x^{(2)} + A_3 x^{(3)} + \dots, \tag{2.2}$$

where $x^{(1)} = x$, $x^{(2)} = x \otimes x$, $x^{(3)} = x \otimes x \otimes x$, etc. and $\otimes$ denotes the Kronecker product. Since $x = 0$ is a stable equilibrium point, A_1 is a stable matrix, i.e. all of its eigenvalues have negative real parts. It is also assumed that each term in the Taylor's expansion is small compared to the previous one.

Now consider the case where a variational parameter α is introduced, i.e. $\dot{x}(t) = f(x(t)) + B\alpha u(t)$ and let the response of the system $x(t)$ be *perturbatively* expanded in a power series in α [4]:

$$x(t) = \alpha x_1(t) + \alpha^2 x_2(t) + \alpha^3 x_3(t) + \dots. \tag{2.3}$$

On comparing terms in the variational parameter α, the following set of n-dimensional differential equations can be derived:

$$\begin{aligned}
\dot{x}_1 &= A_1 x_1 + Bu \\
\dot{x}_2 &= A_1 x_2 + A_2(x_1 \otimes x_1) \\
\dot{x}_3 &= A_1 x_3 + A_2(x_1 \otimes x_2 + x_2 \otimes x_1) + A_3(x_1 \otimes x_1 \otimes x_1) \\
&\vdots
\end{aligned} \tag{2.4}$$

Each n-dimensional equation describes the time evolution of an x_i, where x_i represents the ith order perturbative term in the expansion (2.3). Defining a vector $\underline{x}(t)$:

$$\underline{x}(t) = \begin{bmatrix} x_1 \\ x_2 \\ \vdots \end{bmatrix} \tag{2.5}$$

the system in (2.4) acquires the form:

$$\begin{aligned}
\dot{\underline{x}} &= \underline{A}\underline{x} + \underline{B}\underline{u} \\
\underline{y} &= \underline{C}\underline{x},
\end{aligned} \tag{2.6}$$

where

$$\underline{A} = \begin{bmatrix} A_1 & & & \\ & A_1 & & \\ & & A_1 & \\ & & & \ddots \end{bmatrix} \quad \underline{B} = \begin{bmatrix} B & 0 & 0 & 0 & \dots \\ 0 & A_2 & 0 & 0 & \\ 0 & 0 & A_2 & A_3 & \dots \\ \vdots & \vdots & \vdots & \vdots & \end{bmatrix} \tag{2.7}$$

$$\underline{u}(t) = \begin{bmatrix} u_1 \\ u_2 \\ u_3 \\ \vdots \end{bmatrix} \quad \underline{C} = [C, C, C, \dots], \quad u_1 = u(t), \quad u_2 = x_1 \otimes x_1,$$

$$u_3 = x_1 \otimes x_2 + x_2 \otimes x_1, \quad u_4 = x_1 \otimes x_1 \otimes x_1, \dots$$

The source u_2 for the second equation in (2.4) depends only on the state vector x_1 determined from the first equation and so on. Note that since A_1 is a stable matrix, $\underline{A}$ is also automatically stable. Now, u_2, u_3, etc. are not independent inputs like u_1 and therefore, linear system theory cannot be applied directly to the representation in (2.6). However, subsequent sections will show how the representation may be adapted so that linear system theory and consequently, linear model reduction may be applied to the representation in (2.6).

The well-known bilinear representation (Carleman bilinearization) [4–6] is an alternative approach to approximation of (2.1) for weakly nonlinear systems:

$$\begin{aligned} \dot{\hat{x}}(t) &= \hat{A}\hat{x}(t) + \hat{N}\hat{x}(t)\hat{u}(t) + \hat{B}\hat{u}(t) \\ y(t) &= \hat{C}\hat{x}(t), \end{aligned} \tag{2.8}$$

where

$$\hat{x}(t) = \begin{bmatrix} x^{(1)} \\ x^{(2)} \\ \vdots \end{bmatrix},$$

$\hat{A}$ and $\hat{N}$ are square matrices of dimension $n + n^2 + \cdots + n^K$, $\hat{B}$ and $\hat{C}$ are vectors with $n + n^2 + \cdots + n^K$ components if K terms in the Taylor's series expansion are taken into account. The matrices are defined in [4–6]. For example, for $K = 2$:

$$\hat{A} = \begin{bmatrix} A_1 & A_2 \\ 0 & A_{21} \end{bmatrix}, \quad \hat{N} = \begin{bmatrix} 0 & 0 \\ N & 0 \end{bmatrix}, \quad \hat{B} = \begin{bmatrix} B \\ 0 \end{bmatrix}, \quad \hat{C} = [C, 0],$$

where A_1 and A_2 are the matrices in (2.2), $A_{21} = A_1 \otimes I + I \otimes A_1$, $N = B \otimes I + I \otimes B$, where I is the $n \times n$ identity matrix.

However, the perturbative representation has several advantages over the bilinear representation–namely:

1. The system (2.6) has a simple linear form unlike (2.8).
2. The size of the system (2.6) with K perturbative terms is nK. The size of (2.8) with K terms in the series expansion is $n + n^2 + \cdots + n^K$.
3. There is no need to restrict the input to the perturbative system to guarantee stability. However, a restriction exists on the input to guarantee stability of the bilinear system [5, 6]. A sufficient condition for stability on the interval $[0, T]$ is $|u(t)| \leq K_c$ for all $t \in [0, T]$ where $\hat{A} + \lambda\hat{N}$ is stable for all $\lambda \in [-K_c, K_c]$.

2.3 KRYLOV-Based Model Reduction

The goal of any model reduction technique is to replace the n- dimensional system (2.1) with a system of much smaller dimension $k \ll n$, such that the behaviour of the reduced order system satisfactorily represents the behaviour of the full system. In projection based reduction schemes, a projection matrix, V, is selected such that its columns span the required subspace [4]. The reduced system is then formed from approximating the state vector x with $\hat{x}$ where $\hat{x} = \hat{V}x$. Consider a linear state-space representation:

$$\begin{aligned} \dot{x}(t) &= Ax(t) + Bu(t) \\ y(t) &= Cx(t), \end{aligned} \tag{2.9}$$

The state-space equations for the reduced system are then given as:

$$\dot{\hat{x}}(t) = \hat{A}\hat{x}(t) + \hat{B}\hat{u}(t)$$
$$\hat{y}(t) = \hat{C}\hat{x}(t), \tag{2.10}$$

where

$$\hat{A} = V^t A V, \quad \hat{B} = V^t B, \quad \hat{C} = C V.$$

In Krylov-based methods, the projection matrix is chosen to span the columns of the Krylov subspace

$$K_m((s_0 I - A)^{-1}, (s_0 I - A)^{-1} B) =$$
$$= \{(s_0 I - A)^{-1} B, (s_0 I - A)^{-2} B, \ldots, (s_0 I - A)^{-m} B\}.$$

The rationale for selection of this subspace is that it results in matching the first m moments of the original and reduced systems. Here s_0 is the point in the complex plane about which moments are matched. However, the Krylov-based reduction methods are preferable to direct moment matching techniques as the methods avoid the numerical problems arising in explicit moment matching.

For nonlinear systems, model reduction is not as straightforward as for linear systems. In this contribution, we look at some of the properties of linear systems with a view to adapting the perturbative representation of the nonlinear system so that a reduction strategy similar to that for linear systems can be applied to it.

2.4 Scale Invariance Property

Consider the behaviour of a linear system when $u \to \alpha u$. In this case, the output also changes as $y \to \alpha y$. We term this the scale invariance property which holds for linear systems. The result is that the Krylov-based reduction method is unaffected when $u \to \alpha u$. Similarly, if $x \to \beta x$, the reduction process is unaffected. However, nonlinear systems are not scale invariant. For example, consider the perturbative system under a rescaling of the input. That is consider $u \to \alpha u$. The $\underline{Bu}$ term of (1.6) transforms as:

$$\underline{Bu} \to \alpha \begin{bmatrix} 1 & & & \\ & \alpha & & \\ & & \alpha^2 & \\ & & & \ddots \end{bmatrix} \underline{Bu}. \tag{2.11}$$

It is evident from (2.11) that the scale invariance property does not hold. To enable application of linear theory to (2.6) would require that which is not the case as evident from (2.11). Consequently, linear model reduction techniques may not be applied directly to the perturbative representation and hence, a modification is required. To this end, a parameter μ is introduced with a view to explicitly accounting for the scale dependence of the nonlinear system. The role of μ is to bear the nonlinear properties of the system throughout the reduction process. Consider (2.6) and (2.7). The $\underline{Bu}$ term can be rewritten as:

$$\underline{Bu} = D\underline{BU},$$

where

$$D = \mathrm{diag}\,(1, \mu, \mu^2, \ldots),$$

with

$$\underline{U} = [u_1, \mu^{-1}u_2, \mu^{-2}u_3, \mu^{-2}u_4, \ldots].$$

for any nonzero function μ. If when $u_1(t) \to \alpha u_1(t)$, μ transforms as: $\mu \to \alpha\mu$, then transforms as:

$$\underline{U} \to \alpha\underline{U}. \tag{2.12}$$

It transforms in the same manner as the input to a linear system. The property in (2.12) is very important as it shows that to enable application of linear systems theory to (2.6), then the proper input to (2.6) is actually $\underline{U}$ and not $\underline{u}$.

An estimate for μ may be determined as follows: If $\mu = 0$, then the system in (2.6) is linear. Thus, μ must be proportional to the output due to the nonlinearity $y - y_1$, where $y_1 = Cx_1$ is the output from the linear part of (2.6). For the purposes of practical implementation of the reduction scheme, it is convenient to take μ as a constant parameter. Hence, the following is deemed an appropriate choice for μ:

$$\mu = \frac{|\overline{y - y_1}|}{T|\bar{u}|}, \tag{2.13}$$

where the bar denotes the average value of the waveform over the time interval $[0, T]$ for which the behaviour of the system is under examination, provided $\bar{u} \neq 0$. An exact optimal value for μ for a particular model parameterisation may be chosen from computer simulations for a particular 'test input' that is close to the inputs for which the system is designed. μ is then determined by minimising an error function using the Nelder–Mead algorithm. A suitable error function is the following:

$$err = \frac{\sqrt{\sum (f_{\mathrm{ex}} - f_{\mathrm{red}})^2}}{N}, \tag{2.14}$$

where f_{ex} is the output from the exact model and f_{red} is the output from the reduced model, N is the number of samples taken of f_{ex} to compute the error.

However, for most practical cases, the estimate in (2.13) suffices. Obviously, the reduced model approximates the input–output behaviour of the system locally. No conclusions however, can be drawn about its global behaviour.

2.5 KRYLOV Reduction of Perturbative Representation

The reduction process for the perturbative representation proceeds as follows: Let the approximation in (2.2) involve K terms. The dimension of the system representation in (2.6) is thus NK. Suppose it is required to reduce the system to dimension k. The Krylov space for the first-order response x_1 in (2.4) and (2.6) is formed as $K_1 = (A_1^{-1}, A_1^{-1}B)$ (s_0 is set to zero to simplify the explanation but this is not necessary). An orthogonal projection matrix, V_1 for the first-order system is formed from K_1 i.e. $\hat{x}_1 = V_1 x_1$. Now, the second-order system in (2.4) and (2.6) is formed as:

$$\begin{aligned}\dot{x}_2 &= A_1 x_2 + \mu A_2(\hat{x}_1 \otimes \hat{x}_1) = A_1 x_2 + \mu A_2(V_1 \otimes V_1)(x_1 \otimes x_1)\\ &= A_1 x_2 + \mu A_2(V_1 \otimes V_1)\hat{u}_2 = A_1 x_2 + B_2 \hat{u}_2. \qquad (2.15)\end{aligned}$$

This differs from the standard second-order system such as that presented by Phillips [4]. In the standard version, $\mu = 1$. However, results in section 6 will show that inclusion of the novel proposal for μ achieves greater accuracy.

The Krylov space for the linear system in (2.15) is then formed as $K_2 = (A_2^{-1}, A_2^{-1}B_2)$. An orthogonal projection matrix, V_2, is formed from K_2 and this matrix is used to reduce the second order system. The procedure for reducing the higher order terms in (2.4) and (2.6) , i.e. $x_3, \ldots,$ in the perturbative system is similar.

2.6 Illustrative Example

The circuit employed is the nonlinear ladder shown in Fig. 2.1. The number of nodes in the system is $n = 30$. The ladder in represents a heat flow model [7]. The voltage at the mth node represents the temperature on a rod at a distance proportional to m. The (input) voltage at node 1 represents the heat source. The nonlinearities represent the dependence of the conductivity on the temperature. The average voltage at all nodes is taken as the output and this represents the average temperature of the rod. Varying the circuit parameters corresponds to different spatial or environment

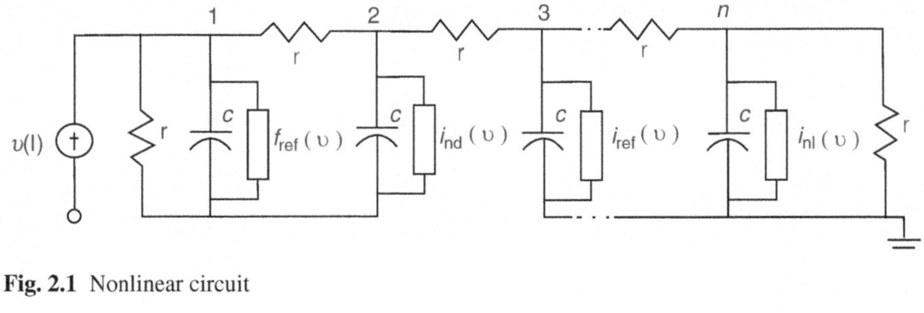

Fig. 2.1 Nonlinear circuit

conditions [7]. The nonlinear resistor introduces a quadratic nonlinearity at each node:

$$i_{\rm nl} = gv^2, \tag{2.16}$$

for $v > 0$. The parameters are $C = r = 1$. The strength of the nonlinearity is varied by varying g.

The dimension of the original state-space is $n = 30$. The perturbative representation (2.6) contains two terms, i.e. $K = 2$. The reduction process is performed from the representation (2.6) of order $nK = 60$ to a representation of order $k = 6$. The value is $\mu = 1.6443$. Fig. 2.2 shows the result for an exponential input e^{-t} from the reduced perturbative model for $g = 0.1$ superimposed on the result from a full nonlinear model of Fig. 2.1. The root mean square error between this result and that computed from a full model is 0.0026. The reduced model is equally accurate for other input signals. In order to confirm the value of inclusion of μ, the root mean square error is 0.0042 when $\mu = 1$.

As a second example, consider the 30-section nonlinear RC ladder shown in Fig. 2.3.

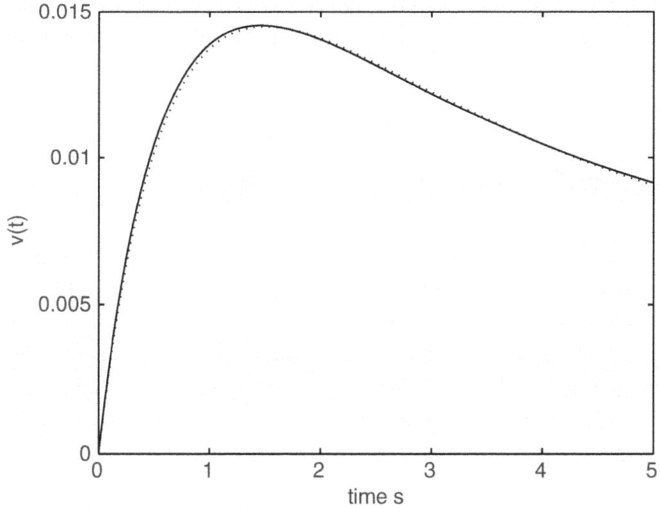

Fig. 2.2 Reduced perturbative model $g = 0.1$ (solid line = full model, dashed line = reduced model)

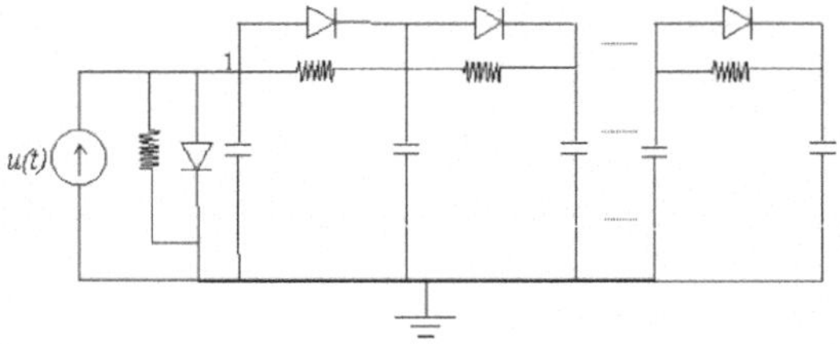

Fig. 2.3 Nonlinear RC ladder

The reduction process described in Section 2.5 is applied. The system is reduced from a dimension of 60 to a dimension of 6. The value of μ is determined from (1.13) as 0.6747. With this value of μ, the RMS error is 0.0037. With the standard approach of [4], the RMS error is 0.0076.

2.7 Conclusions

Krylov model reduction of perturbative representations of weakly nonlinear systems has been addressed (Fig. 2.4). The restriction to weak nonlinear systems arises as the size of the perturbative representation would grow to impractical levels for highly nonlinear systems. This restriction also applies to the bilinear representation and indeed is even greater owing to the larger size of a bilinear representation of the same order. This is the principal advantage of the perturbative representation compared to the bilinear representation–it is the much smaller in size. It is of size (nK) compared to the size of the bilinear representation $(\sim n^K)$. This results in reduced computational cost. The input–output mapping for nonlinear systems depends on inputs and is not scale-independent. To explicitly account for this dependence, a parameter is introduced into the perturbative representation. Results in this contribution indicate that inclusion of the parameter leads to greater accuracy.

Previous work [5, 6] has shown that the same approach also leads to improved accuracy in balanced truncation model reduction.

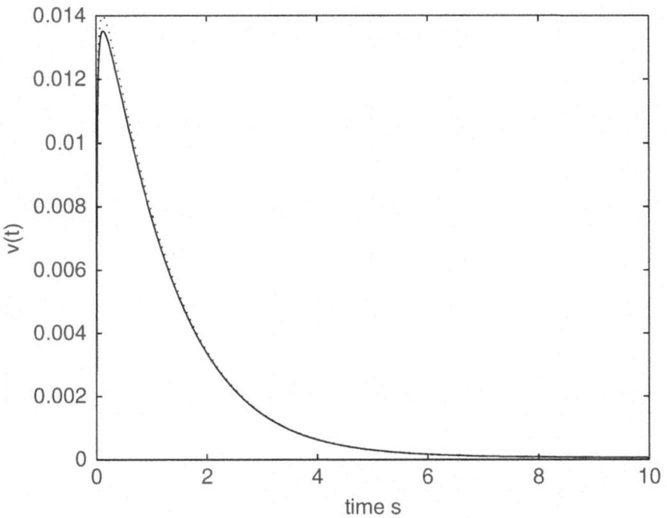

Fig. 2.4 Reduced perturbative model (solid line = full model, dashed line = reduced model)

Acknowledgements This material is based upon works supported by Science Foundation Ireland Science Foundation Ireland under Principal Investigator Grant No. 05/IN.1/I18.

References

1. B. Moore (1981). *Principal Component analysis in linear systems: Controllability, Observability and model reduction*, IEEE Transactions on Automatic Control, **AC-26**, 1, 17–31.
2. A.C. Antoulas, D.C. Sorensen and S. Gugercin (2001). *A survey of model reduction methods for large-scale systems*, Contemporary Mathematics, AMS Publications, Providence, RI.
3. C.A. Beattie and S. Gugercin (2007). *Krylov-based minimization for optimal H2 model reduction*, 46th IEEE Conference on Decision and Control, New Orleans.
4. J.R. Phillips (2003). *Projection-based approaches for model reduction of weakly nonlinear, time-varying systems*, IEEE Transactions on computer-aided design of integrated circuits and systems **22**, 2, 171–187.
5. M. Condon and R. Ivanov (2005a). *Nonlinear systems-Algebraic Gramians and Model Reduction*, COMPEL: The International Journal for Computation and Mathematics in Electrical and Electronic Engineering **24**, 1, 202–219.
6. M. Condon and R. Ivanov (2005b). *Balanced Model Reduction from a Perturbative Representation of Weakly Nonlinear Systems*, NOLTA 2005, Bruges.
7. T. Veijola and L. Costa (1998). *Combined electrical and thermal circuit simulation using APLAC*, Circuit Theory Laboratory Report Series, No. CT-34, Helsinki University of Technology.

Chapter 3
Investigation of a Multizone Drift Doping Based Lateral Bipolar Transistor on Buried Oxide Thick Step

Sajad A. Loan, S. Qureshi, and S. Sundar Kumar Iyer

Abstract A 2D numerical simulation study of a multizone step doped lateral bipolar junction transistor (LBJT) on buried oxide thick step (BOTS) is performed. The combination of multizone doping and BOTS has been used for increasing the breakdown voltage of the device. The steps in doping and step in oxide result in the creation of additional electric field peaks in the collector drift region which increases the uniformity of lateral surface electric field and hence the breakdown voltage. Numerical simulations have revealed that a LBJT with two doping zones and a thick buried oxide results in 200% increase in breakdown voltage than the conventional device. Increasing the number of zones to three from two makes the breakdown voltage 260% higher than the conventional one. An improvement in tradeoff between the breakdown voltage and the on-resistance has been observed in the device. The multizone doping also reduces the kinks in the device characteristics.

Keywords Multizone step · multizone doping · lateral bipolar junction transistor · buried oxide thick step · breakdown voltage

3.1 Introduction

The SOI-BiCMOS is emerging as a promising technology for realization of wireless system-on-chip. This technology offers advantages in terms of reduction of parasitic capacitance, high quality isolation and reduction in crosstalk [1]. However, the problem of compatibility lies in the integration of vertical bipolar device with SOI-CMOS [2]. This problem has been reduced by using lateral bipolar device as an alternative to the vertical device. The lateral bipolar junction transistor (LBJT) offers low parasitic capacitance, promises low power consumption and high breakdown voltage.

S.A. Loan (✉)
Electrical Engineering Department, Indian Institute of Technology, Kanpur, India,
E-mail: sajad@iitk.ac.in

S.I. Ao and L. Gelman (Eds.), *Advances in Electrical Engineering and Computational Science*, Lecture Notes in Electrical Engineering.

To obtain the large breakdown voltage, the lateral surface electric field distribution along the silicon surface must be uniform [3]. Several ideas have been used to enhance the breakdown voltage of the lateral bipolar devices. These include Reduced Surface Field (RESURF) principle [4], fully depleted collector drift region [5], graded drift region [6, 7], semi insulating polysilicon (SIPOS) passivation layers [8]. The concept of extended box has been used to increase the breakdown voltage of lateral Schottky transistor [9]. By using linearly varying doping concentration and linearly varying oxide thickness, the lateral surface component of the electric field in the device can be made perfectly uniform and the maximum breakdown voltage can be obtained. But it is extremely difficult to fabricate a device with linearly varying doping concentration and simultaneously having linearly varying oxide thickness [10, 12].

In this paper, numerical simulation of LBJT with multizone step doped drift region on buried oxide thick step (BOTS) has been performed. we have studied and simulated three types of the devices in this work. In type one, the device is having two doping zones but without BOTS. This is the conventional device. The second type again uses two doping zones but is having BOTS, called as two zone proposed (2ZP) device. The third type uses three doping zones and BOTS, called as three zone proposed (3ZP) device. To increase the breakdown voltage and at the same time retain the ease of fabrication of the device, the linearly varying doping and the linearly varying oxide thickness has been replaced with step profile of both. The multizone step doping and thick step oxide results in increased uniformity in the lateral surface electric field in the drift region, reduction of base-collector junction electric field by using lower doping concentration near the junction and enhancement of collector breakdown voltage. The simulation results using MEDICI [13] has shown that the 2ZP device has a breakdown voltage of 200% higher than the conventional device. A 3ZP device possesses breakdown voltage 260% higher than the conventional one. It has been observed that increasing number of zones further increases breakdown voltage marginally but increases complexity of the device significantly. The high breakdown voltage in the proposed devices can be obtained even at high drift doping concentration. This improves the tradeoff between the breakdown voltage and the on-resistance of the device.

3.2 Ideal, Conventional and Proposed Devices

The four structures of lateral bipolar transistor employed to obtain maximum possible breakdown voltage is shown in Fig. 3.1. Figure 3.1a employs a linearly varying drift doping concentration and a linearly varying buried oxide thickness in a lateral bipolar device. This is an ideal structure for obtaining highest breakdown voltage and best performance in terms of on-resistance and speed of the operation of the device. A highest breakdown voltage is obtained in such a structure even at high drift doping concentration. But, the problem associated with the ideal structure is its fabrication complexity. The designing and fabrication of such a device is extremely

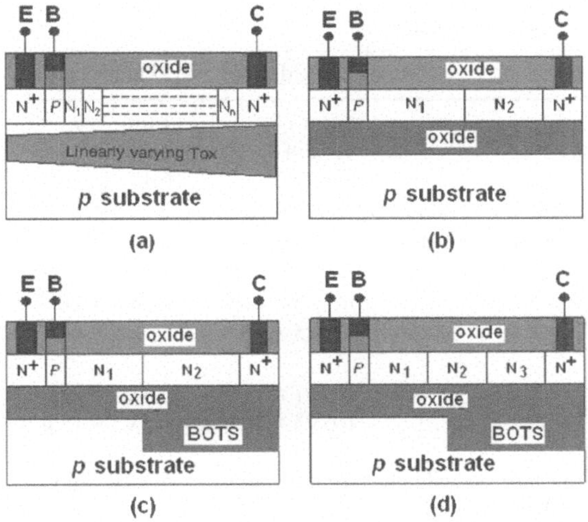

Fig. 3.1 (**a**) An ideal lateral bipolar transistor (**b**) Conventional LBJT (**c**) Two zone proposed (2ZP) LBJT (**d**) Three zone proposed (3ZP) device

difficult as it needs large number of masking and lithographic steps, which renders its fabrication practically impossible. A simple and conventionally used lateral BJT on SOI is shown in Fig. 1b. This structure is easier to fabricate in comparison to the ideal one, but at the cost of increase in on-resistance and the decrease in breakdown voltage. The breakdown voltage is also poor in the conventional structure. The proposed practical alterations to the structure in Fig. 3.1a, that is, 2ZP and 3ZP are shown in Fig. 3.1c and d respectively. These structure retains the advantages of the ideal structure, that is, high breakdown voltage and improved tradeoff between breakdown voltage and on-resistance. The proposed structures are easier to fabricate than the ideal one, but are more complex than the conventional one.

3.3 Simulation Results and Discussion

Extensive device simulations were carried out using the device simulator MEDICI. The various models activated in the simulation are analytic, *prpmob*, *fldmob*, *consrh*, *auger* and *BGN* [13]. The mobility models *prpmob* and *fldmob* specify that the perpendicular and parallel electrical field components are being used. The concentration dependent Shockley-Read-Hall model (*consrh*) and Auger recombination model (*auger*) are activated. Since the device is bipolar, having PN junctions, the band gap narrowing effect has been considered by using *BGN* model. The various device and simulation parameter used are shown in Table 3.1. The current gain and Gummel plot curves for the conventional and 3ZP devices are shown in Fig 3.2. The common emitter current gain of the all devices is chosen to be identical 50) by an appropriate choice of various doping concentrations. This is being done for better comparison of the breakdown voltage. The Gummel plot for the proposed and

Table 3.1 Device and simulation parameters

Parameters	Conv. devices	2ZP devices	3ZP devices
Si film thickness	1 μm	1 μm	1 μm
Oxide thickness	1.5 μm	1.5 μm	1.5 μm
Emitter length	6.4 μm	6.4 μm	6.4 μm
Emitter doping conc.(cm^{-3})(*n* type)	5×10^{19}	5×10^{19}	5×10^{19}
Substrate doping conc.(cm^{-3})(*p* type)	5×10^{13}	5×10^{13}	5×10^{13}
Collector doping conc.(*n* type)(cm^{-3})	$n^+ = 5 \times 10^{19}$ $n_1 = 3 \times 10^{15}$ $n_2 = 5 \times 10^{15}$	$n^+ = 5 \times 10^{19}$ $n_1 = 3 \times 10^{15}$ $n_2 = 5 \times 10^{15}$	$n^+ = 5 \times 10^{19}$ $n_1 = 3 \times 10^{15}$ $n_2 = 5 \times 10^{15}$ $n_3 = 7 \times 10^{15}$
Drift region length	$L_{DN1} = 15$ μm $L_{DN2} = 19$ μm	$L_{DN1} = 15$ μm $L_{DN2} = 19$ μm	$L_{DN1} = 10$ μm $L_{DN2} = 11$ μm $L_{DN3} = 13$ μm
TAUN0, TAUP0 (SRH electron and hole carrier life time coefficients)	5×10^{-6} (s)	5×10^{-6}(s)	5×10^{-6}(s)
VSURF at poly base contact (Surface recombination velocity)	3×10^{6} cm/s	3×10^{6} cm/s	3×10^{6} cm/s
Metal (*Al*) work function	4.17 eV	4.17 eV	4.17 eV

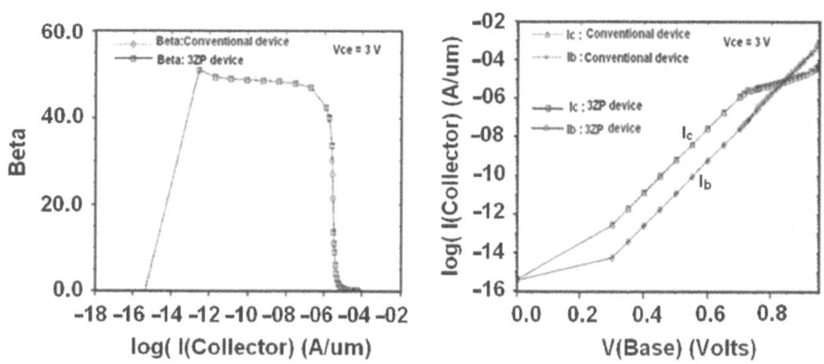

Fig. 3.2 (**a**) Beta versus collector current (**b**) Gummel plots for the the conventional and 3ZP device)

conventional devices are more or less similar, that is why they are overlapping for most of the base voltage range. Since the oxide is fully covering the base region in both the cases, the area of the base remains constant in all cases and hence the base current. The output characteristics of the conventional, 2ZP and 3ZP devices are shown in Fig. 3.3a, b, c respectively. The various parameters used in obtaining these characteristics are shown in Table 3.1. In all structures the buried oxide is 1.5 μm thick and BOTS thickness is varied from 1 to 9 μm.

After comparing the simulation results we have found that the common emitter breakdown voltage with base open (BV_{CEO}) is significantly higher for three zone proposed (3ZP) and two zone proposed (2ZP) devices than the conventional device. The BV_{CEO} of the conventional device is 124 V. Using same device parameters

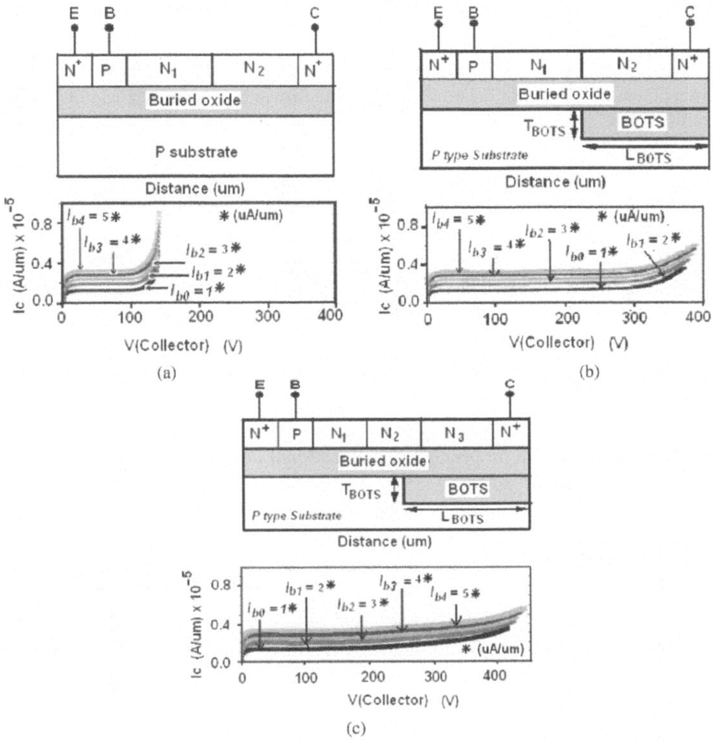

Fig. 3.3 Output characteristics of (**a**) conventional device (**b**) 2ZP device (**c**) 3ZP device

as for the conventional device, the 2ZP device with BOTS length of 34 μm and thickness of 6 μm, the BV_{CEO} is 370 V. Enhancing the number of zones to three, that is, the 3ZP device, the breakdown voltage increases to 440 V. This shows that the breakdown voltage in 3ZP device is 260% higher than the conventional device and more than 20% higher than 2ZP device.

The reason behind this enhancement in breakdown voltage is explained in Fig. 3.4, which shows that the lateral surface electric field is more uniform in the proposed devices than in the conventional device. The conventional device is having least uniformity in the electric field and the 3ZP device is having highest uniformity in the electric field. The enhanced uniformity in surface electric field is due to the presence of steps in doping and step in oxide [10]. These steps result in the generation of additional electric peaks in the collector drift region, which are shown in Fig. 3.4. These generated peaks pull down the peaks at the edges of the collector drift region. Further, the buried oxide thick step (BOTS) reduces the peak electric field at n_2-n^+ junction in 2ZP device and n_3-n^+ in 3ZP device by redistributing applied potential in the collector drift region and in thick oxide. The analytical approach deduced in [12] explains the creation of the electric field component due to BOTS and increase in breakdown voltage.

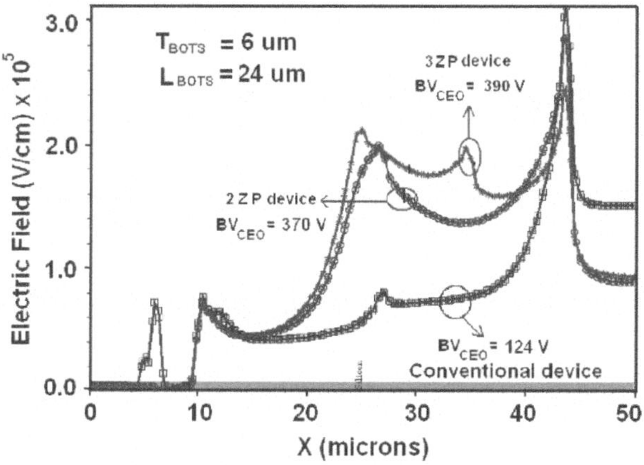

Fig. 3.4 Electric field profile in proposed and conventional devices

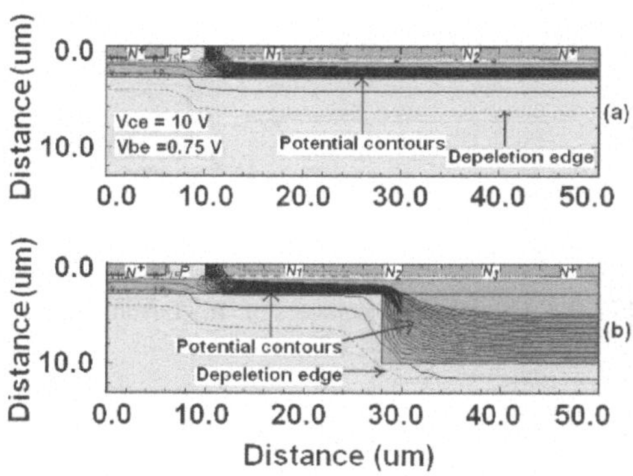

Fig. 3.5 (**a**) Potential contours in the conventional device. (**b**) Potential contours in the proposed device

The potential contours of the conventional and the proposed devices are shown in Fig. 3.5a and b respectively. The crowding of the potential contours along the oxide layer as shown in Fig. 3.5a results in lowering the breakdown voltage in the conventional device. In the proposed device, BOTS helps in reducing the crowding of the potential contours as shown in Fig. 5b. This reduction in crowding of the potential contours in BOTS, makes it possible to apply more voltage to the device before it breaks down.

Figure 6a and b show the electric field profiles at top and bottom of the SOI film respectively. The electric field profile in the oxide layer is shown in Fig. 6c. These figures also show how the location of electric field peaks vary with BOTS length. The increase in breakdown voltage can be attributed to how the electric field peaks are distributed in the device for different values of BOTS length. As can be seen from these figures that there is an optimum length which results in maximum uniformity of the lateral surface electric field and hence the maximum breakdown

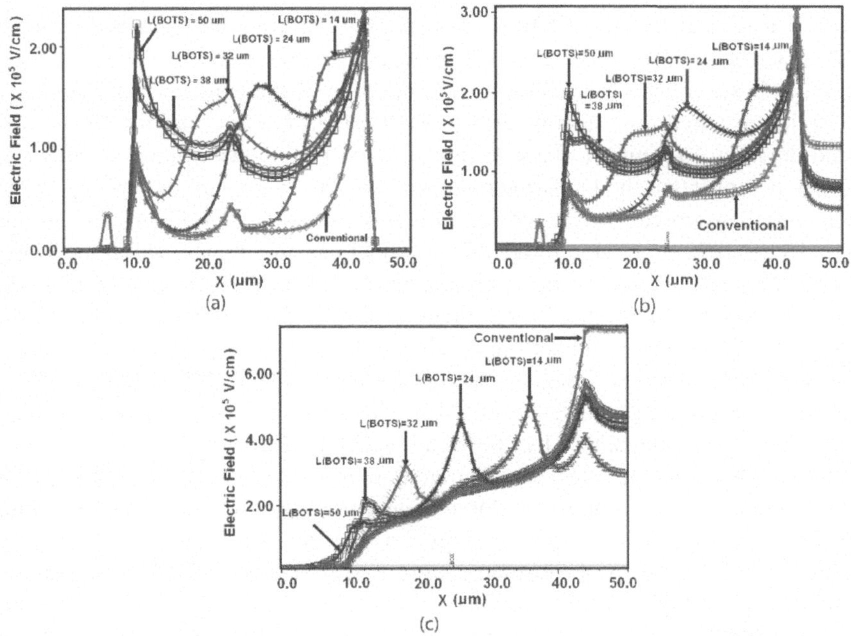

Fig. 3.6 Electric field profile in the proposed device at (**a**) top (**b**) bottom of silicon film (**c**) inside the oxide layer for different values of L_{BOTS}

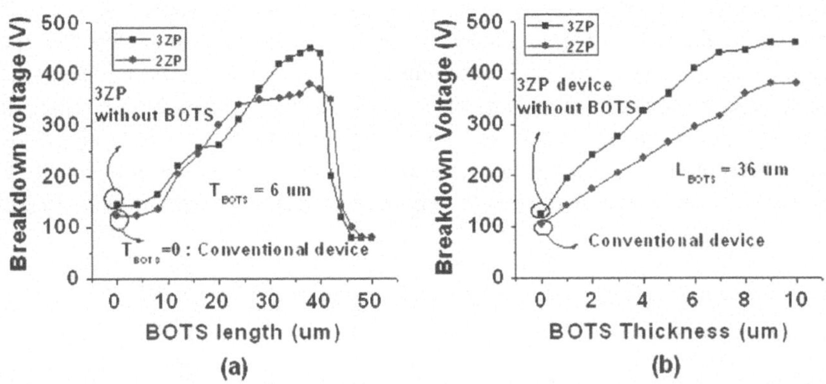

Fig. 3.7 Effect of variation in (**a**) BODS length and (**b**) BODS thickness on breakdown voltage in proposed devices

voltage. It is observed that for L_{BOTS} = 34 µm and T_{BOTS} = 6 µm, a maximum breakdown voltage of 440 and 370 V is obtained in 3ZP and 2ZP devices respectively for specific values of doping concentration in different regions. Figure 3.7 gives the impact of BOTS length and thickness on the breakdown voltage of the proposed device. It is quite obvious from the Fig. 7a that an optimum length of the BOTS is needed to get the maximum breakdown voltage of the device. From this plot it is clear that for the BOTS length of 30–40 μm, the breakdown voltage 2ZP device is >200% higher than that of the conventional device. For 3ZP device, the breakdown voltage is >260% higher than that of the conventional device. At the optimum length, the lateral surface electric field is more uniform in the proposed devices. The impact of oxide thickness on breakdown voltage is shown in Fig. 7b. It shows that the breakdown voltage first increases with increase in BOTS thickness, then subsequently saturates. This is because thick BOTS helps to sustain high electric field to a value lower than the field causing breakdown in the oxide. Since both horizontal and vertical components of the electric field contribute to the avalanche process, increasing BOTS thickness further does not improve breakdown voltage [9]. The simulation results have shown that for T_{BOTS} = 8 μm and L_{BOTS} = 36 μm, the breakdown voltage of the 3ZP device is 445 V, 2ZP device is 380 V and that of the conventional device is 124 V. This shows that the breakdown voltage in the 3ZP device is enhanced by 260% and in 2ZP by 206%.

Figure 3.8 shows the effect of substrate doping on the breakdown voltage. It is observed that there is an optimum doping concentration which gives the maximum breakdown voltage. The effect is explained in Fig. 8b, which shows electric field profile at specified values of substrate doping concentration in a 2ZP and in conventional device. At high substrate doping concentration, the electric field builds up at the n_2-n^+ junction. This electric field peak breaks down the device at low voltage. At low substrate doping, the electric field builds up at the collector-base junction, and ultimately breakdown occurs due to this electric field peak.

Figure 3.9a shows the effect of varying doping concentration in two zones of drift region on breakdown voltage in the two zone proposed device. It is clear that there are optimum values of drift region doping concentration in the two zones, which gives maximum breakdown voltage. The optimum drift region doping concentration in N_1 and N_2 is about 6×10^{15} cm^{-3}. The electric field profile at specified values of doping concentration in two zones N_1 and N_2 is shown in Fig. 3.9b. For $n_1 = 3 \times 10^{16}$ cm^{-3}, the electric field at collector-base junction is high and results

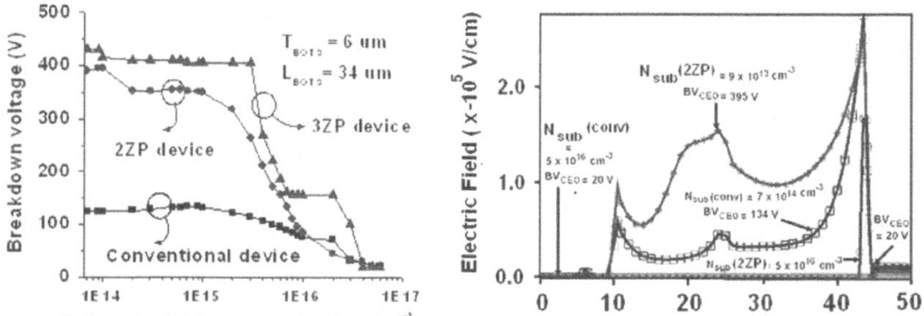

Fig. 3.8 (**a**) Breakdown voltage versus substrate concentration in conventional and proposed devices. (**b**) Electric field profile at specified values of substrate doping concentration in 2ZP and in conventional devices

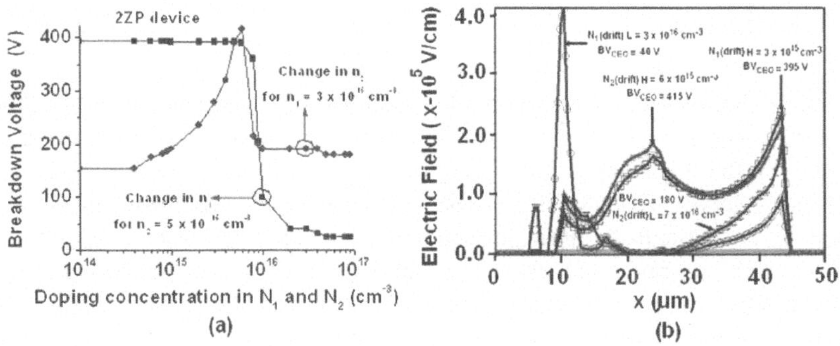

Fig. 3.9 (**a**) Breakdown voltage versus drift doping concentration in 2ZP device. (**b**) Electric field profile at specified values of drift region doping concentration in 2ZP device

Fig. 3.10 Thermal analysis of the proposed and conventional devices

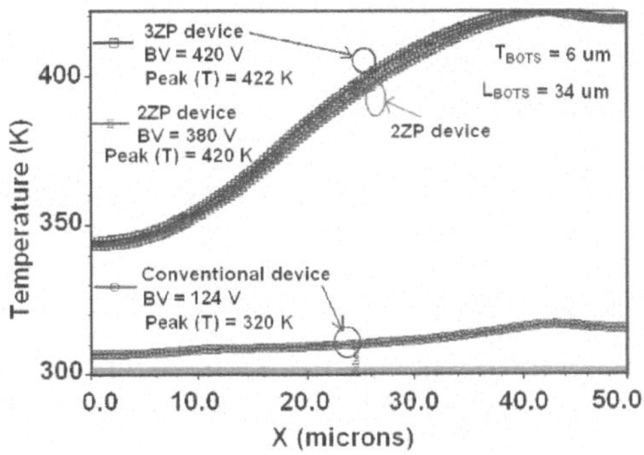

in breaking down the device at 40 V. Similarly, at doping concentration of $n_2 = 7 \times 10^{16}$ cm^{-3}, the electric field peak shifts to n_2-n$^+$ junction and hence breaks down the device at 180 V. The electric field profile is more uniform for optimum values of $n_1 = n_2 = 6 \times 10^{15}$ cm^{-3}.

The thermal analysis of the proposed and the conventional structures has been performed for a base current of $I_b = 1 \times 10^{-16}$ A/μm, L_{BOTS} = 34 μm and T_{BOTS} = 6 μm, as shown in Fig. 3.10. The peak temperature of the conventional device at its breakdown voltage of 124 V is 320 K. In the 2ZP device it rises to 420 K, when operated at its breakdown voltage (BV) of 380 V. The temperature in the 3ZP device is slightly higher than 2ZP device, although having higher breakdown voltage than 2ZP device. The reason behind this temperature enhancement is the presence of thick oxide under collector and operation at high voltage. However, as the breakdown voltage does not increase continuously with increase in oxide thickness, temperature of the device can be kept in safe region by choosing optimum oxide thickness.

3.4 Conclusion

Numerical simulation of a novel lateral bipolar junction transistor on SOI has been performed. The novelty in the device is the combination of two/three zone step doping with the thick step oxide. It is known that the linearly varying drift doping and linearly varying oxide thickness results in maximum breakdown voltage. However, it is very difficult to fabricate such a device. In this paper a device with high breakdown voltage and relatively easy to fabricate has been proposed. The BV_{CEO} has been maximized by choosing appropriate values of T_{BOTS} and L_{BOTS}. The breakdown voltage has been found to be dependent on substrate doping and drift region doping concentration. In this simulation study a more than 260% increase in breakdown voltage has been observed in the three zone proposed device.

References

1. Ning T. H., "Why BiCMOS and SOI BiCMOS" IBM Journal of Res and Dev, vol. 46, no. 2/3, pp. 181–186, Mar./May 2002.
2. I-Shun Michael Sun et al., "Lateral High speed Bipolar transistors on SOI for RF SoC applications" IEEE Transactions on Electron Decives, vol. 52, no. 7, pp. 1376–1383, July 2005.
3. Roy S. D. and Kumar M. J., "Realizing High Voltage Thin Film Lateral Bipolar Transistors on SOI with a Collector Tub" Microelectronics International, 22/1, pp. 3–9, 2005.
4. Huang Y. S. and Baliga B. J., "Extention of resurf principle to dielectrically isolated power devices," Proceedings of the 3rd International Symposium on Power Semiconductor Devices and ICs, pp. 27–30, April 1991.
5. Kumar M. J. and Bhat K. N., "Collector recombination lifetime from the quasi-saturation analysis of high voltage bipolar transistors," IEEE Trans. Electron Devices, vol. 37, no. 11, pp. 172–177, 1995.
6. Zhang S., Sin J. K. O., Lai T. M. L. and KO P. K., "Numerical modeling of linear doping profiles for high-voltage think film SOI devices," IEEE Trans. Electron Devices, vol. 46, no. 5, pp. 1036–1041, 1995.
7. Cao G. J., De Soza M. M. and Narayanan E. M. S., "Resurfed Lateral Bipolar Transistor for High-Voltage, High-Frequency Applications" Proceedings of the 12th International Symposium on Power Semiconductor Devices and ICs. Toulouse, France, pp. 185–187, 2000.
8. Jaume D., Charitat G., Reynes J. M. and Rossel P., "High-Voltage planar devices using filed plate and semi-resistive layers," IEEE Transactions on Electron Devices, vol. 38, no. 7, pp. 1478–1483, July 1991.
9. Kumar M. J. and Roy S. D., "A New High Breakdown Voltage Lateral Schottkey Collector Bipolar transistor on SOI: Design and Analysis" IEEE Transactions Electron Device, vol. 52, no. 11, Nov. 2005.
10. Sunkavalli R et al., "Step drift doping profile for high voltage DI lateral power devices" Proceedings of IEEE International SOI Conference, pp. 139–140, Oct. 1995, 2005.
11. Luo J et al., "A high performance RF LDMOSFET in thin film SOI technology with step drift profile", Solid State Electronics, vol. 47, pp. 1937–1941, 2003.
12. Kim I. J et al., "Breakdown Voltage improvement for thin film SOI Power MOSFET's by a Buried Oxide Step Structures" IEEE Electron Device Letters, vol. 15, no. 5, pp. 148–150, May 1994.
13. TMA MEDICI 4.2 . Technology Modeling Associates Inc. Palo Alto, US 2006.

Chapter 4
Development of a Battery Charging Management Unit for Renewable Power Generation

Yong Yin, Jihong Wang, Xing Luo, and Shen Guo

Abstract To alleviate the intermittences in renewable power generation, use of energy storage is an evitable route of choice. Batteries are nowadays the most realistic means in energy storage. The charging and discharging of batteries affect the battery life time and energy efficiency greatly. At present, batteries have been used to provide electricity for lighting, radios, TVs and communications. The use of a small size wind generators with the assistance of batteries could provide local alternative means to enable more households to have access to electricity. The work reported in the chapter mainly focuses on the development of a battery control unit. The energy conversion modes are analyzed. Then the working principles of chargeable batteries are probed into. Control strategies based on fuzzy logic have been suggested to ensure the optimal performance in battery charging and discharging. A charging/discharging control module using embedded system technology is designed to charge the battery by following the optimal charging/discharging curve of the battery. The whole unit is controlled and monitored by an embedded microprocessor system.

Keywords Renewable power generation · Battery Charging · battery life time · control module · embedded system

4.1 Introduction

Renewable energy refers to sustainable natural resources such as sunlight, wind, wave and tides, which are naturally replenished [1]. Renewable power generation systems have been paid a great attention recently due to the pressure from the global

---

J. Wang (✉)
School of Electronic, Electrical and Computer Engineering, University of Birmingham, Birmingham, B15 2TT, UK,
E-mail: j.h.wang@bham.ac.uk

S.I. Ao and L. Gelman (Eds.), *Advances in Electrical Engineering and Computational Science*, Lecture Notes in Electrical Engineering.

warming. China has abundant inland and offshore wind energy resources, providing potential conditions for various types of capacity, in-grid or off-grid wind stations. According to a new study from an environmental group, China's southern province of Guangdong could support 20 GW of wind generating capacity by 2020, providing as much as 35,000 GWh of clean electrical energy annually, which is equivalent to 17% of Guangdong's total current power demand [2]. By end of year 2005, China had built 59 wind farms with 1,854 wind turbine generators and a 1,266 MW wind power installed capacity [2]. In China, a large number of remote rural or mountainous inhabitants have no access to electricity, but in these areas there are abundant natural energy resources such as wind or solar energy. To alleviate the intermittences in renewable power generation, use of energy storage is an evitable route of choice. Batteries are nowadays the most realistic means in energy storage. The charging and discharging of batteries affect the battery life time and energy efficiency greatly. At present, batteries have been used to provide electricity for lighting, radios, TVs and communications. The use of a small size wind generators with the assistance of batteries could provide local alternative means to enable more households to have access to electricity. Generally, the framework of the energy conversion process in a small-scale renewable energy generation system can be described by Fig. 4.1. The energy conversion process illustrated in Fig. 4.1 can be divided into eight steps.

Many research works such as the modeling of the wind turbine, the control strategies for the three-phase generator have been conducted and the research results were reported in many publications [3, 4]. The work reported in the chapter mainly focuses on the development of a battery control unit. The energy conversion modes are analyzed. Then the working principles of chargeable batteries are probed into. Control strategies based on fuzzy logic have been suggested to ensure the optimal performance in battery charging and discharging. A charging/discharging control module using embedded system technology is designed to charge the battery by following the optimal charging/discharging curve of the battery. The whole unit is controlled and monitored by an embedded microprocessor system.

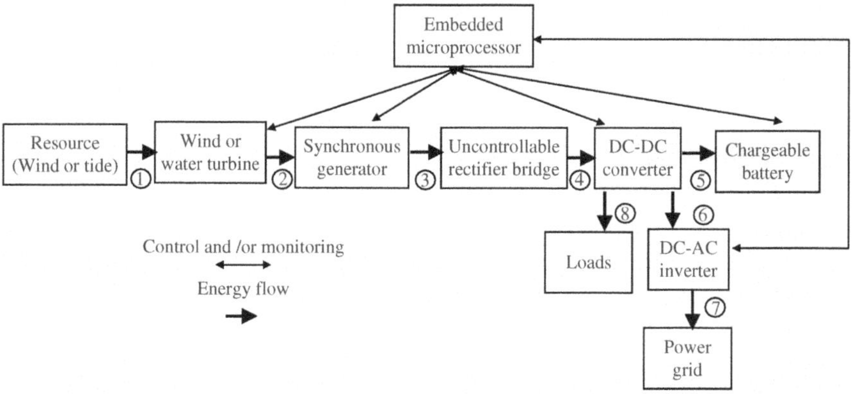

Fig. 4.1 Framework of the renewable energy generation system

4.2 Energy Conversion Modes

If the output electrical power is over the load demand, the surplus is used to charge the battery. Provided that the load does not need all the output power, and the battery is fully charged, the superfluous power is then sent to the main grid. In the system, the output electrical power is provided to the loads with priority. There exist five possibilities with the relationship among P_{output}, P_{load}, $P_{battery}$ and P_{Grid}, and five working modes of energy conversion in the system is formed accordingly [3], in which P_{output}represents the electrical power output, P_{load} the power consumption of the loads, $P_{battery}$the power charged into or discharged from the battery, $P_{Grid}$the power supplied to the main grid. The five modes are:

Mode1: $P_{output} = 0,\ P_{load} = 0,\ P_{battery} = 0,\ P_{Grid} = 0$ – the system is idle.
Mode 2: $P_{output} > P_{load},\ P_{output} - P_{load} < P_{battery},\ P_{Grid} = 0$ – the generated electricity is more than the load demand and the battery is charged to less than its full capacity.
Mode 3: $P_{output} > P_{load},\ P_{output} - P_{load} > P_{battery},\ P_{Grid} > 0$ – the generated electrical power is more than the load demand and the battery is over charged so the extra power is sent to the grid.
Mode 4: $P_{output} < P_{load},\ P_{load} < 0,\ P_{Grid} = 0$ – the generated electrical power is less than the load demand and the battery is discharged.
Mode 5: $P_{output} < P_{load},\ P_{load} = 0,\ P_{Grid} = 0$ – the generated electrical power is much less than the demanded energy supply and the battery is fully discharged.

The battery must be disconnected from the system to avoid over discharge.

From the discussion, it can be seen that the battery works in three statuses: disconnected from the system, charged by the renewable power or discharged to supply power to the loads, as shown in Fig. 4.2. The status of the battery is depended on the working modes of the system, and shifts according to different modes [4].

4.3 Lead Acid Battery

The most commonly used batteries are divided into three categories, alkali battery, nickel iron battery and lead acid battery. The lead acid battery is chosen for our system. The working principles of the lead acid battery can be found in many

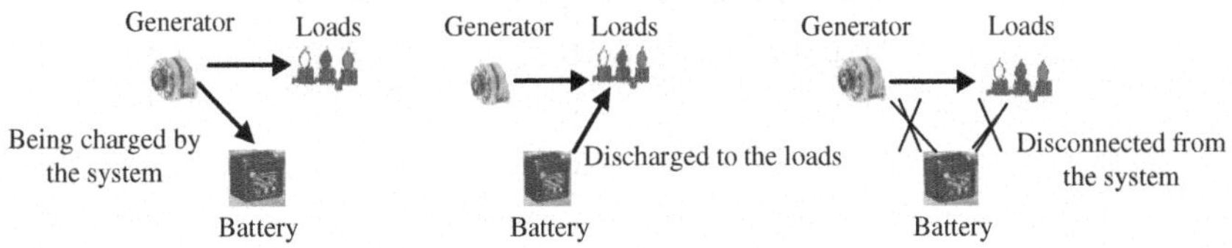

Fig. 4.2 Three working status of the battery

literatures [5, 6]. In 1960s, an America scientist Mass has put forward an optimal charging curve for the battery based on the lowest output gas rate, as shown in Fig. 4.3. If the charging current keeps to the track of the optimal curve, the charging hours can be sharply cut down without any side effect on the capacity and life-span of the battery. The Ampere-hour rule for charging the Lead-acid batteries can be considered as the most efficient charging approach, considering the charging time and the excellent performance provided by this method to maintain the life-span of the battery. One example of the application of the Ampere-hour rule to charge battery is shown in Fig. 4.4. More details in charging method for lead-acid batteries was discussed in reference [5].

When the battery is charged, the charging current I_c, the charging voltage (port voltage) U_c, the potential difference E_b between the positive plate and the negative plate of the battery and the internal resistor R_b of the battery has the following

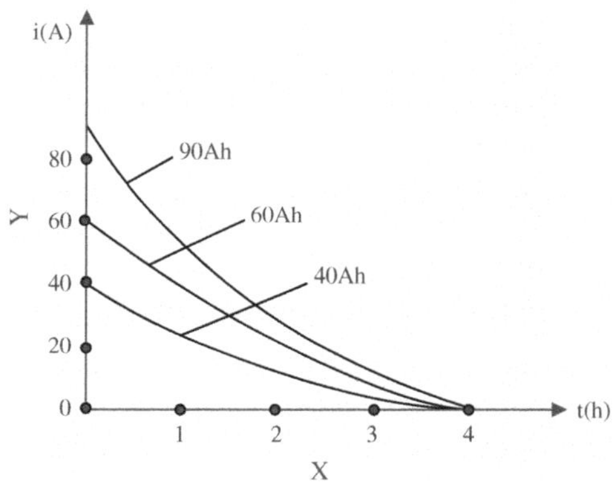

Fig. 4.3 Optimal charging curve of a battery

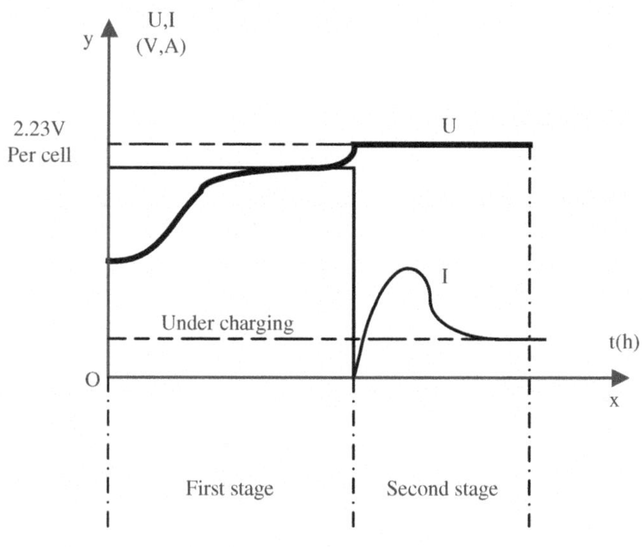

Fig. 4.4 Application of the ampere-hour rule to charge battery

relationship:

$$I_c = \frac{U_c - E_b}{R_b} \tag{4.1}$$

The set point of the charging current is closely related with the capacity of the battery [7]. With the potential difference of the battery gradually increasing, the charging voltage rises with it so that the battery can be charged according to the formula (4.1), but it kept below a certain maximum level as shown in the part one of Fig. 4.4. The second stage starts when the battery voltage reaches a certain level, which is defined by the characteristics of the battery. During this phase, the control variable switches over from current to voltage. The voltage is maintained almost unchanged at a given value, while the current is gradually decreasing until it drops to a value set at the initial stage [8].

4.4 Battery Charging and Discharging Strategies

In general, frequently charging and discharging, overcharging or insufficiently charging a battery will shorten the battery's life time. The problem associated with wind power is when and how the battery should be charged to provide the best energy efficiency for the system and to prolong the battery life time. If the output electrical power is excessive for the consumption of the loads, the surplus will be provided to charge the battery. As shown in Fig. 4.5, from the period t1 to t2 and t7 to t8, the wind power is more than the load consumption, and it lasts a relatively long period of time, so the battery may be charged, while from t3 to t4 or t5 to t6, even though the load can't use up the wind power, it should not change the battery to avoid insufficient or too frequent charging.

To achieve the optimal charging and discharging performance, fuzzy logic control algorithms can be a suitable choice for the management of charging and discharging process. The main advantages of Fuzzy control presents in its feature – including human experiences and decision making into the control logic ([9, 10]).

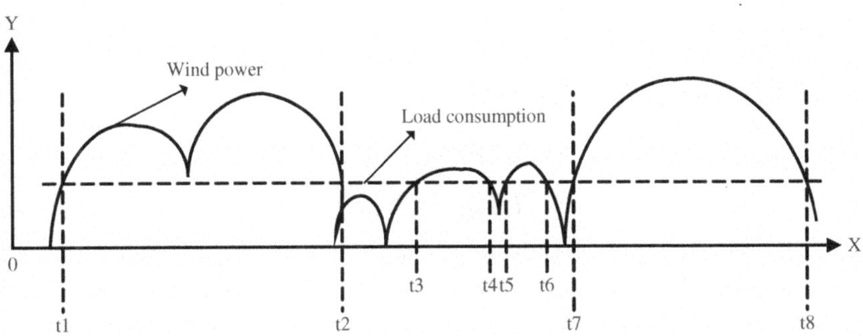

Fig. 4.5 Relationship between the wind power and the load consumption

Fuzzy logic control can be well implemented to a low-cost hardware system and easily updated by amending rules to improve performance or adding new features.

4.4.1 Membership Function Database

Referred to the Ampere-hour charging curve of the battery in Fig. 4.4, the charging process seems infinite. At the beginning of charging process, the charging current is quite large, while it drops quickly with time elapses. During this period, most electrical power has been converted into chemical energy. At the end of the charging process, the current is close to zero and hardly changes. In general, when the charging capacity of the battery reaches to 90% of its rated value, the charging process is considered to complete.

In this system, there are four input variables. $\Delta P = P - Pn$ is the difference of the wind power P and the load consumption Pn; $\Delta P' = d\Delta P/dt'$ represents the changing rate of ΔP; $\Delta T = Tn - T$ the relative temperature of the battery to the surrounding temperature; $\Delta T' = d\Delta T/dt$ the changing rate of ΔT. The output variable is the charging voltage U. So, the control function can be described as:

$$U = F(\Delta P, \Delta P', \Delta T, \Delta T') \tag{4.2}$$

The general fuzzy rules can be represented by (3), which is illustrated in Fig. 4.6.

If (ΔP is ... and $\Delta P'$ is ... and ΔT is ... and $\Delta T'$ is ...)

Then (U is ...).. (4.3)

To reduce the complexity of the controller, an improvement on organization of the fuzzy rules has been considered. According to the control function in (4.2) and (4.3), the fuzzy rules can be organized into two separate parts in formulae (4.4) and (4.5).

$$U1 = F1(\Delta P, P') \tag{4.4}$$

$$U2 = F2(\Delta T, T') \tag{4.5}$$

The fuzzy rules for them can be described as follows.

If (ΔP is ... and $\Delta P'$ is ...) then ($U1$ is...)............................ (4.6)

If (ΔT is ... and $\Delta T'$ is ...) then ($U2$ is...)............................ (4.7)

$$U1 \otimes U2 = U \tag{4.8}$$

The improved fuzzy rules can be demonstrated with the block diagram in Fig. 4.7.

The improved version is much simpler, which can be implemented with two control loops. The outer loop is based on the ΔU and $\Delta U'$, while the inner loop is on the basis of the ΔT and $\Delta T'$. The controller designed by following the above fuzzy rules can be implemented with the structure as shown in Fig. 4.8.

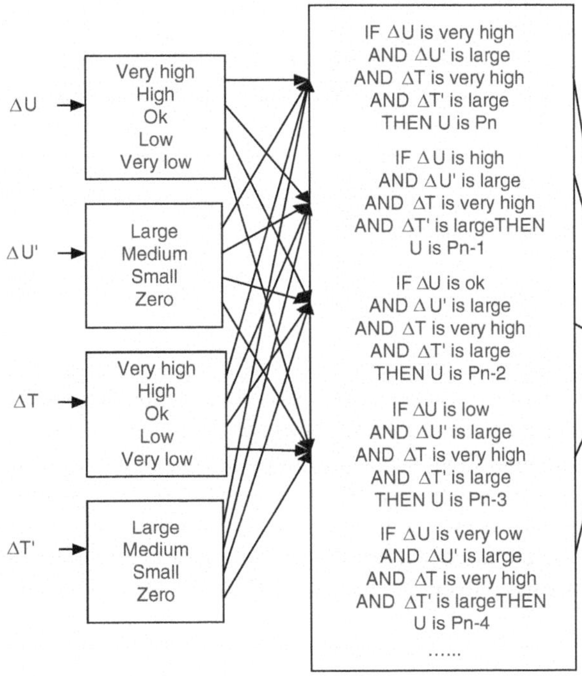

Fig. 4.6 Block diagram of the fuzzy rules

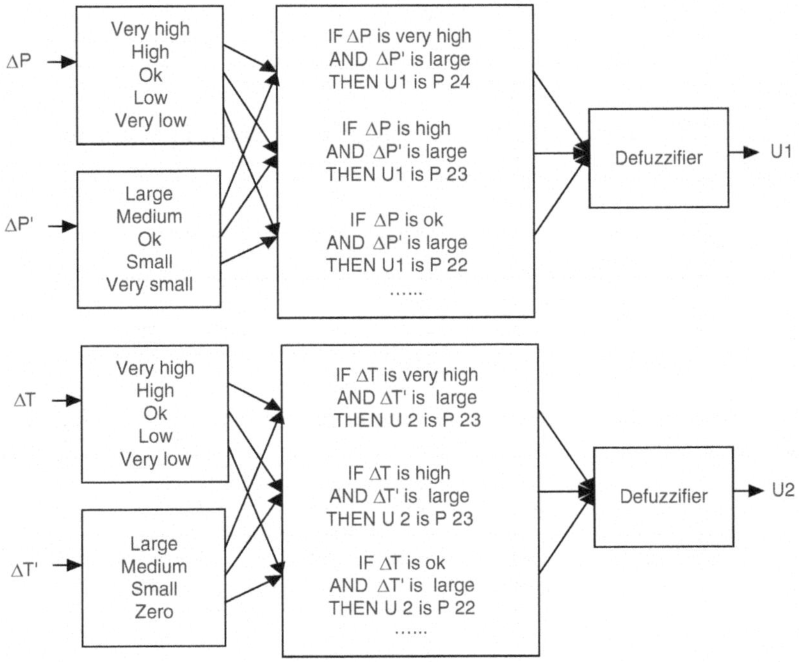

Fig. 4.7 Block diagram of the improved fuzzy rules

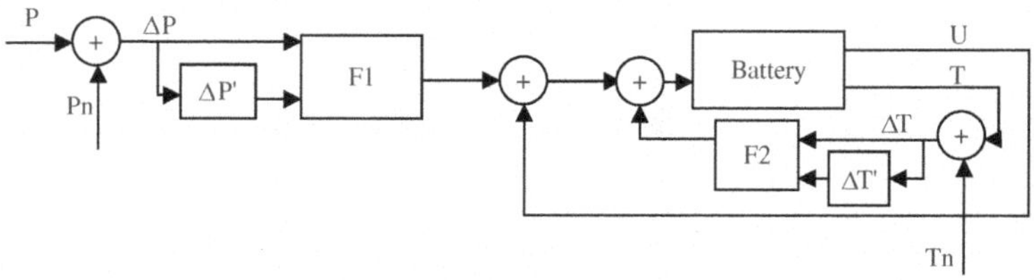

Fig. 4.8 Fuzzy control based on improved rules

4.4.2 Fuzzification

The fuzzification of the system has two separate parts. One is for the outer loop and the other is for the inner loop. As the fuzzification of the outer control loop is much the same as that of the inner loop, we only present the discussion of the outer loop. The input variables for the outer loop is the ΔP and $\Delta P'$. ΔP is positive in the whole charging process. At the same time, it drops gradually in the whole process. As $\Delta P'$ is the differential coefficient of ΔP – to t, it can be positive or negative in the charging process. Given a value set of X in $[0, +1]$, ΔP is labeled by:

PV	PH	PO	PL	PZ
Positive very high	Positive high	Positive ok	Positive low	Positive but close to zero

While $\Delta P'$ is labeled with the following values, defined a value set of X in $[-1, +1]$.

PL	PM	PZ	NM	NL
Positive large	Positive medium	Positive but close to zero	Negative medium	Negative large

The output variable U1 can be positive or negative labeled in five possibilities.

LP	SP	ZE	SN	LN
Large positive	Small positive	Zero	Small negative	Large negative

The fuzzification of ΔP and $\Delta P'$ is shown in Fig. 4.9. The fuzzication of the output *U1* can be demonstrated in Fig. 4.10.

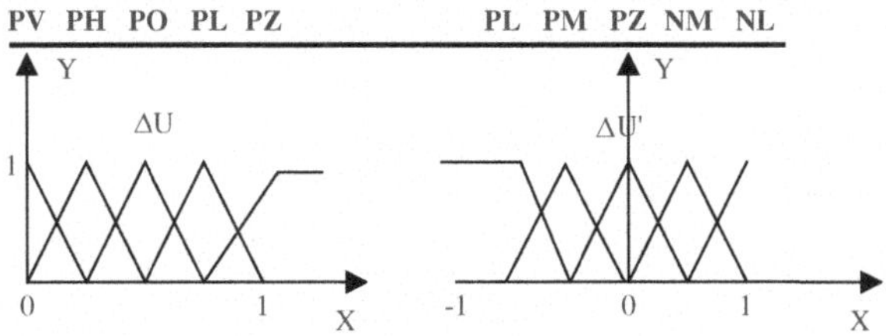

Fig. 4.9 Input variables fuzzification

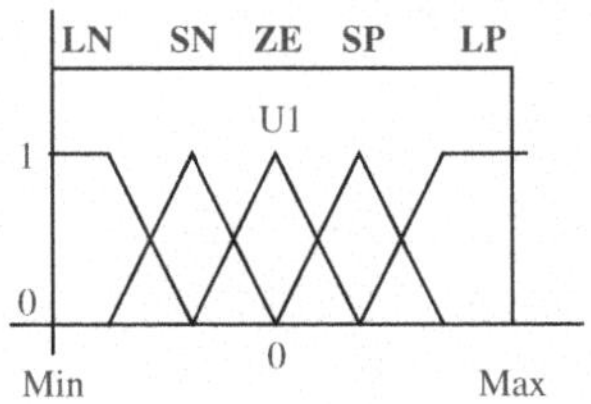

Fig. 4.10 Output variable fuzzification

Table 4.1 Rule base for outer loop

$\Delta P'/\Delta P$	PV	PH	PO	PL	PZ
PL	P24	P23	P22	P21	P20
PM	P19	P18	P17	P16	P15
PZ	P14	P13	P12	P11	P10
NM	P9	P8	P7	P6	P5
NL	P4	P3	P2	P1	P0

Another approach is to label them with the minimum and maximum values of the analogue to digital converter, such as the set of [0, 255] or [−255, 0] with an 8-bit converter (or more details, see [11]).

4.4.3 Fuzzy Rule Evaluation

Fuzzy rule evaluation is the key step for a fuzzy logic control system. This step processes a list of rules from the knowledge base using input values from RAM to produce a list of fuzzy outputs to RAM. There are many mutual methods to evaluate fuzzy rules, among which State Space method is the most popular one due to its close relationship with the transference function of the system [11]. The rule base for outer loop of the system is summarized in Table 4.1.

These rules are expressed in the form of *IF–THEN*, as generally given in formula (4.3). The fuzzy rules are specified as follows:

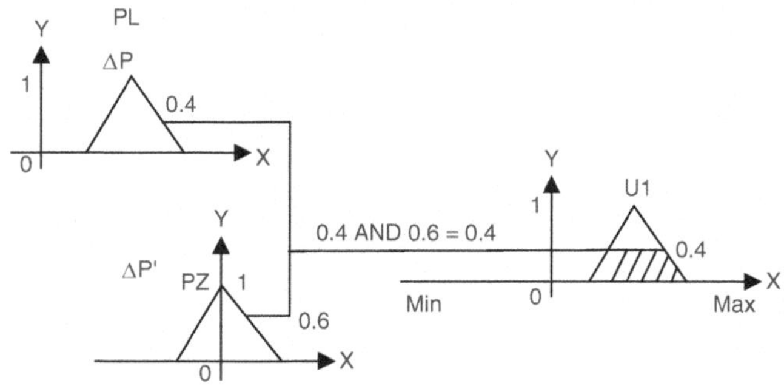

Fig. 4.11 Output variable fuzzification

Rule 1: IF ΔP = PV AND ΔP′ = NL THEN U1 = ZE
Rule 2: IF ΔP = PV AND ΔP′ = NM THEN U1 = SN
Rule 3: IF ΔP = PL AND ΔP′ = PZ THEN U1 = LP
Rule 4: IF ΔP = PV OR ΔP′ = PM THEN U1 = LN
. . . . . .
Rule n: IF ΔP = . . . OR/AND ΔP′ = . . . THEN U1 = . . .

To work out the values of the output P0 to P19 in Table 4.1, the Min–Max method was selected in the rule evaluation process [11]. In the Min–Max rule, the minimum of the two inputs is used as the combined truth-value. Fig. 4.11 displays this method.

Combining with rule3, we can solve out that $P11$ equals 0.4 but belongs to *LP*.

4.4.4 Defuzzification and Simulation

The final step in the fuzzy logic controller is to defuzzify the output, combining the fuzzy output into a final systems output. The Centroid method is adopted, which favours the rule with the output of greatest area [12, 13]. The rule for outputs can be defuzzified using the discrete Centroid computation described in formula (4.9).

$$
\begin{aligned}
Uout &= \frac{\sum_{1}^{4} Si^{*} Fi}{\sum_{1}^{4} Si} \\
&= SUM(i = 1\ to\ 4\ of(\ S(i)^{*}\ F(i)))/SUM(i = 1\ to\ 4\ of\ \ S(i)) \\
&= \frac{S(1)^{*}F(1) + S(2)^{*}F(2) + S(3) * F(3) + S(4)^{*}F(4)}{S(1) + S(2) + S(3) + S(4)}
\end{aligned}
\tag{4.9}
$$

In (4.9), $S(i)$ is the truth value of the result membership function for rule i, and $F(i)$ represents the value while the result membership function (*ZE*) is maximum over the

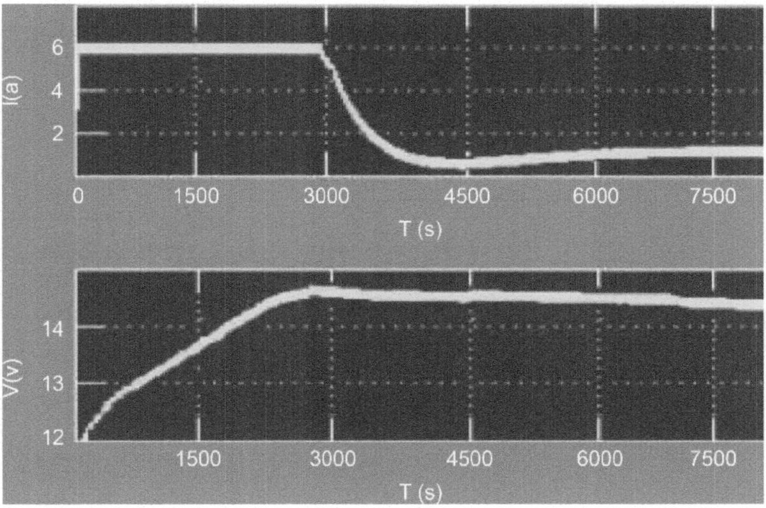

Fig. 4.12 Battery charging current and output port voltage

output variable fuzzy set range. With the fuzzy control rules discussed above, the typical battery charging current and voltage curves are shown in Fig. 4.12.

In the simulation, the setting point for the charging voltage is 14.5 V and that of the charging current to 6 A. The charging process can be separated into two stages. During the first stage, the fuzzy control strategy is implemented to determine the proper start charging time and to prevent it from being over or insufficiently charged. At the beginning, the port voltage Uc starts with a small value, while the current keeps constant to the set value, so the battery will be fully charged. During the second stage, a normal charging strategy is used. The control variable switches over from current to voltage. The voltage is maintained almost unchanged at a given value, while the current is gradually decreasing until it drops to a value set at the initial stage. As can be seen from Fig. 4.12, the performance of the fuzzy controller proves to be very good, and the charging curve is much close to the Ampere-hour curve in Fig. 4.5. Charging time in this method has been reduced to about 30%, comparing to the classical methods. For a fully discharged battery, the required charging time is approximately 2.5 h (9,000 s).

4.5 System Implementation

The performance of the renewable energy generation system, including the battery charging and discharging unit, is controlled and monitored by an embedded microprocessor. In our system, a 32-bit RISC microprocessor from Samsung Company, S3C2410X, is employed. The S3C2410X was developed using an ARM920T core, 0.18 um CMOS standard cells and a memory complier. The S3C2410X provides a

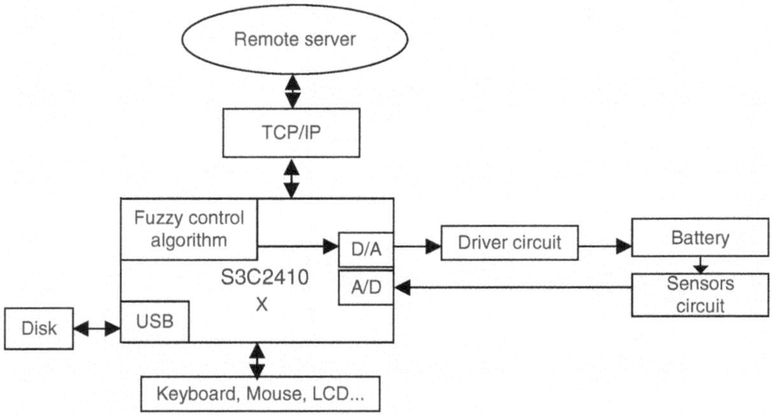

Fig. 4.13 System functional architecture

complete set of common system peripherals, including16 KB Instruction and 16 KB Data Cache, MMU to handle virtual memory management, LCD Controller (STN & TFT), NAND Flash Boot Loader, System Manager (chip select logic and SDRAM Controller), 3-ch UART, 4-ch DMA, 4-ch Timers with PWM, I/O Ports, RTC, 8-ch 10-bit ADC and Touch Screen Interface, IIC-BUS Interface, IIS-BUS Interface, USB Host, USB Device, SD Host & Multi-Media Card Interface, 2-ch SPI and PLL for clock generation and so on [14].

In the battery charging and discharging control unit, the S3C2410X mainly takes charge of three essential functions: charging and discharging control, status surveying and monitoring and data communication, as shown in Fig. 4.13.

The working processes of the system as shown in Fig. 4.13 is: The embedded microprocessor S3C2410X compares the control parameters inputted from Keyboard and those fed back from sensors of the battery. Then the results are processed with fuzzy control algorithm by the embedded microprocessor. After that, corresponding signals, which have been converted from digital instructions into those of analogue form and amplified by the driver, are sent out to control the charging and discharging process. Functions of remote networked monitoring are provided to monitor and survey the status of the battery through networked level, often based on web technology with TCP/IP protocol stack. Components of fault diagnosis expert system can be adopted to employ professional strategies and measures to analyze or judge the existed or potential faults of the battery. Also in the system, history records of the control and monitoring of the battery can be stored to removable disks with USB interface.

The main module is the S3C2410X, which has abundant and powerful peripheral resources and communication ability with very high operational and data processing speed, so various kinds of advanced digital control algorithms can be implanted or emendated into it.

As the charging and discharging process have important relationship with the temperature of the battery, the relationship between the temperature and U_{out} can

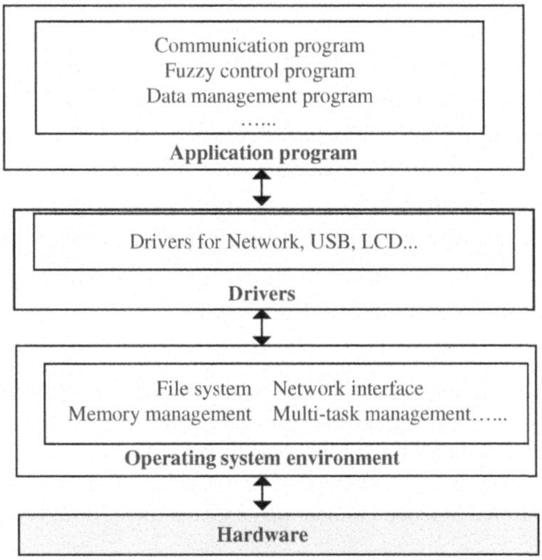

Fig. 4.14 System software architecture

be described with formula (4.10) [15].

$$T(°C) = I_{out} - 273 = \frac{U^*_{out} 10^6}{R3\left(1 + \frac{R5}{R4}\right)} \tag{4.10}$$

The software architecture of the renewable energy generation system can be separated into three parts, namely the operating system environment, hardware driver and application program, as shown in Fig. 4.14.

The operating system environment is based on Embedded Linux, which is transplanted into the S3C2410X microcontroller, providing the basic software environment for file system, network interface, memory management and multi-task schedule etc. to the system. Embedded Linux is a free and open source real-time operating system, and it is programmed with standard C language. It can be tailed according to specific system requirements with some redundant tasks removed and certain enhanced functions added, which guarantees the system's reliability and real-time and powerful performance.

4.6 Conclusions and Future Works

The control process of the battery charging and discharging is non-linear, time-varying with pure time delay, multiple variables and many external disturbances. Many parameters such as the charging rate, the permitted maximum charging current, the internal resistor, the port voltage, the temperature and moisture, change during the charging and discharging process can not be obtained directly, and it is impossible to use traditional control system. A fuzzy control unit for battery charging and discharging in a renewable energy generation system is studied in detail. Simulation based on fuzzy strategies introduced in the paper shows that the control unit has excellent performance in a laboratory environment. Analysis involved in the paper ignores the temperature element in the inner loop of the control unit, but to utilize this unit in the system, changes in the temperature must be taken into account. Future works may involve genetic algorithms or neural network based methods, combined with fuzzy algorithm, to get quicker and accurate system response. Also remote online monitoring and alarming functions should be developed to prevent the battery from destruction.

Acknowledgements The authors would pay their appreciations to The Science Foundation for New Teachers of Ministry of Education of China (No. 20070497059) and The Open Research Projects Supported by The Project Fund (No. 2007A18) of The Hubei Province Key Laboratory of Mechanical Transmission and Manufacturing Engineering, Wuhan University of Science and Technology and The National Science Foundation of China (No. 50675166).

References

1. Renewable energy, last modified on February 2008, Available: http:// en.wikipedia.org/wiki/Renewable_energy#Market_development_of_renewable_heat_energy. Accessed on February 2008.
2. Li, Z. J., China's Wind energy potential appears vast, November, 2005, Available: http://www.worldwatch.org/node/57. Accessed on February 2008.
3. Zhao, Q., Integrated control of load power tracking and battery charging for stand-alone small scale wind power generating systems, University of Inner Mongolia, China, Master's thesis. June 2006.
4. Oneyama, N., Air compressor and its operation control, *The Journal of Energy Conservation Center*, Vol. 50, No. 2, pp. 65–73, 1998.
5. Northern Arizona Wind & Sun, Deep cycle battery FAQ, Available: http://www.windsun.com/Batteries/Battery_FAQ.htm. Accessed on February 2008.
6. Buchmann, I., The lead-acid battery, created on April 2003, last edited on February 2007, Available: http://www.batteryuniversity.com/partone-13.htm. Accessed on February 2008.
7. Kirchev, A., D., Pavlov and B. Monahov., Gas-diffusion approach to the kinetics of oxygen recombination in lead-acid batteries. *Journal of Power Sources*, Vol. 113, pp. 245–254, 2003.
8. Osaka, T. S., M. Rajamäki and T. Momma. Influence of capacity fading on commercial lithium-ion battery impedance. *Journal of Power Sources*. Vol. 119, pp. 929–933, 2003.
9. Fuzzy control system, From Wikipedia, the free encyclopedia, 2007, Available: http://en.wikipedia.org/wiki/Fuzzy_system. Accessed on February 2008.
10. Lee, C. C., Fuzzy logic in control systems: Fuzzy logic controller – part I and II, *IEEE Transactions on Systems, Man and Cybernetics*, Vol. 20, No. 2, pp. 404–435, 1990.
11. Lee, H. D. and S. K. Sul, Fuzzy Logic based Torque control strategy for parallel type HEV, *IEEE Transactions on Industrial Electronics,* Vol. 45, No. 4, pp. 625–632, 1998.
12. Berardinis, L. A., A clear thinking on fuzzy logic, Machine Design, Vol. 64, No. 8, pp. 46–52, 23 April 1992.
13. Cai, B. and D. Konik, Intelligent vehicle active suspension control using fuzzy logic, *IFAC World Congress, Sydney*, Vol. 2, pp. 231–236, 1993.
14. Samsung Electronics, S3C2410X 32-Bit RISC Microprocessor User's Manual, Revision 1.2. Publication Number: 21.2-S3-C2410X-052003, 2003.
15. Analog Devices, Inc., Two-terminal IC temperature transducer, AD590, Available: http://www.analog.com, 1997.

Chapter 5
Optical CDMA Transceiver Architecture: Polarization Modulation with Dual-Balanced Detection

Mohammad Massoud Karbassian and Hooshang Ghafouri-Shiraz

Abstract In this chapter, we propose a novel OCDMA architecture design based on the polarization modulation and dual-balanced detection. The application of optical tapped-delay line (OTDL) as the incoherent OCDMA decoder is also investigated. It is assumed that the signal is degraded by (i) fiber amplifier spontaneous emission (ASE) noise, (ii) electronic receiver noise (iii) photo detectors (PD) shot-noise (iv) and mainly MAI. For optical signature sequences, we have used a coding scheme based on double-padded modified prime code (DPMPC).

Keywords Optical CDMA Transceiver · Polarization Modulation · Dual-Balanced Detection · OCDMA architecture · optical tapped-delay line

5.1 Introduction

Bandwidth-hunger world demands higher bit-rate and ultra-fast services such as video-on-demand and streaming over the Internet protocol e.g. IPTV. Due to tremendous resources of bandwidth and extremely low loss, fiber-optic can be the best physical transmission medium for telecommunications and computer networks. Unlike conventional time- division multiple-access (TDMA) and wavelength-division multiple-access (WDMA) techniques, code-division multiple-access (CDMA) can make full use of the huge bandwidth in the fiber-optic. Meanwhile, it has the potential to support random access protocol, different services with different data-rates and bursty traffics.

Optical CDMA (OCDMA) has been proposed as an access protocol that takes advantage of the excess bandwidth particularly in single-mode fiber-optic (SMF). It provides multi-channel communications over a single frequency band. Accordingly,

---

M.M. Karbassian (✉)
University of Birmingham, Edgbaston, B15 2TT, UK,
E-mails: mmk404@bham.ac.uk,h.ghafouri-shiraz@bham.ac.uk

S.I. Ao and L. Gelman (Eds.), *Advances in Electrical Engineering and Computational Science*, Lecture Notes in Electrical Engineering.

signals must be designed to reduce mutual interferences. This can be achieved by subdividing each data into a number of binary chips. The chip sequences constitute a code that permits a bit-stream broadcast on a network to be selected by means of a correlation process at the receiver destination. A large number of chip sequence signatures can be assigned to different users, in the process of which the set of optical signatures essentially becomes a set of address codes for the network users.

Incoherent OCDMA technique inherently suffers from multiple-access interference (MAI) that requires estimation and removal through cancellation techniques or higher order modulations [1–4]. In the past few years, advances in photonics technology made it possible for incoherent scheme to approach the sensitivity performance of coherent ones. Hence, there has been a considerable shift of interest from coherent to incoherent scheme on the part of both research and industry, due to its simple architecture and cost effectiveness.

Polarization shift keying (PolSK) is the only modulation that takes advantage of vector characteristic of lightwave. It has been experimentally reported [5] that PolSK has insensitive behavior to (i) phase noise, (ii) polarization fluctuations, (iii) self- and cross-phase modulations due to the constant-power envelop of the lightwave. A comprehensive set of results [5, 6] showed that the performance of the binary PolSK is approximately 3 dB better than intensity modulation (on the peak optical power) of other PolSK or equivalent phase shift keying (PSK) modulations. It has very low sensitivity to fiber dispersions in which elevates the system performance accordingly. PolSK encodes the information on a constellation of signal points in the space of the Stokes parameters which are the coordinates of the signal's states of polarization (SOP) over the Poincaré sphere [6]. In general, each signal point corresponds to a given SOP and a given optical power. To perform PolSK detection avoiding optical birefringence compensation, it is necessary to use a receiver that extracts the Stokes parameters of the incoming lightwaves [6]. Additionally, the advantages of PolSK over OCDMA have been reported [7, 8].

In this chapter, we propose a novel OCDMA architecture design based on the polarization modulation and dual-balanced detection. The application of optical tapped-delay line (OTDL) as the incoherent OCDMA decoder is also investigated. It is assumed that the signal is degraded by (i) fiber amplifier spontaneous emission (ASE) noise, (ii) electronic receiver noise (iii) photo detectors (PD) shot-noise (iv) and mainly MAI. For optical signature sequences, we have used a coding scheme based on double-padded modified prime code (DPMPC). It has been extensively introduced and employed in pulse-position modulation (PPM), [2] in overlapping PPM, [4] in coherent heterodyne, [9] and homodyne [10, 11] schemes, and also in frequency modulated OCDMA system, [12] with even novel MAI cancellation technique [13].

5.2 POLSK–OCDMA Transmitter

A generic transformation of the SOP of a fully polarized lightwave, propagating along the z-axis which preserves the degree of polarization, is explained as follows. Let $\bar{E}(t)$ and $\bar{E}'(t)$ be the electromagnetic field vectors before and after the transformation (i.e. modulation) respectively, and given as follows:

$$\bar{E}(t) = (E_x(t)\bar{x} + E_y(t)\bar{y})e^{j\omega t}, \quad \bar{E}'(t) = (E'_x(t)\bar{x}' + E'_y(t)\bar{y}')e^{j\omega t} \tag{5.1}$$

where ω is the optical angular frequency, where $E_x(t)$ and $E_y(t)$ are (x, y)-components of electric filed before transformation and $E'_x(t)$, $E'_y(t)$ are (x, y)-components of electric filed after. Thus, we have:

$$\begin{bmatrix} E'_x(t) \\ E'_y(t) \end{bmatrix} = \mathbf{Q} \begin{bmatrix} E_x(t) \\ E_y(t) \end{bmatrix} \tag{5.2}$$

where $\mathbf{Q}$ is a complex Jones matrix with unit determinant. A subset of Jones matrices, called the set of matrices of birefringence or optical activity that not only preserves the degree of polarizations, but also has the additional feature of preserving two orthogonal fields (according to the Hermitian scalar product) [14] which were orthogonal before the transformation. Matrices of this kind are complex unitary matrices with unit determinant. Throughout this chapter we strictly refer to the $\mathbf{J}j$ as subset of $\mathbf{Q} = [\mathbf{J}_0 \ldots \mathbf{J}_j \ldots \mathbf{J}_{k-1}]$.

By using the Jones representation, the field can be represented by the vector, $\mathbf{J} = [E_x \quad E_y]^T$ and the intensity of the beam can be normalized so that $|E_x|^2 + |E_y|^2 = 1$. Two SOPs represented by $\mathbf{J}_1$ and $\mathbf{J}_2$ are orthogonal if their inner product is zero. Any SOP can be transformed into another by multiplying it by a Mueller matrix [14] which are required for SOP processing e.g. polarizers, rotators and retarders. In PolSK, the angle of one polarization component is switched relative to the other between two angles; therefore, binary data bits are mapped into two Jones vectors. A block diagram of the proposed PolSK–OCDMA transmitter is illustrated in Fig. 5.1. The light source is a highly coherent laser with a fully polarized SOP. If a non-polarized source is used, then a polarizer can be inserted after the laser source. The light beam first passes through the polarization controller that sets the polarization to an angle of 45° for simplicity. Then, the lightwave gets divided through polarization beam splitter (PBS) to become SOP-encoded in PolSK modulator which switches the SOP of the input beam between two orthogonal states

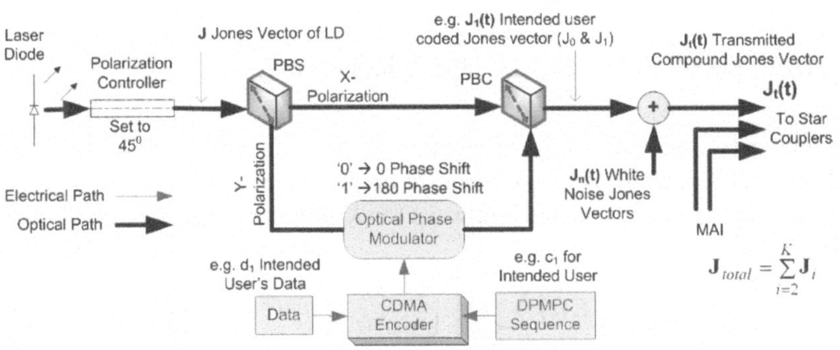

Fig. 5.1 Proposed architecture of incoherent PolSK–OCDMA transmitter

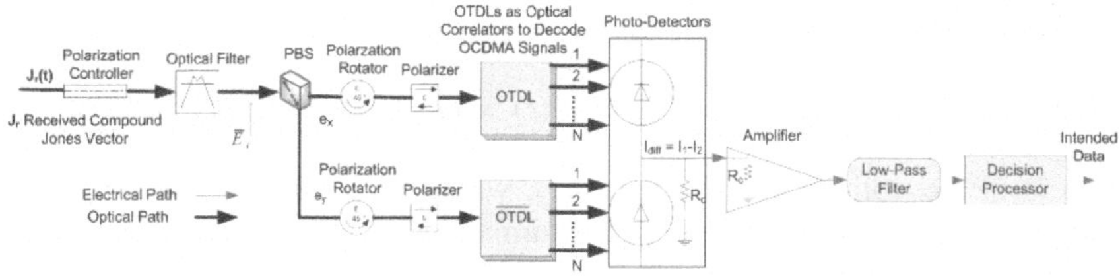

Fig. 5.2 Proposed architecture of incoherent PolSK–OCDMA receiver

(i.e. 0° and 180° at the phase modulator in Fig. 5.1 N times per bit according to an externally supplied code (i.e. DPMPC) that spreads the optical signal into CDMA format. Thereafter, the PolSK–OCDMA modulated signals are combined through polarization beam combiner (PBC) and broadcasted. It is also displayed in Fig. 5.1 that for a K-user system with the first user as the desired one (for example), the *ith* user SOP-encoded signal can be written as:

$$\mathbf{J}_i(t) = \begin{cases} \mathbf{J}_0 & \text{if } d_i(t) \oplus c_i(t) = 0 \\ \mathbf{J}_1 & \text{if } d_i(t) \oplus c_i(t) = 1 \end{cases} \tag{5.3}$$

where $d_i(t)$ is the data signal with symbol duration of T_s, $c_i(t)$ is the N-chip code sequences of DPMPC signal with chip duration of T_c and $d_i(t), c_i(t) \in \{0, 1\}$; $\oplus$ denotes the signal correlation. As the emitted light is initially (linearly) polarized at an angle of 45°, therefore $\mathbf{J}_0 = \frac{1}{\sqrt{2}}[1 \quad 1]^T$ and $\mathbf{J}_1 = \frac{1}{\sqrt{2}}[-1 \quad 1]^T$. In other words, we have [8]:

$$\mathbf{Q} = [\mathbf{J}_0 \quad \mathbf{J}_1] = \frac{1}{\sqrt{2}}\begin{bmatrix} 1 & -1 \\ 1 & 1 \end{bmatrix} \tag{5.4}$$

Therefore, the polarization-modulated signal travels a distance of L km through an optical SMF. Consequently, the SOP-encoded signal undergoes several impairments such as attenuation, dispersion, polarization rotation and fiber nonlinearity. At the receiver end shown in Fig. 5.2, the SOP rotation is compensated by the polarization controller whose function is to insure that the received signal and the optical components at the receiver have the same SOP reference axis.

5.3 Analysis of POLSK–OCDMA Receiver

We previously discussed the configuration of the transmitter. Now we consider the alignment and analysis of the received optical signal. The electric field of the received polarization-modulated lightwave for K number of users can be expressed as [7]:

$$\bar{E}'(t) = \mathrm{Re}\left\{ \bar{E}(t) \sum_{i=1}^{K} \mathbf{Q} \begin{bmatrix} d_i(t) \\ 1 - d_i(t) \end{bmatrix} u_T(t - iT_s).c_i(t) \right\} \tag{5.5}$$

The channel is represented by the Jones matrix $\mathbf{Q}$ and $Re\{.\}$ refers to the real part of complex $\bar{E}'(t)$. Since equipower signal constellations have been considered, both orthogonal components are assumed to be equally attenuated. Thus, these terms can be included in the constant amplitude of the electric field $\bar{E}(t)$, which neglects a loss of orthogonality on the channel. While switching time in SOP (i.e. bit-rate) is much slower than the chip-rate, the elements of the Jones matrix can be understood as time-independent (i.e. $T_c << T_s$). The x-component of the received electric field vector based on $\mathbf{Q} = [\mathbf{J}_0 \quad \mathbf{J}_1]$ (see Eq. (5.4)) is:

$$E'_x(t) = \mathrm{Re}\left\{\bar{E}(t)\sum_{i=1}^{K}[\mathbf{J}_0 d_i(t) + \mathbf{J}_1(1 - d_i(t))]\,u_T(t - iT_s).c_i(t)\right\} \tag{5.6}$$

Thus, orthogonal components of the *ith* user are given as $E_{xi}(t) = \mathbf{J}_0 d_i(t)c_i(t)$ $\bar{E}(t)$ and $E_{yi}(t) = \mathbf{J}_1(1-d_i(t))c_i(t)\bar{E}(t)$ where the (x, y)-components of received modulated signal are [7]:

$$E'_{xi}(t)=\left(\frac{E_{xi}(t)+E_{yi}(t)}{2}+\sum_{i=1}^{K} c_i(t)d_i(t)\frac{E_{xi}(t)-E_{yi}(t)}{2}u_T(t - iT_s)\right)\cos(\varphi_{xi})$$

$$E'_{yi}(t)=\left(\frac{E_{xi}(t)-E_{yi}(t)}{2}+\sum_{i=1}^{K} c_i(t)d_i(t)\frac{E_{xi}(t)+E_{yi}(t)}{2}u_T(t - iT_s)\right)\cos(\varphi_{yi}) \tag{5.7}$$

where φ_{xi} and φ_{yi} describe the frequencies and phases of transmitting lasers in a general form of $\varphi = \omega t + \theta$. Based on the concept of CDMA, the field vectors of all K transmitters are combined and multiplexed over the same channel. Thus, the overall channel field vector can be expressed as:

$$\bar{E}_{Channel} = \sum_{i=1}^{K} \bar{E}'_i(t) \tag{5.8}$$

Figure 5.2 illustrates the application of the OTDLs used as the optical correlator in this incoherent PolSK–OCDMA configuration. The delay coefficients in OTDLs are designed in such a way to make them perform as a CDMA chip-decoder in both branches. Additionally, OTDL in lower branch is set up with complement of code used in upper branch (i.e. shown by $\overline{\textbf{OTDL}}$) to decode other symbol (i.e. '1'). It can be observed from Fig. 5.2 that OTDLs output contain N chip pulses that can be assumed as a parallel circuit of many single PDs so that their currents are added and no interference between the OTDL pulses is possible. The signals are photo-detected in the balanced-detector arrangement to generate the differential electrical current ($I_{diff} = I_1 - I_2$) ready for data-extraction in decision processor unit. The total upper branch current (i.e. x-component) considering all chip currents after photo-detection is then obtained as:

$$I^0 = \Re \int_{t=0}^{T_s} \sum_{n=1}^{N} \left\{ \frac{c(nT_c)+1}{2} \left(\sum_{i=1}^{K} \left(\frac{E_{xi}(t)+E_{yi}(t)}{2} + d_i(t)c_i(t-nT_c)\frac{E_{xi}(t)-E_{yi}(t)}{2} \right) \cos(\varphi_{xi}) \right)^2 \right\} dt \tag{5.9}$$

where $\Re$ is the PD responsivity, $c_i(t-nT_c)$ is the *nth* chip of assigned spreading code of the *ith* user.

By further simplification, it can be modified as:

$$\begin{aligned} I^0 = {} & \frac{\Re}{4} \int_{t=0}^{T_s} \sum_{n=1}^{N} \left\{ \frac{c(nT_c)+1}{2} \left(\sum_{i=1}^{K} \left(E_{xi}^2(t) + E_{yi}^2(t) + d_i(t)c_i(t-nT_c) \right.\right.\right. \\ & \left.\left.\left. \times (E_{xi}^2(t) - E_{yi}^2(t)) \right)(1-\cos 2\varphi_{xi}) \right) \right\} dt \\ & + \frac{\Re}{8} \int_{t=0}^{T_s} \sum_{n=1}^{N} \left\{ \frac{c(nT_c)+1}{2} \left[\sum_{i=1}^{K} \sum_{\substack{j=1 \\ j \neq i}}^{K} \left\{ \left(E_{xi}(t) + E_{yi}(t) \right.\right.\right.\right. \\ & + d_i(t)c_i(t-nT_c) \\ & \times (E_{xi}(t) - E_{yi}(t)) \Big) \left(E_{xj}(t) + E_{yj}(t) + d_j(t)c_j(t-nT_c) \right. \\ & \left.\left.\left.\left. \times (E_{xj}(t) - E_{yj}(t)) \right) \left(\cos(\varphi_{xi} + \varphi_{xj}) + \cos(\varphi_{xi} - \varphi_{xj}) \right) \right\} \right] \right\} dt \end{aligned} \tag{5.10}$$

Since PD frequency response behaves similar to a low-pass filter, both terms of $\cos 2\varphi_{xi}$ in the first element and $\cos(\varphi_{xi} + \varphi_{xj})$ in the second element of Eq. (5.10) are filtered out as they are outside of the PD frequency range. Furthermore, the term $\cos(\varphi_{xi} - \varphi_{xj})$ can also be removed provided that $\varphi_{xi} - \varphi_{xj} >> \omega_c$ where ω_c is the cut-off frequency of the PD. Therefore, the total current of upper branch can be expressed as:

$$I^0 = \frac{\Re}{4} \sum_{n=1}^{N} \left\{ \frac{c(nT_c)+1}{2} \left(\sum_{i=1}^{K} (E_{xi}^2(t) + E_{yi}^2(t) + d_i(t)c_i(t-nT_c)(E_{xi}^2(t) - E_{yi}^2(t))) \right) \right\} \tag{5.11}$$

Then the Stokes parameters are defined as:

$$\begin{aligned} S_i^0 &= E_{xi}^2(t) + E_{yi}^2(t) \\ S_i^1 &= E_{xi}^2(t) - E_{yi}^2(t) \end{aligned} \tag{5.12}$$

where S_i^0 refers to signal intensity part, generated in upper branch of polarization modulator at the transmitter while S_i^1 refers to the linear polarized part, generated in lower branch containing data (see Fig. 5.1. Thus, Eq. (5.11) can be rewritten as:

$$I^0 = \frac{\Re}{4}\sum_{n=1}^{N}\left\{\frac{c(nT_c)+1}{2}\left(\sum_{i=1}^{K}(S_i^0 + d_i(t)c_i(t-nT_c)S_i^1)\right)\right\} \tag{5.13}$$

Similarly the total current of the lower branch (i.e. y-component) can be derived as:

$$I^1 = \frac{\Re}{4}\sum_{n=1}^{N}\left\{\frac{1-c(nT_c)}{2}\left(\sum_{i=1}^{K}(S_i^0 + d_i(t)c_i(t-nT_c)S_i^1)\right)\right\} \tag{5.14}$$

Thus, the balanced-detector output ($I = I^0 - I^1$) is then derived as:

$$I = \frac{\Re}{4}\sum_{n=1}^{N}c(nT_c)\sum_{i=1}^{K}(S_i^0 + d_i(t)c_i(t-nT_c)S_i^1) + n(t) \tag{5.15}$$

where $n(t)$ represents the total filtered Gaussian noise with variance of σ^2 that includes: (i) the PD shot-noise with electric current variance of $\langle i_{av}^2\rangle = 2eiB_o$ where i_{av} is the average photo-current; (ii) optically filtered ASE noise with variance of $\sigma_{ASE}^2 = 2N_0B_0$ where N_0 is the (unilateral) power spectral density (PSD) of the white ASE noise arriving on each polarization and B_0 is the optical filter bandwidth; (iii) electronic receiver noise current (i.e. thermal noise) at the low-pass filter with variance of $\sigma_{LP}^2 = 2k_bTB_{el}/R$. where R is the filter direct-current (dc) equivalent impedance, T the absolute temperature and k_b the Boltzmann constant, B_{el} the filter bandwidth. Thus the overall variance of additive noise of $n(t)$ can be represented as:

$$\sigma_{n(t)}^2 = \langle i_{av}^2\rangle + \sigma_{ASE}^2 + \sigma_{LP}^2 \tag{5.16}$$

By considering the first user as the intended user, then we can modify the differential output current, i.e. Eq. (5.15), as:

$$\begin{aligned} I &= \frac{\Re}{4}\sum_{n=1}^{N}S_1^0c(nT_c) + \frac{\Re}{4}\sum_{n=1}^{N}c(nT_c)c_1(t-nT_c)d_1(t).S_1^1 \\ &+ \frac{\Re}{4}\sum_{i=2}^{K}\sum_{n=1}^{N}c(nT_c)c_i(t-nT_c)d_i(t)S_i^1 + n(t) \end{aligned} \tag{5.17}$$

The first element in Eq. (5.17) is a dc current that needs estimation and removal in the balanced-detector. The second element represents the intended data mixed with its assigned spreading code auto-correlation and polarization while the third element assumes the interference (i.e. MAI) caused by other transmitters and the last one is the noise. Thus, the system SNR can be expressed as:

$$SNR = \frac{\left(\frac{\Re}{4}\sum_{n=1}^{N} c(nT_c)c_1(t - nT_c)d_1(t).S_1^1\right)^2}{\left(\frac{\Re}{4}\sum_{i=2}^{K}\sum_{n=1}^{N} c(nT_c)c_i(t - nT_c)d_i(t)S_i^1\right)^2 + \sigma_{n(t)}^2} \tag{5.18}$$

Both the auto- and cross-correlation of the DPMPC can be expressed respectively as [9–13]:

$$\sum_{n=1}^{N} c(nT_c)c_1(t - nT_c) = P + 2 \tag{5.19}$$

$$X_{li} = \sum_{n=1}^{N} c_l(nT_c)c_i(t - nT_c) \tag{5.20}$$

where P is the prime number, which is set to generate DPMPC sequences.

The cross-correlation, i.e. Eq. (5.20), probability density function (PDF) can be obtained from the independent values of random variable X_{li}. The in-phase cross-correlation value of DPMPC is either *zero* or *one* depending on whether the codes are in the same group or from the different groups [2]. Obviously, the *zero* value does not cause the interference due to perfectly orthogonal sequences, while the *one* value causes the interference which is only among intended user and $(P^2 - P)$ users from the different groups (i.e. P^2 whole sequences and P sequences from the same group of intended user which are orthogonal) [4]. The cross-correlation values are uniformly distributed among interfering users, since the users are independently active. Thus the PDF of w, realization of X_{li}, is:

$$P(w = i) = \frac{i}{P^2 - P} \tag{5.21}$$

where $P(w = i)$ is the probability that w assumes the value i (the number of actively involved users in the transmission). Therefore, by substituting Eqs. (5.19) and (5.21) into Eq. (5.18) and a little further calculation, the system SNR can be simplified as:

$$SNR(K) = \frac{1}{\left(\frac{(K+2)(K-1)}{2(P^2-P)(P+2)}\right)^2 + \frac{16\sigma_{n(t)}^2}{\Re^2 d_1^2(t).S_1^{1^2}(P+2)^2}} \tag{5.22}$$

Note that $SNR(1) = \Re^2 d_1^2(t).S_1^{1^2}(P + 2)^2 \big/ 16\sigma_{n(t)}^2 = E_b/N_0$, where E_b is the energy of one bit, N_0 is the noise PSD, denotes the single-user SNR. Equation (5.22) is one of the main results of this study as it represents the SNR of polarization-modulated optical CDMA system.

5.4 Discussion of Results

BER estimation of binary PolSK modulation has already been evaluated [6, 14]. Here, the numerical results of the BER performance of the proposed transceiver based on the above detailed analysis, resulted the OCDMA system SNR, are demonstrated and discussed.

Figure 5.3 shows the BER of this architecture against the single-user SNR (shown on Figs. by S_{db}). Different trends like 10%, 15%, 20% and 25% of full-load (i.e. $P^2 - P$ interfering users) [2] as the number of simultaneous active users where $P = 19$ have been evaluated in this investigation. As illustrated in Fig. 5.3, the system that can manage 25% of full-load is able to provide $BER = 10^{-9}$ with $S_{db} = 16.5dB$; whereas for $S_{db} = 8.5dB$ the system can support 20% load which is still superior enough to deliver the network services. Furthermore, the system can tolerate 15% load with only $S_{db} = 7$ dB. It is indicated that the delivering network services under these conditions is very power efficient. Although for supporting greater number of users, higher values for P and S_{db} are recommended.

Figure 5.4 also indicates the BER performance against the number of simultaneous users (K) for the discussed architecture. It is observable from Fig. 5.4, when the number of transmissions (i.e. users) increases, the BER also becomes higher due to the growing interferences. The system employed $S_{db} = 14dB$ can tolerate 80 simultaneous users where $P = 19$ which is equal to 24% of full-load. While 73 users (21% of full-load) are guaranteed very consistent communication

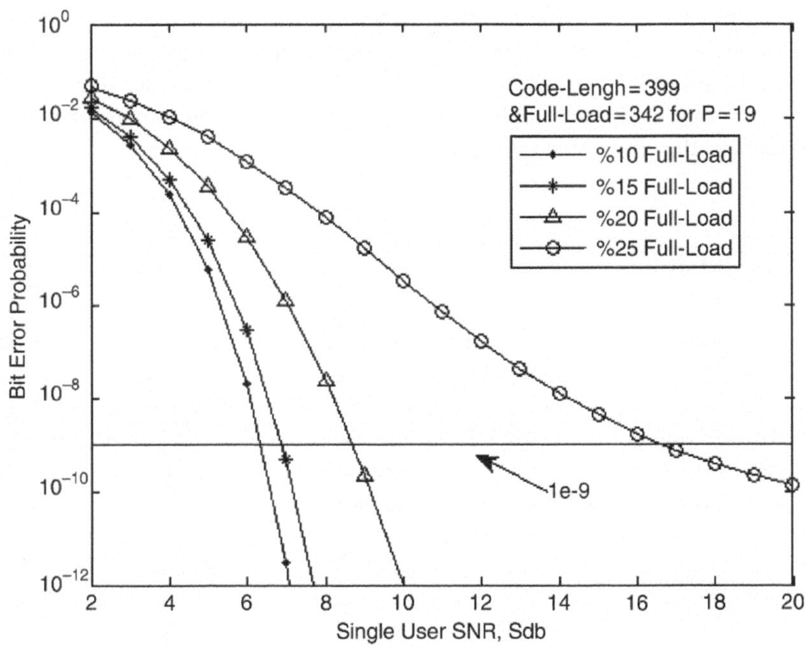

Fig. 5.3 BER performance of the transceiver against single-user SNR, S_{db}

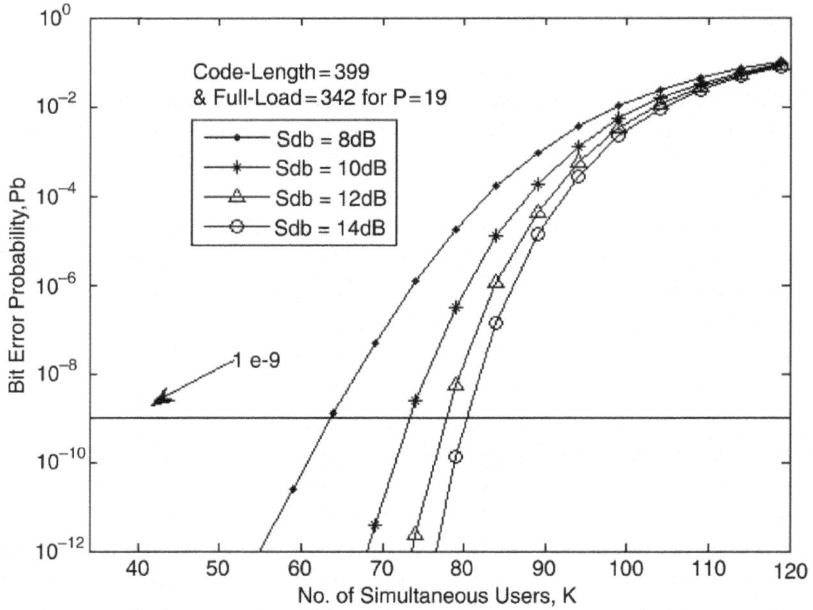

Fig. 5.4 BER performance of the transceiver against the number of simultaneous users, K

link ($BER \leq 10^{-9}$) with only $S_{db} = 10dB$, which refers to cost-effective design with less power consumption.

As a comparison [7, 8], the proposed architecture can tolerate greater number of users with less S_{db}. On the other hand, the results achieved in this analysis are with the code-length of 399 (i.e. for $P = 19$) [2] which is much shorter than analyzed Gold sequences [7, 8]. That implies the proposed structure can provide even higher throughput as the code-length is smaller.

5.5 Conclusion

This chapter has proposed and evaluated novel transceiver architecture of incoherent PolSK–OCDMA with dual-balanced detection. The application of OTDL as the CDMA-decoder has also been investigated. The BER performance of PolSK over OCDMA in cooperation with DPMPC as the spreading code has been demonstrated taking into account the effects of optical ASE noise, electronic receiver noise, PDs shot-noise and mainly the MAI. The results indicated that the architecture can reliably and power-efficiently accommodate great number of simultaneous users, and can be promising for high-data-rate long-haul optical transmission.

References

1. F. Liu, M. M. Karbassian and H. Ghafouri-Shiraz, "Novel family of prime codes for synchronous optical CDMA", J. Opt. Quant. Electron., vol. 39, no. 1, pp. 79–90, 2007
2. M. M. Karbassian and H. Ghafouri-Shiraz, "Fresh prime codes evaluation for synchronous PPM and OPPM signaling for optical CDMA networks", J. Lightw. Tech., vol. 25, no. 6, pp. 1422–1430, 2007
3. H. Ghafouri-Shiraz, M. M. Karbassian, F. Lui, "Multiple access interference cancellation in Manchester-coded synchronous optical PPM-CDMA network", J. Opt. Quant. Electron., vol. 39, no. 9, pp. 723–734, 2007
4. M. M. Karbassian and H. Ghafouri-Shiraz, "Capacity enhancement in synchronous optical overlapping PPM-CDMA network by a novel spreading code", in Proceedings of GlobeCom, pp. 2407–2411, 2007
5. S. Benedetto et al., "Coherent and direct-detection polarization modulation system experiments," in Proceedings of ECOC, 1994
6. S. Benedetto, R. Gaudino and P. Poggiolini, "Direct detection of optical digital transmission based on polarization shift keying modulation", IEEE J. Selected Areas Comms., vol. 13, no. 3, pp. 531–542, 1995
7. K. Iversen, J. Mueckenheim and D. Junghanns, "Performance evaluation of optical CDMA using PolSK-DD to improve bipolar capacity", in SPIE Proceedings, vol. 2450 (Amsterdam), pp. 319–329, 1995
8. N. Tarhuni, T. O. Korhonen and M. Elmusrati, "State of polarization encoding for optical code division multiple access networks", J. Electromagnet. WavesAppl. (JEMWA), vol. 21, no. 10, pp. 1313–1321, 2007
9. M. M. Karbassian and H. Ghafouri-Shiraz, "Performance analysis of heterodyne detected coherent optical CDMA using a novel prime code family", J. Lightw. Technol., vol. 25, no. 10, pp. 3028–3034, 2007
10. M. M. Karbassian and H. Ghafouri-Shiraz, "Phase-modulations analyses in coherent homodyne optical CDMA network using a novel prime code family", in Proceedings of WCE-ICEEE (IAENG), pp. 358–362, 2007
11. M. M. Karbassian and H. Ghafouri-Shiraz, "Performance analysis of unipolar code in different phase modulations in coherent homodyne optical CDMA", J. Eng. Lett. (IAENG), vol. 16, no. 1, pp. 50–55, 2008
12. M. M. Karbassian and H. Ghafouri-Shiraz, "Novel channel interference reduction in optical synchronous FSK-CDMA networks using a data-free reference", J. Lightw. Technol., vol. 26, no. 8, pp. 977–985, 2008
13. M. M. Karbassian and H. Ghafouri-Shiraz, "Frequency-shift keying optical code-division multiple-access system with novel interference cancellation", J. Microw. Opt. Techno. Lett., vol. 50, no. 4, pp. 883–885, 2008
14. S. Benedetto and P. Poggiolini, "Theory of polarization shift keying modulation", IEEE Trans. Comms., vol. 40, no. 4, pp. 708–721, 1992

Chapter 6
Template Based: A Novel STG Based Logic Synthesis for Asynchronous Control Circuits

Sufian Sudeng and Arthit Thongtak

Abstract This article introduces a template based technique in state encoding process to insert additional signals to STG with a small number. According to our method, complete state coding (CSC) property can be satisfied without using state graph tracing. Our method is useful for complicated asynchronous controllers, and also it can guarantee the other relevant properties, such as persistency and consistency. Our process begins with an encoding STG using Petri-net level in order to form a template STG. Then the projection to each non-input signals from an original STG is done. After that, we trace the projection to smaller state space. If the small state space shows conflicts, we have to insert balance signals from template STG. Unbalance signals are inserted after in case the state space still shows conflicts. Finally, we can get the STG with appropriate insertion points which is used to be projected for CSC support on each non-input signals. Asynchronous DMA controller is an example of our proposed method. The final part of this paper is concluded with a complexity comparison between our template based method with state based method and structural encoding method. It shows that the number of iterative signal removal according to our method is less than others.

Keywords Signal Transition Graph (STG) · logic synthesis · asynchronous control circuits · asynchronous DMA controller · template based technique

6.1 Introduction

State encoding is a process in logic synthesis for asynchronous control circuits. It is defined as a given specification to interact between input and non-input signals. The target is to find an encoding for every state of non-input signals in order to

S. Sudeng and A. Thongtak (✉)
Digital System Engineering Laboratory (DSEL), Department of Computer Engineering Faculty of Engineering, Chulalongkorn University, 254 Phyathai Road, Pathumwan, Bangkok, Thailand 10330, E-mails: sovanyy@msn.com, Arthit@cp.eng.chula.ac.th

S.I. Ao and L. Gelman (Eds.), *Advances in Electrical Engineering and Computational Science*, Lecture Notes in Electrical Engineering.

implement the circuits with hazard free. To find a state encoding, several authors have concerned with different ways, such as a logic synthesis using Petri-net unfolding synthesis using state based technique [1–3], synthesis using structural encoding technique, and etc. State based technique is involved several problems. The major problem is the detection of state encoding conflicts. It causes exponential complexity of conflict states with size of specification.

Another problem is the insertion of additional signals to solve the conflict states. Signal insertion must be done in such a way that the new signals are consistent (rising and falling transitions alternate) and their behaviors are hazard free. This method is insufficient for the specification such large scale of signals. Structural encoding method is more suitable than classical state based technique. The structural encoding scheme is Petri-net level encoding. This method encodes with two silent transitions [3, 5] to generate encode STG. Signal removal is begun after encoded STG has been generated. Iteratively removing signal is done with greedy removal from encode STG. In the end, it is produced a reduced STG. The next is to start projection for CSC support for each non-input signals. This method is available for large state encoding. However, it suffers from complexity reduction in greedy removal phase according to the number of iterative signal reduction.

This article proposes a template based method for synthesizing complicated STGs, and it also useful for some large scale STGs. The template based technique not only be useful for complicated and large scale STGs but also solves the reduction complexity of structural encoding technique.

Our template based technique was explained by one complicated example called asynchronous DMA controller that has been used in our asynchronous processor implementation [5].

We introduced three types of synthesis scheme in our explanation. Firstly, the synthesis of asynchronous DMA controller using state based technique. We derived the DMA's STG to state graph and showed the limitation of synthesizable properties which is limited by STG's rule called complete state coding (CSC). The complete state coding meant the state graph must be unique on state space; the asynchronous DMA controller specification is unsatisfied while state based was applied. From the state graph, if we try to insert an additional signal, the complication level is increased and suffers the state explosion.

Secondly, structural encoding technique is applied. The asynchronous DMA controller can be synthesized with this technique, but it shows some complexity in signal removal phase. Finally, template based is introduced. We generate the template STG from original STG using Petri-net level then we trace all non-input signals to small state space and derive it to Karnaugh map, if it shows conflicts, we have to insert some signals from template STG repeatedly, and we can get the final STG called STG with appropriate insertion point.

The final part of this paper is concluded with the complexity comparison between our template based technique, state based technique and structural encoding technique.

6.2 Petri-Net and Signal Transition Graph

To be able to introduce the methods of synthesis of asynchronous circuits in subsequent sections, we will need a more formal definition of an STG. STGs are a particular type of labeled Petri-Nets, while the definition of Petri-net is a net is a triple $N_{df} = (P; T; F)$ such that P and T are disjoint sets of respectively places and transitions, and $F \subseteq (P \times T)[(T \times P)$ is a flow relation. A marking of N is a multi set M of places, i.e., $M : P- > \{0, 1, 2...\}$. The transitions are associated with the changes in the values of binary variables. These variables can be associated with wires, when modeling interfaces between blocks, or with input, output and internal signals in a control circuit [4].

Signal Transition Graph (STG) is a quadruple $I' = (N; M0; Z; \lambda)$,

where

- $\sum = (N; M_0)$ is a Petri net (PN) based on a net $N = (P; T; F)$
- $Z^{\pm}$ is a finite set of binary signals, which generates a finite alphabet $Z^{\pm} = Z^{\pm} \times \{+, -\}$ of signal transitions
- $\lambda : T- > Z^{\pm}$ is a labeling function

An STG inherits the basic operation semantics from the behaviour underlying Petri-Net. In particular, this includes: the rules for transition enabling and firing, the notions of reachable markings, traces, and the temporal relation between transitions. STGs also inherit the various structural and behavioural properties and the corresponding classification of Petri-Nets Namely:

- **Choice place**: A place is called a choice or conflict place if it has more than one output transition.
- **Marked graph and State machine**: A Petri-Net is called a marked graph if each place has exactly one input and one output transition. A Petri-Net is called a state machine if each transition has exactly one input and one output place. Marked graph has no choice. Save state machine has no concurrency.
- **Free-choice**: A choice place is called free-choice if every its output transition has only one input place. A Petri-Net is free-choice if all its choice places are free-choice.
- **Persistency**: A transition t E T is called non-persistence if some reachable marking enables t together with another transition t'. Non-persistency of t with respect to t' is also called a direct conflict between t and t'. A Petri-Net is persistence if it does not contain any non-persistence transition.
- **Boundless and safeness**: A Petri-Net k-bounded if for every reachable marking the number of tokens in any place is not greater than k (a place is called k-bounded if for every reachable marking the number of tokens in it is not greater than k). A Petri-Net is bounded, if there is a finite k for which it is k-bounded. A Petri-Net is safe if it is 1-bounded (1 bounded place is called a safe place).
- **Liveness**: A Petri-Net is live if for every transition t and every reachable marking M there is a firing sequence that leads to a marking M' enabling t.

The properties for synthesizable STGs:

- **Persistency** – Excited signal fire, namely they cannot disable by other transition.
- **Consistency** – Signal strictly alternate between + and −.
- **Complete state coding** – Different marking must represent different states.

6.3 Asynchronous DMA Controller Specification

The Asynchronous processor is a master device on the asynchronous system bus [5], that it is able to initiate read and write transfers to all devices on the bus. The other peripherals are slave devices, which they are only able to response to read or write requests. The DMA controller is required to initiate read and write transfers to memory and also other I/O on the bus, DMA controller also required to be a master device (Fig. 6.1).

As there will be more than one master device in the system, asynchronous system bus specification required the presence of a bus arbiter. The bus arbiter selects which master is to have right to the bus at any one instant in time. An arbiter also exits for the asynchronous processor. Initially the DMA controller is programmed

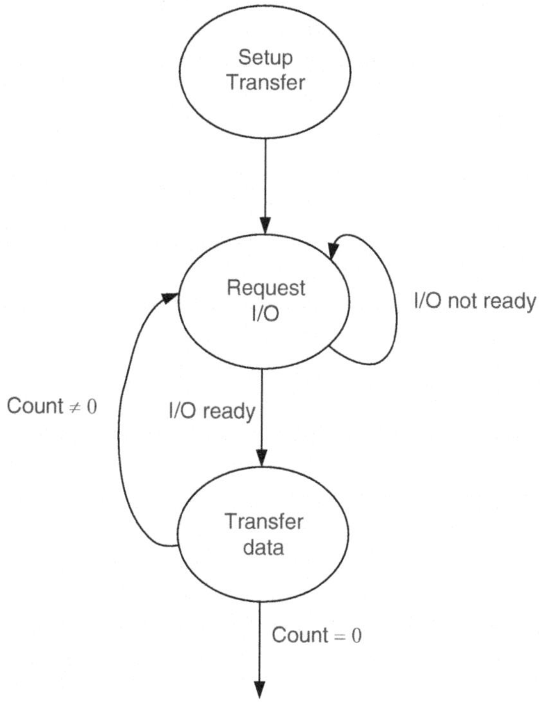

Fig. 6.1 DMA basic

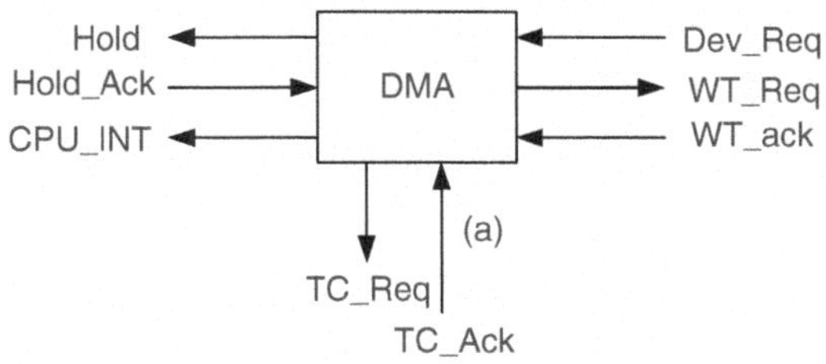

Fig. 6.2 Asynchronous DMA controller specification

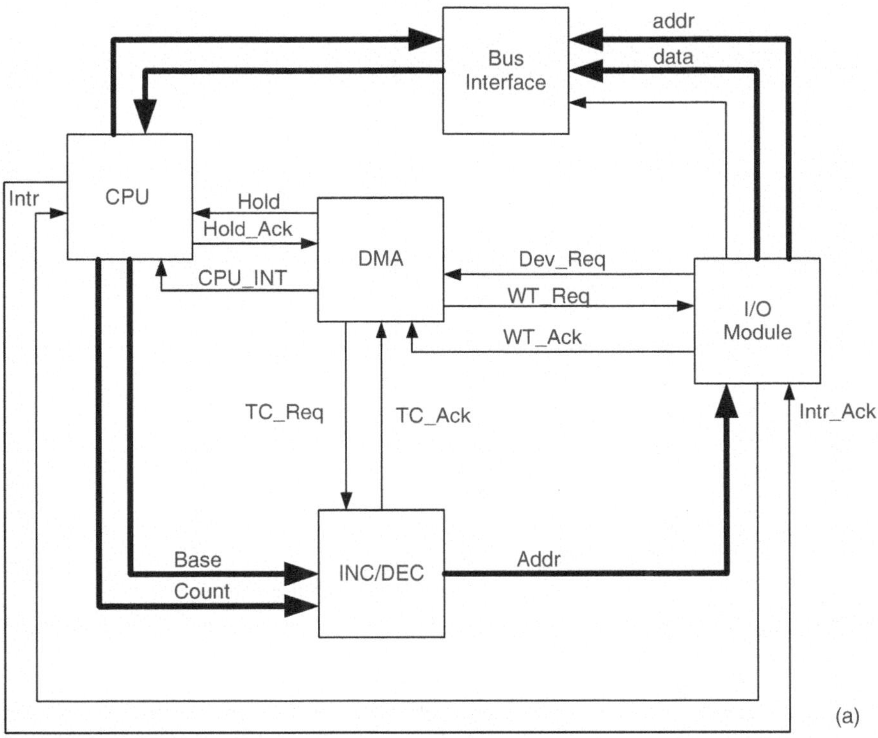

Fig. 6.3 Asynchronous DMA controller architecture

by asynchronous processor. This information is stored in internal DMA registers. Include following information: base address for transfer source, base address in memory to where the data is transferred and size of data transfer. The DMA controller also be slave on the asynchronous system bus in order the asynchronous processor access these registers.

The DMA controller waits for *Dev_req* line to assert and take over the asynchronous system bus from asynchronous processor as shown in Figs. 6.2 and 6.3. It checks its internal registers to obtained details of the transfer. The DMA controller read data from the source, and then writes it out to the memory, until the transfer is complete.

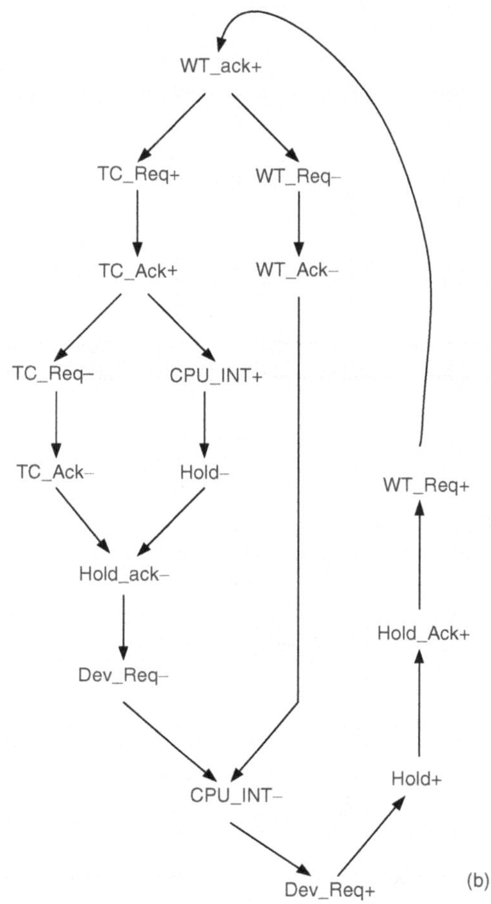

Fig. 6.4 Signal transition graph

The internal architecture is very closely based on the classic DMA controller. The architecture proposed of the DMA controller split into above functional units. The most complex of these units is timing and control unit, which consist of a large and complicated STG which described in the diagram on the Fig. 6.4.

The DMA controller works follow on STG as shown in Fig. 6.4. Finally, DMA releases the usage of asynchronous system bus and activate the *CPU_INT* line in order to indicate the asynchronous processor that transfer was complete.

6.4 Synthesis Using State Based Method

The purpose of this section is to present the synthesis of DMA controller using state-based technique. The key steps in this method are the generation of a state graph, which is a binary encoded reach ability graph of the underlying Petri-net,

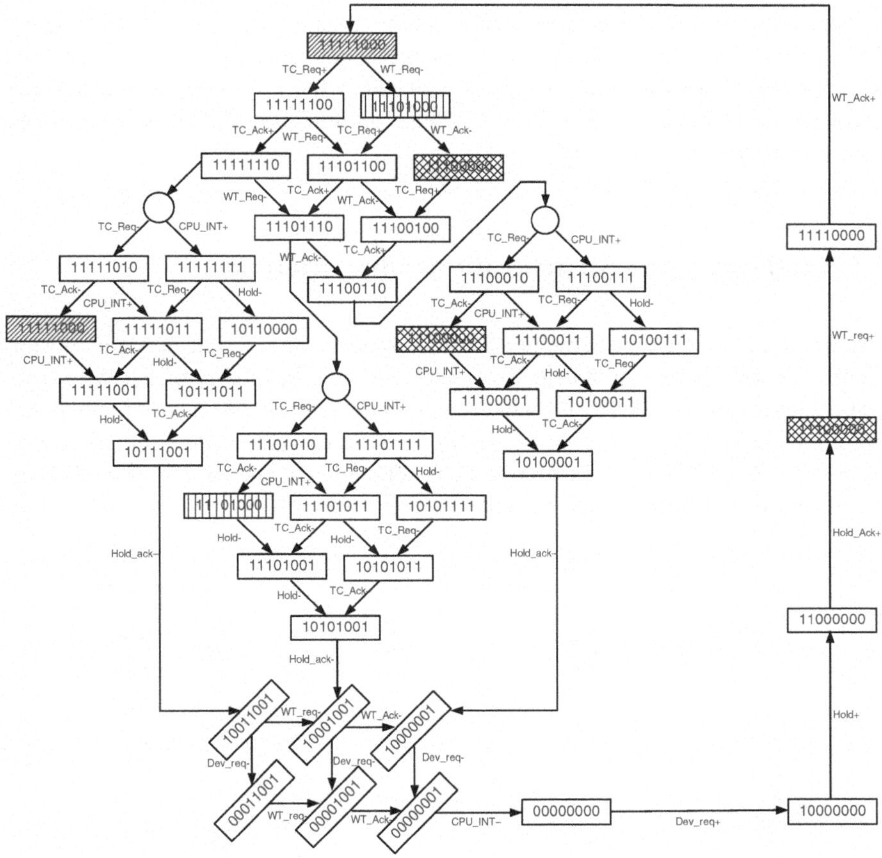

Fig. 6.5 State graph

and deriving boolean equations for the output signals via their next state functions obtained from the state-graph.

The state graph shown in Fig. 6.5 is a state space of DMA controller specification that was derived from Fig. 6.4, the shadowed states are the conflict states, it clearly suffers to state explosion problem of state based method, to solve the explosion problem, an additional states are inserted, from state graph, it hardly finds an insertion of the internal signals due to many of redundant states, it's the major problem of state based technique.

State encoding, at this point, we must have noticed a peculiar situation for the value of the next-state function for signal *that* in two states with the same binary encoding. This binary encoding is assigned to the shadowed states in Fig. 6.5. The binary encoding of the state graph signals alone cannot determine the future behavior. Hence, an ambiguity arises when trying to define the next-state function. This ambiguity is illustrated in the Karnaugh map. When this occurs, the system is said to violate the *Complete State Coding* (CSC) property. Enforcing CSC is one of the most difficult problems in the synthesis of asynchronous circuits.

This section shows a limitation and problem of state based synthesis. However, the solution of state based design is also possible; with inserting internal signals to solve the two conflicting states are disambiguated by the value of CSC, which is the last value in the binary vectors. Now Boolean minimization can be performed and logic equations can be obtained.

6.5 Synthesis Using Structural Encoding Method

The structural encoding scheme provides a way to avoid the state space explosion problem. The main benefit of using structural method is the ability to deal with large highly concurrent specifications, which cannot be tracked by state base method, the structural method for synthesis of asynchronous circuits is the class of mark graph. By transform the STG specification to Petri net, and then encode it to be well form specification as shown in Figs. 6.6 and 6.7.

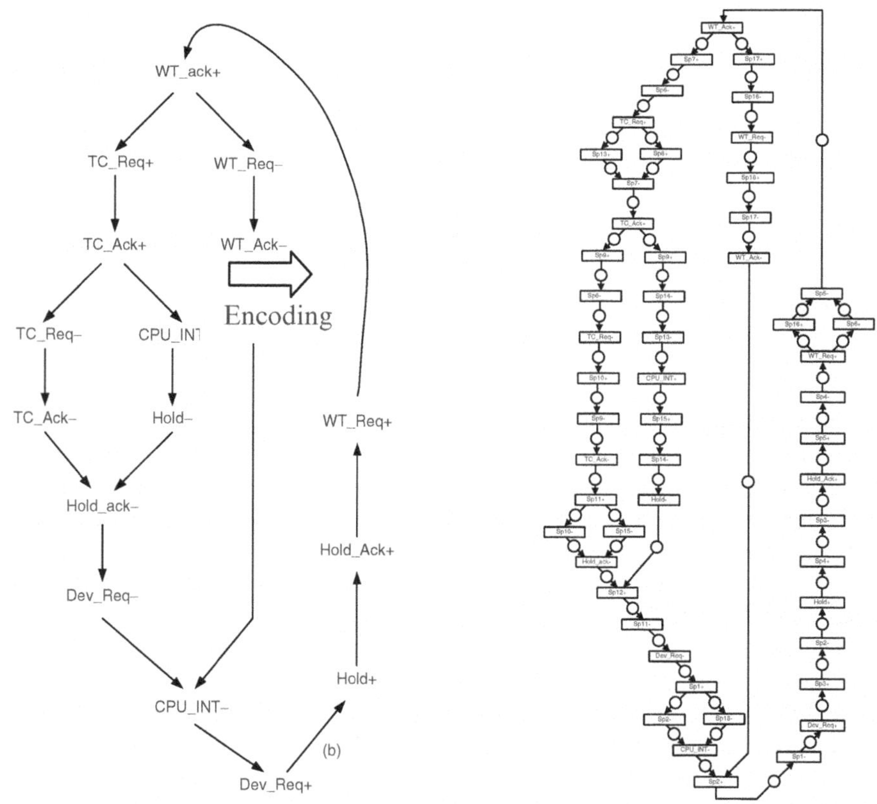

Fig. 6.6 Encoded STG

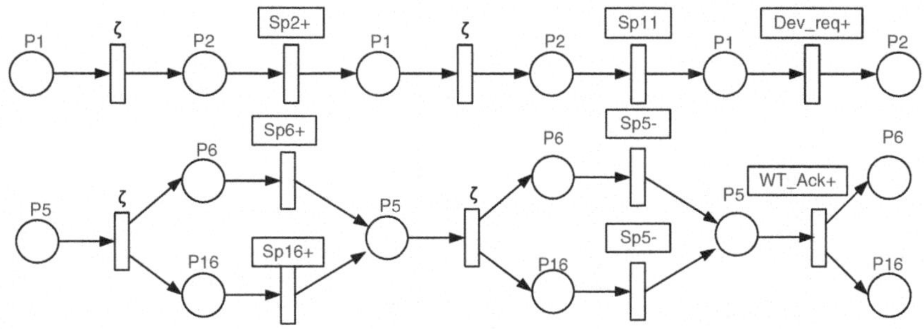

Fig. 6.7 Encoding scheme (*Dev_Req+,WT_Ack+*)

The main idea of structural method is insertion of new signals in the initial specification in a way that unique encoding is guaranteed in the transform specification.

Structural encoding technique, working at level of Petri-net can synthesize a big size of specification. State based is used in the final state of this method, when specification has been decomposing to smaller ones [4].

The synthesis flow is given a consistent STG, encode for all signals resulting an STG contains a new set of signals that ensure unique state and complete state coding property. Depending encoding technique applied, since many of encoding signal may unnecessary to guarantee unique and complete state coding, they are iteratively removed each signal using greedy heuristics until no more signal can be removed without violating unique and complete state coding property. The reduced STG next projected onto different sets of signals to implement each individual output signal. One the reduced STG is reached, it must be computed the complete state support for each non-input signal, applying CSC support algorithm as shown in next section. Afterward the projection of the STG into the CSC support for each non-input signal is performed, finally speed independent synthesis of each projection is performed, provide the small size of the projection, state based techniques can be applied for performing the synthesis. When synthesizing the projection for each non-input signal a, every signal is the projection but a is considered as an input signal. These prevent the logic synthesis to try to solve possible conflicts for the rest of signals (Fig. 6.8).

6.6 Synthesis Using State Based Method

Structural encoding can be synthesizing a big size of specification; state based is used in the final state of this method, when specification has been decomposed to smaller ones. To design with structural encoding method, Petri-net level is applied, begins with encode original STG with two silent transition, the encoding element composes of two transitions and one place, then iterative signals reduction immediately with complete state checking, finally, STG with remaining signals are

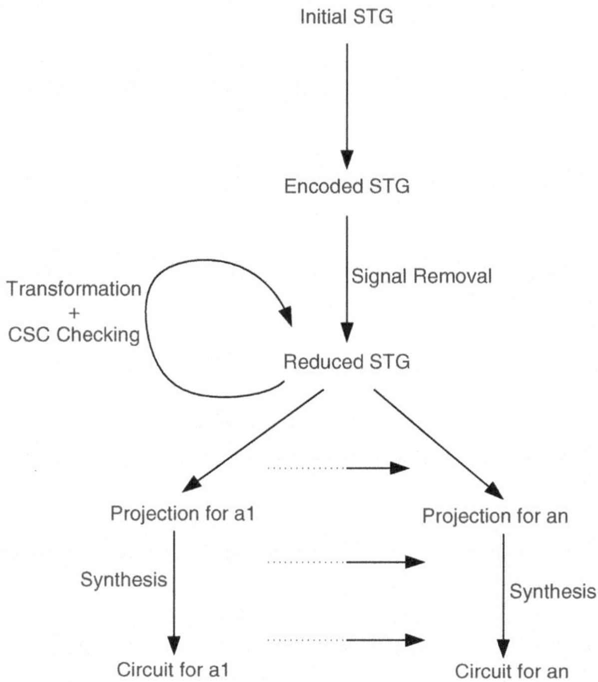

Fig. 6.8 Structural encoding synthesis

produced, and project for all non-inputs signal to get the small size CSC support, apply state based method for each CSC support to get a circuits for each non-input signals, However, structural based still consumed more wasting time in the design phase, especially signal reduction phase, each time of reduction, it must verify the persistence, consistence and complete state coding properties before reached the final reduction.

This work focuses on complexity reduction; our process begins with an encoding STG using Petri-net level in order to form a template STG. Then the projection to each non-input signals from an original STG is done. After that, we trace the projection to smaller state space. If the small state space shows conflicts, we have to insert balance signals from template STG. Unbalance signals are inserted after in case the space still shows conflicts. Finally, we can get the STG with appropriate insertion points which are used to be projected for CSC support on each non-input signal as shown in Fig. 6.9.

Template based method (insertion point calculation)

Let:

a = non input signal

b = balance signal

ub = unbalance signal

Original STG Projection (a)

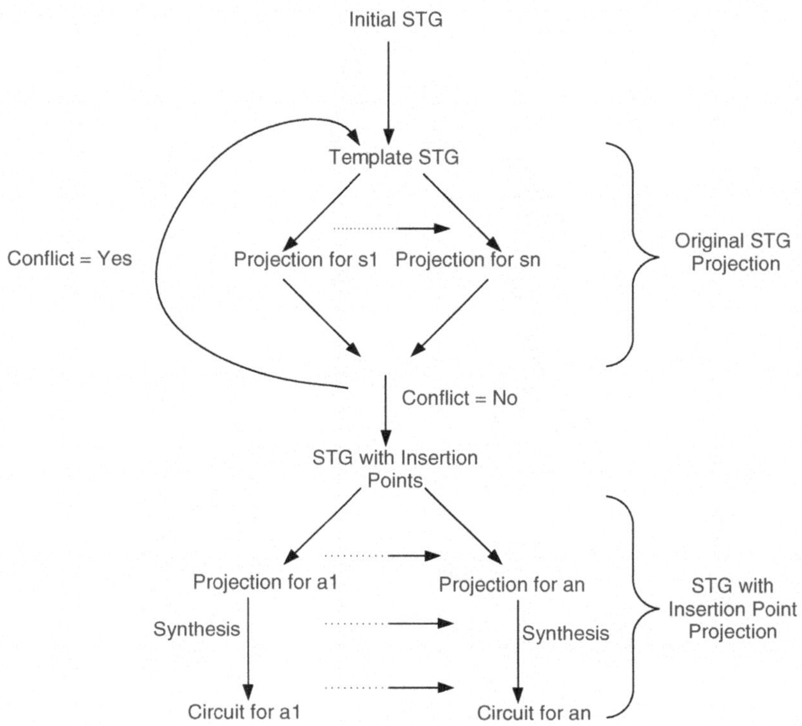

Fig. 6.9 Template based synthesis

While not Conflict do
Trace a;
If Conflict Then
Insert b from template STG;
Else If Still Conflict Then
Insert ub from template STG;
Else
Trace a;
End If;

Figure 6.10 shows the template STG that encoded with the encoding scheme in Fig. 6.7.

After the template based technique was applied. We can get the STG with appropriate insertion signals as shown in Fig. 6.11.

CSC Support calculation:

CSC Support (STG S, a) return CSC support of a
CSC support a = Trigger (a) U {a}
While it's still conflict do
Let b an unbalance signal in z
CSC Support for a = CSC Support for a U {b}

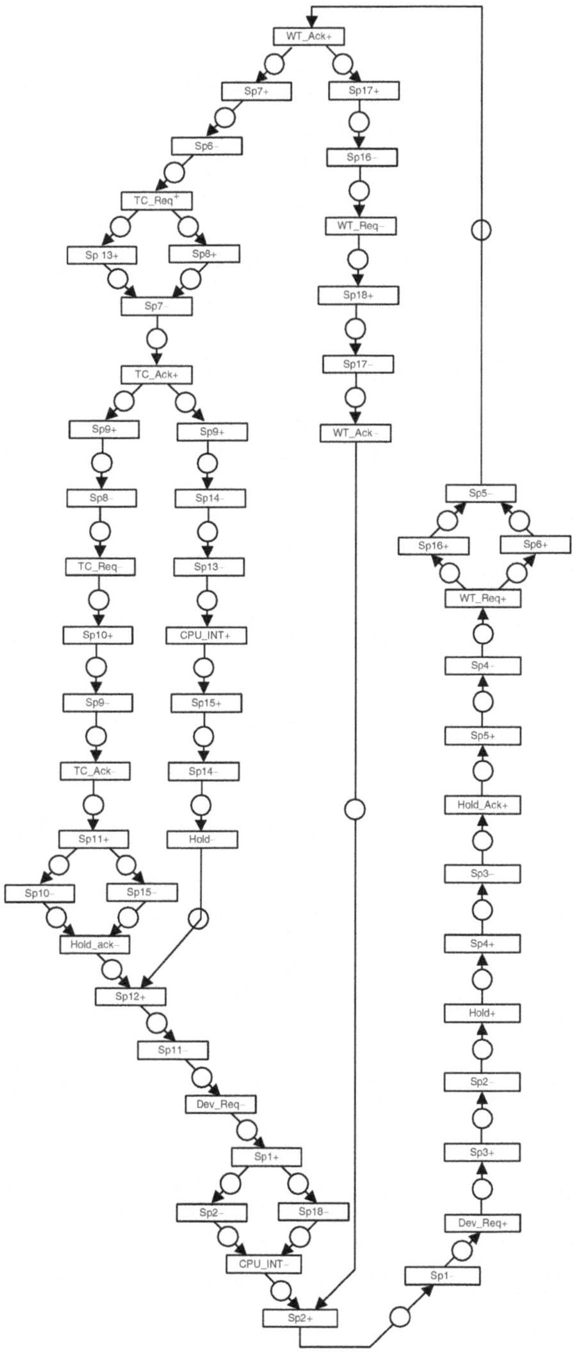

Fig. 6.10 Template STG

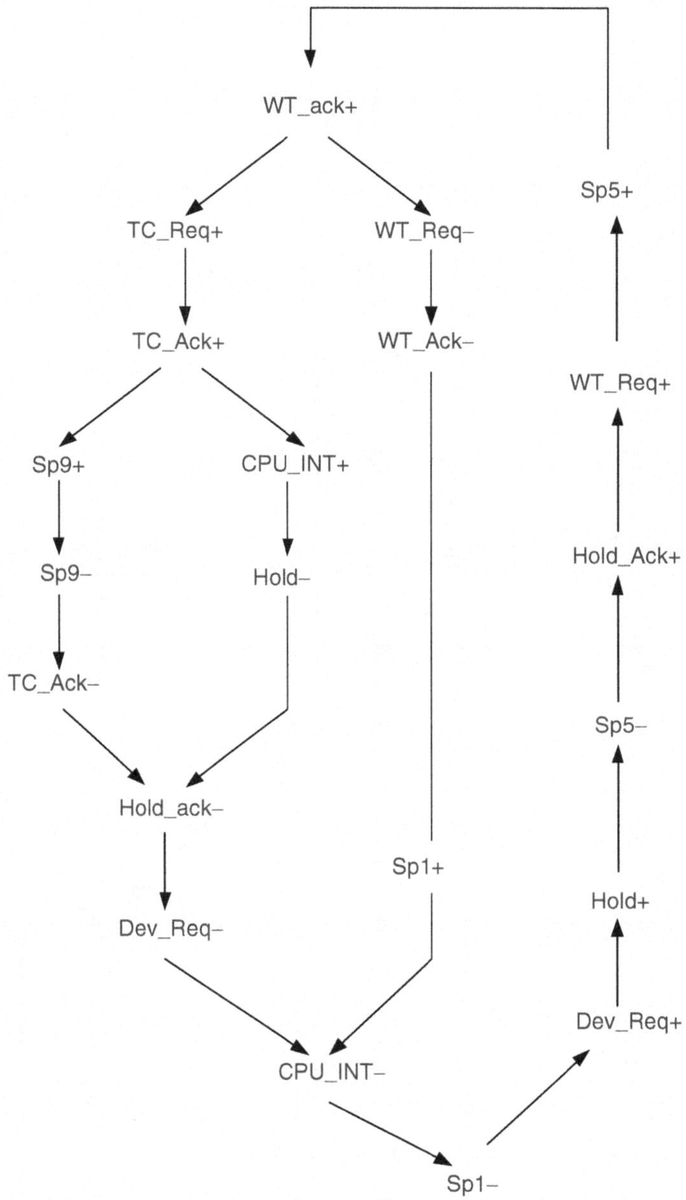

Fig. 6.11 STG with appropriate insertion signals

CSC support calculation for each non-input (output and internal signals) signal is calculated. Finally we got the small state graph which guarantee that it save from state explosion and also satisfied the synthesizing properties (Figs. 6.12 and 6.13).

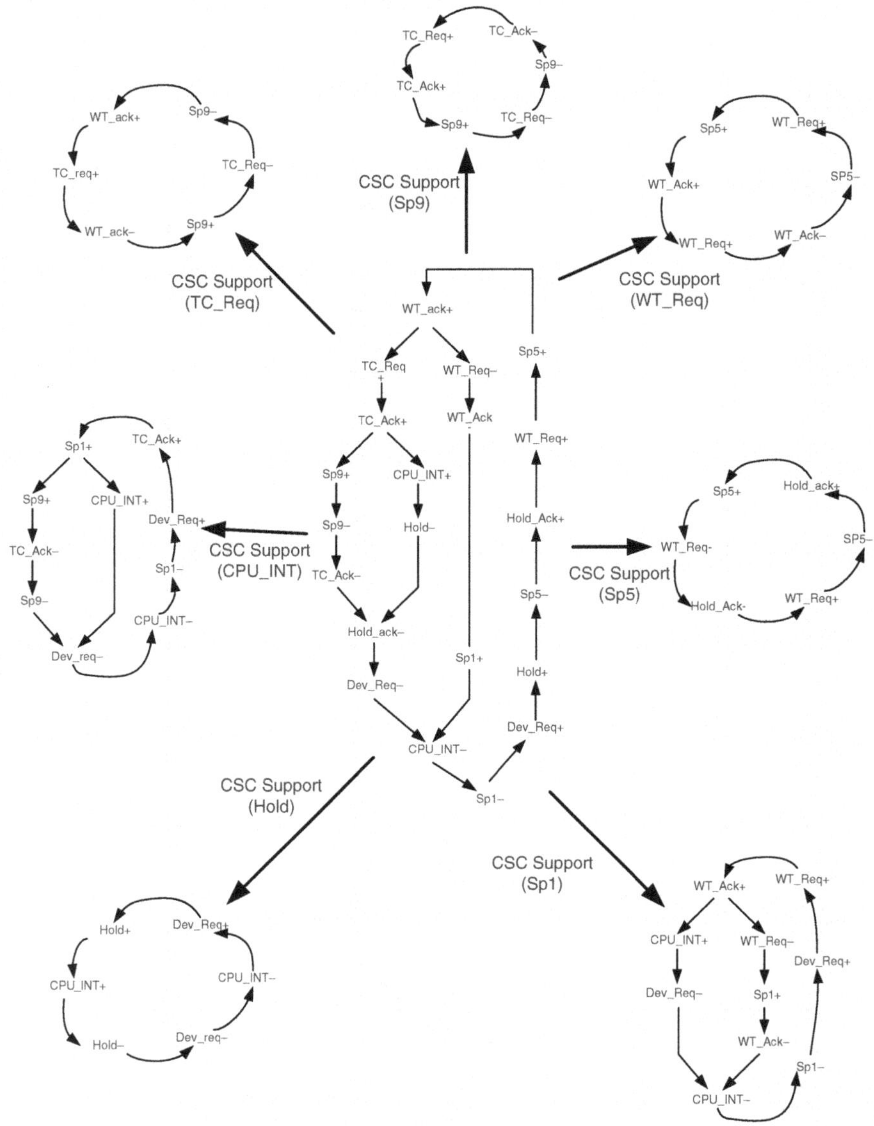

Fig. 6.12 STG with Insertion Point and CSC support calculation

6.7 Result

This section presents complexity comparison between structural encoding method and template based method. The conventional state based method is not shown on a comparison table, it clearly suffers its exponential complexity of conflict state with the size of specification, and others difficult problem. Due to absence of synthesis

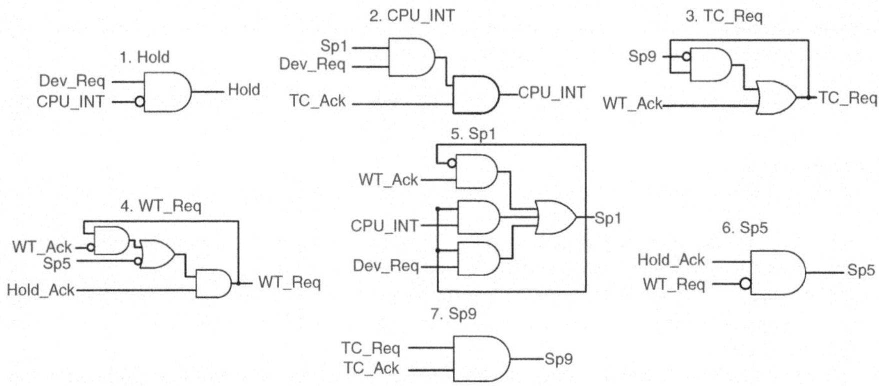

Fig. 6.13 Synthesized circuits from CSC support of Fig.6.12

Table 6.1 Result

Circuits	Signals	Structural encoding		Template based	
		Iteration	Inserted signals	Iteration	Inserted signals
VME bus	6	**10**	2	**6**	2
Asyn DMA	8	**10**	6	**7**	6
Pipeline-Scheduler	14	**26**	2	**18**	2
QDI-bus	16	**28**	4	**20**	3
Pipeline control block	13	**23**	3	**18**	3
Classical DMA	11	**13**	9	**8**	6
PPArbCSC (2,3)	10	**18**	2	**8**	2

tools for asynchronous controller synthesis, all circuits in Table 6.1 were manually synthesis.

6.8 Concluding Remarks

This paper proposes template based technique for reducing additional signals in signal transition graph (STG) based logic synthesis, and solves the reduction complexity of structural encoding method; our method is explained by asynchronous DMA controller specification. The final section showed the comparison between our template based method with state based and structural encoding method. The number of iterative signal removal according to our method is less than others, Roughly speaking, structural encoding is better than classical state based method, and template based method is less complexity than structural encoding.

Acknowledgement The research is supported by TJTTP-OECF (Thailand – Japan Technology Transfer Project – Japanese Overseas Economic Cooperation Fund).

References

1. S.B. Park. Synthesis of Asynchronous VLSI circuit from Signal Transition Graph Specifications, Ph.D. thesis, Tokyo Institute of Technology, 1996.
2. J. Carmona, J. Cortadella. ILP models for the synthesis of asynchronous control circuits. In Proceedings of the International Conference Computer-Aided Design (ICCAD), pp. 818–825, November 2003.
3. J. Carmona and J. Cortadella. State encoding of large asynchronous controllers. In Proceedings of the ACM/IEEE Design Automation Conference, pp. 939–944, July 2006.
4. J. Carmona, J. Cortadella, and E. Pastor. A structural encoding technique for the synthesis of asynchronous circuits. In International Conference on Application of Concurrency to System Design, pp. 157–166, June 2001.
5. S. Sudeng and A. Thongtak FPGA Implementation of Quasi-Delay Insensitive Microprocessor. The World Congress on Engineering and Computer Science (WCECS2007), Clark Kerr Campus, University of California Berkeley, San Francisco, CA, USA on 24–26 October 2007.

Chapter 7
A Comparison of Induction Motor Speed Estimation Using Conventional MRAS and an AI-Based MRAS Parallel System

Chao Yang and Professor John W. Finch

Abstract The Model Reference Adaptive System (MRAS) is probably the most widely applied speed sensorless drive control scheme. This chapter compares induction motor speed estimation using conventional MRAS and AI-based MRAS with Stator Resistance Compensation. A conventional mathematical model based MRAS speed estimation scheme can give a relatively precise speed estimation result, but errors will occur during low frequency operation. Furthermore, it is also very sensitive to machine parameter variations. An AI-based MRAS-based system with a Stator Resistance Compensation model can improve the speed estimation accuracy and is relatively robust to parameter variations even at an extremely low frequency. Simulation results using a validated machine model are used to demonstrate the improved behaviour.

Keywords Dynamic reference model · Model Reference Adaptive System (MRAS) · neural networks · induction motor control

7.1 Introduction

Much effort has been devoted to speed-sensorless induction machine drive schemes, with the Model Reference Adaptive System (MRAS) being the most popular [1]. In a conventional mathematical-model-based MRAS, some state variables are estimated in a reference model (e.g. rotor flux linkage components, ψ_{rd}, ψ_{rq}, or back e.m.f. components, e_d, e_q, etc.) of the induction machine obtained by using measured quantities (e.g. stator currents and perhaps voltages).

These reference model components are then compared with state variables estimated using an adaptive model. The difference between these state variables is then

Prof. J. W. Finch (✉)
School of Electrical, Electronic and Computer Engineering, Newcastle University, Newcastle upon Tyne, NE1 7RU, UK,
E-mail: J.W.Finch@ncl.ac.uk

S.I. Ao and L. Gelman (Eds.), *Advances in Electrical Engineering and Computational Science*, Lecture Notes in Electrical Engineering.

used in an adaptation mechanism, which, for example, outputs the estimated value of the rotor speed ($\hat{\omega}_r$) and adjusts the adaptive model until satisfactory performance is obtained [2–6].

Greater accuracy and robustness can be achieved if the mathematical model is not used at all and instead an AI-based non-linear adaptive model is employed. It is then also possible to eliminate the need of the separate PI controller, since this can be integrated into the tuning mechanism of the AI-based model [7].

However, both the conventional MRAS and AI-based MRAS scheme are easily affected by machine parameter variations, which happen during practical operation [8, 9]. Hence an online stator resistance estimator is applied to the AI-based MRAS scheme which is shown to make the whole scheme more robust using a computer simulation and could make the scheme usable for practical operation [10, 11]. The comparison of schemes presented here is felt to be valuable since much of the literature presents results for the novel approach alone [1].

7.2 Speed Estimation Using Conventional Model Reference Adaptive System

It is possible to estimate the rotor speed by using two estimators (a reference-model-based estimator and an adaptive-model-based one), which independently estimate the rotor flux-linkage components in the stator reference frame ψ_{rd}, ψ_{rq}), and, using the difference between these flux-linkage estimates, drive the speed of the adaptive model to that of the actual speed [12]. The expressions for the rotor flux linkages in the stationary reference frame can be obtained by using the stator voltage equations of the induction machine (in the stationary reference frame). These give (7.1) and (7.2), which are now rearranged for the rotor flux linkages:

$$\psi_{rd} = (L_r/L_m)\left[\int (u_{sD} - R_s i_{sD})dt - L'_s i_{sD}\right] \tag{7.1}$$

$$\psi_{rq} = (L_r/L_m)\left[\int (u_{sQ} - R_s i_{sQ})dt - L'_s i_{sQ}\right] \tag{7.2}$$

These two equations represent a so-called stator voltage model, which does not contain the rotor speed and is therefore a reference model. However, when the rotor voltage equations of the induction machine are expressed in the stationary reference frame, they contain the rotor fluxes and the speed as well. These are the equations of the adaptive model:

$$\hat{\psi}_{rd} = (1/T_r)\int \left(L_m i_{sD} - \hat{\psi}_{rd} - \hat{\omega}_r T_r \hat{\psi}_{rq}\right) dt \tag{7.3}$$

$$\hat{\psi}_{rq} = (1/T_r)\int \left(L_m i_{sQ} - \hat{\psi}_{rq} - \hat{\omega}_r T_r \hat{\psi}_{rd}\right) dt \tag{7.4}$$

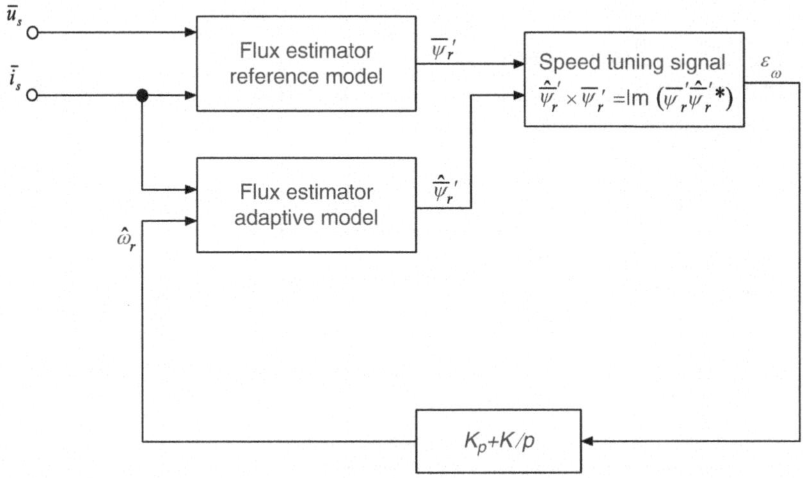

Fig. 7.1 MRAS-based rotor speed observer using rotor flux linkages for the speed tuning signal

The reference and adaptive models are used to estimate the rotor flux linkages and the angular difference of the outputs of the two estimators is used as the speed tuning signal. This tuning signal is the input to a linear controller (PI controller) which outputs the estimated rotor speed as shown in Fig. 7.1. The estimated speed can be expressed as (7.5)

$$\hat{\omega}_r = K_p \varepsilon_\omega + K_i \int \varepsilon_\omega dt \tag{7.5}$$

7.3 Artificial Intelligence-Based Model Reference Adaptive System

The MRAS-based schemes described in the previous section contain a reference model and an adaptive model. Greater accuracy and robustness can be achieved if the mathematical model is partially replaced by a neural network. It is then also possible to eliminate the need of the separate PI controller, since this can be integrated into the tuning mechanism of the neural network-based model.

The neural network-based model can take various forms: it can be an artificial neural network (ANN) or a fuzzy-neural network etc. and there is also the possibility of using different types of speed tuning signals. It is believed that some of these solutions can give high accuracy and are relatively robust to parameter variations even at extremely low stator frequency.

One specific implementation of the ANN-based MRAS speed estimator system which is popular in academic work, shown in Fig. 7.2, which is similar to the conventional MRAS system. In this new model the adaptive model is replaced by a

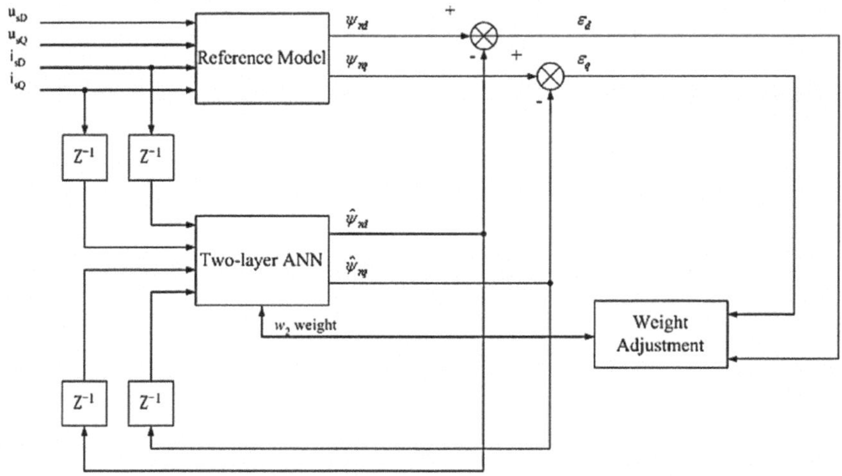

Fig. 7.2 MRAS-based rotor speed estimator containing a two-layer ANN

simple two layer neural network, which gives the whole system a fast response and better accuracy than the conventional MRAS [13, 14].

7.4 MRAS Based Two-Layer ANN Speed Estimator with Dynamic Reference Model

Compared to the conventional MRAS scheme, the MRAS based rotor speed estimator containing a two-layer ANN could give more accurate estimation result and be relatively robust to parameters variations. The two-layer ANN replaces the adjustable model and adaptive mechanism in the conventional MRAS, but the reference model is still necessary for estimation the rotor flux which is used as speed tuning signal. Several machine parameters are used to build the conventional reference model, such as stator resistance (R_s) and stator reluctance (L_s). These parameters may change during the different periods of motor operation. The values of these parameters are fixed in the reference model. So the ANN speed estimator is still sensitive to parameter variations especially during the motor low speed running period.

To solve this problem and make this scheme more independent to the machine parameters a stator resistance estimator is built in the new reference model, in which the stator resistance R_s value could be estimated online. Figure 7.3 shows the total scheme of this neural network based MRAS with a dynamic reference model.

In this new system, both the reference model and adaptive model of the conventional MRAS system are modified for better performance. The whole system can be divided into two main parts, the dynamic reference model part and the neural network part. The dynamic reference part consists of the dynamic reference model derived from Eqs. (7.1) and (7.2), in which the stator resistance R_s is replaced by

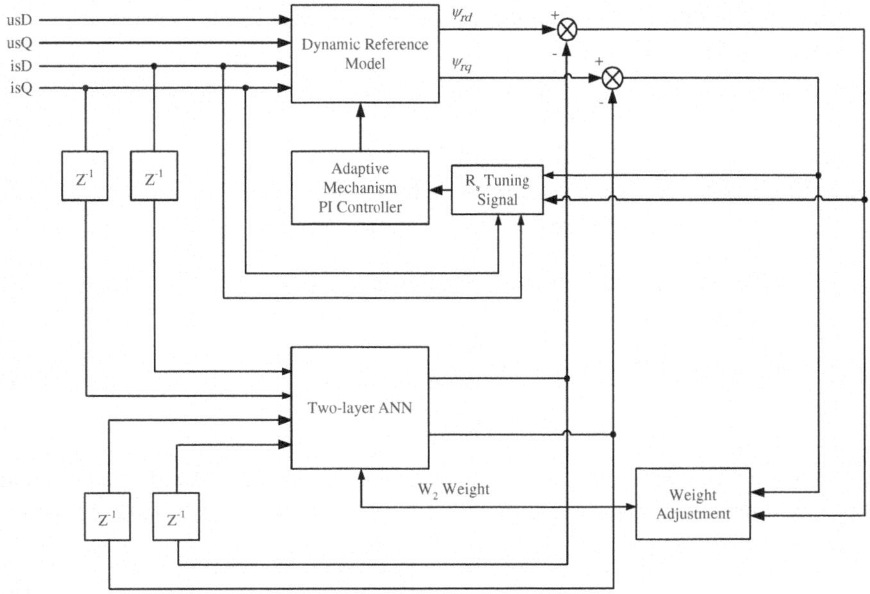

Fig. 7.3 MRAS based ANN speed estimator with dynamic reference model

the online estimated value $\hat{R}_s$ coming from Eqs. (7.6) and (7.7),

$$\hat{R}_s = \left(K_p + \frac{K_i}{p}\right) eR_s \tag{7.6}$$

$$eR_s = i_{sD}(\psi_{rd} - \hat{\psi}_{rd}) + i_{sQ}(\psi_{rq} - \hat{\psi}_{rq}) \tag{7.7}$$

The neural network part contains a simple two-layer network, with only an input layer and an output layer. Adjustable and constant weights are built in the neural network, and the adjustable weights are proportional to the rotor speed. The adjustable weights are changed by using the error between the outputs of the reference model and the adjustable model, since any mismatch between the actual rotor speed and the estimated rotor speed results in an error between the outputs of the reference and adaptive estimators.

To obtain the required weight adjustments in the ANN, the sampled data forms of Eqs. (7.3) and (7.4) are considered. using the backward difference method the sampled data forms of the equations for the rotor flux linkages can be written as (7.8) and (7.9), where T is the sampling time.

$$\begin{aligned}&[\hat{\psi}_{rd}(k) - \hat{\psi}_{rd}(k-1)]/T = -\hat{\psi}_{rd}(k-1)/T_r \\ &-\omega_r\hat{\psi}_{rq}(k-1) + (L_m/T_r)i_{sD}(k-1)\end{aligned} \tag{7.8}$$

$$\begin{aligned}&[\hat{\psi}_{rq}(k) - \hat{\psi}_{rq}(k-1)]/T = -\hat{\psi}_{rq}(k-1)/T_r \\ &-\omega_r\hat{\psi}_{rd}(k-1) + (L_m/T_r)i_{sQ}(k-1)\end{aligned} \tag{7.9}$$

Thus the rotor flux linkages at the kth sampling instant can be obtained from the previous $(k-1)$th values as

$$\begin{aligned}\hat{\psi}_{rd}(k) &= \hat{\psi}_{rd}(k-1)(1-T/T_r)\\ &-\omega_r T\hat{\psi}_{rq}(k-1)+(L_m T/T_r)i_{sD}(k-1)\end{aligned} \tag{7.10}$$

$$\begin{aligned}\hat{\psi}_{rq}(k) &= \hat{\psi}_{rq}(k-1)(1-T/T_r)\\ &-\omega_r T\hat{\psi}_{rd}(k-1)+(L_m T/T_r)i_{sQ}(k-1)\end{aligned} \tag{7.11}$$

Introducing $c = T/T_r$, the following weights are given:

$$\begin{aligned} w_1 &= 1-c\\ w_2 &= \omega_r c T_r = \omega_r T\\ w_3 &= cL_m \end{aligned} \tag{7.12}$$

It can be seen that w_1 and w_3 are constant weights, but w_2 is a variable weight and is proportional to the speed. Thus Eqs. (7.10) and (7.11) take the following forms:

$$\begin{aligned}\hat{\psi}_{rd}(k) &= w_1\hat{\psi}_{rd}(k-1)-w_2\hat{\psi}_{rq}(k-1)\\ &+w_3 i_{sD}(k-1)\end{aligned} \tag{7.13}$$

$$\begin{aligned}\hat{\psi}_{rq}(k) &= w_1\hat{\psi}_{rq}(k-1)+w_2\hat{\psi}_{rd}(k-1)\\ &+w_3 i_{sQ}(k-1)\end{aligned} \tag{7.14}$$

These equations can be visualized by the very simple two-layer ANN shown in Fig. 7.4.

The neural network is training by the backpropagation method, the estimated rotor speed can be obtained from:

$$\begin{aligned}\hat{\omega}_r(k) &= \hat{\omega}_r(k-1)+\Delta w_2(k)/T+(\alpha/T)\Delta w_2(k-1)\\ &= \hat{\omega}_r(k-1)+\eta\{-[\psi_{rd}(k)-\hat{\psi}_{rd}(k)]\hat{\psi}_{rq}(k-1)\\ &+[\psi_{rq}(k)-\hat{\psi}_{rq}(k)]\hat{\psi}_{rd}(k-1)\}/T+(\alpha/T)\Delta w_2(k-1)\end{aligned} \tag{7.15}$$

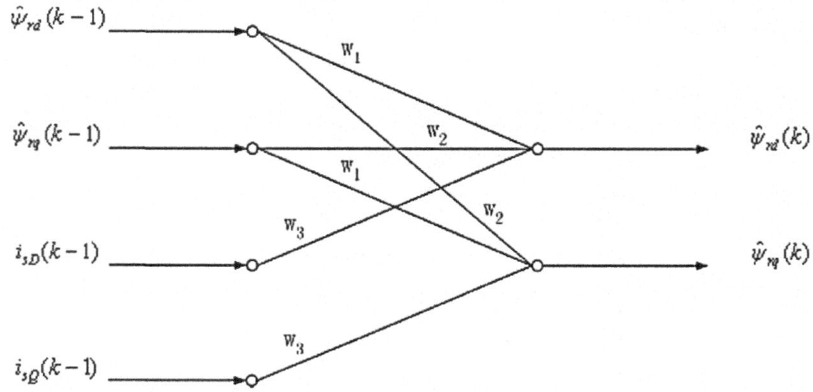

Fig. 7.4 Neural network representation for estimated rotor flux linkages

where η is the learning rate and α is a positive constant called the momentum constant. The inclusion of the momentum term into the weight adjustment mechanism can significantly increase the convergence, which is extremely useful when the ANN shown in Fig. 7.4 is used to estimate in real time the speed of the induction machine.

7.5 Simulation Results and Discussion

To compare the conventional MRAS and the AI-based MRAS with dynamic reference model, simulations are established by using Matlab-Simulink software, based on the standard well established validated 2-axis machine model [6].

Speed estimation results using conventional MRAS and neural network based MRAS are shown in Figs. 7.5 and 7.6 respectively. These results assume the machine parameters are correctly measured and unchanged during operation. Both of the two schemes can give good speed tracking results.

Further simulation were carried out with changed stator resistance to test how much changing parameters would affect the speed estimation results.

In Figs. 7.7 and 7.8, simulations are carried out with the stator resistance changed by a small amount, 2%. Obviously both schemes are still sensitive to parameter variations.

A final simulation for AI-based MRAS with the dynamic reference model is shown in Fig. 7.9. The online estimated stator resistance is displayed in Fig. 7.10. This simulation result shows the effect caused by the stator resistance variation has been considerably improved.

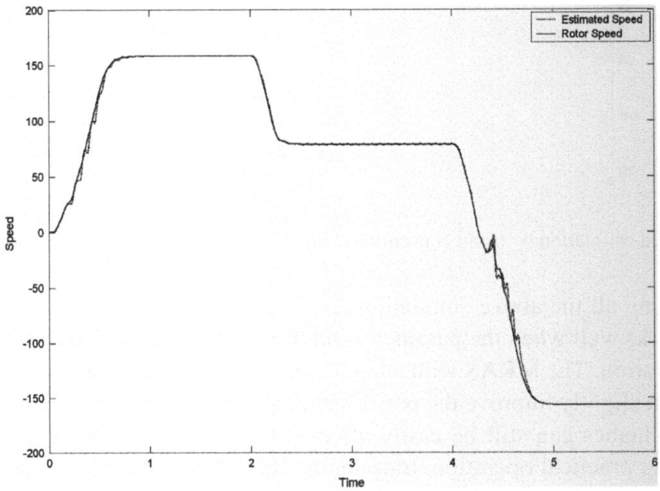

Fig. 7.5 Speed estimation using Conventional MRAS

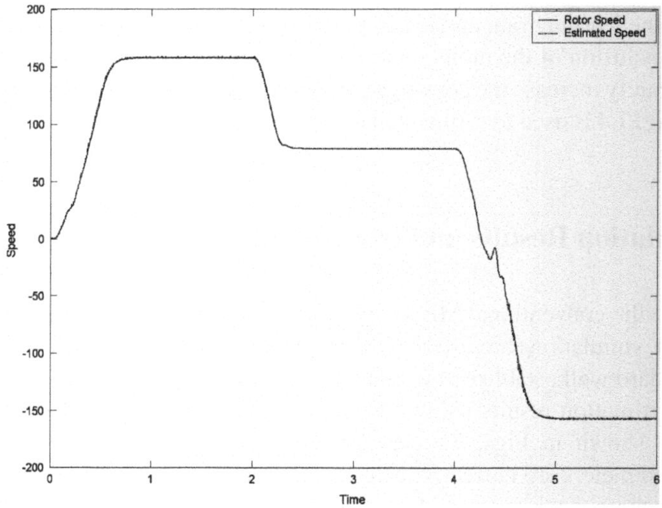

Fig. 7.6 Speed estimation using two-layer ANN MRAS

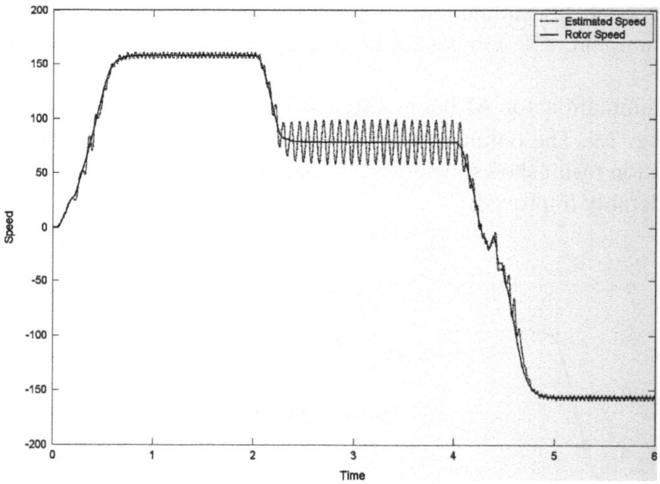

Fig. 7.7 Speed estimation by using conventional MRAS (with stator resistance R_s changed 2%)

Comparing all the above simulation results shows that the conventional MRAS scheme works well when the parameters are precisely measured and do not change during operation. The MRAS with adaptive model replaced by the two-layer neural network can slightly improve the performance when working in the same situation. But both schemes can still be easily affected by parameters variations, which do occur during practical operation. Introducing the stator resistance online estimator gives much improved performance which should enable the scheme to be usable for practical operation.

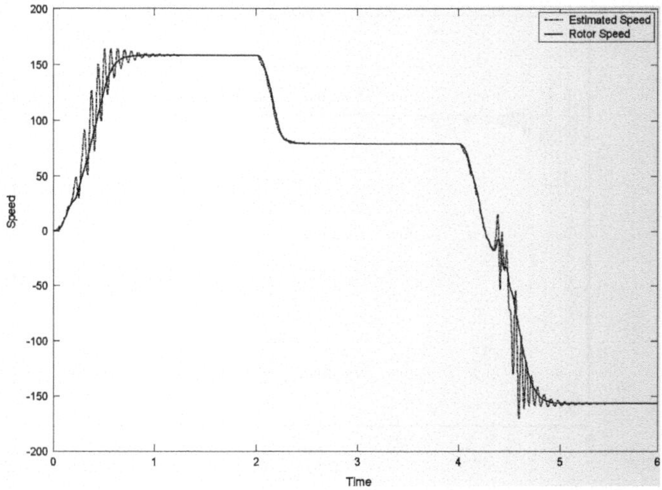

Fig. 7.8 Speed estimation using two-layer ANN MRAS (with stator resistance R_s changed 2%)

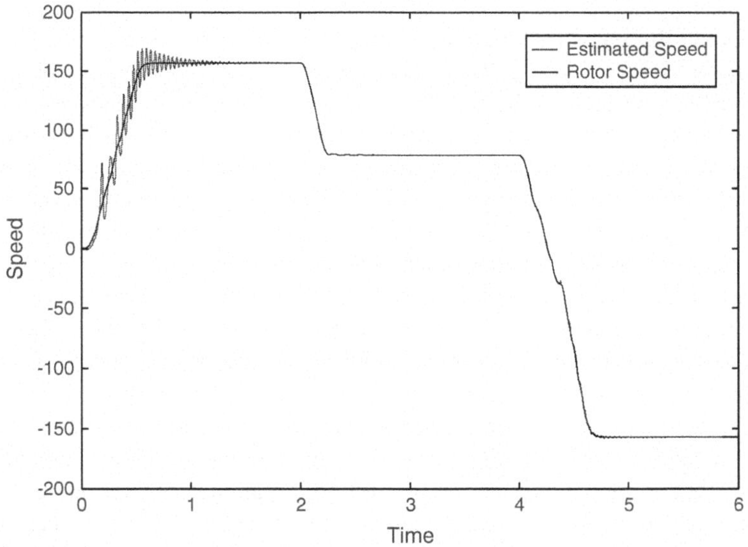

Fig. 7.9 Speed estimation using two-layer MRAS with dynamic reference model

7.6 Conclusion

The main objective of this chapter is to compare conventional MRAS and AI-based MRAS parallel system for induction motor sensorless speed estimation. The conventional MRAS can give good speed estimation in most of the operation period, but errors will occur during low frequency operation mainly caused by machine parameter variations. An AI-based MRAS system can give improved accuracy and

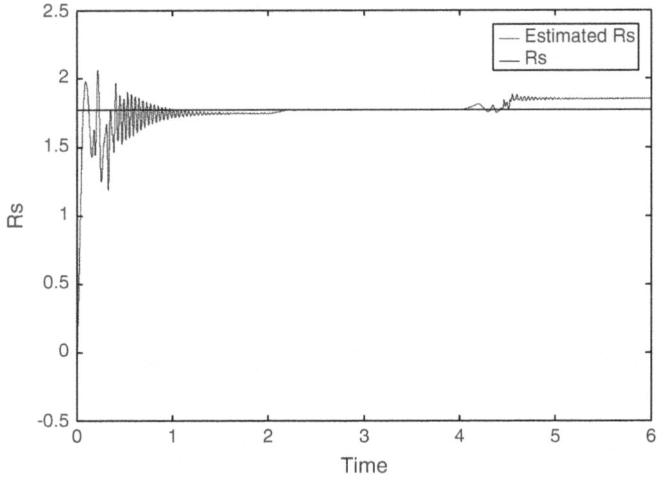

Fig. 7.10 Estimated R_s in the dynamic reference model

bypasses the PI controller tuning problems. The simple structure of the two-layer neural network shown in Fig. 7.4 yields a speed estimation system working online with a fast response. Also the simple two-layer neural network does not require a separate learning stage, since the learning takes place during the on-line speed estimation process. This is mainly due to the fact that the development time of such an estimator is short and the estimator can be made robust to parameter variations and noise. Furthermore, in contrast to most conventional schemes, it can avoid the direct use of a speed-dependent mathematical model of the machine.

However, the Two-layer neural network MRAS lies more in the realm of adaptive control then neural networks. The speed value is not obtained at the output, but as one of the weights. Moreover, only one weight is adjusted in the training. Therefore, it would still be sensitive to parameter variations and system noise.

In the new approach, an online stator resistance estimator is used to compensate the parameter variations. The computer simulation results show that this new approach makes the whole scheme more robust to parameter variations, enhancing the possibility of practical use of the neural network based MRAS scheme. The stator resistance estimator works under an adaptive mechanism (PI controller). Further study could be carried out to replace the PI controller with another simple neural network which could also estimate more machine parameters.

References

1. Finch, J.W. and Giaouris, D., *Controlled AC Electrical Drives*, IEEE Transactions on Industrial Electronics, Feb. 2008, 55, 1, 1–11.
2. Landau, Y.D., *Adaptive Control the Model Reference Approach*, 1979: Marcel Dekker, New York.

3. Vas, P., *Sensorless Vector and Direct Torque Control*, 1998: Oxford University Press, Oxford.
4. Shauder, C., *Adaptive Speed Identification for Vector Control of Induction Motors without Rotational Transducers.* IEEE Transactions on Industry Applications, 1992, 28, 1054–1062.
5. Yang, G. and T. Chin, *Adaptive-Speed Identification Scheme for a Vector-Controlled Speed Sensorless Inverter-Induction Motors.* IEEE Transactions on Industry Applications, 1993, 29, 820–825.
6. Fitzgerald, A.E., C. Kingsley, and S.D. Umans, *Electric Machinery.* 6th ed., 2003: McGraw-Hill International Edition, New York.
7. Vas, P., *Artificial-Intelligence-Based Electrical Machines and Drives.* 1999: Oxford University Press, Oxford.
8. Kumara, I.N.S., *Speed Sensorless Field Oriented Control for Induction Motor Drive.* Ph.D. thesis, 2006, University of Newcastle upon Tyne.
9. Leonhard, W., *Controlled AC Drives, a Successful Transition from Ideas to Industrial Practice*, 1996: Elsevier, Amsterdam, Netherlands.
10. Zhen, L. and L. Xu, *Sensorless Field Orientation Control of Induction Machines Based on a Mutual MRAS Scheme.* IEEE Transactions on Industrial Electronics, 1998, 45, 824–831.
11. Holtz, J. and J. Quan, *Drift-and Parameter-Compensated Flux Estimator for Persistent Zero-Stator-Frequency Operation of Sensorless-Controlled Induction Motors.* IEEE Transactions on Industry Applications, 2003, 39, 1052–1060.
12. Ohtani, T., N. Takada, and K. Tanaka, *Vector Control of Induction Motor without Shaft Encoder.* IEEE Transactions on Industry Applications, 1992, 28, 157–164.
13. Peng, F.Z. and T. Fukao, *Robust Speed Identification for Speed-Sensorless Vector Control of Induction Motors.* IEEE Transactions on Industry Applications, 1994, 30, 945–953.
14. Vasic, V. and S. Vukosavic, *Robust MRAS-Based Algorithm for Stator Resistance and Rotor Speed Identification.* IEEE Power Engineering Review, 2001, 21, 39–41.

Chapter 8
A New Fuzzy-Based Additive Noise Removal Filter

M. Wilscy and Madhu S. Nair

Abstract A digital color image C can be represented in different color space such as RGB, HSV, L*a*b etc. In the proposed method, RGB space is used as the basic color space. Different proportions of red, green and blue light gives a wide range of colors. Colors in RGB space are represented by a 3-D vector with first element being red, the second being green and third being blue, respectively. The general idea in this method is to take into account the fine details of the image such as edges and color component distances, which will be preserved by the filter. The goal of the first filter is to distinguish between local variations due to image structures such as edges. The goal is accomplished by using Euclidean distances between color components instead of differences between the components as done in most of the existing filters. The proposed method uses 2-D distances instead of 3-D distances, and uses three fuzzy rules to calculate weights for the Takagi-Sugeno fuzzy model.

Keywords Additive Noise Removal Filter · Fuzzy-Based · digital color image · fuzzy rules · Image restoration

8.1 Introduction

Images are often degraded by random noise. Noise can occur during image capture, transmission or processing, and may be dependent on or independent of image content. Noise is usually described by its probabilistic characteristics. Gaussian noise is a very good approximation of noise that occurs in many practical cases [1].

The ultimate goal of restoration techniques is to improve an image in some predefined sense. Although there are areas of overlap, image enhancement is largely

M. S. Nair (✉)
Rajagiri School of Computer Science, Rajagiri College of Social Sciences, Kalamassery, Kochi 683104, Kerala, India,
E-mail: madhu_s_nair2001@yahoo.com

S.I. Ao and L. Gelman (Eds.), *Advances in Electrical Engineering and Computational Science*, Lecture Notes in Electrical Engineering.

a subjective process, while image restoration is for the most part an objective process. Restoration attempts to reconstruct or recover an image that has been degraded by using a priori knowledge of the degradation phenomenon [2]. Thus restoration techniques are oriented toward modeling the degradation and applying the inverse process in order to recover the original image. This approach usually involves formulating a criterion of goodness that will yield an optimal estimate of the desired result. By contrast, enhancement techniques basically are heuristic procedures designed to manipulate an image in order to take advantage of the psychophysical aspects of human visual system. For example, histogram equalization is considered an enhancement technique because it is primarily on the pleasing aspects it might present to the viewer, whereas removal of image blur by applying a deblurring function is considered a restoration technique.

Image restoration differs from image enhancement in that the latter is concerned more with accentuation or extraction of image features rather than restoration of degradations. Image restoration problems can be quantified precisely, whereas enhancement criteria are difficult to represent mathematically. Consequently, restoration techniques often depend only on the class or ensemble properties of a data set, whereas image enhancement techniques are much more image dependent. The degradation process is usually modeled as a degradation function that, together with an additive noise term, operates on an input image $f(\mathrm{x},\mathrm{y})$ to produce a degraded image $g(\mathrm{x},\mathrm{y})$. Given $g(\mathrm{x},\mathrm{y})$, some knowledge about the degradation function H, and some knowledge about the additive noise term $\eta(\mathrm{x},\mathrm{y})$, the objective of restoration is to obtain an estimate $\hat{f}(x,y)$ of the original image. The estimate needs to be as close as possible to the original input image and, in general, the more about H and η is known, the closer $\hat{f}(x,y)$ will be to $f(\mathrm{x},\mathrm{y})$.

If H is a linear, position-invariant process, then the degraded image is given in the spatial domain by

$$g(\mathrm{x},\mathrm{y}) = h(\mathrm{x},\mathrm{y})^{*} f(\mathrm{x},\mathrm{y}) + \eta(\mathrm{x},\mathrm{y})$$

where $h(\mathrm{x},\mathrm{y})$ is the spatial representation of the degradation function and the symbol "*" indicates convolution [2].

Image noise reduction has come to specifically mean a process of smoothing noise that has somehow corrupted the image. During image transmission, noise which is usually independent of the image signal occurs. Noise may be additive, where noise and image signal g are independent.

$$f(\mathrm{x},\mathrm{y}) = g(\mathrm{x},\mathrm{y}) + v(\mathrm{x},\mathrm{y})$$

where $f(\mathrm{x},\mathrm{y})$ is the noisy image signal, $g(\mathrm{x},\mathrm{y})$ is the original image signal and $v(\mathrm{x},\mathrm{y})$ is the noise signal which is independent of g [3]. The additive noise image v models an undesirable, unpredictable corruption of g. The process v is called a two-dimensional random process or a random field. The goal of restoration is to recover an image h that resembles g as closely as possible by reducing v. If there is an adequate model for the noise, then the problem of finding h can be posed as the image estimation problem, where h is found as the solution to a statistical optimization problem. The detailed statistics of the noise process v may are unknown. In such cases, a simple linear filter approach can yield acceptable results, if the noise satisfies certain simple assumptions such as zero-mean additive white noise model [4]. Noise may be impulse noise, which is usually characterized by some portion of image pixels that are corrupted, leaving the remaining pixels unchanged. The most common example of the impulse noise is the salt-and-pepper noise removal.

8.2 Conventional Noise Removal Techniques

Noise reduction is the process of removing noise from a signal. Noise reduction techniques are conceptually very similar regardless of the signal being processed, however a priori knowledge of the characteristics of an expected signal can mean the implementations of these techniques vary greatly depending on the type of signal. Although linear image enhancement tools are often adequate in many applications, significant advantages in image enhancement can be attained if non-linear techniques are applied [3]. Non-linear methods effectively preserve edges and details of images, whereas methods using linear operators tend to blur and distort them. Additionally, non-linear image enhancement tools are less susceptible to noise.

One method to remove noise is to use linear filters by convolving the original image with a mask. The Gaussian mask comprises elements determined by a Gaussian function. It gives the image a blurred appearance if the standard deviation of the mask is high, and has the effect of smearing out the value of a single pixel over an area of the image. Averaging sets each pixel to the average value of itself and its nearby neighbors. Averaging tends to blur an image, because pixel intensity values which are significantly higher or lower than the surrounding neighborhood would smear across the area. Conservative smoothing is another noise reduction technique that is explicitly designed to remove *noise spikes* (e.g. salt and pepper noise) and is, therefore, less effective at removing *additive noise* (e.g. Gaussian noise) from an image [5].

Additive noise is generally more difficult to remove from images than impulse noise because a value from a certain distribution is added to each image pixel, for example, a Gaussian distribution. A huge amount of wavelet based methods [6] are available to achieve a good noise reduction (for the additive noise type), while preserving the significant image details.. The wavelet denoising procedure usually consists of shrinking the wavelet coefficients, that is, the coefficients that contain primarily noise should be reduced to negligible values, while the ones containing a significant noise-free component should be reduced less. A common shrinkage approach is the application of simple thresholding nonlinearities to the empirical wavelet coefficients [7, 8]. Shrinkage estimators can also result from a Bayesian approach, in which a prior distribution of the noise-free data (e.g., Laplacian [4], generalized Gaussian [9–11]) is integrated in the denoising scheme.

Fuzzy set theory and fuzzy logic offer us powerful tools to represent and process human knowledge represented as fuzzy if-then rules. Several fuzzy filters for noise reduction have already been developed, e.g., the iterative fuzzy control based filters from [12], the GOA filter [13, 14], and so on. Most of these state-of-the-art methods are mainly developed for the reduction of fat-tailed noise like impulse noise. Nevertheless, most of the current fuzzy techniques do not produce convincing results for additive noise [15, 16]. Another shortcoming of the current methods is that most of these filters are especially developed for grayscale images. It is, of course, possible to extend these filters to color images by applying them on each color component separately, independent of the other components. However, this introduces many artifacts, especially on edge or texture elements.

A new fuzzy method proposed by Stefan Schulte, Valérie De Witte, and Etienne E. Kerre, is a simple fuzzy technique [17] for filtering color images corrupted with narrow-tailed and medium narrow-tailed noise (e.g., Gaussian noise) without introducing the above mentioned artifacts. Their method outperforms the conventional filter as well as other fuzzy noise filters. In this paper, we are presenting a modified version of the fuzzy approach proposed by Stefan Schulte, et al. [17], which uses a Gaussian combination membership function to yield a better result, compared to the conventional filters as well as the recently developed advanced fuzzy filters.

8.3 Proposed Fuzzy Noise Removal Filter

A digital color image C can be represented in different color space such as RGB, HSV, L*a*b etc. In the proposed method, RGB space is used as the basic color space. Different proportions of red, green and blue light gives a wide range of colors. Colors in RGB space are represented by a 3-D vector with first element being red, the second being green and third being blue, respectively. These three primary color components are quantized in the range 0 to $2^m - 1$, where $m = 8$. A color image C can be represented by a 2-D array of vectors where (i, j) defines a position in C called pixel and $C_{i,j,1}$, $C_{i,j,2}$, and $C_{i,j,3}$, denotes the red, green and blue components, respectively.

8.3.1 Fuzzy Sub-filter I

The general idea in this method is to take into account the fine details of the image such as edges and color component distances, which will be preserved by the filter. The goal of the first filter is to distinguish between local variations due to image structures such as edges. The goal is accomplished by using Euclidean distances between color components instead of differences between the components as done in most of the existing filters. The proposed method uses 2-D distances instead of 3-D distances (distance between three color components red, green and blue), that is, the distance between red-green (*rg*) and red-blue (*rb*) of the neighbourhood centered at (i, j) is used to filter the red component [4]. Similarly, the distance between RG and green-blue (*gb*) is used to filter the green component and the distance between *rb* and *gb* is used to filter the blue component, respectively. The method uses three fuzzy rules to calculate weights for the Takagi-Sugeno fuzzy model [10].

The current image pixel at position (i, j) is processed using a window size of $(2K + 1) \times (2K + 1)$ to obtain the modified color components. To each of the pixels in the window certain weights are then assigned namely $W_{k,l}$, where $k, l \in \{-1, 0, 1\}$. $W_{i+k,j+l,1}$, $W_{i+k,j+l,2}$, and $W_{i+k,j+l,3}$ denotes the weights for the red, green and blue component at position (i + k, j + l), respectively. These weights are assigned according to the following three fuzzy rules. Let DT (a, b) represents the distance between the parameters a and b, and NG(y) represents the neighbourhood of the parameter y. In this case, y represents a pixel with the neighbourhood given by a 3×3 window. The three fuzzy rules can be represented as follows:

1. If DT(*rg*, NG(*rg*)) is *SMALL* AND DT(*rb*, NG(*rb*)) is *SMALL* THEN the weight $W_{k,l,1}$ is *LARGE*
2. IF DT(*rg*, NG(*rg*)) is *SMALL* AND DT(*gb*, NG(*gb*)) is *SMALL* THEN the weight $W_{k,l,2}$ is *LARGE*
3. IF DT(*rb*, NG(*rb*)) is *SMALL* AND DT(*gb*, NG(*gb*)) is *SMALL* THEN the weight $W_{k,l,3}$ is *LARGE*

In the above fuzzy rules DIST represents the Euclidean distance.

$$\mathrm{DT}(rg,\ \mathrm{NG}(rg)) = [(C_{i+k,j+l,1} - C_{i,j,1})^2 + (C_{i+k,j+l,2} - C_{i,j,2})^2]^{1/2}$$
$$\mathrm{DT}(rb,\ \mathrm{NG}(rb)) = [(C_{i+k,j+l,1} - C_{i,j,1})^2 + (C_{i+k,j+l,3} - C_{i,j,3})^2]^{1/2}$$
$$\mathrm{DT}(gb,\ \mathrm{NG}(gb)) = [(C_{i+k,j+l,2} - C_{i,j,2})^2 + (C_{i+k,j+l,3} - C_{i,j,3})^2]^{1/2}$$

Fuzzy sets are commonly represented by membership functions from which the corresponding membership degrees are derived. Membership degrees between zero and one indicate the uncertainty that whether the distance is small or not. In the proposed approach, a new membership function SMALL has been used which incorporates a two-sided composite of two different Gaussian curves. The Gaussian membership function depends on two parameters σ and c as given by

$$f(x, \sigma, c) = e^{\frac{-(x-c)^2}{2\sigma^2}}$$

The membership function *gauss2mf*() is a combination of two of these two parameters. The first function, specified by $\sigma 1$ and c1, determines the shape of the leftmost curve. The second function specified by $\sigma 2$ and c2 determines the shape of the rightmost curve. Whenever $c1 < c2$, the *gauss2mf*() reaches a maximum value of 1. Otherwise, the maximum value is less than one. The membership function SMALL is then defined as

$$\mu_S(x) = \mathrm{gauss2mf}(x,\ [\sigma_x, C_x, \sigma_x, 0])$$

where σ_x is the standard deviation of the distance measure and C_x is the mean of the distance measure, respectively.

In the above fuzzy rules, the intersection of two fuzzy sets is involved. The intersection of two fuzzy sets A and B is generally specified by a binary mapping T, which aggregates two membership functions as follows:

$\mu_{A\cap B}(y) = T(\mu_A(y), \mu_B(y))$, where μ_A and μ_B are the membership functions for the two fuzzy sets A and B, respectively. The fuzzy intersection operator, known as triangular norms (T-norms), used in this paper is the algebraic product T-norms. For example, the antecedent of Fuzzy rule 1 is:

$$\mu_{SMALL}(DT(rg, NG(rg))).\ \mu_{SMALL}(DT(rb, NG(rb)))$$

The above obtained value, called the activation degree of the fuzzy rule 1, is used to obtain the corresponding weight. So the weights $W_{i+k,j+l,1}$, $W_{i+k,j+l,2}$, and $W_{i+k,j+l,3}$ are calculated as follows:

$$W_{i+k,j+l,1} = \mu_{SMALL}(DT(rg, NG(rg))).\ \mu_{SMALL}(DT(rb, NG(rb)))$$
$$W_{i+k,j+l,2} = \mu_{SMALL}(DT(rg, NG(rg))).\ \mu_{SMALL}(DT(gb, NG(gb)))$$
$$W_{i+k,j+l,3} = \mu_{SMALL}(DT(rb, NG(rb))).\ \mu_{SMALL}(DT(gb, NG(gb)))$$

The output of the Fuzzy Sub-filter I, denoted as F1, is then given by:

$$F1_{i,j,1} = \frac{\sum_{k=-K}^{+K} \sum_{l=-K}^{+K} W_{i+k,j+1,1} . C_{i+k,j+1,1}}{\sum_{k=-K}^{+K} \sum_{l=-K}^{+K} W_{i+k,j+1,1}}$$

$$F1_{i,j,2} = \frac{\sum_{k=-K}^{+K} \sum_{l=-K}^{+K} W_{i+k,j+1,2} . C_{i+k,j+1,2}}{\sum_{k=-K}^{+K} \sum_{l=-K}^{+K} W_{i+k,j+1,2}}$$

$$F1_{i,j,3} = \frac{\sum_{k=-K}^{+K} \sum_{l=-K}^{+K} W_{i+k,j+1,3} . C_{i+k,j+1,3}}{\sum_{k=-K}^{+K} \sum_{l=-K}^{+K} W_{i+k,j+1,3}}$$

where $F1_{i,j,1}$, $F1_{i,j,2}$ and $F1_{i,j,3}$ denotes the red, green and blue components of the Fuzzy sub-filter I output image respectively.

8.3.2 *Fuzzy Sub-filter II*

The second sub-filter is used as a complementary filter to the first one. The goal of this sub-filter is to improve the first method by reducing the noise in the color components differences without destroying the fine details of the image. In this step, the local differences in the red, green and blue environment are calculated separately. These differences are then combined to calculate the local estimation of the central pixel. In this step also, a window of size (2L + 1) × (2L + 1) is used centered at (i, j) to filter the current image pixel at that position. The local differences for each element of the window for the three color components are calculated as follows:

$$\mathrm{DR}_{k,l} = \mathrm{F1}_{i+k,j+l,1} - \mathrm{F1}_{i,j,1}$$
$$\mathrm{DG}_{k,l} = \mathrm{F1}_{i+k,j+l,2} - \mathrm{F1}_{i,j,2}$$
$$\mathrm{DB}_{k,l} = \mathrm{F1}_{i+k,j+l,3} - \mathrm{F1}_{i,j,3}$$

where k, l $\in \{-1, 0, +1\}$. The correction term $\boldsymbol{\varepsilon}$ is calculated as follows:

$$\boldsymbol{\varepsilon}_{k,l} = (1/3).(\mathrm{DR}_{k,l} + \mathrm{DG}_{k,l} + \mathrm{DB}_{k,l})$$
$$\text{for } k, l \in \{-L, \ldots, 0, \ldots, +L\}.$$

The output of the Fuzzy sub-filter 2, denoted as FS2, is then given by

$$\mathrm{F2}_{i,j,1} = \frac{\sum_{k=-L}^{+L} \sum_{l=-L}^{+L} (F1_{i+k,j+l,1} - \varepsilon_{k,l})}{(2L+1)^2}$$

$$\mathrm{F2}_{i,j,2} = \frac{\sum_{k=-L}^{+L} \sum_{l=-L}^{+L} (F1_{i+k,j+l,2} - \varepsilon_{k,l})}{(2L+1)^2}$$

$$\mathrm{F2}_{i,j,3} = \frac{\sum_{k=-L}^{+L} \sum_{l=-L}^{+L} (F1_{i+k,j+l,3} - \varepsilon_{k,l})}{(2L+1)^2}$$

where $\mathrm{F2}_{i,j,1}$, $\mathrm{F2}_{i,j,2}$ and $\mathrm{F2}_{i,j,3}$ denotes the red, green and blue components of the output image respectively.

8.4 Results and Discussion

The performance of the discussed filter has been evaluated and compared with conventional filters dealing with additive noise, using MATLAB software tool. As a measure of objective similarity between a filtered image and the original one, we use the peak signal-to-noise ratio (PSNR) in decibels (dB).

$$\mathrm{PSNR}(img, org) = 10 \log_{10}(\mathrm{S}^2/\mathrm{MSE}(img, org))$$

This similarity measure is based on another measure, namely the mean-square error (MSE).

$$\text{MSE}\,(img, org) = \frac{\sum_{c=1}^{3}\sum_{i=1}^{N}\sum_{j=1}^{M}\left[org\,(i,j,c) - img\,(i,j,c)\right]^2}{3.N.M}$$

where *org* is the original color image, *img* is the filtered color image of size $N.M$, and S is the maximum possible intensity value (with m-bit integer values, S will be $2^{\mathrm{m}} - 1$). The standard color images used in this paper are House and Gantrycrane images. The original image, the noisy image (original image corrupted with Gaussian noise with a selected σ value) and restored images using mean filter, median filter, fuzzy method of [17] and the modified fuzzy method of the above mentioned standard color images along with their corresponding PSNR values are shown in Figs. 8.1 and 8.2. From experimental results, it has been found that our

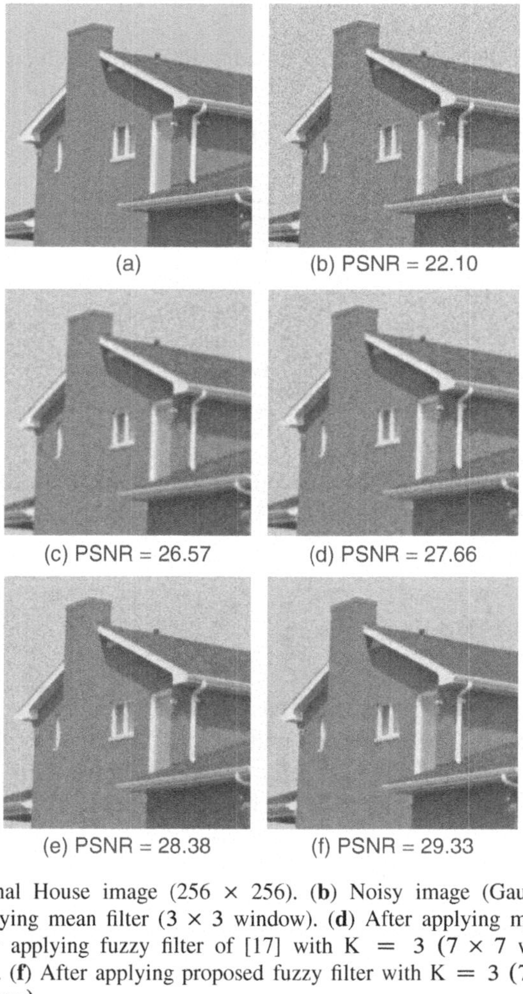

Fig. 8.1 (**a**) Original House image (256 × 256). (**b**) Noisy image (Gaussian noise, σ = 20). (**c**) After applying mean filter (3 × 3 window). (**d**) After applying median filter (3 × 3 window). (**e**) After applying fuzzy filter of [17] with K = 3 (7 × 7 window) and L = 2 (5 × 5 window). (**f**) After applying proposed fuzzy filter with K = 3 (7 × 7 window) and L = 2 (5 × 5 window)

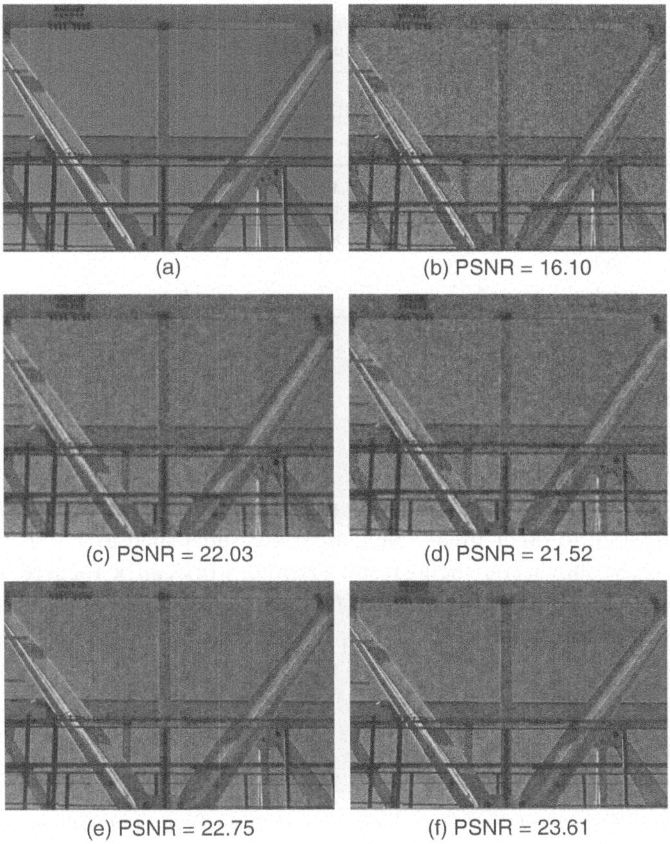

Fig. 8.2 (**a**) Original Gantrycrane image (400 × 264). (**b**) Noisy image (Gaussian noise, σ = 40). (**c**) After applying mean filter (3 × 3 window). (**d**) After applying median filter (3 × 3 window). (**e**) After applying fuzzy filter of [17] with K = 3 (7 × 7 window) and L = 2 (5 × 5 window). (**f**) After applying proposed fuzzy filter with K = 3 (7 × 7 window) and L = 2 (5 × 5 window)

proposed method receives the best numerical and visual performance for low levels and higher levels of additive noise, by appropriately selecting window size for the two fuzzy sub-filters. Numerical results that illustrate the denoising capability of the proposed method, modified method and conventional methods are pictured in Table 8.1. Table 8.1 shows the PSNRs for the colored House image that were corrupted with Gaussian noise for $\sigma = 5, 10, 20, 30$ and 40. The window size for different filters is appropriately chosen to give better PSNR value. The PSNR value of the noisy image and the best performing filter were shown bold.

Table 8.1 Comparative results in PSNR of different filtering methods for various distortions of Gaussian noise for the (256×256) colored House image

	PSNR (dB)				
	$\sigma = 5$	$\sigma = 10$	$\sigma = 20$	$\sigma = 30$	$\sigma = 40$
Noisy	**34.13**	**28.12**	**22.10**	**18.57**	**16.08**
Mean	28.05	27.70	26.57	25.13	23.72
Median	32.31	30.81	27.66	25.02	22.95
Proposed fuzzy method	34.12	31.79	28.38	25.85	23.76
Modified fuzzy method	**34.22**	**32.77**	**29.33**	**26.51**	**24.18**

8.5 Conclusion

A fuzzy filter for restoring color images corrupted with additive noise is proposed in this paper. The proposed filter is efficient and produces better restoration of color images compared to other filters. Numerical measures such as PSNR and visual observation have shown convincing results. Further work can be focused on the construction of other fuzzy filtering methods for color images to suppress multiplicative noise such as speckle noise.

References

1. A. K. Jain, "*Fundamentals of Digital Image Processing*," Pearson Education, Prentice Hall, Englewood Cliffs, NJ, 1989, pp. 233–356.
2. R. C. Gonzalez and R. E. Woods, "*Digital Image Processing*," 2nd Ed., Pearson Education, Prentice-Hall, Englewood Cliffs, NJ, 2002, pp. 147–163.
3. G. R. Arce, J. L. Paredes and J. Mullan, "Nonlinear Filtering for Image Enhancement," *Handbook of Image and Video Processing*, Academic, London, 2006, pp. 81–100.
4. M. Hansen and B. Yu, "Wavelet thresholding via mdl for natural images," *IEEE Transactions on Inference Theory*, Vol. 46, No. 5, pp. 1778–1788, August 2000.
5. http://homepages.inf.ed.ac.uk/rbf/HIPR2/csmooth.htm
6. H. L. Resnikoff and R. O. Wells, "Wavelet Analysis: The Scalable Structure of Information," Springer-Verlag, New York, 1998.
7. D. Donoho, "Denoising by soft-thresholding," *IEEE Transactions on Inference Theory*, Vol. 41, No. 3, pp. 613–627, May 1995.
8. S. Chang, B. Yu, and M. Vetterli, "Adaptive wavelet thresholding for image denoising and compression," *IEEE Transactions on Image Processing*, Vol. 9, No. 9, pp. 1532–1546, September 2000.
9. E. Simoncelli and E. Adelson, "Noise removal via Bayesian wavelet coring," in *Proceedings of IEEE International Conference on Image Processing*, 1996, pp. 379–382.
10. P. Moulin and J. Liu, "Analysis of multi-resolution image denoising schemes using generalized Gaussian and complexity priors," *IEEE Transactions on Inference Theory*, Vol. 45, No. 3, pp. 909–919, April 1999.
11. A. Pižurica, W. Philips, I. Lemahieu, and M. Acheroy, "A joint inter and intra-scale statistical model for Bayesian wavelet based image denoising," *IEEE Transactions on Image Processing*, Vol. 11, No. 5, pp. 545–557, May 2002.

12. F. Farbiz and M. B. Menhaj, "A fuzzy logic control based approach for image filtering," in *Fuzzy Technical Image Processing*, E. E. Kerre and M. Nachtegael, Eds., 1st ed. Physica Verlag, Heidelberg, Germany, 2000, Vol. 52, pp. 194–221.
13. D. Van De Ville, M. Nachtegael, D. Van der Weken, W. Philips, I. Lemahieu, and E. E. Kerre, "A new fuzzy filter for Gaussian noise reduction," in *Proceedings of SPIE Visual Communications and Image Processing*, 2001, pp. 1–9.
14. D. Van De Ville, M. Nachtegael, D. Van der Weken, E. E. Kerre, and W. Philips, "Noise reduction by fuzzy image filtering," *IEEE Transactions on Fuzzy Systems*, Vol. 11, No. 8, pp. 429–436, August 2003.
15. M. Nachtegael, S. Schulte, D. Van der Weken, V. De Witte, and E. E. Kerre, "Fuzzy filters for noise reduction: The case of Gaussian noise," in *Proceedings of IEEE International Conference on Fuzzy Systems*, 2005, pp. 201–206.
16. S. Schulte, B. Huysmans, A. Pižurica, E. E. Kerre, and W. Philips, "A new fuzzy-based wavelet shrinkage image denoising technique," *Lecture Notes in Computer Science*, Vol. 4179, pp. 12–23, 2006.
17. S. Schulte, V. De Witte, and E. E. Kerre, "A Fuzzy noise reduction method for color images," *IEEE Transaction on Image Processing,* Vol. 16, No. 5, May 2007, pp. 1425–1436.
18. A. C. Bovik and S. T. Acton, "Basic Linear Filtering with Application to Image Enhancement," *Handbook of Image and Video Processing*, Academic, New York, 2006, pp. 71–79.
19. B. Vidakovic, "Non-linear wavelet shrinkage with Bayes rules and Bayes factors," *Journal of the American Statistical Association*, Vol. 93, pp. 173–179, 1998.
20. H. Chipman, E. Kolaczyk, and R. McCulloch, "Adaptive Bayesian wavelet shrinkage," *Journal of the American Statistical Association*, Vol. 92, pp. 1413–1421, 1997.
21. L. Sendur and I. W. Selesnick, "Bivariate shrinkage functions for wavelet-based denoising exploiting inter-scale dependency," *IEEE Transactions on Signal Processing*, Vol. 50, No. 11, pp. 2744–2756, November 2002.
22. M. Crouse, R. Nowak, and R. Baranuik, "Wavelet-based statistical signal processing using hidden Markov models," *IEEE Transactions on Signal Processing*, Vol. 46, No. 4, pp. 886–902, April 1998.
23. J. Romberg, H. Choi, and R. Baraniuk, "Bayesian tree-structured image modeling using wavelet-domain hidden Markov models," *IEEE Transactions on Image Processing*, Vol. 10, No. 7, pp. 1056–1068, July 2001.
24. M. Malfait and D. Roose, "Wavelet-based image denoising using a Markov random field *a priori* models," *IEEE Transactions on Image Processing*, Vol. 6, No. 4, pp. 549–565, April 1997.

Chapter 9
Enhancement of Weather Degraded Color Images and Video Sequences Using Wavelet Fusion

Jisha John and M. Wilscy

Abstract The image in outdoor scene is degraded by optical scattering of light which produces additional lightness present in some parts of the image. In this work a method for enhancing visibility and maintaining color fidelity is proposed using wavelet fusion. This method mainly consists of three phases. Given an input image, first phase is to apply a contrast correction using the depth information. Here we compute the value of airlight present in the image by optimizing a cost function. The second phase consists of finding an approximate airlight value by using the intensity information of YIQ color model. Contrast restoration in the first and second phase is performed by removing the airlight from the image and applying depth information. The third and final phase of the proposed method consists of a wavelet fusion method to get a resultant image which has considerable visibility improvement and also maintains the color fidelity.

Keywords Weather Degraded · Color Images · Video Sequences · Wavelet Fusion · Contrast restoration · Enhancing visibility

9.1 Introduction

One of the major reasons for accidents in air, on sea and on the road is the poor visibility due to presence of fog or mist in the atmosphere. During winter, visibility is worse, sometimes up to few feet only. Under such conditions, light reaching the human eye is extremely scattered by constituents of atmosphere like fog, haze and aerosols and the image is severely degraded. Images taken under such bad weather conditions suffer from degradation and severe contrast loss. The loss of image quality is a nuisance in many imaging applications. For example, in underwater imaging

J. John (✉)
Mtech student, Department of Computer Science, University of Kerala, Karyavattom – 695581, Trivandrum, Kerala, India,
E-mail: jisha.json@yahoo.com

S.I. Ao and L. Gelman (Eds.), *Advances in Electrical Engineering and Computational Science*, Lecture Notes in Electrical Engineering.

in murky water, the detection of artifacts becomes difficult due to poor image quality. Hence, imaging must be performed at close range and this usually results in a long time required to inspect a small area. Another example is in the navigation of surface ships and aircraft in bad weather. In weather conditions such as fog, visibility is low and navigation is more difficult, dangerous and slow.

The image in outdoor scene is degraded by optical scattering of light which produces additional lightness present in some parts of the image, an effect that has been referred to as "atmospheric background radiation" [1, 2] or "airlight" [3, 4]. This results in degradation of image contrast, as well as alteration of scene color, which finally leads to a poor visual perception of the image. Contrast enhancement methods fall into two groups: non-model-based and model-based.

Non-model-based methods analyze and process the image based solely on the information from the image. The most commonly used non-model-based methods are histogram equalization and its variations [5–8]. For color images, histogram equalization can be applied to R, G, B color channels separately but this leads to undesirable change in hue. Better results are obtained by first converting the image to the Hue, Saturation, Intensity color space and then applying histogram equalization to the intensity component only [9]. However, even this method does not fully maintain color fidelity.

There are also other non-model-based methods like unsharp masking [10], approaches based on the Retinex theory [11–13] and wavelet-based methods [14, 15]. Generally, all non-model-based methods have a problem with maintaining color fidelity. They also distort clear images, which is an important limitation for fully automatic operation. Model-based methods use physical models to predict the pattern of image degradation and then restore image contrast with appropriate compensations. They provide better image rendition but usually require extra information about the imaging system or the imaging environment.

In [16] John P Oakley and Hong Bu have suggested a method of enhancement by correcting contrast loss by maintaining the color fidelity. In this method it is assumed that if the distance between a camera position and all points of a scene represented by an image generated by the camera is approximately constant, the airlight will be uniform within the image. But in most real-time situations this assumption is not valid. This method gives good contrast restoration but does not provide much visibility enhancement. To enhance the visibility R.T. Tan et al. [17] have proposed a visibility enhancement method which makes use of color and intensity information. Visibility is greatly improved in the resulting images but color fidelity is not maintained. Hence in situations were the naturality of the image is important this method cannot be used.

In this work a method for enhancing visibility and maintaining color fidelity is proposed using wavelet fusion. This method mainly consists of three phases. Given an input image, first phase is to apply a contrast correction using the depth information. Here we compute the value of airlight present in the image by optimizing a cost function. The second phase consists of finding an approximate airlight value by using the intensity information of YIQ color model. Contrast restoration in the first and second phase is performed by removing the airlight from the image and applying depth information. The third and final phase of the proposed method consists of a wavelet fusion method to get a resultant image which has considerable visibility improvement and also maintains the color fidelity.

The rest of the paper is organized as follows: In Section 9.2 we will discuss the atmospheric scattering models concentrating on the airlight model which forms the basis of this method. In Section 9.3, the contrast correction is given which forms the first phase of this work. In Section 9.4 the approximate airlight estimation is discussed. Section 9.5 introduces the third phase were a wavelet fusion is applied to get the enhanced image. Section 9.6 explains how the method can be applied on video sequences. To show the effectiveness of the proposed method, performance analysis is done in Section 9.7 with the help of a contrast improvement index and sharpness measure. Section 9.8 includes the experimental results and discussion.

9.2 Atmospheric Scattering Models

Scattering of light by physical media has been one of the main topics of research in the atmospheric optics and astronomy communities. In general, the exact nature of scattering is highly complex and depends on the types, orientations, sizes, and distributions of particles constituting the media, as well as wavelengths, polarization states, and directions of the incident light [1,2]. Here, we focus on one of the airlight model, which form the basis of our work.

While observing an extensive landscape, we quickly notice that the scene points appear progressively lighter as our attention shifts from the foreground toward the horizon. This phenomenon, known as *airlight* results from the scattering of environmental light toward the observer, by the atmospheric particles within the observer's cone of vision. The environmental illumination can have several sources, including, direct sunlight, diffuse skylight and light reflected by the ground. While attenuation causes scene radiance to decrease with path length, airlight increases with path length. It therefore causes the apparent brightness of a scene point to increase with depth. The, the irradiance due to airlight is given by

$$E(x) = I_{\infty}\rho(x)e^{-\beta d(x)} + I_{\infty}\left(1 - e^{-\beta d(x)}\right). \tag{9.1}$$

The first term in the equation represents the direct transmission, while the second term represents the airlight. E is the image intensity, x is the spatial location, I_{∞} is the atmospheric environmental light, which is assumed to be globally constant and ρ is the normalized radiance of a scene point, which is the function of the scene point reflectance, normalized sky illumination spectrum, and the spectral response of the camera. β is the atmospheric attenuation coefficient. d is the distance between the object and the observer. It is the second term in Eq. (9.1) or airlight that causes degradation in the image taken under bad weather conditions and hence all contrast restoration methods are aimed at removing this additional lightness from the image.

9.3 Contrast Correction

In Simple Contrast Loss, the degradation can be described by the applying

$$y = m(x - \lambda) \tag{9.2}$$

to each pixel of the image, where

'x' is the distorted pixel value
'λ' is an estimate of the airlight-contributed part of the pixel value
'm' is a scaling parameter
and 'y' is a modified pixel value.

To estimate the airlight in an image the normalized brightness value needs to be known

9.3.1 *Normalized Brightness and Airlight*

The normalized brightness, B_K, is defined by:

$$B_k = \frac{\rho_k}{\bar{\rho}_k}, k = 1, 2, \ldots\ldots.K \tag{9.3}$$

were k is the pixel position, ρ_k is the value of the image at pixel position k, and $\bar{\rho}_k$ is the output of a spatial low-pass filter at pixel position k. K is the total number of pixels in the image. The type of spatial low-pass filter is not critical. The Gaussian-shaped kernel is used here but other shapes provide similar results. The Gaussian shape is preferred since it introduces the least spurious structure. Also, an efficient recursive implementation of the Gaussian kernel may be used to reduce computational effort. For natural images, B_k has a near-symmetric and near-Gaussian distribution with a mean close to unity.

9.3.2 *Correcting Contrast*

In order to perform contrast correction, an airlight estimate is required. An algorithm is used for estimating the level of this airlight given the assumption that it is constant throughout the image. Airlight estimation is done by finding the minimum value of a cost function Eq. (9.4) that is a scaled version of the standard deviation of the normalized brightness is given by;

$$S_{gm}(\lambda) = STD\left\{\frac{p'_k - \lambda}{\bar{p}'_k - \lambda} : k = 1, 2, \ldots.K\right\}.GM\left\{\bar{p}_k - \lambda : k = 1, 2, \ldots.K\right\} \tag{9.4}$$

Where $GM\{.\}$ denotes the geometric mean. The geometric mean can also be written as

$$GM\{x_k : k = 1, 2......K\} = \exp\left\{\frac{1}{k}\sum_{k=1}^{K} \ln(x_k)\right\} \tag{9.5}$$

Another possible variation on the cost function is to use sample variance in Eq. (9.4) rather than sample standard deviation, in which case the scaling factor must be squared.

$$S(\lambda) = \frac{1}{k}\sum_{k=1}^{K}\left(\frac{\rho_k - \bar{\rho}_k}{\bar{\rho}_k - \lambda}\right)^2 . \exp\left\{\frac{1}{k}\sum_{k=1}^{K} \ln(\bar{\rho}_k - \lambda)^2\right\} \tag{9.6}$$

Obtain the optimum value of λ which minimizes the cost function by calculating,

$$\hat{c} = \arg\min\left\{S_{gm}\ (\lambda)\right\} \tag{9.7}$$

This is done using a standard optimization algorithm. This estimated value of λ represents the airlight present in the image. From this computation we can rewrite Eq. (9.1) as

$$\mathrm{E(x)} = \mathrm{I}_\infty \rho(x)\mathrm{e}^{-\beta \mathrm{d(x)}} + \lambda \tag{9.8}$$

Hence the enhanced image

$$\mathrm{I}_\infty \rho(\mathrm{x}) = (\mathrm{E(x)} - \lambda)\mathrm{e}^{-\beta \mathrm{d(x)}} \tag{9.9}$$

From Eq. (9.1) we can again write

$$\lambda = \mathrm{I}_\infty\left(1 - \mathrm{e}^{-\beta \mathrm{d(x)}}\right) \tag{9.10}$$

So, in order to estimate $\mathrm{e}^{\beta \mathrm{d(x)}}$ we can rewrite Eq. (9.10) as

$$\lambda = \left(I_\infty^r + I_\infty^g + I_\infty^r\right)\left(1 - \mathrm{e}^{-\beta \mathrm{d(x)}}\right), \text{ Hence}$$

$$\mathrm{e}^{\beta \mathrm{d(x)}} = \frac{1}{1 - \frac{\lambda}{(I_\infty^r + I_\infty^g + I_\infty^{r)}}} \tag{9.11}$$

Where $(I_\infty^r + I_\infty^g + I_\infty^r)$, namely, the environmental light, is assumed to be the largest intensity in the image. λ is found by optimizing the cost function Eq. (9.4) and depth information is obtained from Eq. (9.11). Thus Eq. (9.9) gives the contrast corrected image. Section IV and V describes how visibility enhancement can be achieved.

9.4 Intensity Based Enhancement

To accomplish the goal of intensity enhancement, airlight (λ) is computed based on the intensity value of YIQ color model [17] which is defined as

$$Y = 0.257 * E_r + 0.504 * E_g + 0.098 * E_b \tag{9.12}$$

where E_r, E_g, E_b represents the r, g and b color channels respectively.

It is assumed that the value of Y is the value of λ. However the values of λ are approximated values and thus to create a better approximation, these values are diffused by using Gaussian blur. Depth information, $e^{\beta d(x)}$ is computed as described in the earlier section by Eq. (9.11). Enhanced image is obtained as in Eq. (9.9).This resultant image contains all the detailed information present in the image.

9.5 Wavelet Fusion

The first phase described in Section 9.3 results in an image maintaining the color fidelity but the visibility is not enhanced particularly in scenes were the distribution of airlight is not uniform. The second phase uses an approximate airlight estimation method which results in an image with enhanced visibility but the color fidelity is not maintained. In the third phase a novel fusion method is used which helps in extracting the useful information from both images and hence obtaining an image with enhanced visibility at the same time maintaining the color fidelity. The daubechies wavelet is used here. The two images obtained as described in Section 9.3 and 9.4 are decomposed by using Daubechies wavelet method. The wavelet decomposition is done using shift invariant wavelet transform. The four images obtained per image after decomposition are coefficients extracted from the given image.

The first image is actually approximate coefficients displayed while the second image is formed when horizontal coefficients are displayed. The third image is formed when vertical coefficients are displayed. And the final image comes when diagonal coefficients are displayed. These coefficients are obtained by the following process. The image is actually passed through some sets of filters then these images are obtained. The image is passed through two low pass filters one aligned vertically and one aligned horizontally.

If image is passed through two filters, one low pass aligned horizontally and a high pass filter aligned vertically the vertical coefficients are obtained. Vertical coefficients and are obtained from high pass filter aligned horizontally and low pass filter aligned vertically. And the final image is for diagonal coefficients which are obtained with both high pass filters aligned horizontally and vertically. After obtaining the wavelet bands of the two images merge the coefficients by obtaining the mean value of the approximate coefficients and maximum value from the detailed coefficients. The resultant image is an enhanced image which contains the maximum details and also maintains the color fidelity. video enhancement.

9.6 Video Enhancement

There are only few existing methods to enhance weather degraded video sequences. In order to apply the enhancement process on video sequence, two methods can be adopted. Each of the frames can be separately processed or the method can be applied to the background and foreground pixels separately.

In the first method the computation time is more since the whole method is applied on each frame. In the approach followed in this paper the background and foreground pixels are separated and processed separately. The computation time is much less and the visibility of the resultant video sequence is found to increase considerably while color fidelity is maintained. In this method, first a static background frame is estimated. Then, analyze each of the frames to detect the motion pixels using any standard methodology [7] and apply morphological operations to remove all noise pixels. The motion pixels are then bounded by boxes so that each of the frames will be divided into sub images containing motion pixels. The background frame is processed using the wavelet fusion method. The estimated lightness parameter in step 1 is utilized globally for the entire video sequence and hence need not be computed for the sub images. Wavelet fusion is applied to each of the sub images using the estimated lightness and approximate lightness parameters. Finally the sub images are merged with the background image to form the enhanced video sequence. This method saves a lot of computation time since the processing is done only on the sub images of each frame and the estimated lightness parameter in step 1 of wavelet fusion method is computed only for the background image.

9.7 Performance Analysis

There is a lack of methodology to assess the performances of the methods or to compare them with one another. Unlike image quality assessment or image restoration areas, there is no easy way to have a reference image, which makes the problem not straightforward to solve.

For performance analysis a contrast improvement index is used here as proposed in [18]. This measure helps in comparing the contrast of foggy and restored image and hence analyzing the efficiency of the proposed method. However, it does not rate the fidelity of the contrast restoration method. To achieve such an objective, the same scene without fog must be grabbed and compared with restored image. The contrast improvement index is given by

$$\mathrm{CI} = \frac{C_{processed}}{C_{Original}} \tag{9.13}$$

where C is the average value of the local contrast measured with a 3*3 window as:

$$\mathrm{C} = \frac{\max - \min}{\max + \min} \tag{9.14}$$

In order to evaluate the effectiveness of the resultant image a well-known benchmark-image sharpness measure, the tenengrad criterion [19, 20] can be used. The tenengrad criterion is based on gradient $\nabla I(x, y)$ at each pixel (x, y), were the partial derivatives are obtained by a high-pass filter, eg., the sobel operator, with the convolution kernels i_x *and* i_y. The gradient magnitude is given by

$$S(x, y) = \sqrt{(i_x * I(x, y))^2 + (i_y * I(x, y))^2} \tag{9.15}$$

And the tenengrad criteria is formulated as

$$\text{TEN} = \sum_x \sum_y S(x, y)^2, \text{ for S(x, y > T)} \tag{9.16}$$

were T is the threshold. The image quality is usually considered higher if its tenengrad value is larger.

The tenengrad values (TEN) of all images given below has been calculated and listed in corresponding figure, captions. It is noted that images processed using the wavelet fusion method described above gives significantly larger tenengrad values, which indicate the effectiveness of this method. This result agrees with the visual evaluation of human eye.

9.8 Results and Discussion

The performance of the proposed method has been evaluated and compared with conventional methods of contrast enhancement using MATLAB software tool. The performance is analyzed using the measures described above. As a measure of objective similarity between a contrast restored image and the original one, the mean-squared error (MSE) is used.

$$MSE(img,org) = \frac{\sum_{c=1}^{3} \sum_{i=1}^{N} \sum_{j=1}^{M} [org\,(i, j, c) - img\,(i, j, c)]^2}{3.N.M} \tag{9.17}$$

were *org* is the original color image, *img* is the restored color image of size $N.M$

The color images used in this paper are Trees, Traffic and Aerialview images of size 256 × 256. The original image, the noisy image (original image degraded by fog) and restored images using visibility enhancement method and proposed wavelet fusion method along with their corresponding tenengrad, contrast improvement index values and mean-squared error values are shown in Fig. 9.1. From experimental results, it has been found that the proposed method has good contrast improvement and visibility (measured by the sharpness measure-tenengrad). It also maintains color fidelity which is shown using the low mean-squared error compared to other method.

(a) (b) TEN = 224626 (c)TEN = 225000,CI = 2.6356 MSE = 2.0401 e +004 (d)TEN = 225000,CI = 1.1555 MSE = 9.4580 e+ 003

Fig. 9.1 (**a**) Original Trees image (256 × 256). (**b**) Foggy image. (**c**) After applying method of visibility enhancement. (**d**) After applying proposed method

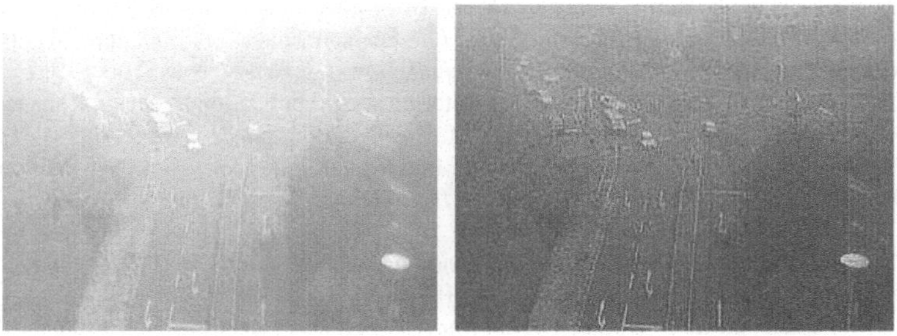

(a) TEN = 225017 (b) TEN = 2 30371, CI = 1.4743

Fig. 9.2 (**a**) Frame #1 Traffic video sequence. (**b**) After applying proposed method

(a) TEN = 173397 (b) TEN = 184135 CI = 4.7925

Fig. 9.3 (**a**) Degraded Aerialview image. (**b**) After applying proposed method

Figures 9.2 and 9.3 gives the foggy Traffic image and the Aerialview image degraded by mist. The method proposed is not restricted to uniform suspension of airlight and hence is applicable for all real-time scenarios.

9.9 Conclusion

In this paper the problem of restoring the contrast of atmospherically degraded images is addressed. This method uses the shift invariant discrete wavelet transform to decompose the images into various levels and then the frames are reconstructed by coefficient based fusion. It provides excellent color fidelity as well as visibility enhancement and hence is applicable for all real time operations. The method can be efficiently applied on video sequences also. Future research will be focused on contrast restoration of images degraded by dense fog or mist.

References

1. S. Chandrasekhar, *Radiative Transfer*. Dover, New York, 1960. Raefal C. Gonzalez, Richard E. Woods, *"Digital Image Processing," 2nd Ed.*, Pearson Education, 2002, pp. 147–163.
2. H. C. Van De Hulst, *Light Scattering by Small Particles*. Wiley, New York, 1957. Alan C. Bovik and Scott T. Acton, "Basic Linear Filtering with Application to Image Enhancement," *Handbook of Image and Video Processing, Academic,* 2006, pp. 71–79.
3. J. P. Oakley and B. L. Satherley, "Improving image quality in poor visibility conditions using a physical model for contrast degradation," *IEEE Transactions on Image Processing*, vol. 7, no. 2, pp. 167–179, February 1998.
4. S. K. Nayar and S. G. Narasimhan, "Vision in bad weather," in *Proceedings of IEEE International Conference Computer Vision*, 1999, vol. 2, pp. 820–827.
5. R. C. Gonzalez and R. E. Woods, *Digital Image Processing*. Reading, MA: Addison-Wesley, 1993.
6. S. M. Pizer et al., "Adaptive histogram equalization and its variations," *Computer Vision, Graphics, and Image Processing*, vol. 39, pp. 355–368, 1987.
7. K. Zuiderveld, "Contrast limited adaptive histogram equalization," in *Graphics Gems IV*, P. Heckbert, Ed. New York: Academic, 1994, ch. VIII.5, pp. 474–485.
8. J. A. Stark, "Adaptive image contrast enhancement using generalizations of histogram equalization," *IEEE Transactions on Image Processing*, vol. 9, no. 5, pp. 889–896, May 2000.
9. J. J. Rodriguez and C. C. Yang, "High-resolution histogram modification of color images," *Graphical Models and Image Processing*, vol. 57, no. 5, pp. 432–440, September 1995.
10. A. Polesel, G. Ramponi, and V. J. Mathews, "Image enhancement via adaptive unsharp masking," *IEEE Transactions on Image Processing*, vol. 9, no. 3, pp. 505–510, March 2000.
11. D. J. Jobson, Z. Rahman, and G. A. Woodell, "Properties and performance of a center/surround retinex," *IEEE Transactions on Image Processing*, vol. 6, no. 3, pp. 451–462, March 1997.
12. D. J. Jobson, Z. Rahman, and G. A. Woodell, "A multi-scale retinex for bridging the gap between color images and the human observation of scenes," *IEEE Transactions on Image Processing*, vol. 6, no. 7, pp. 965–976, July 1997.
13. Z. Rahman, D. J. Jobson, and G. A. Woodell, "Retinex processing for automatic image enhancement," *Journal of Electronic Imaging*, vol. 13, no. 1, pp. 100–110, January 2004.
14. P. Scheunders, "A multivalued image wavelet representation based on multiscale fundamental forms," *IEEE Transactions on Image Processing*, vol. 10, no. 5, pp. 568–575, May 2002.
15. L. Grewe, and R. R. Brooks, "Atmospheric attenuation reduction through multi-sensor fusion," *Proceedings of SPIE*, vol. 3376, pp. 102–109, 1998.
16. J. P. Oakley, and H. Bu, "Correction of simple contrast loss in color images," *IEEE Transactions on Image Processing*, vol. 16, no. 2, February 2007.
17. R. T. Tan, N. Pettersson Lars Petersson, "Visibility enhancement of roads with foggy or hazy scenes," *Proceedings of the 2007 IEEE Intelligent Vehicles Symposium Istanbul*, Turkey, June 13–15, 2007.

18. P. Zeng, H. Dong, J. Chi, X. Xu, School of Information Science and Engineering, Northeastern University. Shenyang, China, *Proceedings of the 2004 IEEE International Conference on Robotics and Biomimetics*, August 22–26, 2004, Shenyang, China.
19. E. P. Krotkov, *Active Computer vision by Cooperative focus and stereo*. Springer-Verlag, New York, 1989.
20. A. Buerkle, F. Schmoeckel, M. Kiefer, B. P. Amavasai, F. Caparrelli, A. N. Selvan, and J. R. Travis, "Vision-based closed-loop control of mobile micro robots for micro handling tasks," *Proceedings of SPIE*, vol. 4568, Microrobotics and Micro assembly 111, pp. 187–198, 2001.

Chapter 10
A GA-Assisted Brain Fiber Tracking Algorithm for DT-MRI Data

L.M. San-José-Revuelta

Abstract This work deals with the problem of fiber tracking in diffusion tensor (DT) fields acquired via magnetic resonance (MR) imaging. Specifically, we focus on tuning-up a previously developed probabilistic tracking algorithm by making use of a genetic algorithm which helps to optimize most of the adjustable parameters of the tracking algorithm. Since the adjustment of these parameters constitutes a hard NP-complete problem, traditionally, this task has been heuristically approached. In previous work, we have already developed a multilayer neural network that was successfully applied to this issue. Though robustness was one of its major advantages, the complexity of the whole algorithm constituted its main drawback. In order to avoid this situation, here we have explored the possibility of using a computationally simpler method based on a micro-genetic algorithm. This strategy is shown to outperform the NN-based scheme, leading to a more robust, efficient and human independent tracking scheme. The tracking of white matter fibers in the human brain will improve the diagnosis and treatment of many neuronal diseases.

Keywords Diffusion tensor magnetic resonance imaging · fiber tracking · genetic algorithm

10.1 Introduction

The Diffusion Tensor Magnetic Resonance Imaging (DT-MRI) technique measures the diffusion of hydrogen atoms within water molecules in 3D space. Since in cerebral white matter most random motion of water molecules are restricted by axonal membranes and myelin sheets, diffusion anisotropy allows depiction of directional anisotropy within neural fiber structures, [1, 2]. The DT-MRI technique

L.M. San-José-Revuelta
E.T.S.I. Telecomunicación, University of Valladolid, 47011 Valladolid, Spain,
E-mail: lsanjose@tel.uva.es

S.I. Ao and L. Gelman (Eds.), *Advances in Electrical Engineering and Computational Science*, Lecture Notes in Electrical Engineering.

has raised great interest in the neuro-science community for a better understanding of the fiber tract anatomy of the human brain. Among the many applications that arise from tractography we find: brain surgery (knowing the extension of the fiber bundles could minimize the functional damage to the patient), white matter visualization using fiber pathways (for a better understanding of brain anatomy) and inference of connectivity between different parts of the brain (useful for functional and morphological research of the brain).

Most of DT-MRI visualization techniques focus on the integration of sample points along fiber trajectories, [3], using only the principal eigenvector of the diffusion ellipsoid as an estimate of the predominant direction of water diffusion, [2]. Several approaches have already been developed, some of them include the *Runge-Kutta* approach, the *multiple diffusion tensor* approach, the *tensorline* approach and the *exhaustive search* approach, to name a few.

Though, the *in vivo* visualization of fiber tracts opens up new perspectives for neurological research, these algorithms may depict some fiber tracts which do not exist in reality or miss to visualize important connectivity features (e.g. crossing or branching structures) mainly due to both some deficiencies in these methods and several shortcomings inherent in datasets. In order to avoid misinterpretations, the viewer must be provided with some information on the uncertainty of every depicted fiber and of its presence in a certain location. In [4], we proposed an estimation algorithm that takes into account the whole information provided by the diffusion matrix, i.e., it does not only consider the principal eigenvector direction but the complete 3D information about the certainty of continuing the path through every possible future direction. An improved version of this algorithm was developed in [5]. This article included two main aspects: (i) a procedure that on-line adapts the number of *offspring paths* emerging from the actual voxel, to the degree of anisotropy observed in its proximity (this strategy was proved to enhance the estimation robustness in areas where multiple fibers cross while keeping complexity to a moderate level), and (ii) an initial version of a neural network (NN) for adjusting the parameters of the algorithm in a user-directed training stage. Subsequent work, [6], studied with more detailed the architecture of the neural network and numerically evaluated its tracking capability, robustness and computational load when being used with both synthetic and real DT-MR images. This work showed that in many cases, such as real images with low SNR, a huge computational load was required.

In this work we propose to use an evolutionary computation-based approach for tuning-up the parameters of the tracking algorithm instead of using the neural network scheme. The main aim is to adjust the parameters with a less complex procedure and to obtain a robust and efficient tracking algorithm. The required human intervention time should also be reduced. Specifically, we propose a *genetic algorithm* for this optimization task. Numerical results will prove that this approach leads to similar and even better convergence results while offering much lower computational requirements.

10.2 Brief Tracking Algorithm Description

The basic version of the algorithm here used is described in [4]. Thus, this section just presents a summary of the method, with special emphasis on the new aspects. The algorithm uses probabilistic criteria and iterates over several points in the analyzed volume (the points given by the highest probabilities in the previous iteration). The process starts in a user-selected seed voxel, V_0, and, at every iteration, it evaluates a set of parameters related to the central voxel of a cubic structure consisting of $3 \times 3 \times 3 = 27$ voxels, similar to that shown in Fig. 32.3, left.

The central point, V_c, (No. 14 in the figure) represents the last point of the tract being analyzed. In the first iteration, $V_c = V_0$. Obviously, there exist 26 possible directions to take for the next iteration in order to select the next point of the tract. Once V_c is selected, the previous point and all those points exceeding the limits of the MR volume are also removed from the list of possible destination points (*valid points*).

10.2.1 Basic Concepts

First, a measure $P_i, i \in \{validpoints\}$, is evaluated based on the probability of going from voxel V_c to voxel V_i. This probability takes into account the eigenvalues and eigenvectors available at point V_c from the DT-MR image diffusion matrix. In order to calculate this probability, the information shown in Fig. 32.3, right, is used.

The table shows, for every voxel shown in Fig. 32.3, left, the changes that must occur in indices (m, n, p), when a pathway goes from voxel V_c to voxel V_i. For instance: when going from point No. 14 to point No. 17, coordinates m and n increase by 1, and p remains the same. This is represented in the table with "$\pi_m \pi_n \pi_p = (+ + 0)$". With this information, the probability of each possible destination V_i can be calculated taking into account the projection of each of the eigenvectors to each of the directions defined in triplet $\pi_m \pi_n \pi_p$. Besides, each projection is weighted by the corresponding eigenvalue λ. Thus, in the previous example, P_i should be calculated as $P_i = V_{1y}\lambda_1 + V_{2y}\lambda_2 + V_{3y}\lambda_3 + V_{1z}\lambda_1 +$

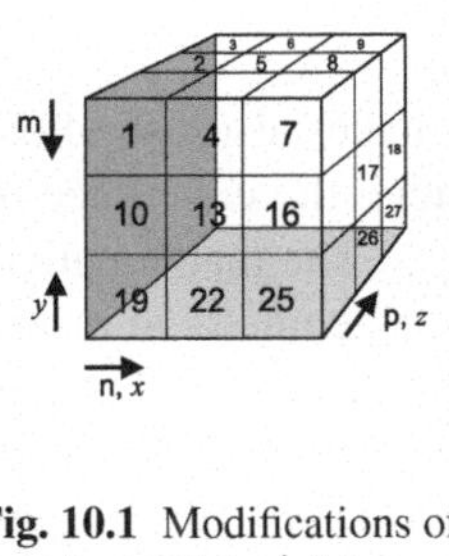

	mnp	$\pi_m\pi_n\pi_p$		mnp	$\pi_m\pi_n\pi_p$		mnp	$\pi_m\pi_n\pi_p$
1°)	111	(---)	10°)	211	(0--)	19°)	311	(+--)
2°)	112	(--0)	11°)	212	(0-0)	20°)	312	(+-0)
3°)	113	(--+)	12°)	213	(0-+)	21°)	313	(+-+)
4°)	121	(-0-)	13°)	221	(00-)	22°)	321	(+0-)
5°)	122	(-00)	14°)	222	(000)	23°)	322	(+00)
6°)	123	(-0+)	15°)	223	(00+)	24°)	323	(+0+)
7°)	131	(-+-)	16°)	231	(0+-)	25°)	331	(++-)
8°)	132	(-+0)	17°)	232	(0+0)	26°)	332	(++0)
9°)	133	(-++)	18°)	233	(0++)	27°)	333	(+++)

Fig. 10.1 Modifications of indices (m, n, p) when moving from V_c to the neighboring voxel V_i, $1 \leq i \leq 27, i \neq 14$

$V_{2z}\lambda_2 + V_{3z}\lambda_3$, where $V_{j\alpha}$ represents the α-component of eigenvector j, $1 \leq j \leq 3$, $\alpha \in \{x, y, z\}$. In the general case we have,

$$P_i = \sum_{\alpha \in \{x,y,z\}} \chi_\alpha \sum_{j=1}^{3} V_{j,\alpha}\lambda_j \tag{10.1}$$

with χ_x, χ_y, χ_z being zero if π_m, π_n, π_p are zero, respectively, and equal to 1 otherwise.

The axes reference criterion for the (x, y, z) vector components is also shown in Fig. 32.3, left. Note that, for this calculus, the sign "−" in the triplet is equivalent to sign "+". In order to properly calculate P_i, it must be weighed by 0.33 if there are no zeros in triplet i, and by 0.5 if there is one zero.

10.2.2 Anisotropy and Local Probability

The following anisotropy index is used in the algorithm:

$$\text{fa} = \sqrt{\frac{(\lambda_1 - \lambda_2)^2 + (\lambda_2 - \lambda_3)^2 + (\lambda_1 - \lambda_3)^2}{2(\lambda_1^2 + \lambda_2^2 + \lambda_3^2)}}, \tag{10.2}$$

where $\lambda_1 \geq \lambda_2 \geq \lambda_3$. When both fa($V_c$) and fa($V_i$) do not exceed a certain threshold, then point V_i is eliminated as a possible destination point.

Taking into account both P_i and the anisotropy given by Eq. (10.2), the local probability of voxel i is defined as

$$P_i' = a \cdot \mu_1 \cdot \text{fa}(V_i) + (1 - a) \cdot \mu_2 \cdot P_i, \qquad 0 < a < 1 \tag{10.3}$$

where parameter a allows the user to give a higher relative weight to either the anisotropy or the local probability, and μ_1 and μ_2 are scaling factors (normally, 1 and 1,000, respectively). The set of values P_i' is properly normalized so that they can be interpreted as probabilities.

10.2.3 Eigenvectors and Direction Considerations

Besides these considerations, the final probability of voxel i makes also use of the so-called *smoothness parameters* (described in [7]) which judge the coherence of fiber directions among the trajectories passing through voxel V_c. The mathematical expressions of these four parameters, $\{sp_i\}_{i=1}^4$, as well as their geometrical meaning, is explained in [6]. They measure the angles between the directions that join successive path points, as well as the angles between these directions and the eigenvectors associated to the largest eigenvalues found in those voxels. sp_2, sp_3 and sp_4 are used to maintain the local directional coherence of the estimated tract and avoid the trajectory to follow unlikely pathways, [7]. The threshold for sp_1 is set such that the tracking direction could be moved forward consistently and smoothly, preventing the computed path from sharp transitions.

Next, the following parameter is calculated for every valid point whose smoothness parameters satisfy the four corresponding threshold conditions,

$$P_i^{''} = b(\xi_1 sp_1 + \xi_2 sp_2 + \xi_3 sp_3 + \xi_4 sp_4) + (1-b)P_i^{'} \tag{10.4}$$

where, ξ_1, ξ_2, ξ_3 and ξ_4 are the corresponding weights of the smoothness parameters (normally, 0.25), and b stands for a weighting factor.

10.2.4 Path Probabilities

Probabilities P_i'' can be recursively accumulated, yielding the probability of the path generated by the successive values of V_c,

$$P_p(k) = P_i^{'''} \cdot P_p(k-1) \tag{10.5}$$

with k being the iteration number, and $P_i^{'''} = P_i^{''} / \sum_i P_i^{''}$.

At the end of the visualization stage, every estimated path is plotted with a color that depends on its probability P_p.

10.2.5 Final Criterion and Pool of "Future Seeds"

A pool of voxels is formed by selecting, at the end of each iteration, the s best voxels according to Eq. (10.4). The first voxel of the pool becomes the central voxel V_c at next iteration, expanding, this way, the current pathway.

As proposed in [6], the value of s is adjusted depending on the degree of anisotropy found in current voxel V_c and its surroundings. When this anisotropy is high, it means that a high directivity exists in that zone, and the probability that V_c belongs to a region where fibers cross is really low. Consequently, s takes a small value $(1, 2$ or $3)$. On the other hand, if V_c is found to be situated in a region of high anisotropy, the probabilities of having fibers crossing or branching is higher. In this case, it is interesting to explore various paths starting in V_c. This can be achieved by setting parameter s to a higher value.

10.2.6 Parameters to Be Estimated

In this work we propose to use an genetic algorithm for adjusting the parameters of the algorithm $(a, b, \mu_1, \mu_2, \xi_1, \xi_2, \xi_3, \xi_4)$, instead of using the complex and time consuming NN proposed in [5,6]. This adjustment is necessary when the algorithm is applied to a different area of the brain (fiber bundles) or even to the same portion but having been scanned with under different conditions. In these cases, the volume of interest will have a different smoothness and anisotropy characterization.

10.3 Proposed GA for Parameter Estimation

Genetic algorithms (GAs) represent a set of potential solutions (*population*) with a predetermined encoding rule. At every iteration, each potential solution (*chromosome*) is associated to a figure of merit, or *fitness* value, in accordance to its proximity to the optimal solution. Considering the tracking problem described in Section 10.2, the goal is to estimate the set of algorithm's parameters and thresholds $\Omega = (a, b, \mu_1, \mu_2, \xi_1, \xi_2, \xi_3, \xi_4)$.

10.3.1 Estimation Procedure

When the proposed estimation strategy is used, the user is requested to manually draw a sample fiber path as well as to compare this fiber path to those estimated by the GA during its first stages. Specifically, the steps for the estimation of the parameters are: (i) the user manually draws a sample fiber path, $\mathbf{r}_u$, (ii) the GA starts with a randomly generated population of $n_p = 10$ individuals $\{\mathbf{u}_i\}_{i=1}^{n_p}$, each of them being a possible binary representation of parameters Ω, (iii) the tracking algorithm of Section 10.2 is applied n_p times, each of them with the set of parameters represented by each GA's individual. This way, n_p different paths $\mathbf{r}_i$ are obtained, (iv) every path $\mathbf{r}_i$ is compared with $\mathbf{r}_u$ and given a fitness value λ_i, (v) iterate the GA during $n_g = 25$ generations and then go to step (ii).

Every time that the fiber paths are obtained at step (ii) the user must compare them to his sample $\mathbf{r}_u$ and, in case he finds that a tract $\mathbf{r}_j$, $1 \leq j \leq n_p$, is better than his first estimation $\mathbf{r}_u$, then $\mathbf{r}_j$ becomes the new reference path $\mathbf{r}_u$. At the end, solution Ω is obtained from the encoding of the fittest individual.

Though this scheme seems initially complicated, experiments show that a few iterations lead to sets Ω that allow to obtain good tracking results. The user doesn't have to assign too many fitness values or to perform many comparisons. The extremely reduced size of the population and the low number of generations per GA execution, lead to moderately short training periods.

10.3.2 GA Description

The initial set of potential solutions, $\mathcal{P}[0]$ (population at $k = 0$), is randomly generated – in our specific application, the user drawn tract could be included in the initial population. Let us denote $\mathcal{P}[k] = \{\mathbf{u}_i\}_{i=1}^{n_p}$ to the population at iteration k.

As previously mentioned, the population size n_p has been fixed to a low value exploiting, this way, the properties of the so called *micro genetic algorithms* (μ-GAs).

10.3.3 Genetic Operators

In order to perform step (v) in subsection *estimation procedure*, individuals are modified by making use of the *genetic operators*, mainly *mutation* and *crossover*, to a subset of individuals selected using the *roulette wheel* selection scheme, [8]. This selection procedure is based on a biased random selection, where the fittest individuals have a higher probability of being selected than their weaker counterparts. This way, the probability of any individual to be selected is $P_i(k) = \lambda_i(k)/\sum_{j=1}^{n_p} \lambda_j(k)$, with $\lambda_i(k)$ being the fitness of the ith individual in the population during the kth iteration of the algorithm.

The mutation operator modifies specific individuals with probability p_m, changing the value of some concrete positions in the encoding of $\mathbf{u}_i$. Both the specific positions and the new values are randomly generated, and mutation is effectively performed with probability p_m. Notice that this genetic operator promotes the exploration of different areas of the solutions space.

On the other hand, the crossover operator requires two operands (*parents*) to produce two new individuals (*offspring*). These new individuals are created when merging parents by crossing them at specific internal points. This operation is performed with probability p_c. Since parents are selected from those individuals having a higher fitness, the small variations introduced within these individuals are intended to also generate high fit individuals.

10.3.4 Elitism and Entropy Dependent Operators

The proposed GA also implements an elitism strategy: The best individual in the population is preserved and directly introduced in the new population. Nevertheless, a duplicate of it can also be used as input for the genetic operators.

This elitist model of the genetic algorithm presents some convergence advantages over the standard GA. In fact, using Markov chain modeling, it has been proved that GAs are guaranteed to asymptotically converge to the global optimum – with any choice of the initial population – if an elitist strategy is used, where at least the best chromosome at each generation is always maintained in the population, [9]. However, Bhandari et al. [9] provided the proof that no finite stopping time can guarantee the optimal solution, though, in practice, the GA process must terminate after a finite number of iterations with a high probability that the process has achieved the global optimal solution.

Following the ideas described in [10], the crossover and mutation probabilities depend on the Shannon entropy of the population (excluding the elite) fitness, which is calculated as

$$\mathcal{H}(\mathcal{P}[k]) = -\sum_{i=1}^{n_p} \lambda_i^*(k) \log \lambda_i^*(k) \tag{10.6}$$

with $\lambda_i^*(k)$ being the normalized fitness of individual $\mathbf{u}_i$, i.e., $\lambda_i^*(k) = \lambda_i(k)/\sum_{i=1}^{n_p} \lambda_i(k)$. When all the fitness values are very similar, with small dispersion, $\mathcal{H}(\mathcal{P}[k])$ becomes high and p_c is decreased – it is not worthwhile wasting time merging very similar individuals. This way, exploration is boosted, while, conversely, exploitation decreases. On the other hand, when this entropy is small, there exists a high diversity within the population, a fact that can be exploited in order to increase the horizontal sense of search. Following a similar reasoning, the probability of mutation is increased when the entropy is high, so as to augment the diversity of the population and escape from local suboptimal solutions (exploitation decreases, exploration becomes higher). Therefore, we have that probabilities p_m and p_c are directly/inversselly proportional to the population fitness entropy, respectively.

Some exponentially dependence on time k must also be included in the model – making use of exponential functions – in order to relax (decrease), along time, the degree of dependence of the genetic operators' probabilities with the dispersion measure. This avoids abandoning good solution estimates when very low fitness individuals are sporadically created, specially when most of the population individuals have converged to the global optimum.

As a consequence of these entropy dependent genetic operators, the resulting complexity of the GA is notably decreased since crossover is applied with a very low probability (and only on individuals not belonging to the elite), and the diversity control allows the algorithm to work properly with a much smaller population size, [10].

10.4 Numerical Results

In order to evaluate the proposed algorithm (parameter tuning + tracking), both synthetic and real DT-MR images have been used. For the sake of comparison, we have used the same test images as in [6].

10.4.1 Synthetic Images

Figure 6 in [6] shows four different synthetic DT-MRI data defined in a $50 \times 50 \times 50$ grid (we will refer to them as "cross", "earth", "log" and "star" as in previous work). To make the simulated field more realistic, Rician noise – [11] – was added in the diffusion weighted images which were calculated from the Stejskal-Tanner equation using the gradient sequence in [12], and a b-value of 1,000.

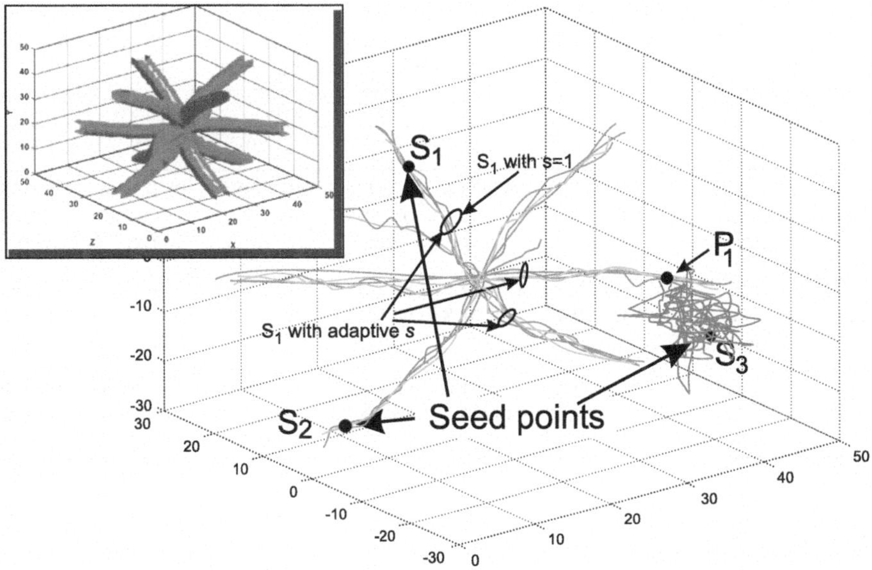

Fig. 10.2 Tracking results for the "star" synthetic DT-MR image. Black: seed points. Blue: fiber paths obtained using adjustment with NN, Green: paths using estimation with the proposed AG. Red: extrinsic voxels. Initial seeds $V_0 = \{S_1, S_2, S_3\}$. Top left: original synthetic image

Satisfactory tracing results for the first three cases can be found in [4], where a simpler algorithm was used. For the sake of brevity, in this paper we have worked with the most complex case, the *star* (Fig. 10.2, top left). This image consists of six orthogonal sine half-waves, each of them with an arbitrary radius. Under this scenario the diffusion field experiments variations with the three coordinate axes and there exists a crossing region. Three different tracking results are shown in Fig. 10.2, each of them for a different seed $V_0 = \{S_1, S_2, S_3\}$. Blue tracts were obtained with an algorithm were parameters were estimated with a NN – [6] – while green ones correspond to the estimation using the proposed GA.

It can be seen that, in both cases, the path estimates pass through isotropic zones where different fiber bundles cross. It is also appreciated how both methods differentiate between the totally isotropic zones extrinsic to the tracts and the fiber bundles.

The differentiation between voxels belonging to a fiber or to a very isotropic area, respectively, is attained by mapping the path probabilities given by Eq. (10.5) into a color scale and classifying them according to some fixed thresholds. Notice that seeds S_1 and S_2 belong to the intrinsic volume (voxels with a very high anisotropy). In this case both methods move through the most probable direction following the main direction of the star in each situation. When extrinsic point S_3 is selected as seed, the algorithms explore in the neighboring voxels until they find a voxel with a high anisotropy value (point P_1). Once P_1 is found, the tracking algorithm proceeds as in the case of S_1 and S_2. Fig. 10.2 shows how the algorithm finds the proper

Table 10.1 Convergence performance for different SNRs values. Cell values represent percentage of right convergence for two configurations of the algorithm: $s = 1/s = 4$. Each cell shows: top: NN-estimation, middle: GA-estimation, bottom: Bayesian tracking, [13]

		SNR (dB)					
Image	Method	5	10	15	20	25	30
	NN	78.3/**82.8**	89.7/**93.6**	92.1/**94.3**	98.3/**98.7**	99.0/**99.0**	100/**100**
Cross	AG	81.0/**84.9**	93.1/**94.1**	94.2/**95.7**	98.4/**98.3**	99.0/**100**	100/**100**
	Friman, 2005	76.8	89.0	90.7	97.0	100	100
	NN	77.7/**76.2**	88.6/**87.5**	89.9/**89.0**	98.2/**98.2**	99.0/**99.0**	100/**100**
Earth	AG	81.5/**79.0**	88.9/**91.8**	92.8/**93.3**	98.4/**98.5**	99.0/**100**	100/**100**
	Friman, 2005	74.4	83.2	85.0	97.3	99.2	100
	NN	71.0/**69.7**	82.1/**81.0**	86.1/**85.5**	96.0/**95.8**	98.0/**97.8**	100/**100**
Log	AG	73.5/**74.2**	84.6/**84.2**	89.1/**87.0**	96.7/**97.0**	97.8/**98.0**	100/**100**
	Friman 2005	68.8	78.3	85.2	96.0	98.0	100

fiber path whatever (extrinsic or intrinsic) seed voxel is chosen, for both methods of parameters' estimation.

Next, the robustness of the tracking algorithm with both parameter estimation methods is now studied. For the sake of brevity, these experiments were run with parameter s kept constant during the fiber tract estimation (see Section 10.2).

The convergence performance for different SNRs is shown in Table 10.1. The first row in each cell corresponds to tracking results when parameters where estimated using the NN, the second contains the results when the proposed GA is used for this estimation, and the third one shows the values obtained with a slightly modified version of the Bayesian method proposed in [13].

It can be seen that both algorithms (with NN- and AG-based adjustment) converge properly within a wide range of SNRs, with the AG version showing a convergence gain of about 3–6% in all cases. The percentage values for the "cross" and the "earth" test images are very close, while for the "log" case both algorithms exhibit a slightly lower convergence. Comparing our methods with the Bayesian approach, we see that the proposed tracking algorithm performs slightly better when the SNR is low, while the three methods tend to similar results with high SNRs.

Analyzing the simulations of the synthetic images considered, it is seen that convergence results improve whenever the MR image contains branching or crossing areas – as it is the case in real DT-MR images. This is the case of our "cross" image. For this image, the convergence results are improved $\sim 5\%$ when parameter s is modified according to the anisotropy. Besides, for these studied cases, we see that the influence of the procedure that adapts s is higher for low SNRs.

10.4.2 Real Images

The proposed tracking algorithm has also been applied to real DT-MR images. Specifically, we have selected the *corpus callosum* of the brain (see Fig. 10.3).

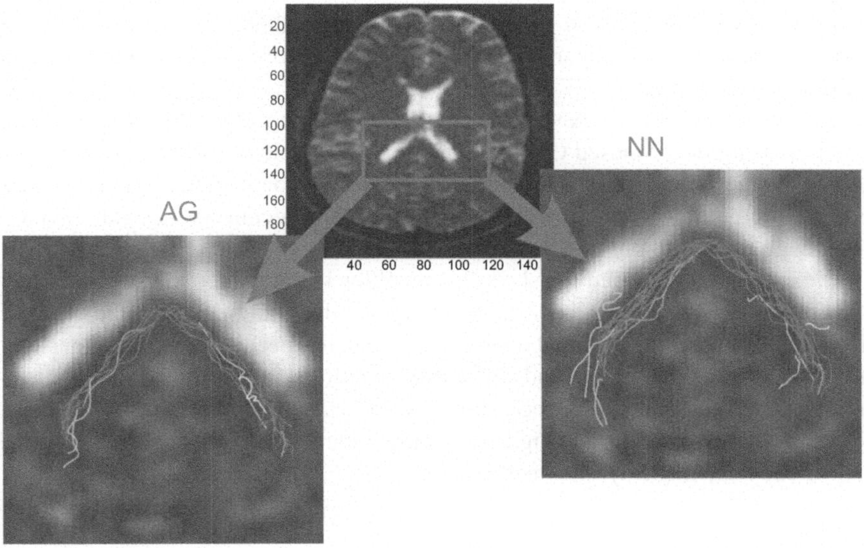

Fig. 10.3 Tracking results for the *corpus callosum* area of the human brain. *Left:* tracts obtained with the tracking algorithm tuned-up with the proposed GA, *Right:* parameter estimation with NN

Simulation results show that whichever parameters tuning-up method is used, the algorithm is able to follow the main fiber bundle directions without getting out of the area of interest. Fig. 10.3 shows some bundles of properly estimated tracts. Red/green color indicates high/low certainty.

10.4.3 Final Remarks

The proposed parameter estimation procedure is useful when the volume being analyzed varies. For instance, with just 5–10 training iterations (repetitions of the procedure described in 10.3.1), in synthetic images, or 8–16, in real images, the parameters of the algorithm are fine-tuned so as to get satisfactory results. Note that these training times are: (i) always inferior to those required by the NN-based method proposed in [5, 6], (ii) always greatly inferior to the time required to heuristically adjust the parameters, (iii) only required when the scanning conditions vary.

10.5 Conclusions

The work here presented expands upon previous work of the author on fiber tracking algorithms to be used with DT-MR images. Previous papers presented and improved the basic probabilistic tracking algorithm [4] and developed a novel multilayer neural network that helps to tune-up the tracking method, [5]. In this paper we have presented an Evolutionary Computation-based algorithm that outperforms the neural network approach.

Numerical simulations have shown that the tracking algorithm that has been tuned-up using the proposed GA-based method is capable of estimating fiber tracts both in synthetic and real images. The robustness and convergence have been studied for different image qualities (SNRs). Results show a convergence gain of about 3–6% with respect to our previous work in [5, 6].

The experiments carried out show that an efficient parameter adjustment in conjunction with precise rules to manage and update the pool of future seeds lead to: (i) a better use of computational resources, (ii) a better performance in regions with crossing or branching fibers, and (iii) a minimization of the required human intervention time. The method has been tested with synthetic and real DT-MR images with satisfactory results, showing better computational and convergence properties than already existing Bayesian methods.

References

1. Bjornemo, M. and Brun, A. (2002). White matter fiber tracking diffusion tensor MRI. *Master's Thesis, Linkoping University.*
2. Ehricke, H. H., Klose, U. and Grodd, U. (2006). Visualizing MR diffusion tensor fields by dynamic fiber tracking and uncertainty mapping. *Computers & Graphics*, 30:255–264.
3. Mori, S., van Zijl, P.C.M. (2002). Fiber tracking: principles and strategies – a technical review. *Nuclear Magnetic Resonance in Biomedicine*, 15:468–480.
4. San-José-Revuelta, L. M., Martín-Fernández, M., and Alberola-López, C. (2007). A new proposal for 3D fiber tracking in synthetic diffusion tensor magnetic resonance images. *Proceedings IEEE International Symposium on Signal Processing and Its Applications, Sharjah, United Arab Emirates.*
5. San-José-Revuelta, L. M., Martín-Fernández, M., and Alberola-López, C. (2007). Neural-network assisted fiber tracking of synthetic and white matter DT-MR images. *Proceedings International Conference on Signal and Image Engineering, ICSIE 2007, London, United Kingdom* I:618–623.
6. San-José-Revuelta, L. M., Martín-Fernández, M., and Alberola-López, C. (2008). Efficient tracking of MR tensor fields using a multilayer neural network. *IAENG International Journal of Computer Science* 35:129–139.
7. Kang, N. et al. (2005). White matter fiber tractography via anisotropic diffusion simulation in the human brain. *IEEE Trans. on Medical Imaging*, 24:1127–1137.
8. Mitchell, M. (1996). *An Introduction to Genetic Algorithms*. MIT, Cambridge, MA.
9. Bhandari, D., Murthy, C.A, and Pal, S.K. Genetic algorithm with elitist model and its convergence. *International Journal on Pattern Recognition and Artificial Intelligence*, 10:731–747.
10. San-José-Revuelta, L. M. (2005). Entropy-guided micro-genetic algorithm for multiuser detection in CDMA communications. *Signal Processing*, 85:1572–1587.
11. Gudbjartsson, H., Patz, S. (1995). The Rician Distribution of Noisy MRI Data. *Magnetic Resonance in Medicine*, 34:910–914.
12. Westin, C.-F. et al. (2002). Processing and visualization for diffusion tensor MRI. *Medical Image Analysis*, 6:93–108.
13. Friman, O., Westin, C.-F. (2005). Uncertainty in white matter fiber tractography *Proceedings of the MICCAI 2005, LNCS 3749*, 107–114.

Chapter 11
A Bridge-Ship Collision Avoidance System Based on FLIR Image Sequences

Jun Liu, Hong Wei, Xi-Yue Huang, Nai-Shuai He, and Ke Li

Abstract In this paper, a forward-looking infrared (FLIR) video surveillance system is presented for collision avoidance of moving ships to bridge piers. An image preprocessing algorithm is proposed to reduce clutter background by multi-scale fractal analysis, in which the blanket method is used for fractal feature computation. Then, the moving ship detection algorithm is developed from image differentials of the fractal feature in the region of surveillance between regularly interval frames. When the moving ships are detected in region of surveillance, the device for safety alert is triggered. Experimental results have shown that the approach is feasible and effective. It has achieved real-time and reliable alert to avoid collisions of moving ships to bridge piers.

Keywords Bridge-ship collision avoidance · infrared image · fractal · target detection

11.1 Introduction

The fact that more and more ships are built while their size becomes bigger and bigger is introducing the high risk of collision between bridge and ship in inland waterways. Incidences of ship-bridge collision mainly cause six types of results, i.e. damage of bridge, people casualty, damage of ship and goods, economical loss, social loss and environmental loss. A large amount of statistical analysis indicates that one of main reasons resulting in ship-bridge collision is the execrable natural environment such as poorly visible conditions, floods, etc. [1, 2].

J. Liu (✉)
Navigation & Guidance Lab, College of Automatic, University of Chongqing,
Chongqing, 400030, China,
E-mail: h.wei@reading.ac.uk

S.I. Ao and L. Gelman (Eds.), *Advances in Electrical Engineering and Computational Science*, Lecture Notes in Electrical Engineering.

Mainly, there are two existing strategies to avoid bridge-ship collision at present [3, 4]. One is a passive strategy in which fixed islands or safeguard surroundings are built around bridge piers. The shortages of the passive method are: it could not avoid ship damage from a collision; the costs are normally high; and it becomes less effective with constant increase of ship size. The other is an active strategy that uses radar or video images to monitor moving ships by measuring their course for collision estimation. Compared with the passive method, the active method avoids damage of both bridge and ship and its costs are low. However radar is difficult to detect course changes immediately due to its low short-term accuracy, and the high noise level makes radar sometimes hardly detect any objects from a clutter background. Sensors for visible light do not work well under poorly illuminated conditions such as fog, mist and night. In contrast, infrared sensors are capable of adapting weather and light changes during a day. Moreover, the FLIR images overcome the problems that radar has, i.e. they have high short-term angle accuracy.

In design, the first consideration of the FLIR surveillance system is its robustness for detecting moving ships. The main difficulties are: (1) low thermal contrast between the detected object and its surroundings; (2) relatively low signal to noise ratio (*SNR*) under the weak thermal contrast; and (3) insufficient geometric, spatial distribution and statistical information for small targets [5].

Motion detection in a surveillance video sequence captured by a fixed camera can be achieved by many existing algorithms, e.g. frame difference, background estimation, optical flow method, and statistical learning method [6]. The most common method is the frame difference method for the reason that it has a great detection speed and low computation cost. However the detection accuracy by using this method is strongly affected by background lighting variation between frames. The most complex algorithm is the optical flow method, by which the computation cost is high. The statistical learning method needs training samples which may not be available in most cases, and its computation cost is also relatively high. The background estimation method is extremely sensitive to the changes of the lighting condition in which the background is established.

In the FLIR surveillance video sequence used for moving ship detection, a background normally consists of various information, such as sky, the surface of river, water waves, large floating objects (non-detected objects) in a flooding season, etc. In many cases, ships and background in FLIR images are visually merged together. It is very difficult to detect the targets (moving ships) using normal methods mentioned above.

The new FLIR video surveillance system for bridge-ship collision avoidance is proposed in Section 11.2. Section 11.3 presents the novel infrared image preprocessing algorithm using multi-scale fractal analysis based on the blanket method. The moving ship detection algorithm in the region of surveillance is also developed in Section 11.3. Section 11.4 demonstrates the experimental results with detailed discussion and analysis. Finally, conclusions and future work are presented in Section 11.5.

11.2 The FLIR Video Surveillance System

In the system, a pan-tilt is fixed on bridge pier, and FLIR camera is installed on the pan-tilt. The visual region of FLIR i.e. the region of surveillance, can be adjusted by the pan-tilt, and is configured according to real conditions. The FLIR camera links to a personal computer (*PC*) through a frame grabber. When images are captured, the image processing program in *PC*'s memory is used to detect the moving ships. When the moving ships are detected in the region of surveillance, the device for safety alert is started. The ship driver could be alarmed if necessary, and he/she would take maneuvers to avoid ship-bridge collision. The flowchart of the system and sketch map of installation is depicted in Fig. 11.1.

Large amount of experiments carried out in the Yangtse River have proved that the minimum pre-warning distance between bridge pier and ship to avoid collision is 500 m in inland waterway, and the valid distance for moving ship detection is from 800 to 2,000 m when the uncooled infrared *FPA* (focal plane arrays) thermal imaging camera is used. Therefore, this type of camera is suitable for the application. The camera resolution is 320 × 240 pixels. There are three ways designed to trigger the pre-warning signal, i.e. automatically broadcast the pre-recorded voice through very high frequency (*VHF*), automatically broadcast the pre-recorded voice through loudspeaker, and automatically turn on the assistant lighting system.

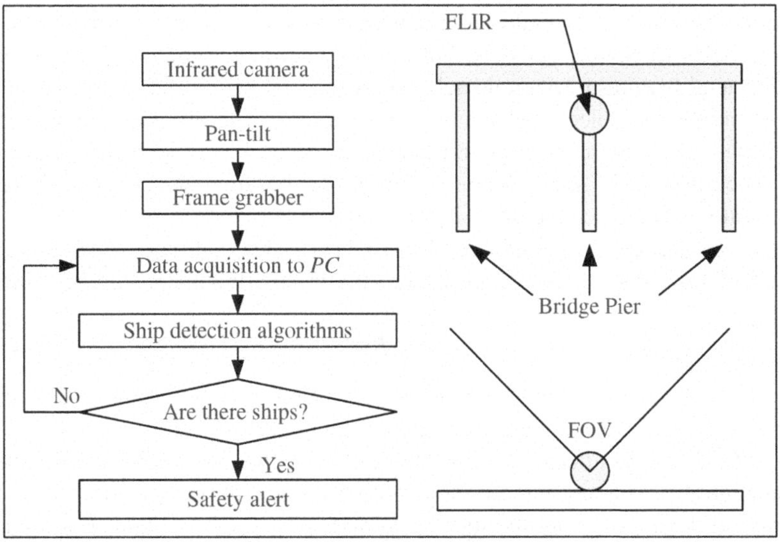

Fig. 11.1 Flowchart of the system framework and sketch of installation

11.3 The Detection Algorithm for Moving Ships

In order to detect moving ships in an FLIR image from complicated background along the inland waterway, a novel detection algorithm is developed. It consists of four main stages: extracting the region of interest, i.e. the region of surveillance (*ROS*); calculating the multi-scale fractal feature; generating the binary image based on the fractal feature; detecting moving ships by a frame difference method. The algorithm is schematically demonstrated in Fig. 11.2.

11.3.1 Extracting the ROS

The *ROS* is defined based on various conditions in real parts of the inland waterway. Consequently, the *ROS* appears as a part of region in the original image. The image analysis and processing is focused on this region only. This excludes unwanted regions to reduce computation cost in further processing.

11.3.2 Calculating the Multi-scale Fractal Feature

In practice, ships in FLIR images are treated as man-made objects in contrast to natural background. A fractal model can well describe complex surface structure characteristics for natural objects, but not for man-made objects [7, 8]. To some extent, fractal features of natural background keep relatively stable, but fractal features of man-made objects behave obviously variety. Therefore, the fluctuation of fractal features distinguishes natural and man-made objects with the variation of scale. The multi-scale fractal feature is proposed to reduce interference of natural background and enhance the intensity of ships.

Many researchers [9, 10] adopted Mandelbrot's idea and extended it to surface area calculation. For a grey image surface, the fractal dimension can be estimated as in Eq. (11.1).

$$A(\varepsilon) = K\varepsilon^{2-D} \tag{11.1}$$

where D is the fractal dimension, K is a constant, ε is the scale with value of $0, 1, \cdots, \varepsilon_{\max}$, and $A(\varepsilon)$ is the surface area of image in the scale ε.

Let $f(x, y)$a grey image, the image could be viewed as a hilly terrain surface where height from the normal ground is proportional to the image grey value. Then all points at distance ε from the surface on both sides create a blanket of thickness 2ε. The estimated surface area is the volume of the blanket divided by 2ε. For different scale $\varepsilon(\varepsilon \geq 0)$, the blanket area can be iteratively estimated. The upper surface $u(x, y, \varepsilon)$ and the lower surface $b(x, y, \varepsilon)$ are expressed in Eqs. (11.3) and (11.4), respectively.

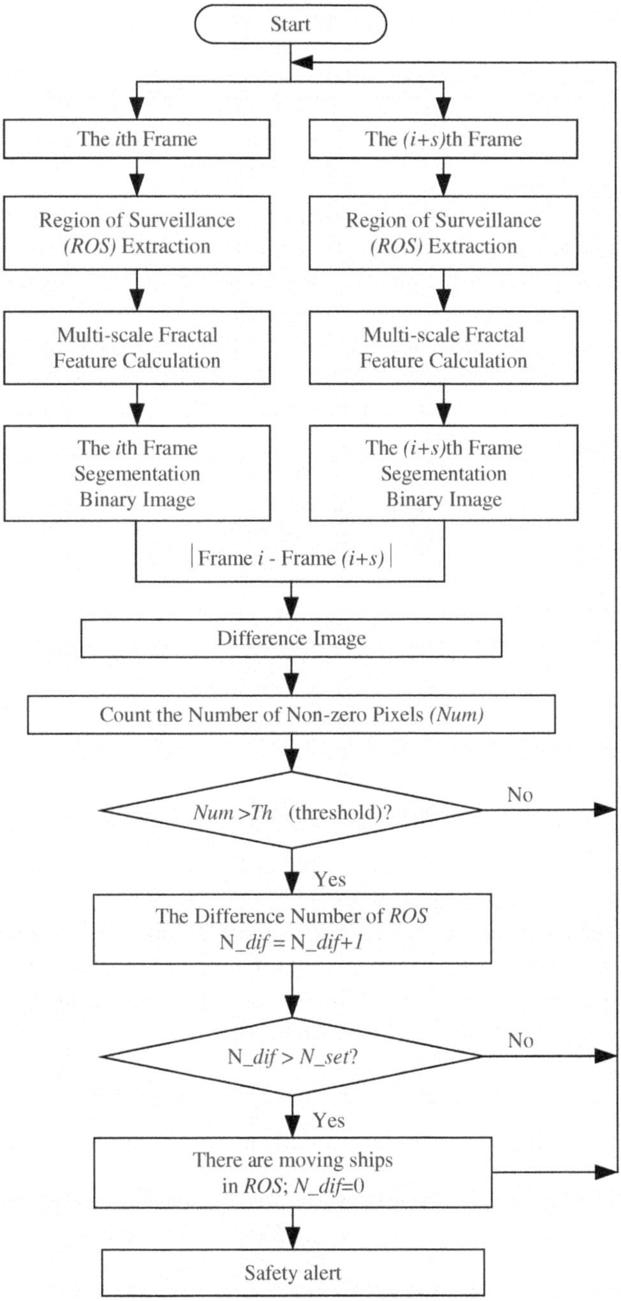

Fig. 11.2 The detection algorithm

$$u(x, y, 0) = b(x, y, 0) = f(x, y) \quad \varepsilon = 0 \tag{11.2}$$

$$u(x, y, \varepsilon) = \max\{u(x, y, \varepsilon - 1), \max_{|(m,n)-(x,y)|\leq 1} u(m, n, \varepsilon - 1)\} \tag{11.3}$$

$$b(x, y, \varepsilon) = \min\{b(x, y, \varepsilon - 1), \min_{|(m,n)-(x,y)|\leq 1} b(m, n, \varepsilon - 1)\} \tag{11.4}$$

where, $\varepsilon = 0, 1, \cdots, \varepsilon_{\max}$, the image point (m, n) with distance less than one from (x, y) is the four neighbors of (x, y). To assure the blanket of the surface for scale ε includes all the points of the blanket for scale $\varepsilon - 1$, (m, n) are chosen to be the eight neighbors of pixel (x, y) in the experiments.

$\varepsilon_{\max}$ is the maximum scale when the fractal features are calculated. $\varepsilon_{\max} \in N$, $\varepsilon_{\max} \geq 2$.

At scale ε, the volume between the upper surface and the lower surface is calculated by Eq. (11.5):

$$V(x, y, \varepsilon) = \sum_{k=x-\varepsilon}^{x+\varepsilon} \sum_{m=y-\varepsilon}^{y+\varepsilon} [u(k, m, \varepsilon) - b(k, m, \varepsilon)] \tag{11.5}$$

Then, the estimate of the surface area at (x, y) and ε can be obtained by Eq. (11.6):

$$A(x, y, \varepsilon) = \frac{V(x, y, \varepsilon)}{2\varepsilon} \tag{11.6}$$

Taking the logarithm of both sides in Eq. (11.1), we have:

$$\log A(x, y, \varepsilon) = (2 - D)\log(\varepsilon) + \log K \tag{11.7}$$

Linearly fitting $\log A(x, y, \varepsilon)$ against scale ε in Eq. (11.7), fractal dimension D can be obtained as a constant for all scales ε.

For the constant K in Eq. (11.1), also named D-dimensional area [12], which characterizes the roughness of surface, i.e. different surfaces have different K values. From this point of view, K is not a constant for the variation of scale ε. When two scales $\varepsilon_1, \varepsilon_2$ are used in Eq. (11.7), we have:

$$\log A(x, y, \varepsilon_1) = (2 - D)\log(\varepsilon_1) + \log K \tag{11.8}$$

$$\log A(x, y, \varepsilon_2) = (2 - D)\log(\varepsilon_2) + \log K \tag{11.9}$$

In Eqs. (11.8) and (11.9), fractal dimension D is a constant for scale ε_1 and ε_2, and let $\varepsilon_1 = \varepsilon, \varepsilon_2 = \varepsilon + 1, \varepsilon = 0, 1, 2, \cdots \varepsilon_{\max}$, $K(x, y, \varepsilon)$ is derived as

$$K(x, y, \varepsilon) = \exp\left(\frac{\log A(x, y, \varepsilon)\log(\varepsilon + 1) - \log A(x, y, \varepsilon + 1)\log(\varepsilon)}{\log(\varepsilon + 1) - \log(\varepsilon)}\right) \tag{11.10}$$

From Eq. (11.10), the D-dimension area $K(x, y, \varepsilon)$ can be calculated from surface area $A(x, y, \varepsilon)$ at point (x, y) along with scale ε. We use a new function $C(x, y)$ to measure the deviation of $K(x, y, \varepsilon)$ against scale ε, as presented in Eq. (11.11).

$$C(x, y) = \sum_{\varepsilon=2}^{\varepsilon_{\max}} \left[K(x, y, \varepsilon) - \frac{1}{\varepsilon_{\max} - 1} \sum_{\varepsilon=2}^{\varepsilon_{\max}} K(x, y, \varepsilon) \right]^2 \tag{11.11}$$

$C(x, y)$ is the fractal feature used for ship detection in the algorithm. $C(x, y)$ appears high value for man-made objects, and much lower value for natural background.

11.3.3 Segmenting the Fractal Feature Image

$C(x, y)$ provides sufficient information to discriminate natural background and ships. The simplest OSTU segmentation method [11] is used to segment the $C(x, y)$ image. In the resulting binary image, pixel value 255 represents ships and other man-made objects.

11.3.4 Detecting Moving Ships

In the process of moving ship detection, a difference between two binary images generated from segmentation of $C(x, y)$ is used. A group of pixels with non-zero values represent the difference. Based on the fact that the FLIR camera is fixed on a bridge pier, the process is summarized as follows.

1. Generate two binary images $C_i(x, y)$ and $C_{i+s}(x, y)$ by segmentation of two frames with interval of s, $f_i(x, y)$ and $f_{i+s}(x, y)$. In practice, the interval s is chosen as five to ten frames.
2. Calculate the difference between image $C_i(x, y)$ and $C_{i+s}(x, y)$, and obtain $D(x, y)$.
3. To count the number of non-zero pixels in $D(x, y)$, and record it as *Num*.
4. If *Num* is larger than the threshold (*th*) which is experimentally obtained, denote the number of pixels in *ROS* have changed once, then the difference number of *ROS* (*N-dif*) add one.
5. If *N-dif* is larger than a pre-set value *N_set* from experiments, denote that moving ships are detected in the *ROS*. In practice, the value *N_set* is set as 2 or 3, which is effective in reducing the false alarm ratio.

11.4 Experimental Results and Discussion

The testing experiments were carried out in the Yangtse River in Chongqing city, China. An FLIR camera was mounted on a bridge pier. A Celeron1.5 Ghz *PC* was connected with the camera through a frame grabber, the frame size was 320 × 240, and the frame rate was 30*fps*. The parameter settings in the algorithm were, the frame interval as ten frames, the value of threshold (*th*) as 5, the value of *N_set* as 2, and the value of $\varepsilon_{\max}$ as 4. A group of testing results is demonstrated in Fig. 11.3. The average processing time for each step in the algorithm is shown in Table 11.1.

Observations have indicated that the speed of moving ships is from 20 to 30 km/h. The *ROS* defines the distance between moving ships and bridge pier as 800–2,000 m. Therefore the time during which a ship driver takes action to avoid collision to a bridge pier after altered is between 96 and 360 s. From Table 11.1, it is clearly shown that the FLIR surveillance system takes about one s to complete a process, due to the value *N_set* as 2 or 3. It is satisfactory to the application with a real-time manner.

Comparative experiments were also carried out for system performance analysis in terms of reliability and effectiveness. The frame difference method was implemented to be compared with the proposed method. FLIR frames were carefully selected for this comparison. Weather conditions and time of a day were taken into account when 400 frames with 286 moving ships involved were chosen as the testing set. Two parameters were introduced as the criterion for the performance, i.e. false alarm ratio (*FAR*) and missed alarm ratio (*MAR*). The comparative results are shown in Table 11.2. Some typical experimental results are demonstrated in Fig. 11.4.

From the results in Table 11.2, it can be seen that the proposed method for bridge-ship collision avoidance is superior to the frame difference method in the criterion of both false alarm ratio and missed alarm ratio. From the results of the testing set and Fig. 11.4, the system is capable of adapting weather and light changes during a day.

It is worth to mention that while the FLIR system is mounted on bridge deck, the performance of surveillance system is impaired by vibration caused by moving vehicles on the bridge. Therefore, the FLIR system is mounted on a bridge pier in practice.

11.5 Conclusion

This paper presented a novel FLIR video surveillance system for bridge-ship collision avoidance by using the multi-scale fractal feature, by which moving ships have successfully been separated from the complex background in inland waterway images. The proposed algorithm for moving ship detection has achieved the real-time performance within the *ROS* in FLIR video sequences. Experimental results have proved that the developed FLIR video surveillance system is efficient in detecting moving ships to alert possible bridge-ship collisions. Its wide adaptability and

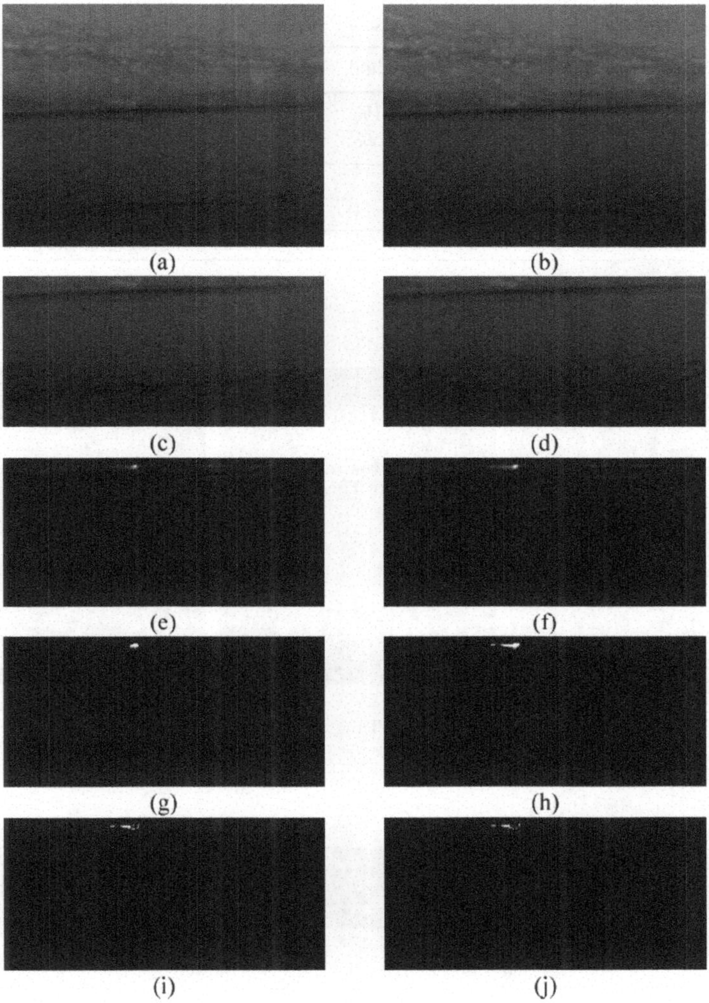

Fig. 11.3 Testing results. (a) The 1st original frame. (b) The 11th original frame. (c) The *ROS* image extracted from (a). (d) The *ROS* image extracted from (b). (e) $C(x, y)$ of the 1st frame. (f) $C(x, y)$ of the 11th frame. (g) Segmenting binary image from (e). (h) Segmenting binary from (f). (i) $D(x, y)$ between the 11th and the 1st frames, *Num* = 32. (j) between the 21st and the 11th frames, *Num* = 25

Table 11.1 Average processing time for each step

Steps	Time (ms)
Extracting the region of surveillance	5
Calculating $C(x, y)$	331
Segmenting $C(x, y)$ image	22
Detecting moving ships based on frame difference	25
Other steps	10
Total	393

Table 11.2 Comparative performance analysis

Frame difference	method	The proposed method
FAR	23.1%	Lower than 1%
MAR	17.6%	Lower than 1%

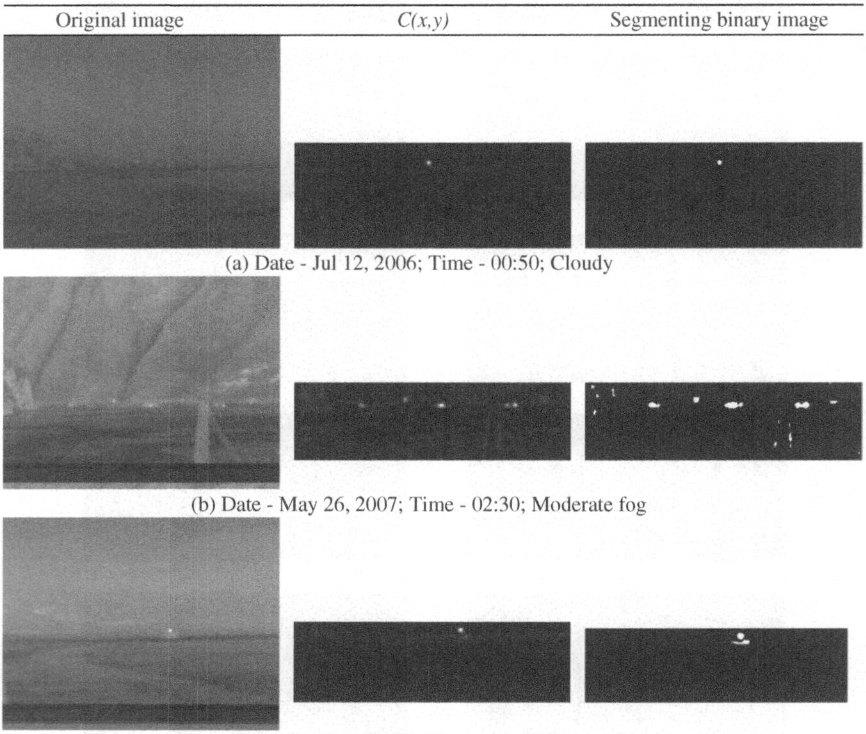

(a) Date - Jul 12, 2006; Time - 00:50; Cloudy

(b) Date - May 26, 2007; Time - 02:30; Moderate fog

(c) Date - Jul 11, 2006; Time - 21:20; Shower

Fig. 11.4 Testing results

high reliability fro weather and light changes have been proved in long period testing in Yangtse River, Chongqing city, China.

Our future work will be extended to two venues. First, the investigation for tracking multiple ships will be carried out based on the ship detection results. Relative tracking algorithms will be developed to address the issue in a clutter background. Second, a data fusion of radar and infrared sensors may be explored to improve system performance in making use of the complement of data from different sensors.

Acknowledgement The authors would like to thank the Maritime Safety Administration of Chongqing for providing experimental sites. This research is partially supported by the China Scholarship Council (CSC).

References

1. Dai, T. D., Lie, W., and Liu, W. L., 1993, The analysis of ship-bridge collision in main waterway of the Yangtze River, *Navigation of China*, 4:44–47.
2. Van Manen, S. E., 2001, Ship collisions due to the presence of bridge, International Navigatio Association (PIANC), Brussels, Report of WG 19.
3. Zhu, Q. Y., 2006, Pier anticollision system based on image processing, Master's thesis, Department of Information Engineering, Wuhan University of Technology, Wuhan.
4. Wu, J., 2004, Development of ship-bridge collision analysis, *Journal of Guangdong Communication Polytechnic*, 4:60–64.
5. Liu, J., Huang, X. Y., Chen, Y., and He, N. S., 2007, Target recognition of FLIR images on radial basis function neural network, in *Proceedings of Advances in Neural Networks, ISNN 2007, 4th International Symposium on Neural Networks*, Nanjing, pp. 772–777.
6. Zhan, C. H., Duan, X. H., Xu, S. Y., Song, Z., and Luo, M., 2007, An improved moving object detection algorithm based on frame difference and edge detection, in *Proceedings of the Fourth International Conference on Image and Graphics*, Chengdu, pp. 519–523.
7. Mandelbrot, B. B., 1982, *The fractal geometry of nature*. New York: W.H. Freeman, pp. 1–24.
8. Pentland, A., 1984, Fractal-based description of natural scenes, *IEEE Transactions on Pattern Analysis and Machine Intelligence*, 6:661–674.
9. Peleg, S., Naor, J., Hartley, R., and Avnir, D., 1984, Multiple resolution texture analysis and classification, *IEEE Transactions on Pattern Analysis and Machine Intelligence*, 6:518–523.
10. Zhang, H., Liu, X. L., Li, J. W., and Zhu, Z. F., 2006, The study of detecting for IR weak and small targets based on fractal features, In *Lecture Notes in Computer Science, Advances in Multimedia Modeling*. Springer, Heidelberg/Berlin, pp. 296–303.
11. Ostu, N., 1979, A threshold selection method from gray-level histograms, *IEEE Transactions System, Man and Cybernetics*, SMC-9:62–66.
12. Li, J. and Zhang, T. X., 1996, Target detection based on multiscale fractal parameter change, *Journal of Data Acquisition & Processing*, 11(3):218–221.

Chapter 12
A Gray Level Feature Detector and Its Hardware Architecture

Neeta Nain, Rajesh Kumar, and Bhavitavya Bhadviya

Abstract This chapter describes a fast real time gray scale based feature point detector and its hardware architecture for *FPGA* based realization. The implementation is based on a new efficient technique which is fusion of both affine transformation invariance and robustness to noise. The novelty of the proposed approach lies in its highly accurate localization and realization only in terms of addition, subtraction and logic operations. The algorithm is designed to keep the high throughput requirements of today's feature point detectors and applications in Silicon. The proposed implementation is highly modular with custom scalability to fit devices like *FPGAs* etc, with different resource capacity. The implementation can be ported to any real time vision processing systems where power and speed are of utmost concern.

Keywords Feature Detector · gray scale · Hardware Architecture · affine transformation invariance · FPGA

12.1 Introduction

A one dimensional change in intensity can be classified as an edge and it can be approximated using gradient or first order derivative of the image. Similarly feature points (FPs) are the sudden changes in two dimensions or second order properties of a surface, which can be approximated with second order derivatives. Since second order derivatives are noise sensitive, and hence the proposed FP detector uses only first order derivatives to approximate the second order derivative. FPs are locations in the image where the signal changes significantly in two-dimensions. Examples include corners and junctions as well as locations where the texture varies

N. Nain (✉)
Associate Professor, Department of Computer Engineering,
Malaviya National Institute of Technology Jaipur-302017, India,
E-mail: neetanain@yahoo.com

S.I. Ao and L. Gelman (Eds.), *Advances in Electrical Engineering and Computational Science*, Lecture Notes in Electrical Engineering.

significantly. A large number of feature point detectors like [1–7] etc. are reported in literature which are mainly classified into two categories: template based and geometry based. Template based approach is to build a set of corner templates and determine the similarity between the templates and all the sub windows of the gray level image. Though simple these techniques are time consuming due to multiple iterations. Geometry based relies on measuring differential geometry features of corners. They are of two kinds: boundary based and gray level based. Boundary based, first extract the boundary as a chain code, and then search for significant turnings at boundary. These techniques suffer from high algorithm complexity, as multiple steps are needed also the corner accuracy is dependent on the boundary extraction technique used. Gray-level based methods, directly operate on the gray level image. The feature detector proposed in this work belongs to this last category.

The chapter is organized as Section 12.2 describes the proposed feature point detection algorithm. Its hardware realization is explained in Section 12.3. The synthesis results and its analysis is given in Section 12.4 followed by related discussion and conclusions in Section 12.5.

12.2 Feature Point Detection Model

A true FP can be defined as a sudden and significant $2D$ change in intensity generally classified as corners [2, 3]. Alternatively FP can also be defined as a point where multiple regions of different intensity meet and convey significant information change in $2D$. As the number of regions meeting at a point increases, the measure of entropy also increases at that point which is defined as Information Content(IC) of the FP in this chapter. Such points are commonly referred as interest points [5, 8]. The proposed technique detects both corners and interest points as FPs. With the aim of satisfying the universal criterion of localization, consistency, accuracy and noise immunity in a real time environment a short but effective five step algorithm is proposed. Where first $1D$ changes in intensity are detected using a very simple 2×2 difference mask unlike the usual convolution masks of sizes varying from 3×3 to 7×7, the complete algorithm is as follows:

1. **Apply Difference Mask** – The complex convolution is replaced with simple differences between the pixel intensity values. The use of the difference masks gives rise to four orthogonal difference values: H_1, H_2, V_1 and V_2 as shown in Fig. 12.1. These are the four critical difference values that responds positively to intensity changes. Each of these can be defined as:

$$H_1 = | I_{(i,j)} - I_{(i,j+1)} | \tag{12.1}$$

$$H_2 = | I_{(i+1,j)} - I_{(i+1,j+1)} | \tag{12.2}$$

$$V_1 = | I_{(i,j)} - I_{(i+1,j)} | \tag{12.3}$$

$$V_2 = | I_{(i,j+1)} - I_{(i+1,j+1)} | \tag{12.4}$$

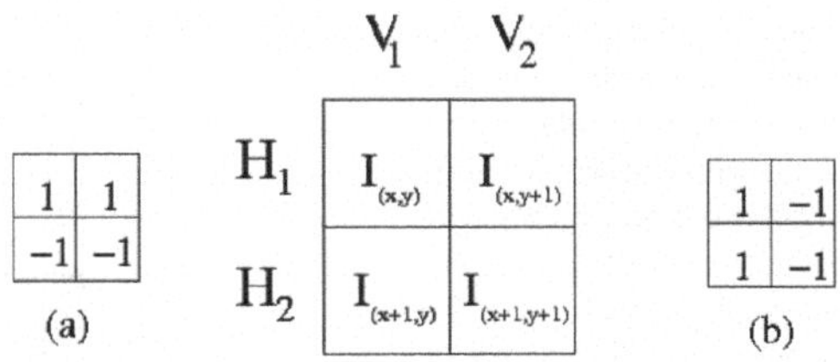

Fig. 12.1 The 2 × 2 difference operator mask (**a**) and (**b**) are 90° rotated to each other

Where $I(i, j)$ is the pixel under consideration. The response of the 2 × 2 difference operator can be classified as, If $(H_1 = H_2 = V_1 = V_2 = 0)$ then it is a constant region; If $((H_1 \text{ AND } H_2 > P_1) \text{ OR } (V_1 \text{ AND } V_2 > P_1))$ then it is a region of only $1D$ change; If $((H_1 \text{ OR } H_2 \text{ AND } V_1 \text{ OR } V_2) > P_1)$ then the region contains FPs (detected as $2D$ change). Here P_1 is the threshold value for FP detection which can be varied as per the interest of feature detection. For example, $P_1 = 1$ will detect all intensity variations as FPs, and $P_1 = 255$ will detect only step changes (black-to-white or white-to-black) as FPs. The difference operator detects candidate points for feature detection. As we are interested only in the $3rd$ category of responses for FP detection hence each pixel satisfying this criteria is further processed to determine whether it is a true FP or not.

2. **Apply the Pseudo Gaussian Mask to the Points Satisfying the First Difference Operator –** In real case scenario we receive images which are blurred due to several types of noise and thus FPs and also edges are not well defined. To overcome these problems and increase the noise immunity of our algorithm we propose a Pseudo-Gaussian kernel which is derived from a 5 × 5 Gaussian kernel having $\sigma = 1.3$. It is further modified and normalized so as to have all factors some exponent of 2 to accomplish operations like multiplication or division(in the Gaussian) by simple shift of the bits during calculations. Due to constrained time and performance intensive requirement for real time applications the Gaussian smoothing is applied only around those pixels which are part of the region containing FPs. Additionally due to the nature of our difference kernel we apply only a partial of the entire Gaussian kernel. This partial implementation as shown in Fig. 12.2 reduces the Gaussian averaging overhead by 75% but produces the desired noise smoothing on the set of 2 × 2 pixels under consideration. The new Gaussian averaged values of the four pixels under consideration (where we get the third category of responses) are calculated as:

$$I^{'}_{(i,j)} = G_k * I_{(i,j)}; \quad I^{'}_{(i,j+1)} = G_l * I_{(i,j+1)} \tag{12.5}$$

$$I^{'}_{(i+1,j)} = G_m * I_{(i+1,j)}; \; I^{'}_{(i+1,j+1)} = G_n * I_{(i+1,j+1)} \tag{12.6}$$

where G_k, G_l, G_m, G_n are as shown in Fig. 12.2.

3. **Calculate the Second Difference Operator with Threshold Parameter P_2 –** To approximate second order differences the difference operator is re-applied on the above new Gaussian averaged values of the four pixels under consideration and the four orthogonal differences H_1, H_2, V_1 and V_2 are updated using a

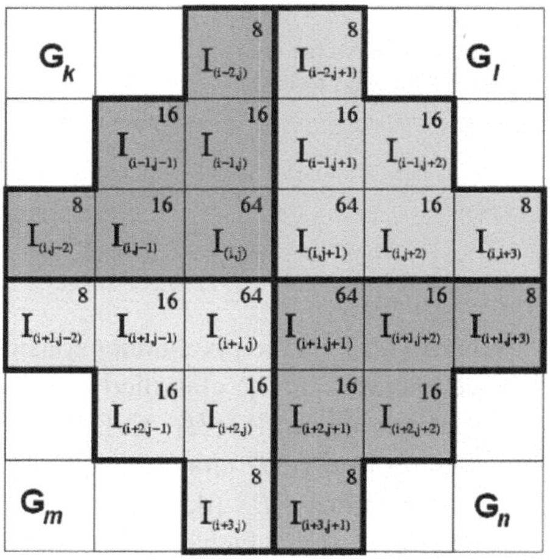

Fig. 12.2 Pseudo-Gaussian mask. Each of the four central pixel(weight = 64) is averaged with the neighboring similar colored pixels with their respective weights(in top right corner). Note the weights are all a power of 2 and that the total weight of the partial mask is also $64+2*16+16+(2*8) = 128$ (a power of 2)

second threshold parameter P_2. The use of P_2 is encouraged to avoid missing weak FPs in the presence of noise. A low P_1 will detect most of the FPs and P_2 gives control over the noise interference with the desired FPs. The pseudo Gaussian kernel is significant in many ways and is one the principle part of the proposed technique. In addition to noise removal it is amplifying the change in intensity at the meeting point of regions, in contrast to smoothing the meeting point. Consider a set of pixels where four different gray scale regions meet. As the values of each pixel become averaged by pixels having the same pixel intensity the difference mask in the next step responds very strongly to this point. This is unlike applying a standard kernel wherein the pixel is averaged with all neighboring pixels and may lead to a very low or no response at all if the nearby regions are close to each other in gray scale values.

4. **Compute Information Content of the** FP **–** The measure of entropy or distinctiveness of the detected FP is computed as the Information Content(IC) of the FP. For the set of four pixels that form a part of the zero crossings (satisfy the second difference operator) the $IC_{(i,j)}$ of that FP is increased by a constant. For example, we can add a 12 to its IC, a 7 to the IC of its 4-connected neighbors and a 5 to the IC of its diagonally connected neighbors (any multiple of 12, 7 and 5 can be used). This results in a averaged and normalized IC as 60, when all 8-connected neighbors satisfy the second difference operator including itself as maximum entropy at the current pixel. The more significantly distinctive the FP is, the larger is the average IC. For the best accuracy we pin-point all FPs with $IC > 52$ in the nxn neighborhood (where n depends upon the localization accuracy required) as the most prominent FPs (as shown in red), where at least 6 pixels from different regions in its neighborhood are contributing to this FP. Similarly the less dominant are shown in blue with $44 < IC < 52$, in green for $32 < IC < 44$, and in white for $22 < IC < 32$, where at least 5, 4, and 3 pixels from different regions are contributing to the FP respectively.

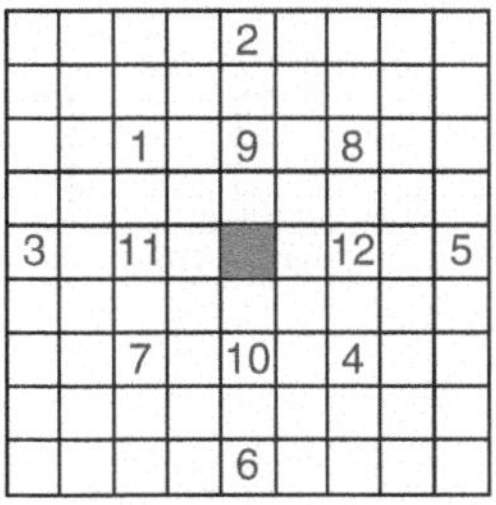

Fig. 12.3 Mask used for removing the false positives

5. **Eliminate False Positives –** The orthogonal difference masks will respond strongly to all diagonal edges (false positive FPs). To avoid these we eliminate all the candidate pixels that are part of the diagonal edges and are only 2-pixel connected in the direction of the diagonal. A 4×4 mask as shown in Fig. 12.3 is used to eliminate the false positives. To understand this, we have defined two logical functions: $Inrange2(Px_1, Px_2)$ where in Px_1 and Px_2 are the points of concern as shown in the mask (numbered $1 - 12$). $Inrange2(Px_1, Px_2)$ is true if and only If $|(Px_1 - Px_2)| < P_2$. Similarly we define $Inrange3(Px_1, Px_2, Px_3)$, which is true if and only If $|(Px_1 - Px_2)|$ AND $|(Px_2 - Px_3)| < P_2$. We eliminate all pixels that return true for only $Inrange3(x, y, z)$ in the same direction and no $Inrange2()$ is true other than $Inrange2(x, y)$ and $Inrange2(y, z)$. These are the non maxima FPs which are part of a diagonal edge and needs to be eliminated.

12.3 Hardware Implementation

The feature detector is implemented on *FPGA*. Two different implementation approaches have been used for realizing the proposed hardware as explained in the following sub Sections 12.3.1 and 12.3.2. In the first approach the entire algorithm is completely implemented in hardware with custom scalability to fit *FPGA*'s with different resource capacity. In the second approach it is implemented using the concept of *SOPC*.

12.3.1 Architecture 1: A Complete Hardwired Approach

The overall architecture is depicted in the block diagram as shown in Fig. 12.4. Two *RAM* blocks are used where *RAM* block 1 stores the original image with each pixel represented in gray scale value of 8 bits. Whereas *RAM* block 2 stores the resultant image after false positive removal and also stores the IC of each pixel. *RAM* block 2 is twice the size of the input image. *RAM* block 2 also feeds the Non-Maxima suppression module with the resultant IC for suppression of weak features. It is best to have the fastest possible memory for reducing the processing time of the

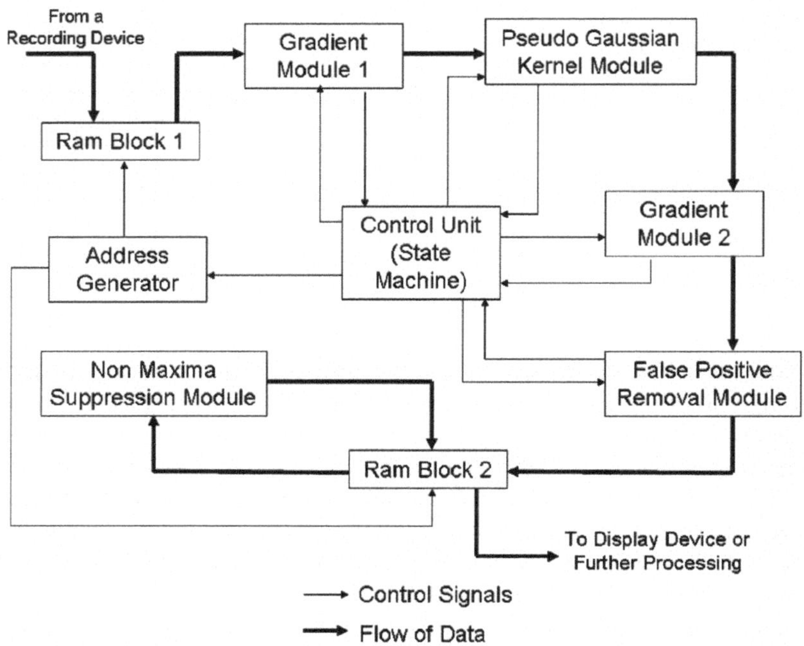

Fig. 12.4 Block diagram of the Proposed Architecture 1

algorithm as reading and writing from the *RAM* is the most expensive operation in our entire algorithm.

Address Generator module generates the memory addressing sequence for both the *RAM* modules. It is capable of generating individual addresses depending upon the control signals from the Control Block.

Control Block is the most important module as it synchronizes the various blocks. It is made of a State Machine converted to its *VHDL* entity. It controls the read/write sequences necessary to proceed further. It takes as input the results of each block compares them to its present state and decides where to proceed. It is capable of clearing the module data and assigning the threshold values P_1 and P_2 for comparison operation.

Gradient Module 1 uses 4 comparators and 4 subtractors along with other logic elements to calculate the Gradient mask. This module is again reused as Gradient Module 2, the only difference being in the threshold value which is P_2 instead of P_1 as used in the $1st$ module. This requires 4 read cycles to obtain the pixel information necessary for calculations.

Pseudo Gaussian Kernel Module is called only if a pixel satisfies the Gradient Module 1, otherwise processing of all the following modules are skipped and the address generator moves to the next pixel. This step requires another 20 read cycles to obtain the total of 24 pixel values compulsory for its mask calculation and obtaining the updated values for the next gradient module. It has been determined experimentally (on real images) that on an average only 40% of the entire image pixels actually enters this module. This module uses only left shift operations and 20 adders(8 bit) to approximates the convolution and multiplication.

This is followed by the Gradient Module 2 in which apart from mask calculations the *IC* is also computed. It is a background module which adds specific value to the *IC* map and writes it into the *RAM* module 2 taking a total of 9 write cycles.

All the pixels that satisfy the Gradient Module 2 are then processed into the False Positive Removal Module which requires another 7 read cycles, to obtain the unread pixels required for its processing. It further requires around 16 comparators and 16 subtractors to formulate the input for its Conditional Block.

Once the entire image is over, the Control Unit starts the Non-Maxima Suppression module which reads the *IC* of each pixel and determines its significance. It uses 3 comparators which compares the *IC* of the current pixel with 3 constant values in a step manner and colors the pixel accordingly. It uses 2 Latches(8 bit wide) that prevents unwanted comparing operations. The module first works on all the column pixels of the current row and then increments the row number for further processing. Maximum column and row number for a module can be specified in the Control Unit Registers. We have used a common shared register having 31 8-bit values, this register stores the pixel values as and when they are read from the *RAM* for the current iteration. Once the iteration is completed, instead of clearing all the values, we shift the values to new positions for the next iteration. This saves the common read values for the next iteration. This on an average reduces the memory overhead by 50%. Figure 12.5 shows the Low level block diagram of two of the various modules depicting the actual mapping of the *VHDL* into schematic entities. The top level module is a schematic and hence each *VHDL* module was converted into its schematic before adding to the top level module.

12.3.2 Architecture 2: A Hardware Software Co-design

The complete hardwired approach is suitable for use in *ASIC* design or a system with a vision co-processor. It would be difficult to be used in an existing embedded

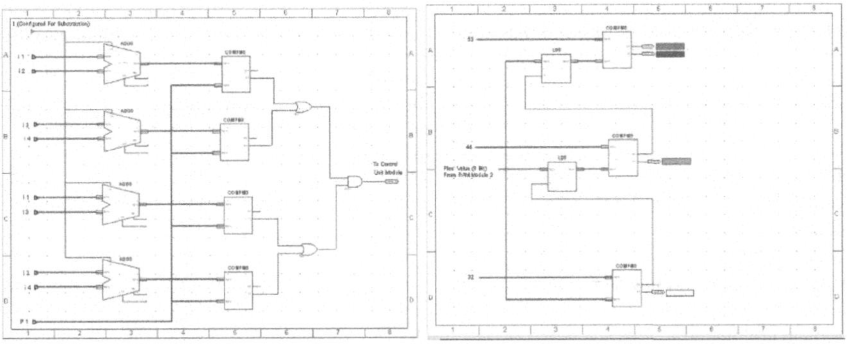

Fig. 12.5 Internal structure of module 1 (**a**) Gradient module, (**b**) partial non-maxima suppression module

applications. For full filling this usability we took another approach. The system was designed as an *SOPC* with *NIOS II* as the core processor, and then the algorithm's logic was added by means of Custom Instructions. The *NIOS II* processor was then programmed in *C* code for executing on the [11] and [12] systems.

12.4 Synthesis Results and Analysis

The Compilation, verification and Simulation of the architecture is done for both *Altera's Cyclone II* and *FPGA* device (*EP2C20*). The entire verification process was carried on 480 x 640 sized images. The timing analysis post synthesis and simulation resulted in a throughput speed of about 500 images/s (assuming a minimum average delay of 6.69 ns). The worst case scenario is evaluated by testing an image with alternate black and white pixels and assigning P_1 and P_2 both to 1, thus ensuring that each and every pixel enters all the modules and time taken is maximum. This resulted in the worst case timing delay of 19 ns, making our implementation execute 200 images/s. Further the complete implementation requires a tightly coupled (or a dedicated) *RAM* having a total of 1,500 KB of free space. Its synthesis resulted in the implementation summary as enlisted in Table 12.1. Table 12.1 also depicts the minimum timing latency for each module. Assuming the serial nature of the iteration for single feature detection module the minimum latency post synthesis turns out to be 18.2 ns. Thus a 480×640 size image would take 5.3 ns to execute. Although in real time applications it never takes that much as only 40% of the total pixels reach the second module, from which only 20% reach the Gradient module 2 while only 10% of the total image pixels actually reach the false positive removal module. Further this results in only 10% of all the pixels being finally fed into the Non-maxima suppression module. Re-calculating the modified minimum delay it comes out as:

$$Delay = 3.08 * 4/21 + 2.02 + 0.4 * (3.08 * 5/21 + 4.12 + 2.02) \\ + 0.2 * (4.98) + 0.1 * (3.99) = 6.69\ ns$$

Table 12.1 Resource utilization summary and timing analysis for each of the individual modules

Module name	Utilization (slices) (out of 18742)	Utilization percentage	Latency (ns)
Address generator	16	0.00	3.08
Control unit	16	0.00	0
Gradient module	112	0.60	2.02
Gaussian module	133	0.71	4.12
False positive removal module	168	0.90	4.98
Non maxima suppression module	48	1.32	3.99
Complete architecture	493	2.63	20.21

Table 12.2 Compilation report of the *SOPC* system

Resource type	Utilized	Available	Percentage
LUTs	5,278	18,752	28
Total pins	287	315	91
Memory bits	79,360	239,616	33
PLLs	2	4	50
NIOS II system code	72 KB	8,192 KB	0.9
Run time memory usage	1,872 KB	8,704 KB	21

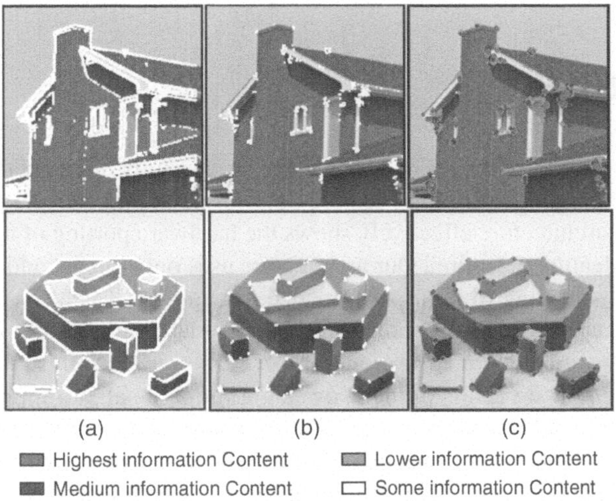

Fig. 12.6 Output on two of the test images after (**a**) First difference operator, (**b**) Second difference operator (**c**) False positive removal

The *SOPC* system with the updated *NIOS II* core (Supplemented with our Custom Instructions) was compiled and executed on the $DE1$ board. Table 12.2 depicts a detailed description of the system's resource utilization. The simulated result of the architecture on two of the test images is shown in Fig. 12.6. Comparing our implementation results with two of the most popular FP detection algorithms by [5, 8] as shown in Table 12.3 ([9] implementation is yet to be completed and thus we have only included its estimated resource utilization) we can conclude that our architecture can be regarded as one of the most resource effective and high performing implementation of a feature detector in hardware. Further a frequency of 50 MHz ensures that the power requirement for this implementation is very competitive. Thus for real time applications, we can use Architecture 1 for high throughput requirements, as it is suitable for the highest resolution video and vision processing systems. Whereas Architecture 2 can be added to an existing embedded system and can process $QCIF$ and CIF frames at a rate of upto 25 frames/s.

Table 12.3 Comparison with two other popular feature detectors implemented in hardware

Implementation name	Resource utilization-slices	Maximum frequency-MHz	Throughput (images/s)
[9]	78,728	—	—
[10]	685	58.72	120
Our Algorithm-Architecture 1	493	50	500
Our Algorithm-Architecture 2	5,278	100	30

12.5 Conclusions

We proposed a true real time *FP* detector and successfully depicted the possibility of embedding a high performance feature detector in an embedded Vision System. The proposed architecture effectively shows the hardware porting of our own novel algorithm for feature detection. Our architecture uses only integer add, subtract and shift operations leading to high performance throughput. It is also modular in its entirety and can be scaled depending upon the resource availability of the *FPGA* device. The above resource utilization numbers can be used for estimating the utilization for the scaled version of the above algorithm. The algorithm incurs at most 28 subtractions(or additions) at any pixel, also assuming the worst case scenario of highest noise content in the images only 80% of all the pixels will pass the first step for further processing, giving an upper limit of average subtractions per pixel to 22. No expensive multiplication or division operations are involved in our algorithm making it one of the most suitable contender for real time implementations. Further emphasizing the claim that it is an excellent contender for use in hardware implementations and embedded systems alike. Comparing the algorithms code-size to [7], it is a mere 5 KB (100 lines of Code) instead of a 145 KB (3,500 lines of code) of Rostenś. Thus it will find direct use in embedded systems having a very low memory footprint. Apart from the performance gain, as every pixel operation is dependent only on its surrounding 5×5 pixels independent of the rest of the image, so coarse level parallelization is very much possible. As a future work the proposed technique can be parallelized and pipelined to obtain higher throughput.

References

1. Moravec, H P (1977). In *Towards Automatic Visual Obstacle Avoidance*, page 584. Proceedings of the 5th International Joint Conference on Artificial Intelligence.
2. Harris, C. and Stephens, M. (1988). In *A Combined Corner and Edge Detector*, volume 23, pages 147–151. Proceedings of 4th Alvey Vision Conference, Manchester.

3. Mokhtarian, F and Soumela, R (1998). In *Robust Image Corner Detection Through Curvature Scale Space*, volume 20, pages 1376–1381. IEEE Transcations on pattern Analysis and Machine Intelligence.
4. Lowe, D G (2004). In *Distinctive Image Features from Scale-Invariant Keypoints*, volume 60, pages 91–110. International Journal of Computer Vision.
5. Smith, Stephen M. and Brady, J. Michael (1997). Susan – a new approach to low level image processing. *International Journal for Computer Vision*, 23(1):45–78.
6. Trajkovic, M and Hedley, M (1998). Fast corner detection. *International Journal for Image and Vision*, 16(2):75–87.
7. Rosten, Edward and Drummond, Tom (2006). In *Machine Learning for high-speed Corner Detection*, volume 17, pages 211–224. ECCV.
8. C Schmid, Roger Mohr and Bauckhage, C (2004). In *Comparing and Evaluating Interest Points*, volume 655, pages 63–86. INRIA Rhone, Montbonnot, France, Europe.
9. Cabani, Cristina and Maclean, W. James (2006). In *A Proposed Pipelined Architecture for FPGA Based Affine Invariant Feature Detectors*, pages 112–116. IEEE Conference on Computer Vision and Pattern Recognition.
10. Cesar Torris-Huitzil, Miguel Arias-Estrada (2000). In *An FPGA Architecture for High Speed Edge and Corner Detection*, pages 112–116. IEEE Transactions in Computing.

Chapter 13
SVD and DWT-SVD Domain Robust Watermarking using Differential Evolution Algorithm

Veysel Aslantas

Abstract This study aims to develop two optimal watermarking techniques based on SVD and DWT-SVD domain for grey-scale images. The first one embeds the watermark by modifying the singular values of the host image with multiple SFs. In the second approach, a hybrid algorithm is proposed based on DWT and SVD. Having decomposed the host image into four bands, SVD is applied to each band. Then, the same watermark is embedded by modifying the singular values of each band with different SFs. Various combinations of SFs are possible, and it is difficult to obtain optimal solutions by trial and error. Thus, in order to achieve the highest possible transparency and robustness, optimization of the scaling factors is necessary. This work employs DE to obtain optimum SFs. DE can search for multiple solutions simultaneously over a wide range, and an optimum solution can be gained by combining the obtained results appropriately. Experimental results of developed techniques show both the significant improvement in transparency and the robustness under attacks.

Keywords Watermarking · Differential Evolution Algorithm · grey-scale image · singular values · hybrid algorithm

13.1 Introduction

In the past decade, there have been great developments in computer and communication technology that have enabled digital multimedia contents such as audio, image and video to be reproduced and distributed with ease. Besides these advantages are provided by the technology, it also enables illegal operations in these materials such as duplication, modification and forgery. Due to that, protection of multimedia data

V. Aslantas
Erciyes University, Department of Computer Engineering, 38039 Melikgazi, Kayseri, Turkey,
E-mail: aslantas@erciyes.edu.tr

S.I. Ao and L. Gelman (Eds.), *Advances in Electrical Engineering and Computational Science*, Lecture Notes in Electrical Engineering.

has become a very crucial and urgent matter for data owners and service providers. Several methods have been proposed in order to provide protection against copyright forgery, misuse or violation. Among the methods, digital watermarking techniques are the most commonly used.

The principal objective of digital watermarking is to embed data called a watermark (tag, label or digital signal) into a multimedia object with the aim of broadcast monitoring, access control, copyright protection etc. The object may be an image or video or audio. Watermarking techniques can be divided into various categories according to visibility, permanency and domain [1, 2].

In relation to the visibility, digital watermarks can be of two types: visible or invisible watermarks. Visible watermarks can easily be detected by observation such as an embossed logo superimposed upon the image indicating the ownership of the image. Although the owner of the multimedia object can be identified with no computation, watermarks can be removed or destroyed with ease. On the contrary, invisible watermarks are designed to be imperceptible (transparent) and are inserted on the unknown places in the host data. The watermarked data should look similar to the original one, and should not cause suspicion by others. If it is used in an illegal way, the embedded watermark can then be utilized as a proof for showing the ownership of the multimedia object. The invisible watermarks are more secure and robust than the visible watermarks and are the subject of this chapter.

In the class of invisible watermarks, one may further categorise techniques according to permanency as fragile, semi-fragile and robust. Fragile watermarks are embedded in such a way that slight changes to the watermarked image would destroy the watermark. Hence, they are mainly used with the purpose of authentication. Semi-fragile watermarks are designed to tolerate some degree of modification to the watermarked image, for instance, the addition of quantization noise from lossy compression. Robust watermarks are designed to resist intentional or unintentional image modifications called attacks that attempt to destroy or remove the watermark. Such attacks include filtering, geometric transformations, noise addition, lossy compression, etc. In general, this kind of watermarks is employed for copyrights protection and ownership verification.

Depending upon the domain in which the watermark is embedded, the techniques can be grouped into two classes: spatial and frequency-domain techniques. In spatial domain, the pixel values of the cover image are modified for embedding the watermark. On the other hand, the watermark is embedded in transform domain by means of modulating the coefficients of the transformed host image according to watermark. Most of the transform domain watermarking techniques developed with the use of Discrete Fourier Transform (DFT), Discrete Cosine Transform (DCT), and Discrete Wavelet Transform (DWT). Although spatial domain methods are less complex as no transform is used, transform domain watermarking techniques are more robust in comparison to spatial domain methods against attacks.

Two important and conflicting requirements in watermarking are perceptual transparency and robustness. If significant modifications are introduced to the host image either in spatial or transform domain, greater robustness, in general, can be obtained. However, such modifications are distinguishable and hence do not satisfy the requirement of transparency (invisibility). The design of an optimal watermarking for a given application always involves a trade-off between these competing criteria. Thus, image watermarking can be formulated as an optimization task. Artificial intelligence techniques such as genetic algorithm, differential evolution, particle swarm, Fuzzy logic, neural networks, etc. have been employed to solve the problem optimally in spatial domain [3], DCT [4–9], DWT [10–12] and SVD [12–14].

In general, a watermarked image can be obtained by adding the watermark multiplied by a scaling factor (SF) to the host image either in spatial or transform domain. SF is used to control the watermark strength. The smaller the SF, the better the image quality (transparency) but the weaker the robustness. In contrast, the larger the SF, the stronger the robustness but the more the degradation of the quality of the host image. Moreover, a single SF may not be applicable for modifying all the values of host image because different spectral components may exhibit different tolerance to modification. Therefore, multiple SFs need to be used for adapting to them to reduce visual artifacts [13].

Singular value decomposition is one of the most popular methods of linear algebra with various applications in signal processing including watermarking [12–23]. This study aims to develop two optimal watermarking techniques based on SVD and DWT-SVD domain for grey-scale images. The first one embeds the watermark by modifying the singular values of the host image with multiple SFs. In the second approach, a hybrid algorithm is proposed based on DWT and SVD. Having decomposed the host image into four bands, SVD is applied to each band. Then, the same watermark is embedded by modifying the singular values of each band with different SFs. Various combinations of SFs are possible, and it is difficult to obtain optimal solutions by trial and error. Thus, in order to achieve the highest possible transparency and robustness, optimization of the scaling factors is necessary. This work employs DE to obtain optimum SFs. DE can search for multiple solutions simultaneously over a wide range, and an optimum solution can be gained by combining the obtained results appropriately. Experimental results of developed techniques show both the significant improvement in transparency and the robustness under attacks.

13.2 SVD Domain and DWT-SVD Domain Watermarking Techniques

SVD-Domain Watermarking: SVD decomposes an $N \times N$ matrix A into a product of 3 matrices $A = USV^T$ where U and V^T are $N \times N$ orthogonal matrices. S is an $N \times N$ diagonal matrix. The elements of S are only nonzero on the diagonal and are called the SVs of A. The steps of SVD based watermarking algorithm is described in Table 13.1 [15]:

DWT-SVD Domain watermarking: DWT transforms a signal from spatial domain to frequency domain. The transformation is done using wavelet filters such as Haar Wavelet Filter, Daubechies Orthogonal Filters and Daubechies Bi-Orthogonal Filters. Each of these filters decomposes the image into several

Table 13.1 SVD based watermarking scheme

Watermark embedding	Watermark extracting
1. Apply SVD to the host image (I): $I = USV^T$	1. Apply SVD to the watermarked (possibly distorted) image (I_W^*): $I_W^* = U^* S_W^* V^{*T}$
2. Modify the S with the watermark $S_M = S + kW$	2. Compute possibly corrupted S_M^*: $S_M^* = U_W S_W^* V_W^T$
3. Apply SVD to the S_M: $S_M = U_W S_W V_W^T$	3. Extract the watermark (possibly distorted) image (W^*): $W^* = (S_M^* - S)/k$
4. Compute watermarked image (I_W): $I_W = US_W V^T$	

Table 13.2 DWT-SVD based watermarking scheme

Watermark embedding	Watermark extracting
1. Apply DWT to decompose the host image I into 4 subbands	1. Decompose the watermarked (possibly distorted) image I_{dW} into 4 subbands (LL, HL, LH and HH) using DWT
2. Compute SVD of each subband $I^k = U^k S^k V^{kT}$, $k = 1, 2, 3, 4$ where k denotes subbands, and λ_i^k, $i = 1, \ldots, N$ are the SVs of S^k	2. Compute SVD of each subband image: $I_{dW}^k = U_{dW}^k S_{dW}^k V_{dW}^k$, $k = 1, 2, 3, 4$
3. Apply SVD to the watermark: $W = U_W S_W V_W$ where $\lambda_{Wi} i = 1, \ldots, N$ are the singular values of S_W	3. Obtain possible corrupted singular values of the watermark from each subband: ${\lambda_{dWi}}^k = ({\lambda_{di}}^k - {\lambda_i}^k)/\alpha_k$,
4. Modify the SVs of each subband with the SVs of the watermark: $\lambda_i^{*k} = \lambda_i^k + \alpha_k \lambda_{wi}$	4. Using the SVs, calculate the four watermarks (possibly distorted): $W_k^* = U_W {S_{dW}}^k V_W^T$, $k = 1, 2, 3, 4$
5. Obtain the modified coefficients of the each subbands: $I^{*k} = U^k S^{*k} V^{kT}$	
6. Compute the watermarked image by applying the inverse DWT using	

frequencies. In two dimensional DWT, each level of decomposition produces four subbands of the image identified as LL1, HL1, LH1 and HH1 (L: i.e., low, H: i.e., high). The LL1 band is termed as approximate subband while the others i.e. HL1, LH1, HH1 are termed as detailed subbands. The low pass band (LL1) can further be decomposed to obtain another level of decomposition. The process continues until the desired number of levels determined by the application is reached. The watermarking algorithm initially decomposes the host image into subbands. Afterwards, it determines the SVs of each subband and modifies these SVs with the same watermark by scaling with different scaling factors. The steps of DWT-SVD domain watermarking scheme is given in Table 13.2 [16].

13.3 Optimal Watermarkings using DE

The reminder of this section explains the way that DE is employed to optimize watermark embedding process with respect to the two conflicting requirements: transparency and robustness to different attacks.

An NxN host image (or subband) can have N SVs that may reveal different tolerance to modification. Prior knowledge may not be available about the sensitivity of the image with respect to various values of the SFs so an algorithm needs to be employed to compute the optimum scaling factors that produce maximum robustness and transparency. Thereby, DE is employed in this work to achieve this objective.

13.3.1 Basic Concepts of DE

Differential Evolution (DE) [24] is a novel population-based heuristic related to evolutionary computation. It works with real-valued parameters and is fairly fast and reasonably robust. It is capable of handling nonlinear, non-differentiable and multimodal continuous space functions. Like other evolutionary algorithms, DE also utilizes similar operators such as the initial population generation, crossover, mutation and selection but differs from those algorithms such as Genetic Algorithms and Evolutionary Strategies, where perturbation occurs in accordance with a random quantity, DE employs weighted differences between solution vectors to perturb the population. For each generation, the operators are continually repeated until the best vector representing the optimum solution is produced or some predefined termination criterion, such as maximum generation number or when there is no improvement in the fitness function is reached. The operators of DE algorithm are briefly described as follows:

Initial Population: DE starts with a population of NP solution vectors as is the case with other evolutionary algorithms. The population consists of real valued vectors with dimension D which equals to the number of design parameters. Each vector forms a candidate solution to optimization problem. If a priori knowledge is not available about the problem, the first population of solutions can be generated randomly as follows:

$$x_{i,G} = x_{i(L)} + rand_i[0,1]\left(x_{i(H)} - x_{i(L)}\right) \tag{13.1}$$

where $x_{i,(L)}$ and $x_{i,(H)}$ are the lower and higher boundaries of D-dimensional vector $x_i = \{x_{j,i}\} = \{x_{1,i}, x_{2,i}, \ldots, x_{D,i}\}^{\mathrm{T}}$, respectively.

Mutation: The main operator of DE is the mutation operator that is employed to expand the search space. For each target vector $x_{i,G}$, a mutant vector $v_{i,G+1}$ is generated by the combination of vectors randomly chosen from the current population at generation G as:

$$v_{i,G+1} = x_{r1,G} + F(x_{r2,G} - x_{r3,G}) \tag{13.2}$$

where i, r_1, r_2, and r_3 are mutually different indexes selected from the current generation $\{1, 2, \ldots, NP\}$. As a result, the population size must be greater than 3. $F \in [0, 2]$ is a user-defined real constant called step size which control the perturbation and improve convergence by scaling the difference vector. Smaller values for F result in faster convergence in the generated population and larger values in higher diversity.

Crossover: This operator combines successful solutions of the previous generation in order to maintain diversity in the population. By considering the elements of the mutated vector, $v_{i,G+1}$, and the elements of the target vector, $x_{i,G}$, the trial vector $u_{i,G+1}$ is produced as:

$$u_{j,i,G+1} = \begin{cases} v_{j,i,G+1} & \text{if } rand_{j,i} \leq CR \text{ or } j = I_{rand} \\ x_{j,i,G} & \text{if } rand_{j,i} > CR \text{ and } j \neq I_{rand} \end{cases} \tag{13.3}$$

where $j = 1, 2, \ldots, D$; $rand_{j,i} \in [0; 1]$ is the random number; $CR \in [0; 1]$ is user-defined crossover constant and $I_{rand} \in [1, 2, \ldots, D]$ is a randomly chosen index. I_{rand} ensures that the trial vector gets at least one parameter from the mutant vector, i.e., $v_{i,G+1} \neq x_{i,G}$.

Selection: The selection operator chooses the vectors that are going to compose the population in the next generation by employing the so called "greedy" criterion. If the fitness value of the trial vector is better than or equal to fitness value of the target vector, the latter is replaced by the former otherwise the latter is preserved in the population of the next generation.

$$x_{i,G+1} = \begin{cases} u_{i,G+1} & \text{if } f(u_{i,G+1}) \leq f(x_{i,G}) \\ x_{i,G} & \text{otherwise} \end{cases} \quad i = 1, 2, \ldots, NP \tag{13.4}$$

13.3.2 Optimisation of Scaling Factors

In order to accomplish the optimal performance of digital image watermarking techniques, the proposed methods employ the DE algorithm to search for optimal parameters that are the scaling factors. By applying DE algorithm, the SFs (k) are obtained for optimal watermarking depending on both the robustness and the transparency requirements under certain attacks at each generation of optimisation process. Figure 13.1 illustrates the flowchart of the developed DE based watermarking techniques in this study [12–14].

In the application of DE, each member vector in the population represents a possible solution to the problem and hence consists of a set of SFs. At the beginning of the optimization process, DE utilizes randomly generated initial solutions produced by random number generator. By employing each vector (the values of SFs), the watermarked images of the current generation are computed according to the watermark embedding schemes described in Tables 13.1 and 13.2 in Section 13.2.

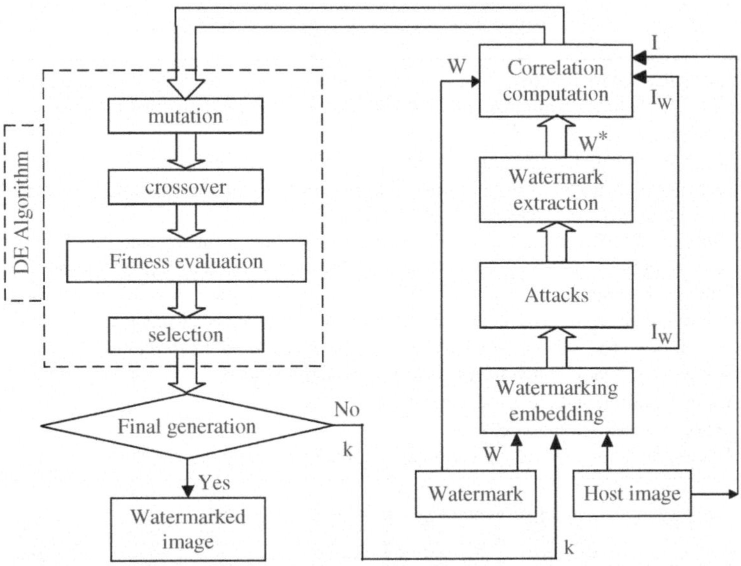

Fig. 13.1 Block diagram of DE-based watermark embedding scheme

In practice, different watermarking applications may require robustness to different attacks. Hence, the level of robustness of the watermark differs from application to application. The robustness of the developed watermarking schemes is evaluated with respect to the commonly employed attacks in literature. They include average filtering (AV), rotation (RT), sharpening (SH), and resizing (RS). As can be seen from Fig. 13.1, the developed system is very flexible. Therefore, the other attacking methods can easily be included to it or replaced with those used in this work [4, 13].

By employing the extraction procedures given in Section 13.2, the watermarks are extracted from the attacked watermarked images. The correlation values are computed between the host and watermarked images ($corr_I$) and between the original watermark and the extracted watermarks ($corr_W$). These values are then used to compute the fitness of a solution in the population of DE at each generation. Having assigned the fitness values for all vectors in the population, the vectors producing the better fitness values are selected for the next generation. By employing the operators of DE (mutation, crossover, and selection), the population of the next generation is then generated from these vectors. The above process on the population is repeated until the predefined criterion is fulfilled.

The fitness function designed for SVD based method is formulated as:

$$f_i = \left[1 \Big/ \left(\frac{1}{t} \sum_{i=1}^{t} corr_w \left(W, W_i^* \right) \right) - corr_I \left(I, I_W \right) \right]^{-1} \tag{13.5}$$

where f_i and t are the fitness value of the ith solution and the number of attacking methods, respectively. $corr_I()$ and $corr_W()$ are related to transparency and

Table 13.3 Outlines of the developed DE based watermarking algorithms

1. Define the fitness function, numbers of variables (D) and the values for population size (*NP*), control parameters (F and *CR*), and number of generations (or any other terminating criteria)
2. Generate an initial population of potential solutions at random
3. Calculate watermarked images using the solutions in the population by means of embedding process (Tables 13.1 and 13.2)
4. Apply the attack functions upon the watermarked images one by one
5. Extract out the watermarks from the attacked images using the extraction procedure (Tables 13.1 and 13.2)
6. Obtain the NC values between the host and each watermarked images
7. Compute the NC values between the watermark and the extracted ones
8. Evaluate the fitness value for each corresponding solution
9. Apply mutation and crossover operators
10. Apply the selection process
   Repeat Steps 3–11 until a predefined termination criteria are satisfied

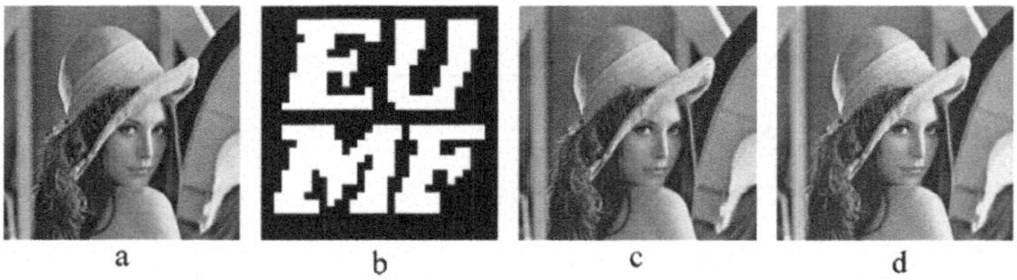

Fig. 13.2 (**a**) Host image, (**b**) watermark, (**c**) SVD domain watermarked image, (**d**) DWT-SVD domain watermarked image

robustness measure, respectively. In DWT-SVD domain method, $corr_W()$ in Eq. (13.5) is replaced with $\max(corr_W)$ which represents the maximum correlation value produced by the best quality watermark among the watermarks extracted from the four subbands. Table 13.3 illustrates the outlines of the developed DE based watermarking algorithms: The steps given in the table are the same for both of the techniques provided that the watermark embedding and extracting stages must be done according to the procedures of each technique explained in Section 13.2.

13.4 Results

To evaluate the performances of the proposed schemes, several experiments were conducted using the 256×256 Lena (host) and 32×32 watermark images that are illustrated in Fig. 13.2a and b. The attacks employed in the fitness evaluation process were: AV (3×3), SH (3×3), RS (bicubic: $256 \rightarrow 128 \rightarrow 256$), and RT (30°). The control parameters of DE employed for SVD and DWT-SVD domain methods are listed in Table 13.4.

Table 13.4 Control parameters used for SVD and DWT-SVD domain methods

Control parameters	SVD domain	DWT-SVD domain
NP	150	16
F	0.6	0.5
CR	0.8	0.6
Number of parameters (D = SFs)	32	4
Maximum generation number	400	200
Mutation method	DE/best/1/exp	DE/best/1/exp

In the representation scheme of solutions, each string consisted of 32 parameters for SVD domain and 4 parameters for DWT-SVD domain techniques. Every parameter represented a possible SF. Table 13.5 shows the optimal SVD domain SFs obtained. In DWT-SVD domain, the optimized SFs for LL, HL, LH and HH subbands are 0.873, 0.083, 0.092 and −0.012, respectively. By employing the SFs obtained, the watermarked images of both techniques are computed and are shown in Fig. 13.2c and d. As can be observed from those figures, the watermarked images are not distinguishable from the original one. In addition to the attacks used in the optimisation process, the effectiveness of the algorithms were also tested to cope with the Gaussian noise (GN) (mean = 0, variance = 0.001), 30×30 pixel translation (TR), cropping (CR) on left half and JPEG compression (50:1) attacks [12–14]. Distorted images after attacks and corresponding watermarks extracted by both techniques are illustrated in Fig. 13.3. The extracted watermarks by DWT-SVD given in the figure show the best quality watermarks obtained from the 4 subbands which are also indicated in the figure in parenthesis.

The results obtained are given in Tables 13.6 and 13.7 by comparing the correlation values. The same experiments were also carried out with single SF as shown in the tables [15, 16]. As can be seen from the tables, the larger the SF, the more the distortion of the quality of the host image (transparency) and the stronger the robustness. In contrast, the smaller the SF, the better the image quality and the weaker the robustness. On the other hand, results of the multiple SFs obtained by DE show both the significant improvement in transparency and the robustness under attacks.

13.5 Conclusions

In this chapter, two optimal watermarking schemes based on SVD and DWT-SVD domain are presented. As different images (or subbands) may have different singular values, the proposed techniques provide a general solution for determining the optimum values of the scaling factors for each image/subband. Differential

Table 13.5 SVD domain SFs obtained by DE (NV = number of variables)

NV	1	2	3	4	5	6	7	8	9	10	11
SFs	−51.05	242.21	1.65	0.71	0.76	−1.11	−0.97	−2.30	−0.96	1.08	0.85
NV	**12**	**13**	**14**	**15**	**16**	**17**	**18**	**19**	**20**	**21**	**22**
SFs	0.95	−0.77	−0.63	7.90	−13.61	−1.13	−0.59	−1.22	0.77	−0.99	−1.10
NV	**23**	**24**	**25**	**26**	**27**	**28**	**29**	**30**	**31**	**32**	
SFs	−0.85	−1.22	0.62	1.07	−1.15	−1.24	−20.31	−19.48	27.54	−7.75	

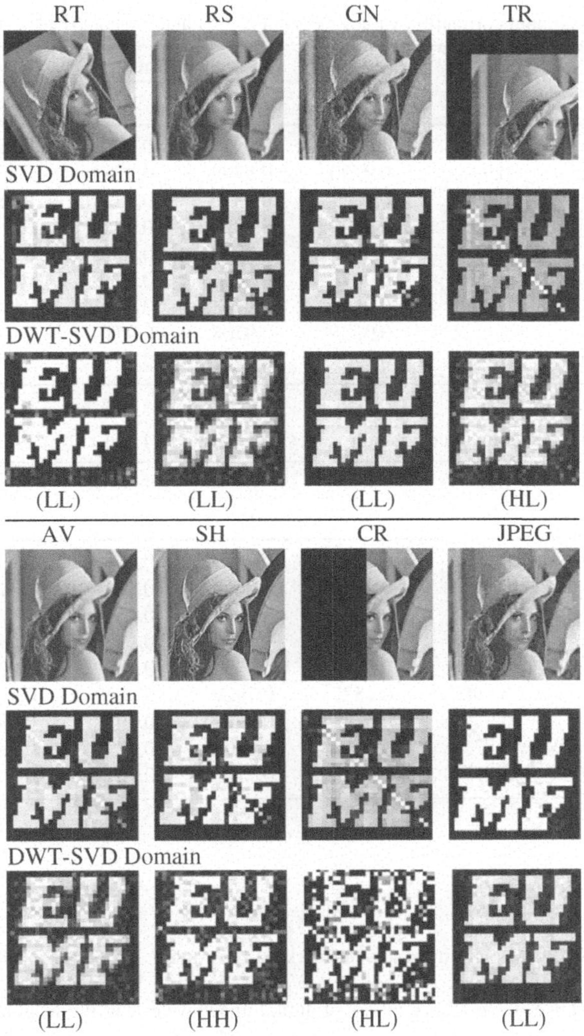

Fig. 13.3 Distorted images and corresponding extracted watermarks

evolution algorithm is adopted in the techniques in order to achieve the highest possible robustness without degrading image quality. Experimental results show the feasibility of the developed techniques and their superiority over the pure SVD and DWT-SVD domain techniques.

Table 13.6 SVD domain correlation values ($corr_I(I, I^*)$ and $corr_W(W, W^*)$)

		DE	Single SF				
		SFs	0.1	0.3	0.5	0.7	0.9
$corr_I$		0.9995	1.0000	0.9996	0.9989	0.9986	0.9981
	RT	0.9965	0.9790	0.9799	0.9783	0.9703	0.9741
	RS	0.9947	0.9827	0.9820	0.9820	0.9838	0.9840
	AV	0.9963	0.9829	0.9829	0.9829	0.9846	0.9847
	SH	0.9843	0.9673	0.9728	0.9713	0.9710	0.9733
$corr_W$	GN	0.9944	0.9678	0.9744	0.9749	0.9774	0.9837
	TR	0.9809	0.9596	0.9708	0.9773	0.9776	0.9804
	CR	0.9815	0.9625	0.9708	0.9710	0.9751	0.9784
	JPEG	0.9996	0.9763	0.9922	0.9973	0.9975	0.9988

Table 13.7 DWT-SVD domain correlation values ($corr_I(I, I^*)$ and $corr_W(W, W^*)$)

		DE	Single SF				
		SFs	0.005	0.01	0.05	0.1	0.5
$corr_I$		0.9990	0.9999	0.9999	0.9996	0.9985	0.9750
	RT	0.9910	0.2512	0.2693	0.3551	0.4121	0.6851
	RS	0.9815	0.2916	0.3040	0.4574	0.6227	0.9619
	AV	0.9868	0.5281	0.5154	0.4353	0.6176	0.9780
	SH	0.9840	0.8213	0.8332	0.8859	0.9227	0.9761
$corr_W$	GN	0.9982	0.7527	0.7929	0.9859	0.9950	0.9933
	TR	0.9896	0.5733	0.9298	0.9978	0.9994	0.9996
	CR	0.8339	0.6942	0.6999	0.7297	0.7998	0.9755
	JPEG	0.9974	0.6320	0.7105	0.9315	0.9634	0.9971

References

1. Hartung F, Kutter M. Multimedia watermarking techniques. Proceedings of IEEE 1999;87: 1079–1107.
2. Potdar VM, Han S, Chang E. A survey of digital image watermarking techniques. IEEE third International Conference on Industrial Informatics, INDIN'05, 2005, pp. 709–716.
3. Zhang F, Zhang H. Applications of a neural network to watermarking capacity of digital image. Neurocomputing 2005;67:345–349.
4. Shieh CS, Huang HC, Wang FH, Pan JS. Genetic watermarking based on transform-domain techniques. Pattern Recognition 2004;37:555–565.
5. Shih FY, Wu YT. Enhancement of image watermark retrieval based on genetic algorithm. Journal of Visual Communication and Image Representation 2005;16:115–133.
6. Chang YL, Sun KT, Chen YH. ART2-based genetic watermarking. 19th International Conference on Advanced Information Networking and Applications, 2006, pp. 729–734.
7. Aslantas V, Ozer S, Ozturk S. A novel clonal selection algorithm based fragile watermarking method. LNCS 2007; 4628:358–369.
8. Khan A, Mirza A. Genetic perceptual shaping: utilizing cover image and conceivable attack information using genetic programming. Information Fusion, 2007;8:354–365.
9. Aslantas V, Ozer S, Ozturk S. A novel fragile watermarking based on particle swarm optimization. IEEE International Conference on Multimedia and Expo, 2008, pp. 269–272.

10. Kumsawat P, Attakitmongcol K, Srikaew A. A new approach for optimisation in image watermarking by using genetic algorithms. IEEE Transactions on Signal Process, 2005;53: 4707–4719.
11. Lee D, Kim T, Lee S, Paik J. Genetic algorithm-based watermarking in discrete wavelet transform domain. LNCS 2006; 4113:709–716.
12. Aslantas V; Dogan A, Ozturk S. DWT-SVD based image watermarking using Particle Swarm Optimizer. IEEE International Conference on Multimedia and Expo, 2008, pp. 241–244.
13. Aslantas V. A singular-value decomposition-based image watermarking using genetic algorithm, International Journal of Electronics and Communications (AEU), 2008;62:386–394.
14. Aslantas V. Optimal SVD based robust watermarking using differential evolution algorithm. World Congress of Engineering, London, UK, 2008, pp. 629–631.
15. Liu R, Tan T. An SVD-based watermarking scheme for protecting rightful ownership. IEEE Transactions on Multimedia, 2002;4:121–128.
16. Ganic E, Eskicioglu AM. Robust DWT-SVD domain image watermarking: embedding data in all frequencies. ACM multimedia and security workshop, Magdeburg, Germany, 2004, pp. 166–174.
17. Huang F, Guan ZH. A hybrid SVD-DCT watermarking method based on LPSNR. Pattern Recognition Letter, 2004;25:1769–1775.
18. Chang CC, Tsai P, Lin CC. SVD-based digital image watermarking scheme. Pattern Recognition Letter, 2005;26:1577–1586.
19. Bao P, Ma X. Image adaptive watermarking using wavelet domain singular value decomposition. IEEE Transactions on Circuits Systems Video Technology, 2005;15:96–102.
20. Shieh JM, Lou DC, Chang MC. A semi-blind digital watermarking scheme based on singular value decomposition. Computer Standards Interfaces 2006;28:428–440.
21. Chung, KL, Yang, WN, Huang, YH, Wu, ST, Hsu, YC. On SVD-based watermarking algorithm. Applied Mathematics and Computation 2007;188:54–57.
22. Chang CC, Hu YS, Lin CC. A digital watermarking scheme based on singular value decomposition. ESCAPE 2007, pp. 82–93.
23. Mohan BC, Kumar SS. A robust image watermarking scheme using singular value decomposition. Journal of Multimedia 2008;3:7–15.
24. Storn R, Price K. Differential evolution – A simple and efficient heuristic for global optimization over continuous spaces. Journal of Global Optimization, 1997;11:341–359.

Chapter 14
Design and Performance Evaluation of a Prototype Large Ring PET Scanner

M. Monjur Ahasan and David J. Parker

Abstract Positron emission tomography (PET) is a powerful radiotracer imaging technique. Over the last 2 decades, PET has been widely applied as a non-invasive tool for in-vivo diagnosis of human diseases. The macroPET system described here was constructed as a prototype to demonstrate the feasibility of performing PET scans on a large scale. It was created by reconfiguring components from an original ECAT 951 system, which had 32 detector modules ("buckets") mounted in two rings with an inner diameter of 100 and a 10.8 cm axial field of view. In the macroPET system the 32 buckets are mounted in a single ring with an inner diameter of 2.34 m. This paper presents the macroPET design and initial measurements of its characteristics.

Keywords Positron emission tomography · radiotracer imaging · macroPET design · detector module · PET scanner

14.1 Introduction

Positron emission tomography (PET) is a powerful radiotracer imaging technique, in which the distribution of radiotracer is measured by detecting pairs of back-to-back gamma rays produced in positron-electron annihilation. Over the last 2 decades, PET has been widely applied as a non-invasive tool for in-vivo diagnosis of human diseases [1–3]. Considerable work has also been done on developing small diameter PET scanners for small animal imaging [4–6]. At Birmingham, the technique is also used for studying engineering processes [7].

We believe that PET also has considerable potential in the area of large animal veterinary medicine. However, a PET scanner designed to accommodate human

M.M. Ahasan (✉)
School of Physics and Astronomy, The University of Birmingham, Birmingham B15 2TT, UK,
E-mail: mma532@bham.ac.uk

S.I. Ao and L. Gelman (Eds.), *Advances in Electrical Engineering and Computational Science*, Lecture Notes in Electrical Engineering.

subjects will not be suitable for large animals such as horses. The macroPET system described here was constructed as a prototype to demonstrate the feasibility of performing PET scans on a large scale. It was created by reconfiguring components from an original ECAT 951 system [8], which had 32 detector modules ("buckets") mounted in two rings with an inner diameter of 100 and a 10.8 cm axial field of view. In the macroPET system the 32 buckets are mounted in a single ring with an inner diameter of 2.34 m. This paper presents the macroPET design and initial measurements of its characteristics. To the best of our knowledge, this is the first attempt to construct and test a PET scanner with such a large diameter.

14.2 The Macropet Design

The ECAT 951 scanner, manufactured by CTI Inc, is based on bismuth germanate (BGO) block detectors. Each block consists of a BGO crystal $50 \times 60 \times 30\,\mathrm{mm}^3$, backed by four photomultiplier tubes (PMTs). The front of the crystal is cut into an 8×8 array of crystal elements, each $6.25 \times 6.75\,\mathrm{mm}^2$. By comparing the light intensity detected in the four PMTs, a γ-ray interaction can be unambiguously assigned to an individual crystal element. The blocks are grouped into buckets each comprising four detector blocks together with their associated electronics (preamplifiers and discriminators) under the control of a microprocessor. Data from all the buckets are fed to a set of coincidence processors which identify events in which a pair of 511 keV γ-rays have been detected in two buckets within the resolving time of the system (6 ns).

In the original ECAT 951 the 32 buckets were mounted in two adjacent rings, each with an inner diameter of 100 cm. The 8,192 individual detector elements were thus arranged in 16 rings, each of 512 elements. For each bucket, coincidences could be detected with any of the seven opposing buckets in the same ring or in the other ring. The resulting field of view was a cylinder approximately 60 cm in diameter.

For macroPET (Fig. 14.1), the detector blocks have been remounted in a single horizontal ring of 128 blocks. Because the blocks are tapered to fit tightly together in the original 64 block rings, in the new arrangement there are gaps of approximately 7.5 mm between the front faces of adjacent blocks. The inner diameter of the single ring is 234 cm. For convenience, the blocks are mounted on eight separate aluminium base plates, each corresponding to a 45° segment of the ring. The blocks are connected in fours to the original sets of bucket electronics, with bucket controllers from the two original rings alternating around the new single ring. Overlapping of controllers is made possible by displacing alternate controllers vertically.

By alternating the buckets from the original rings, the original coincidence combinations of buckets are appropriate in the new geometry, and enable imaging to be performed over a field of view approximately 125 cm in diameter. The eight rings of detector elements span an axial (vertical) field of view of 5.4 cm, which

Fig. 14.1 A photography of the macroPET scanner

is divided into 15 image planes (eight direct planes and seven cross planes) with a plane separation of 3.375 mm.

The initial results reported here were obtained by accepting data with a maximum ring difference of 3, as is generally the case in acquisition of "2D" PET data. Purpose-built septa were not constructed for macroPET, but in an attempt to investigate whether such septa would be valuable some studies were performed with the original ECAT 951 septa mounted inside the macroPET ring. This limits the FOV to 82 cm diameter. Apart from these septa, macroPET has no shielding against out of field activity.

Events within the energy window from 250 to 850 keV were accepted from the system. A delayed timing window was used to record the distribution of random coincidences. Data were initially recorded in histograms (prompt minus delayed events) appropriate to the image planes of the original geometry and were then rebinned into the correct sinograms for the new geometry. In this rebinning the gaps between blocks were allowed for by treating each large ring as if it consisted of 1,152 (128 $\times$ 9) detector elements with every ninth detector absent, and interpolation was used to complete the missing sinogram elements. Arc correction of the projection data was achieved by performing a further rebinning with linear weighting. To compensate for the difference in efficiency of different detector elements, a normalisation was applied based on the response to a central uniform phantom.

Simple 2D filtered backprojection has been used to reconstruct all the images presented here, using a Hamming filter of cutoff frequency 0.4 and a 256 by 256 matrix with zoom factor from 1 to 4. Results measured with septa in place are reported in brackets.

14.3 Performance Evaluation

14.3.1 Spatial Resolution

The transaxial spatial resolution of the system, with or without septa, was measured using data from one or more 68-Ge line sources mounted parallel to the scanner axis. In each case, a Gaussian fit was performed on the profile across the central plane of the reconstructed images (Fig. 14.2), and the full-width at half-maximum (FWHM) was calculated as FWHM $= 2\sqrt{2\ln 2}\sigma$, where σ is the standard deviation of the fitted Gaussian function. The pixel size in the reconstructed images was determined experimentally from the image obtained using two 68-Ge line sources separated by a known distance (40 cm apart).

On the central axis, the transaxial resolution (FWHM) was determined as 11.2 mm (11.7 mm). The resolution degrades slightly as the source is moved off axis. At 40 cm off-centre, values of 12.0 mm (11.4 mm) were measured, and at 50 cm off-centre the no-septa resolution was measured as 12.1 mm.

The axial resolution of the system was measured using data from an 18-F point source positioned at the scanner centre (no-septa mode), and fitting a Gaussian function to the profile across the 15 image planes (Fig. 14.3). This central point source gave a transaxial resolution of 10.8 mm and an axial resolution of 9.3 mm.

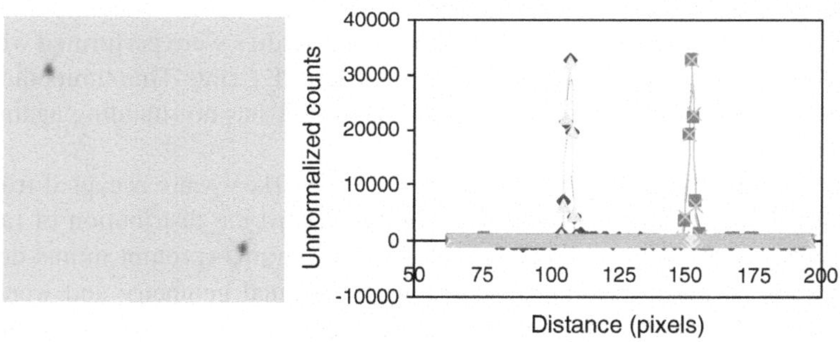

Fig. 14.2 Reconstructed image for two 68-Ge line sources and its profile

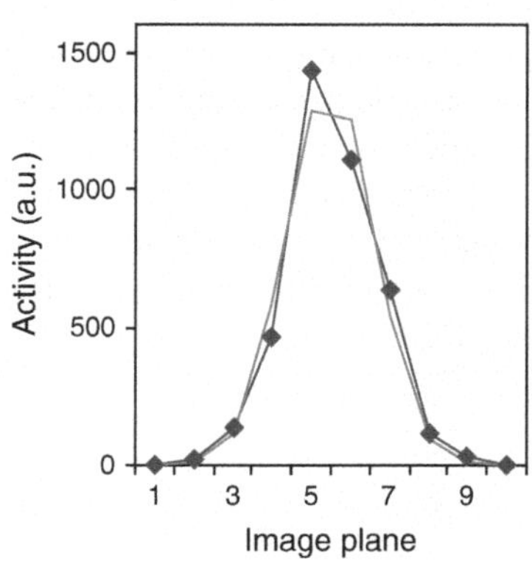

Fig. 14.3 The axial image profile for a central 18-F point source

These resolution values are all significantly larger than the spatial resolution quoted for the original ECAT 951 system (around 6 mm). Much of the difference can be attributed to the effect of acollinearity of the two photons emitted in electron/positron annihilation. The mean angular deviation from collinearity is around 0.4° [9] and the resultant effect on resolution is 0.0022D, where D is the ring diameter [10]. This is expected to contribute around 5.1 mm to the FWHM of images measured using a detector ring of diameter 2.34 m. It is also possible that errors in positioning individual blocks in the large ring contribute to the poor spatial resolution. Nevertheless, the scanner has been shown to give acceptable spatial resolution over a FOV of at least 1 m diameter.

14.3.2 Sensitivity

The sensitivity of the scanner was determined from the count rate measured using bare sources of known activity in both acquisition modes. To avoid problems of dead-time etc., these measurements were performed using sources of relatively low activity. A 14 cm long 68-Ge line source of total activity 200 kBq was mounted along the scanner axis, so that the activity in each 3.375 mm slice was approximately 4.8 kBq, and the count rate in each image plane was determined. In another measurement, an 18-F point source with an activity of 15 MBq was mounted at the centre of the field of view, and the total count rate (all planes) was found. The sensitivity calculations were corrected for the positron branching ratios of 18-F (0.967) and 68-Ge (0.89).

Figure 14.4 shows the axial sensitivity profiles determined using a 68-Ge line source on the scanner axis. The profiles are approximately as expected based on the number of detector rings which can contribute to each image plane, using a maximum ring difference of 3 (four combinations in each cross plane from 4 to 12, three combinations in each direct plane from 3 to 13, and fewer combinations in the end planes). The septa reduce the sensitivity by approximately 55% in each plane.

The absolute values of sensitivity without septa (approximately 1.87 and 1.39 cps/kBq for cross and direct plane respectively) are significantly lower (by

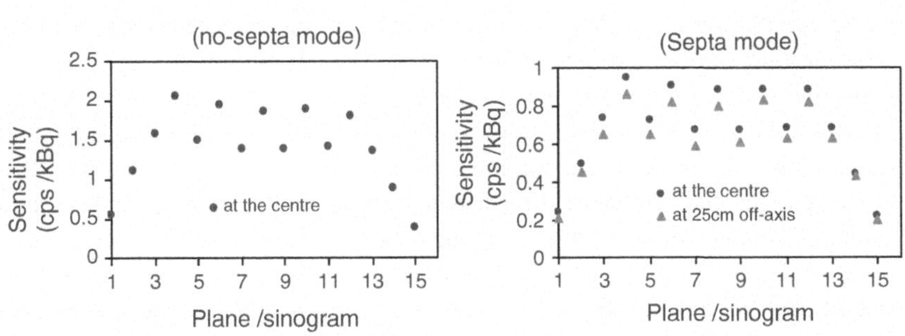

Fig. 14.4 Axial sensitivity profile for a line source positioned along the scanner axis

approximately 35%) than expected from a simple theoretical calculation, assuming a detector efficiency of 0.5 and neglecting the gaps between blocks. In a detector ring of diameter 234 cm the 128 gaps, each 7.5 mm wide, represent 6.5% of the solid angle. Overall, we might expect these gaps to reduce the sensitivity for detecting a pair of γ-rays by approximately 13%. This is insufficient to account for the observed discrepancy.

The sensitivity measured using a central point source was 2.16 (0.80) cps/kBq, which is closer to the expected value. This may suggest an error in determining the activity of the line source.

The sensitivity of a PET scanner is often quoted in terms of the count rate achieved using a cylindrical phantom uniformly filled with activity. Results have been reported using cylinders of various sizes [11, 12]. For macroPET using a 20 cm diameter cylinder filled with 18-F solution, sensitivity values of approximately 900 (400) cps kBq^{-1} ml^{-1} (averaged over all planes) were measured. Naturally, this is significantly lower than the equivalent sensitivity values quoted for clinical PET scanners [6, 13].

The total sensitivity for the original ECAT 951, measured using a 20 cm diameter phantom in 2D mode, was quoted as 110 cps μCi^{-1} ml^{-1} (4,070 cps kBq^{-1} ml^{-1}). In general the sensitivity of a scanner is proportional to Z^2/D (where Z is the axial length and D the ring diameter), since the activity within each slice is proportional to Z and the solid angle subtended by the detector ring is proportional to Z/D. Compared with the ECAT 951, macroPET has Z reduced by a factor $1/2$ and D increased by a factor 2.34, so we expect the sensitivity to be reduced overall by approximately a factor 0.107, giving a figure of 435 cps kBq^{-1} ml^{-1}. The observed sensitivity of 400 cps kBq^{-1} ml^{-1} (with septa) is entirely consistent with this. Actually we would expect the sensitivity to be somewhat reduced because of the gaps between blocks.

14.3.3 Scatter Fraction

Detection of Compton scattered events is a major source of background in PET, especially when imaging large objects without any collimating septa. To find the effect of septa on scattered photons, the scatter fraction was measured in no-septa and septa acquisition modes. As is conventional, the scatter fraction was measured for a 68-Ge line source mounted at the centre of a 20 cm diameter cylinder filled with inactive water. Figure 14.5 shows the profile across the resulting sinograms. The scatter fraction was then determined from each profile by estimating the number of counts within the peak, and using the equation $SF = S/(T + S)$. For the 20 cm diameter cylinder, the scatter fraction was measured as 26% (17%). To investigate the effect of scatter within a large animal, measurements were also performed in no-septa mode using three 68-Ge line sources mounted inside a large rectangular water tank ($62 \times 38\,cm^2$). The three sources were mounted vertically at intervals of 25 cm along the long axis of the tank, with the middle source at the centre of the

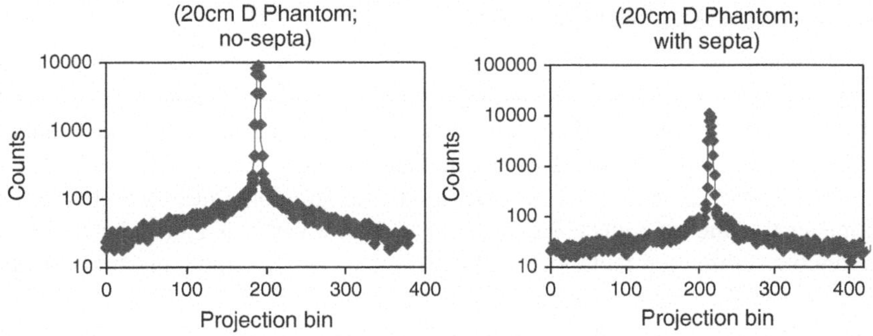

Fig. 14.5 Profile across the sinogram measured for a 68-Ge line source at the centre of a 20 cm diameter cylinder of water without septa (*left*) and with septa (*right*)

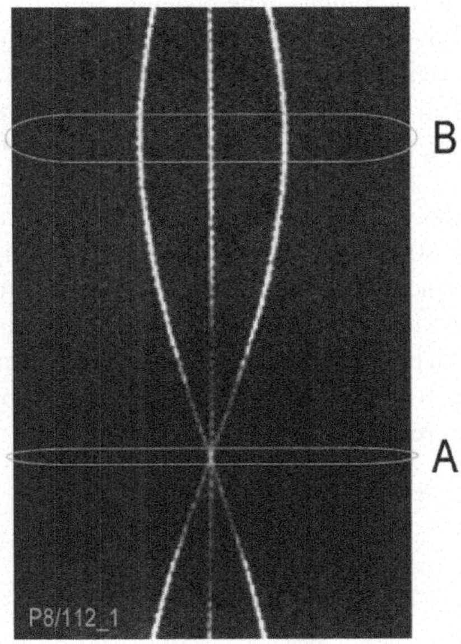

Fig. 14.6 Central plane (plane 8) sinogram measured from three 68-Ge line sources inside a large ($62 \times 38\,\text{cm}^2$) water tank (no-septa mode)

tank, and coincidence events were acquired for 66 h. Figure 14.6 shows the resulting sinogram.

Values for scatter fraction were obtained by analysing the profiles at angles A (looking directly along the long axis of the water tank, with the three sources in line) and B (looking along the short axis). SF values of approximately 63% and 38% were obtained along A (62 cm long) and B (38 cm wide) respectively. This measurement was repeated with septa in plane but the results were inconclusive because the count rate was very low and there was significant influence from out-of-field activity.

14.3.4 Count Rate Performance

Count rate performance is evaluated by calculating the noise equivalent count rate (NEC) which takes into account the statistical noise due to scatter and random events. The general formula [14], to calculate the NEC rate is $NEC = T^2/(T + S + kR)$, where T, S and R are the true, scatter and random rates respectively. The parameter k is the randoms correction factor, with a value of 1 or 2 depending on the randoms correction method. A value of 2 is used when randoms are measured using a delayed coincidence window as in the present work.

NEC rates are usually quoted using a 20 cm diameter cylindrical phantom with an axial length of 20 cm. For macroPET, a similar phantom was used but it was only filled to a depth of around 5 cm to avoid the effect of out-of-field of activity due to the lack of any shielding on the current scanner. NEC rates were measured in both no-septa and septa acquisition mode. The phantom was filled with 1,500 ml water containing an initial activity of 425 kBq/ml (1.1 MBq/ml) of 18-F, and was then placed in the centre of the scanner. A set of 10 min scans was acquired every half hour for almost 12 h. The contribution of scattered events was estimated assuming a scatter fraction of 26% (17%). The noise equivalent count rates were calculated from the NEC formula.

Figure 14.7 shows the various components of the count rate in both acquisition modes, as a function of source activity, and also the resulting NEC rate. Although the scanner delivers coincidence rates of up to 100 kcps, the NEC rate peaks at around 30 kcps in the no-septa case due to the significant contributions of random and scattered events. The peak NEC rate was also measured as 28 kcps in septa configuration. The NEC performance would be even poorer in both acquisition modes if out of field activity were present.

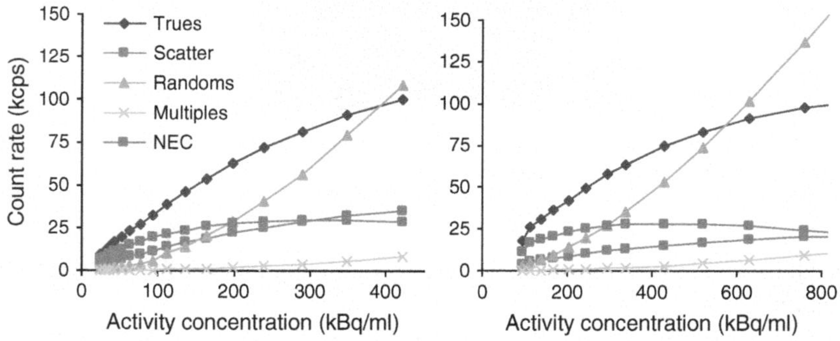

Fig. 14.7 Count rates for a 20 cm diameter phantom: without septa (*left*), with septa (*right*)

14.3.5 Imaging Studies

14.3.5.1 Jaszczak Phantom

The Jaszczak SPECT phantom is often used for assessing the image quality in nuclear medicine images. It consists of a cylinder with perspex inserts which create cold lesions when the phantom is filled with radioactive solution. In the present work a Jaszczak phantom with internal diameter 18.6 cm was used, containing six different sets of cylindrical rods with diameters 4.8, 6.4, 7.9, 9.5, 11.1, 12.7 mm.

The images reported here were obtained in no-septa mode by filling only the lower half of the phantom with approximately 2.5 l of water, and adding approximately 200 MBq of 18-F. Data were acquired for 600 s per image, with the phantom mounted at the centre of the field of view and at different radial offsets. Each image was reconstructed using filtered backprojection and incorporating the calculated attenuation correction for a water-filled cylinder.

Due to the curved nature of the PET geometry, lines of response (LORs) near the periphery of the field of view are more closely spaced than those near the centre. To avoid distortion due to this effect, the sinogram data were rebinned before reconstruction (a process referred to as "arc correction"). Figure 14.8 shows the centre plane image from the phantom mounted with its axis displaced by 45 cm from the centre of the field of view, (a) before arc correction and attenuation correction, and (b) after applying these corrections. It can be seen that the distortion due to changing LOR separation is removed by the arc correction process, and a reliable image is obtained out to a radius of at least 55 cm. In this image, only the two largest sets of cold rods (12.7 and 11.1 mm diameter) are clearly visible. In other images measured with the phantom at the centre of the FOV the 9.5 mm rods can also be seen.

14.3.5.2 Uniform Phantom

Images of a central uniform 20 cm diameter cylindrical phantom filled with 18-F solution to the depth of the FOV were used to assess the uniformity and noise level

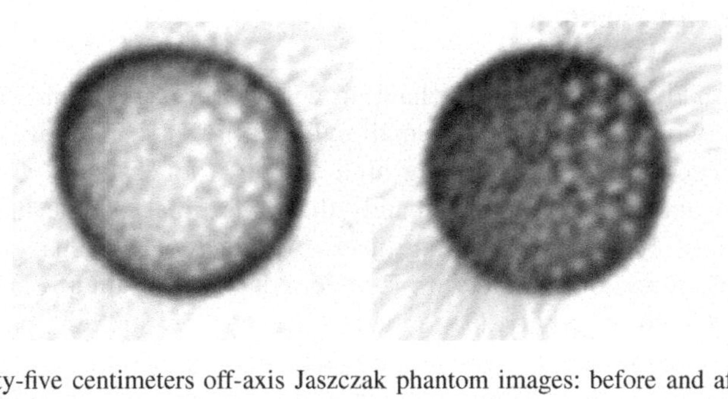

Fig. 14.8 Forty-five centimeters off-axis Jaszczak phantom images: before and after attenuation and arc correction

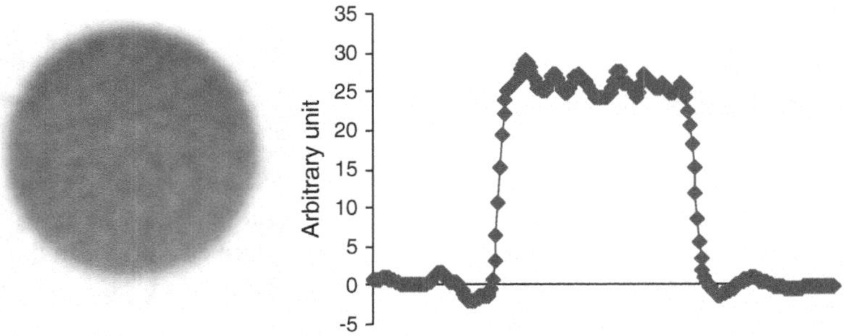

Fig. 14.9 Uniform phantom image (slice 3) and the corresponding horizontal profile across the centre of the image

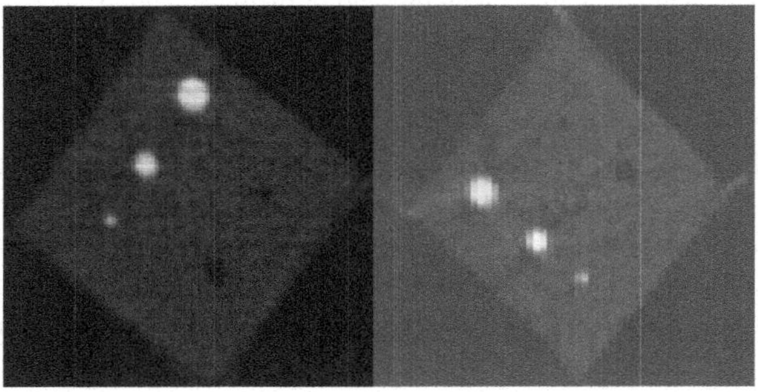

Fig. 14.10 Contrast phantom image (central slice): no-septa (*left*) and septa (*right*)

in no-septa mode. Figure 14.9 shows the image (slice 3) for a measurement in which 150 M counts were obtained in 600 s. Also shown is the horizontal profile across this image. From the ratio of standard deviation (σ) to the mean pixel counts (μ) within a circular region of interest (ROI) covering the central part of the phantom in the reconstructed image, the coefficient of variation (CV) was calculated as 5.9%.

14.3.5.3 Contrast Phantom

A laboratory-made 35.7 cm square phantom with warm background embedded by three hot and two cold lesions (five small cylinders) was scanned in no-septa and septa mode. The cylinders containing high activity (hot lesions) were 4, 2, and 1 cm diameter and all have approximately the same uptake value in each mode. The diameters of cylinders containing inactive water (cold lesions) were 3 and 1 cm. The phantom was only filled with F-18 to a depth of around 5 cm to avoid the effect of out-of-field of activity due to the lack of any side shielding on the system.

Figure 14.10 shows the image slice for the square phantom generated without and with septa. For both configurations, three hot and two cold lesions are visible in each case. Contrast recovery coefficient (CRC) [15] values close to unity can be achieved for the 2 and 4 cm hot lesions and for the 3 cm cold lesion. For both the 1 cm hot and cold lesions the CRC is around 0.3.The 1 cm hot lesion is more prominent than the 1 cm cold lesion.

14.4 Conclusions

After design, setup and calibration of the macroPET system, initial results of its performance characteristics were obtained. The spatial resolution is around 11 mm, significantly poorer than for a standard scanner, probably because of the significant contribution from acollinearity. Without septa, the sensitivity measured for a 20 cm diameter phantom is 900 cps/kBq/ml and the peak NEC rate is 30 kcps. The septa used (which were not designed specifically for this scanner) reduce the scatter fraction by around 35% at the cost of reducing the sensitivity by 55%.

The main weakness of the prototype scanner is the lack of side shielding, so that it is sensitive to out-of-field activity. Most of the measurements reported here were obtained with the activity confined axially to the field of view. Out of field activity will contribute significantly to the randoms rate. Also, the scatter fraction grows dramatically as larger objects are imaged. The septa used have only a limited effect in blocking out-of-field activity (and can actually make the situation worse as they reduce the true rate more than the background rate). To date all the images in this work have been reconstructed using simple 2D FBP. In the future it may be appropriate to use 3D image reconstruction and/or iterative techniques.

With these improvements, the results so far indicate that a geometry like macroPET offers a practical solution for large animal imaging.

References

1. M. E. Phelps. Molecular imaging with positron emission tomography, *Annu. Rev. Nucl. Part. Sci.* 52, 303–338 (2002).
2. C. Das, R. Kumar, Balakrishnan, B. Vijay, M. Chawla, and A. Malhotra. Disseminated tuberculosis masquerading as metastatic breast carcinoma on PET-CT, *Clin. Nucl. Med.* 33(5), 359–361 (2008).
3. R. Myers, S. Hume, P. Bloomfield, and T. Jones. Radio-imaging in small animals, *Psychopharmacol* 13(4), 352–357 (1999).
4. A. P. Jeavons, R. A. Chandler, C. A. R. Dettmar. A 3D HIDAC-PET camera with submillimetre resolutionnfor imaging small animals, *IEEE Trans. Nucl. Sci.* 46, 468–473 (1999).
5. J. Missimer, Z. Madi, M. Honer, C. Keller, A. Schubiger, and S. M. Ametamey. Performance evaluation of the 16-module quad-HIDAC small animal PET camera, *Phy. Med. Biol.* 49(10), 2069–2081 (2004).
6. W. W. Moses, P. R. G. Virador, S. E. Derenzo, R. H. Huesman, and T. F. Budinger. Design of a high-resolution, high-sensitivity PET camera for human brains and small animals, *IEEE Trans. Nucl. Sci.* 44(4), 1487–1491 (1997).

7. A. Sadarmomtaz, D. J. Parker, and L. G. Byars. Modification of a medical PET scanner for PEPT studies, *Nucl. Instr. Meth. A* 573, 91–94 (2007).
8. R. D. Badawi, M. P. Miller, D. L. Bailey, and P. K Marsden. Randoms variance reduction in 3D PET, *Phys. Med. Biol.* 44, 941–954 (1999).
9. S. DeBenedetti, C. E. Cowan, W. R. Konneker, and H. Primakoff. On the angular distribution of two-photon annihilation radiation, *Phys. Rev.* 77(2), 205–212 (1950).
10. C. S. Levin and E. J. Hoffman. Calculation of positron range and its effect on the fundamental limit of positron emission tomography system spatial resolution, *Phys. Med. Bio.* 44, 781–799 (1999).
11. V. Bettinardi, M. Danna, A. Savi, M. Lecchi, I. Castiglioni, M. C. Gilardi, H. Bammer, G. Lucignani, and F. Fazio. Performance evaluation of the new whole-body PET/CT scanner: Discovery ST, *E. J. Nucl. Med. Mol. Imag.* 31(6), 867–881 (2004).
12. T. DeGrado, T. Turkington, J. Williams, C. Stearns, and J. Hoffman. Performance characteristics of a whole-body PET scanner, *J. Nucl. Med.* 35, 1398–1406 (1994).
13. T. K Lewellen, S. G. Kohlmyer, R. S. Miyaoka, M. S. Kaplan, C. W. Stearns, and S. F. Schubert. Investigation of the Performance of the General Electric ADVANCE Positron Emission Tomograph in 3D Mode, *IEEE Trans. Nucl. Sci.* 43(4), 2199–2206 (1996).
14. S. C. Strother, M. E. Casey, and E. J. Hoffmann. Measuring PET scanner sensitivity: relating count rates to image signal-to-noise ratios using noise equivalent counts, *IEEE Trans. Nucl. Sci.* 37, 783–788 (1990).
15. M. E. Daube-Witherspoon, J. S. Karp, M. E. Casey, F. P. DiFilippo, H. Hines, G. Muehllehner, V. Simcic, C. W. Stearns, L. E. Adam, S. Kohlmyer, and V. Sossi. PET performance measurements using the NEMA NU 2-2001 standard, *J. Nucl. Med.* 43(10), 1398–1409 (2002).

Chapter 15
Robust Wavelet-Based Video Watermarking Using Edge Detection

Tamás Polyák and Gábor Fehér

Abstract In this paper a robust video watermarking method is presented, which embeds data in the wavelet domain using edge detection. The algorithm uses the luminance values around the edges where changes are less noticeable for the human visual system. Watermark is embedded in an additive way using spread spectrum noise. The complexity of the method is low. The algorithms used for wavelet transformation and edge detection are fast and simple. Results show that the watermark signal is imperceptible, and the method is robust against low bitrate lossy compressions, like H.264.

Keywords video watermarking, wavelet, edge detection, low bitrate lossy compression

15.1 Introduction

Digital video streaming is a more and more popular service among content providers. These streams can be displayed on various types of devices: computer, PDA etc. The content can be easily distributed through the internet, but digital contents carry a big security risk, i.e. the copying and reproduction of the content is very easy and effortless. Even users without special knowledge can easily share the downloaded content with other people.

To avoid illegitimate access, digital content is often encrypted, and travels in an encrypted form to the consumer. Although encryption secures the content on the way to the consumer, during playback it must be decrypted, and this is the point where it exposed to illegal copying. In these cases watermarks can be used to

T. Polyák (✉)
Budapest University of Technology and Economics, Department of Telecommunications and Media Informatics, Magyar tudósok krt. 2., 1117 Budapest, Hungary, BME TMIT,
E-mail: tpolyak@tmit.bme.hu

S.I. Ao and L. Gelman (Eds.), *Advances in Electrical Engineering and Computational Science*, Lecture Notes in Electrical Engineering.

give some extra protection to the content, since watermarking can embed additional information in the digital content.

Watermarks are embedded in the digital content in an imperceptible way. Watermarks can carry some information about the content: owner of the content, metadata, QoS parameters etc. This protection does not eliminate the need for encryption, but is a supplemental technique to secure the content, or store additional data. These watermarks are called robust watermarks because they must survive transformations of the underlying content e.g. lossy compression.

Digital watermarking algorithms are generally divided into two groups: algorithms that hide data in the spatial domain. It means that information is embedded by modifying the pixel values directly [1–3], and algorithms that use a transform domain for data hiding. Discrete cosine transform (DCT) and discrete wavelet transform (DWT) are often used at transform domain watermarking [4–7]. These watermarking techniques modify coefficients of the given transform domain. Wavelet transform is commonly used because it has many advantages over DCT transform. It is closer to the human visual system (HVS), instead of processing 8×8 pixel blocks it processes the whole frame. It divides the frame into four parts with different frequencies (LL, LH, HL and HH) and directions.

Some watermarking techniques embed data in object borders, and other perceptually important areas. It has several advantages: the HVS is less sensitive to changes made in these components, and modern lossy compression algorithms leave them relatively untouched to maintain high video quality. These properties make these regions ideal for watermarking. The detection of suitable regions can be realized in the uncompressed domain. Pröfrock, Schlauweg and Müller use the Normed Centre of Gravity (NCG) algorithm to find the blocks that contain object borders or other significant changes in the video [8]. Ellinas presents an algorithm that embeds data in images using four level DWT and edge detection. This algorithm also modifies the surroundings of the edges; it is accomplished using a morphological dilatation with a structuring element of 9×9.

The proposed algorithm is designed for watermarking low resolution (CIF and QCIF) video streams. During the design of a video watermarking algorithm complexity has to be taken into count. A trade-off has to be made between complexity, quality loss, and robustness. The algorithm uses the wavelet domain for data hiding. It modifies the wavelet coefficients that belong to object borders. Visible artifacts will not appear on the source video. The suitable coefficients are selected by an edge detection algorithm. Watermark is inserted in an additive way that a spread spectrum pseudorandom noise is added to the luminance pane of the middle frequency components.

15.2 The Watermark Embedding Process

Figure 15.1 shows the process of watermark embedding. First the input video signal ($X_{i,j}$) is transformed using a fast n-level ($n = 1, 2$ or 3) DWT transform, the Haar wavelet [9].

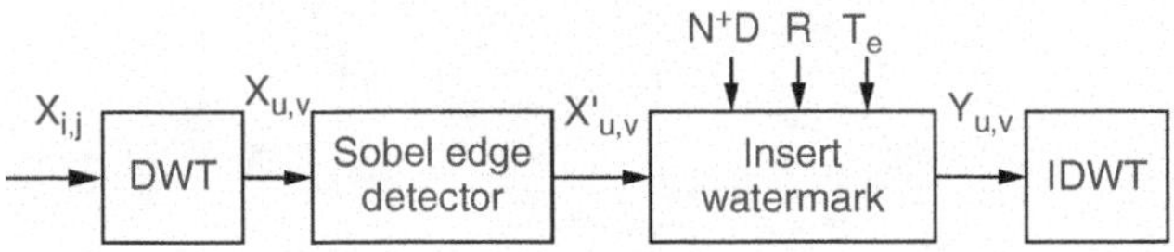

Fig. 15.1 Block diagram of the watermark embedding process

Fig. 15.2 Video frame after two-level wavelet transformation

The transformation produces a low frequency, low resolution approximation subband, and 3*n detail subbands. Watermark is inserted into the middle frequency components (HL and LH) of each level. Figure 15.2 shows the video frame after a two-level wavelet decomposition

The transformed video then gets through an edge detecting filter, where edge detection is applied on the middle frequency components. Edge detection is performed using the Sobel edge detector [10].

The Sobel operator performs a 2-D spatial gradient measurement on an image. It is used to find the approximate absolute gradient magnitude at each point in an input grayscale image. The Sobel edge detector uses a pair of 3 × 3 convolution masks, one estimating the gradient in the x-direction (G_x) and the other estimating the gradient in the y-direction (G_y). A convolution mask is usually much smaller than the actual image. As a result, the mask is slid over the image, manipulating a square of pixels at a time. The two kernels are shown in (15.1).

$$G_x = \begin{bmatrix} -1 & 0 & 1 \\ -2 & 0 & 2 \\ -1 & 0 & 1 \end{bmatrix}, \quad G_y = \begin{bmatrix} -1 & -2 & -1 \\ 0 & 0 & 0 \\ 1 & 2 & 1 \end{bmatrix}, \tag{15.1}$$

The magnitude of the gradient (G) is then calculated using (15.2)

$$|G| = \sqrt{G_x^2 + G_y^2} \tag{15.2}$$

Fig. 15.3 Video frame with Sobel edge detection in the middle frequency DWT coefficients

To improve performance an approximate magnitude is used (15.3)

$$|G| = |G_x| + |G_y| \tag{15.3}$$

Figure 15.3 shows the frame after edge detection in the middle frequencies.

During watermark insertion the edges with value greater than a given threshold (T_e), and the pixels around them in a given radius (R) are selected. The value of the radius is different at the different levels of the transformed image. Radius of two pixels is used at the first level, and radius of one pixel is used at the second and third level considering that the higher level coefficients contain lower frequency components, which affect the quality of the video.

Data is embedded into the 2*n selected middle frequency areas by adding a pseudo random spread spectrum noise.

First, data is extended with a chip rate cr to improve robustness.

$$d_i = D_x, \quad where\, i = x \cdot cr, ..., (x+1) \cdot cr - 1 \tag{15.4}$$

Insertion is done in an additive way using (15.5).

$$Y_{u,v} = X'_{u,v} + \alpha \cdot N_i \cdot d_i \tag{15.5}$$

where $Y_{u,v}$ are the modified wavelet coefficients, $X'_{u,v}$ are the selected wavelet coefficients for watermarking, α is the strength of the watermark (the value which the luminance values are modified by), N_i is the pseudorandom Gaussian noise consisting of -1 and 1, d_i is the data to embed. The data also consists of -1 and 1 according to 0 and 1.

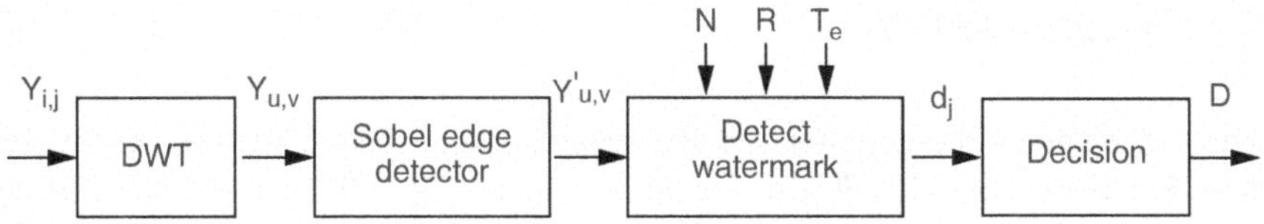

Fig. 15.4 Block diagram of the watermark detection process

The pseudorandom noise N_i is calculated from a seed, and from the u and v coordinates of the selected coefficients. This seed is used as the key for the watermarking method.

The same data are embedded into consequent frames to improve robustness.

15.3 The Watermark Detection Process

The process of the watermark detection is also made in the wavelet domain. After the n level DWT the middle frequency blocks get through an edge detection. Figure 15.4 shows the block scheme of the watermark detecting process.

The embedded data are extracted by correlating the pseudorandom noise N_i with the area containing the detected edges and their surroundings. The detection is blind. Assuming that the value of $Y_{u,v}$ coefficients is almost equal,

$$d_j = \frac{1}{K}\sum_{u,v} Y_{u,v} \cdot N_i, \tag{15.6}$$

can be used, where K is the number of suitable coefficients and d_j is the extracted value. The embedded bit is calculated the following way:

$$\begin{aligned} \delta &= 0, \quad if\; d < -T_w \\ \delta &= 1, \quad if\; d > T_w \end{aligned}, \tag{15.7}$$

where T_w is a threshold used to decrease the number of false results.

Its value is:

$$T_w = \frac{\alpha}{2} \tag{15.8}$$

Because the embedded information changes only after nine frames, simple voting is applied on the detected bits.

15.4 Experimental Results

The watermarking algorithm was tested on four video streams: "Mobile", "Bus" and "Flower", which are highly detailed streams, and "Foreman", which contains more smooth regions. The size of all video streams is 352×288 pixels. Video quality was evaluated using two metrics: PSNR and SSIM [11, 12].

15.4.1 Quality Results

Figure 15.5 shows the original and the watermarked version of the same video frame. For better visibility it also contains cropped, 20 × 20 pixels blocks of the two frames, which contain significant edges (situated at the top of the bus on the right side of the frame). It can be seen that the noise caused by watermark signal is imperceptible.

Figure 15.6 shows the PSNR and SSIM error maps between the original and the watermarked frame. The error maps are converted to be negative for better visibility. It can be seen that watermark is inserted around the edges. The PSNR error map shows that numerous pixels have been altered (PSNR = 40.31 dB), but the SSIM error map shows a better quality (SSIM = 0.9952).

Table 15.1 presents the objective quality values of the tested videos using two level DWT.

It can be seen that the video quality depends on the content. If the content contains more edges or textured components, then more noise is added to it. Although it improves robustness of the watermark, the objective quality of the video becomes worse.

Table 15.2 shows the objective quality values of the test sequence "Bus" after embedding watermark into different number of DWT levels. At the first level $R = 2$, at the second and the third levels $R = 1$ was used as radius to select the pixels to modify.

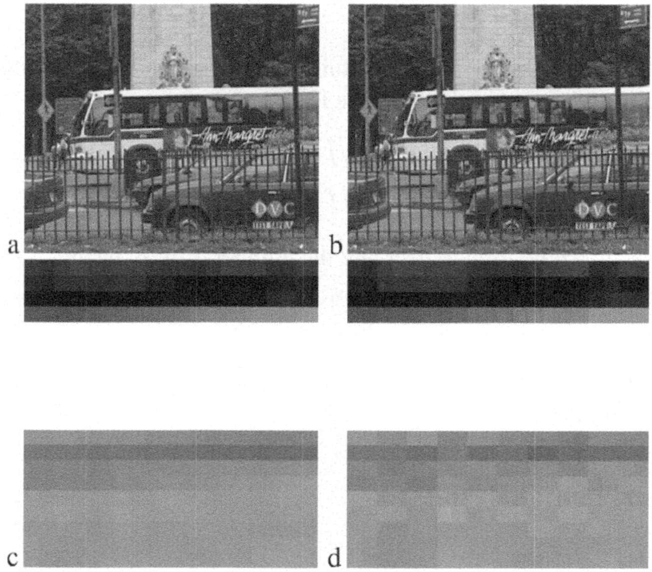

Fig. 15.5 Original (**a**) and watermarked (**b**) video frame and blocks of 20 × 20 pixels from the original (**c**) and watermarked (**d**) video frame using two level DWT

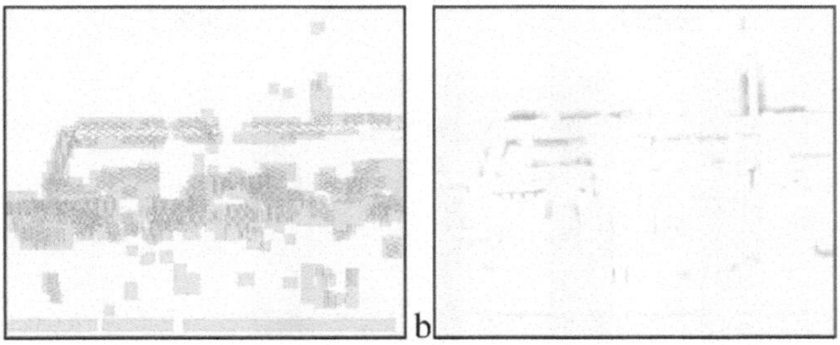

a b

Fig. 15.6 Error maps between original and watermarked frame using two-level DWT (converted to negative for better visibility): (**a**) PSNR; (**b**) SSIM

Table 15.1 Average PSNR and SSIM values of watermarked test sequences

Videos	Mobile	Foreman	Bus	Flower
PSNR [dB]	35.29	45.17	37.37	35.27
SSIM	0.9842	0.9969	0.9905	0.9905

Table 15.2 Average PSNR and SSIM values after embedding data using different levels of DWT

DWT level	1	2	3
PSNR [dB]	40.90	37.37	35.88
SSIM	0.9967	0.9905	0.9818

As expected, the more DWT levels used the more pixels are modified, this results worse video quality.

15.4.2 Robustness of the Algorithm

Robustness of the algorithm has been tested against lossy compression. The H.264/AVC with Baseline profile was used, this codec is widely used for low bitrate video coding.

First, test data of 128 bits were embedded in the video streams; 8 bits were embedded at every frame using a watermarking strength (α) of 5. Then the watermarked videos have been compressed using the H264/AVC codec with different bitrates. Finally, the accordance was measured between the original and the extracted data produced by the detector.

Accordance of 100% means that every embedded bit can be correctly identified, while 50% means that watermark signal can not been retrieved, the embedded and the extracted data are uncorrelated.

The algorithm was tested using one-, two- and three-level DWT.

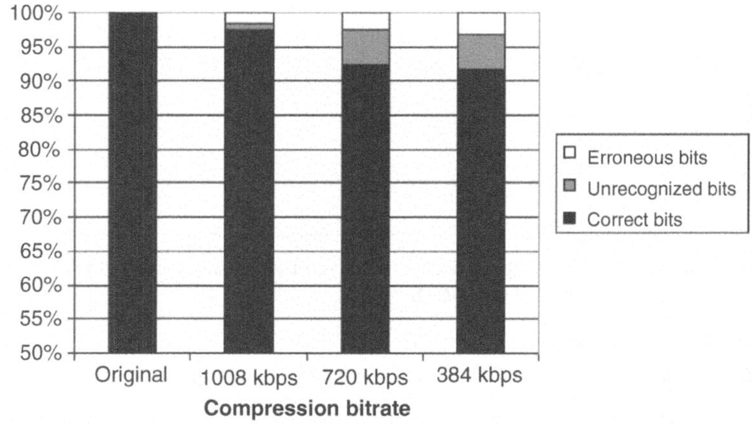

Fig. 15.7 Robustness results against H.264/AVC lossy compression, 1-level DWT

15.4.2.1 Robustness Results Using One-Level DWT

Figure 15.7 shows the test results of the watermarking algorithm with one-level DWT after H.264/AVC compression.

"Correct bits" means the number of bits that can be detected correctly.

"Erroneous bits" is the number of false bits. The embedded bit was 0 and the detected was 1 and vice versa.

"Unrecognized bits" is the number of bits that the detecting algorithm can not decide whether the embedded bit was 0 or 1. Results show that the watermark can be extracted even from videos compressed using low bitrate H.264 compression.

15.4.2.2 Robustness Results Using Two-Level DWT

Figure 15.8 shows the test results of the watermarking algorithm with two-level DWT after H.264/AVC compression

"Correct bits", "Erroneous bits" and "Unrecognized bits" mean the same.

Using two levels produces better results: even at bitrate of 384 kbps the 94% of the embedded data can be extracted.

The number of false bits also has reduced.

15.4.2.3 Robustness Results Using Three-Level DWT

Figure 15.9 shows the test results of the watermarking algorithm with three-level DWT after H.264/AVC compression

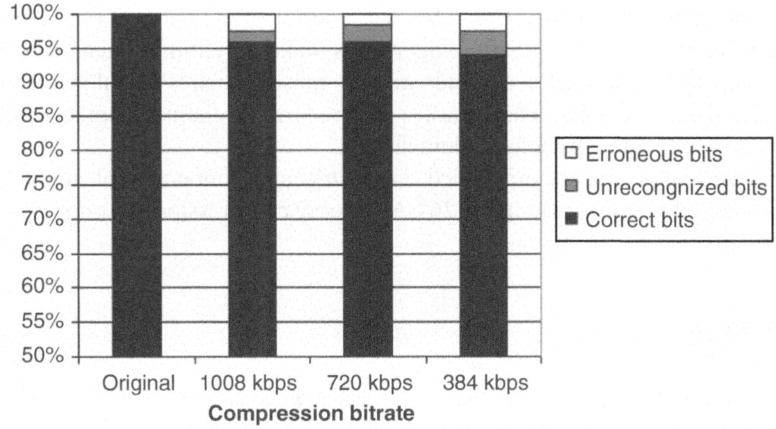

Fig. 15.8 Robustness results against H.264/AVC lossy compression, 2-level DWT

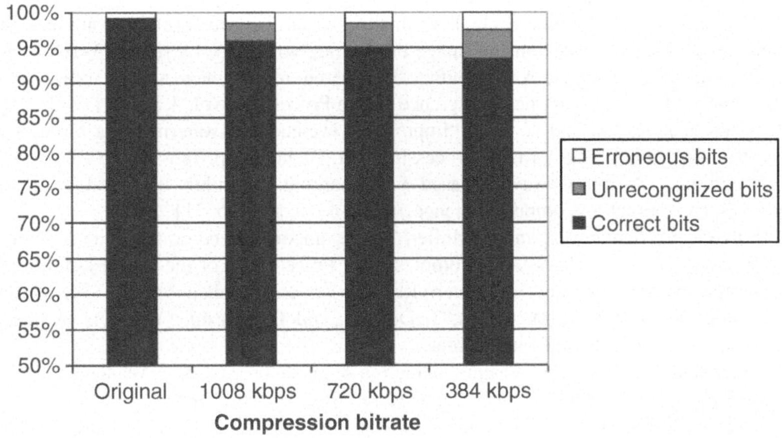

Fig. 15.9 Robustness results against H.264/AVC lossy compression, 3-level DWT

"Correct bits", "Erroneous bits" and "Unrecognized bits" mean the same. It can be seen that using the third level of DWT does not improve the robustness of the algorithm.

15.5 Conclusion

In this paper a novel video watermarking algorithm is presented. Data embedding is made in the wavelet domain. It uses the Sobel edge detector for finding the appropriate areas for watermarking – these areas are the significant edges of the frames and their surroundings. The HVS is less sensitive to modifications on middle and high frequencies. Compression algorithms make only minor changes to these areas. The watermark itself is a pseudo random noise, which is calculated from the input data and a seed value. Watermark is detected by correlating the pixel values of the selected area with a pseudo random noise.

Test results show that the embedded watermark can be imperceptible to the HVS, and the method is resistant to the H.264/AVC lossy compression algorithms.

References

1. R. Schyndel, A. Tirkel, and C. Osborne, "A digital watermark," IEEE Proceedings of the International Conference Image Processing, vol. 2, pp. 86–90, 1994.
2. W. Bender, D. Gruhl, N. Morimoto, and A. Lu, "Techniques for data hiding," IBM Systems Journal, vol. 35, no. 3–4, pp. 313–336, 1996.
3. R. B. Wolfgang and E. J. Delp, "A watermark for digital images," in IEEE Proceedings of the International Conference Image Processing, vol. 3, pp. 219–222, 1996.
4. I. J. Cox, J. Kilian, T. Leighton, and T. Shamoon, "Secure spread spectrum watermarking for multimedia," IEEE Transactions on Image Processing, vol. 6, no. 12, pp. 1673–1687, 1997.
5. M. D. Swanson, B. Zhu, and A. H. Tewfik, "Transparent robust image watermarking," in IEEE Proceedings of the International Conference Image Processing, vol. 3, pp. 211–214, 1996.
6. M. Barni, F. Bartolini, and A. Piva, "Improved wavelet-based watermarking through pixel-wise masking," IEEE Trans. Image Processing, vol. 10, no. 5, pp. 783–791, 2001.
7. J. N. Ellinas, "A robust wavelet-based watermarking algorithm using edge detection," International Journal of Computer Science, vol. 2, no. 3, pp. 206–211, 2007.
8. D. Pröfrock, M. Schlauweg, and E. Müller, "A new uncompressed-domain video watermarking approach robust to H.264/AVC compression," Proceedings of the IASTED International Conference on Signal Processing, Pattern Recognition, and Applications, pp. 99–104, 2006.
9. P. Schrder, W. Sweldens, M. Cohen, T. DeRose, and D. Salesin, "Wavelets in Computer Graphics," SIGGRAPH 96 Course Notes.
10. M. Nixon and A. S. Aguado, "Feature extraction & image processing," Academic, 2007.
11. S. Winkler, "Digital Video Quality: Vision Models and Metrics," Wiley, 2005.
12. Z. Wang, A. C. Bovik, H. R. Sheikh, and E. P. Simoncelli, "Image quality assessment: From error visibility to structural similarity," IEEE Transactions on Image Processing, vol. 13, no. 4, pp. 600–612, 2004.

Chapter 16
High Speed Soft Computing Based Circuit for Edges Detection in Images

Nashaat M. Hussein and Angel Barriga

Abstract In this chapter a technique for detecting edges in images is presented. The technique is based on applying soft computing techniques such as fuzzy logic and Lukasiewicz algebra operator. The utility of this technique is related to the simplicity of the operations for edge calculation that makes it very suitable for low cost hardware implementation and high processing speed.

Keywords Edge detection · image · soft computing · fuzzy logic · Lukasiewicz algebra operator

16.1 Introduction

Many image processing techniques requires the detection of edges. An edge is defined as a sharp change in the luminosity intensity between a pixel and another adjacent one. Most of the edge detection techniques can be grouped in two categories: gradient based techniques and Laplacian based methods. The techniques based on gradient use the first derivative of the image and look for the maximum and the minimum of this derivative. Examples of this type of strategies are: the Canny method [1], Sobel method, Roberts method [2], Prewitt method [3], etc. On the other hand the techniques based on Laplacian look for the cross by zero of the second derivative of the image. An example of this type of techniques is the zero-crossing method [4].

Normally the edge extraction mechanisms are implemented executing the corresponding software realisation on a processor. Nevertheless in applications that demand constrained response times (real time applications) the specific hardware implementation is required. The main drawback of implementing the edge detec-

N. M. Hussein and A. Barriga
Instituto de Microelectrónica de Sevilla (CNM-CSIC)/University of Seville, Spain
E-mails: nashaat@imse.cnm.es, barriga@imse.cnm.es

S.I. Ao and L. Gelman (Eds.), *Advances in Electrical Engineering and Computational Science*, Lecture Notes in Electrical Engineering.

tion techniques in hardware is the high complexity of the existing algorithms. In this chapter a technique for detecting edges in images is presented. The technique is based on applying soft computing techniques such as fuzzy logic and Lukasiewicz algebra operator. The utility of this technique is related to the simplicity of the operations for edge calculation [5]that makes it very suitable for low cost hardware implementation and high processing speed.

16.2 Edge Detection Algorithm

The process of edge detection in an image consists of a sequence of stages. The first stage receives the input image and applies a filter in order to eliminate noise. Then a binary image is obtained applying a threshold in order to classify the pixels of the image under two categories, black and white. Finally, in the last stage the detection of edges is performed.

16.2.1 The Filter Stage

The filter stage allows to eliminate noise patterns. The target of the filter step consists in eliminating all those points that do not provide any type of information of interest. The noise corresponds to nonwished information that appears in the image. It comes principally from the capture sensor (quantisation noise) and from the transmission of the image (fault on transmitting the information bits). Basically we consider two types of noise: Gaussian and impulsive (salt & peppers). The Gaussian noise has its origin in differences of gains of the sensor, noise in the digitalization, etc. The impulsive noise is characterized by some arbitrary values of pixels that are detectable because they are very different from its neighbours. A way to eliminate these types of noise is by means of a low pass filter. This filter makes smoothed of the image replacing high and low values by average values.

The filter used in the proposed edge detection system is based on the bounded-sum Lukasiewicz's operator. This operator comes from multi-valued Lukasiewicz algebra and is defined as:

$$BoundedSum(x, y) = \min(1, x + y) \tag{16.1}$$

where $x, y \in [0, 1]$. The main advantage of applying this operator comes from the simplicity of the hardware realisation as it is seen in Section 16.3.

The Lukasiewicz's bounded sum filter performs the smoothing of the image and is suitable for salt & peppers noise as well as Gaussian. Figure 16.1 shows the effect of applying this type of filter.

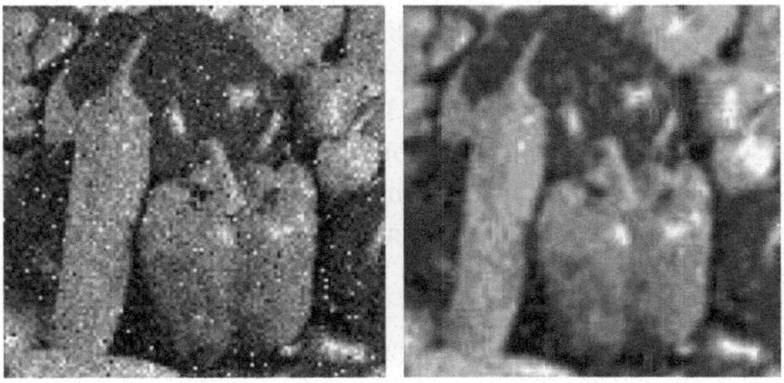

Fig. 16.1 (**a**) Input image with salt & peppers noise, (**b**) Lukasiewicz's bounded sum filter output

The application of the filter has been made using a mask based on a 3 × 3 array. For the pixel x_{ij} the weighted mask is applied to obtain the new value y_{ij} as is shown in the following expression:

$$y_{ij} = \min(1, \frac{1}{8} \sum_{k=-1}^{1} \sum_{l=-1}^{1} x_{i+k,j+l}) \tag{16.2}$$

16.2.2 The Threshold Stage

The techniques based in thresholding an image allow classifying the pixels in two categories (black and white). This transformation is made to establish a distinction between the objects of the image and the background. The way of generating this binary image is making the comparison of the values of the pixels with a threshold T.

$$y_{i,j} = \begin{cases} 0 & if\ x_{i,j} < T \\ 255 & if\ x_{i,j} > T \end{cases} \tag{16.3}$$

where $x_{i,j}$ is a pixel of the original image and $y_{i,j}$ is the pixel corresponding to the binary image. In the case of a monochrome image in that the pixels are codified with 8 bits the range of values that the pixels take corresponds to the range between 0 and 255. It is usual to express the above mentioned range with normalized values between 0 and 1.

A basic technique for threshold calculation is based on the frequency of grey level. In this case, for an image with n pixels and n_i pixels with the grey level i, the threshold T is calculated by means of the following expression:

$$T = \sum_{i=1}^{255} p_i i \ with\ p_i = n_i/n \ and \sum_{i=1}^{255} p_i = 1 \tag{16.4}$$

where p_i represents the grey level frequency (also known as the probability of the grey level).

In 1965 Zadeh [6] proposed the fuzzy logic as a reasoning mechanism that uses linguistic terms. The fuzzy logic is based on the fuzzy set theory in which an element can belong to several sets with different membership degree. This is opposed with the classic set theory in which an element belongs or not to a certain set. Thus a fuzzy set A is defined as

$$A = \{(x, \mu(x)) \,|x \in X\} \tag{16.5}$$

where x is an object of the set of objects X and $\mu(x)$ is the membership degree of the element x to the set A. In the classic set theory $\mu(x)$ takes values 0 or 1 whereas in the fuzzy set theory $\mu(x)$ belongs to the range of values between 0 and 1.

An advantageous aspect of applying fuzzy logic in the calculation of the threshold T is that the calculation mechanism improves the processing time since it only requires processing the image once and allows calculating in a direct way the value of the threshold.

The fuzzy system has an input that receives the pixel that is going to be evaluated and an output that corresponds to the result of the fuzzy inference. Once read the image the output shows the value of threshold T. Basically the operation that makes the fuzzy system corresponds to the calculation of the centre of gravity of the histogram of the image with the following expression:

$$T = \sum_{i=1}^{M} \sum_{j=1}^{R} \alpha_{ij} c_{ij} \Bigg/ \sum_{i=1}^{M} \sum_{j=1}^{R} \alpha_{ij} \tag{16.6}$$

where T is the threshold, M is the number of pixels of the image, R is the number of rules of the fuzzy system, c is the consequent of each rule and α is the activation degree of the rule.

The knowledge base of the fuzzy system contains both the membership functions and the rule base. It is made a partition of the universe of discourse of the histogram in a set of N equally distributed membership functions. Figure 16.2a shows a partition example for $N = 9$. Triangular membership functions have been used since they are easier for hardware implementation. These functions have an overlapping degree of 2 what allows to limit the number of active rules. The membership functions of the consequent are singletons equally distributed in the universe of discourse of the histogram. The use of singleton-type membership functions for the consequent allows applying simplified defuzzification methods such as the Fuzzy Mean. This defuzzification method can be interpreted as one in which each rule proposes a conclusion with a "strength" defined by its grade of activation. The overall action of several rules is obtained by calculating the average of the different conclusions weighted by their grades of activation. These characteristics of processing based on active rules and simplified defuzzification method allows a low cost and high speed hardware implementation.

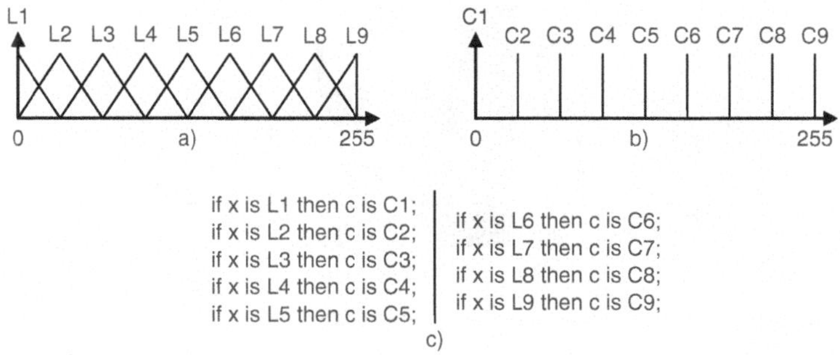

Fig. 16.2 Membership functions for $N = 9$, (**a**) antecedent, (**b**) consequent, (**c**) rulebase

The rule base of the system of Fig. 16.2c used the membership functions previously defined. The knowledge base (membership functions and the rule base) is common for any images, for that reason the values can store in a ROM memory.

It is possible to optimize the expression shown in Eq. (16.6) if the system is normalized. In this case the sum extending to the rule base of the grades of activation of the consequent takes value 1 ($\sum_{j=1}^{R} \alpha_{ij} = 1$). Then Eq. (16.6) transforms in:

$$T = \frac{1}{M} \sum_{i=1}^{M} \sum_{j=1}^{R} \alpha_{ij} c_{ij} \tag{16.7}$$

For each pixel the system makes the inference in agreement with the rule base of Fig. 16.2c. The output of the system accumulates the result corresponding to the numerator of Eq. (16.7). The final output is generated with the last pixel of the image.

16.2.3 The Edge Detection Stage

The input image for the edge detection is a binary image in which pixels take value black (0) or white (255). In this case the edges appear when a change between black and white takes place between two consecutive pixels. The edge generation consists of determining if each pixel has neighbours with different values. Since the image is binary every pixel is codified with a bit (black = 0 and white = 1). This operation of edge detection is obtained calculating the *xor* logic operation between neighbouring pixels using a 3 × 3 mask. Using the 3 × 3 mask it is possible to refine the edge generation detecting the orientation of the edges. Figure 16.3c shows an example of applying the *xor* operator on the binary image. Figure 16.4 shows the obtained results when making the edge detection on a set of test images.

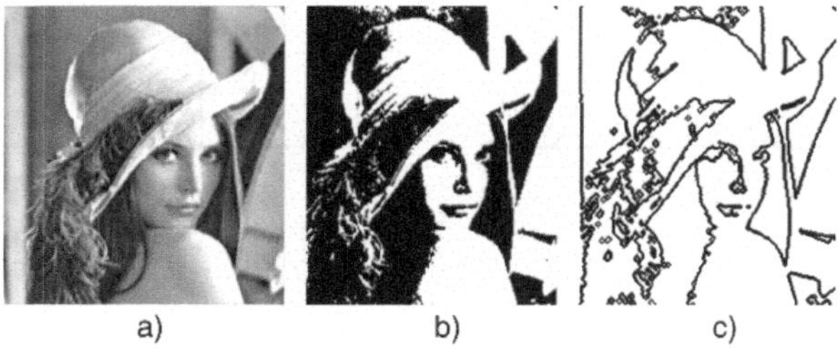

Fig. 16.3 (**a**) Lena's image, (**b**) binary image, (**c**) edge detection

Fig. 16.4 Test images and edge detection results

16.3 Hardware Implementation

The circuit that realizes the edge detection has been implemented on a low cost FPGA device of the Xilinx Spartan3 family. This circuit receives the image from a vision sensor based on a CMOS chip that gives a CIF resolution (352 × 288). The image is stored in a double port RAM memory. The data memory width is 32 bits. This allows to read two words simultaneously.

Figure 16.5 shows the block diagram of the system [5]. The image is captured by the camera. The camera provides an 8 bits pixel in each clock cycle. Then the pixel is stored in the RAM memory under the control of the memory control circuit. During the reading of the image the threshold is calculated since the threshold generation circuit receives the pixel coming from the camera. As soon as the image is stored in the memory the threshold generation circuit produce the threshold value of the image.

Next, the edge detection circuit initiates its operation reading from the memory eight pixels in each clock cycle (2 words of 32 bits). This way the edge detection circuit is able to provide four output data in parallel that are stored in the external memory. Each data corresponds to a pixel of the edge image. This image is binary

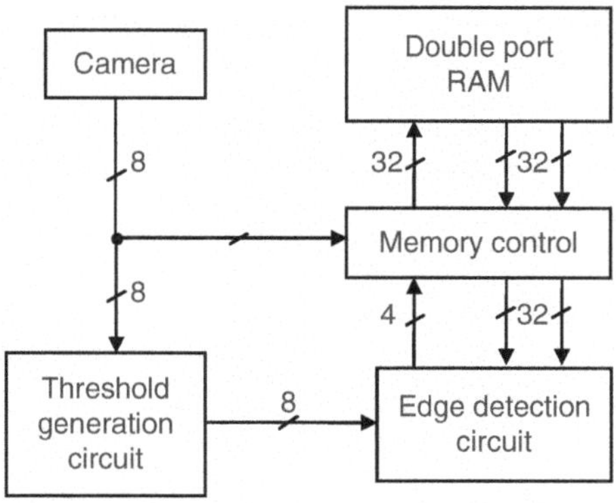

Fig. 16.5 Block diagram of the system

reason why only a bit is needed to represent the value of the pixel (0 if edge or 1 if background). The new image of the edges is stored in the above mentioned memory.

16.3.1 The Threshold Generation Circuit

In order to implement the fuzzy system circuit that generates the threshold the design methodology is based on the hardware description language VHDL as a way to describe and model the system at high level [7]. To achieve a behavioural modelling of the fuzzy inference module a VHDL description style will be used. In this description the system structure (fuzzy sets, rule base) and the operator description (connectives, fuzzy operations) are defined separately. This makes it possible to describe independently both the fuzzy system structure and the processing algorithm. The fuzzy system description must be synthesizable in order to generate the hardware realizations.

Figure 16.6 shows the VHDL architecture of the fuzzy system. The rule base is described in the architecture body. It contains a rule base structure with the nine rules of Fig. 16.2. Each rule can be divided into two components: the antecedent of the rule and the consequent. The antecedent is an expression of the input variables related to their linguistic values. The consequent sets the linguistic value of the rule output.

The processing mechanisms of the fuzzy operation *is* (=) and the inference then (*rule*(,)) are not defined in the VHDL description. Only the structure of the rulebase is defined. Such a description is a high level description because it does not assume any specific implementation criteria. It only describes the knowledge base in terms of a behavioral rule base.

Linguistic labels represent a range of values within the universe of discourse of input and output variables. These labels can be described by functions in order to compute the membership degree of a certain input value. Membership functions

```
architecture knowledge_base of threshold is
 signal R: consec;
 -- MF for x
 constant L1: triangle:=((0,   0,  31),(0, 32));
 constant L2: triangle:=((0,  31,  63),(32,32));
      . . .
 constant L9: triangle:=((223,255,255),(32,0));
 --MF for z
 constant C1: integer := 0;  constant C2: integer := 31;
      . . .
 constant C9: integer := 255;
begin
   R(1)<=rule((x=L1),C1); R(2)<=rule((x=L2),C2);
   R(3)<=rule((x=L3),C3); R(4)<=rule((x=L4),C4);
   R(5)<=rule((x=L5),C5); R(6)<=rule((x=L6),C6);
   R(7)<=rule((x=L7),C7); R(8)<=rule((x=L8),C8);
   R(9)<=rule((x=L9),C9);
   Zout<=defuzz(R);
end knowledge_base;
```

Fig. 16.6 VHDL architecture of the knowledge base

associated to a linguistic label can be triangular or trapezoidal. The declarative part of the architecture of Fig. 16.8 shows the definition of such membership functions.

The data type triangle is defined in a VHDL package. This data type contains the definitions of the points that define a triangle as well as the slopes. On the other hand the rule base expresses the knowledge of Fig. 16.2. The function *rule* is also defined in the VHDL package. This function makes the inference of the rule. The set of rules is evaluated concurrently since the signal assignments in the architecture body are concurrent. The operator " = " has been redefined taking advantage of the functions overload properties in VHDL.

The functions used in the description of the fuzzy system have been described in agreement with the restrictions of VHDL for synthesis. This has allowed to generate the circuit that implements the fuzzy system using conventional circuit synthesis tools. The fuzzy inference module is a combinational circuit that makes the fuzzy inference.

The circuit can operate at a frequency of 137 MHz which would allow that the processing of an CIF image (352×288 pixels) will carry out in 0.7 ms. In case of an VGA image ($1,024 \times 768$) require a time of 5.7 ms to calculate the threshold.

16.3.2 Design of the Edge Detection System

The edge detection algorithm basically is constituted by three stages. In the first stage the Lukasiewicz bounded-sum is performed. After the filter stage a thresholding step is applied. This gives rise to a black and white monochrome image. In the third stage the edges of the image are obtained. For it the final value of each pixel is

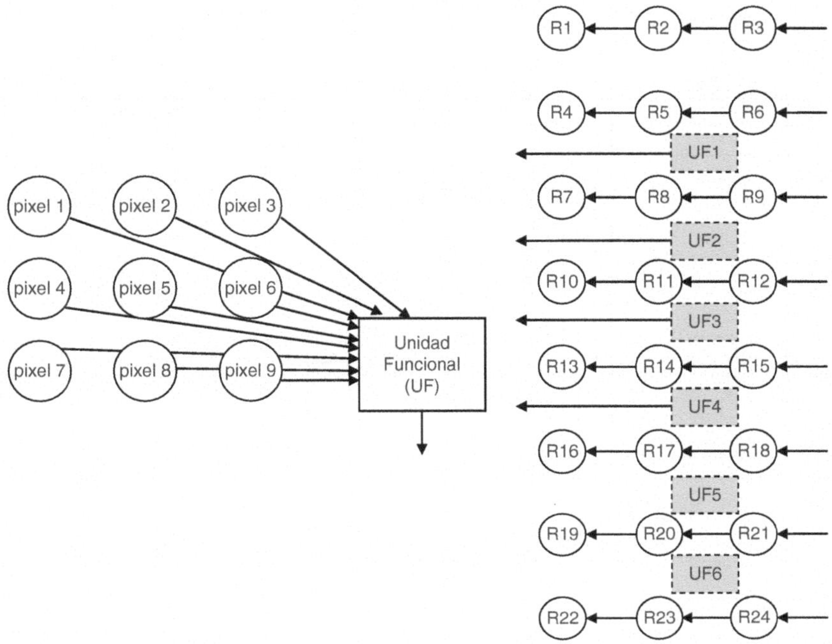

Fig. 16.7 Block diagram of the (**a**) 3 × 3 architecture and (**b**) 8 × 3 architecture

evaluated. Only those pixels that are around the target pixel are of interest (a 3 × 3 mask). By this reason, if in the surroundings of a pixel the value is the same (all white or all black) then it means no edge and the output value associates the above mentioned pixel with the background of the image. When it is detected that some value of the surroundings of the pixel changes indicates that the pixel at issue is in an edge, reason why the black value is assigned to it.

Figure 16.7a shows the basic processing scheme to calculate the value of one pixel. Pixels 1–9 correspond to the 3 × 3 mask register that moves through the image. The Functional Unit (FU) makes the processing of the data stored in the mask registers.

In order to improve the image processing time the mask was spread to an 8 × 3 matrix as is shown in Fig. 16.7b. Each Functional Unit (FU) operates on a 3 × 3 mask in agreement with the scheme of Fig. 16.7a. The data are stored in the input registers (R3, R6, R9, ...) and they move in each clock cycle to their interconnected neighbours registers. In the third clock cycle the mask registers contain the data of the corresponding image pixels. Then the functional units operate with the mask data and generate the outputs. In each clock cycle the mask advances a column in the image. Pixels enter by the right and shift from one stage to another outgoing at the left side. It is a systolic architecture that follows a linear topology and allows processing several pixels in parallel.

The system receives two input data of 32 bits (these mean eight pixels). These data come from a double port memory that stores the image. The circuit also receives as input data the threshold (T) that has been calculated previously. The circuit

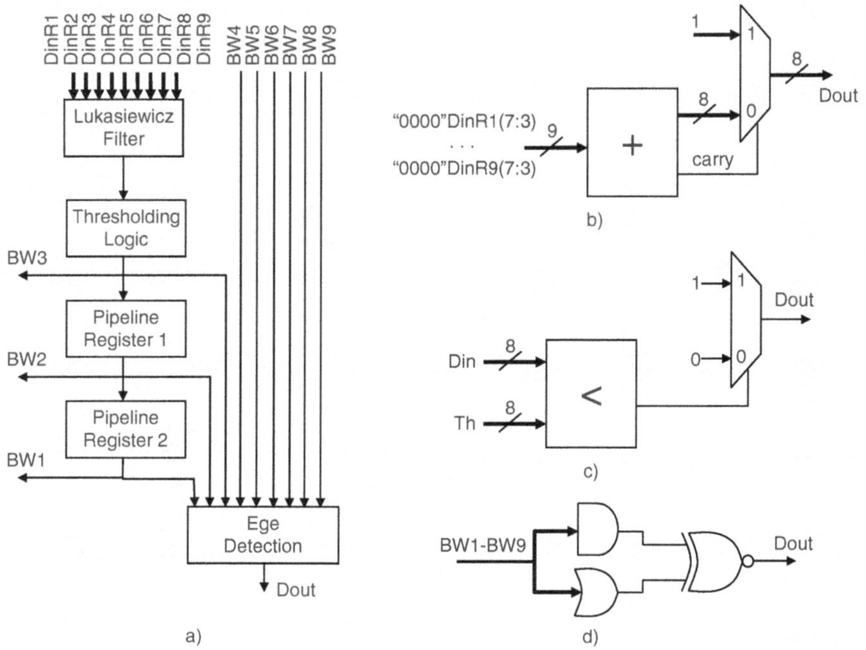

Fig. 16.8 (**a**) Functional Unit (FU) circuit schematic, (**b**) Lukasiewicz filter, (**c**) thresholding logic circuit, (**d**) edge detection circuit

generates as output the 4 bits corresponding to the output values of the processed pixels stored in R5, R8, R11 and R14.

The functional unit operates on the 3×3 mask and generates the output value corresponding to the evaluated element of the mask. A block diagram of a functional unit is shown in Fig. 16.8a. The circuit consists of two pipeline stages so that the data has a latency of two clock cycles. The first stage is the image filter. Then threshold T is applied. The edge detector, in the output stage, operates on the binary mask (black and white image).

Figure 16.8 shows the circuits corresponding to the different blocks of the functional unit (FU). As we can observe in Fig. 16.8b the filter based on Lukasiewicz's bounded sum receives the data stored in registers R1 to R9. These data are scaled by the factor 0.125 entailing division by 8, which signifies a displacement of three places to the left. The sum of the data is compared (using the carry as control signal) with value 1. The threshold logic (Fig.16.8c) compares the pixel with the threshold value. The output is a binary image (black and white) and only therefore requires one bit. Finally, the edge detection circuit receives a 3×3 binary image. It carries out the *xor* operation of the bits. If all the bits of the mask are equal the output pixel is in the background, whereas if some bit is different the output is an edge pixel.

Figure 16.9 shows the chronogram of the circuit. It can be observed that the operation of the system begins with the falling edge of signal *CS*. Whenever a row begins two clock cycles are needed to initialize the mask registers. In the third clock

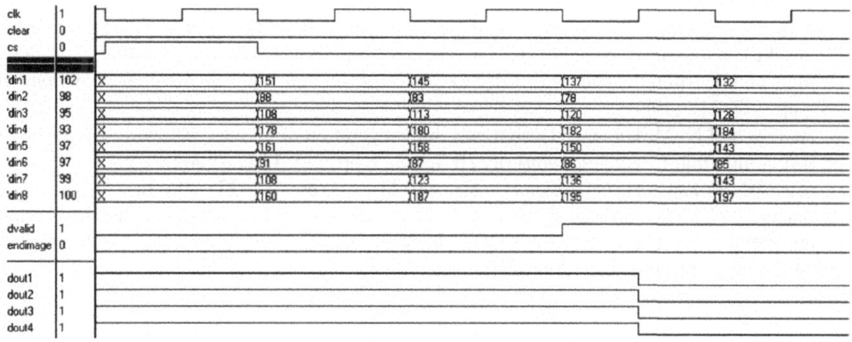

Fig. 16.9 Chronogram of the circuit

cycle *Dvalid* signal take value 1, indicating a valid output. Input data are provided in each clock cycle. Once *Dvalid* has been activated the output data in the following cycles is also valid since *Dvalid* = 1 (data of the following columns being processed in successive cycles).

The circuit occupies an area of 293 slices on an FPGA of the Spartan3 Xilinx family that supposes a cost of 5,361 equivalent gates. It can operate at a 77 MHz frequency, although the development board used has a 50 MHz clock. This supposes that the required time to process an image is 0.335 ms what allows to process 2,985 images per second.

16.4 Conclusion

The hardware implementation of an edge extraction system has been described. The edge detection algorithm is based on soft computing technique. Thus the calculation of the threshold that allows to obtain a binary image is realized by means of a fuzzy logic inference mechanism. The conditioning of the image for its later processing is based on a filter that uses Lukasiewicz's bounded sum. The main advantage of this edge detection technique is that it allows a very efficient hardware implementation in terms of cost and speed. This makes it specially indicated in applications that require a real time processing.

References

1. J. F. Canny, A computational approach to edge detection, *IEEE Transactions on Pattern Analysis and Machine Intelligence*, 679–698 (1986).
2. L. G. Roberts, Machine perception of three-dimensional solids, in J. T. Tippet et al., editor, Optical and Electro-Optical Information Processing (MIT, Cambridge, MA, 1965), pp. 159–197.

3. J. M. S. Prewitt, Object enhancement and extraction, in A. Rosenfeld and B. S. Lipkin, editors, Picture Processing and Psychophysics (Academic, New York, 1970), pp. 75–149.
4. D. Marr and E. Hildreth, Theory of Edge Detection, *Proceedings of the Royal Society London*, 207, pp. 187–217 (1980).
5. A. Barriga and N. M. Hussein, Design and implementation of an edge detection circuit based on soft computing, *Communications of SIWN*, vol. 3, pp. 135–139 (2008).
6. L. A. Zadeh, Fuzzy Sets, *Information and Control*, vol. 8, pp. 338–353 (1965).
7. A. Barriga, S. Sánchez Solano, P. Brox, A. Cabrera, and I. Baturone, Modelling and Implementation of Fuzzy Systems based on VHDL, *International Journal of Approximate Reasoning*, vol. 41, issue 2, pp. 164–278 (2006).

Chapter 17
A Comparison Between 3D OSEM and FBP Image Reconstruction Algorithms in SPECT

Khalid S. Alzimami, Salem A. Sassi, and Nicholas M. Spyrou

Abstract In this chapter, we will only concern our selves with the two most widely used algorithms in SPECT reconstruction, namely the FBP and Ordered-Subsets Expectation Maximization (OSEM) algorithms. This chapter describes the basic principles of these two algorithms, and summarises SPECT image attenuation and scatter correction methods. Finally, it presents a study evaluating the volumetric iterative reconstruction algorithm OSEM 3-D, and compares its performance with the FBP algorithm.

Keywords Image Reconstruction · Single Photon Emission Computed Tomography · Filtered Back Projection · ordered-subsets expectation maximization · image attenuation

17.1 Introduction

Single Photon Emission Computed Tomography (SPECT) is a nuclear medicine imaging technique used to visualize the distribution of radioactive tracer uptake in the patient [1]. In SPECT, the radio-pharmaceutical emits single gamma (γ) ray photons isotropically. A gamma camera system is used to acquire a 2-D image of the distribution of the radiotracer uptake within the object of interest. These images often called views or projections and are acquired at discrete angles around the patient [2].

Mathematical algorithms are then used to reconstruct 3-D images of selected planes within the object from the 2-D projection data. This process is known as Emission Computed Tomography (ECT) [3]. Images produced by SPECT result

K.S. Alzimami (✉)
Department of Physics, University of Surrey, Guildford, Surrey GU2 7XH, UK,
E-mail: K.Alzimami@surrey.ac.uk

S.I. Ao and L. Gelman (Eds.), *Advances in Electrical Engineering and Computational Science*, Lecture Notes in Electrical Engineering.

in a significant improvement of object contrast, and can allow more accurate quantification of radiotracer uptake at specific locations within the body [1].

Even though, the first tomographic imaging systems used iterative reconstruction methods, the analytical Filtered Back Projection (FBP) method is still the most widely used method in the reconstruction of SPECT images. However, with the rapid increase in the processing speed of modern computers, interest in the development and implementation of iterative reconstruction methods in emission tomography modalities has grown rapidly and a plethora of iterative reconstruction algorithms have been developed. These include algebraic methods like the algebraic reconstruction technique (ART) [4], simultaneous iterative reconstruction technique (SIRT) and iterative least-squares technique (ILST), as well as Statistical reconstruction methods like maximum likelihood expectation maximization (MLEM) [5], ordered-subsets expectation maximization (OSEM) [6], maximum likelihood algorithm (RAMLA), maximum a posteriori (MAP) algorithms and gradient and conjugate gradient (CG) algorithms [7].

Many of these methods have been implemented in SPECT imaging systems. In this chapter, however, we will only concern our selves with the two most widely used algorithms in SPECT reconstruction, namely the FBP and Ordered-Subsets Expectation Maximization (OSEM) algorithms. This chapter describes the basic principles of these two algorithms, and summarises SPECT image attenuation and scatter correction methods. Finally, it presents a study evaluating the volumetric iterative reconstruction algorithm OSEM 3-D, and compares its performance with the FBP algorithm for the reconstruction of $^{99}Tc^{m}$ SPECT imaging.

17.2 Background

The purpose of tomographic reconstruction is to obtain axial cross-sections of an object from projections of that object. In 1917 Radon showed the reconstruction of a 2-D object using an infinite number of its 1-D projections [8]. Modified realizations of this method have been adopted to reconstruct 3-D SPECT images from the acquired 2-D projections. These methods have been used widely in SPECT image reconstruction due to their computational speed and simple implementation.

Historically, all image reconstruction methods have aimed to reduce the problem of 3-D image reconstruction into a 2-D problem by dividing the 2-D projection data into 1-D profiles to produce cross-sectional images of the object. These algorithms are often referred to as 2-D reconstruction algorithms. Recently, due to the massive increase in processing speeds of modern computers and the availability of cheap high capacity memory chips, a new generation of image reconstruction algorithms has evolved. These algorithms have permitted the full 3-D reconstruction of the projection data.

In order to correct the SPECT images accurately for photon crosstalk between trans-axial slices, one needs fully 3D reconstruction. Unlike 2-D reconstruction, slice by slice, full 3-D reconstruction uses a large matrix making it possible for the photons that are detected in the out-of-slice projection pixels to be accounted for [9]. Although the computational burden of full 3-D iterative reconstruction is somewhat high, the accuracy is significantly improved and there are noteworthy improvements of signal-to-noise ratio (SNR) in comparison to 2D reconstruction [10]. A detailed description of the 3-D reconstruction problem is given in [9, 11, 12].

In general, image reconstruction in SPECT can be divided into two main approaches. The first approach is based on direct analytical methods, while the second is based on algebraic and statistical criteria and iterative algorithms.

17.2.1 Analytical Reconstruction Algorithms: Filtered Backprojection

Filtered back projection (FBP) is the most commonly used reconstruction algorithm in SPECT imaging. This is principally due to the simple concept of the algorithm and relatively quick processing time. FBP can be viewed as a two-step process: filtering of the data and back projection of the filtered projection data. There is extensive literature on these reconstruction algorithms and reviews of FBP and its applications to SPECT [1, 4, 13, 14].

17.2.1.1 Filtering

3-D SPECT images are created by back projecting planar 2-D views of the object into image space along the line of response [15]. This process will result in blurred images and reconstruction star like artefacts. To overcome this problem, a ramp filter is applied to the projection data, which in turn increases the magnitude of the high frequency components of the image including statistical noise. Further filtration is therefore required to temper this effect. This can be done by applying an image smoothing kernel, but because convolution is a computationally intensive task, the data is first transferred to the frequency domain and projection data filtration is performed in the frequency (Fourier) domain [1, 2].

There are basically three types of reconstruction filters, these can be categorised as low pass, high pass and band pass filters. While high pass filters allow high frequency information including image noise to pass, these usually referred to as image enhancement filters. Band-pass filter allows only data with frequencies within the pass band to be retained, while suppressing or eliminating all other frequencies [15, 16].

Reconstruction filters can be described by two important parameters: the "cut-off frequency" and "order or power" of the filter function. The cut-off frequency determines the filter rolling off to an infinitely low gain whereas the shape of the curve is determined by the order of the filter function. The location of the cut-off frequency determines how the filter will affect both image noise and resolution.

17.2.1.2 Back Projection

The main reconstruction step involves the backprojection of acquired data into the image domain. For simplicity, the position of photon detection within the gamma camera head is assumed to be perpendicular the location photon emission. So by smearing back the number of counts at each point in projection profiles into the image space, a 3-D image of radioactive tracer distribution can be built. Regions with higher concentration of back projected lines (ray sums) and greater number of counts form a reconstructed image of radiotracer concentration within the object. This process is done in spatial (image) domain unlike image filtration, which is normally carried out in the frequency domain [1, 13, 14].

17.2.2 Iterative Reconstruction Algorithms

There is increasing recognition that iterative reconstruction plays a key role in improving the quality of reconstructed images and may improve the accuracy of SPECT image quantification, particularly where attenuation is non homogeneous or where a more exact model of the emission and detection processes is required [17, 18]. Iterative reconstruction algorithms, such as OSEM have become a clinically practical alternative to FBP [19].

The iterative reconstruction algorithm begins with a preliminary estimate of the object source distribution. This is done by assuming a homogenous object distribution or created using a single measured projection. A set of projection data is estimated from the initial estimate using a mathematical projector that models the imaging process. Differences between estimated projection data and measured projection data are calculated for every projection angle. Using an algorithm derived from specific statistical criteria, these differences are used to update the initial image estimate. The updated image estimate is then used to recalculate a new set of estimated projection data that are again compared with the measured projection data. The number of iterations is either limited by the user or is controlled by pre-specified conversion criteria [9].

The iterative reconstruction techniques can be classified into three groups, defined mainly by the underlying assumptions regarding the nature of the data, whether statistical or non statistical. These are:

- Algebraic reconstruction algorithms
- Statistical algorithms assuming Gaussian noise and
- Statistical algorithms assuming Poisson noise

17.2.2.1 Ordered Subsets Expectation Maximization (OSEM)

The ML-EM algorithm has proven to be effective, but also to be very slow for daily use. Splitting up the measured datasets into different subsets and using only one subset for each iteration speeds up the algorithm by a factor equal to the number of subsets. The method is known as Ordered Subsets Expectation Maximization (OSEM) [6].

In SPECT, the sequential processing of ordered subsets is natural, as projection data is collected separately for each projection angle. In other words, counts on single projections can form successive subsets. With OSEM, the careful selection of subset ordering can greatly accelerate convergence. However, the lack of convergence of OSEM algorithms is of theoretical importance since in practice, the algorithm is terminated after only a few iterations [20].

In recent years many attempts have been made to improve the quality of the 3D OSEM algorithms by incorporating corrections for the major image-degrading factors in the projection and backprojection operations of the iterative steps. Specifically, several commercial algorithms (such as Flash 3-D by Siemens [21], Astonish 3-D by Philips [22], HOSEM by HERMES [23] and Wide-Beam Reconstruction by UltraSPECT [24]) have been developed to improve the SNR of the image by modelling Poisson noise that results from low counts as well as implementing resolution recovery algorithms to restore resolution degradation due to collimator spread function and source detector distance in the case of WBR.

17.2.3 Physical Factors Affecting Quantitative and Qualitative Accuracy

Several physical factors, the most important being photon attenuation, Compton scatter and spatially varying collimator response, degrade the qualitative and quantitative accuracy of SPECT images. Theoretically, the number of counts in reconstructed SPECT images should be directly proportional to the absolute concentration and distribution of the radiotracer within the imaged object. It has been reported that the presence of scatter and attenuation in the images limits the accuracy of quantification of activity. The uncertainty could be as high as 50–100% if the scatter and attenuation are not corrected [25].

17.2.3.1 Attenuation Correction

Attenuation causes inconsistent projection data that can increase or decrease counts in the image, especially near hot regions. It also causes shadows or regions of diminished activity, within the patient. Thus, attenuation may potentially affect tumour detectablility or artificially enhance a noise blob [26].

AC methods can be classified as: (a) constant attenuation coefficient (μ), known as the Chang method [27]; or (b) variable μ, or transmission source method. The most widely used AC method is the Chang method, but it can only be used in the brain or abdomen regions as they can be considered essentially uniform. Obviously, the Chang method will only work well where μ is, in fact, approximately constant.

However, this is not the case when the attenuating medium is non-uniform (e.g. cardiac studies). Therefore, AC requires a variable attenuation coefficient dependent on the spatial location of the pixel in the object (i.e. patient or phantom) [28].

17.2.3.2 Scatter Correction

Scattered photons limit resolution and act as a modifier to the effective attenuation coefficients in a depth dependent way. Moreover, the presence of scattered photons in SPECT projection data leads to reduced contrast and loss of the accuracy of quantification in reconstructed images [25].

In order to compensate for the effects of scatter, several scatter compensation techniques have been developed. These techniques can be categorised into two main groups; subtraction based and reconstruction based scatter correction. In the subtraction-based approach the scatter component is estimated and subtracted from the projection data prior to reconstruction. Typically the scatter estimate is produced from multiple energy window acquisition. Kadrmas et al. have illustrated the advantages of using multiple energy windows when imaging isotopes that have multiple emission peaks [29].

In the reconstruction-based approach, the scatter is incorporated in the algorithm model. The main advantage of this approach is that the noise increase associated with the scatter subtraction method is avoided. This is because there is no explicit subtraction of scatter counts. A number of studies have shown that iterative reconstruction with accurate modelling of scatter is superior to pre-reconstruction scatter subtraction [25]. However, the major setback to model based reconstruction is that it is not obvious how the scatter originating from radio-nuclides in object regions out of the field of view (FOV) sources is compensated [10].

17.2.3.3 Distance-Dependent Resolution Compensation

The distance-dependent acceptance of photons through the holes of the collimator is the main contributor to the degradation of the spatial resolution in SPECT. 3D modelling of the collimator line spread function (LSF) utilizing the collimator parameters and assumed distance between collimator and source is used to compensate for collimator and distance dependent SPECT image blurring. This resolution recovery is achieved by applying stationary or non-stationary deconvolution filters to the acquired projection data prior to image reconstruction or to the reconstructed data. Deconvolution methods using Wiener, Metz, or power spectrum equalisation filtering have been used in 2-D by ignoring the interslice blurring, but the 3-D filtering gives better results. These methods have been used mainly in conjunction with the FBP reconstruction technique [30].

In an Alternative approach, a point spread function (PSF) of the camera system is modelled at each point of the reconstructed area for each camera projection, and is included into the projection/back-projection operator in the iterative reconstruction process. This modelling may be made in two dimensions, by ignoring the interslice blur. However, for best results calculations should be done in 3-D [30, 31].

Both deconvolution and modelling methods improve contrast of the reconstructed images. However, the main drawback of deconvolution methods is that they may also substantially increase high frequency noise. To compensate for this consequence one needs to apply a low pass filter, which, in turn, degrades image resolution. The modelling method on the other hand has a regularising effect on the reconstructed image. Compared to reconstruction without blurring compensation, high frequency noise is reduced, while contrast and resolution are improved [30,31].

17.3 A Comparison of 3D OSEM with FBP Image Reconstruction Algorithms

17.3.1 Methods

The performance of 3-D OSEM, with and without AC, was characterized with respect to subset and iteration number for a constant pixel size and the same amount of post reconstruction filtering. Due to space constraints, more details about methods and data analysis are given in [32].

17.3.2 Results and Discussions

As shown in Fig. 17.1, for the same number of subsets, there is a linear relationship between noise termed as variation coefficient and the number of iterations. The graph also shows a linear relationship between noise and number of subsets for the same number of iterations.

This linear relationship ($R^2 \approx 1$) leads to a predictable and accurate characterization of noise. Also, Fig. 17.1 shows that the effect of subset and iteration number is additive over noise, in accordance with OSEM theory [33].

As shown in Fig. 17.2, the number of iterations needed to reach a constant FWHM was approximately 17 for the periphery and 22 for the centre of the FOV. These results agree with those reported by Yokoi et al. and Kohli et al. [34, 35]. Kohli et al. concluded that the reason for the slow convergence in the central region was the effect of attenuation [35]. However this is not right as simulations by Yokoi et al. assumed no attenuating medium for a point source [34]. According to Yokoi et al. the reason of the slow convergence is that the SPECT resolution is originally better at the periphery than at the centre [34]. Another reasonable explanation of the slow convergence could be due to the fact that many more ray sums contribute to the formation of objects located at the centre of the image than in the periphery [1].

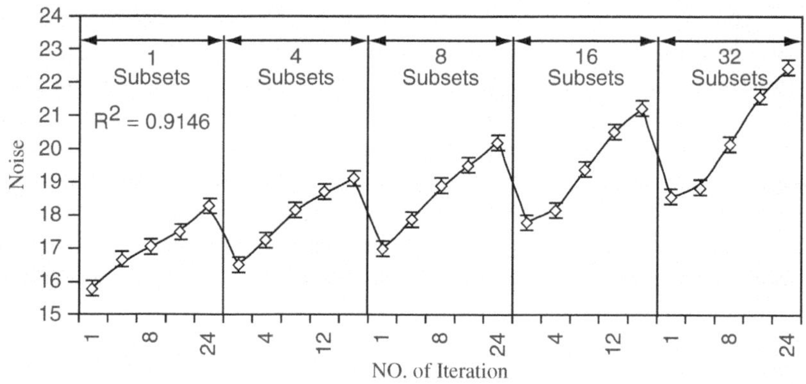

Fig. 17.1 Relationship between noise and number of iteration at different number of subsets

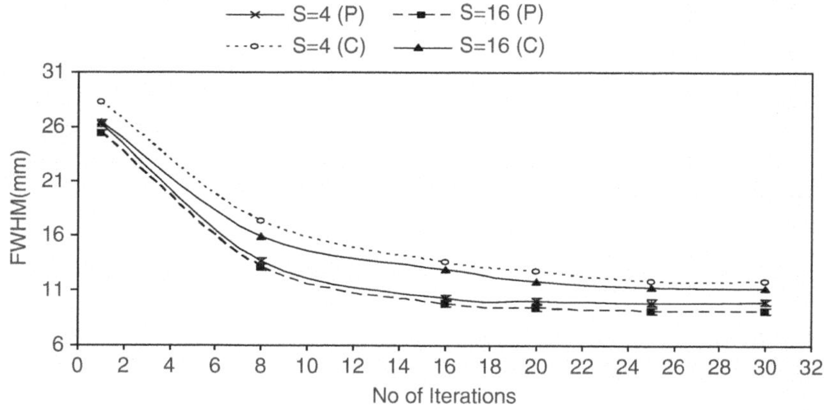

Fig. 17.2 Variation of the spatial resolution (FVHM) with the number of iterations for a range of subsets at the centre (C) and the periphery (P)

Figure 17.3 demonstrates a statistically significant ($P < 0.01$) improvement in spatial resolution using 3-D OSEM compared to FBP. This could be attributed to the resolution recovery properties of Flash 3D [21].

Figure 17.4 illustrates that 3D OSEM is the best choice for low count statistics study as 3D OSEM is significantly ($P < 0.01$) better than FBP in terms of noise. This could be due to the fact that IA takes the Poisson nature of the data into account, and unlike conventional 1-D or 2-D reconstruction methods like FBP and 2-D OSEM respectively, OSEM 3-D allows the reconstruction of all slices simultaneously, thus improving count statists [9].

As shown in Fig. 17.5, there was no significant difference ($P > 0.01$) in image contrast between FBP and 3D OSEM without applying AC. However, with AC 3-D OSEM image contrast has improved significantly in comparison to FBP with AC. This means that 3D OSEM with AC may extend the range of detectable lesion sizes and contrast levels, and may provide clarity in situations in which FBP is

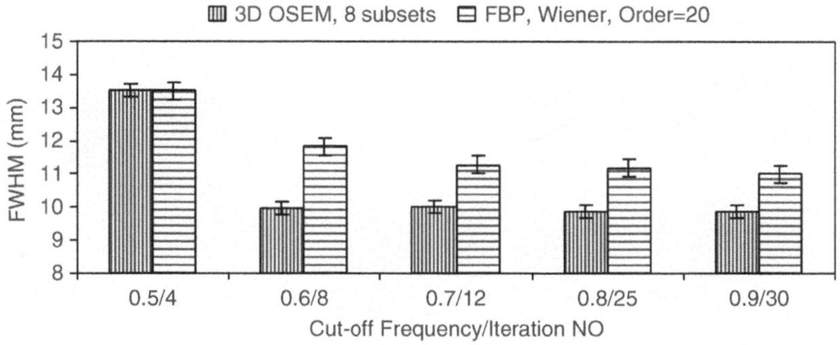

Fig. 17.3 Spatial resolution versus number of iteration in 3D OSEM with eight subsets and cut-off frequency in FBP with Wiener filter with order of 20

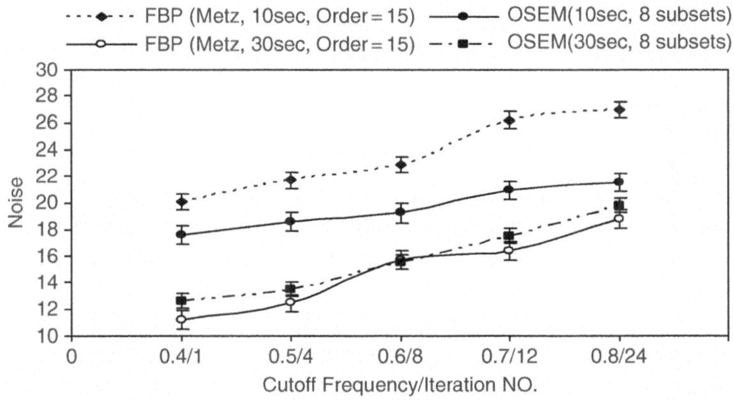

Fig. 17.4 Estimated noise versus number of iteration in 3D OSEM and cut-off frequency in FBP with Metz filter with order of 15 for 10 s per projection angle

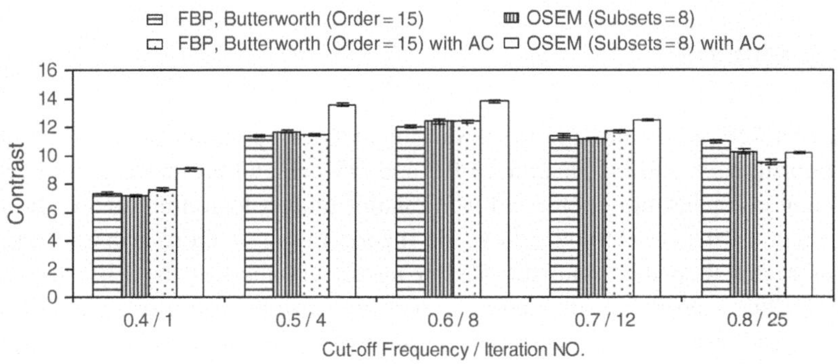

Fig. 17.5 Measured contrast with and without AC versus number of iteration in 3D OSEM and cut-off frequency in FBP with Butterworth filter with order of 15

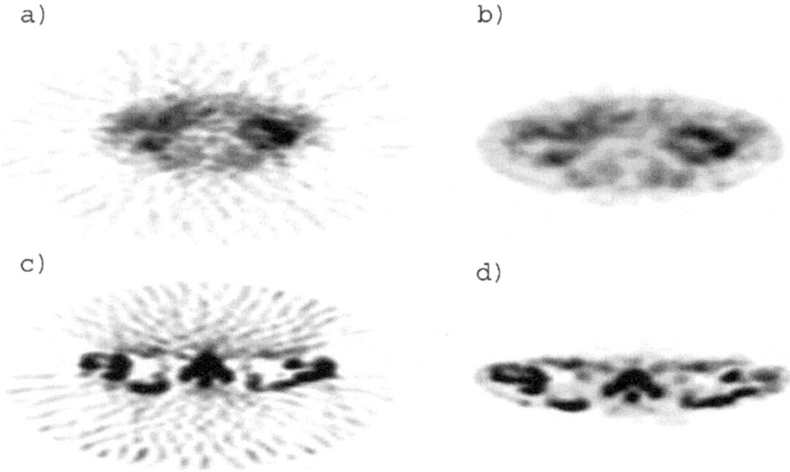

Fig. 17.6 Reconstructed images of bone scan (above): (**a**) using FBP – Wiener filter (Cut-off: 0.7/Order: 15); and (**b**) using 3D OSEM (Subsets: 8/Iterations: 16) and reconstructed images of adrenal scan (bottom); (**c**) using FBP – Metz filter (Cut-off: 0.7/Order: 15); and (**d**) using 3D OSEM (Subsets: 8/Iterations: 12)

ambiguous. It must be acknowledged that Chang AC is only approximate; therefore, the benefit gained by the AC may not be as great as could be gained with a more accurate correction method such as one based on a CT scan.

To demonstrate the performance of 3-D OSEM vs. FBP in clinical studies, Fig. 17.6 shows SPECT bone and adrenal clinical scans reconstructed with 3D OSEM and FBP with an optimised set of parameters. Visual assessment of these images suggests that the superiority of 3D OSEM in terms of spatial resolution and contrast.

17.4 Summary

We have described the basic principles of the two most commonly used reconstruction algorithms in SPECT imaging: FBP and OSEM, and presented a brief review of current compensation methods for the major image-degrading effects. Iterative image reconstruction methods allow the incorporation of more accurate imaging models rather than the simplistic Radon model assumed in the FBP algorithm. These include scatter and attenuation corrections as well as collimator and distance response and more realistic statistical noise models. In our study we have demonstrated the superior performance of 3-D OSEM compared to FBP particularly for low count statistics studies, including improved image contrast and spatial resolution. Using 3-D OSEM with suitable AC may improve lesion detectability due to the significant improvement of image contrast. Indeed, 3-D iterative reconstruction algorithms are likely to replace the FBP technique for most SPECT and PET clinical applications. However, for the full potential of these methods to be fully realised, more exact image compensation methods need to be developed and optimal image reconstruction parameters need to be used. The full impact of these methods on quantitative SPECT imaging is yet to be assessed. Finally, with development of new faster and more accurate 3-D and possibly 4D iterative algorithms, the future of SPECT image reconstruction is certain to be based on iterative techniques rather than any analytical methods.

References

1. S. R. Cherry, J. A. Sorenson and M. E. Phelps, *Physics in Nuclear Medicine* (Saunders, Philadelphia, PA, 2003).
2. H. H. Barrett, Perspectives on SPECT. *SPIE* **671**, 178–183 (1986).
3. K. Kouris, N. M. Spyrou and D. F. Jackson, *Imaging with Ionizing Radiation* (Surrey University Press, London, 1982).
4. G. T. Herman, *Image Reconstruction from Projections: the Fundamental of Computerised Tomography* (Academic, New York, 1980).
5. L. A. Shepp and Y. Vardi, Maximum likelihood reconstruction for emission tomography, *IEEE Transactions on Medical Imaging* **1**(2), 113–122 (1982).
6. H. M. Hudson and R. S. Larkin, Accelerated image reconstruction using ordered subsets of projection data, *IEEE Transactions on Medical Imaging* **13**(4), 601–609 (1994).
7. B. F. Hutton, J. Nuyts and H. Zaidi, *Iterative Reconstruction Methods*, edited by H. Zaidi (Springer, New York, 2005) pp. 107–140.
8. J. Radon, Uber die bestimmung von funktionen durch ihre integral-werte langs gewisser mannigfaltigkeiten, *Ber Verh Sachs Akad Wiss* **67**, 226 (1917).
9. P. Gilbert, Iterative methods for the three-dimensional reconstruction of an object from projections. *Journal of Theoretical Biology* **36**(1), 105–117 (1972).
10. F. J. Beekman, W. A. M. de Jong, and S. van Geloven, Efficient Fully 3-D Iterative SPECT reconstruction with Monte Carlo-based scatter compensation, *IEEE Transactions on Medical Imaging* **21**(8), 867–877 (2002).
11. M. Defrise, D. W. Townsend, R. Clack, Three-dimensional image reconstruction from complete projections, *Physics in Medicine and Biology* **34**(5), 573–587 (1987).
12. D. R. Gilland, R. J. Jaszcak, T. Riauka, R. E. Coleman., Approximate 3D iterative reconstruction for SPECT, *Medical Physics* **24**(9), 1421–1429 (1997).
13. M. W. Tsui and E. C. Frey, *Analytical Image Reconstruction Methods in Emisssion Computed Tomography*, edited by H. Zaidi (Springer, New York, 2005) pp. 82–106.
14. G. L. Zeng, Image reconstruction: a tutorial, *Computerized Medical Imaging and Graphics* **25**(2), 97–103 (2001).
15. W. R. Hendee and E. R. Ritenour, *Medical Imaging Physics* (Wiley, New York, 2002).
16. D. R. Gilland, B. M. W. Tsui, W. H. McCartney, J. R. Perry, and J. Berg, Determination of the optimum filter fuction for SPECT imaging, *Journal of Nuclear Medicine* **29**(5), 643–650 (1988).
17. S. Vandenberghe, Y. D'Asseler, R. Van de Walle, T. Kauppinen, M. Koole, L. Bouwens, K. Van Laere, I. Lemahieu and R. A. Dierckx, Iterative reconstruction algorithms in nuclear medicine, *Computerized Medical Imaging and Graphics* **25**, 105–111 (2001).
18. P. P. Bruyant, Analytic and iterative reconstruction algorithms in SPECT, *Journal of Nuclear Medicine* **43**(10), 1343–1358 (2002).
19. W. Koch, C. Hamann, J. Welsch, G. Pöpperl, P. E. Radau, and K. Tatsch, Is iterative reconstruction an alternative to filtered backprojection in routine processing of dopamine transporter SPECT studies?, *Journal of Nuclear Medicine* **46**(11), 1804–1811 (2005).

20. R. L. Byrne, Editorial recent developments in iterative image reconstruction for PET and SPECT, *IEEE Transactions on Medical Imaging* **19**(4), 257–260 (2000).
21. Flash 3D and e-soft, 2008, Siemens Medical Solutions (September 4, 2008); http://www.medical.siemens.com/siemens
22. Astonish 3-D, 2008, Philips (September 4, 2008); http://www.medical.philips.com
23. HOSEM, 2008, HERMES (September 4, 2008); http://www.hermesmedical.com
24. WBR, 2008, UltraSPECT (September 4, 2008); http://www.ultraspect.com
25. H. Zaidi, Scatter modeling and compensation in emission totmography. *European Journal of Nuclear Medicine and Molecular Imaging* **31**(5), 761–782 (2004).
26. T. Kauppinen, M. O. Koskinen, S. Alenius, E. Vanninen and J. T. Kuikka, Improvement of brain perfusion SPET using iterative reconstruction with scatter and non-uniform attenuation correction, *European Journal of Nuclear Medicine* **27**(9), 1380–1386 (2000).
27. L. T. Chang, A method for attenuation correction in radionuclide computed tomography, *IEEE Transactions on Nuclear Science* **NS-25**, 638–643 (1978).
28. M. G. Erwin, SPECT in the year 2000: basic principles, *Journal of Nuclear Medicine* **28**(4), 233–244 (2000).
29. D. J. Kadrmas, E. C. Frey, and B. M. W. Tsui, Application of reconstruction-based scatter compensation to thallium-201 SPECT: implementations for reduced reconstructed image noise. *IEEE Transactions on Medical Imaging* **17**(3), 325–333 (1998).
30. B. F. Hutton and Y. H. Lau, Application of distance-dependent resolution compensation and post-reconstruction filtering for myocardial SPECT, *Physics in Medicine and Biology* **43**, 1679–1693 (1998).
31. P. Gantet, P. Payoux, P. Celler, P. Majorel, D. Gourion, D. Noll and J. P. Esquerré, Iterative three-dimensional expectation maximization restoration of single photon emission computed tomography images: application in striatal imaging. *Medical Physics* **33**(1), 52–59 (2006).
32. 32. K. Alzimami, S. Sassi, and N. M. Spyrou, Optimization and comparison of 3D-OSEM with FBP SPECT imaging, The 2008 *International Conference of Signal and Image Engineering Proceeding* **I**, London July 2008, 632–636.
33. M. Brambilla, B. Cannillo, M. Dominietto, L. Leva, C. Secco and E. Inglese, Characterization of ordered-subsets expectation maximization with 3D post-reconstruction Gauss filtering and comparison with filtered backprojection in ^{99m}Tc SPECT, *Annals of Nuclear Medicine* **19**(2), 75–82 (2005).
34. T. Yokoi, H. Shinohara and H. Onishi, Performance evaluation of OSEM reconstruction algorithm incorporating three-dimensional distance-dependent resolution compensation for brain SPECT: A simulation study, *Annals of Nuclear Medicine* **16**(1), 11–18 (2002).
35. V. Kohli, M. King, S. J. Glick, T. S. Pan, Comparison of frequency-distance relationship and Gaussian-diffusionbased method of compensation for distance-dependent spatial resolution in SPECT imaging, *Physics in Medicine and Biology* **43**, 1025–1037 (1998).

Chapter 18
Performance Improvement of Wireless MAC Using Non-Cooperative Games

Debarshi Kumar Sanyal, Matangini Chattopadhyay, and Samiran Chattopadhyay

Abstract This chapter illustrates the use of non-cooperative games to optimize the performance of protocols for contention-prone shared medium access in wireless networks. Specifically, it proposes and analyses a set of new utility functions conforming to a recently proposed generic game-theoretic model of contention control in wireless networks [5]. The functions ensure that the game has a unique non-trivial Nash equilibrium. Simulations carried out for IEEE 802.11 style networks indicate that the aggregate throughput of the network at the Medium Access Control (MAC) layer has improved while the collision overhead is reduced.

Keywords Wireless networks · medium access control · game theory · nash equilibrium · throughput

18.1 Introduction

In most wireless networks, including ad-hoc and sensor networks, access to the shared wireless medium is random-access based. In absence of a central regulating authority, it is only natural that multiple wireless nodes will attempt to access the medium simultaneously, resulting in packet collisions. IEEE 802.11 (hereafter briefly called 802.11) is the most commonly followed standard in wireless networks. It defines a distributed mechanism called the distributed coordination function (DCF) to resolve contention among the contending nodes. It involves channel sensing to assess the state of the shared medium and adjusting the channel access probability accordingly to minimize chances of collision. In 802.11, the channel

---

D.K. Sanyal (✉)
Senior Member of Technical Staff, Interra Systems (India) Pvt. Ltd., Salt Lake Electronics Complex, DN-53, Kolkata-700 091, India (West Bengal)
E-mail: debarshi@cal.interrasystems.com

S.I. Ao and L. Gelman (Eds.), *Advances in Electrical Engineering and Computational Science*, Lecture Notes in Electrical Engineering.

access probability is determined by a contention window (CW) maintained by a node. DCF uses a binary feedback signal (collision or no collision) and sharply updates CW using binary exponential backoff to adapt to the contention.

However, various studies [1–4] report that this contention control scheme results in drastic reduction in throughput with increasing number of contending nodes. Every new transmission begins with the same channel access probability and thus the history of contention level in the network is not used. Moreover, doubling the contention window for every failed transmission only reduces the chances of transmission by the nodes that were already unsuccessful.

These inferences indicate that instead of using drastic update strategies, it is better to steadily guide the system to a state that is optimal with respect to the current contention level in the network. Since this state is optimal, it is sustainable and has high throughput, greater fairness and sparing collisions. This requires an analytical framework where these desirable features can be mathematically characterized. Towards this end, we follow [5–8] and model the nodes as selfish players in a non-cooperative game [9]. Unlike common studies [10], [11], we do not reverse-engineer DCF as a game but use game theory as a tool for optimization. We define utility functions reflecting the gain from channel access and the loss from collision. The strategy set of each node is the set of allowable channel access probabilities of the node. We next use gentle update strategies that drive the network to its Nash equilibrium from which no player has an incentive to unilaterally deviate. This characterizes the desired stable operating point or the optimal state of the network. A continuous feedback signal is used to assess the contention level.

The main contribution of this work is to propose new utility functions that allow a large range of channel access probabilities (thus ensuring good performance in a wide spectrum of network sizes) and result in a unique non-trivial Nash equilibrium. Non-triviality and uniqueness ensure that the equilibrium is efficient and leads to high short-term fairness. As we show through extensive simulations, the resulting design is able to provide far better results than DCF used in 802.11. Throughput gets higher, and collision overhead and time slots wasted in collisions are drastically reduced in the new design. The nodes do not need to know the total number of nodes in the network nor is any message passing necessary. A completely distributed mechanism that allows each node to behave selfishly to maximize its own payoff is used to ensure a socially beneficial state. The chapter is organized as follows: Section 18.2 provides a background of this study, journeying through the recent past of game-theoretic models of medium access control in wireless networks. We also point out the differentiation and novelty of our work as regards related studies in the literature. Section 18.3 presents the proposed game model and its many properties. Section 18.4 investigates distributed mechanisms to achieve the equilibrium of the game in real networks. The model is evaluated via simulations in Section 18.5.

After a brief discussion in Section 18.6 conclusions follow in Section 18.7.

18.2 Background

Game theory [9] provides a tool to study situations in which there is a set of rational decision makers who take specific actions that have mutual, possibly conflicting, consequences. A game models an interaction among parties (called the *players*) who are rational decision makers and have possibly conflicting objectives. Each player has a set of actions called its *strategies* and with each strategy is associated a *payoff function*. Rationality demands each player maximize its own payoff function. Non-cooperative games are commonly used to model problems in medium access control in telecommunication networks. In such a game, the solution concept is a notion called a stable set or *Nash equilibrium* that identifies a set of strategies for all the participating players, from which no player has an incentive to unilaterally deviate as any unilateral deviation will not result in an increase in payoff of the deviating player.

In [12] the authors model the nodes in an Aloha network as selfish players in a non-cooperative Aloha game. They assume that the number of backlogged users is always globally known. This limitation is addressed in [13] where only the total number of users is known. More recent research includes modeling the DCF in 802.11 in a game-theoretic framework [11], and reverse engineering exponential backoff as a strategy [10] of non-cooperative games.

In contradistinction to these works, we do not explicitly reverse-engineer 802.11 or DCF into a game model but use ideas from non-cooperative game theory to optimize the performance of 802.11. We consider each node in a wireless network as having a set of strategies which are its channel access probabilities. Unlike common practices in optimization of 802.11 like [4], we do not assume that each node knows the number of users in the network. This is more reflective of practical situations where such knowledge is difficult to acquire. Non game-theoretic approaches that do not depend on the nodes' knowing the network size include the ones presented in [2] and [3]. We use selfishness to achieve optimized system-wide performance. In [5, 6, 8] the authors use game theory to study and optimize 802.11. Although they remain our motivation, their illustrative utility functions are different from ours. Their utility functions are well-behaved in a much more restricted domain than ours. Moreover, as we show, our utility functions give far better results for large network sizes. Supermodular games for contention control are presented in [7].

18.3 Game-Theoretic Model of Medium Access Control

The system we consider is a network of N nodes that are all able to hear one another. To facilitate analysis we will adopt a description of our access mechanism in terms of channel access probabilities. It has been shown in [1] that in a saturated regime the constant channel access probability p relates to the corresponding contention window cw according to

$$p = \frac{2}{cw + 1} \tag{18.1}$$

Now we turn to the game model. We define the game G as a 3-tuple $G =< N, \{S_i\}_{i \in N}, \{u_i\}_{i \in N} >$, where N is a set of players (wireless nodes), player $i \in N$, each player having a strategy set $S_i = \{p_i | p_i \in [v_i, w_i]\}$ with $0 < v_i < w_i < 1$ and payoff u_i. The strategy set of each player is the set of its channel access probabilities. Note that it is a generalization to a continuous space of the simpler strategy set {wait, transmit} denoting the two deterministic actions that a node can perform. Denote the strategy profile of all nodes by $\mathbf{p} = (p_1, p_2, \ldots, p_{|N|})$. The payoff is naturally of the form: $u_i = U_i(p_i) - p_i q_i(\mathbf{p})$. Here $U_i(p_i)$ is the utility function denoting the gain from channel access and $q_i(\mathbf{p})$ is the conditional channel probabilty given by

$$q_i(\mathbf{p}) = 1 - \prod_{j \in N - \{i\}} (1 - p_j) \tag{18.2}$$

Thus $p_i q_i(\mathbf{p})$ is the cost of channel access.
We propose a novel utility function:

$$U_i(p_i) = \frac{p_i(\ln w_i - \ln p_i + 1)}{\ln r_i} \tag{18.3}$$

where p_i = channel access probability $(0 < v_i \leq p_i \leq w_i < 1)$, $r_i = w_i / v_i > 1$. In non-cooperative games, the final outcome is described by investigating its Nash equilibrium (NE). The strategy profile of all nodes but i is denoted by $\mathbf{p}_{-i} = (p_1, p_2, \ldots, p_{i-1}, p_{i+1}, \ldots, p_{|N|})$. Then the NE is defined as the strategy profile $\mathbf{p}^* = (p_1^*, p_2^*, \ldots, p_{|N|}^*)$ with the property that for every player i and every strategy p_i of player i, $u_i(p_i^*, \mathbf{p}_{-i}^*) \geq u_i(p_i, \mathbf{p}_{-i}^*)$. So no player gains by unilaterally changing its strategy.

Note in passing that our utility function does not correspond to a physical quantity like throughput, delay, or the like but is a mathematical expression formulated to satisfy the desirable property of unique non-trivial NE of the game. The rest of the paper proves these characteristics and shows that it leads to performance far superior to DCF in contention control.

Theorem 18.1. *The game G has an NE.*

Proof. The strategy set $S_i = [v_i, w_i]$ $(0 < v_i < w_i < 1)$ of each player is a nonempty, convex and compact subset of the Euclidean space. Further u_i is continuous. Since

$$\frac{\partial^2 u_i}{\partial p_i^2} = \frac{-1}{p_i \ln r_i} < 0$$

and $p_i \in S_i$, u_i is quasi-concave in p_i. Hence the game has an NE.
An NE is non-trivial if the following condition holds for all nodes $i \in N$:

$$U_i'(p_i^*) = q_i(\mathbf{p}^*) \tag{18.4}$$

and trivial otherwise.

Theorem 18.2. *The game G has a non-trivial NE.*

Proof. Let us define $f_i(\mathbf{p})$ as:

$$f_i(\mathbf{p}) = (U_i^{'})^{-1}(q_i(\mathbf{p})) \tag{18.5}$$

Since

$$U_i^{'}(p_i) = \frac{\ln(w_i/p_i)}{\ln r_i} \tag{18.6}$$

is one-to-one in S_i, the inverse function

$$(U_i^{'})^{-1}(q_i) = w_i r_i^{-q_i}$$

exists, which is continuous and decreasing in q_i. Now observe, $(U_i^{'})^{-1}(0) = w_i$ and $(U_i^{'})^{-1}(1) = v_i$. Thus, $f_i(\mathbf{p})$ maps any $\mathbf{p} \in S_1 \times S_2 \times \cdots \times S_{|N|}$ into a point $p_i \in [v_i, w_i]$. Hence the vector function $f(\mathbf{p}) = (f_1(\mathbf{p}), f_2(\mathbf{p}), \ldots, f_{|N|}(\mathbf{p}))$ maps the set $S_1 \times S_2 \times \cdots \times S_{|N|}$ into itself. Hence, by Brouwer's fixed point theorem, there is a fixed point of $f(\mathbf{p})$ in $\times_{i \in N} S_i$, that is, $f_i(\mathbf{p}^*) = p_i^*$ for each $i \in N$. It immediately follows from the definition of $f_i(\mathbf{p})$ in (18.5) above that at this fixed point, the condition (18.4) for non-trivial NE holds. In other words, G has a non-trivial NE.

The existence of non-trivial NE signifies that the channel access probability of any node in NE is not at the boundary of strategy space. This reduces unfair sharing of the wireless channel among the contending nodes.

Theorem 18.3. *The NE is unique for all p_i in $[v_i, w_i]$ for all $i \in N$ if $v_i > \frac{w_i}{e^{1/w_i - 1}}$.*

Proof. First consider the function:

$$\varphi_i(p_i) = (1 - p_i)(1 - U_i^{'}(p_i))$$

Putting the value of $U_i(p_i)$ and differentiating with respect to p_i,

$$\varphi_i^{'}(p_i) = \frac{1}{\ln r_i}(\frac{1}{p_i} - 1 - \ln(\frac{p_i}{v_i}))$$

If $v_i > \frac{p_i}{e^{\frac{1}{p_i}-1}}$, then, $\ln v_i + (\frac{1}{p_i} - 1) > \ln p_i$, which means $\varphi_i^{'}(p_i) > 0$. Since $\frac{p_i}{e^{\frac{1}{p_i}-1}}$ is strictly increasing in p_i, if $v_i > \frac{w_i}{e^{1/w_i - 1}}$, we have $\varphi_i^{'}(p_i) > 0$. For the rest of the proof, assume that the above relation between v_i and w_i holds. Recollect from condition (18.4) that at non-trivial NE,

$$U_i^{'}(p_i^*) = q_i(\mathbf{p}^*) = 1 - \prod_{j \in N - \{i\}} (1 - p_j^*)$$

making $\varphi_i(p_i)$ identical for all nodes i. Now assume for the sake of contradiction, that there exist at least two distinct non-trivial NE $\mathbf{x}^*$ and $\mathbf{y}^*$ of the game G. Let

$\varphi_i(x_i^*) = \sigma_1$ and $\varphi_i(y_i^*) = \sigma_2$ for some node i. Since φ_i is strictly increasing, $\sigma_1 \neq \sigma_2$. Assume without loss in generality, $\sigma_1 > \sigma_2$. Then $x_i^* > y_i^*$ for all $i \in N$. From definition of $q_i(\mathbf{p})$, since $q_i(\mathbf{p})$ is increasing in $\mathbf{p}$, $U_i^{'}(x_i^*) = q_i(\mathbf{x}^*) > q_i(\mathbf{y}^*) = U_i^{'}(y_i^*)$. But $U_i^{''}(p_i) = -1/(p_i \ln r_i)$ showing that $U_i^{'}(p_i)$ is decreasing, thus a contradiction. Hence $\mathbf{x}^* = \mathbf{y}^*$ proving the uniqueness of the NE.

Theorem 18.4. *Suppose all the players in G have identical strategy sets. Then at the unique non-trivial NE of G, all the players choose the same strategies.*

Proof. Since all players have the same strategy sets, for any two players i and j, $w_i = w_j$ and $v_i = v_j$. Now suppose that the unique non-trivial NE is $\mathbf{p}^*$. Let there exist p_i^* and p_j^* in $\mathbf{p}^*$ corresponding to nodes i and j such that $p_i^* \neq p_j^*$. Denote the set of strategies in some order of all nodes except i and j as $\mathbf{p}_{-ij}^*$ so that the NE can be described as an ordered set $(\mathbf{p}_{-ij}^*, p_i^*, p_j^*)$. By symmetry, every permutation of the NE must be an NE. Interchange the strategies of nodes i and j to get a new NE $(\mathbf{p}_{-ij}^*, p_j^*, p_i^*)$. But this contradicts the uniqueness of the NE, violating Theorem 18.3. Hence, $p_i^* = p_j^*$. i and j being arbitrary, it follows that $p_i^* = p_j^*$ for all $i, j \in N$.

This has a very important and useful consequence: the wireless channel is accessed by the nodes with identical access probabilities at the equilibrium point thus resulting in fair sharing of the resource among the contenders, and consequently short-term fairness.

18.4 Distributed Mechanisms to Achieve Nash Equilibrium

The non-cooperative game G is played repeatedly by the nodes, that is, it is a multi-stage game. A stage may be a single transmission or a sequence of K transmissions for a fixed $K > 1$.

Suppose that each node after playing a round observes the cumulative effect (in the sense of conditional collisional probability) of the actions of all players in the previous round. This knowledge may now be used by the node to select its strategy for the next round. Based on how this knowledge is used, there are many common techniques to push the system to its Nash equilibrium. Two more popular ones are:

- Best response: This is the most obvious mechanism where at each stage every player chooses the strategy that maximizes its payoff given the actions of all other players in the previous round:

$$p_i(t+1) = \arg \max_{p_i(t) \in [v_i, w_i]} (U_i(p_i(t)) - p_i(t) q_i(\mathbf{p}(t)))$$

  for each node $i \in N$.
- Gradient play: In this case, each player adjusts its persistent probability gradually in a gradient direction suggested by the observations of the effect of other

players' actions. Mathematically,

$$p_i(t+1) = p_i(t) + f_i(p_i(t))(U_i'(p_i(t) - q_i(\mathbf{p}(t)))$$

for each node $i \in N$ where step size $f_i(.) > 0$ can be a function of the strategy of player i. It has an intuitively appealing interpretation: if at any stage of the game, the marginal utility $U_i'(p_i(t))$ exceeds the "price of contention" $q_i(p(t))$, the persistence probability is increased, but if the price is greater, the persistence probability is reduced.

To eliminate the need to know the current strategies of all other players, a node needs to estimate $q_i(\mathbf{p})$ in an indirect way. A simple method is to track the number of packet transmissions (nt) and the number of packet losses (nl) over a period of time and calculate $q_i = nl/nt$. A more sophisticated approach [3] involves observing the number of idle slots which follows a geometric distribution with mean $I = p^{idle}/(1 - p^{idle})$ where p^{idle} = probability of a slot being idle = $\prod_{i \in N}(1 - p_i)$. Thus by observing I, q_i can be estimated as $q_i = 1 - (I/(I+1))/(1 - p_i)$ allowing a completely distributed update mechanism. See [5] for a thorough discussion of the convergence of these procedures.

18.5 Performance Evaluation

In this section, we report results from numerical simulations performed to analyze the model with our utility functions. The simulation results are obtained in a saturation regime. The performance is compared with the standard DCF in basic access 802.11b. We choose to measure the channel access probabilities, aggregate throughput of all nodes, collision overhead incurred and the average number of time slots wasted in collisions per successful transmission.
The saturation throughput S_i of a node i in an 802.11 network is given by the following expression in Bianchi's model [1]:

$$S_i = \frac{p_i^{suc} p^{trans} \wp}{p^{idle}\sigma + p^{trans} p_i^{suc}\tau + p^{col}\tau^{col}}$$

where $\wp$ = packet payload (assumed constant), p^{trans} = probability of a transmission by any node = $1 - \prod_{i \in N}(1 - p_i)$, p_i^{suc} = conditional success probability of node i = probability of a successful transmission by the node given that there is a transmission = $\frac{p_i \prod_{j \in N-\{i\}}(1-p_j)}{p_{tr}}$, p^{idle} = probability of a slot being idle = $\prod_{i \in N}(1 - p_i)$, and finally $p^{col} = 1 - p^{idle} - \sum_{i \in N} p_i^{suc}$.
τ is the slot time, τ^{suc} is the average duration of a slot in which there is a successful transmission occurs and τ^{col} is the average duration of a slot in which a collision occurs.

In our experiments, the slot time is 20 µs, SIFS = 10 µs, DIFS = 20 µs, basic rate = 1 Mbps, data rate = 1 Mbps, data rate = 11 Mbps, propagation delay = 1 µs,

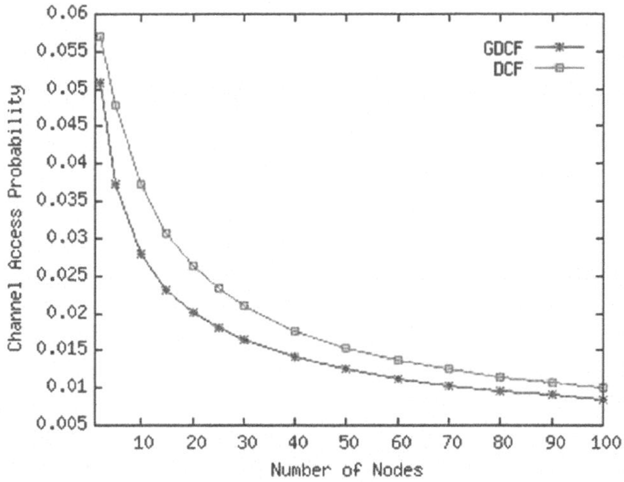

Fig. 18.1 Network size vs. channel access probabilities

PHY header = 24 bytes, MAC header = 34 bytes, ACK = 14 bytes and the packet payload is fixed at 1,500 bytes. We assume all nodes use the same parameters for the utility function. To compare with DCF, we set bounds of the contention window to powers of 2 (specifically 32 and 1,024) and derive the bounds of the strategy sets using (18.1). This gives $w_i = 2/33 = 0.06061$ and $v_i = 2/1025 = 0.00195$. Note that in this range, the condition of Theorem 18.3 is satisfied since $v_i > \frac{w_i}{e^{1/w_i - 1}} \approx 1 \times 10^{-8}$. In the following sub-sections, we call our game-theoretic distributed coordination function GDCF.

18.5.1 Channel Access Probabilities

Figure 18.1 shows the variation of the probability with which a node accesses the wireless channel in the saturation regime in GDCF and DCF as the network size increases. In both protocols, the domain of the probability is the set [0.06061, 0.00195] since contention window varies from 32 to 1,024. Recollect from Section 18.3 that the probability with which any given node in GDCF accesses the channel at the NE is identical for all nodes. We observe that the channel access probabilities are always observed to be higher in DCF than in GDCF. This has an important consequence as will be evident in the ensuing discussion.

18.5.2 Aggregate Throughput

The aggregate throughput (see Fig. 18.2) which measures the total throughput of all nodes under saturation conditions is higher in DCF when the network size is small but it gets far lower than GDCF as the network size grows. This is because the channel access probabilities are higher in DCF. In a small network, collisions are few and

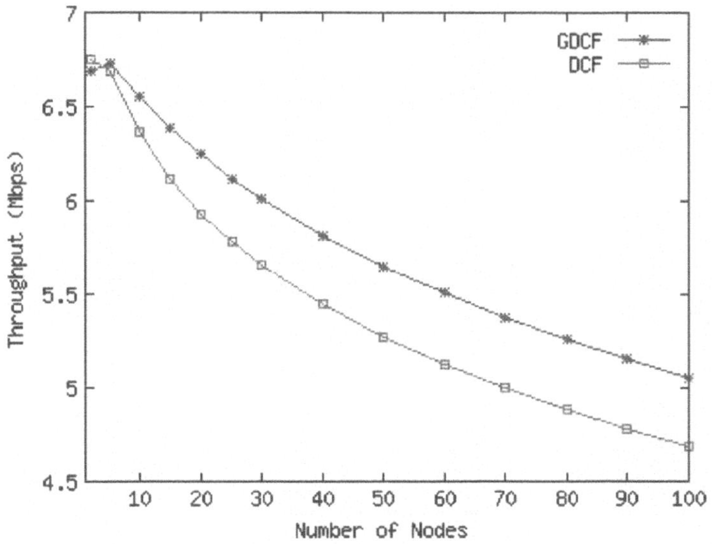

Fig. 18.2 Variation of aggregate throughput with network size

so higher channel access probabilities translate to higher throughput. But as network size grows, collisions increase if all nodes aggressively access the channel, triggering a drastic reduction in throughput. Since channel access probabilities are lower in GDCF, it supersedes DCF as network size grows. Consequently, large beds of wireless nodes achieve much greater throughput if they run GDCF instead of DCF.

18.5.3 Collision Overhead

We observe from Fig. 18.3 that the conditional collision probability is higher in case of DCF than in GDCF. This is expected as the channel access probabilities are lower in case of GDCF resulting in reduced medium contention even when the number of nodes increases. Indeed GDCF is designed with the aim of keeping the *price of contention* low.

18.5.4 Slots Wasted in Collisions

In 802.11, a large number of slots are wasted due to collisions. This seriously affects the performance and lifetime of the power-constrained devices in ad-hoc and sensor networks. Figure 18.4 makes clear that the number of slots wasted in collisions for every successful transmission is much higher in DCF than in GDCF, making the later attractive in these networks.

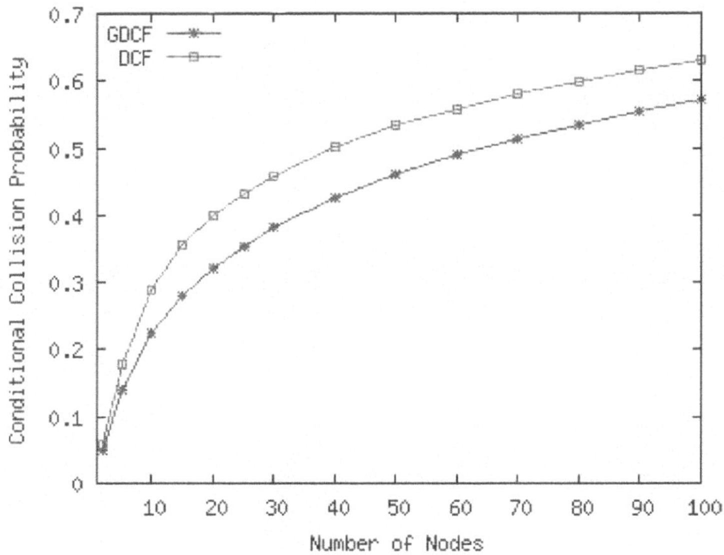

Fig. 18.3 Variation of conditional collision probability with network size

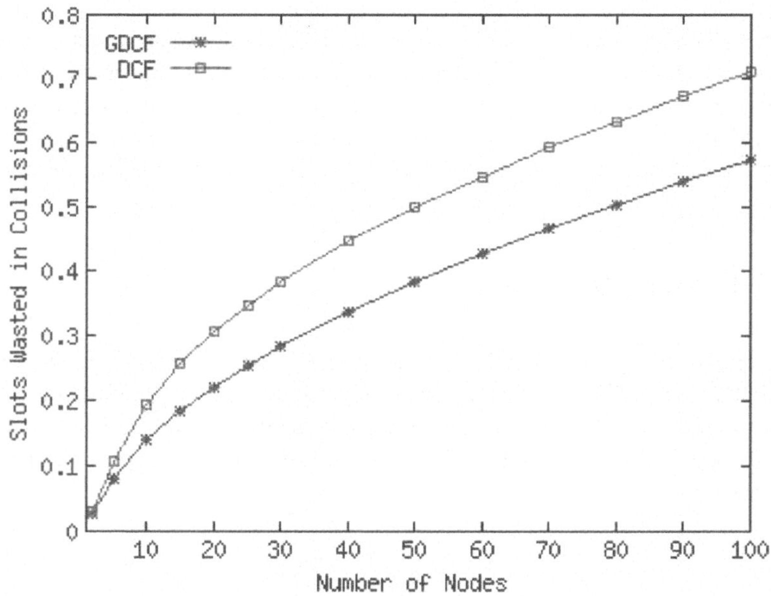

Fig. 18.4 Number of slots wasted in collisions per successful transmission

Table 18.1 Brief comparison of utility functions

Number of nodes	Throughput		Conditional collision probability	
	GDCF	U_Chen	GDCF	U_Chen
80	5.26	5.13	0.48	0.56
90	5.15	4.93	0.55	0.59
100	5.06	4.73	0.57	0.63

18.6 Discussion

The motivation for this work is [5]. So we make a brief comparison with it in this section. Our utility function produces unique non-trivial NE for a wide range of v_i and w_i while the one in [5] has more strict constraints on v_i once w_i is specified. This makes our utility functions admit a much larger game space and hence is superior for widely varying network sizes since the channel access probability should be very small for very large network sizes and high for small network sizes. Table 18.1 presents a performance comparison as regards throughput and collision for large networks. Here, as in our experiments, we vary contention windows from 32 to 1,024 as is the default for DSSS PHY in 802.11 standard while in [5] (call it U_Chen) the window ceiling is limited to 256 if the base is 32 and only powers of 2 are chosen. As expected, with these settings, our utility function exhibits particularly better performance for large network sizes.

18.7 Conclusion

This chapter describes an elegant game-theoretic model aimed at optimizing the aggregate throughput and distributing it fairly among the contending nodes in an 802.11 network, thus inducing a desirable state of the entire network. It also drastically brings down the collision overhead, reducing energy loss due to failed transmissions. We believe it contends as a serious and superior alternative to the traditional distributed coordination function in 802.11 MAC. The game-theoretic model uses selfish behavior of the participating players to steer the network to desirable system-wide characteristics. Our utility functions assure the existence of unique, non-trivial Nash equilibrium where all nodes possess identical channel access probabilities. This results in fairly distributed high throughput with low collision overhead. Rigorous analysis and simulations are used to prove the ideas. In future we intend to simulate the algorithm in settings with bit errors to gather more performance measures. Further, we intend to investigate other utility functions with similar characteristics and explore if this design can be used in real networks. We also intend to analyse the more generic features of the model.

References

1. Bianchi, G. (2000). "Performance analysis of the IEEE 802.11 distributed coordination function", *IEEE Journal on Selected Areas in Communication*, 18(3):535–547.
2. Cali, F., M. Conti, and E. Gregori. (2000). "Dynamic IEEE 802.11: design, modeling and performance evaluation", *IEEE Journal on Selected Areas in Communication*, 18(9): 1774–1786.
3. Heusse, M., F. Rousseau, R. Guillier, and A. Dula. (Aug 2005). "Idle sense: an optimal access method for high throughput and fairness in rate-diverse wireless LANs", *Proceedings of ACM Sigcomm*.
4. Toledo, A. L., T. Vercauteren, X. Wang. (Sep, 2006). "Adaptive optimization of IEEE 802.11 DCF based on Bayesian estimation of the number of competing terminals", *IEEE Transactions on Mobile Computing*, 5(9):1283–1296.
5. Chen, L., S. H. Low, and J. C. Doyle. (2006). "Random access game and medium access control design", *Caltech CDS, Technical Report*.
6. Chen, L., S. H. Low, and J. C. Doyle. (2007). "Contention Control: A game-theoretic approach", *Proceedings of IEEE Conference on Decision and Control*.
7. Chen, L., T. Cui, S. H. Low, and J. C. Doyle. (2007). "A game-theoretic model for medium access control", *Proceedings of WICON*.
8. Cui, T., L. Chen, S. H. Low, and J. C. Doyle. (2007). "Game-theoretic framework for medium access control", *Caltech CDS, Technical Report*.
9. Fudenburg, D., and J. Tirole. (1991). "Game Theory", MIT, Cambridge, MA.
10. Tang, A., J. W. Lee, J. Huang, M. Chiang, and A. R. Calderbank. (2006). "Reverse engineering MAC", *Proceedings of WiOpt*.
11. Xiao, Y., X. Shan, and Y. Ren. (Mar 2005). "Game theory models for IEEE 802.11 DCF in wireless ad-hoc networks", *IEEE Communications Magazine*, 43(3):22–26.
12. MacKenzie, A. B., and S. B. Wicker. (Oct, 2001). "Selfish users in aloha: a game-theoretic approach", *Proceedings of IEEE Vehicular Technology Conference*.
13. Altman, E., R. E. Azouzi, and T. Jimenez. (2002). "Slotted Aloha as a stochastic game with partial information", *Proceedings of WiOpt*.

Chapter 19
Performance Evaluation of Mobile Ad Hoc Networking Protocols

Nadia Qasim, Fatin Said, and Hamid Aghvami

Abstract In this paper performance evaluations of mobile ad hoc network's protocols are made with its quality of service factors. It is seen that mobile ad hoc networks will be an integral part of next generation networks because of its flexibility, infrastructure less nature, ease of maintenance, auto configuration, self administration capabilities, and cost effectiveness. This research paper shows comparative evaluation within mobile ad hoc networks' routing protocols from reactive, proactive and hybrid categories. We have comprehensively analyzed the results of simulation for mobile ad hoc routing protocols over the performance metrics of packet delivery ratio, end to end delay, media access delay and throughput for optimized link state routing, temporary ordered routing algorithm and ad hoc on demand distance vector protocol. In mobile ad hoc networks, mobile nodes must collaborate with each other in order to interconnect, organize the dynamic topology as mobility cause route change and establish communication over wireless links. The simulation results showed the lead of proactive over reactive and hybrid protocols in routing traffic for dynamic changing topology. Proactive protocol, optimized link state routing, a protocol for building link tables for ad-hoc networks, can transmit traffic more rapidly though involve less processing speed in packet forwarding.

Keywords mobile ad hoc network · Performance Evaluation · routing protocol · Proactive protocol · packet forwarding

19.1 Introduction

Mobile ad hoc networks (MANETs) are rapidly evolving as an important area of mobile communication. MANETs are infrastructure less and wireless. It has several

N. Qasim (✉)
The Centre for Telecommunications Research, King's College London, Strand, WC2R 2LS, London, UK,
E-mail: nadia.qasim@kcl.ac.uk

S.I. Ao and L. Gelman (Eds.), *Advances in Electrical Engineering and Computational Science*, Lecture Notes in Electrical Engineering.

Fig. 19.1 Mobile ad hoc networks

routers, which are free to move arbitrarily and perform management. MANETs as shown in Fig. 19.1. It has characteristics that network topology changes very rapidly and unpredictably where many mobile nodes (MNs) can move to and from a wireless network without any fixed access point, consequently routers and hosts move, hence topology is dynamic.

MANETs has to support multi hop paths for mobile nodes to communicate with each other and can have multiple hops over wireless links; therefore its' connection point to the internet may also change. If mobile nodes are within the communication range of each other than source node can send message to the destination node otherwise it can send through intermediate node. Now-a-days MANETs have robust and efficient operation in mobile wireless networks as it can include routing functionality into MNs which is more than just mobile hosts and reduces the routing overhead and saves energy for other nodes. Hence, MANETs are very useful when infrastructure is not available [1], unfeasible, or expensive because it can be rapidly deployable, without prior planning. Mostly mobile ad hoc networks are used in military communication by soldiers, automated battlefields, emergency management teams to rescue [1], search by police or fire fighters, replacement of fixed infrastructure in case of earthquake, floods, fire etc., quicker access to patient's data from hospital database about record, status, diagnosis during emergency situations, remote sensors for weather, sports stadiums, mobile offices, electronic payments from anywhere, voting systems [2], vehicular computing, education systems with set-up of virtual classrooms, conference meetings, peer to peer file sharing systems [2].

Major challenges in mobile ad hoc networks are routing of packets with frequently MNs movement, there are resource issues like power and storage and there are also wireless communication issues. As mobile ad hoc network consists of wireless hosts that may move often. Movement of hosts results in a change in routes. In this paper we have used routing protocols from reactive, proactive and hybrid categories to make evaluation.

19.2 Mobile Ad Hoc Network's Routing Protocols

Mobile ad hoc network's routing protocols are characteristically subdivided into three main categories. These are proactive routing protocols, reactive on-demand routing protocols and hybrid routing protocols. Each category has many protocols and some of these protocols are shown in Fig. 19.2.

Proactive routing protocols maintain regular and up to date routing information about each node in the network by propagating route updation at fixed time intervals throughout the network, when there is a change in network topology. As the routing information is usually maintained in tables, so these protocols are also called *table-driven* protocols i.e. ad hoc on demand distance vector protocol (AODV), dynamic source routing (DSR), admission control enabled on-demand routing (ACOR) and associativity based routing (ABR). Reactive routing protocols establish the route to a destination only when there is a demand for it, so these protocols are also called *on demand* protocols i.e., destination sequenced distance vector (DSDV), optimized link state routing (OLSR), wireless routing protocol (WRP) and cluster head gateway switch routing (CGSR). When a source wants to send to a destination, it uses the route discovery mechanisms to find the path to the destinations by initiating route request. When a route has been established, then route remains valid till the destination is reachable or when the route is expired. Hybrid routing protocols is the combination of both proactive and reactive routing protocols i.e. temporary ordered routing algorithm (TORA), zone routing protocol (ZRP), hazy sighted link state (HSLS) and orderone routing protocol (OOPR). Proactive and reactive algorithms are used to route packets. The route is established with proactive routes and uses reactive flooding for new MNs. In this paper, we have compared MANETs routing protocols from reactive, proactive and hybrid categories, as we have used randomly one protocol from each categories as from reactive AODV, proactive OLSR, hybrid TORA.

Ad hoc on demand distance vector protocol is reactive protocol and construct route on demand and aims to reduce routing load [3]. It uses a table driven routing framework and destination sequence numbers for routing packets to destination mobile node (MN) and has location independent algorithm. It sends messages only

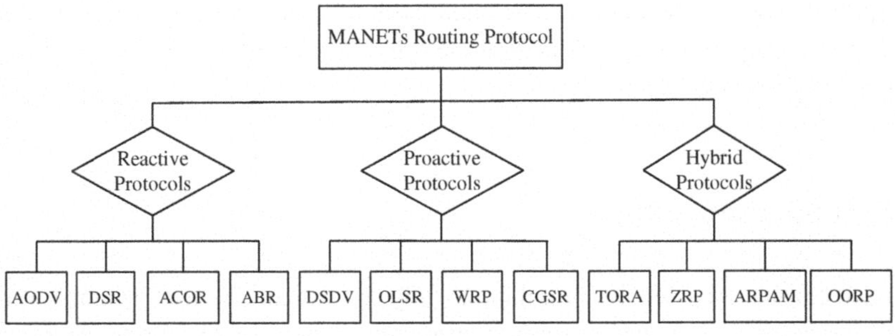

Fig. 19.2 Mobile ad hoc network's routing protocols

when it was demanded and has bi-directional route from the source to destination. When it has packets to send from source to destinations MN then it floods the network with route request (RREQ) packets. All MNs that receive the RREQ checks its routing table to find out that if it is the destination node or if it has fresh route to the destination then it unicast route reply (RREP) which is routed back on a temporary reverse route generated by RREQ from source node, or else it re-broadcast RREQ.

Optimized link state routing is a proactive routing protocol [4]. In which each node periodically broadcasts its routing table allowing each node to build a global view of the network topology. The periodic nature of the protocol creates a large amount of overhead. In order to reduce overhead it limits the number of MN that can forward network wide traffic and for this purpose it uses *multi point relays* (MPRs) which is responsible for forwarding routing messages and optimization for controlled flooding and operations. Each node independently elects a group of MPRs from its one hop neighbors. All MNs periodically broadcast a list of its MPR selectors instead of the whole list of neighbors. MPRs are also used to form a route from MN to destination node and perform route calculation. All MNs maintain the routing table that contains routes to all reachable destination nodes. OLSR does not notify the source immediately after detecting a broken link.

Temporary ordered routing algorithm is hybrid protocol, which is distributed and routers only maintain information about adjacent routers [5]. During reactive operation, sources initiate the establishment of routes to a given destination on demand. Where in dynamic networks, it is efficient with relatively sparse traffic patterns; as it does not have to maintain routes at all the time. It does not continuously execute a shortest path computation and the metric used to establish the routing structure does not represent a distance. TORA maintains multiple routes to the destination when topology changes frequently. It consists of link reversal of the *Directed Acyclic Graph* (ACG). It uses internet MANET encapsulation protocol (IMEP) for link status and neighbor connectivity sensing. IMEP provide reliable, in-order delivery of all routing control messages from a node to all of its neighbors, and notification to the routing protocol whenever a link neighbors is created or broken. As TORA is for multihop networks which is considered to minimize the communication overhead associated with adapting to network topological changes by localization of algorithmic reaction. Moreover, it is bandwidth efficient and highly adaptive and quick in route repair during link failure and providing multiple routes to destination node in wireless networks.

19.3 Simulation Setup

We have conducted extensive simulation study to evaluate the performance of different mobile ad hoc networks routing protocols reactive AODV, proactive OLSR, hybrid TORA. We have used OPNET 14.0 simulator to carry out simulation study

[6], which is used for network modeling and simulation results as it has fastest event simulation engine. In our *Mobility Model, MNs* in the simulation area move according to random waypoint model [1]. *Radio Network Interface* used are the physical radio characteristics of each mobile node's network interface, such as the antenna gain, transmit power, and receiver sensitivity, were chosen to approximate the direct sequence spread spectrum radio [2]. *Media Access Control* is the distribution coordination function (DCF) of IEEE 802.1 1b, which was used for underlying MAC layer [2]. Default values are used for MAC layer parameters. For *Network Traffic,* in order to compare simulation results for performance of each routing protocol, communication model used for network traffic sources is FTP. For *Traffic Configuration*, all experiments have one data flow between a source node to a sink node consisting of TCP file transfer session and TCP transmits with the highest achievable rate. TCP is used to study the effect of congestion control and reliable delivery [7].

19.4 Simulation Environment

It consists of 50 wireless nodes which were placed uniformly and forming mobile ad hoc network, moving about over a $1,000 \times 1,000$ m area for 900 s of simulation time [8].

The AODV simulation parameters used are the same as in [8] except the active route timeout which was set to 30 s, the TORA parameters we used are similar to those in [9]; moreover, the OLSR parameters we used are similar to those in [8]. In OLSR's Hello interval and TC interval were set to 2 and 5 s respectively, its neighbor hold time was 6 s, and entries in topology table expire after 15 s. In OLSR scenario the MNs in the network are grouped in clusters and each cluster has MPR. The transmission range is 300 m. The MPRs are selected in each cluster, which are the MNs which have high willingness parameter. We have five clusters each having its own five MPRs that move towards stable state. MPR of MNs in the network sends topology control messages periodically. The numbers of MPR in network are directly proportional to the number of TC traffic sent. Each MN sends periodically Hello messages in the network that consists of list of neighbors and node movement's changes [6] (Tables 19.1–19.3).

Table 19.1 Constants used in the AODV simulation

Route discovery parameter	Gratuitous reply
Active route timeout	30s
Hello interval	Uniform (10,10.1)
Allowed hello loss	10
TTL parameter	2s

Table 19.2 Constant of IMEP used in the TORA simulation

Beacon periods	3s
Max beacon timer	9s
Max tries	Three attempts

Table 19.3 Constants used in the OLSR simulation

Hello interval	2s
TC interval	5s
Neighbor hold time	6s
Topology table entries expire time	15s

In our simulation study, performance evaluations are made using following parameters:

- *Throughput* is the total number of packets received by the destination.
- *End to End Delay* is the average end to end delay of data packets from senders to receivers.
- *Media Access Delay* is the media transfer delay for multimedia and real time traffics' data packets from senders to receivers.
- *Packet delivery ratio* (PDR) is ratio between the number of packets received by the TCP sink at the final destination and number of packets generated by the traffic sources. Moreover, it is the ratio of the number of data packets received by the destination node to the number of data packets sent by the source MN [10].

There have been previous papers [2, 7–9, 11] to provide a comparative analysis of routing protocols for ad hoc networks, although those simulations used substantially different input parameters than ours. Yet, there has been no comprehensive evaluation study done to compare the performance based on categories of routing protocols, which are reactive, proactive and hybrid routing protocols. Specifically, the total simulation time was 900 s over which the performance statistics are collected. Another important difference between our study and previous studies was that we aim to evaluate the varying state behavior of routing protocols from different categories.

19.5 Simulation Results

When the mobile ad hoc network simulations were run than result shows that all MNs were in range of each other, no data packets experience collisions in presence of ftp traffic load and all MNs were capable of sending packets. Hence, it shows that carrier sense and back off mechanisms of the 802.11b were working precisely. All results were obtained by averaging over ten random mobility scenarios of mobile ad hoc networks.

The most of events were simulated by OLSR which are 143,571,00. Consequently, average number of events simulated by TORA and AODV are 199,354,5 and 229,537 respectively. On the other hand, high simulation speed for most of events simulated per seconds was observed in TORA routing protocol simulation runs that was 544,829 events per second, than it was in AODV and OLSR for about 398,557 and 232,943 events per seconds. These statistics shows that proactive protocol can simulate millions of more event than reactive and hybrid protocols.

19.5.1 Throughput

Throughput which is the number of routing packets received successfully by each routing protocol was shown in Fig. 19.3a. When comparing the routing throughput packets received by each of the protocols, OLSR has the high throughput. Throughput is a measure of effectiveness of a routing protocol. OLSR receives about 1,950,000 routing packets at start of simulation time, then fluctuates for 60 s and gradually becomes stable around 1,600,200 data packets received. In Fig. 19.3b AODV and TORA are plotted on the different scales to best show the effects of varying throughput. TORA's throughput increases IMEP's neighbor discovery mechanism, which requires each node to transmit at least one Hello packet per BEACON period (3 s). For 900 s of simulations with 50 MNs, this results in a maximum throughput of 1, 4,500 packets. In reactive protocol AODV as the number of sources nodes increases than the number of routing packets receive increases to 8,000 packets as it maintains cache of routes in routing table to destination and unicast reply by reversing route generated by source node or re broadcast route request. Delivery of broadcast packets are not reliable at receiver as there cannot be reservation for the wireless medium at the receivers before transmitting a broadcast packet by exchange of request to send or clear to send (RTS/CTS) packets. The source nodes generate packets and broadcast packets which are received by many MNs, so number of packets received is much higher than the number of packets sent. This difference does not exist in wired networks and shows fundamental limitation of wireless networks. Overall, proactive routing protocol has highest throughput in MANETs.

19.5.2 End to End Delays

Figure 19.3c shows that OLSR has lowest steady end to end delays which are about 0.0004 s. Further on, the end to end delay start to rise and fall abruptly in AODV and TORA therefore ends up in less end to end delays in AODV as compare to TORA that is around on average 0.0015 s and 0.0032 respectively. TORA had higher delays because of network congestion. As created loop where the number of routing packets sent caused MAC layer collisions, and data, Hello and ACK packets were lost that resulted in assuming that links to neighbors was broken by IMEP. Therefore, TORA reacted to these link failures by sending more UPDATEs, in turn that created

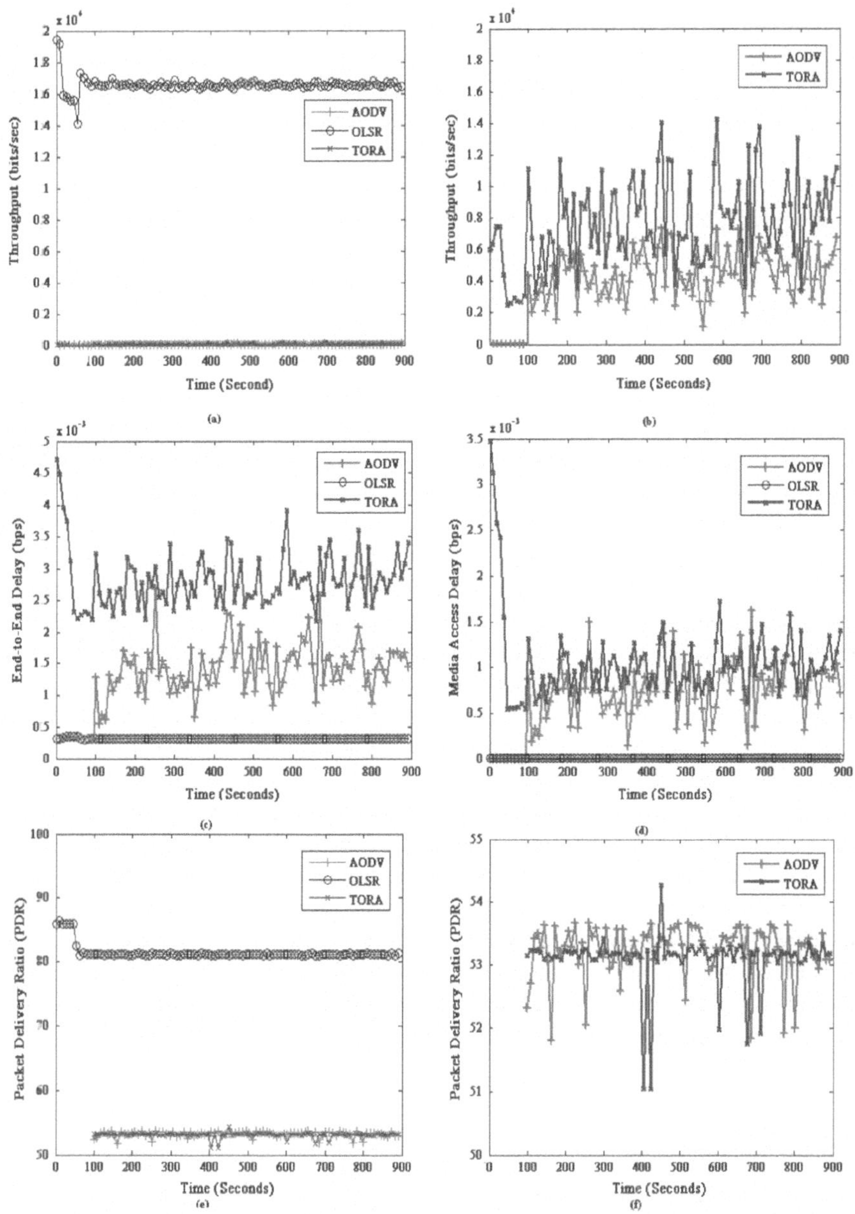

Fig. 19.3 Comparison between the MANETs routing protocol's simulation results: (**a**) Throughput, (**b**) Throughput for TORA and AODV, (**c**) End to End Delay, (**d**) Media Access Delay, (**e**) Packet Delivery Ratio, (**f**) Packet Delivery Ratio for AODV and TORA

Table 19.4 Simulation results over simulation time of 900s

Protocols	Average number of events simulated	Average speed
AODV	229,576	398,557 events/s
TORA	199,354,5	544,829 events/s
OLSR	143,571,00	232,943 events/s

more congestion as failure to receive an ACK from retransmitted UPDATEs was considered as link failure indication.

19.5.3 Media Access Delay

In Fig. 19.3d we plotted media access delay which is very important for multimedia and real time traffic; furthermore it is vital for any application where data is processed online. Media access delay was low for OLSR that is around 0.0001 s. However, the media access delay for AODV and TORA fluctuates more frequently but AODV fluctuates more frequently above and below its mean while TORA mainly around its mean, thus in both case fluctuation is higher and more frequent as compared to OLSR that remains steady over 900s of simulation time (Table 19.4).

19.5.4 Packet Delivery Ratio

The fraction of the originated application data packets by each protocol was able to deliver at varying time as shown in Fig. 19.3e. As packet delivery ratio shows both the completeness and correctness of the routing protocol and also the measure of efficiency. We used packet delivery rate as the ratio of number of data packets received at the sink node to the number of data packets transmitted by source nodes having a route in its routing table after a successful route discovery. For all protocols packet delivery ratio is independent of offered traffic load, where routing protocols OLSR, AODV, TORA delivering about 81%, 53.6% and 53.1% of the packets in all cases. OLSR provides better packet delivery rate than all other routing protocols, on the other hand AODV has higher delivery ratio as compared to TORA. AODV and TORA takes time for computing route to send data packets as these protocols constructs route on demand whereas OLSR is an proactive routing protocol and uses routing table to send data packets at once. As packet delivery ratio indicates the loss rate that can be seen by the transport protocols that effects the maximum throughput that the network can handle. OLSR have MRPs for each cluster which maintains routes for the group of destination, packets that the MAC layer is unable to deliver are dropped since there are no alternate routes. In Fig. 19.3f, we have used different scales of axes to show results of packet delivery ratio visibly for reactive and hybrid protocol. In AODV and TORA graph starts after 100 s because AODV and TORA takes time for computing route to receive data packets on destination as these protocols constructs route on demand whereas OLSR is a proactive routing protocol and uses routing table to send data packets at once. TORA in 50 MN wireless networks delivered around 53% of data packets over simulation time, TORA fall short to converge because of increased congestion. In TORA mainly data packets were dropped because of short lived routing loops, which are part of its link reversal process.

When the packet of next hop and previous hop are same then more data packets were dropped because the packets were looped until time to live expires or when loop exited; moreover, data packets which were in loops interfered by broadcast UPDATE packet from neighbor MNs which in turn can resolve routing loop. It was observed that packet delivery ratio was less in TORA than AODV. Moreover, routing protocols have differed in how much protocols can deliver packets to destination MN.

19.6 Conclusion

The mobile nodes' mobility management is key area since mobility causes route change and frequent changes in network topology, therefore effective routing has to be performed immediately. This paper makes contributions in two areas. Firstly, this paper compared the performance of reactive ad hoc on demand distance vector protocol; proactive optimized link state routing protocol and hybrid temporary ordered routing algorithm protocol in mobile ad hoc networks under ftp traffic. Secondly, we have presented the comprehensive results of packet delivery ratio, throughput, media access delay, and end to end delay over mobile ad hoc networks of 50 MNs moving about and communicating with each other. The simulation results were presented for a range of node mobility at varying time. OLSR performs quite predictably, delivering virtually most data packets at node mobility. In [8] also shows that OLSR shows the best performance in terms of data delivery ratio and end-to-end delay. TORA, although did not perform adequate in our simulation runs in terms of routing packet delivery ratio, delivered over fifty three percentage of the packets. Since the network was unable to handle all of the traffic generated by the routing protocol and a significant fraction of data packets were dropped. As well as in [9] shows that the relative performance of TORA was decisively dependent on the network size, and average rate of topological changes; TORA can perform well in small network size but TORA's performance decreases when network size increases to 50 nodes. On the other hand, AODV performed better than TORA in most performance metrics with response to frequent topology changes. Finally, the overall performance of OLSR was very good when MNs movement was changing over varying time. We have analyzed that all routing protocol successfully delivers data when subjected to different network stresses and topology changes. Moreover, mathematical analysis and simulation results both show that optimized link state routing protocol, from proactive protocol category, is a very effective, efficient route discovery protocol for MANETs.

References

1. H. Pucha, S. M. Das, Y. C. Hu, "*The Performance Impact of Traffic Patterns on Routing Protocols in Mobile Ad Hoc Networks*", Journal (COMNET), vol. 51(12), pp. 3595–3616, August 2007.
2. J. Broch, D. A. Maltz, D. B. Johnson, Y. C. Hu, J. Jetcheva, "*A Performance Comparison of Multi-Hop Wireless Ad-Hoc Network Routing Protocols*," Proceedings of the 4th ACM/IEEE International Conference on Mobile Computing and Networking MOBICOM'98, pp. 85–97, Texas, October 1998.
3. C. E. Perkins, E. M. Royer, I. D. Chakeres, "*Ad hoc On-Demand Distance Vector (AODV) Routing Protocol*", draft-perkins-manet-aodvbis-00.txt, October 2003.
4. T. Clausen, C. Dearlove, P. Jacquet, "*The Optimized Link State Routing Protocol version 2*", MANET Working Group, Available: http://www.ietf.org/internet-drafts/draft-ietf-manet-olsrv2-05.txt, February 2008.
5. V. Park, S. Corson, "*Temporally-Ordered Routing Algorithm (TORA) Version 1*", Internet draft, IETF MANET working group, Available: http://tools.ietf.org/id/draft-ietf-manet-tora-spec-04.txt, July 2001.
6. Opnet.com (2008), The OPNET Modeler 14.0, Available: http://www.opnet.com.
7. E. Nordstrom, P. Gunningberg, C. Rohner, O. Wibling, "*A Comprehensive Comparison of MANET Routing Protocols in Simulation, Emulation and the Real World*", Uppsala University, pp. 1–12, 2006.
8. C. Mbarushimana, A. Shahrabi, "*Comparative Study of Reactive and Proactive Routing Protocols Performance in Mobile Ad Hoc Networks*", 21st International Conference on Advanced Information Networking and Applications Workshops (AINAW'07), IEEE Computer Society, March 2007.
9. S. R. Das, R. Castaneda, J. Yan, R. Sengupta, "*Comparative Performance Evaluation of Routing Protocols for Mobile Ad hoc Networks*", Proceedings of the International Conference on Computer Communications and Networks, pp. 153–161, 1998.
10. G. Ivascu, S. Pierre, A. Quintero, "*QoS Support based on a Mobile Routing Backbone for Ad Hoc Wireless Networks*", Proceedings of the International Conference on Wireless Communications and Mobile Computing IWCMC'06, pp. 121–126, Vancouver, Canada, July 2006.
11. V. D. Park, M. S. Corson, "*A Performance Comparison of TORA and Ideal Link State Routing*", Proceedings of IEEE Symposium on Computers and Communication, June 1998.

Chapter 20
IEEE 802.11E Block Acknowledgement Policies

O. Cabral, A. Segarra, and F. J. Velez

Abstract Optimization of IEEE 802.11e MAC protocol performance is addressed by modifying several parameters left open in the standard, like block size and acknowledgement policies in order to improve the channel efficiency. The use of small block sizes leads to a high overhead caused by the negotiation on the other hand, the use of large block sizes causes long delays, which can affect negatively real-time applications (or delay sensitive applications). An event driven simulator was developed, and results with a single service and several services running simultaneously were extracted. By using the Block Acknowledgement (BA) procedure, for video and background traffics in a single service situation, the capacity was improved in the case when the number of stations is equal or higher than 16 and 12, respectively. However, for lower values of the number of stations, the use of BA leads to a slightly worst system performance. In a scenario with mixture of services the most advised block size is 12 (less delay in a highly loaded scenario). The number of supported user (total) increases from 30 to 35.

Keywords IEEE 802.11e · simulator · block acknowledgement · quality of service

20.1 Introduction

Recent years have seen an immense growth in the popularity of wireless applications that require high throughput. To support such growth, standardization bodies such as the IEEE 802.11 have formed task groups to investigate and standardize features providing increased quality of service (QoS) and higher throughputs. One of these extensions is the acknowledgement (ACK) policy feature included in the ratified IEEE 802.11e amendment for QoS support [3], which is the focus of this work. In

O. Cabral (✉)
Instituto de Telecomunicacoes, DEM-UBI Calcada Fonte do Lameiro 6201-001 Covilha Portugal,
E-mail: orlando@ubi.pt

S.I. Ao and L. Gelman (Eds.), *Advances in Electrical Engineering and Computational Science*, Lecture Notes in Electrical Engineering.

particular, we investigate the policy regarding the block size for a video application. The Block Acknowledgement (BA) procedure improves system throughput results by reducing the amount of overhead required by a station to acknowledge a burst of received traffic [1,2]. It acknowledges a block of packets by a single ACK, instead of using several ACKs, one for each packet, saving the Arbitration Inter-frame Spacing (AIFS) period, the backoff counter time, and the acknowledgement time. The number of frames that can be transmitted within a block is called block size. It is limited and is not specified in the standard. In this chapter, to find the most suitable block size we have tried several block sizes and several loaded scenarios with and without mixture of services. This chapter is organised as follows. In Section 20.2, a brief introduction to the IEEE 802.11e standard is presented along with the main directive lines of the BA procedure. Section 20.3, gives the description of the state of the art. Section 20.4 defines the problem and the scenario, including details on traffic parameters. Section 20.5 gives the simulation results obtained for several scenarios with and without the use of the BA procedure, with and without mixtures of traffic. Conclusions are given in Section 20.6, as well as suggestions for further work.

20.2 IEEE 802.11e and Block Acknowledgement Description

20.2.1 IEEE 802.11e User Priorities and Access Categories

The so-called enhanced distributed channel access (EDCA) provides differentiated, distributed access to the medium for Quality Stations (terminals that support IEEE 802.11e) using four access categories (ACs) voice (VO), video (VI), best effort (BE), and background (BK). This differentiation is achieved by mapping the traffic to four queues that correspond to the four AC. The traffic prioritisation is performed by varying the amount of time a station queue senses the channel to be idle before backoff or transmission, or the length of the contention window to be used for the backoff, or the duration a station queue may transmit after it acquires the channel. Each AC contends to access the medium with a single CSMA instance, [3] [1]. Each queue has an Arbitration Inter-frame Spacing (AIFS) period preceding the next backoff contention window (see Fig. 20.1).

20.2.2 Block Acknowledgement

The BA mechanism (see Fig. 20.2) improves channel efficiency by aggregating several acknowledgements into one frame [4,5]. There are two types of BA mechanisms: immediate and delayed. Immediate BA is suitable for high-bandwidth, low-latency traffic while the delayed BA is suitable for applications that tolerate moderate latency. The QSTA with data to send using the BA mechanism is referred

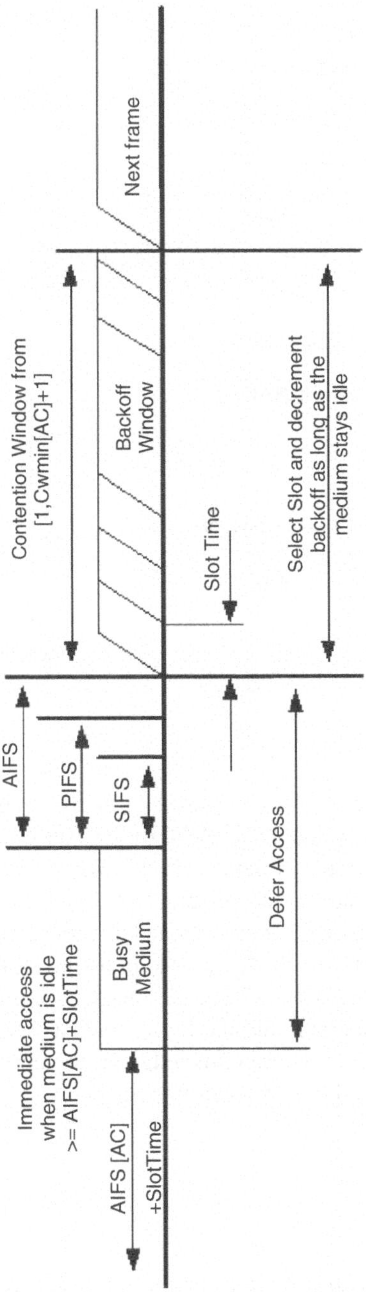

Fig. 20.1 Timing relationship in EDCA

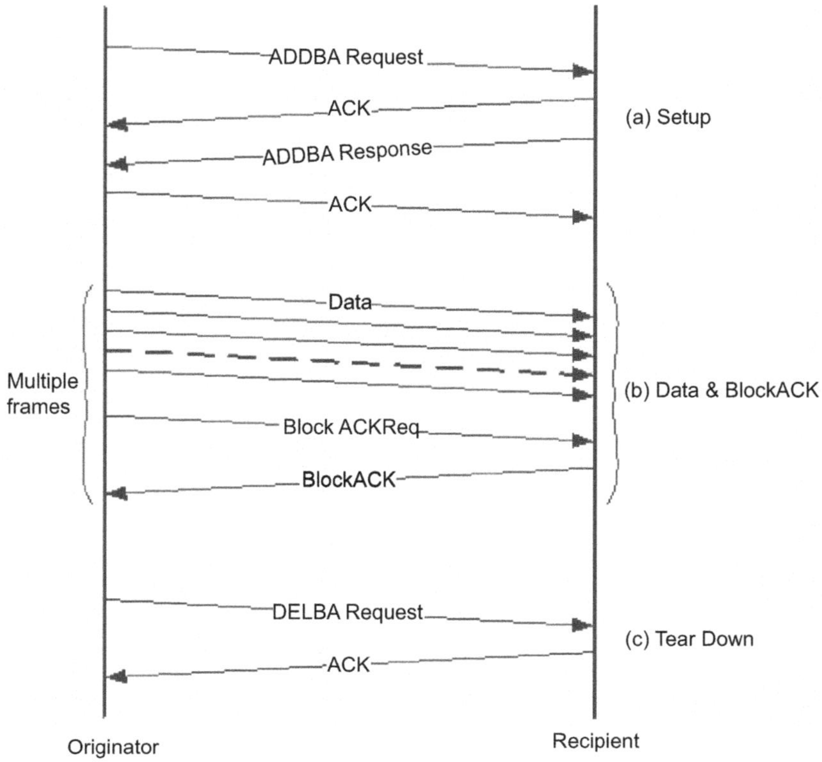

Fig. 20.2 Block ACK sequence

to as the originator, and the receiver of that data as the recipient. The BA mechanism is initialized by an exchange of Add Block Acknowledgment (ADDBA) Request/Response frames. After initialization, blocks of QoS data frames can be transmitted from the originator to the recipient. A block may be started within a polled TXOP or by winning EDCA contention. The number of frames in the block is limited, and the amount of state that is to be kept by the recipient is bounded. The MAC Packet Data Units (MPDUs) within the block of frames are acknowledged by a BA control frame, which is requested by a BlockACKReq control frame. The response to the BlockACKReq will acknowledge all the correctly received frames and request the badly received to be transmitted again.

20.3 State of the Art

Analytical frameworks to model BA have been published [6-8] but the results are not based on a realistic approach to the problem nor account for the achievable QoS, because the use of several service classes with different priorities (the base for EDCA) is not considered at all. The existing theoretical approaches [6-8] do not consider the hidden terminal problem, assume that the buffer is always full, nor assume a multi-rate scenario. In [6,7], an analytical framework was presented to model an ad-hoc network under the standard IEEE 802.11e with the BA procedure for a completely simplified scenario. The hidden/exposed terminal problem is one fundamental issue in WLANs but most of the existing analytical models either assume it does not exist, or do not consider the EDCA features of IEEE 802.11e (or do not account for the delay or any other QoS metric) [6,7]. Results presented in [6,7] express the block size as a function of the goodput in saturation conditions. Results show that the block size should be as high as possible, which can be misleading when we consider QoS. In [8] an analytical approach to model the BA in IEEE 802.11e was proposed without accounting for the hidden terminal problem. The multi-rate feature in the same environment was also not considered. Further, the packet loss due to errors in the channel was not taken in consideration. Finally, the use of EDCA procedures, like several virtual queues, for the several classes of service are also not considered, i.e., this work does not consider the IEEE 802.11e standard at all. From the simulation approaches presented in the literature the one that is most similar to the work proposed here was proposed in [9]. In [9] several combinations for the block size are presented where a scheduler based on the delay and amount of data in the buffer is proposed. The work presented here is an improvement of this approach and provides a more extensive study on the block size while considering use-perceived QoS.

20.4 System, Scenario and Assumptions

A cellular WiFi system was considered where each cell has a set of N+1 IEEE 802.11e stations communicating through the same wireless channel. While station 0 is the QoS Access Points (QAP), the other N are QoS stations (QSTA). Each station has four buffers whose size depends on the kind of service being dealt in order to guarantee a given value for the goodput (payload of the packet at MAC level). These buffers will be filled with a MAC Service Data Units (MSDU) generated that characterises the service being dealt in the given buffer. If the MSDU is bigger than a fragmentation threshold, it will be fragmented. In order to cope with service quality the packet transmission follows the Enhanced Distributed Channel Access (EDCA) IEEE 802.11e MAC procedure. Due to collisions or interference a packet may not be correctly received. The interference issues are addressed by using a radio propagation model. Each packet exits the buffer only after the reception of an acknowledgement, or if it has suffered more than a given collision threshold. The users are assumed to be static, and are distributed uniformly in a square area of 2,500 square meter. The physical layer specification assumed in this work is the IEEE 802.11a standard 802, 1999, [10], that defines an Orthogonal Frequency Division Multiplexing (OFDM) based PHY layer that operates in the 5 GHz frequency bands, being able to achieve bit-rates as high as 54 Mbps. The physical

Table 20.1 MAC and PHY parameters

Slot time	0.009 ms
ACK size	112 bit
SIFS	0.016 ms
DIFS	0.034 ms
RTS threshold	4,000 bit
RTS size	160 bit
CTS size	112 bit
CWmin	31 slots
CWmax	1,280 slots
Collisions threshold	7
Fragmentation threshold	8,192
Simulation time	100 s

Table 20.2 Traffic parameters

AC	Voice (VO)	Video (VI)	Background (BK)
Packet size	1,280 bit	10,240 bit	18,430 bit
Packet interval	20 ms	10 ms	12.5 ms
Usage	50%	30%	20%
Symmetry	Symmetric	Asymmetric (downlink)	Asymmetric (downlink)

and MAC parameters are presented in Table 20.1. The use of the Request/Clear-to-send (RTS/CTS) procedure is implemented only if the packet size is larger than a given threshold, RTS threshold in Table 20.1.

In the algorithm proposed in [10] the sender chooses the bit-rate that achieves the highest throughput taking into account the SINR estimate. More details on physical layer implementation used in the simulator can be found in [11]. Three types of traffic sources were chosen, namely high priority voice (VO), medium priority video (VI), and low priority FTP data as background traffic (BK). The traffic sources parameters are presented in Table 20.2 as well as the Access Categories (AC) of each type of traffic.

We implemented the BA procedure with minor modification to the existing features like TXOP and RTS/CTS as already explained and without disturbing the overall TXOP and RTS/CTS procedure. When a TXOP is gained, the transmission starts and the origin will know if a BA procedure is implemented with this destination. If this is true, it will not wait for an acknowledgement, but just a SIFS and start the next transmission for the same destination or not depending on which is the next packet in the buffer. Figure 20.3 presents this procedure for destination 1 that has the BA procedure implemented and for destination 2 without the BA. At the beginning of a transmission for a user with active BA procedure, if the block size threshold is reached, a block ACK request packet is added to the buffer to the top-1 place in the queue to be transmitted as the next in line packet.

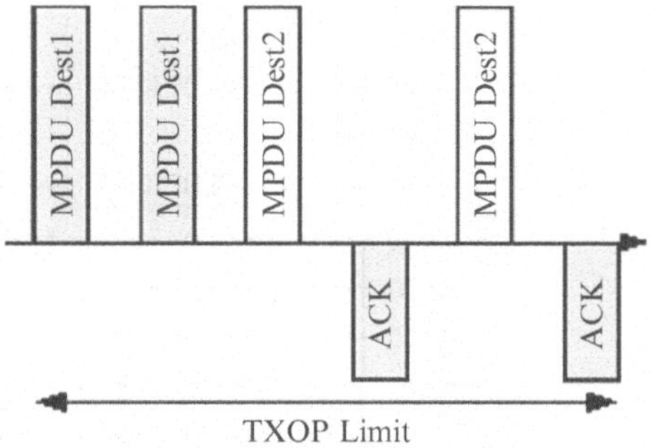

Fig. 20.3 Illustration of the BA procedure within a TXOP

20.5 Results

20.5.1 Block Acknowledgement with Standalone Services

The objective of the simulations was to investigate the gain of the BA procedure for a single service and for a mixture of services. In terms of grade of service, the voice application supports delays up to 30 ms, the video application supports delays up to 300 ms, while the delay for background applications can be up to 500 ms [13]. As previously mentioned, the BA mechanism improves the channel efficiency by aggregating several acknowledgments into one frame. To investigate the proposed BA procedure for the stand-alone service a BA procedure with a block size of 16 fragments was adopted. This procedure is simulated only for BK and VI traffic. For VO application it is certain that BA is not a solution because of the large delays that occur when the buffer is filled with 16 packets (i.e., the delay of the first one would be 320 ms, application). Figure 20.4 presents the delay for BK and VI traffic. It starts to increase around 12 stations for BK and at 16 stations for VI and increases more for a higher number of stations. As expected, the delay is lower when the BA procedure is used. The improvement (reduction) for BK traffic is 300 ms on the average after 12 stations, while for VI traffic it is 420 ms on the average after 16 stations. The improvement in the goodput is of 2 and 2.2 Mb/s in average, after 12 stations, for BK traffic and after 16 stations for VI, respectively [12].

20.5.2 Block Acknowledgement with Mixtures of Applications

This part of the work explores which should be the policy regarding the block size with mixtures of applications. For the purpose we have investigated the BA policy for a video service, an application that is delay-sensitive. Additionally, we have investigated the BA block size for background service that is not delay sensitive. Simulations were performed for 100 s and around 40 times for each number of stations, 15, 20, 25, 30, 35, 40, 45 and each block size, 4, 8, 12, 16. The users started the negotiation to initiate BA procedure in the beginning of the simulation

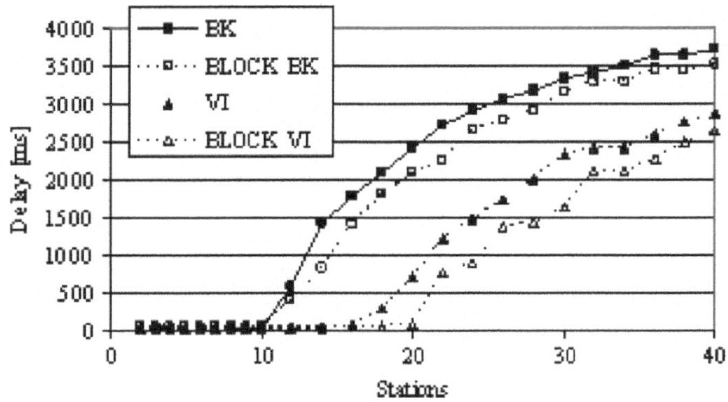

Fig. 20.4 Delay for BK and VI with and without Block ACK

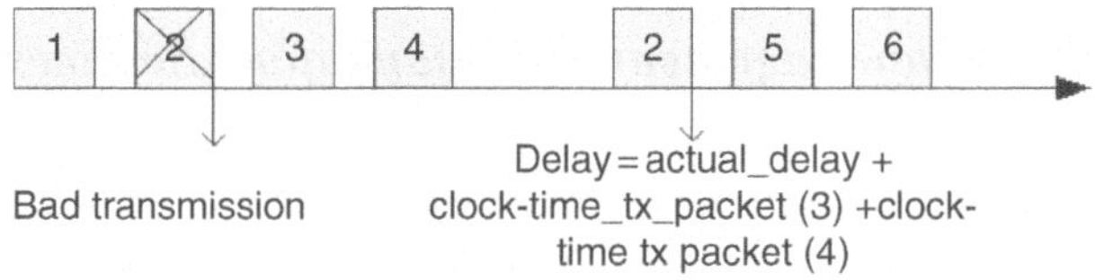

Fig. 20.5 Procedure to count delay

so all the packets of video and background traffic were transmitted within the BA procedure. The use of the BA procedure in a scenario with a mixture of applications running at the same time was further studied. On the one hand, the overhead caused by the negotiation to establish the BA procedure, and to maintain the BA causes bad performance when a small block size is used. On the other hand, the packet losses caused by the voice users with higher priority in the system, influences the overall QoS on the system and mainly on the video service since we considered in this simulations, that if a packet is not correctly received at a given time the following packets sent within this block will have the delay of the badly sent packet after transmitted added to their delay, Fig. 20.5.

The impact of packet retransmissions in the delay therefore increased. Figure 20.6 presents the delay for each access category. For BK traffic the delay starts to increase considerably with more than 20 stations (12 in the standard procedure). This service class will transmit very rarely since the voice and video traffic will always be scheduled first. For video traffic the delay starts to increase in the 35 stations. In contrast with BK and VI, VO applications present a negligible delay.

The delay impacts the most the video application. Figure 20.7 shows the results for various block sizes for transmitted video application. Regardless on the number of stations, the BA procedure reduces the delay. Certain sensitivity is observed for a block size of 16, which is the size that gives the highest delay for a total of 1–25 stations. For a larger number of station the best block size is 16 as smaller block sizes are not that efficient in decreasing the delays.

When 30 stations are being served in the system the results for the delay with BA is near 5 ms for all block sizes, while without BA the values of the delay

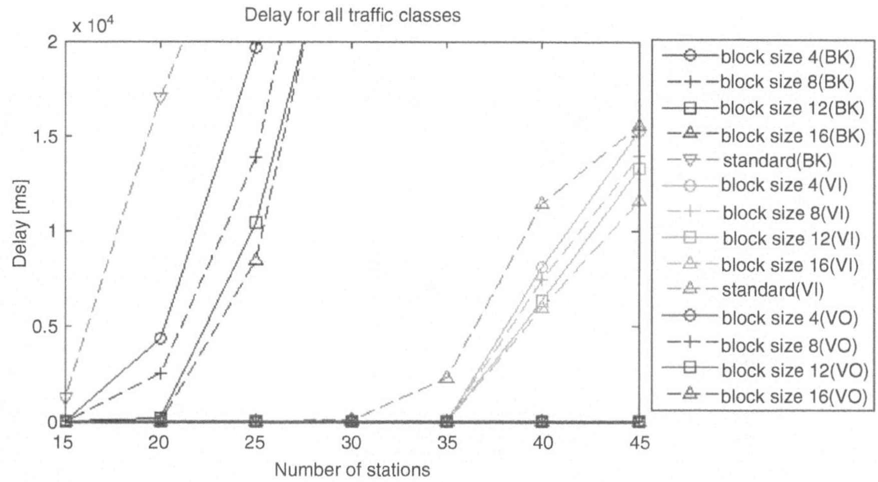

Fig. 20.6 Delay for all services with Block ACK implemented for VI and BK traffic

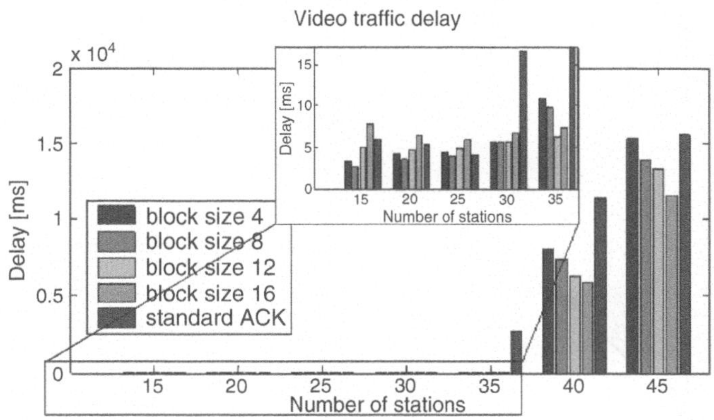

Fig. 20.7 Delay for video traffic

goes up to 80 ms. For 35 stations the difference is even higher. A minimum near 5 ms is obtained for block size 12 while for delay without BA the delay is near 2.3 s. Results for 40 and 45 advise the use of block size 16, although the network is already overloaded (and the delay will just keep growing up to infinity). When less then 40 stations are present the confidence intervals are very small for all the buffers. When there are 40 or more stations the confidence intervals increase. One can therefore extrapolate that there are stations that manage to transmit and have a small delay while others transmit from time to time, causing very high delays. The behaviour observed in Fig. 20.7 occurs due to the overhead caused by the blockACK request, and the delays caused by bad receptions affected mostly the block size 16, but providing lower delays for higher loaded scenarios. The solution is neither to use a large block size (as large as 16) nor a small block size (as low as 4). The

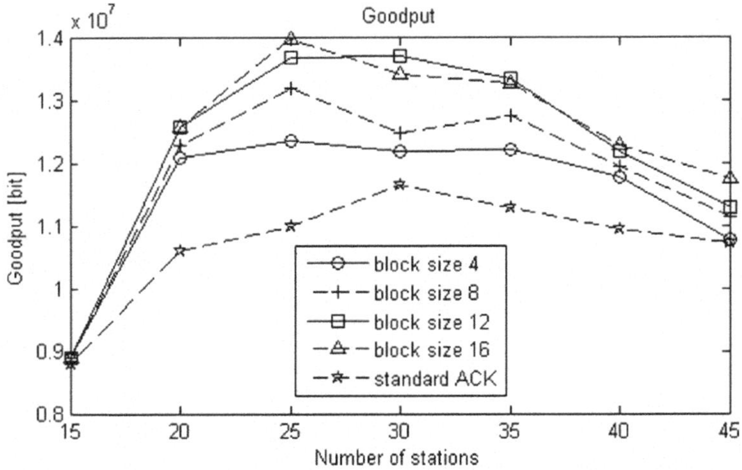

Fig. 20.8 Total goodput with BA implemented for video and background traffic

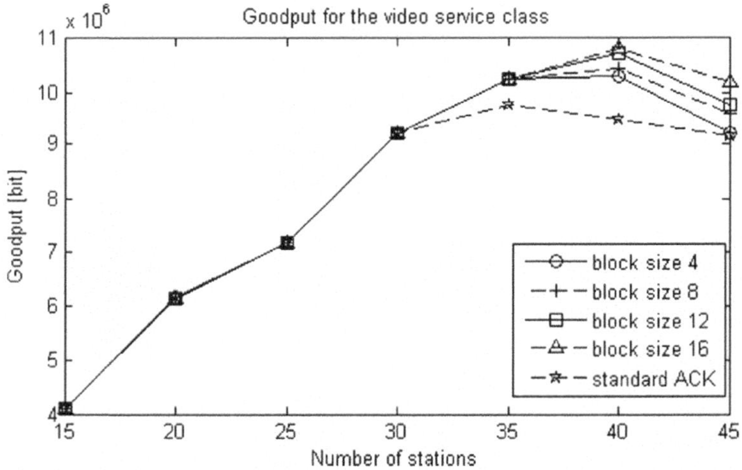

Fig. 20.9 Video goodput with Block ACK implemented for video traffic

former increases the delay causing problems to the application using these packets (for more than 35 stations), while the latter causes some unnecessary overhead by requesting very often block ACK requests (for less than 35 stations). The advised block size shall be 12 since it is the one that provides lower delay in a scenario where the load is still manageable, at least for the case where BA is used.

Figure 20.8 presents the results of the goodput in the system in the downlink. Without the BA the maximum goodput achieved is near 11 Mbit/s, while with BA is near 14 Mbit/s for block sizes 12 and 16. The decreasing behaviour after 25 stations occurs due to the high number of collision, and as the background traffic has low priority ends up not being transmitted giving its turn to voice and video users. As voice traffic provides higher overhead the resulting goodput is lower.

Figure 20.9 presents the goodput only for the video service class. Only after 30 stations the achieved throughput is different when using and not using BA. The highest goodput is found for block size 16, and is more than 10% higher relatively to the case of standard ACK. The increasing behaviour of the goodput is verified up to 40 stations.

20.6 Conclusions

This work investigated the BA size as a policy to decrease delays and ensure a stable system. The investigation was based on tuning several parameters and investigated the effect for a procedure with and without BA. Several policies were investigated. Future work will test policies that provide access to the medium, that ensure some degree of service, based on the channel SINR, delays, bit error rate, can be tested. Although, for lower values of the number of stations, the use of BA leads to a slightly worst system performance, the BA procedure provides an improvement in highly loaded scenarios. The improvement is of 2 and 2.2 Mb/s in average, for BK traffic and VI traffic, respectively. In a scenario with mixture of services the most advised block size is 12 since it is the one that provides lower delays in a highly loaded scenario while the users are still within the capacity of the AP. The number of supported user increases from 30 to 35. Note that 35 stations is the total number of VO plus VI and BK user.

Acknowledgements This work was partially funded by CROSSNET (Portuguese Foundation for Science and Technology, FCT, POSC project with FEDER funding), by IST-UNITE, by the FCT PhD grant SFRH/BD/28517/2006, by Fundação Calouste Gulbenkian, and by "Projecto de Re-equipamento Científico" REEQ/1201/EEI/2005 (a Portuguese Foundation for Science and Technology project).

References

1. A. R. Prasad, and N. R. Prasad, (2005). *802.11 WLANs and IP Networking*. Artech House, Boston, MA.
2. J. L. Mukerjee, R. ă Dillinger, M. ăMohyeldin and Schulz, E.ă (2003). Investigation of radio resource scheduling in wlans coupled with 3g cellular network. *IEEE Communications Magazine*, 41(6):108–115.
3. T. Li, Q. Ni, T. Turletti and Y. Xiao, (2005). Performance analysis of the ieee 802.11e block ack scheme in a noisy channel. In *IEEE BroadNets 2005 – The Second International Conference on Broadband Networks*, Boston, MA.
4. T. Li, Q. Ni and Y. Xiao, (2006). Investigation of the block ack scheme in wireless ad-hoc networks. *Wiley jornal of Wireless Communications and Mobile Computing*, 6(6):877–888.
5. I. Tinnirello, and S. Choi, (2005). Efficiency analysis of burst transmissions with block ack in contention-based 802.11e wlans. In *ICC 2005 – IEEE International Conference on Communications*, Seoul, Korea.

6. V. Scarpa, G. Convertino, S. Oliva and C. Parata, (2005). Advanced scheduling and link adaptation techniques for block acknowledgement. In *7th IFIP International Conference on Mobile and Wireless Communications Networks (MWCN 2005)*, Marrakech, Morocco.
7. Ni, Quiang (2005). Performance analysis and enhancements for ieee 802.11e wireless networks. *IEEE Networks*, 19(4):21–27.
8. A. Grilo, (2004). *Quality of Service in IP-based WLANs*. Ph.D. thesis, Instituto Superior Tcnico, Technical University of Lisbon, Lisbon, Portugal.
9. O. Cabral, A. Segarra and F. J. Velez, (2008). Event-driven simulation for ieee 802.11e optimization. *IAENG International Journal of Computer Science*, 35(1):161–173.

Chapter 21
Routing in a Custom-Made IEEE 802.11E Simulator

João Miguel Ferro and Fernando J. Velez

Abstract The custom-made IEEE 802.11E simulator in this chapter is an evolution of the previous one. To provide support for multi-hop environment it was necessary to implement the routing algorithm above the already existing Medium Access Control (MAC) plus Physical (PHY) layers. After the random placement of all the stations in the field, the simulator determines which can communicate directly. With this data, the selected routing algorithm determines the shortest path from a station to all the others. For these initial tests we chose the well-known Dijkstra's algorithm. In this work, however, we present no cross-layer at all, and the routing table is not modified during the simulations. The engine starts collecting events at the beginning of the simulation. For the end user, this simulator allows for simulating a network with an unrestricted number of nodes, in any configuration type. By using the chosen routing protocol, it supports connections to every reachable station from its neighbours, allowing for a message to reach any destination.

Keywords IEEE 802.11E Simulator · Custom-Made · Routing · multi-hop environment · network simulation

21.1 Introduction

Wireless networks are gaining more and more importance in our world. Cellular phones with GPRS/UMTS, Wi-Fi, and WiMAX networks are very common these days, and they share a common feature: they require some sort of backbone infrastructure in order to allow for packets from different communication peers to reach each other. For example, if someone makes a phone call, the conversation will always pass from the cell phone to the operators' infrastructure, and then to the

---

J. M. Ferro (✉)
Instituto de Telecomunicações, DEM-UBI, Portugal,
E-mail: ferro@lx.it.pt

S.I. Ao and L. Gelman (Eds.), *Advances in Electrical Engineering and Computational Science*, Lecture Notes in Electrical Engineering.

receivers phone, even if they are both in the same building. Resources would be certainly saved if somehow the cell phones could connect directly to each other.

In an ad-hoc network, all the participants (also called nodes) can communicate directly with their neighbours. Two nodes are considered neighbours if their communication devices can reach each other. Nodes wanting to communicate to others that are not neighbours will simply send a message to another node which is located nearer the destination. As so, a centralised infrastructure to establish the connectivity is not required, since each node will determine by itself to where it should forward its data. The specification IEEE 802.11 refers to this type of network as Independent Basic Service Set (IBSS).

This chapter is organized as follows. Section 21.2 presents some background information about the previous simulator and about IEEE 802.11e. Section 21.3 describes the features of the new simulator highlighting the modifications we performed. Section 21.4 presents the results of the initial simulations, while Section 21.5 presents conclusions and suggestions for future work.

21.2 Previous Work

This simulator is based on the one developed at the Instituto de Telecomunicações, Laboratório da Covilhã, as part of the IST-UNITE project. The purpose of that previous version was to create a trustworthy simulator to be used in the context of IST-UNITE, allowing the study of the interoperability between Wi-Fi and High Speed Downlink Packet Access (HSDPA), [1]. This involved the creation of an HSDPA time-driven simulator and a Wi-Fi event-driven simulator that may be able to run separately or together for the coexistence scenario. In the latter case, there is communication between the two simulators which will run separately in a synchronous way.

The Wi-Fi simulator was initially built only for the infrastructure mode, i.e., an architecture with one access point (AP) and several wireless stations (STAs) attached to it. It can simulate traffic from one STA to the AP and vice-versa, and calculates the throughput (total in the simulation and per unit of time), packet delay (total in the simulation and average), lost packets, retransmissions, and collisions. This computation is oriented to the support of the Quality of Service (QoS), i.e., it implements IEEE 802.11e with four access categories. As a consequence, these metrics are calculated for each traffic type: voice (VO), video (VI), background (BK) and best-effort (BE).

The IEEE 802.11e standard was conceived to implement Quality of Service (QoS) in IEEE 802.11a/b/g. QoS refers to the way resources are reserved for certain applications, guaranteeing that the most urgent data flows will have higher priority levels. The IEEE 802.11 Medium Access Control (MAC) coordination functions are the Distributed Coordination Function (DCF) and the Point Coordination Function (PCF). They establish the timing and sequence for each station to access the shared medium. The latter is only used in the infrastructure mode, with the AP acting as the coordinator. It provides some basic QoS support, but since it is defined as optional and is more complex and costly to implement, only a few APs support it. The former does not present QoS guarantees at all. To ensure the QoS performance, IEEE 802.11e introduced the Hybrid Coordination Function (HCF), which defines the HCF Controlled Channel Access (HCCA) and Enhanced Distributed Channel Access (EDCA). In EDCA, a Transmission Opportunity (TXOP) is won according to the traffic type, scheduling firstly higher priority traffic, like voice and video. The HCCA is more complex, but allows greater precision in terms of QoS. All these enhancements are fully explained by Prasad [2].

In the EDCA simulator each station (and AP) has four buffers, one for each traffic type. By using an event-driven approach, for each new packet arrived at a machine a new event is generated. The simulator engine uses its internal clock to pick up the next event and pass the corresponding packet to the developed MAC + PHY layers. A more comprehensive description of the lower layers of the simulator can be found in [3].

21.3 Overview

The current simulator is an evolution of the previous one. To provide support for multi-hop environment it was necessary to implement the routing algorithm above the already existing Medium Access Control (MAC) plus Physical (PHY) layers, as it is presented in Fig. 21.1.

After the random placement of all the stations in the field, the simulator determines which can communicate directly. With this data, the selected routing algorithm determines the shortest path from a station to all the others. For these initial tests we chose the well-known Dijkstra's algorithm, [4], which calculates the least cost path from one node to all the remaining ones. This characteristic allows the routing calculation to take place before the simulation starts, making the routing table available to all nodes since time instant $t = 0$ s. Besides this initial calculation, the routing table can be accessed and modified at any time, allowing the use of cross-layer design (i.e., gather information from physical and MAC layers and use it in the routing algorithm) for optimisation purposes. In this work, however, we present no cross-layer at all, and the routing table is not modified during the simulations.

The engine starts collecting events at the beginning of the simulation. When a packet arrives at a wireless machine, a new simulation module verifies if the packet has another destination by checking the two additional fields now included in the packet header: besides the "payload", "origin station", and "destination station", among others, each packet now includes the "initial origin" and the "final destination" fields. If it is found that the destination is another node, the routing table is accessed to determine the next hop, and the packet is added to the machine buffer with the new destination stamped.

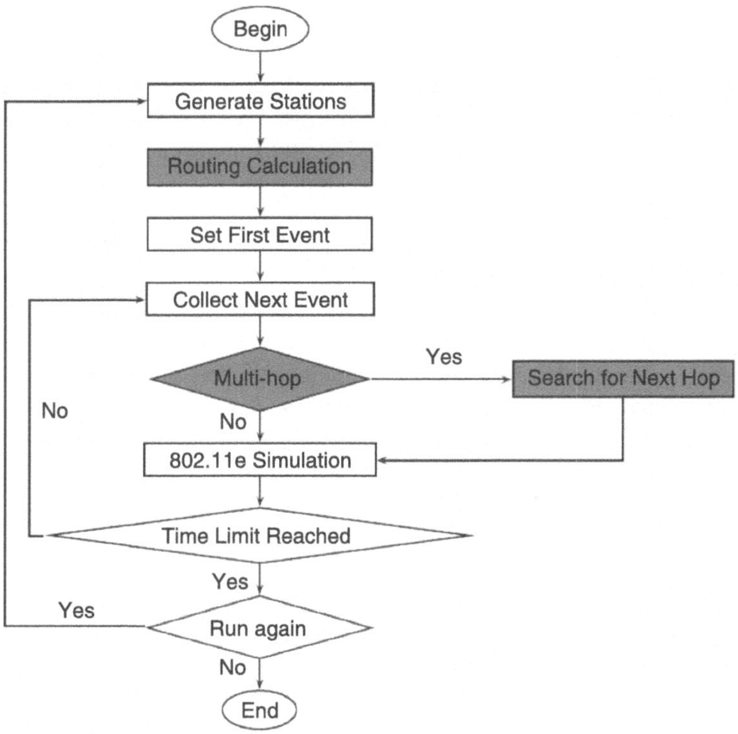

Fig. 21.1 Multi-hop simulator flowchart, modules added to the previous version are displayed in grey

The simulation will run until the time limit is reached, and will be repeated during a pre-determined number of times. The final output of the simulator is obtained as an average of the results for these simulations.

For the end user, this simulator allows for simulating a network with an unrestricted number of nodes, in any configuration type. By using the chosen routing protocol, it supports connections to every reachable station from its neighbours, allowing for a message to reach any destination. Stations are randomly placed in a field and even if a station is isolated the simulator will still run (unless this station is the source/final destination of a packet, which will terminate the simulation). Several parameters can be tuned by the user to its needs, some of them are presented in Table 21.1.

21.4 Results

In order to verify if the modified simulator was able to simulate a multi-hop environment, we ran a few tests. In the first one we tried to verify if the simulation time was going to affect the end-to-end delay. For this purpose, we considered a fixed

Table 21.1 Some values that can be modified by the user

Parameter	Default value	Description
Side	60 m	Side of the square where the stations are placed
Range	35.365 m	Transmission range of each station
Simulation Time	10,000 ms	Period of time to be simulated
Payload XX	– bits	Length of each packet of the type XX
Inter Arrival XX	– ms	Interval of time to generate a new packet of the type XX

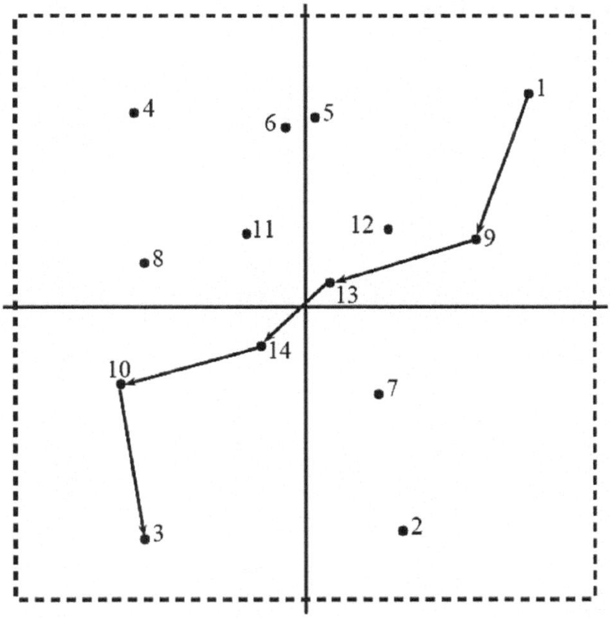

Fig. 21.2 Video traffic being transmitted from station 1 to station 3

configuration of 14 stations, with a video stream being generated from station 1 to station 3, Fig. 21.2.

The following parameters were used:

- Video traffic starts at: $t = 1$ ms
- New packet generated each: 10 ms
- Simulation ends at: $t = 0.1/0.5/1/5/10/15$ s
- Video payload size: 10,240 bits (each packet will be sent in three fragments)

Figure 21.3 presents the average delay for each simulation time. As one can observe, the delay is constant with the simulation time, despite the slight differences for the shortest simulations.

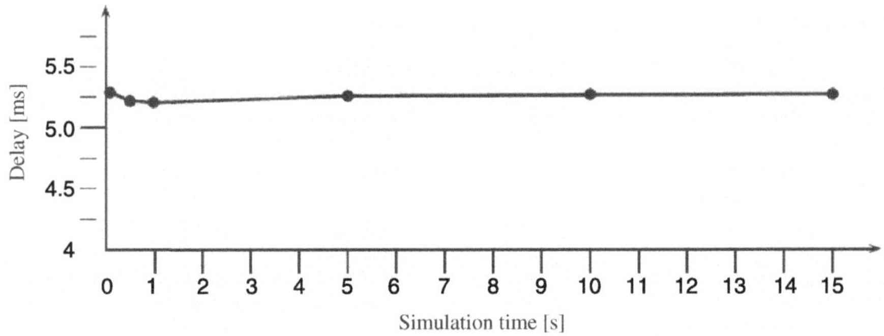

Fig. 21.3 Delay versus simulation time

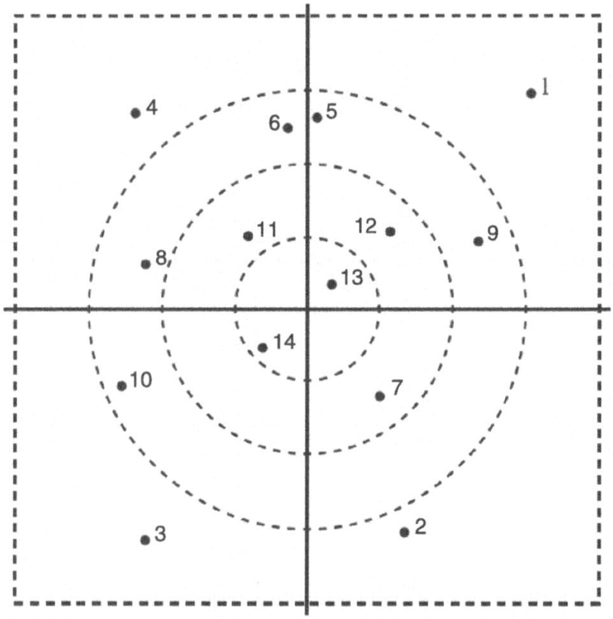

Fig. 21.4 Station deployment for the tests

After this initial test, 10 s was assumed as a reasonable simulation time, and the subsequent tests were made with this value for the duration.

Besides this, further tests were made to evaluate the influence of multiple streams in the delay and packet loss. For these tests we used the same deployment of stations as in the previous test, Fig. 21.4.

The inner circle has a radius of 15 m, the next one 30 m, and the outer one 45 m. The square has $120 \times 120\,m^2$. In the next set of tests we added sequentially video streams between the following stations:

A. From station 1–3
B. From station 4 to 2

C. From station 5 to 6
D. From station 8 to 9

For each of these streams the routing algorithm established the following paths:

A. 1 → 9 → 13 → 14 → 10 → 3
B. 4 → 11 → 14 → 7 → 2
C. 5 → 6
D. 8 → 14 → 13 → 9

Figure 21.5 presents the results obtained for the delay of video stream from station 1 to station 3. By increasing the number of stations that generate packets, the end-to-end delay (latency) increases.

Another interesting metric is the number of lost packets, which is presented as the darker areas in Fig. 21.6. When the number of packets to be transmitted increases, there are more collisions and packet losses.

One interesting characteristic of this simulator is the possibility of simulating four different types of traffic and their mixtures. By using this feature we produced a new set of tests, keeping the same topology but modifying the type of packets in each stream:

A. Video stream from station 1–3
B. Background traffic from station 4–2
C. Voice traffic from station 5–6

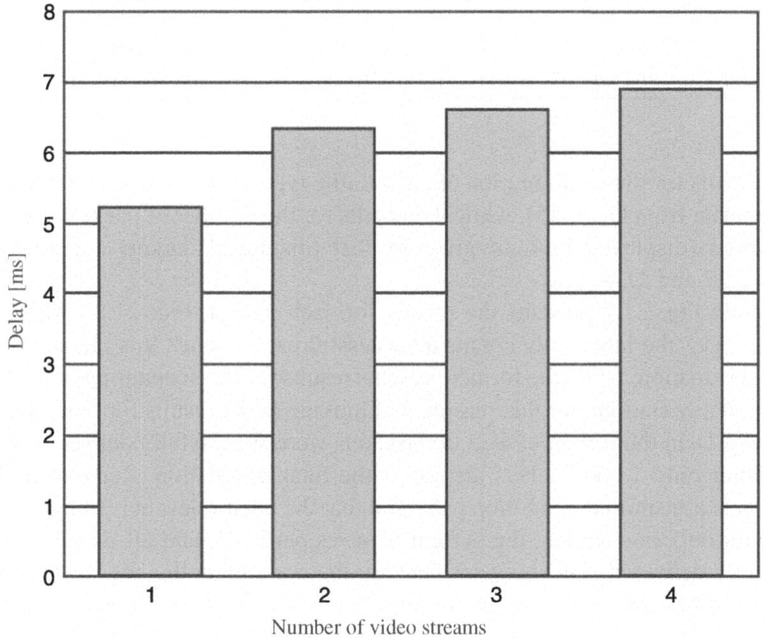

Fig. 21.5 End-to-end delay for video stream 1

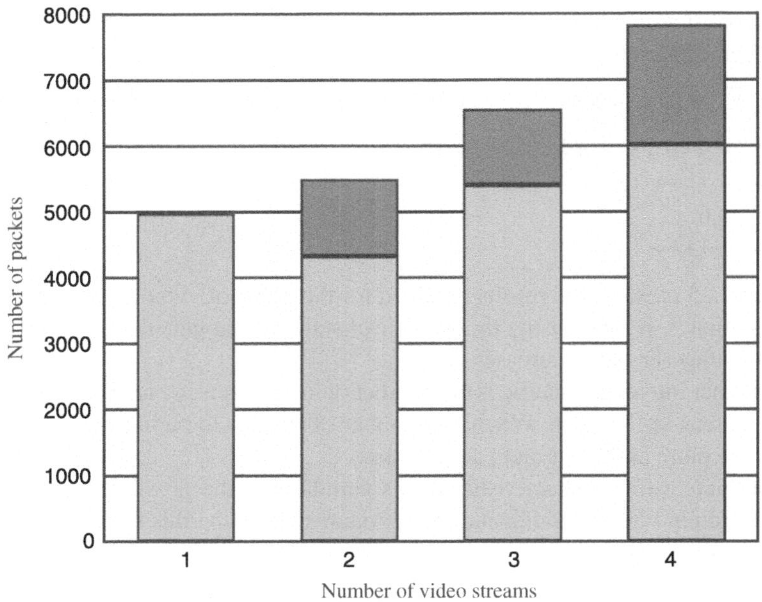

Fig. 21.6 Total of packets received/lost

Table 21.2 Configuration of each traffic type

Type	Packet size (bits]	Period (ms]
Video	10,240	10
Voice	1,280	20 (bi-directional)
Background	18,430	12.5

The details for the configuration of each traffic type are presented in Table 21.2 and were taken from Qiang [5], while the results for the number of packets successfully delivered (displayed in light grey) or lost (displayed darker) are presented in Figs. 21.7 and 21.8.

While Fig. 21.7 presents the results for each packet, Fig. 21.8 addresses each stream, i.e., the latter only counts a successful packet when this one arrives at the final destination, while the former presents results for the accounting of packets that arrive at any station. For this reason, and looking at the results for the video stream alone (VI), in the former case, 5,000 packets were successfully delivered, while for the latter only 1,000 packets arrived at the final destination. Remember that this stream is transmitted in a 5-hop path, and that the most relevant result is the latter.

With only one stream, the system behaves perfectly, and all of the packets are correctly delivered. This happens because there are no collisions, since the period is longer than the end-to-end delay for every string. However, when more than one type of traffic is being generated simultaneously, the packets start to collide and some of them are considered lost (current policy establishes that a station will drop

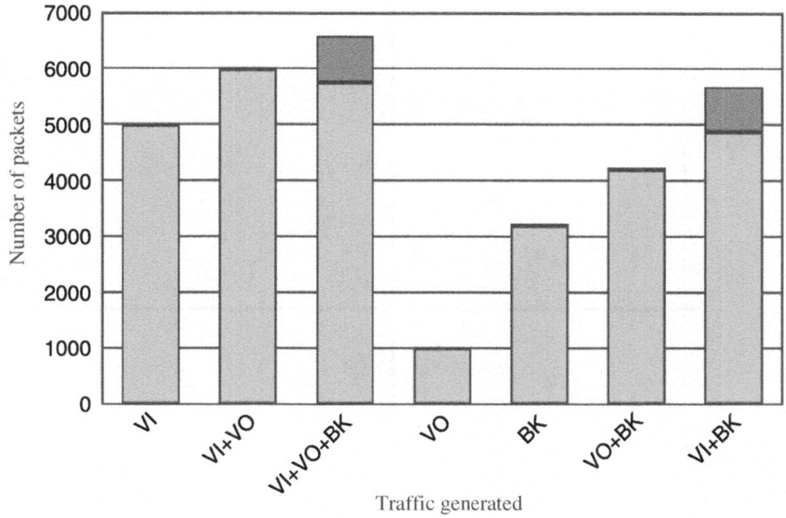

Fig. 21.7 Total of packets received/lost

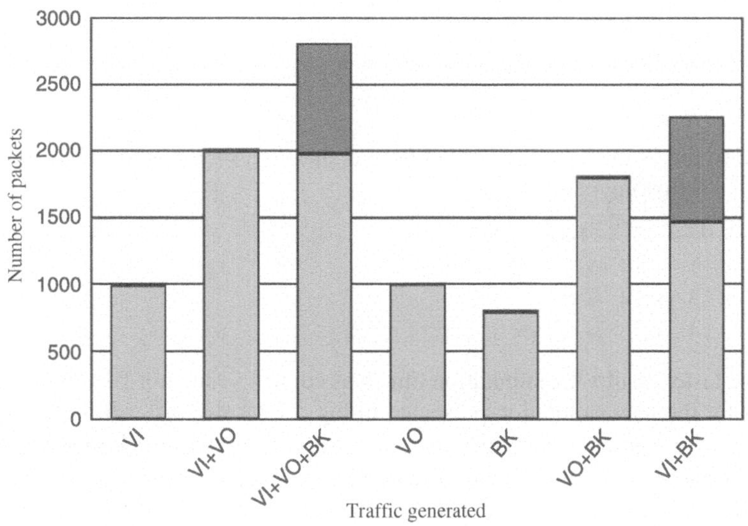

Fig. 21.8 Total of packets lost/received at destination

a packet after it experiences two collisions). The extension of this problem is larger for background traffic, which can be explained by the largest size of each packet, which will require the fragmentation into four smaller packets. For voice traffic, the delay is very low and it seems that it does not affect the remaining traffic at all. This is due to the small packets generated by this traffic type (no fragmentation required), and because these stations are in line-of-sight, so no multi-hop is needed.

For the next tests, a new deployment was made changing the position of station 6, while keeping all the others in the same position, Fig. 21.9.

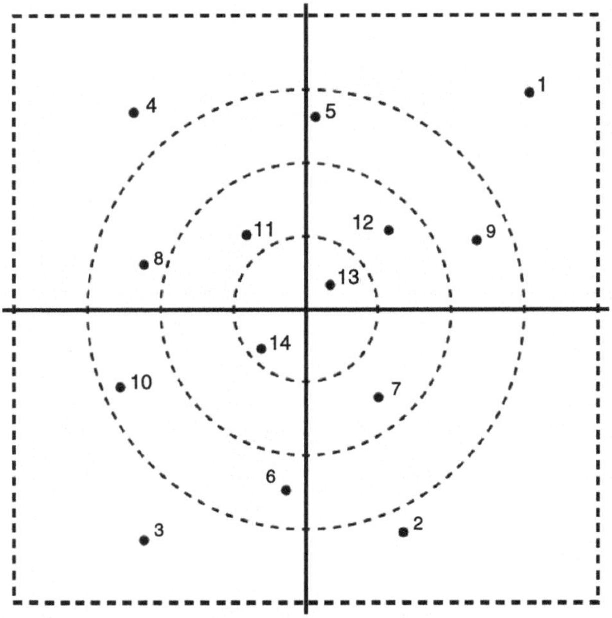

Fig. 21.9 Station deployment for the next set of tests

The traffic is the same generated before, and as so the routing algorithm computed the following paths:

A. $1 \rightarrow 9 \rightarrow 13 \rightarrow 14 \rightarrow 10 \rightarrow 3$
B. $4 \rightarrow 11 \rightarrow 14 \rightarrow 7 \rightarrow 2$
C. $5 \rightarrow 13 \rightarrow 14 \rightarrow 6$
D. $8 \rightarrow 14 \rightarrow 13 \rightarrow 9$

To have faster results the simulation time was cut to 1 s (and not 10 s anymore). The results for the new simulation presented in Fig. 21.10 also differ from the previous because a new policy establishes that a packet will be discarded after experiences eight collisions (previously it was discarded after just two collisions). By increasing the collision limit, one should expect that the number of packets lost will be reduced (for each packet it will be given eight chances to be transmitted, four times more than previously). Comparing the new results with the previous (Fig. 21.8), one can see that in fact the number of packets lost was reduced, while the number of packets successfully delivered increased.

21.5 Conclusions

In this chapter we presented the potentialities of a custom-made IEEE 802.11e simulator, with multi-hop capabilities. Based on a previous work that developed a MAC plus PHY layers simulator for this standard, we added above these layers a routing

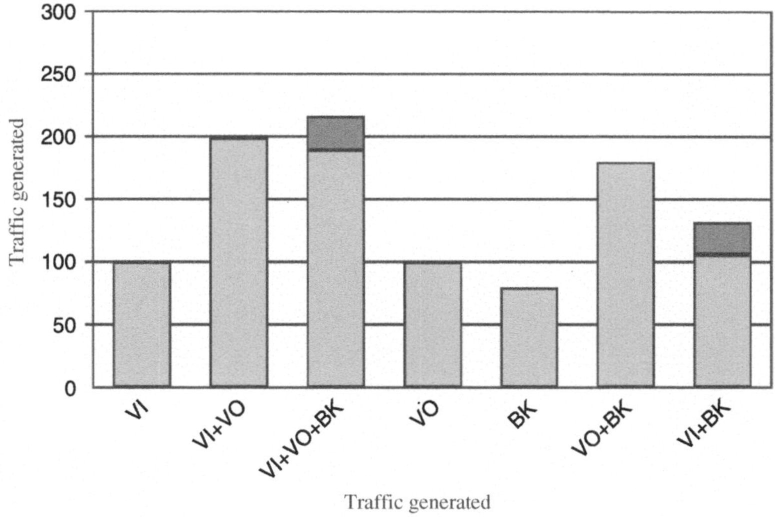

Fig. 21.10 Total of packets lost (grey)/received (light grey) at destination

algorithm in order to allow the simulation of traffic among stations that can not communicate directly. We performed some initial tests to verify if the simulator was performing as expected, and the results were encouraging. A few improvements for this simulator are being considered, as well as the release of its source code to the scientific community for research purposes. Another possibility currently under study, is to adapt the simulator to allow simulations in the field of Wireless Sensor Networks (WSN) – another research the authors of this chapter are interested in. The simulator was built using the MAC and PHY layers of the standard IEEE 802.11e, and it is intended to replace it by using a MAC layer specification developed specifically for WSN.

References

1. H. Holma and A. Toskala, in: *WCDMA for UMTS: HSPA Evolution and LTE*, edited by Harri Holma and Antti Toskala (Wiley, Chichester, 2007).
2. A. R Prasad and N. R. Prasad, in *802.11 WLANs and IP Networking*, edited by Anand R. Prasad and Neeli R. Prasad (Artech House, London, 2005).
3. O. Cabral and F. J. Velez, Event-Driven Simulation for IEEE 802.11e Optimization, *IAENG International Journal of Computer Science* **35**(1), 161–173 (2008).
4. E. W. Dijkstra, A note on two problems in connexion with graphs, *Numerische Mathematik* **1**, 269–271 (1959).
5. N. Qiang, Performance analysis and enhancements for IEEE 802.11e wireless networks, *IEEE Network* **19**(4), 21–27 (2005).

Chapter 22
Web-Based Management of Distributed Services

George Oikonomou and Theodore Apostolopoulos

Abstract This paper presents WebDMF, a Web-based Framework for the Management of Distributed services. It is based on the Web-based Enterprise Management (WBEM) family of standards and introduces a middleware layer of entities called "Representatives". WebDMF can be integrated with existing WBEM infrastructures and is not limited to monitoring. On the contrary, it is capable of actively modifying the run-time parameters of a managed service. Due to its abstract design, it is suitable for the management of a variety of distributed services, such as grids and content delivery networks. The paper includes a discussion on WebDMF's design, implementation and advantages. We also present experiments on an emulated network topology as an indication of the framework's viability.

Keywords Common information model · distributed services management · web-based enterprise management · WebDMF

22.1 Introduction

Legacy management approaches such as the Simple Network Management Protocol (SNMP) [1], target single nodes and are mainly used for the management of devices. The current paradigm of highly decentralized, distributed systems poses new challenges and increased complexity in the area of network and service management. There is need for solutions that are better-suited for software and services. Such solutions should also take into account the managed system's distributed nature.

G. Oikonomou (✉)
Department of Informatics, Athens University of Economics and Business, Athens, Greece
E-mail: geo@aueb.gr

S.I. Ao and L. Gelman (Eds.), *Advances in Electrical Engineering and Computational Science*, Lecture Notes in Electrical Engineering.

We present the design, implementation and evaluation of WebDMF, a Web-based Management Framework for Distributed services. It uses standard web technologies: Its core is based on the Web-based Enterprise Management (WBEM) family of specifications [2–4]. It is not limited to monitoring but is also capable of modifying the run-time parameters of a managed service. Due to its abstract design it has wide scope and is suitable for the management of a variety of services. We have particularly studied its application for the management of grids [5] and content delivery networks [6]. This paper is a revision of our previous work, originally published in [7].

In this context, the paper's contribution is as follows:

- It proposes a framework for the management of distributed services, called WebDMF. We demonstrate how WebDMF can be integrated with existing WBEM infrastructures and, in doing so, it can achieve its goal without need for modifications to the managed service.
- It provides indications for the viability of the approach through a preliminary performance evaluation.

Section 22.2 briefly outlines existing efforts in the field of distributed service management. In Section 22.3 we describe WebDMF's architectural design. Implementation details and preliminary evaluation results are presented in Section 22.4. Finally, in Section 22.5 we discuss our conclusions.

22.2 Related Work

Existing research and development efforts investigate techniques for the management of distributed applications and services. The Open Grid Forum's GMA [8] and the Relational GMA (R-GMA) [9] focus on monitoring grids. MonALISA [10] and the CODE toolkit [11] have wider scope but still only perform monitoring.

There are some proposals that can go beyond monitoring. The Unified Grid Management and Data Architecture (UGanDA) [12] contains an infrastructure manager called MAGI. MAGI has many features but is limited to the management of UGanDA deployments. MRF is a Multi-layer resource Reconfiguration Framework for grid computing [13]. It has been implemented on a grid-enabled Distributed Shared Memory (DSM) system called Teamster-G [14].

Ganglia is another noteworthy proposal [15]. It performs monitoring of distributed environments (mainly computer clusters) by adopting a hierarchical approach and by breaking down a cluster into "federations". Astrolabe [16] partitions a distributed system into non overlapping zones. Each zone has a "representative" that is chosen automatically among the zone's nodes through a gossip protocol. Users can monitor a zone's operational parameters by specifying aggregation functions. Lastly, UPGRADE-CDN [17] introduces "AMonitor", a monitoring framework for content delivery networks. AMonitor uses state-of-the-art technologies but is limited to monitoring.

Thorough research reveals that, existing management systems with many features tend to target specific distributed applications. On the other hand, generic approaches with wider scope offer less features and are often limited to monitoring. Furthermore, many current research efforts are based on proprietary technologies and are limited in terms of interoperability and ease of integration with existing infrastructures. Lastly, there are questions about the scalability of some approaches.

Compared to those necessary efforts, the framework discussed in this paper has wide scope while maintaining a rich set of features, including the ability to perform modifications on the managed service. It is based on open standards and is easy to integrate with existing WBEM infrastructures.

22.3 WebDMF: A Web-Based Management Framework for Distributed Services

WebDMF stands for Web-based Distributed Management Framework. It treats a distributed system as a number of nodes that are interconnected over a network and share resources to provide services to the end user. The proposed framework's aim is to provide management facilities for the nodes. Through those, we achieve management of the entire deployment.

WebDMF's architecture is based on the "Web-based Enterprise Management" (WBEM) family of specifications, published and maintained by the Distributed Management Task Force (DMTF). WBEM adopts the client-server paradigm and specifies a method for the modeling of management data, called the Core Information Model (CIM) [2]. A WBEM server is also called a CIM Object Manager (CIMOM) and interacts with clients through a set of well-defined request-response HTTP packets, as specified in "CIM Operations over HTTP" [4]. The payload of those packets is CIM data encoded in XML [3].

WebDMF nodes function as WBEM entities; clients, servers or both, depending on their role in the deployment. The framework's design introduces a middleware layer of entities that we call "Management Representatives". They act as peers and form a management overlay network, which can be integrated with an existing WBEM-based management infrastructure. Representatives act as intermediaries between existing WBEM clients and CIM Object Managers. In our work, we use the terms "Management" and "Service" node when referring to those entities. Figure 22.1 displays the three management entities mentioned above, forming a very simple topology.

This resembles the "Manager of Managers" (MoM) approach. However, in MoM there is no direct communication between domain managers. In WebDMF, representatives are aware of the existence of their peers. Therefore, it adopts the "Distributed Management" approach. By distributing management over several nodes throughout the network, we can increase reliability, robustness and performance, while network communication and computation costs decrease [18].

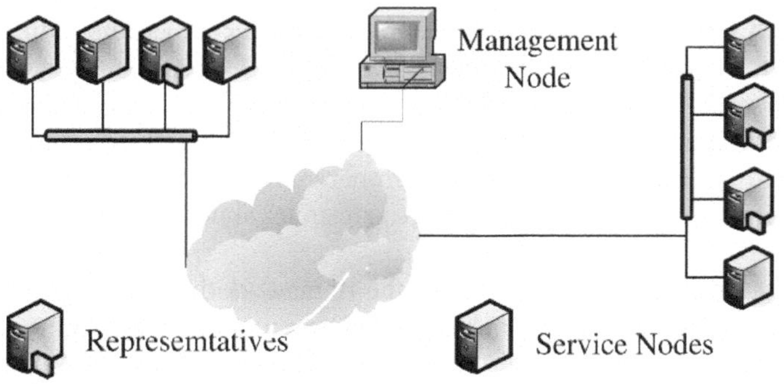

Fig. 22.1 WebDMF entities

22.3.1 Management and Service Nodes

A "Management Node" is a typical WBEM client. It is used to monitor and configure various operational parameters of the distributed service. Any existing WBEM client software can be used without modifications.

A "Service Node" is the term used when referring to any node – member of the distributed service. For instance, in the case of a computational grid, the term can describe an execution host. Similarly, in a content delivery network a service node is the term used to refer to an intermediate relay node or a node hosting content. The role of a node in a particular distributed deployment does not affect the design of our framework.

Typically, a Service Node executes an instance of the (distributed) managed service. As displayed in Fig. 22.2a, a WebDMF request is received by the CIMOM on the Service Node. A WBEM provider, specifically written for the service, handles the execution of the management operation. The existence of such a provider is a requirement. In other words, the distributed service must be manageable through WBEM. Alternatively, a service may be manageable through SNMP, as shown in Fig. 22.2b. In this scenario, the node may still participate in WebDMF deployments but some functional restrictions will apply.

22.3.2 Management Representative

The framework introduces an entity called the "Management Representative". This entity initially receives a request from a WBEM client (management node) and performs management operations on the relevant service nodes. After a series of message exchanges, it will respond to the initial request. A representative is more than a simple 'proxy' that receives and forwards requests. It performs a number of other operations including the following:

- Exchanges messages with other representatives regarding the state of the system as a whole

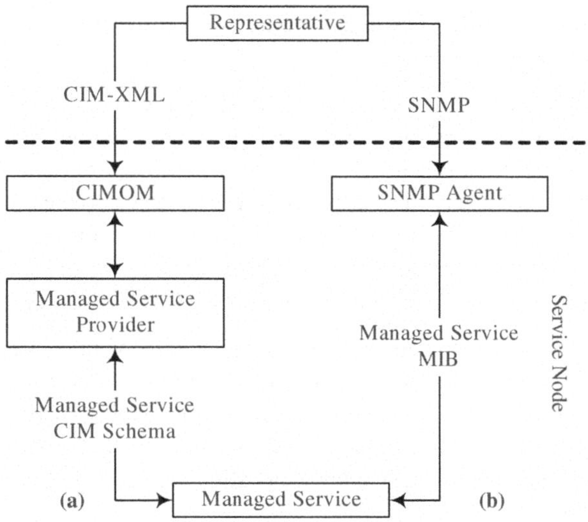

Fig. 22.2 Two methods of communication between the WebDMF representative and service nodes

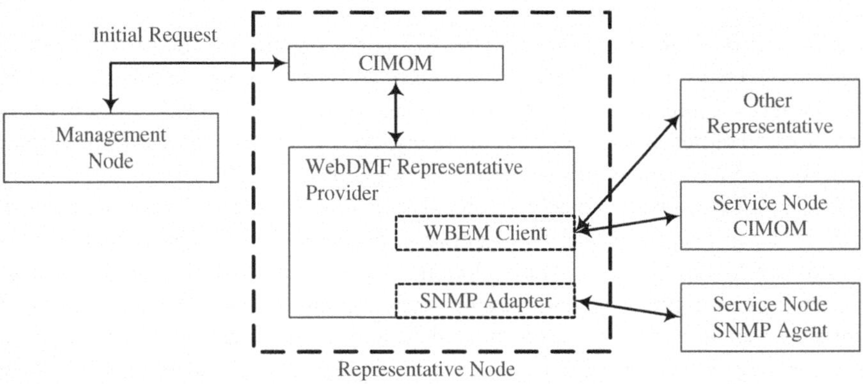

Fig. 22.3 WebDMF representative and its building blocks

- Keeps a record of Service Nodes that participate in the deployment
- Redirects requests to other representatives

An initial request does not state explicitly which service nodes are involved in the management task. The functionality implemented in the representative is partly defined by its ability to determine the destination of intermediate messages. All intermediate message exchange is transparent to the management node as well as to the end user.

In order to achieve the above functionality, a representative is further split into building blocks, as shown in Fig. 22.3. It can act as a WBEM server as well as a client. Initial requests are received by the CIMOM on the representative. They are delegated to the WebDMF provider module for further processing. The module performs the following functions:

- Determines whether the request can be served locally.
- If the node can not directly serve the request then it selects the appropriate representative and forwards it.
- If the request can be served locally, the representative creates a list of service nodes that should be contacted and issues intermediate requests.
- It processes intermediate responses and generates the final response.
- It maintains information about the distributed system's topology.

In some situations, a service node does not support WBEM but is only manageable through SNMP. In this case, the representative attempts to perform the operation using SNMP methods. This is based on a set of WBEM to SNMP mapping rules. There are limitations since it is not possible to map all methods. However, even under limitations, the legacy service node can still participate in the deployment.

22.3.3 Domains

In a WebDMF deployment, a representative is responsible for the management of a group of service nodes, either on its own or in cooperation with other peers. We use the term "Domain" when referring to such groups. The relationship between domains and representatives is many-to-many. Thus, a representative may be responsible for the management of more than one domain.

Domains are organized in a hierarchical structure. The hierarchy's top level (root node of the tree) corresponds to the entire deployment. The exact rationale behind the domain hierarchy of each individual deployment can be based on a variety of criteria. For example, a system may be separated into domains based on the geographical location of nodes or on the structure of an organization. In any case, the hierarchy can affect the performance, ease of management and scalability of the system. The domain hierarchical structure and the relationships between domains and nodes are depicted in the class diagram in Fig. 22.4.

22.3.4 CIM Schemas and Operations

WebDMF defines two categories of management operations: (i) Horizontal (Category A) and (ii) Vertical (Category B).

Horizontal Operations enable management of the WebDMF overlay network itself. Those functions can, for example, be used to perform topology changes. The message exchange that takes place does not involve Service Nodes. Therefore, the managed service is not affected in any way. When a service node joins the network, it registers itself with a representative. This can be performed by a script on the

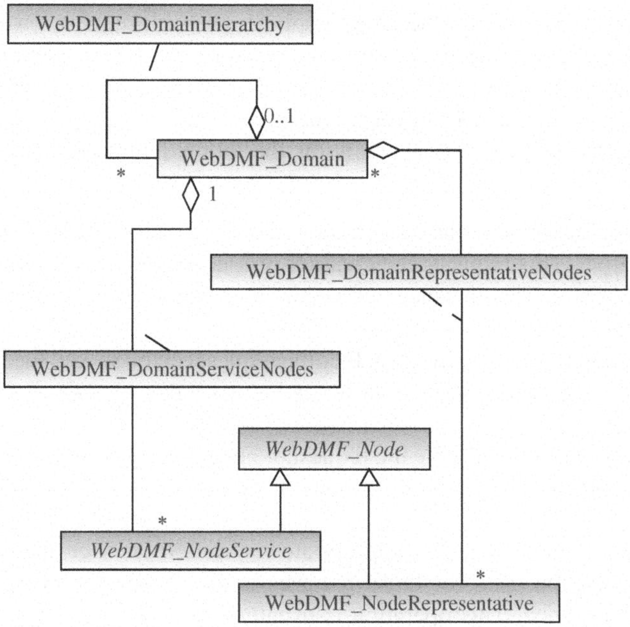

Fig. 22.4 Domains and nodes – the WebDMF core schema

service node itself or manually by a user. However, there is no automated discovery mechanism for new service nodes.

The registration operation is very lightweight. However, its "semi-manual" nature makes WebDMF more suitable for deployments with relatively infrequent topology changes. It is less suitable for systems with very frequent and abrupt topology changes (disconnecting nodes, roaming nodes), such as wireless, ad-hoc networks.

Vertical operations read and modify the CIM schema on Service Nodes, thus achieving management of the target application. Typical examples include:

- Setting new values on CIM objects of many service nodes
- Reading operational parameters from service nodes and reporting an aggregate (e.g. sum or average)

In line with the above, we have designed two CIM Schemas for WebDMF, the core schema ("WebDMF_Core") and the request factory. They both reside on representative repositories.

The former schema models the deployment's logical topology, as discussed earlier. The class diagram presented in Fig. 22.4 is the actual diagram representing WebDMF_Core. Horizontal functions correspond to WBEM operations on instances of classes declared in this schema.

The latter schema corresponds to vertical functions. Users can call WBEM methods on instances of this schema. In doing so, they can define management operations that they wish to perform on the target application. Each request towards the distributed deployment is treated as a managed resource itself. For example, users can create a new request. They can execute it periodically and read the results. They can modify it, re-execute it and finally delete it.

In order to complete a vertical operation, the following message exchange takes place:

- The management node sends a `CreateInstance()` WBEM message to any representative. This requests the creation of a new instance for class `WebDMF_RequestWBEM`. This instance defines the management operation that needs to be performed on service nodes.
- The representative determines whether the request can be served locally. If not, it chooses the appropriate representative and issues a new `CreateInstance()` request.
- If the representative can serve the request, it generates a list of service nodes that must be contacted, based on the values of properties of the recently generated instance.
- The representative sends the appropriate requests to service nodes. The type of CIM operation used for those requests is also based on values of properties in the instance. This operation is usually a WBEM `GetInstance()` or a `ModifyInstance()`.
- After all service nodes have been contacted and responses have been sent, the instance on the first representative contains results. It remains available to the user for potential modification and/or re-execution. All other intermediate instances are deleted.

Request factory classes are generic. They are not related in any way with the CIM schema of the managed application. This makes WebDMF appropriate for the management of a wide variety of services. Furthermore, no re-configuration is needed with each new release of the target service.

22.4 Implementation and Performance Evaluation

22.4.1 Implementation Details

The WebDMF representative is implemented as a single shared object library file (.so). It is comprised of a set of WBEM providers. Each one of them implements CIM management operations for a class of the WebDMF schemas.

The interface between the CIMOM and providers complies with the Common Manageability Programming Interface (CMPI). Providers themselves are written in C++. This does not break CIMOM independence, as described in [19].

The representative was developed on Linux 2.6.20 machines. We used gcc 4.1.2 and version 2.17.50 of binutils. Testing took place using version 2.7.0 of the Open Pegasus CIMOM.

22.4.2 Performance Evaluation

In order to evaluate WebDMF, we installed a testbed environment using ModelNet [20]. Results presented here have been obtained from actual code execution on an emulated network topology and are used as an indication of the solution's viability. They also bring out one of WebDMF's key features, the ability to perform changes to the managed service.

The topology emulated by ModelNet represents a wide-area network. It consists of 250 virtual nodes situated in three LANs with each LAN having its own gateway to the WAN. The three gateways are interconnected via a backbone network, with high bandwidth, low delay links. We have also installed two WebDMF representatives (nodes R1 and R2).

In this scenario, the distributed managed service is a content delivery network implemented with the OpenCDN (oCDN) open source software [21]. Service nodes correspond to oCDN Origin nodes. Each of those registers with an oCDN centralized control entity called "Request Routing and Distribution Management" (RRDM). We have also designed a CIM schema and the relevant WBEM providers that enable management of OpenCDN nodes. Figure 22.5 portrays the emulated topology and test scenario. For clarity reasons, we omit service nodes residing in other domains.

In our experiment, we wish to perform a change in the 200 service nodes residing in domain "ccslab.aueb.gr". To be more specific, we wish to set them to register with a different RRDM. This involves changing properties of the oCDN_Origin instance on the service node's CIMOM.

The client forms a WBEM `CreateInstance()` request for class `WebDMF_RequestWBEM` of the request factory. It is initially sent to the WebDMF Representative (R1). The request is forwarded to R2. R2 sends `ModifyInstance()` requests to the 200 service nodes. R2 reports to R1. R1 sends the final response to the client.

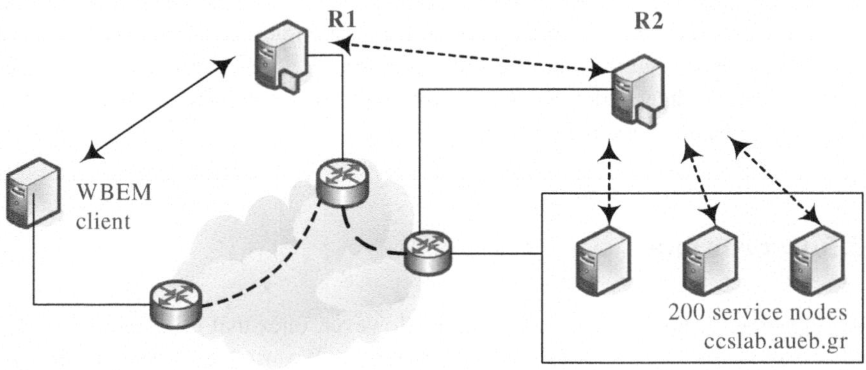

Fig. 22.5 Emulated topology and test scenario

Table 22.1 Evaluation results

Metrics		Values
Repetitions	N	200
Central tendency	Arithmetic mean	3.851944
	Median	3.869485
Dispersion	Variance	0.005281
	Standard deviation	0.072670
Quartiles	Q0 (min)	3.710042
	Q1	3.797917
	Q3	3.910481
	Q4 (max)	3.997615
95% Confidence interval for the mean		
	From	3.841873
	To	3.862016

The above experiment was repeated 200 times. Table 22.1 summarizes the results with times measured in seconds. Consider the fact that each repetition involves 204 request-response exchanges among various nodes. Furthermore, consider that packets crossing the network are of a small size (a few bytes). The total execution time includes the following:

- Communication delays during request–response exchanges. This includes TCP connection setup for all WBEM message exchanges.
- Processing overheads on R1 and R2. This is imposed by WebDMF's functionality.
- Processing at the service nodes to calculate the requested value and generate a response.

The absolute value of the average completion time may seem rather high. However, in general terms, processing times are minimal compared to TCP connection setup and message exchange. With that in mind, we can see that each of the 204 request-responses completes in less than 20 ms on average. This is normal.

After 200 repetitions we observe low statistical dispersion (variance and standard deviation). This indicates that the measured values are not widely spread around the mean. We draw the same conclusion by estimating a 95% confidence interval for the mean. This indicates that the same experiment will complete in the same time under similar network load conditions.

22.5 Conclusions

Existing monitoring solutions are generic. However, ones that offer more features and the ability to "set" tend to have narrow scope. We wanted to design a framework that would be generic enough and suitable for a wide variety of services. At the same time, it should not be limited to monitoring. WebDMF achieves that by detaching the details of the managed service from the representative logic. Management functions for a specific service are realised by WBEM providers on the service nodes. Representatives unify those on a deployment scale. WebDMF has some other noteworthy features:

- It is based on WBEM. This is a family of open standards. WBEM has been considered adequate for the management of applications, as opposed to other approaches (e.g. SNMP) that focus on the management of devices.
- It provides interoperability with existing WBEM-based management infrastructures without need for modifications.

References

1. W. Stallings, *SNMP, SNMPv2, SNMPv3, RMON 1 and 2*. Addison Wesley, Redwood City, CA, 1999.
2. *CIM Infrastructure Specification*, DMTF Standard DSP0004, 2005.
3. *Representation of CIM in XML*, DMTF Standard DSP0201, 2007.
4. *CIM Operations over HTTP*, DMTF Standard DSP0200, 2007.
5. G. Oikonomou, and T. Apostolopoulos, Using a web-based framework to manage grid deployments, in *Proceedings of The 2008 International Conference on Grid Computing and Applications* (part of WORLDCOMP 08), Las Vegas, 2008, pp. 10–16.
6. G. Oikonomou, and T. Apostolopoulos, Web-based management of content delivery networks, in *Proceedings of 19th IFIP/IEEE International Workshop on Distributed Systems: Operations and Management (DSOM). Managing Large Scale Service Deployment (MANWEEK 08)*, Samos, Greece, 2008, pp. 42–54.
7. G. Oikonomou, and T. Apostolopoulos, WebDMF: A web-based management framework for distributed services, in *Proceedings of The 2008 International Conference of Parallel and Distributed Computing* (part of WCE 08), Vol. I, London, 2008, pp. 593–598.
8. *A Grid Monitoring Architecture*, Open grid Forum GFD.7, 2002.
9. W. Cooke, et al., The Relational Grid Monitoring Architecture: Mediating Information about the Grid, *Journal of Grid Computing*, **2**(4), 2004, 323–339.
10. I. C. Legrand, H. B. Newman, R. Voicu, C. Cirstoiu, C. Grigoras, M. Toarta, and C. Dobre, MonALISA: An Agent based, Dynamic Service System to Monitor, Control and Optimize Grid based Applications, in *Proceedings of Computing in High Energy and Nuclear Physics (CHEP)*, Interlaken, Switzerland, 2004.
11. W. Smith, A System for Monitoring and Management of Computational Grids, in *Proceedings of International Conference on Parallel Processing (ICPP '02)*, 2002, p. 55.
12. K. Gor, D. Ra, S. Ali, L. Alves, N. Arurkar, I. Gupta, A. Chakrabarti, A. Sharma, and S. Sengupta, Scalable enterprise level workflow and infrastructure management in a grid computing environment, in *Proceedings of Fifth IEEE International Symposium on Cluster Computing and the Grid (CCGrid '05)*, Cardiff, UK, 2005, pp. 661–667.
13. P.-C. Chen, J.-B. Chang, T.-Y. Liang, C.-K. Shieh, and Y.-C. Zhuang, A multi-layer resource reconfiguration framework for grid computing, in *Proceedings of 4th International Workshop on Middleware for Grid Computing (MGC '06)*, Melbourne, Australia, 2006, p. 13.
14. T.-Y. Liang, C.-Y. Wu, J.-B. Chang, and C.-K. Shieh, Teamster-G: A grid-enabled software DSM system, in *Proceedings of Fifth IEEE International Symposium on Cluster Computing and the Grid (CCGrid '05)*, Cardiff, UK, 2005, pp. 905–912.
15. M. L. Massie, B. N. Chun, and D. E. Culler, The ganglia distributed monitoring system: Design, implementation, and experience, *Parallel Computing*, **30**, 817–840, 2004.
16. K. P. Birman, R. V. Renesse, and W. Vogels, Navigating in the storm: Using astrolabe to adaptively configure web services and their clients, *Cluster Computing*, **9**, 127–139, 2006.

17. G. Fortino, and W. Russo, Using p2p, grid and agent technologies for the development of content distribution networks, *Future Generation Computer Systems*, **24**, 180–190, 2008.
18. M. Kahani, and P. H. W. Beadle, Decentralised approaches for network management, *ACM SIGCOMM Computer Communication Review*, **27**(3), 36–47, 1997.
19. *Common Manageability Programming Interface*, The Open Group, C061, 2006.
20. A. Vahdat, K. Yocum, K. Walsh, P. Mahadevan, D. Kostic, J. Chase, and D. Becker, Scalability and Accuracy in a Large-Scale Network Emulator, in *Proceedings of 5th Symposium on Operating Systems Design and Implementation (OSDI)*, December 2002.
21. OpenCDN Project [Online]. Available: http://labtel.ing.uniroma1.it/opencdn/

Chapter 23
Image Index Based Digital Watermarking Technique for Ownership Claim and Buyer Fingerprinting

Sarabjeet S. Bedi, Rabia Bano, and Shekhar Verma

Abstract Digital Image Watermarking is a technique for inserting information into an image that can be later extracted for a variety of purpose including identification, ownership authentication, and copyright protections. The goal of this paper is to design a secure watermarking scheme in spatial domain to identify the true buyer and to protect the ownership copyright of digital still images. The scheme inserts a binary watermark in an original image that serves as a fingerprint for a specific buyer. Utilizing the one-way property of the hash function, it generates the unique buyer fingerprint. An image index concept has been used to split the image randomly into disjoint blocks. A threshold is defined for manipulation of pixel intensities of these blocks. The amount of manipulation in pixel intensities absolutely depends upon the buyer fingerprint. The recovery of the watermark not only identifies the buyer but also protects the owner's copyright. The extracted buyer fingerprint identifies the buyer of the original image. The scheme has the ability to trace the buyer involved in forgery. The experimental results show that the watermark can be retrieved from the attacked watermarked image even if the opponent has the complete information of watermarking scheme.

Keywords Buyer fingerprint · cryptographic hash function · Digital Watermarking · Image key

23.1 Introduction

Information security aspects come into role when it is necessary or desirable to protect information as it is being shared during transmission or storage from an immediate future opponent who may present a threat to confidentiality, authenticity,

S. S. Bedi (✉)
MJP Rohilkhand University, Bareilly (UP), India-243006
E-mail: dearbedi@gmail.com

S.I. Ao and L. Gelman (Eds.), *Advances in Electrical Engineering and Computational Science*, Lecture Notes in Electrical Engineering.

integrity, access control, and availability. The need for information security has been termed as security attack, mechanism, and services [1]. The various data hiding techniques like cryptography, stenography, digital signatures, finger printing, have been developed to address the information security issues but fail to provide the complete solutions to protect the intellectual property rights of digital multimedia data. The existing basket of technologies like cryptography secures the multimedia data only during storage or transmission and not while it is being consumed [2]. Digital Watermarking provides an answer to this limitation as the watermark continues to be in the data throughout its usage.

Digital Watermarking (DWM) is the process of embedding information into digital multimedia contents such that the embedded information (watermark) can be extracted later [3]. The extracted information can be used for the protection of intellectual property rights i.e. for establishing ownership right, ensuring authorized access and content authentication. The watermarking system can be implemented using either of two general approaches. One approach is to transform the original image into its frequency domain representation and embed the watermark data therein. The second is to directly treat the spatial domain data of the host image to embed the watermark.

According to Hartung [4] most proposed watermark method utilize the spatial domain, this may be due to simplicity and efficiency. The spatial domain method is about embedding watermark information directly into image pixels proposed by [5]. These techniques embed the watermark in the LSB plane for perceptual transparency which is relatively easy to implement but their significant disadvantages includes the ease of bypassing the security they provide [5, 6] and the inability to lossy compression the image without damaging the watermark.

The methods [6, 7] extended the work to improve robustness and localization in their technique, in which watermark is embedded by adding a bipolar M-Sequence in the spatial domain and detection is via a modified correlation detector. But these schemes were not very much capable to protect the watermark and also not resist with lossy compression.

Regarding security and content authentication a new method [8] introduce the concept of hash function, in which author insert the binary watermark into the LSB of original image using one-way hash function. The technique is too sensitive since the watermark is embedded into the LSB plane of the image and algorithm also does not very resist with lossy compression. Thus the limitations of spatial domain methods are that, in general they are not robust to common geometric distortion and have a low resistance to JPEG lossy compression. Therefore a scheme is required to fulfill the existing gap in the use of watermarking and cryptographic techniques together.

In this paper, the robust and secure digital invisible watermarking scheme in spatial domain is proposed. The proposed scheme combines the advantages of cryptographic concept and imperceptibility feature of digital image watermarking. The security and perceptual transparency are achieved by using cryptographic one-way hash function and computation of threshold value respectively. For the robustness the proposed technique does not depend upon perceptually significant regions; rather it utilizes the concept of image key and buyer fingerprint generator. The unique binary sequence serves as the buyer authenticator of a particular image. The recovery of watermark also protects the copyrights.

The rest of the paper is organized as follows. Section 23.2 describes the concept of generation of image key and buyer fingerprint. Section 23.3 explains the proposed watermarking scheme with watermark insertion and extraction process. Experimental results with discussion and conclusion are given in Section 23.4 and Section 23.5 respectively.

23.2 Image Key and Buyer Fingerprint

The image key and generation of buyer fingerprint used in proposed scheme are described as below.

23.2.1 Image Key

An Image key, I for any grayscale image, I_m of size $X \times Y$ pixels (let, $X = 2^x$ and $Y = 2^y$) is of the same size as of image. The original image is spatially divided into $z = 2^n (= 2^x = 2^y)$ disjoint blocks. Each block is denoted by an index k, for $1 \leq k \leq 2^n$ and for every block $B(k)$, $B_i(k) \cap B_j(k) = \varnothing$ for $1 \leq k \leq 2^n, i \neq j$. The length of blocks which contains different number of locations may vary from each other. Each location for two-dimensional image key, I is represented as (i, j), where i corresponds to row for $1 \leq i \leq 2^n$ and j corresponds to column for $1 \leq j \leq 2^n$. The image key, I store the index numbers of image pixels. As the indexes are random, so it is not possible for attacker to guess the image key. The pixel value of blocks in the image is modified corresponding to the indexes of the image key. Now even if the attacker knows the factor by which manipulation is done, he/she will not be able to locate the pixels whose values are modified. The image key is stored with the owner of the image and is used during the extraction of the watermark.

23.2.2 Buyer Fingerprint

The Buyer fingerprint, F is a binary sequence of length 2^n and will be equal to the number of blocks. Each location of buyer fingerprint is denoted by index, k for $1 \leq k \leq 2^n$. The unique buyer fingerprint, F is generated using cryptographic hash function. A cryptographic hash function $H(S) = \{f_1, f_2, \ldots, f_p\}$ where S is string of data of arbitrary length, f_i is binary output bits of the hash function, and p is a size of the output bit string, has the two important properties [1]. First property it is computationally infeasible to find any input which maps to pre-specified output.

Second is computationally infeasible to find any two distinct inputs that map to same output referred as collision-resistant. The generation of Buyer fingerprint is discussed in Section 23.3.

23.3 Proposed Technique

23.3.1 Basic Idea

In this watermarking scheme, the original image, I_m is divided into blocks based on the image key. The intensity value of pixels in each block is modified depending upon the bit of watermark to get the watermarked image. Only those pixels are modulated whose value is greater than the threshold, $T(k)$ for $1 \leq k \leq 2^n$. The motivation for selecting the threshold is to increase the perceptual transparency as it filters out the pixels having intensity values less than threshold. The threshold is calculated for each block. The arithmetic mean of pixels for each group is calculated separately which is taken as the threshold. In the extraction phase watermarked blocks are compared with the original image block and the buyer fingerprint is generated.

23.3.2 Generation of Watermark

The watermark is generated through the creation of unique Buyer fingerprint, F of an original image using one-way hash function. The Buyer fingerprint is created as $F = H(X, Y, I, M)$ where X is an image height, Y is an image width, I is an image key and M is a MSB array of block. The two parameters image key and MSB array of block makes the Buyer fingerprint unique. The MSB of pixels at index, k are summed together to form the k_{th} element of array M. The image key, I store the index numbers of block of image pixels. As the indexes are generated randomly, so it is not possible for the attacker to guess the image key. The generated Buyer fingerprint is of length 2^n and shall be equal to the number of blocks of original image.

23.3.3 Insertion of Watermark

The original image, I_m is spatially divided into $z = 2^n (=2^x = 2^y)$ blocks. The arithmetic mean of each block is calculated which is taken as threshold for that block. As the length of the generated Buyer fingerprint is equal to the number of blocks of original image. Therefore for each bit of watermark the intensities of the pixels of correspond indexed blocks shall be modified. This modification is based on threshold value of the specific block. If the watermark bit is '1', then the pixels having intensity value greater than threshold are increased by a factor, α. Where as for the watermark bit ‘0’, no modification is recommended. The detailed algorithm of watermark insertion is as given in the following steps:

1. For $1 \leq k \leq 2^n$
   (a) Let $T(k)$ be the threshold of the block having index k.
   (b) Suppose $F(k)$ is the k_{th} bit of watermark.
2. For $1 \leq i \leq 2^x, 1 \leq j \leq 2^y$
   (a) Let $I_m(i, j)$ be pixel intensity of original image at location (i, j).
   (b) Assume that (i, j) belongs to the block $B(k)$.
   (c) If $F(k) = 0$, then $W_m(i, j) = I_m(i, j)$.
   (d) If $F(k) = 1$, then
   (e) If $I_m(i, j) > T(k)$, then $W_m(i, j) = I_m(i, j) + \alpha$ and $d(k) = d(k) + \alpha$.

The factor α is taken as positive throughout the insertion. The value of α is chosen so as to maintain the fidelity. The larger values will degrade the quality of the image. From step 2e of algorithm it is clear that the factor $d(k)$ records the increase in the value of intensities for each block.

The watermark insertion process is described in the block diagram as shown in Fig. 23.1. The buyer signature, threshold and *image key* are the required input attributes for watermark insertion.

23.3.4 Extraction of Watermark

This section illustrates the extraction of watermark form watermarked image, W_m. The extraction procedure requires original image, and image key. The algorithm for the extraction of watermark is as given in the following steps:

1. For $1 \leq k \leq 2^n$

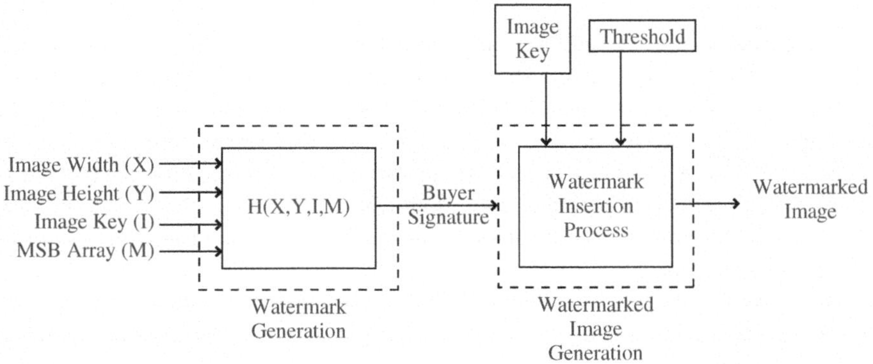

Fig. 23.1 Watermark insertion process

a. Set values $s(k) = 0$.
b. Set bit values $b(k) = 0$.
c. Let $T'(k)$ be the new threshold of the group having index k.

2. For $1 \leq i \leq 2^x, 1 \leq j \leq 2^y$

   If $|W_m(i, j) - I_m(i, j)| > |\alpha|$, then
   $W_m(i, j) = I_m(i, j) + \alpha$.

3. If (i, j) belongs to the block $B(k)$ and $W_m(i, j) > T'$, then

   $$s(k) = s(k) + (W_m(i, j) - I_m(i, j)).$$

4. If βs(k) $\leq d(k)$ and $s(k) \neq 0$, then $b(k) = 1$

   Else $b(k) = 0$

5. The retrieved watermark is $b(k)$.

The value of α is used to reduce the watermarked pixel intensities (may be attacked) to optimum value. As the value of α is known, the limits of pixel values can be determined after watermarking. This fact can be used to rectify the values of attacked pixels. The value of α is utilized for the calculation of the new value of threshold in step 2 of algorithm 2. For the smaller value α there will smaller change in the value of threshold. The difference of watermarked and original pixel is found for every block in step 3 of algorithm 2. The damping factor $\beta(<1)$ decreases the value of $s(k)$ in step 4 of algorithm 2 so as to satisfy the inequality. The value of $s(k) = 0$ for unaltered watermarked image. The extracted watermark, b is obtained which is equal to inserted watermark; F as there is exact bit-wise match.

The Fig. 23.2 describes the process of extraction of the watermark using the original image, the watermarked image, the threshold and the *image key*. The *buyer signature* output of this phase is used to verify the ownership claim.

23.4 Results and Discussion

The proposed scheme is applied on different grayscale original images of size 128×128 pixels. The unique watermark of 128 bits and the unique image key of size 128×128 pixels are generated. The threshold is computed for every block and the watermark is inserted in the spatial domain. The different watermarked images of size 128×128 pixels are obtained for different buyer fingerprint. The common image processing operations like modification, low pass filter, medium pass filter, cropping, combination of rotation and scaling and compression are imposed on watermarked image.

The Normalized Cross Correlation (NCC) values between original image and the attacked watermarked image is computed to demonstrate the robustness and fidelity of the proposed scheme.

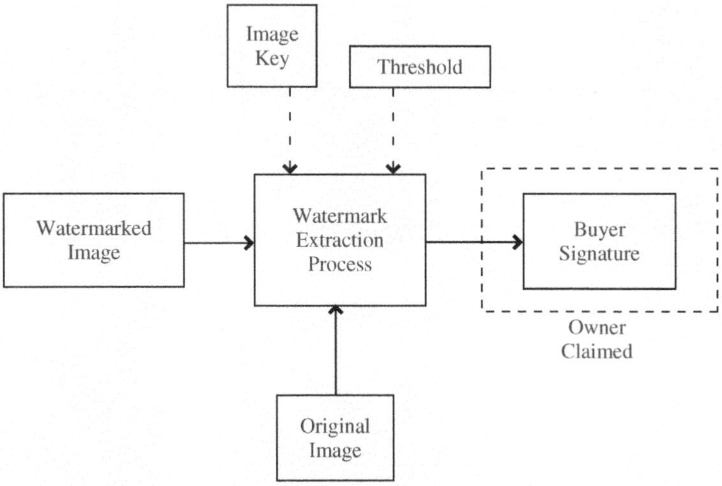

Fig. 23.2 Extraction of the watermark

Table 23.1 NCC values of images for some common image processing operations

Attacks image	Scaling	Scaling + rotation	Compression	MPF	LPF	Modify	Cropping
Lena	0.9999	0.9878	0.9943	0.98461	0.96294	0.68474	0.66389
Cameraman	0.9999	0.9878	0.99523	0.9502	0.93504	0.71265	0.88474

The simulation results have been produced on various sets of images. The original gray-scale image of "Lena" of size 128 × 128 pixels is taken as shown in Fig. 23.3a. Unique image key of size 128 × 128, and MSB array of 128 bits are taken as input to MD5 hash function. The 128-bit string buyer fingerprint is generated and inserted in original image which produced the watermarked image of size 128 × 128 as shown in Fig. 23.3b. The result shows that the watermark is invisible. The NCC value between the original image and the watermarked image is 0.99998. However the watermark is extracted from watermarked image. The exact bit-wise match between extracted watermark and the inserted buyer fingerprint identifies the true buyer of the original image.

The effect of some attacks on the watermarked image is also shown in Table 23.1 and Fig. 23.1. Table 23.2 shows the bit-wise match of inserted buyer fingerprint and extracted buyer fingerprint. The graph of the resultant values have been plotted as shown in Fig. 23.4.

In the low pass filter attack, a mask of 3 × 3 is used. The NCC value of 0.96294 is obtained between the original and the modified watermarked image whereas it is 0.98461 for median-pass filter attack. The modified watermarked image is shown in Fig. 23.3c. An exact match of 128 bits is obtained for both the filtering operations as illustrated in Table 23.2. The watermarked image is scaled to twice of its size as shown in Fig. 23.3d and the measured value of NCC is 0.9999. For the combined

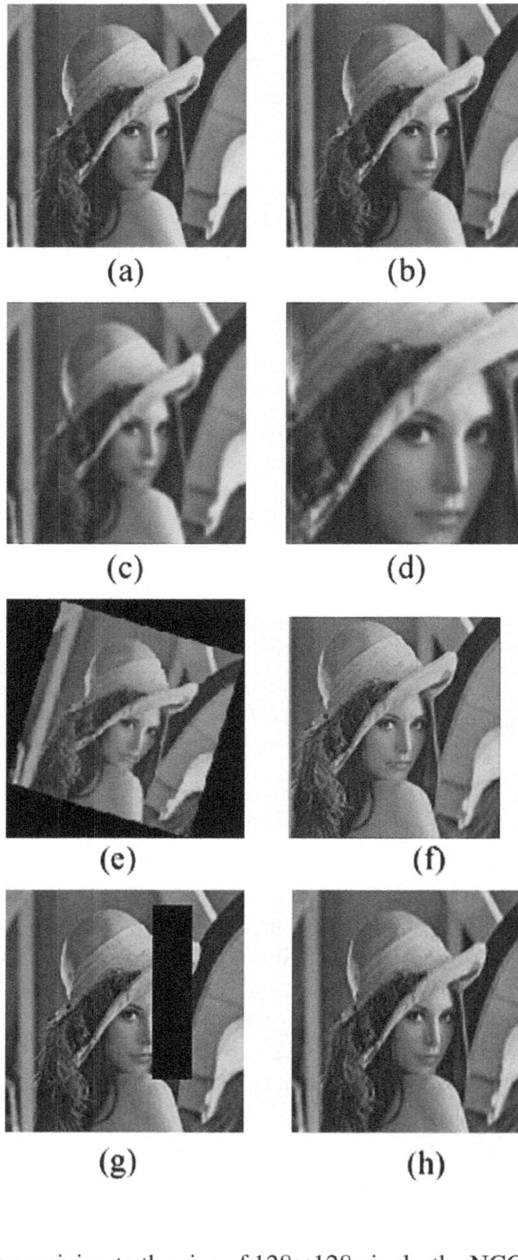

Fig. 23.3 (**a**) Original Lena image. (**b**) Watermarked Lena image. (**c**)–(**h**) Results of watermarked images with some attacks: (**c**) low pass filtered; (**d**) scaled; (**e**) rotated (17°); (**f**) cropped from top and left border; (**g**) random modification; (**h**) JPEG lossy compression (quality factor = 90)

attack of rotation of 17° followed by resizing to the size of 128×128 pixels, the NCC value between original and modified watermarked image is 0.9878 (Fig. 23.3e). The 126–127 bits are recovered for scaling and combined attack. The cropped and randomly modified images are shown in Fig. 23.3f–g. In case of modification, the watermarked image has been tampered at specific locations by changing the pixel values. In case of severe manipulation to pixel intensities, a bit-wise match of 120–

Table 23.2 BIT-wise match between inserted and extracted buyer fingerprint (LENA IMAGE)

Attacks β	Scaling	Scaling + rotation	Compression	MPF	LPF	Modify	Cropping
1	127	126	128	128	128	120	123
0.9	128	128	128	128	128	128	128

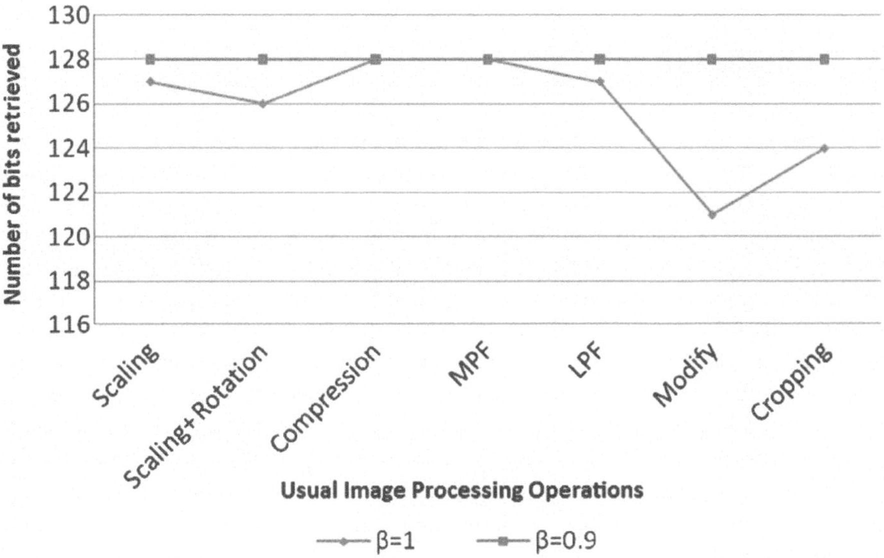

Fig. 23.4 Wise match between the inserted and extracted buyer signature for the image, "Lena"

126 bits is obtained. With the use of damping factor of 0.9, exact 128 bits is obtained for the buyer fingerprint and the value of NCC is 0.68474. In case of cropping the NCC value becomes 0.66389. In a rigorously cropped image 3–4 bits of inserted Buyer fingerprint are lost which can be recovered by using a damping factor of 0.9. The robustness against lossy JPEG compression with quality factor 90 is demonstrated in Fig. 23.3h and the NCC value 0.9843 is obtained with all the 128 bits of Buyer fingerprint are recovered.

The NCC values in Table 23.1 shows that scaling attack and the combined attack of scaling and rotation as the pixel values are not changed intensely at the center of the image, so there is not much distortion in the NCC values that are obtained. In the Low Pass Filter (LPF) attack only the pixels having lower intensities are retrievable, so depending on the image if the LPF's threshold for a block of the image is higher than the threshold assumed in the watermark insertion algorithm then the NCC value is higher than that obtained if the threshold of the LPF is lower than the one assumed. In the Medium Pass Filter (MPF) attack only the pixels having medium intensities are retrievable, so depending on the image if the threshold assumed in the watermark insertion algorithm lies towards the lower range of the MPF's thresholds for a block of the image then the NCC value is higher than that obtained if the assumed threshold lies towards the higher threshold of the MPF. In the modification and cropping attacks as some of the pixels are tampered badly the original pixel values cannot be identified. Therefore results demonstrates that proposed scheme is more robust to geometric attacks and compression, whereas robust to modification and cropping.

23.5 Conclusion

The proposed watermarking technique is for copyright protection and buyer fingerprinting. The image key and the unique watermark are the key features for sustaining the security of algorithm. The watermark generated uses the property of the hash function which is an important feature in making the watermark cryptographically secure. When the positive value of α is used the watermark increased the intensity of an image block. An attacker who is known to the watermarking process might intentionally utilize this fact and try to remove the watermark. For this purpose the attacker must know the image key. The attacker cannot guess the secret image key as the image key has been kept secret and the indexes have been generated randomly. The watermark extraction algorithm rectifies the pixel values in case the attacker increases or decreases the pixel values. An invalid image key would be unsuccessful in forging the watermark. The technique survives with common image transformations as well as intentional attacks, maintaining the objective of buyer fingerprinting and ownership claim.

References

1. W. Stallings, "Cryptography and Network Security: Principles and Practices," *Pearson Education, Inc.*, NJ, 3rd Ed., 2005, ISBN 81–7808–902–5.
2. S. S. Bedi and S. Verma, "Digital Watermarking Technology: A Demiurgic Wisecrack Towards Information Security Issues," *Invertis Journal of Science and Technology*, vol. 1, no. 1, pp. 32–42, 2007.
3. A. Kejariwal, "Watermarking," Magazine of IEEE Potentials, October/November, 2003, pp. 37–40.
4. F. Hartung and M. Kutter, "Multimedia Watermarking Techniques," *Proceedings of IEEE*, vol. 87, no. 7, pp. 1079–1106, July 1999.
5. M. Yeung and F. Mintzer, "Invisible watermarking for image verification," *Journal of Electric Imaging*, vol. 7, no. 3, pp. 578–591, July 1998.
6. R. Wolfgang and E. Delp, "Fragile watermarking using the VW2D watermark," *Proceedings of the IS & T/SPIE Conference on Security and Watermarking of Multimedia Contents*, pp. 204–213, San Jose, CA, January 1999.
7. J. Fridrich, "Image watermarking for temper detection," *Proceedings of the IEEE International Conference on Image Processing*, vol. 2, pp. 404–408, Chicago, IL, October 1998.
8. P. W. Wong and N. Memon, "Secret and Public key Image Watermarking Schemes for Image Authentication and Ownership Verification," *IEEE Transaction on Image Processing*, vol. 10, no. 10, October 2001.

Chapter 24
Reverse Engineering: EDOWA Worm Analysis and Classification

Madihah Mohd Saudi, Emran Mohd Tamil, Andrea J Cullen, Mike E Woodward, and Mohd Yamani Idna Idris

Abstract Worms have become a real threat for computer users for the past few years. Worm is more prevalent today than ever before, and both home users and system administrators need to be on the alert to protect their network or company against attacks. It is coming out so fast these days that even the most accurate scanners cannot track all of the new ones. Indeed until now there is no specific way to classify the worm. To understand the threats posed by the worms, this research had been carried out. In this paper the researchers proposed a new way to classify the worms which later is used as the basis to build up a system which is called as the EDOWA system to detect worms attack. Details on how the new worm of classification which is called as EDOWA worm classification is produced are explained in this paper. Hopefully this new worm classification can be used as the basis model to produce a system either to detect or defend organization from worms attack.

Keywords Classification · worm analysis · payload · worm classification

24.1 Introduction

Computer worm can caused millions dollars of damage by infecting hundreds and thousands of host in a very short period of time. A computer worm is a computer program or a small piece of software that has the ability to copy itself from machine to machine. It uses computer networks and security holes to replicate itself. Computer worm is classified as a highly threat to the information technology world. McCarthy [1] defines computer worm as a standalone malicious code program that copies itself across networks. Meanwhile Nachenberg [2] stated that the computer worm is a program that is designed to copy itself from one computer to another, dominate some network medium such as through email. The computer worm would

M. M. Saudi (✉)
Faculty Science & Technology, Universiti Sains Islam Malaysia (USIM), Nilai, Malaysia
E-mail: madihah@usim.edu.my

S.I. Ao and L. Gelman (Eds.), *Advances in Electrical Engineering and Computational Science*, Lecture Notes in Electrical Engineering.

infect as many machines as possible on the network. The prototypical computer worm infects a target system only once; after the initial infection, the worm attempts to spread to other machines on the network.

While there are thousands of variations of computer worms, the classification of computer worm can be done via several ways Nazario [3] proposed a function structure framework that consist six components. The components are reconnaissance capabilities, specific attack capabilities, a command interface, communication capabilities, intelligence capabilities and unused attack capabilities. The framework mainly predicts the future research on network worms. Another form of classification of computer worm Weaver et al. [4], they classified computer worm into five major classifications: target discovery, carrier, activation, payload and attackers. As for Kienzle et al. [5], they classified computer worm into three basic classes by propagation strategies. The computer worms are classified into e-mail worm, windows file-sharing worm and traditional worm.

24.2 Method of Testing

In order to produce a new worm classification, the researchers' had conducted few testing and researches. A worm code analysis laboratory is build to test and analyze the worm. A controlled laboratory environment is built to conduct the testing. This laboratory is not connected to the real network. Three machines were used and connected in LAN using a hub. Figure 24.1 below illustrates the laboratory.

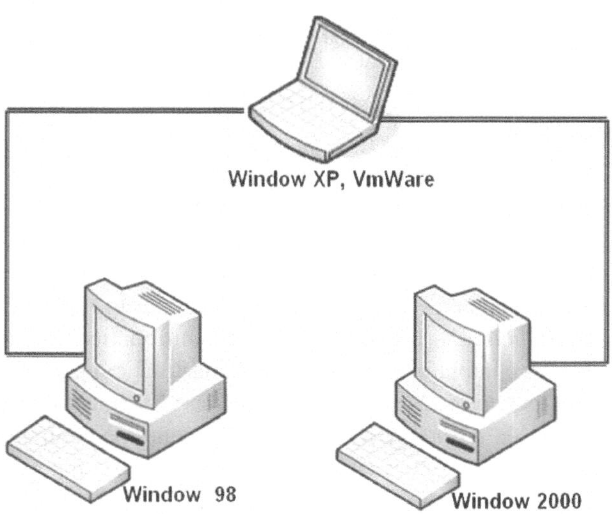

Fig. 24.1 Worm code analysis lab architecture

24.2.1 *Worm Analysis Process*

For a new computer worm, the main purpose of analyzing it is to know the intention of the code. As for computer worm that has been released in real time, the worm analysis process is for verification of what the worm source code intention and to verify as what has been published in anti virus website or CERT website or it might be a new worm feature. The analysis techniques can be divided into two techniques, either the static analysis or dynamic analysis. Before loading the computer worm specimen into the machine, researchers make sure all the preparation and the verification has been done. While conducting the analysis, all the process is document in writing. A written record of analytic techniques and the computer worm action is useful in understanding how the computer worm works, tracing through its function in a repeatable fashion and improving the worm analyst skills.

24.2.2 *Loading Specimen*

When all the preparation and the verification have been done, the lab is disconnected from any production network. The USB thumb drive is used to transfer the computer worm specimen onto the lab system.

Once the specimen already placed in the lab system, the analysis can be carried out. To determine the purpose and the capabilities of this piece of code, researchers can used either the static analysis or dynamic analysis.

24.2.2.1 Static Analysis

Static analysis is also known as white box analysis. It involves analyzing and understanding source code. If only binary code available, the binary code has to be compiled to get the source code. White box analysis is very effective in finding programming errors and implementation errors in software and to know the flow of the program. The static analysis looks at the files associated with the computer worm on the hard drive without running the program. With static analysis, the general idea of the characteristics and purpose of the computer worm can be analyzed. The static analysis phase involves antivirus checking with research, analyzing the strings, looking for scripts, conducting binary analysis, disassembling and reverses compiling.

Antivirus Checking with Research

When computer worm specimen already copied to the testing machine, researchers will run the antivirus to check if the antivirus installed detects anything. If the antivirus detects the computer worm, check the name of the computer worm and search it in any antivirus website for further information. If the computer worm is in compressed or archived form, researchers will open the archive to get its content. The researchers need to verify if the information available from the antivirus website is correct.

String Analysis

The strings that extracted from the computer worm could help the researchers to know more about the computer worm characteristics. A few tools such as TDS3 and Strings.exe (from Sysinternal) were used to extract the strings. The information that could be retrieved from the extracted strings are consist of the worm specimen's name, user dialog, password for backdoors, URLs associated with the malware, email address of the attacker, help or command-line options, libraries, function calls and other executables used by the malware.

Looking for Script

The language written for the computer worm can be identified based on strings extracted from it. The following Table 24.1 gives some clues:

Disassemble Code

Disassemble and debugger is used to convert a raw binary executable into assembly language for further analysis. Researchers use the tools that have been listed in Appendix A to disassemble and debug the computer worm.

24.2.2.2 Dynamic Analysis

Dynamic analysis involves executing the computer worm and watching its actions. The computer worm is activated on a controlled laboratory system.

Table 24.1 Scripting language

Scripting language	Identifying characteristics inside the file	File's common suffix
Bourne shell scripting language	Starts with line !#/bin/sh	.sh
Perl	Start with line !#usr/bin/perl	.pl, .perl
JavaScript	Includes the word javascript or JavaScript, especially in the form <Script language = "JavaScript">	.js, .html, .htm
Visual basic script (VBScript)	Includes the word VBScript, or characters vb scattered throughout the file	.vbs, .html, .htm

Monitoring File Activities

Most computer worms read from or write to the file system. It might attempts to write files, altering existed programs, adding new files or append itself to the file system. By using tool such as Filemon all actions associated with opening, reading, writing, closing and deleting files can be monitored.

Monitoring Process

The monitoring tool such as Prcview v3.7.3.1 or Process Explorer, displays each running program on a machine, showing the details of what each process is doing. With this kind of tool the files, registry keys and all of the DLLs that each process has loaded can be monitored. For each running process, the tool displays its owner, its individual privileges, its priority and its environment variables.

Monitoring Network Activities

From a remote machine which will be in the same LAN with the infected testing machine, the port scanner, Nmap program and a sniffer will be installed. The port scanner and Nmap program are used to monitor the listening port. A sniffer will be installed to sniff the worm traffic. All of the related tools like Ethereal, NeWT and TDS-3 use the sniffer. Using the sniffer, details of individual packets and all packets transmitted across the LAN can be monitored. As for the local network monitoring tool (TDIMon), it will monitor and records all request to use the network interface and show how the worm grabbed the network resources and used them.

The computer worm might have placed the network interface in promiscuous (broadcast) mode which allowed it to sniff all packets from LAN. To determine if the infected machine in promiscuous mode state of the interface, run the Promiscdetect.exe tool.

Monitoring Registry Access

The registry need to be monitored as the registry is the hierarchical database containing the configuration of the operating system and most programs installed on the machine. The monitoring registry access can be done by using the Regmon.

24.3 EDOWA Classification

After the analysis done in the laboratory, it leads researchers to produce a new classification for the EDOWA system. A proposal of the classification of worm is made. This classification is based on several factors: infection, activation, payload,

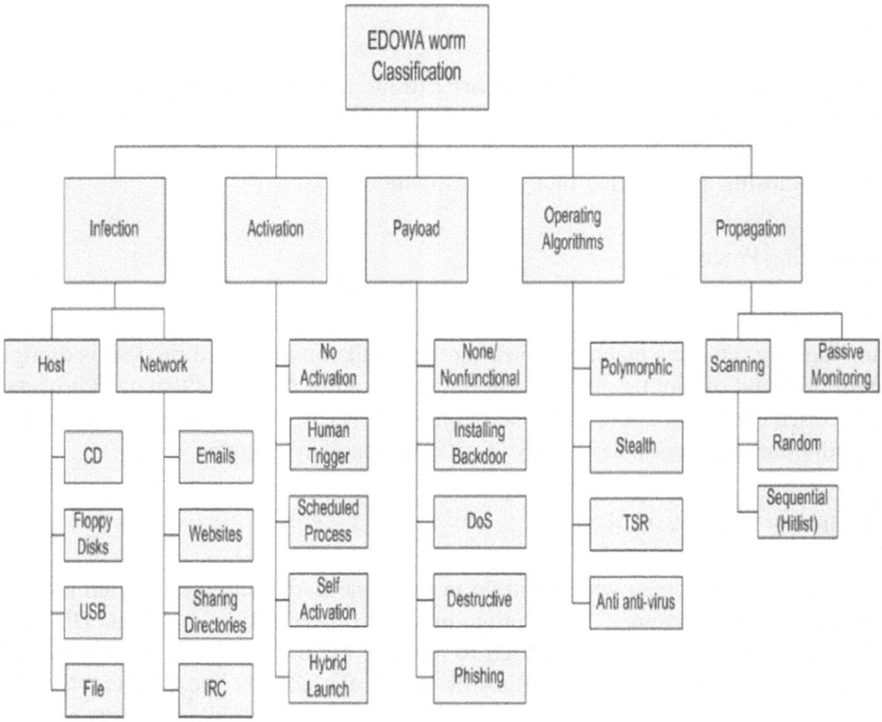

Fig. 24.2 EDOWA worm classification

operating algorithms and propagation. Figure 24.2 is an overview of the EDOWA classification.

Infection is the first step before a worm infects a computer. The activation is a mechanism that will activate a worm. Payload is a code that carries a destructive mechanism of a worm. Operating algorithms is a techniques used to avoid detection. Finally, the propagation mechanisms are how the worm spread to reproduce.

Sections below will elaborate more details on the EDOWA classification.

24.3.1 Infection

Infection is the phase on how a computer gets infected by a worm. From the research of eight classifications that is available, we found out that only one research made infection as the first phase. Albanese et al. [6] says that infection refers to how a worm gains initial control of a system. Worms rely on two general methods to infect a host. Either they exploit an error in software running on a system, or they are the result of some action taken by a user. Here we proposed two techniques:

24.3.1.1 Host

Host is a mechanisms needed by the worm copy itself to a new systems that are not yet been infected. It cannot propagate autonomously across the network. Host computer worms where the original terminates itself after launching a copy on another host so there is only one copy of the worm running somewhere on the network at any given moment. It requires human help to move from one machine to another. CD, Floppy Disks, USB (thumb-drive and external hard disk) and File are the most common host available now.

24.3.1.2 Network

Network is the fastest way in moving worm. It consists of multiple parts, each running on different machines and possibly performing different actions also using the network for several communication purposes. Propagating from one machine to another is only one of those purposes. It can infect a computer without human interaction. Most simply copy themselves to every computer with which the host computer can share data. Most Windows networks allow machines within defined subgroups to exchange data freely, making it easier for a worm to propagate itself.

24.3.2 Activation

For this classification, activation is defined as a trigger mechanism of a worm. This phase is where the worm enters the host once it found a machine. According to Nazario [3] in his book "Defense and Detection Strategies against Internet Worms", stated these are used to launch an attack against an identified target system.

24.3.2.1 No Activation

Worm with no activation will just stay in the computer doing nothing. It just used up some hard disk space.

24.3.2.2 Human Trigger

Human trigger is the slowest activation mechanisms. Usually this approach use worm that propagates using emails. The social engineering technique is used to attract user to click on the file to activate the worm [7]. According to Christoffersen et al. [8], some worms are activated when the user performs some activity, like resetting the machine, logging onto the system and thereby running the login scripts or executing a remotely infected file. Evidently, such worms do not spread very rapidly.

24.3.2.3 Schedule Process

Based on Weaver et al. [4], the second fastest worms activate is by using scheduled system processes. Schedule process is an activation that is based on specific time and date. Many desktop operating systems and applications include auto-updater programs that periodically download, install and run software updates.

24.3.2.4 Self Activation

The worms that are fastest activated are able to initiate their own execution by exploiting vulnerabilities in services that are always on and available (e.g., Code Red [9] exploiting IIS Web servers) or in the libraries that the services use (e.g., XDR [10]). Such worms either attach themselves to running services or execute other commands using the permissions associated with the attacked service.

24.3.2.5 Hybrid Launch

Hybrid Launch uses the combination of two or more activation mechanism to launch a worm. ExploreZip [2] is an example of a hybrid-launch worm. ExploreZip send an e-mail that required a user to launch the infected attachment to gain control of the system. Once running on the computer system, ExploreZip would automatically spread itself to other computers over the peer-to-peer network. These targeted machines would then become infected on the next reboot without any known user intervention.

24.3.3 Payload

For this classification, payload is defined as a destructive mechanism of a worm. A payload is code designed to do more than spread the worm. Many worms have been created which are only designed to spread, and don't attempt to alter the systems they pass through.

24.3.3.1 No Payload

Worm with no payload does not do any harm to the computer system. This kind of worm will just propagate without infecting any destructive mechanisms to the computer.

24.3.3.2 Installing Backdoor

Backdoor is a term used to describe a secret or undocumented means of getting into a computer system. Many programs have backdoors placed by the programmer to allow them to gain access to troubleshoot or change the program. Some backdoors are placed by hackers once they gain access to allow themselves an easier way in next time or in case their original entrance is discovered. Example of backdoor attack is the worm called Blaster [11] that used the backdoor mechanism to transfer the worm payload to newly infected systems.

24.3.3.3 Denial of Services

A denial of service (DoS) attack floods a network with an overwhelming amount of traffic, slowing its response time for legitimate traffic or grinding it to a halt completely. The more common attacks use built-in "features" of the TCP/IP protocol to create exponential amounts of network traffic. Example of DoS attack is the well-known worm called Code Red [9] was programmed to unleash a denial-of-service attack on the Whitehouse.gov that targeting the actual Whitehouse.gov IP address.

24.3.3.4 Destructive

This will do harm to the machine or the host. According to Shannon et al. [12], Witty worm deletes a randomly chosen section of the hard drive, which, over time, renders the machine unusable.

24.3.3.5 Phishing

Phishing [13] is a criminal activity using social engineering techniques. Phishers attempt to fraudulently acquire sensitive information, such as usernames, passwords and credit card details, by masquerading as a trustworthy entity in an electronic communication. eBay and PayPal are two of the most targeted companies, and online banks are also common targets. Phishing is typically carried out by email or instant messaging, and often directs users to give details at a website, although phone contact has been used as well. Attempts to deal with the growing number of reported phishing incidents include legislation, user training, and technical measures.

24.3.4 Operating Algorithms

Operating algorithms is defined as a detecting techniques used by worms to avoid detection. Among the eight classifications that are available, we found out that only one research have operating algorithms in their classification. Albanese et al. [6] classified it as survival. Operating algorithms are the mathematical and logical ways that a worm attempts to avoid detection. It can be categorized as:

24.3.4.1 Polymorphic

A polymorphic worm is a worm that changes all part of their code each time they replicate, this can avoid scanning software. Kruegel et al. [14] paper defined polymorphic worms as a worm that is able to change their binary representation as part of the spreading process. It can be achieved by using self-encryption mechanisms or semantics-preserving code manipulation techniques. As a consequence, copies of a polymorphic worm might no longer share a common invariant substring of sufficient length and the existing systems will not recognize the network streams containing the worm copies as the manifestation of a worm outbreak.

24.3.4.2 Stealth

Stealth worm use a concealment mechanisms. It spread slow, evokes no irregular communication pattern and spread in an approach that makes detection hard. Cheetancheri [15] stated in his thesis that the goal of stealth worm is to spread to as many hosts as possible without being detected. However, once such a worm has been detected, manual means of mitigation are possible.

24.3.4.3 Terminate and Stay Resident

Terminate and stay resident (TSR) worm exploit a variety of techniques to remain resident in memory once their code has been executed and their host program has terminated. These worms are resident or indirect worm, known as such because they stay resident in memory, and indirectly find files to infect as they are referenced by the user.

24.3.4.4 Anti Anti-virus

Anti anti-virus will corrupt the anti-virus software by trying to delete or change the anti-virus programs and data files so the anti-virus does not function properly. According to Nachenberg [16], anti anti-virus or are usually called retroviruses, are computer viruses that attack anti-virus software to prevent themselves from being detected. Retroviruses delete anti-virus definition files, disable memory resident anti-virus protection and attempt to disable anti-virus software in any number of ways.

24.3.5 Propagation

Propagation is defined as a worm that spread itself to another host or network. After researching the propagation issue, we strongly believe that there are two ways for a worm to reproduce itself: scanning and passive.

24.3.5.1 Scanning

Scanning is a method used by worms to find their victims. We strongly agree with the method proposed by Weaver et al. [4]. There are two possible scanning method that is random scanning and sequential scanning.

Random Scanning

It is the most popular method where the worm simply picks a random IP address somewhere in the Internet Address space and then tries to connect to it and infect it. Example of a random scanning worm is the Blaster [11] that picks a random number to determine whether to use the local address it just generated or a completely random one.

Sequential Scanning (Hitlist)

The worm releaser scans the network in advance and develops a complete hitlist of all vulnerable systems on the network. According to Staniford [17], the worm carries this address list with it, and spreads out through the list.

Passive

Worms using a passive monitoring technique are not actively searching for new victims. Instead, they are waiting for new targets to contact them or rely on the user to discover new targets. Christoffersen et al. [8] says passive worms tend to have a slow propagation rate, they are often difficult to detect because they generate modest anomalous reconnaissance traffic.

24.4 Conclusion

This new classification is produced based on the research and testing that have been done in the laboratory. The classifications are divided into five main categories: Infections, Activation, Payload, Operating Algorithms and Propagation. Efficient

Detection of Worm Attack (EDOWA) system is produced based on this classification. This EDOWA system is not discussed in this paper. Hopely this paper can be used as the basis model for worm classification and can be used for other upcoming research.

References

1. L. McCarthy, "Own Your Space: Keep Yourself and Your Stuff Safe Online (Book)," Addison-Wesley, Boston, MA, 2006.
2. C. Nachenberg, "Computer Parasitology," *Proceedings of the Ninth International Virus Bulletin Conference*, September/October 1999, pp. 1–25.
3. J. Nazario, "Defense and Detection Strategies against Internet Worms" (BOOK), *Artech House Inc.*, 2003. Or a paper entitles "The Future of Internet Worm" by Nazario, J., Anderson, J., Wash, R., and Connelly, C. Crimelabs Research, Norwood, USA, 2001.
4. N. Weaver, V. Paxson, S. Staniford and R. Cunningham, "A Taxonomy of Computer Worms," *Proceedings of the ACM CCS Workshop on Rapid Malcode (WORM)*, pp. 11–18, 2003.
5. D.M. Kienzle and M.C. Elder, "Recent Worms: A Survey and Trends," *Proceedings of the ACM CCS Workshop on Rapid Malcode (WORM)*, pp. 1–10, 2003.
6. D.J. Albanese, M.J. Wiacek, C.M. Salter and J.A. Six, "The Case for Using Layered Defenses to Stop Worms (Report style)," UNCLASSIFIED-NSA Report, pp. 10–22, 2004.
7. C.C. Zou, D. Towsley and W. Gong, "Email worm modeling and defense," *Computer Communications and Networks, ICCCN 2004*, pp. 409–414, 2004.
8. D. Christoffersen and B.J. Mauland, "Worm Detection Using Honeypots (Thesis or Dissertation style)," Master dissertation, Norwegian University of Science and Technology, June 2006.
9. H. Berghel, "The Code Red Worm: Malicious software knows no bounds," *Communication of the ACM*, vol. 44, no. 12, pp. 15–19, 2001.
10. CERT. CERT Advisory CA-2002–25 Integer Overflow in XDR Library, http://www.cert.org/advisories/ca-2002–25.html
11. M. Bailey, E. Cooke, F. Jahanian, D. Watson and J. Nazario, "The Blaster Worm: Then and Now," *IEEE Security & Privacy*, vol. 3, no. 4, pp. 26–31, 2005.
12. C. Shannon and D. Moore, "The Spread of the Witty Worm," *IEEE Security & Privacy*, vol. 2, no. 4, pp. 36–50, 2004.
13. A. Tsow, "Phishing With Consumer Electronics: Malicious Home Routers," *15th International World Wide Web Conference (WWW2006)*, Edinburgh, Scotland, May 2006.
14. C. Kruegel, E. Kirda, D. Mutz, W. Robertson and G. Vigna, "Polymorphic Worm detection using structural information of executables," *8th International Symposium on Recent Advances in Intrusion Detection (RAID)*, 2005.
15. S.G. Cheetancheri, "Modeling a computer worm defense system (Thesis or Dissertation style)," Master dissertation, University of California, 1998.
16. C. Nachenberg, "The Evolving Virus Threat," *23rd NISSC Proceedings*, Baltimore, Maryland, 2000.
17. S. Staniford, President of Silicon Defense. "The Worm FAQ: Frequently Asked Questions on Worms and Worm Containment," *The Worm Information Center*, 2003

Chapter 25
Reconfigurable Hardware Implementation of a GPS-Based Vehicle Tracking System

Adnan Yaqzan, Issam Damaj, and Rached Zantout

Abstract In this chapter, we build on a recently produced VTS (The Aram Locator) offering a SOC replacement of the microcontroller-based implementation. Although the microcontroller-based system has acceptable performance and cost, an FPGA-based system can promise less cost and a more cohesive architecture that would save processing time and speeds up system interaction. The performance of the proposed implementations is evaluated on different FPGAs. The suggested designs show enhancement in terms of speed, functionality, and cost.

Keywords Hardware Implementation · Vehicle Tracking System · GPS-Based · Reconfigurable · FPGA

25.1 Introduction

After a great evolution, reconfigurable systems fill the flexibility, performance and power dissipation gap between the application specific systems implemented with hardwired Application Specific Integrated Circuits (*ASICs*) and systems based on standard programmable microprocessors. Reconfigurable systems enable extensive exploitation of computing resources. The reconfiguration of resources in different parallel topologies allows for a good matching with the inherent intrinsic parallelism of an algorithm or a specific operation. The reconfigurable systems are thus very well-suited in the implementation of various data-stream, data-parallel, and other applications.

The introduction of a new paradigm in hardware design called Reconfigurable Computing (*RC*) offers to solve any problem by changing the hardware configurations to offer the performance of dedicated circuits. Reconfigurable computing

I. Damaj (✉)
Electrical and Computer Eng'g Dept, Dhofar University, P.O. Box 2509, 211 Salalah, Oman
E-mail: i_damaj@du.edu.om

S.I. Ao and L. Gelman (Eds.), *Advances in Electrical Engineering and Computational Science*, Lecture Notes in Electrical Engineering.

enables mapping software into hardware with the ability to reconfigure its connections to reflect the software being run. The ability to completely reprogram the computer's hardware implies that this new architecture provides immense scope for emulating different computer architectures [1,2].

The progression of field programmable gate arrays (*FPGA*s) *RCs* has evolved to a point where *SOC* designs can be built on a single device. The number of gates and features has increased dramatically to compete with capabilities that have traditionally been offered through *ASIC* devices only. *FPGA* devices have made a significant move in terms of resources and performance. The contemporary *FPGA*s have come to provide platform solutions that are easily customizable for system connectivity, digital signal processing (*DSP*), and data processing applications. Due to the importance of platform solutions, leading *FPGA* vendors are coming up with easy to-use design development tools [3,4].

As the complexity of *FPGA*-based designs grows, a need for a more efficient and flexible design methodology is required. Nowadays, hardware implementations have become more challenging, and the device densities have increased at a pace that such flows have become cumbersome and outdated. The need for a more innovative and higher-level design flow that directly incorporates model simulation with hardware implementation is needed. One of the modern tools (also used in the proposed research) is *Quartus II*, started with *Altera. Quartus II* is a compiler, simulator, analyzer and synthesizer with a great capability of verification and is chosen to be used for this implementation. It can build the verification file from the input/output specification done by the user. *Quartus II* design software provides a complete, multiplatform design environment that easily adapts to your specific design needs. It is a comprehensive environment for system-on-a-programmable-chip (*SOPC*) [5,6].

An application that needs real-time, fast, and reliable data processing is GPS-based vehicle tracking. In this chapter, we build on a recently produced *VTS* (The Aram Locator) offering a *SOC* replacement of the microcontroller-based implementation. Although the microcontroller-based system has acceptable performance and cost, an *FPGA*-based system can promise less cost and a more cohesive architecture that would save processing time and speeds up system interaction.

This chapter is organized so that Section 25.2 presents the currently available existing microprocessor-based *VTS* and its proposed update. Section 25.3 presents different designs and implementations with different levels of integration. In Section 25.4, performance analysis and evaluation of results are presented. Section 25.5 concludes the chapter and sheds light by summarizing the achievements described in the chapter and suggesting future research.

25.2 Upgrading the *Aram* Locator *GPS* System

One recently implemented *VTS* is the Aram Locator [5,7]. It consists of two main parts, the Base Station (*BS*) and the Mobile Unit (*MU*). The *BS* consists of a *PIC* Microcontroller based hardware connected to the serial port of a computer. The *MU*

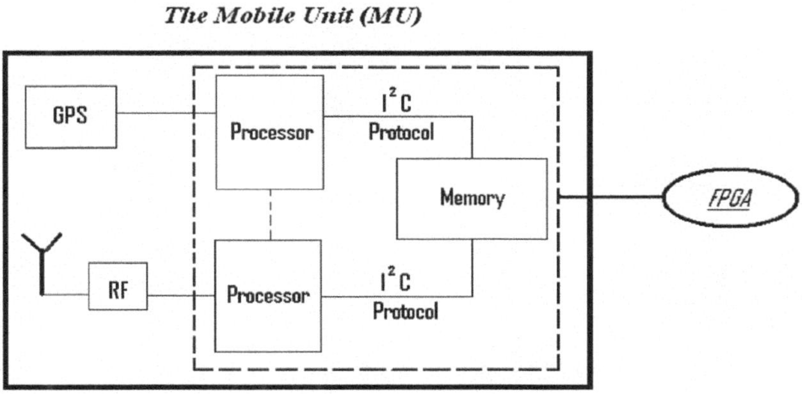

Fig. 25.1 Modules of the FPGA system

is a self-contained *PIC* Microcontroller based hardware and a *GPS* module. The latter would keep track of all the positions traversed by the vehicle and records them in its memory. The system has a great storage capacity, and could perform a significant recording with a considerable sampling rate. The mobile unit (*MU*) of the addressed *Aram Locator* consists of two communicating microcontrollers interfaced with memory. There is also a *GPS* unit and *RF* transceiver (simply sketched in Fig. 25.1) [7–9].

The processor of the *Aram* follows a sequential procedure of different parallel components. For instance, although two processes $P1$ and $P2$ hardware both exist parallel, these processes run sequentially. An interrupt, sent by the I2C controller, is used to activate one of the processes ($P1$ or $P2$). But the inner part of the processes contain several parallel operations like bit assignments and selectors (corresponding to if-statements). In Fig. 25.2, the state diagram describes the behavior of a system, some part of a system, or an individual object (See Fig. 25.3).

25.3 The *FPGA*-Based *Aram* System

The microcontrollers make use of the same memory using a specific protocol. The system is performing properly and has a big demand in the market. However, *FPGAs* promise a better design, with a more cohesive architecture that would save processing time and speeds up system interaction.

The two microcontrollers along with memory would be incorporated into or better supported with a high-density *PLD*. This will transform the hard slow interface between them into a faster and reliable programmable interconnects, and therefore makes future updates simpler. This design estimated to save a considerable percentage of the overall cost of one working unit. For a large number of demands, there would be a significant impact on production and profit. Hence, *PLDs* such as *FPGAs* are the way, for a better design in terms of upgradeability and speed, and it

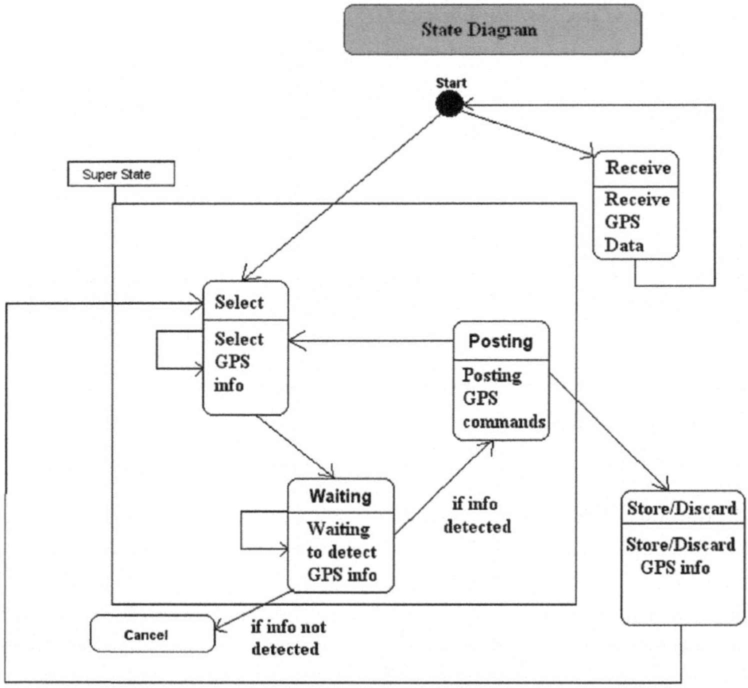

Fig. 25.2 The state diagram of the FPGA system

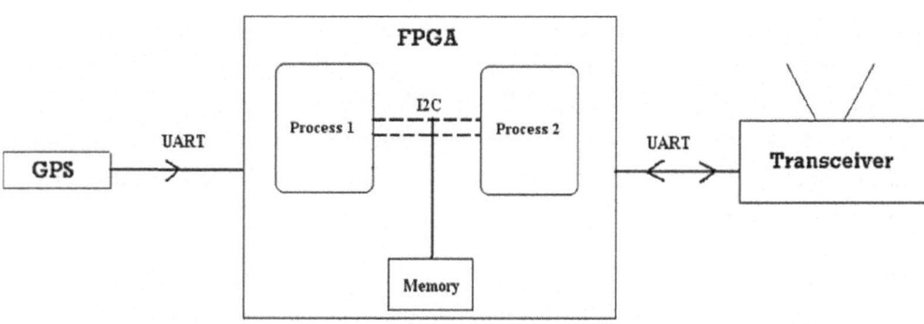

Fig. 25.3 Block diagram of the FPGA system

is a promising advancement for the production cost and revenue. The block diagram of the *FPGA* system is depicted in Fig. 25.3.

Hiding the detailed architecture of the underlying *FPGA*, The proposed system is of two communicating processes, $P1$ and $P2$, along with a shared memory. In addition to the *FPGA*-based system, the *GPS* antenna and the mobile unit play significant roles. The memory block of the microcontroller-based design is replaced by hardware entity controlled by the $I2C$.

Table 25.1 The parameters sent by the GPS

NMEA	Description
HPGGA	Global positioning system fixed data
GPGLL	Geographic position-latitude/longitude
GPGSA	GNSS DOP and active satellites
GPGSV	GNSS satellites in view
GPRMC	Recommended minimum specific GNSS data
GPVTG	Course over ground and ground speed
GPMSS	Radio-beacon signal-to-noise ratio, signal strength, frequency, etc.
GPZDA	PPS timing message (synchronized to PPS)

25.3.1 The Intersystem Process 1

This process has to deal with the message received from the *GPS*. The default communication parameters for *NMEA* (the used protocol) output are 9,600 bps baud rate, eight data bits, stop bit, and no parity. The message includes information messages as shown in Table 25.1.

$GPGGA,161229.487,3723.2475,N,12158.3416,W,1,07,1.0,9.0,M,,,,0000*18
$GPGLL,... $GPGSA,... $GPGSV,... $GPGSV,...
$GPRMC,161229.487,A,3723.2475,N,12158.3416,W,0.13,309.62,120598,*10,
$GPVTG,... $GPMSS,... $GPZDA,...

From these *GPS* commands, only necessary information is selected (i.e. longitude, latitude, speed, date, and time). The data needed is found within the commands *RMC* and *GGA*; others are of minor importance to the *FPGA*. The position of the needed information is located as follows:

$GPRMC: < **time** > , < validity > , < **latitude** > , latitude hemisphere, < **longitude** > , longitude hemisphere, < **speed** > , < course over ground > , < date > , magnetic variation, check sum [7]

$GPGGA, < **date** > , latitude, latitude hemisphere, longitude, longitude hemisphere, < GPS quality > , < # of satellites > , horizontal dilution, < altitude > , Geoidal height, DGPS data age, Differential reference, station Identity (*ID*), and check sum. This information is stored in memory for every position traversed. Finally and when the *VTU* reaches its base station (*BS*), a large number of positions is downloaded to indicate the route covered by the vehicle during a time period and with a certain download speed.

Initially, a flag *C* is cleared to indicate that there's no yet correct reception of data. The first state is "*Wait for GPS Parameters*", as mentioned in the flow chart, there's a continuous reception until consecutive appearance of the *ASCII* codes of ***"R, M, C"*** or ***"GGA"*** comes in the sequence. For a correct reception of data, *C* is set (i.e. *C* = "1"), indicating a correct reception of data, and consequently make the corresponding selection of parameters and saves them in memory. When data storing ends, there is a wait state for the *I2C* interrupt to stop *P*1 and start *P*2, *P*2 download the saved data to the base station (*BS*). It is noted that a large number of vehicles might be in the area of coverage, and all could ask for reserving the channel with the base station; however, there are some predefined priorities that are distributed among the vehicles and therefore assures an organized way of communication. This is simply achieved by adjusting the time after which the unit sends its *ID* when it just receives the word "free".

25.3.2 The Intersystem Process 2

The Base station is continuously sending the word "free", and all units within the range are waiting to receive it and acquire communication with the transceiver. If the unit received the word "free", it sends its *ID* number, otherwise it resumes waiting. It waits for acknowledge, if Acknowledge is not received, the unit sends its *ID* number and waits for feedback. If still no acknowledgement, the communication process terminates, going back to the first step. If acknowledge is received, process 2 sends Interrupt to process 1, the latter responds and stops writing to memory.

Process 2 is then capable of downloading information to the base station. When data is transmitted, the unit sends the number of points transmitted, to be compared with those received by the base station. If they didn't match, the unit repeats downloading its information all over again. Otherwise, if the unit receives successful download, it terminates the process and turns off.

Initially, the circuit shown in Fig. 25.4 is off. After car ignition, current passes through *D*1, and continues its way towards the transistor. This causes the relay to switch and supports an output voltage of 12 V. The circuit (C^*) is now powered and could start its functionality. Using the two regulators, it becomes feasible to provide an adequate voltage to the *FPGA*, which in turn navigates the whole switching technique of the system. In other words, the *FPGA* adapts itself so that it can either put a zero or 5 V at the side connecting *D*2. For the 5 V, the circuit is all on, and the vehicle is in normal functionality. When data download ends, the *FPGA* perceives that, and changes the whole circuit into an idle one, and waits for another car ignition. So, it is well known now that the *FPGA* will be the controller of the behavior of the *VTS* system.

25.3.3 The Memory Block

The suggested memory blocks are addressed by a 12-bit address bus and stores 8-bit data elements. This means that the memory can store up to 4 KB of data. The memory controller navigates the proper memory addressing. Multiplexers are distributed along with the controller to make the selection of the addressed memory location and do the corresponding operation (See Fig. 25.5).

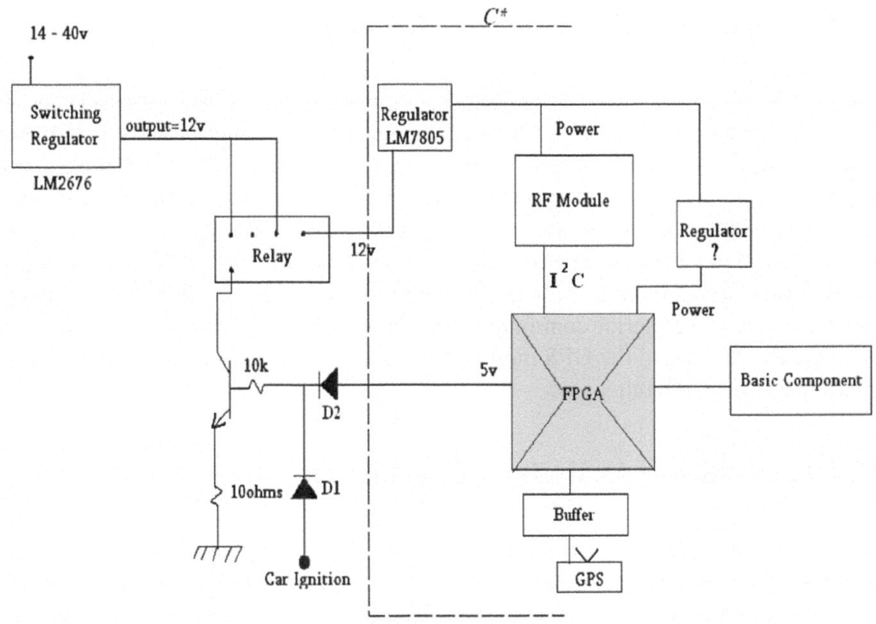

Fig. 25.4 The electric components driving the FPGA

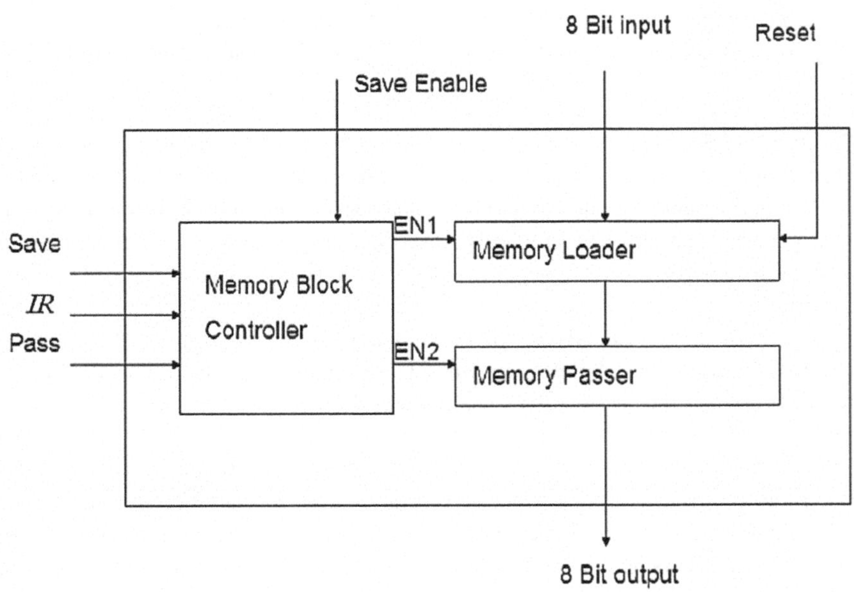

Fig. 25.5 The memory block of the *FPGA* system

25.3.4 Communication Protocols: I2C and UART

The *I2C* bus is a serial, two-wire interface, popularly used in many systems because of its low overhead. It is used as the interface of process 1 and process 2 with the shared memory. It makes sure that only one process is active at a time, with good reliability in communication. Therefore, it writes data read from the *GPS* during process 1, and reads from memory to output the traversed positions into the base station. The Universal Asynchronous Receiver Transmitter (*UART*) is the most widely used serial data communication circuit ever. *UART* allows full duplex communication over serial communication links as RS232. The *UART* is used to interface *Process* 1 and the GPS module from one side, and *Process* 2 and the Base Station (*BS*) from the other side.

25.4 Performance Analysis and Evaluation

Three different systems are to be tested for the *FPGA* implementation. The suggested systems gradually add more parts to the designed FPGA implementation till we reach a complete stand alone system. The three suggested integrations are as follows:

First Integration: Process 1, process 2, and I2C
Second Integration: Process 1, process 2, I2C, and memory
Third integration: Process 1, process 2, I2C, UART, and memory (standalone FPGA).

According to the local market cost, around 9.6% could be saved per unit if the FPGA-based all-in-one system is adopted. For the kind of memory implemented in the system, the vehicle cannot store many locations, so the Vehicle Tracking System (*VTS*) is to be used within a small city. If the sampling rate is to be set one reading every 2 min, one could get a general but not very specific overview of the tracks traversed.

The vehicle tracking needs 344 bits of data to store the five important parameters (longitude, latitude, speed, date, and time). Consequently, this would need 43 memory locations. In other words, the system needs 43 locations for one reading lapsed 2 min.

With 4,096 memory locations, the system makes 95.25 readings, meaning that the vehicle should come back to its base station every 3 h and 10 min to download its information. This would be 4 h and 45 min if the rate is one reading every 3 min. This is not very satisfactory but is tolerated as long as the intended area has a small range. However, this is one problem subject to upgradeability. One solution is to use an *FPGA* with a large memory.

Table 25.2 shows the results of simulation done on integration of parts (modules) forming the *FPGA*-based system. Each integrated system is tested on the component, integration, and system levels.

Table 25.2 Results summary taken from the syntheses of integrations on STRATIX EP1S10F484C5. The execution time in some cases varies according to "RMC", "GGA", or "free"

Integration	% Area in logic elem.	Prop. delay (ns)	Execution time (ns)	Max. op. freq. (MHz)
First	18	118.77	Varies	8.149
Second	50	116.00	Varies	8.62
Third	50	94.47	Varies	10.58

Table 25.3 The readings obtained from the integrations compiled on STRATIX EP1S10F484C5

Integration	Size (bits)	Number of clock cycles	Prop. delay (ns)	Speed of processing (μ*s*)
First	1,448	181	118.773	21.497
Second	1,448	181	116.003	20.996
Third	1,448	181	94.474	17.099

The first design used 1,910 logic elements out of 10,570 (*STRATIX EP1S10 F484C5*, 175.47 MHz), and a maximum operating frequency of 8.149 *MHz*, leaving 82% free space of the *FPGA* capacity. However, after adding memory to the integration, the number of logic elements increased to 5,303, with 50% usage of the capacity. The propagation delay decreased slightly inside the *FPGA*, The decrease in propagation delay means that the optimizer found a better way to reduce its critical path.

Similar results are shown when the *UART* part is added (standalone *FPGA*), with an improvement in propagation delay. Although the number of logic elements has increased, it contributed to better interaction among the parts and raised the operating frequency to 10.58 MHz. Therefore, integration of parts has enhanced the delay with an expected increase in number of logic elements positively affects the processing speed when propagation finds its paths among combinations of gates. Suppose that the *GPS* message ***"M"*** received by the *UART* has come in the following sequence:

$GPGGA,161229.487,3723.2475,N,12158.3416,W,1,07,1.0,9.0,M,,,,0000*18
$GPGLL,3723.2475,N,12158.3416,W,161229.487,A*2C
$GPRMC,161229.487,A,3723.2475,N,12158.3416,W,0.13,309.62,120598,*10

"***M***" is to be tested on the three obtained integrations (Table 25.3), taking into account that the *GPS* parameters needed for the application come as represented in Section 25.3. The system takes selective parameters according to their position in the sequence, and checks if they correspond to the desired information. Every character is represented by its 8-bit *ASCII* representation. Table 25.3 shows an exact interpretation of data when received from the *GPS* and processed via several modules of the integration. After integrations have been inspected, the proposed system synthesized on different *FPGA*s, and the results appear in Table 25.4 [7, 8].

Table 25.4 Syntheses of the VTS on different FPGAs. The execution time in some cases varies according to "RMC", "GGA", or "free"

FPGA	Logic area in logic elements	Prop. delay (ns)	Exec. time (ns)	Max. operating frequency (MHz)
STRATIX EP1S10F484C5	50%	94.474	Varies	10.58
STRATIX-II EP2S15F484C3	36%	151.296	Varies	6.609
MAX3000A EPM3032ALC44–4	Doesn't fit	NA	NA	NA
Cyclone EP1C6F256C6	90%	90.371	Varies	11.06
FLEX6000 EPF6016TI144–3	Doesn't fit	NA	NA	NA
APEXII EP2A15B724C7	30%	181.418	Varies	5.512

From the readings of Table 25.4, the following could be concluded testing the all-in-one system:

- *STRATIX EP1S10F484C5* (175.47 MHz) has enough number of logic elements, and the capacity taken by the project is one half its total capacity. The propagation delay is 94.474 ns, thus, the system runs with a frequency of 10.58 MHz.
- *STRATIX-II EP2S15F484C3* (420 MHz) has a larger number of logic elements, so the project took lesser capacity (36%), and caused more propagation delay.
- The project fits 90% in *Cyclone EP1C6F256C6* (405 MHz), but with minimum propagation delay of 90.371 ns and thus 11.065 MHz operating frequency.
- *APEXII EP2A15B724C7* (150 MHz) has the largest capacity among the listed devices allocating 30% only for the *ARAM* project with a largest propagation delay (181.418 ns) and minimum frequency (5.512 MHz).

25.5 Conclusion

In this chapter, we have presented an alternative design of an existing modern *GPS*-based *VTS* using *FPGAs*. The performance of the proposed implementations is evaluated on different *FPGAs*. The suggested designs show enhancement in terms of speed, functionality, and cost. Future work includes refining the proposed designs in order to eliminate the sequential alternation of the two main internal processes, and investigating larger buffering by providing more memory elements and using state-of-art *FPGAs*.

References

1. C. Ashraf, Y. Arjumand, An Approach to Task Allocation for Dynamic Scheduling in Reconfigurable Computing Systems, International Multitopic Conference, 2005, pp. 1–6.
2. V. Subramanian, J. Tront, S. Midkiff, C. Bostian, A Configurable Architecture for High Speed Communication Systems, Military and Aerospace Applications of Programmable Logic Devices (MAPLD) International Conference, Vol. 3, pp. E11 1–9, Sept. 2002.
3. I. Damaj, "Parallel Algorithms Development for Programmable Logic Devices", Advances in Engineering Software, Elsevier Science, 2006. Issue 9, Vol. 37, pp. 561–582.

4. V. Subramanian, J. Tront, C. Bostian, S. Midkiff, Design and Implementation of a Configurable Platform for Embedded Communication Systems, IPDPS, 2003, p. 189.
5. M. Shanblatt, B. Foulds, A Simulink-to-FPGA Implementation Tool for Enhanced Design Flow, International Conference on Microelectronic Systems Education, 2005, pp. 89–90.
6. F. Vahid, T. Givargis, Embedded System Design, A Unified Hardware/Software Introduction (Wiley, New York, 2001).
7. Z. Osman, M. Jrab, S. Midani, R. Zantout, Implementation of a System for Offline Tracking Using GPS, Mediterranean Microwave Symposium, 2003.
8. NMEA Reference Manual SiRF Technology, Inc. 148 East Brokaw Road San Jose, CA 95112, USA; http://www.nmea.org
9. R. Kühne, R. Schäfer, J. Mikat, K. Thiessenhusen, U. Böttger, S. Lorkowski, New Approaches for Traffic Management in Metropolitan Areas, Proceedings of the IEEE Conference on ITS, 2003.

Chapter 26
Unknown Malicious Identification

Ying-xu. Lai and Zeng-hui. Liu

Abstract The detection of unknown malicious executables is beyond the capability of many existing detection approaches. Machine learning or data mining methods can identify new or unknown malicious executables with some degree of success. Feature set is a key to apply data mining or machine learning to successfully detect malicious executables. In this paper, we present an approach that conducts an exhaustive feature search on a set of malicious executables and strives to obviate over-fitting. To improve the performance of Bayesian classifier, we present a novel algorithm called Half Increment Naïve Bayes (HIB), which selects the features by carrying an evolutional search. We also evaluate the predictive power of a classifier, and we show that our classifier yields high detection rates and learning speed.

Keywords Unknown malicious detection · Half Increment Naïve Bayes · classification

26.1 Introduction

As network-based computer systems play increasingly vital roles in modern society, a serious security risk is the propagation of malicious executables. Malicious executables include viruses, Trojan horses, worms, back doors, spyware, Java attack applets, dangerous ActiveX and attack scripts. Identifying malicious executables quickly is an important goal, as they can cause significant damage in a short time. Consequently detecting the presence of malicious executables on a given host is a crucial component of any defense mechanism.

Traditional malicious executables detection solutions use signature-based methods, in that they use case-specific features extracted from malicious executables in

---

Y.-x. Lai (✉)
College of Computer Science, Beijing University of Technology, Beijing 100124, China
E-mail: laiyingxu@bjut.edu.cn

S.I. Ao and L. Gelman (Eds.), *Advances in Electrical Engineering and Computational Science*, Lecture Notes in Electrical Engineering.

order to detect those same instances in the future [1]. Security products such as virus scanners are examples of such application. While this method yields excellent detection rates for existing and previously encountered malicious executables, it lacks the capacity to efficiently detect new unseen instances or variants. Due to detect malicious accurately is a NP problem [2, 3], heuristic scanners attempt to compensate for this lacuna by using more general features from viral code, such as structural or behavioral patterns [4]. Although proved to be highly effective in detecting unknown malicious executables, this process still requires human intervention.

Recently, attempts to use machine learning and data mining for the purpose of identifying new or unknown malicious executables have emerged. Schultz et al. examined how data mining methods can be applied to malicious executables detection [5] and built a binary filter that can be integrated with email servers. Kolter et al. used data mining methods, such as Naïve Bayes, J48 and SVM to detect malicious executables [6]. Their results have improved the performance of these methods. Bayes or improved Bayes algorithm has the capability of unknown malicious detection, but it spends more time to study. A new improved algorithm (half-increment Bayes algorithm) is proposed in this paper.

In this paper, we are interested in applying data mining methods to malicious executables detection, and in particular to the problem of feature selection. Two main contributions will be made through this paper. We will show how to choose features which are most representative properties. Furthermore, we propose a new improved algorithm and will show that our method achieve high learning speed and high detection rates, even on completely new, previously unseen malicious executables.

The rest of this paper is organized as follows: Section 26.2 is a brief discussion of related works. Section 26.3 gives a brief description of Bayesian algorithm. Section 26.4 presents details of our methods to obtain high learning speed. Section 26.5 shows the experiment results. Lastly, we state our conclusions in Section 26.6.

26.2 Related Work

At IBM, Kephart et al. [7] proposed the use of Neural Networks to detect boot sector malicious binaries. Using a Neural Network classifier with all bytes from the boot sector malicious code as input, it had shown that 80–85% of unknown boot sector malicious programs can be successfully identified with low false positive rate (<1%). The approach for detecting boot-sector virus had incorporated into IBM's Anti-Virus software. Later, Arnold et al. [8, 9] applied the same techniques to win32 binaries. Motivated by the success of data mining techniques in network intrusion system [10, 11], Schultz et al. [5] proposed several data mining techniques to detect different types of malicious programs, such as RIPPER, Naïve Bayes and Multi-Naïve Bayes. The authors collected 4,301 programs for the Windows operating system and used MacAfee Virus Scan to label each as either malicious or benign. The authors concluded that the voting naïve-Bayesian classifier outperformed all other methods. In a companion paper [12] the authors developed an Unix mail filter that detect malicious Windows executables based on the above work. Kolter et al. [6] also used data mining methods, such as Naïve Bayes, J48 and SVM to detect malicious codes. The authors gathered 1,971 benign and 1,651 malicious codes and encoded each as a training example using n-grams of byte codes as features, boosted decision trees outperformed other methods with an area under the ROC curve of 0.996 (applied detectors to 291 malicious executables discovered, and boosted decision trees achieved a TP rate of 0.98 for a desired FP rate of 0.05). Zhang et al. [13] used SVM and BP neural network to virus detection, the D-S theory of evidence was used to combine the contribution of each individual classifier to give the final decision. It showed that the combination approach improves the performance of the individual classifier significantly. Zhang et al. [14, 15] established methods based on fuzzy pattern and K-nearest neighbor recognition applying to detect malicious executables for the first time.

26.3 Naïve Bayesian and Application

The goal of our work was to improve a standard data mining technique to compute accurate detectors for new binaries. We gathered a large set of programs from public sources and separated the problem into two classes: malicious and benign executables. We split the dataset into two subsets: the training set and the test set. The data mining algorithms used the training set while generating the rule sets. We used a test set to check the accuracy of the classifiers over unseen examples.

In a data mining framework, features are properties extracted from each example in the data set – such as strings or byte sequences. A data mining classifier trained with features can use to distinguish between benign and malicious programs. We used strings that are extracted from the malicious and benign executables in the data set as features.

We propose an exhaustive search for strings. Typically there are many "disorder-words" when files are read as ASCII code, like "autoupdate.exe 垗尰膜岞 ?software*abbbbbb 膜岃 download", etc. The string selection program extracts consecutive printable characters from files. To avoid yielding a large number of unwanted features, characters are filtered before they are recorded. Our feature selection involves an extraction step followed by an elimination step.

26.3.1 String Extraction and Elimination

In string extraction step, we scan files and read character one by one, record consecutive printable characters (like English letter, number, symbolic and etc.), and construct lists of all the strings. The length of the string is specified as a number of characters. The shorter the length, the more likely the feature is to have general relevance in the dataset. But a short length will yield a larger number of features.

Extracted strings from an executable are not very robust as features because some strings are unmeaning, like "abbbbbb", so we select a subset of strings as feature set by an eliminate step.

Many have noted that the need for a feature filter is to make use of conventional learning methods [13,14], to improve generalization performance, and to avoid over-fitting. Following the recommendation of those, the glossary filter criterion is used in the paper to select a subset of strings.

The glossary is a computer glossary which includes 7,336 items:

- Computer words, such as function name, API name
- Abbreviations on computer networks, like "QQ", "msn", etc
- Postfixs, like ".dll", presenting dynamic link libraries

26.3.2 Naïve Bayes

The naive Bayes classifier computes the likelihood that a program is malicious given the features that are contained in the program. We treat each executable's features as a text document and classified based on that. Specifically, we want to compute the class of a program given that the program contains a set of features F. We define C to be a random variable over the set of classes: benign, and malicious executables. That is, we want to compute $P(C|F)$, the probability that a program is in a certain class given the program contains the set of features F. We apply Bayes rule and express the probability as:

$$P\left(C|F\right) = \frac{P\left(F|C\right) \times P\left(C\right)}{P\left(F\right)} \tag{26.1}$$

To use the naïve Bayes rule we assume that the features occur independently from one another. If the features of a program F include the features $F_1, F_2, F_3 \ldots, F_n$, then Eq. (26.1) becomes:

$$P\left(C|F\right) = \frac{\prod_{i=1}^{n} P\left(F_i|C\right) \times P\left(C\right)}{\prod_{j=1}^{n} P\left(F_j\right)} \tag{26.2}$$

Each $P(F_i|C)$ is the frequency that string F_i occurs in a program of class C. $P(C)$ is the proportion of the class C in the entire set of programs.

The output of the classifier is the highest probability class for a given set of strings. Since the denominator of Eq. (26.1) is the same for all classes we take the maximum class over all classes C of the probability of each class computed in Eq. (26.2) to get:

$$\text{Most Likely Class} = \max_{C}\left(P(C)\prod_{i=1}^{n} P(F_i|C)\right) \tag{26.3}$$

In Eq. (26.3), we use max_C to denote the function that returns the class with the highest probability. Most Likely Class is the class in C with the highest probability and hence the most likely classification of the example with features F.

26.4 Increment Naïve Bayes

26.4.1 Naïve Bayes

To train the classifier, we record how many programs in each class contained each unique feature. We use this information to classify a new program into an appropriate class. We first use feature extraction to determine the features contained in the program. Then we apply Eq. (26.3) to compute the most likely class for the program.

The Naïve Bayes algorithm requires a table of all features to compute its probabilities. This method requires a machine with one gigabyte of RAM, because the size of the binary data was too large to fit into memory.

To update the classifier, when new programs are added to the training set, we update feature set at first, and apply Eq. (26.3) to compute the most likely class for the program again. So it is time-consuming to update the classifier by NB algorithm.

26.4.2 Multi-naïve Bayes

To correct NB algorithm problem the training set is divided to smaller pieces that would fit in memory. For each set we train a Naïve Bayes classifier. Each classifier gives a probability of a class C given a set of strings F which the Multi-Naïve Bayes uses to generate a probability for class C given F over all the classifiers.

For each classifier, the probabilities in the rules for the different classifiers may be different because the underlying data the each classifier is trained on is different. The prediction of the Multi-Naïve Bayes algorithm is the product of the predictions of the underlying Naïve Bayes classifier.

$$P(C|F) = \prod_{k=1}^{n} P_k(C|F) \tag{26.4}$$

When new programs are added to the training set, these new programs as a subset and train a Naïve Bayes classifier over the subproblem. Based on the Eq. (26.4), we update the probability. The Multi-Naïve Bayes algorithm does not need to compute its probabilities over all the training set, but the accuracy of the classifier will be worsen.

26.4.3 Half Increment Bayes (HIB)

The above NB and MNB algorithms obtain feature set over all training set or subset at first. Once the final feature set was obtained, we represent our positive and zero data in the feature space by using, for each feature, "1" or "0" to indicate whether or not the feature is present a given executable file. The probability for a string occurring in a class is the total number of times it occurred in that class's training set divided by the total number of times that the string occurred over the entire training set.

We can derive a method purely from the NB algorithm for increment update. In our method, feature set is increased with studying of classifier. That is, composed there are k_1 string features extracted from the first sample, so k_1 strings are elements of set F. If there are S_2 strings extracted from the second sample, k_2 elements are not found in F, these elements should be added in feature set F. Set F will includes $k_1 + k_2$ elements. Classifier is trained based on the evolutional feature set.

Claim 1 *We can obtain the class-conditional probability* $P\left(F^{(n+1)}\middle|C\right)$ *over* $n + 1$ *samples by that of former* n *samples and the* $(n + 1)th$ *sample, that is* $P\left(F_i^{(n+1)}\middle|C_j\right) = \frac{\left[P\left(F_i^{(n)}\middle|C_j\right)\times n + P\left(x^{(n+1)}\middle|C_j\right)\right]}{n+1}$. *Where,* $P\left(F^{(n+1)}\middle|C\right)$ *is the class-conditional probability over* $n + 1$ *samples,* $P\left(F^{(n)}\middle|C\right)$ *is class-conditional probability over n samples,* $P\left(x^{(n+1)}\middle|C\right)$ *is class-conditional probability over the* $(n + 1)th$ *sample.*

Proof. Composed there are a features obtained from n samples, that is $F^{(n)} = \{F_1, F_2..., F_a\}$, then

$$P\left(F^{(n)}\middle|C\right) = \prod_{i=1}^{a} P(F_i|C_j)(j = 1, 2) \tag{26.5}$$

$$P(F_i|C_j) = \frac{P(F_iC_j)}{P(C_j)} = \frac{count(F = F_i \wedge C = C_j)}{count(C = C_j)} \tag{26.6}$$

Where, $count(F = F_i \wedge C = C_j)$ is sample number for $F = F_i$ and class $C = C_j$, $count(C = C_j)$ is sample number for class $C = C_j$.

If there are n samples with $C = C_j$ in training set, then the class-conditional probability $P\left(F^{(n)}\middle|C\right)$ for n samples is:

$$P\left(F_i^{(n)}\middle|C_j\right) = \frac{count((F = F_i^{(n)}) \wedge (C = C_j))}{n} \tag{26.7}$$

If the $(n + 1)$th sample for class $C = C_j$ was added, there are two cases.

Case 1. the strings in the $(n + 1)$th sample are all found in F, then

$$P\left(F_i^{(n+1)}\middle|C_j\right) = \frac{count((F = F_i^{(n)}) \wedge (C = C_j))}{n+1} \tag{26.8}$$

Case 2. there are b strings in the $(n+1)$th sample are not found in F, then these strings are added in F, that is $F^{(n+1)} = \{F_1, F_2 \ldots, F_a, F_{a+1}, \cdots, F_{a+b}\}$. For those new b probabilities, due to $P\left(F_i^{(n)}\middle|C_j\right) = 0 \quad a < i \leq a + b$, then

$$P\left(F_i^{(n+1)}\middle|C_j\right) = \frac{1}{n+1} \quad a < i \leq a + b \tag{26.9}$$

We rewrite Eqs. (26.8) and (26.9) as Eq. (26.10):

$$P\left(F_i^{(n+1)}\middle|C_j\right) = \frac{\left[P\left(F_i^{(n)}\middle|C_j\right) \times n + P\left(x^{(n+1)}|C_j\right)\right]}{n+1} \tag{26.10}$$

Therefore, information of $n+1$ samples can obtained from those of former n samples and the $(n+1)$th sample.

Claim 2 *For NB and HIB, based on same training set, same feature sets can be obtained by same string extraction and elimination methods, that is* $F_N = F_{\mathrm{H}}$. *Where* F_{H} *is feature set obtained by half increment algorithm,* F_{N} *is feature set obtained by naïve bayes algorithm.*

Proof. Composed there are a features obtained from n samples, that is $F_N = \{F_1, F_2 \ldots, F_a\}$, Set F_N is made of strings that extracted from n samples and eliminated based on computer dictionary.

F_{H} is increased as training samples. When all of training samples are all extracted, Set F_H is made of strings that extracted from n samples and eliminated based on computer dictionary. So $F_{\mathrm{N}} = F_{\mathrm{H}}$.

Based on Claim 1 and 2, we can obtain Theory 1.

Theory 1 *Most Likely Classes computed by half increment classification and Naïve Bayes classification are same, that is* $C_1 = C_2$. *Where* C_1 *is the most likely class obtained by naïve Bayes, and* C_2 *is the most likely class obtained by half-increment classification.*

Proof. Composed a features were obtained from n samples, that is $F_N = \{F_1, F_2 \ldots, F_a\}$, Based on Eq. (26.3),

$$C_1 = \max\left(P(C)\prod_{i=1}^{a} P(F_i|C)\right) \tag{26.11}$$

When a new sample is added, $F_N = \{F_1, F_2 \ldots, F_a, F_{a+1}, \cdots, F_{a+b}\}$ the most likely class computed by naïve Bayes is

$$C_1 = \arg\max P(C) \times \prod_{i=1}^{a+b} P(F_i|C) \tag{26.12}$$

As $F_N = F_H$, based on Eq. (26.10)

$$C_2 = \arg\ \max P(C) \times \prod_{i=1}^{a+b} P(F_i | C) \tag{26.13}$$

Therefore, the $C_1 = C_2$.

26.4.4 Complexity

Based on former algorithm, time-consuming of NB and HIB are made of two parts:

1. Extract unrepeated strings from samples
2. Fix on feature set and training set, build up classifier

For step (1), time-consuming of two algorithms are same. But for step (2), they are different.

For HIB, feature set is increased with studying of classifier. That is, composed there are S_1 strings in the first sample, and k_1 strings are unrepeated, so k_1 strings are elements of set F. If there are S_2 strings in the second sample, the time-consuming of computing whether k_1 elements are found in S_2 or not is $T_{H2} = O(S_2 \times k_1)$. If k_2 elements are not found in F, these elements should be added in feature set F. Set F will includes $k_1 + k_2$ elements. For n samples, the all time-consuming of fix on set F is

$$T_H = O(S_2 \times k_1 + S_3 \times (k_1 + k_2) + \ldots + S_n \times (k_1 + k_2 + \ldots + k_r)) \tag{26.14}$$

For NB, the set F are obtained before classifier study. Composed there are K elements in F, the time consuming to indicate whether or not the feature is present a given executable file is

$$T_N = O(K \times (S_1 + S_2 + \ldots + S_n)) \tag{26.15}$$

Based on Claim 2, $\boldsymbol{F}_{\mathbf{N}}=\boldsymbol{F}_{\mathbf{H}}$, $K = k_1 + k_2 + \ldots + k_r$.

Based on Eq. (26.14),

$$T_H = O(K \times (S_2 + S_3 + \ldots + S_n) - S_2 \times (k_2 + \ldots + k_r) - \ldots - S_{n-1} \times k_r) \tag{26.16}$$

If $k_2 = k_3 = \ldots = k_r = 0$, or $K = k_1$, $T_H = T_N$, the time-consuming of two algorithm are same. But malicious executables include viruses, Trojan horses, worms, back doors, spyware, Java attack applets, dangerous ActiveX and attack scripts, a sample cannot include all feature elements.

26.5 Experimental Results

26.5.1 Experimental Design

Our experiment are carried out on a dataset of 2,995 examples consisting of 995 previously labeled malicious executables and 2,000 benign executables, collected from desktop computers using various versions of the Windows operating system. The malicious executables are taken from an anti-Virus software company.

For each run, we extract strings from the executables in the training and testing sets. We select the most relevant features from the training data, apply elimination method, and use the resulting classifier to rate the examples in the test set.

26.5.2 Performance Analysis

26.5.2.1 Time-Consuming

In our experiments, we used VC++ implementation of the NB, MNB and HIB classifiers. In this section, we evaluate the performance of NB, MNB and HIB three algorithms in comparison.

For HIB algorithm, increasing rule of feature elements is based on order of studied samples. Figure 26.1 is the curve between number of studied samples and number of feature elements. In Fig. 26.1, the slope of the curve is steep at some points. When classifier begins to study some type sample, the slope of the curve is steep, that is, the number of feature elements increase quickly. After classifier has studied some samples, the slope of the curve is smooth.

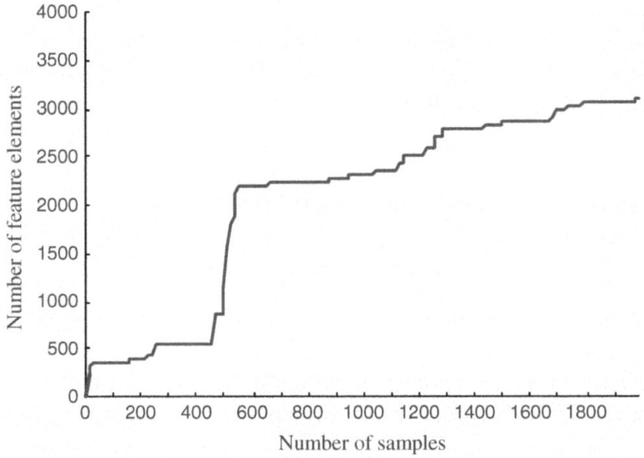

Fig. 26.1 Curve between number of feature elements and number of samples

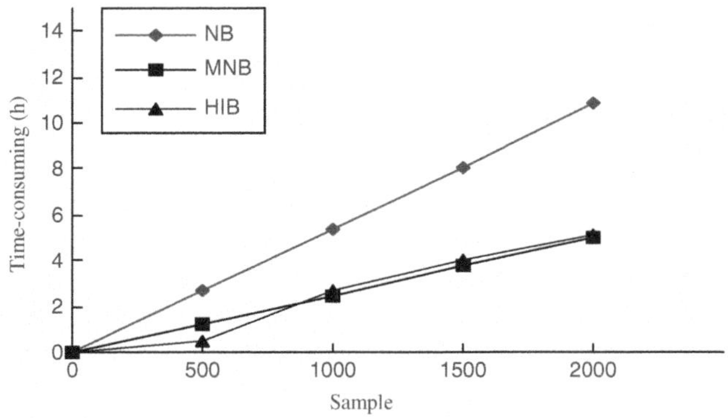

Fig. 26.2 NB, MNB and HIB curves of time-consuming

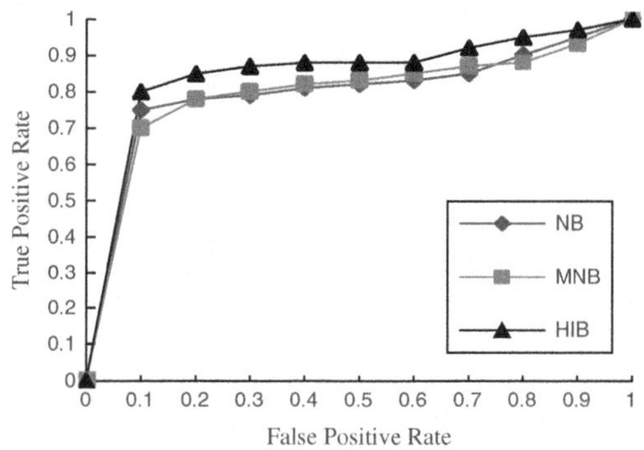

Fig. 26.3 NB, MNB and HIB ROC

Figure 26.2 is the curve of three algorithm's time-consuming. The slope of NB algorithm is initially much steeper than the HIB and MB algorithms. HIB algorithm has better efficiency.

The ROC curves in Fig. 26.3 show a more quickly growth than the NB and MNB until the false positive rate climbed above 4%. Then the three algorithms converged for false positive rates greater then 6% with a detection rate greater then 95%.

26.5.2.2 Comparison Against Benchmark Model

To evaluate our system we are interested in several quantities:

1. True Positives (TP), the number of malicious executable examples classified as malicious code
2. True Negatives (TN), the number of benign programs classified as benign
3. False Positives (FP), the number of benign programs classified as malicious code
4. False Negatives (FN), the number of malicious codes classified as benign

The Overall Accuracy is defined as $\frac{TP+TN}{TP+TN+FP+FN}$.

Table 26.1 Experimental results using our method and traditional methods

Method	Feature	Classifier	Accuracy (%)
Our method	Strings	NB	96.8
Our method	Strings	MNB	93.3
Our method	Strings	HIB	97.9
Schultz	String	NB	97.11
Schultz	Bytes	MNB	96.88
Kolter	4-gram	SVM	93
Henchiri	8-gram	J48	93.65

We compare our method with a model used in previous research [7, 15]. The results, displayed in Table 26.1, indicate that a virus classifier can be made more accurate by using features representative of general viral properties, as generated by our feature search method. With up to 97.9% overall accuracy, our system outperforms NB and MNB algorithms and achieves better results than some of the leading research in the field, which performs at 97.11%.

26.6 Conclusion

The naïve Bayes classifier is widely used in many classification tasks because its performance is competitive with state-of-the-art classifiers, it is simple to implement, and it possesses fast execution speed. In this paper, we discussed the problem of how to classify a set of query vectors from the same unknown class with the naïve Bayes classifier. Then, we propose the method HIB algorithm and compare it with naïve Bayes and multi-naïve Bayes. The experimental results show that HIB algorithm can take advantage of the prior information, can work well on this task. Finally, HIB algorithm was compared with a model used in previous research [7,15]. Experimental results reveal that, HIB can reach a higher level of accuracies as 97.9%. HIB's execution speed is much faster than MNB and NB, and HIB has low implementation cost. Hence, we suggest that HIB is useful in the domain of unknown malicious recognition and may be applied to other application.

References

1. G. McGraw, G. Morrosett, Attacking malicious code: a report to the infosec research council, *IEEE Transactions on Software*, vol. 2, Aug. 1987, pp. 740–741.
2. F. Cohen, Computer Viruses Theory and Experiments, *Computers & Security*, vol. 6, Jan. 1987, pp. 22–35.
3. D. Spinellis, Reliable Identification of Bounded-Length Viruses Is NP Complete, *IEEE Transactions on Information Theory*, vol. 49, Jan. 2003, pp. 280–284.
4. MacAfee. Homepage-MacAfee.com. Online Publication, 2000. http://www.macafee.com

5. M.G. Schcltz, E. Eskin, E. Zadok, S.J. Stolfo, Data mining methods for detection of new malicious executables, *In: Proceedings of IEEE Symposium in Security and Privacy*, 2001.
6. J.Z. Kolter, M.A. Mloof, Learning to detect malicious executables in the wild, *In: Proceedings of the 10th ACM SIGKDD International Conference on Knowledge Discovery and Data Mining*, ACM Press, New York, 2004, pp. 470–478.
7. J.O. Kepahart, G.B. Sorkin, W.C. Arnold, D.M. Chess, G.J. Teauro, S.R. White, Biologically inspired defenses against computer viruses. *In: Proceedings of IJCAI'95*, Montreal, 1995. pp. 985–996.
8. W. Arnold, G. Tesauro, Automatically generated Win32 heuristic virus detection, *In: Proceedings of the 2000 International Virus Bulletin Conference,* 2000.
9. G. Tesauro, J.O. Kephart and G.B. Sorkin, Neural networks for computer virus recognition, *IEEE Expert*, vol. 11, Apr. 1996, pp. 5–6.
10. W. Lee, S. Stolfo, K. Mok, A Data Mining Framework for Building Intrusion Detection Models, *IEEE Symposium on Security and Privacy*, 1999.
11. W. Lee, S.J. Stolfo, P.K. Chan, Learning patterns from UNIX processes execution traces for intrusion detection, *AAAI Workshop on AI Approaches to Fraud Detection and Risk Management,* AAAI Press, 1997, pp. 50–56.
12. M.G. Schcltz, E. Eskin, E. Zadok, M. Bhattacharyya, S.J. Stolfo, MEF: Malicious Email filter, A Unix mail filter that detects malicious windows executables, *In: Proceeding of USENIX Annual Technical Conference,* 2001.
13. B. Zhang, J. Yin, J. Hao, Unknown computer virus detection based on multi-naive Bayes algorithm. *Computer Engineering*, vol. 32, Oct. 2006, pp. 18–21.
14. B. Zhang, J. Yin, J. Hao, Using fuzzy pattern recognition to detect unknown malicious executables code, *Lecture Notes in Computer Science,* vol. 36, Mar. 2005, pp. 629–634.
15. B. Zhang, J. Yin, D. Zhang, J. Hao, Unknown computer virus detection based on K-nearest neighbor algorithm, *Computer Engineering and Applications*, vol. 6, 2005, pp. 7–10.

Chapter 27
Understanding Programming Language Semantics for the Real World

Trong Wu

Abstract Computer is used virtually everywhere everyday in the world. Before 1990s computer systems are generally used for mathematics, engineering, and business computations. In this period, mainly use FOTRAN, COBOL, and PL/1 for computation on mainframe systems. In the last two decades scientists found that the natural world is complicated that has overwhelmed with data that required more sophisticated computation facilities and better languages for computation and that created new computing disciplines. This paper addresses the understanding of programming language semantics that will help user in selection of programming language features for various applications needs and that can help programmers in designing reliable, accurate, efficient, and user-friendly software systems for the real world.

Keywords User-defined types · exception handling · multitasking · communication between program units · and user-defined precision

27.1 Introduction

It is a chaotic world, it is a systematic world; it is a confused age, it is illuminate age; it is disordered society, it is organized society. Everyday we are struggling in a complex, intricate, and difficult environment for our life. Computer science can give one abilities to solve a category of complicated, obscure, and oppressive problems. We can take the advantage of computer speed and large volume of storage spaces to add our ability for solving complex problems in the world. This is a very challenge task and interesting assignment to youngsters. It attracts many of them to study computer science.

T. Wu
Department of Computer Science at Southern Illinois University Edwardsville, IL 62026, USA
e-mail: twu@siue.edu

S.I. Ao and L. Gelman (Eds.), *Advances in Electrical Engineering and Computational Science*, Lecture Notes in Electrical Engineering.

In October 1962, Purdue University established the first department of computer science in the United States [1]. Since then Computer Science education has become an integral discipline in colleges and universities across the country. Initially, students in physics, engineering, and mathematics were advised to take one course in the *FORTRAN* language while students in the school of business particularly in Management Information Systems ware required to take a course in the *COBOL* language. Students who majored in Computer Science were required to enroll in *FORTRAN*, *COBOL,* and *PL*/1 courses. It was thought that this would prepare them potentially to work in industrial firms or business companies after completion of their degrees.

Since then, computer software and its use have spread to nearly every aspect of peoples' lives from work to play. Today almost every one uses computer and software everywhere. Most of the equipment and devices, which use software are designed and manufactured by engineers. For these applications, users need not only robust hardware, but also reliable software.

Today, the most serious problems with software products are *expense*, *ease of use*, and *reliability* [2]. Computer scientists not only need to study computer mechanisms and how to increase the productivity and efficiencies of a computer system, but also need to design computer systems, write programs, and execute them. The former is the *state of the practice of software engineering* and the latter is the *state of the art*. Unfortunately, there exists a gap between the former and the latter. The task for computer scientists is to eliminate or narrow this gap. To implement this, one should thoroughly analyze system requirements, carefully design the system and programs, and then perform various tests including unit tests and system tests. Finally computer scientists need to deliver systems and make sure that they meet their requirements and fit in their environments [3].

However, computer scientists need a good language tool to implement these software engineering principles. Human languages such as English, Chinese, Spanish, Russian, etc. are all ambiguous languages. None of them can be used to communicate with a computer system. Therefore, mathematicians, computer scientists, and linguists must work hard to develop *semantically unambiguous* languages, starting by designing *grammar rules* and then developing the *sentential structures* of the language

Nevertheless, most engineers have had only one computing course in *FORTRAN* or in the *C* language or both while they were in college and perhaps they believe that *FORTRAN* or the *C* language is enough for their engineering application needs. In fact, the *FORTRAN* language lacks user-defined types that limit its application domain. Moreover, both the *FORTRAN* and *C* languages do not have *predefined accuracy features*, *friendly concurrency facilities*, and *effective exception handling methods*. Today, neither *FORTRAN* language nor the *C language* remains adequate to program in this complex and chaotic world for designing and implementing more sophisticated and reliable systems for biological, physical, and other natural related systems. Therefore, we need understand programming Language semantics thoroughly so that we can deal with natural world computation.

This paper describes some necessary elements in the understanding programming language semantics that is important for the computation of the natural world. It requires *computation more accurate*, *more reliable*, *more precise*, *more efficient*, and *friendlier*. Therefore, we should consider the *extends of application domain*, the application of *accuracy* and *efficient* facilities, the use of *modularization* and *communication*, the *utilization of parallelism* or *concurrency*, and *applying exception handling* and for *reliability critical projects* in this real world.

27.2 Extension of the Applications to a Real World

A department store usually consists of tens' of thousands merchandises; it is very difficult to manage them *properly*, *efficiently*, and *cost-effectively*. One of the most effective ways to manage such department stores is to partitioning all the merchandises into departments such as *man's ware*, *lady's ware*, *house ware*, *juniors*, *sports*, *electronics*, *hardware*, *shores*, *bath and bed*, *pharmacy*, etc. Any merchandise in a department store must belong to one and only one department, and department-to-department are discriminates to each other. No merchandises can belong to two distinct departments at the same time. Each department consists of a set of similar or the same kind merchandises and a set of operation rules and policies for manage these merchandises. For example, *hardware* and *electronic* departments have different return or refund policies; some *electronic items* are not allowed customers to test or to tryout. *Pharmacy department* has its operational procedures or rules; medications are classified into *prescription* and *non-prescription*. For any prescription medication is strictly required a medical doctor's signed prescription [4].

In a special season of the year, like Christmas time, most department stores want to make some additional business for this splendor season; they create a new department called "*Santa*;" it is for children to take pictures with the *Santa*. Likewise, in the springtime they create a new department called "*Plants*" by selling saplings, flowers, seedling, soils, seeds, fertilizer, rocks, and other goods for the garden [4].

Today real world projects can be very large and complicated for applications in newly developed domains. These projects can include *hardware projects*, *software projects*, and *firmware projects*. Many engineering projects need to take years or tens' of years to complete them such as the *Yangtze River Three Gorges Dam project* that was launched in 1993 and the water level in the reservoir will reach to 175 m in 2009, when the project is finally completed [5]; the *Airbus A380 project* that was started in 2000 for worldwide market campaign and made its maiden flight in 2005 [6]; and the *Boeing 777 program* was launched in October 1990 and the first 777–300 was delivered to Cathay Pacific Airways in June 1998 [7]. These projects involved thousands designers, hundreds subcontractors, and tens of thousands manufactures.

A computer system commonly manages a large amount of data objects. In structure it is similar to a *department store*, with its thousands of items of merchandise, or *a large engineering project*, with hundreds of subcontractors. Therefore, we need to group *data objects* into *types*. *Types* are discriminates to each other, unless a *type* conversion function is applied that can converts from one *type* to another. *Data objects* in a computer system are analogous to *merchandise* in a department store or *devices*, *parts*, and *equipment* in an engineering project. The *association rules* used to manage *merchandise*, *parts*, *devices*, and *equipments* are similar to *respect operations* of *types* in a computer system. To create a new department in a department store is to expand its business; much like creating a *new type* in a computer system enlarges its application domain [4] to *multidisciplinary areas* and to the control of *more complex physical systems*. These new types are called *user-defined types*.

Programming language likes *FORTRAN IV* have only *basic types* and do not allow users to define their own types. Therefore, its application domain is so limited. However, the *Ada* and the *C++* programming languages do provide *user-defined* type capability; and their application domain is not limited [4]. Today, these two languages are considered general purpose programming languages. Hence, *user-defined* types are vital for this complex and chaotic world in computing.

27.3 Modularization and Communication in the Chaotic World

Modularity partitions a complicated problem into sub-problems, and we implement them as *sub-programs*. A *sub-program* is a unit of a program declared out of line and invoked via calls. The purposes of a *subprogram* are many folds:

1. Result of *modular design* of program
2. *Factoring of common logic* that occurs several places in a program
3. *Parameterized* calls allow operating on different objects at different times, and
4. Simplifying and easing the complexity of the problem

In general, there two distinct subprogram types exist in commonly used programming languages.

1. **Procedures**. It represents computational segments, and its results may be passed back via parameters or side effects. These are used wherever statements are permitted.
2. **Functions**. It returns values of a designated type. These are used anywhere an expression is accepted.

Communication between the main program and subprograms or from one subprogram to another occurs via *parameter passing*. Each programming language defines its own *parameter passing* mechanisms. There are five *parameter-passing* mechanisms in current commonly used programming languages such as parameter passing by *value*, by *reference*, by *name*, by *value-result*, and by *copy rules* that including *copy- in*, *copy-out*, and *copy-in out* rules [8, 9].

Among these five parameter-passing mechanisms, *parameter passing by value*, *parameter passing by reference*, and *parameter passing by name* can be defined mathematically below: For a parameter pass, in evaluating a variable *name* to get a *value* by finding the *location* associated with the *name* and extracting the *value* from the *location*. Let **names, locations**, and **values** be three sets, we define two mappings:

$$\rho : \textbf{names} \rightarrow \textbf{locations}, \text{ and}$$
$$\sigma : \textbf{locations} \rightarrow \textbf{values}$$

Differences in the parameter passing mechanism are defined by when ρ and σ are applied.

1. Parameter passing by **value**
   ρ and σ both applied at point of call, argument completely evaluated at point of call.
2. Parameter passing by **reference** (location)
   Location is determined at the point of call and location is bound as the value of parameter. ρ is applied at point of call and σ is applied with every reference to the parameter.
3. Parameter passing by **name**
   At time of call, neither ρ nor the σ is applied. ρ and σ are applied with every reference to the parameter.

These three parameter-passing mechanisms formed hierarchical structures; the diagram is given in Fig. 27.1. For parameter *passing by value*, both ρ and σ are applied at *calling time*; for parameter *passing by reference*, ρ is applied at *calling time* and σ is applied at reference time; for parameter *passing by name* both ρ and σ are applied with every reference to the parameter.

The *ALGOL*, *C*, *C++*, and *Pascal* languages provide parameter **passing by value** mechanism; and it is a convenient and effective method for enforcing write protection. The *ALGOL*, *C*, *C++*, and *Pascal* languages also implement parameter **passing by reference**. This eliminates duplication of memory. But, there are disadvantages in the parameter passing by **reference**. First, it will likely be slower because one additional level memory addressing is needed compared to parameter passing by **value.** Second, if only one-way communication to the called subprogram is required, unexpected and erroneous changes may occur in the actual parameter.

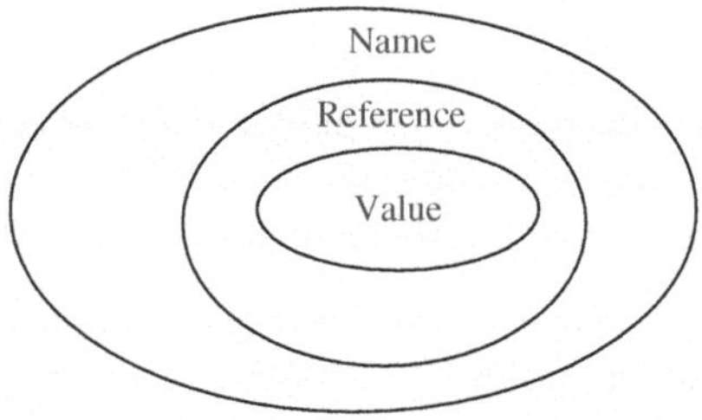

Fig. 27.1 Hierarchical structures of parameter passing

Finally, parameter passing by **reference** can create **aliases**, which are harmful to *readability* and *reliability*. They also make program verification difficult.

The *ALGOL* programming language by default provides parameter passing by **name**. When implementing parameter passing by **name**, the system will create a *run-time* subprogram to evaluate the expression in the calling unit of the program and return the result to the called unit of the program. Therefore, it requires some additional overhead to implement such a *run-time subprogram*. In the *ALGOL* programming language, if one wants parameter passing by **value**, he or she must declare with the word "**value**" for that variable in the actual parameter. *ALGOL* treats parameter passing by **reference**, as a special case of parameter passing by **name**; therefore, the programmer does not need to specify anything. We can define a hierarchical structure for parameter passing by ***value*, reference**, and **name** in the *ALGOL* language.

In the programming language *PL/*1, for actual parameters with a single variable, *PL/*1 uses parameter passing by **value**. However, for a constant or an expression as an argument in the calling statement, *PL/*1 refers to a dummy formal parameter in the called subprogram and it implements a default value or parameter passing by **value.**

In most *FORTRAN* implementations before *FORTRAN* 77, parameters were passed by reference. In later implementations parameter passing by **value-result** has been used commonly.

For the *Ada* language, parameter passing has three modes: mode **in**, mode **out**, and mode **in out**. These are different from parameter passing by value and by reference [10–13].

Mode **in**

- This is the default mode (i.e., **in** may be omitted).
- The actual parameter is **copied** into a local variable. The actual parameter must have a defined value at the point of call.
- The actual parameter may be an expression of compatible type.

Mode **out**

- The result is **copied** into the actual parameter upon exit.
- The actual parameter may be a variable of compatible type.
- The actual parameter need not have a value upon entry.

Mode **in out**

- The value of the actual parameter is **copied** into a local variable upon entry.
- The value of local parameter is **copied** into the actual parameter upon exist.
- The actual parameter must be a variable with a defined value upon entry.

These parameter-passing mechanisms are serving communication facilities between main program and its subprograms or between one subprogram and another. The purpose of engineering computing is to solve problems in the complicated world by means of *granulation* [14, 15], *organization*, and *causation*. *Granulation* subdivides the problem into a set of more manageable sub-problems. *Granulation* is an effective tool to modularize the original problem and to write it into subprograms. From a programmer viewpoint, communication between the main program and subprograms and the communication from one subprogram to another is the *causation* within the program. This paper emphasizes using features of programming languages for engineering computation. It is worth noting that the *Ada* programming language is designed for *embedded systems*, *safety-critical software*, and *large projects* that require high *portability*, *reliability*, and *maintainability*. For example, over 99% of the aviation software in the *Boeing* 777 airplane uses the *Ada* language [16]. It is not surprisingly; the *Ada* language was the first object-oriented design programming language to be accepted as an International Standard.

Today, we design software systems to meet complex application requirements. Almost all the activities in human society, the biological world, physical systems, and engineering projects are *concurrent* or *parallel*; and purely *sequential* activities are special cases. Therefore, *concurrency* reflects the nature of designing software projects. In the next section, we will address the *multitasking* features in the *Ada* programming language [17, 18].

27.4 Parallel or Concurrency Is the Nature of the World

Among all commonly used programming languages, the *Ada* language has the most complete and best features for **multitasking. Multitasking** permits a programmer to partition a big job into many parallel **tasks** [1, 13]. Other programming languages like the *C* and *C++* programming languages can only apply some predefined functions. Thus, they are lack of flexibility and limit their applicability.

A **task** is a unit of computation that can be scheduled independently and in parallel with other such units. An ordinary *Ada* program can be thought of as a **single task**; in fact, it would be called the **main task**. Other tasks must be declared in the **main task** (as **subtasks**) or be defined in a **package** [7, 10, 13, 19]. Several independent *tasks* are often to be executed *simultaneously* in an application. *Ada tasks* can be executed in true *parallelism* or with apparent concurrency simulated by *interleaved execution*. *Ada tasks* can be assigned relative **priorities** and the underlying operating system can schedule them accordingly. A *task* is terminated when its execution ends. A *task* can be declared in *packages*, *subprograms*, *blocks*, or *other tasks*. *All tasks* or *sub-tasks* must terminate before the declaring *subprogram*, *block*, or *task* can be terminated.

A **task** may want to communicate with other tasks. Because the execution speed of tasks cannot be guaranteed, a method for synchronization is needed. To do this, the *Ada* language requires the user to declare **entry** and **accept** statements in two respective tasks engaged in communication. This mechanism provides for *task* interaction and is called a **rendezvous** in the *Ada* language [19].

The *Ada* language also gives an optional scheduling called a **priority** that is associated with a given *task*. A **priority** expresses relative urgency of the *task* execution. An expression of a **priority** is an integer in a given defined range. A numerically smaller value for **priority** indicates lower level of urgency. The **priority** of a *task*, if defined, must be static. If two *tasks* with no **priorities** or two *tasks* of equal priority exist, they will be scheduled in an *arbitrary order*. If two tasks of different priorities are both eligible for execution, they could sensibly be executed on the same processor. A lower priority task cannot execute while a higher priority *task* waits. The *Ada* language forbids time-sliced execution scheduling for *tasks* with explicitly specified priorities. If two *tasks* of prescribed priorities are engaged in a **rendezvous**, the **rendezvous** is executed with the higher of the two priorities. If only one *task* has a defined priority, the **rendezvous** is executed at least at that priority.

A *task* may **delay** its own execution or put itself to *sleep* and *not use processing resources* while waiting for an event to occur by a *delay statement*. The **delay** statement is employed for this purpose. Zero or negative values have no effect. The *smallest delay time* is **20 ms** or **0.020 s**. The m*aximum delay* duration is up to **86,400 s** or **24 h**. The duration only specifies minimum delay; the task may be executed any time thereafter, if the processor is available at that time.

The *Ada* language also provides a **select statement**. There are three forms of select statements. Selective wait, conditional **entry call**, and **timed entry** call. A **selective wait** may include (1) a terminate alternative, (2) one or more **delay alternatives**, or (3) an **else** part, but only one of these possibilities is legal. A task may designate a family (an array) of *entries* by a single name. They can be declared as:

Entry request (0..10) (reqcode: integer);
    Entry alarm (level); – where type level
                                        – must be discrete
An accept statement may name an index entry
    Accept request (reqcode: integer) (0) **do** ...
Accept request (1) (reqcode: integer) **do** ...
    Accept alarm (level);

An entry family allows an accepting task to select entry calls to the same function deterministically.

Among all commonly used programming languages, the *Ada* language is the unique one that provides *multitasking* features at the programming level, and it is very important useful feature for *modeling and simulating* of *real-time and concurrent events* in programming.

The goals of computing are *reliability*, *efficiency*, *accuracy*, and *ease of use*. From the programming point of view, to provide reliable computation is to prevent or eliminate *overflow*, *underflow*, and other *unexpected conditions* so that a program can be executed safely, completely, and efficiently. *Efficient computation* requires an effective computational algorithm for the given problem using proper programming language features for that computation. For *accurate computation*, one should consider problem solving capability, accuracy features, and parallel computation abilities in a given programming language. For *ease of use*, the software engineer should put himself in the situation of the user. A software engineer should remember that users have a job to be done, and they want the computer system to do the job with a minimum of effort. In the next section, we will discuss issues of **accuracy** and **efficiency** of the *Ada* language in *numerical computation capability* [8].

27.5 Accuracy and Efficiency Are Required by This Sophicificate World

The area of *numerical computation* is the backbone of computer science and all engineering disciplines. *Numerical computation* is critical to real world engineering applications. For example, on November 10, 1999, the U.S. National Aeronautics and Space Administration (*NASA*) reported that the *Mars Climate Orbiter Team* found:

> *The 'root cause' of the loss of the spacecraft was the failed translation of English units into metric units in a segment of ground-based, navigation-related mission software as NASA previously announced* [20].

This example indicates that numerical computation and software design are crucial tasks in an engineering project. The goal of *numerical computation* is to reach to a sufficient level of accuracy for a particular application. Designing software for efficient computation is another challenge.

For *engineering applications*, we need to deal with many of *numerical computations*. Among most commonly used programming languages, the *Ada* language has the best numerical computation capability. From the **precision** aspect, the *Ada* language allows a user to define his own accuracy requirement. This section will address the *Ada* language numerical computation capability [8]. To deal this, we should consider the following four criteria: problem solving capability, accuracy of computation, execution time for solving problems, and the capability of parallelism

1. **Problem solving capability**: The *Ada* language provides user-defined types and separate compilation. The former supports programmers solving a wide range of engineering problems and the latter permits development of large software systems. The *Ada* language provides data abstraction and *exception handling* that support information hiding and encapsulation for writing a reliable program. In the real world, many engineering projects consist of concurrent or parallel activities in their physical entities. *Ada multitasking* meets this requirement [21, 22]. In fact, multiprocessor computer systems are now available, thus simulating a truly parallel system becomes possible.
2. **Precision and accuracy**: The *Ada* language's real number types are subdivided into *float-point types* and *fixed-point types*. *Float-point type* have values are numbers
with the format, $\pm.dd..d \times 10^{\pm dd}$. *Fixed-point types* have values with the formats $\pm dd.ddd$, $\pm dddd.0$ or $\pm 0.00ddd$ [1, 11, 13].

For the *float-point number types*, **model numbers** other than zero, the numbers that can be represented exactly by a given computer, are of the form:

$$sign \times mantissa \times (radix \times \times exponent)$$

In this form, *sign* is either $+1$ or -1; *mantissa* is expressed in a number base given by *radix* and *exponent* is an integer. The *Ada* language allows the user to specify the number of significant decimal digits needed. A floating-point type declaration with or without the optional range constraint is shown:

$$\textbf{type } T \textbf{ is digit } D[\textbf{range } L..R];$$

In addition, most *Ada* compilers provide the types *long_float* and *long_long_float* (used in package standard) and *f_float*, *d_float*, *g_float*, and *h_float* (used in package system) [22]. The size and the precision of each of the *Ada* floating-point types are given as follows:

Type	Size(bits)	Precision (digits)
f_float	32	6
d_float	64	9
g_float	64	15
h_float	128	33

The goal of computation is accuracy. Higher accuracy will provide more reliability in the real-time environment. Sometimes, a single precision or a double precision of floating point numbers in *FORTRAN* 77 [21] is not enough for solving some critical problems. In the *Ada* language one may use the floating point number type: *long_long_float* (*h_float*) by declaring digit 33 to use 128 bits for floating point numbers provided by Vax *Ada* [23] to provide a precision of 33 decimal digits accuracy, and the range of exponent is about from 10^{-134} to 10^{+134} or -448 to +448 of base 2 [1,11,13,24] for the range. The author has employed this special accuracy feature in the computation of *hypergeometric distribution function* [25,26].

For the *fixed-point types*, the **model numbers** are in this form:

$$sign \times mantissa \times small$$

The sign is either $+1$ or -1; *mantissa* is a positive integer; small is a certain positive real number. *Model numbers* are defined by a fixed-point constraint, the number small is chosen as the largest power of two that is not greater than the **delta** of a fixed accuracy definition. The *Ada* language permits the user to determine a possible range and an error bound which is called **delta** for computational needs. Examples are the follows:

$$\text{Overhead has a } \textbf{delta} \text{ of } 0.01;$$

$$\text{Overhead has a } \textbf{range} - 10E5\text{–}1.0E5;$$

These indicate *small* is 0.0078125 which is 2^{-7} and model numbers are $-12,800,000 \times small$ to $+12,800,000 \times small$. The predetermined range provides a reliable programming environment. The user assigned error bound delta guarantees an accurate computation. These **floating-point number** and **fixed-point number**

types not only provide good features for real-time critical computations, but also give extra reliability and accuracy for general numerical computations.

3. **The *Ada* for parallel computation**: The author has used *exception handlings* and *tasks* [17, 19] for computation of a division of a product of factorials and another product of factorials in the computation of the *hypergeometric distribution function* [26, 27]. *Exception handling* is used to prevent an *overflow* or *underflow* of multiplications and divisions, respectively. The tasks are used to compute the numerator and denominator concurrently. In addition, *tasks* and *exception handling* working together can minimize the number of divisions and maximize the number of integer multiplications in both of the numerator and denominator, reduce round off errors, and obtain the maximum accuracy. When both products in the numerator and denominator have reached a maximum before an overflow occurs, both *task one* and *task two* stop temporarily and invoke *task three* to perform a division of the products that have been obtained in the numerator and denominator before an overflow occurs. After *task three* completes its job, *task one* and *task two* resume their computation and repeat this procedure until the final result is obtained. These **tasks** work together and guarantee that the result of this computation will be the most accurate and the time for the computation is reasonable. The author has performed these computations on a single processor machine, so the parallelism is a logical parallelism. If one has a multiprocessor machine, he can perform an actual parallelism, tasking and exception *handling* can easily be employed in the computation of the *hypergeometric distribution function* and some computation results and required time for this problem are given in [19], along with those for the computation of the *multinomial distribution function,* multivariate *hypergeometric distribution function*, and other comparable functions. We conclude here that it is not possible to carry out the computation without using these *Ada* special features.
4. **Execution time**: In the 1990s, compiler technology was inadequate to support many *Ada* features. In a real-time system, the response time for *multitasking* features was seemed not fast enough. Therefore, speed was an important criterion for choosing an *Ada* compiler for *real-time* applications. However, the second generation of *Ada* compilers has doubled the speed of the first generation compilers. Today, compilers are fast enough to support all *Ada* features. Currently, *Ada* compilers are available for supercomputers, mainframe computers, minicomputers, and personal computers at reasonable prices.

In running an application, a program *crash* is a disaster. If it is not possible to prevent a *crash* from occurring, the programmer should provide some mechanisms to handle it, to eliminate it, or to minimize its damage. In the next section, we will address *exception-handling* features for these purposes.

27.6 Exception Handling Is the Safe Guard in the Dangerous World

An **exception** is an *out-of-the-ordinary* condition, usually representing a *fault state* that can cause a program crash. The *Ada* language provides for *detection* and *handling* of such conditions [1,11,25]. It is raised implicitly from an operation of integer overflow during the evaluation of an expression, or assigning a negative value to a type with positive data item. It is also raised explicitly as a result of checking when a determinant is found to be zero during matrix inversion or a *stack* is found to be empty during "*pop*" operation. It is not always possible or practical to avoid all exceptions. Allowing exceptions to occur without providing a means to handle the condition could lead to an *unreliable program* or a *crash*, however. Therefore, *exception handling* is very important in designing a programming language [12,25]. The *Ada* language provides five predefined exceptions. They are:

Constraint_error

- It possibly the most frequently used exception in *Ada* programs.
- Constraint_error is raised when a range constraint is violated during assignment or a discriminant value is altered for a constrained type.

Numeric_error

- When an arithmetic operation cannot deliver the correct result
- Overflow or underflow condition occurs

Storage_error

- It is raised if out of memory during the *elaboration* of a data object
- During a subprogram execution, creation of a *new-access type* object

Tasking_error

- It is raised during *inter-tasking communication*

Program_error

- A program attempts to enter an unelaborated procedure, e.g. a *forward* is not declared
  The *Ada* language encourages its user to define and raise his exceptions in order to meet the specific purposes needed. An **exception handler** is a segment of subprogram or block that is entered when the exception is raised. It is declared at the end of a block or subprogram body. If an *exception* is raised and there is no handler for it in the unit, then the execution of the unit is abandoned and the *exception* is *propagated dynamically*, as follows:

- If the unit is a block, the *exception* is raised at the end of the block in the containing unit.

- If the unit is a *subprogram*, the *exception* is raised at the point of call rather than the statically surrounding units; hence the term dynamic is applied to the propagation rule.
- If the unit is the *main program*, program execution is terminated with an appropriate (and nasty) diagnostic message from the run time support environment.

The *Ada* language strongly encourages its user to define and use his own exceptions even if they manifest in the form of predefined exceptions at the outset, rather than depend on predefined ones. The following is an example that shows the *exception, handler*, and *propagation* working together for the computation of factorial function.

```
Function Power (n, k: natural) return natural is
    Function Power (n, k: natural) return natural is
        begin                                   – inner
            if k < 1 then
                return 1;
            else
                return n * Power(n, k−1);
        end if;
    end Power;                                  – inner
begin                                           – outer
    return Power(n, k);
    Exception
        when numeric_error = >
    return natural'last;
end Power;                                      – outer
```

The advantage of this segment of code is that when an overflow condition occurs, the inner **Power** function will exit. Once the outer function returns the **natural'last**, it is not possible to get back to the exact point of the exception.

```
Function Power (n, k: natural) return natural is
    begin
        if k < 1 then
            return 1;
        else
            return n * Power (n, k−1);
        end if;
        Exception
            when numeric_error = >
                return natural'last;
    end Power;
```

If the function had been written as above, each execution of the function would encounter an exception when an overflow occurs. An undesired example is given as follows for comparison. This function will continuously raise exceptions when the first *overflow* is detected and at the end of all the recursive calls.

Designing a programming language for application to all human activities and their needs is not an easy task. However, the language we want for engineering computation must be a reliable one. The *Ada* language is a language that provides complete **exception-handling** facilities. For a program to be considered *reliable*, it must be operate sensibly even when presented with improper input data as well as a software or hardware malfunction. A truly reliable program must monitor itself during execution and take some appropriate actions when a computation is entering an unexpected state. This is the purpose of an **exception handler**.

27.7 Conclusion

In this paper we have discussed the goal of engineering computing as *accuracy*, *efficiency*, *reliability*, and *ease of use*. To accomplish the goals of engineering computation is not easy; it involves human intelligence, knowledge of variety of programming languages, and various programming facilities. In this paper, we have examined many programming language features that are critically important for engineering applications. Some of these features such as *subprograms*, *user-defined types*, and *basic numerical computation facilities* are provided by most commonly used languages like *FORTRAN* 90, the *C*, the *C++*, *Pascal,* and *Ada* languages. This paper has highlighted the use of the *Ada* programming language's *strong typed feature*, *predefined exception handling*, *user defined exception handling*, and *user-defined types for developing reliable programs*. All of these good features in the *Ada* programming languages are unique among all commonly used programming languages. In particular, the use of the *Ada* language's **delta**, **digits**, and **model numbers** for designing engineering projects, which require accurate critical numerical computation, are very important. For the *Ada* language, most of their compilers allow users to have a 128-*bit floating-point number* or 33 *significant digits* for numerical computation. Programmers may specify required accuracies through **digits** and **delta clauses** for *floating-point numbers* and **fixed-point numbers**, respectively. In addition, *Ada's exception handling* can prevent *overflow* or *underflow* during the execution of programs and *multitasking* can perform *parallel computations*. Therefore, from the software reliability point of view, the *Ada* language is better than *FORTRAN*, the *C*, the *C++* and languages in *numerical computation* for engineering applications.

Today, most engineers have only *FORTRAN* or the *C/C++* programming languages for computing. Perhaps some of them have used **Mathematica**, **Mathlab**, or **Maple** for computation. All these mathematical packages have one or more of following drawbacks. Each mathematical software package has its own input and output formats; these formats might not compatible with an engineering project. Each of these software packages has its predefined accuracy, which might not meet the needs of engineering projects. All of these mathematical packages are designed for classroom teaching or laboratory research computations; *efficiency* is not critically important for these purposes. Moreover, these mathematical packages are not for *real-time* or *embedded systems* applications; they lack *exception-handling* capabilities. To do this, each engineering student needs to take at least two courses in computing, one is to learn the basic computation capabilities, a *FORTRAN* or the $C/C{+}{+}$ course will serve for this purpose, the other is to study how to build *efficient*, *accurate*, *reliable*, and *ease of use* software systems to satisfy all engineering domain needs. However, the instructor must have knowledge about engineering experiences in real world, and backgrounds in inter-disciplinary applications in order to qualify for this purpose and lead students to design, implement, and handle engineering projects for the challenges of the real world.

References

1. Demillo, R. and Rice, J. eds. *Studies in Computer Science: In Honor of Samuel D. Conte*, Plenum, New York, 1994.
2. Myers, G. J. *Software Reliability*, Wiley, New York, 1976.
3. Pfleeger, S. L. *Software Engineering, the Production of Quality Software*, 2nd edition, Macmillan, New York, 1991.
4. Wu, T. Some tips in computer science, a talk given at University of Illinois at Springfield, Dec. 2, 2004.
5. *China Daily,* CHINAdaily.com.cn, October 28, 2006.
6. *The Aviation Book, A visual encyclopedia of the history of aircraft*, www. Chronicale-Books.com
7. Boeing 777, globalsecurity.org/military/systems/ aircraft/b777.html
8. Pratt, T. W. and Zelkowitz, M. V. *Programming Languages, Design and Implementation*, 3rd edition, Prentice-Hall, Englewood Cliffs, NJ, 1996.
9. Sebesta, R. W. *Concepts of Programming Languages*, 6th edition, Addison-Wesley, Reading, MA, 2003.
10. Barnes, J. G. P. *Programming in Ada*, 3rd edition, Addison-Wesley, Reading, MA, 1989.
11. Barnes, J. *Programming in Ada 95*, Addison-Wesley, Reading, MA, 1995.
12. Smedema, C. H., et al. *The Programming languages Pascal, Modula, Chill,* and *Ada*, Prentice-Hall, Englewood Cliffs, NJ, 1983.
13. Watt, D. A., et al. *Ada Language and Methodology*, Prentice-Hall, Englewood Cliffs, NJ, 1987.
14. Zadeh, L. A. Fuzzy sets and information granularity, In M. M. Gupta, P. K. Ragade, R. R. Yager, eds., *Advances in Fuzzy Set Theory and Applications*, North-Holland, Amsterdam, pp. 3–18, 1979.
15. Zadeh, L. A. Towards a theory of fuzzy information granulation and its centrality in human reasoning and fuzzy logic, *Fuzzy Sets and Systems*. 19, 111–117, 1997.
16. Ada information clearance house, the web site: www.adaic.org/atwork/boeing.html
17. Booch, G. *Software engineering with Ada*, 2nd edition, Benjamin/Cummings Publishing, Reading, MA, 1987.
18. Habermann, A. N. and Perry, D. E. *Ada for Experienced Programmers*, Addison-Wesley, Reading, MA, 1983.
19. Welsh, J. and Lister, A. A Comparative Study of Task Communication in *Ada*, *Software Practice and Experience*. 11, 257–290, 1981.
20. Mas project http://mars.jpl.nasa
21. Parnas, D. L. Is *Ada* Too Big? (letter) *Communications of ACM*. 29, 1155–1155, 1984.
22. Wichmann, B. A. Is *Ada* Too Big? A Designer Answers the Critics, *Communications of ACM*. 29, 1155–1156, 1984.
23. Struble, G. *Assembler Language Programming*, The IBM System/370 Family, Addison-Wesley, Reading, MA, 1984.

24. *Vax Ada Language Reference Manual*, Digital Equipment Corporation, Maynard, MA, 1985.
25. Wu, T. Built-in reliability in *the Ada* programming language, *Proceedings of IEEE 1990 National Aerospace and Electronics Conference*, pp. 599–602, 1990.
26. Wu, T. An Accurate Computation of the Hypergeometric Distribution Function, *ACM Transactions on Mathematical Software*. 19 (1), 33–43, 1993.
27. Wu, T. *Ada* programming language for Numerical Computation, *Proceedings of IEEE 1995 National Aerospace and Electronics Conference*, pp. 853–859, 1995.
28. Joint Task Force for Computing Curricula 2004, *Overview Report*. ACM and IEEE-CS, 2004.

Chapter 28
Analysis of Overhead Control Mechanisms in Mobile AD HOC Networks

S. Gowrishankar, T.G. Basavaraju, and SubirKumarSarkar

Abstract In this chapter a description of several techniques that are proposed for minimizing the routing overhead in ad hoc routing protocols is discussed. Different algorithms are classified into several categories such as clustering, hierarchical, header compression and Internet connectivity to mobile ad hoc networks based on main objective of minimizing the routing overheads. Clearly, the selection of this area in this paper is highly subjective. Besides, the routing overhead minimizing schemes discussed in this chapter, there are dozen of research schemes that are currently the focus of the research community. With this tutorial, readers can have a comprehensive understanding of different schemes that are employed to reduce the routing overhead.

Keywords Routing overhead · minimizing · Mobile AD HOC Network · performance comparison

28.1 Introduction

Mobile Ad hoc networks are autonomous systems formed by mobile nodes without any infrastructure support. Routing in MANET is challenging because of the dynamic nature of the network topology. Fixed network routing protocols can assume that all the routers have a sufficient description of the underlying network, either through global or decentralized routing tables. However, dynamic wireless networks do not easily admit such topology state knowledge. The inherent randomness of the topology ensures that a minimum overhead is associated with the routing in MANET and is greater than that of fixed networks. It is therefore of interest to

S. Gowrishankar (✉)
Department of Electronics and Telecommunication Engineering, Jadavpur University, Kolkata 700032, West Bengal, India,
E-mail: gowrishankarsnath@acm.org

S.I. Ao and L. Gelman (Eds.), *Advances in Electrical Engineering and Computational Science*, Lecture Notes in Electrical Engineering.

know how small the routing overhead can be made for a given routing strategy and random topology [1].

To evaluate the performance of routing protocols in MANET, several performance metrics such as packet delivery ratio, average end-to-end delay and routing overhead are commonly used. Among these metrics routing overhead is the important one as it determines the scalability of a routing protocol. Routing Overhead means how many extra messages were used to achieve the acceptance rate of improvement.

To evaluate the routing overhead in mobile ad hoc routing protocols different methods are followed. They are (a) simulations, (b) physical experiments, and (c) theoretical analysis [2].

In simulations a controlled environment is provided to test and debug many of the routing protocols. Therefore, most of the literature [3–5] evaluates the control overhead in routing protocols using software simulators like NS-2 [6], Glomosim [7], Qualnet [8] and OPNET [9]. However, simulations are not foolproof method and it may fail to accurately reveal some critical behaviors of routing protocols, as most of the simulation experiments are based on simplified assumptions.

Physical experiments evaluate the performance of routing protocols by implementing them in real environment. Some of the papers in the literature evaluate routing overhead in real physical environment [10, 11]. But, physical experiments are much more difficult and time consuming to be carried out.

Analysis of routing overhead from a theoretical point of view provides a deeper understanding of advantages, limitations and tradeoffs found in the routing protocols in MANET. Some of the papers in literature [12–15] have evaluated routing protocols in MANET from theoretical analysis perspective.

28.2 Theoretical Analysis of Overhead in Hierarchical Routing Scheme

Traditionally the routing schemes for ad hoc networks are classified into proactive and reactive routing protocols. Proactive protocols like DSDV [16] and OLSR [17] maintain routing information about the available paths in the network even if these paths are not currently used. The drawback of such paths is that it may occupy a significant part of the available bandwidth. Reactive routing protocols like DSR, TORA [5] and AODV [18] maintain only the routes that are currently available. However, when the topology of network changes frequently, they still generate large amount of control traffic.

Therefore, the properties of frequent route breakage and unpredictable topology changes in MANET make many of the routing protocols inherently not scalable with respect to the number of nodes and the control overhead. In order to provide routing scalability a hierarchy of layers is usually imposed on the network. Scalability issues are handled hierarchically in ad hoc networks. Many hierarchical routing algorithms are adopted for routing in ad hoc wireless networks. For e.g., cluster based routing and the dominating set based routing.

Sucec and Marsic provide a formal analysis of the routing overhead i.e., they provide a theoretical upper bound on the communication overhead incurred by the clustering algorithms that adopt the hierarchical routing schemes. There can be many scalability performance metrics like hierarchical path length, least hop path length and routing table storage overhead. Among these metrics, control overhead per node (Ψ) is the most important one because of scarce wireless link capacity, which has severe performance limitation. The control overhead Ψ is expressed as a function of |V|, where V is the set of network nodes. It is shown that with reasonable assumptions, the average overhead generated per node per second is only polylogarithmic in the node count i.e., $\Psi = O\left(\log^2 |V|\right)$ bits per second per node [19].

Communication overhead in hierarchically organized networks may result from the following phenomenon: (a) Hello Protocols; (b) level-k cluster formation and cluster maintenance messaging, k $\in \{1, 2 \ldots L\}$, where L is the number of levels in the clustered hierarchy; (c) flooding of the cluster topology updates to cluster members; (d) addressing information required in Datagram headers; (e) location management events due to changes in the clustered hierarchy and due to node mobility between clusters; (f) hand off or transfer of location management data; (g) Location query events. Total communication overhead per node ψ in hierarchically organized networks is the sum of the above contributing elements.

The control overhead and network throughput under a cluster based hierarchical routing scheme is discussed in [20]. The authors claim that when the routing overhead is minimized in a hierarchical design then there is a loss in the throughput from the same hierarchical design. A strict hierarchical routing is assumed which is not based on any specific routing protocol. In MANET, hierarchical routing protocols do not require every node to know the entire topology information. Only a few nodes called the cluster head nodes need to know about the topology information and all the other nodes can simply send their packets to these cluster heads.

Hierarchical routing protocols reduce the routing overhead, as lesser nodes need to know the topology information of an ad hoc network. The throughput of ad hoc network with hierarchical routing scheme is smaller by a factor of $O\left(\frac{N2}{N1}\right)$, where N2 is the number of cluster head nodes and N1 is the number of all the nodes in the network. Hence, the authors claim that there is a tradeoff between the gain from the routing overhead and the loss in the throughput from the hierarchical design of the ad hoc routing protocols.

The control overhead in a hierarchical routing scheme can be due to packet transmissions per node per second (ϕ), due to the maintenance of routing tables as well as due to the address management or location management. Therefore the overhead ϕ required by hierarchical routing is a polylogarithmic function of the network node count (N) i.e., $\Phi = \Theta\left(\log^2 |N|\right)$ packet transmissions per node per second. In this equation, overhead due to hierarchical cluster formation and location management are identified [15].

28.3 Theoretical Analysis and Overhead Minimizing Techniques for AD HOC Networks Using Clustering Mechanisms

The concept of dividing the geographical regions into small zones has been presented as clustering in the literature.

Clustering basically transforms a physical network into a virtual network of interconnected clusters or group of nodes. These clusters are dominated by clusterheads (CH) and connected by gateways or border terminals as shown in Fig. 28.1. Any node can be CH, if it has the necessary functionality such as the processing and transmission power. The node registered with the nearest CH becomes the member of that cluster. By partitioning a network into different clusters both the storage and communication overhead can be reduced significantly.

Different clustering algorithms may use different clustering schemes but generally three different types of control messages are needed: (a) Beacon messages also known as Hello messages are used to learn about the identities of the neighboring nodes, (b) cluster messages are used to adapt to cluster changes and to update the role of a node, (c) route messages are used to learn about the possible route changes in the network [21].

The various types of control messages overhead are (a) *Hello Overhead*: To reduce the hello overhead messages; the frequency of hello messages generated by a node to learn about its neighboring node when a new link is formed should be at least equal to the link generation rate. The link generation between any two nodes can be notified by sending the hello messages and each of the nodes can hear the hello message sent by the other node. (b) *Cluster message overhead due to link break between cluster members and their cluster heads*: This event causes the node to change its cluster or become a cluster head when it has no neighboring clustering heads. The cluster members send the cluster messages due to this type of link changes. To minimize the control message overhead the ratio of such link breaks to total link breaks should be equal to the ratio of links between the cluster members and cluster heads divided by the total number of links in the entire network. (c) *Cluster message overhead due to link generation between two cluster heads*: When

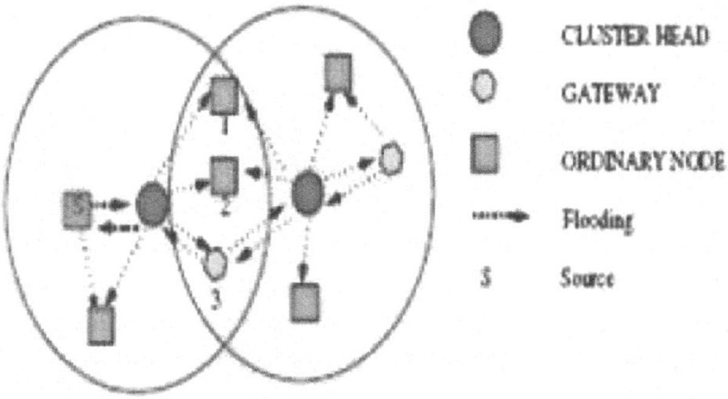

Fig. 28.1 Route establishment in clustering mechanism

a link is generated between two cluster heads, one of the cluster heads needs to give up its cluster head role, which is decided by the clustering algorithm. Every time a link between two cluster heads appears, the number of cluster messages generated is same as the number of nodes in the cluster that needs to undergo reclustering. (*d*) *Routing overhead:* When a particular node in the cluster should be updated with the route to other nodes in the cluster, the routing storage overhead is proportional to the size of the cluster.

An analytical study on routing overhead of two level cluster based routing protocols for mobile ad hoc networks is done in [2]. Routing protocols can be summarized into generic proactive routing protocol and a generic reactive routing protocol. It's generic because there may be some different strategy employed for each of the groups, but the underlying nature of the routing is similar.

In two level cluster based routing scheme, the routing is divided into two separate parts, i.e., routing among different clusters (i.e., intercluster routing) and routing within a cluster (i.e., intracluster routing). Since there are two types of routing schemes i.e., proactive and reactive which can be employed in intercluster routing and intracluster routing, there are totally four types of two level hierarchical routing schemes with different combinations of them. Hence we have proactive to proactive, reactive to reactive, proactive to reactive and reactive to proactive routing scheme.

When a proactive scheme is adapted for intercluster routing each cluster head periodically collects its local cluster topology and then broadcasts it to its direct neighbor cluster head via gateways. When a reactive routing scheme is used for inter cluster routing, a route request for a route to the destination node that is in another cluster will be broadcasted among cluster heads. When a proactive routing scheme is utilized for intracluster routing, each node broadcasts its local node topology information, so the route to the destination within the same cluster will be available when needed. When a reactive routing scheme is employed for intracluster routing, a route request to the destination will be flooded within the same cluster.

Thus a proactive to proactive routing scheme will work as a proactive routing protocol with a hierarchical structure. The proactive to proactive routing scheme produces overhead due to periodical topology maintenance of

$$\mathrm{O}\left(\frac{1}{n}N^2 + \frac{1}{kN_c}N^2\right)$$

where n is the total number of clusters in the network, N is the network size, k is the cluster radius, N_c is the cluster size.

A reactive to reactive routing protocol operates as a purely reactive routing protocol with a hierarchical structure. Reactive to reactive routing protocol yields a routing overhead due to route discovery of

$$\mathrm{O}\left(\frac{1}{k}N^2\right).$$

In a proactive to reactive routing scheme the cluster heads periodically exchange topological information and a node always sends a route request to its cluster head where there is no available route to an expected destination. Then the cluster head will send a route reply packet to the node, which indicates that the destination is within the local cluster or contains a route to the destination node, which is in another cluster. Proactive to reactive routing protocol have a basic routing overhead due to topology maintenance, cluster maintenance and route discovery which is found to be

$$\mathrm{O}\left(\frac{1}{kN_c}N^2 + \frac{r}{nk}N^2\right) \text{ where r is average route lifetime.}$$

In a reactive to proactive routing scheme each node in the cluster will periodically broadcast local node topology information within the cluster. Thus, when the destination is within the cluster, the route is immediately available. Otherwise, the node will send a route request packet to its cluster head, which will be broadcasted among the cluster heads. Reactive to proactive routing protocol have a basic routing overhead due to cluster maintenance and route discovery, which is approximately equal to

$$\mathrm{O}\left(\frac{1}{n}N^2 + \frac{1}{k}N^2\right).$$

A mathematical framework for quantifying the overhead of a cluster based routing protocol (D-hop max min) is investigated by Wu and Abouzeid [13]. The authors provide a relationship between routing overhead and route request traffic pattern. From a routing protocol perspective, 'traffic', could be defined as the pattern by which the source destination pairs are chosen. The choice of a source destination pair depends on the number of hops along the shortest path between them. Also the network topology is modeled as a regular two-dimensional grid of unreliable nodes. It is assumed that an infinite number of nodes are located at the intersections of a regular grid. The transmission range of each node is limited such that a node can directly communicate with its four immediate neighbors only. It is reported that the clustering does not change the traffic requirement for infinite scalability compared to flat protocols, but reduces the overhead by a factor of

$$\mathrm{o}\left(\frac{1}{m}\right) \text{ where } M \text{ is the cluster size.}$$

Wu and Abouzeid show that the routing overhead can be attributed to events like (a) route discovery, (b) route maintenance, and (c) cluster maintenance.

Route discovery is the mechanism where by a node i wishing to send a packet to the destination j obtains a route to j. When a source node i wants to send a packet to the destination node j, it first sends a route request (RREQ) packet to its cluster head along the shortest path. The route reply (RREP) packet travels across the shortest path back to the cluster head that initiated the RREQ packet. So the route discovery event involves an RREP and RREQ processes. The overhead for RREQ is generally higher than the RREP since it may involve flooding at the cluster head level. Therefore, the minimum average overhead of finding a new route is

$$\frac{f(k-3)\left(6M+6+\frac{3}{M}\right)+f(k-2)\left(4M^2+2\right)-f(k-1)M\left(M^2-1\right)}{3\left(2M^2+2M+1\right)f(k-1)}$$

Where, M is the radius of the cluster and changing the value of k controls the traffic pattern.

Route maintenance is the mechanism by which a node i is notified that a link along an active path has broken such that it can no longer reach the destination node j through that route. When route maintenance indicates a link is broken, i may invoke route discovery again to find a new route for subsequent packets to j. In cluster based routing, the neighboring node sends a RERR packet to its cluster head rather than the source node itself. The cluster head could patch a path locally without informing the source node, if the failed node is not the destination node. This is called local repair. In this case, the path is locally fixed. Also, the RERR packet sent from a neighboring node to the cluster head is considered as the cluster maintenance overhead. Therefore the minimum overhead required for route maintenance is $4C\ (f\ (k-2) - f\ (k-1))$ where, $k > 3$ and C is a constant.

Clustering incurs cluster maintenance overhead, which is the amount of control packet needed to maintain the cluster membership information. The membership in each cluster changes overtime in response to node mobility, node failure or new node arrival. The average cluster maintenance overhead is

$$\frac{4\left(M-1\right)M\left(M+1\right)}{3\left(2M^2+2M+1\right)}+2M$$ Where, M is the radius of the cluster.

The work done in [14] by Zhou, provide a scalability analysis of the routing overheads with regard to the number of nodes and the cluster size. Both the interior routing overhead within the cluster and the exterior routing overhead across the clusters are considered. The routing protocol includes a mechanism for detecting, collecting and distributing the network topology changes. The process of detecting, collecting and distributing the network topology changes contribute to a total routing overhead R_t. The routing overhead can be separated into interior routing overhead (R_i) and the exterior routing overhead (R_e). The interior routing overhead R_i is the bit rate needed to maintain the local detailed topology. This includes the overhead of detecting the link status changes by sending "HELLO" message, updating the cluster head about the changes in link status and maintaining the shortest path between the regular nodes to their cluster head.

Exterior routing overhead (R_e) is the bit rate needed to maintain the global ownership topology, which includes the overheads of the distributing the local ownership topologies among the cluster heads. Hence $R_t = R_i + R_e$

28.4 Minimizing Overhead in AD HOC Networks by Header Compression

In literature it has been studied that [22] approximately half of the packets sent across the Internet are 80 bytes long or less. This percentage has increased over the last few years in part due to widespread use of real time multimedia applications. The multimedia application's packet size is usually smaller in size and these small packets must be added with many protocol headers, while traveling through the networks. In Ipv4 networks there can be at least 28 bytes (UDP) or 40 bytes (TCP) overheads per packet. These overheads consume much of the bandwidth, which is very limited in wireless links. Small packets and relatively larger header size translates into poor line efficiency. Line efficiency can be defined as the fraction of the transmitted data that is not considered overhead. Figure 28.2 shows some of the common header chains and size of each component within the chain.

Ad hoc networks create additional challenges such as context initialization overhead and packet reordering issues associated with node mobility. The dynamic nature of ad hoc networks has a negative impact on header compression efficiency.

A context is established by first sending a packet with full uncompressed header that provides a common knowledge between the sender and receiver about the static field values as well as the initial values for dynamic fields. This stage is known as context initialization. Then the subsequent compressed headers are interpreted and decompressed according to a previously established context. Every packet contains a context label. Here the context label indicates the context in which the headers are compressed or decompressed.

A novel hop-by-hop context initialization algorithm is proposed in [23] that depends on the routing information to reduce the overhead associated with the context initialization of IP headers and uses a stateless compression method to reduce the overhead associated with the control messages. Context initialization of IP headers is done on a hop-by-hop basis because the headers need to be examined in an uncompressed state at each of the intermediate nodes. The context initialization

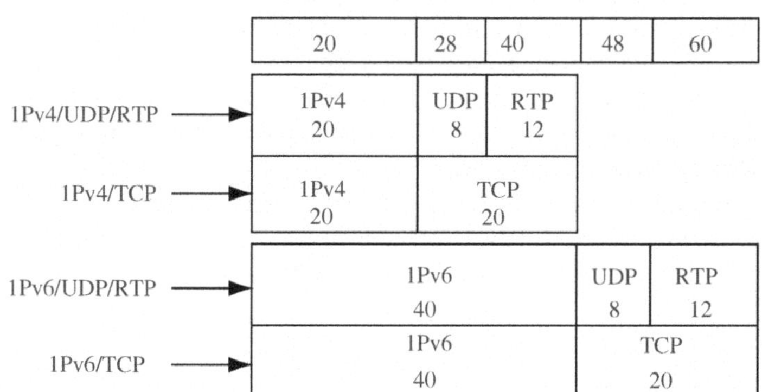

Fig. 28.2 Common header chains and their sizes

overhead is reduced by caching the address information that is transmitted in the routing messages, in order to reduce the size of the context initialization headers.

Also a stateless header compression is proposed. It is stateless because the state of the context is fixed and it does not change with time. Header compression improves the line efficiency by exploiting the redundancies between the header fields within the same packet or consecutive packets belonging to the same stream.

The overall result is the reduced overhead, increased network capacity and line efficiency even in the presence of rapid path fluctuations.

An ad hoc robust header compression protocol has been proposed (ARHC) in [24]. ARHC protocol can be used to compress UDP, TCP and raw IP headers in ad hoc network. The mechanism of ARHC is that when the first packet of a session arrives, the compressor generates a unique number called context ID, which indexes the quintuplet (source address, destination address, source port, destination port, protocol) and all the constant fields. Compressor then records the context id, quintuplet and all the constant fields. Then the compressor will send the full packet header along with the context ID. Upon receiving the very first packet the decompressor records this information. When the subsequent packets arrive later, the compressor and decompressor act as follows. The compressor will remove the constant fields and the quintuplet from the header and transmits only the context ID. The decompressor then retrieves the quintuplets and the constant fields from the context tables indexed by the context ID, thereby restoring the original header.

28.5 Minimizing Overhead for AD HOC Networks Connected to Internet

Today Internet has become the backbone of the wired and wireless communication. Also mobile computing is gaining in popularity. In order to meet the rapid growing demand of mobile computing, many of the researchers are interested in the integration of MANET with the Internet.

When a mobile node in MANET wants to exchange packets with the Internet, first the node must be assigned a global IP address and then the available Internet gateways has to be discovered to connect to the Internet as shown in Fig. 28.3. But, this is achieved at the cost of higher control overhead.

For gateway discovery, a node depends on periodic gateway advertisement. To make efficient use of this periodic advertisement, it is necessary to limit the advertisement flooding area. For gateway discovery, a node depends on periodic gateway advertisement. To make efficient use of this periodic advertisement, it is necessary to limit the advertisement flooding area. A complete adaptive scheme to discover IG in an efficient manner for AODV is given in [13]. In this approach both the periodic advertisement and adaptive advertisement schemes are used. At a relatively long interval each gateway sends the periodic advertisement messages. Periodic advertisements performed at a widely spaced interval do not generate a great deal of overhead but still provides the mobile nodes with a good chance of finding the short-

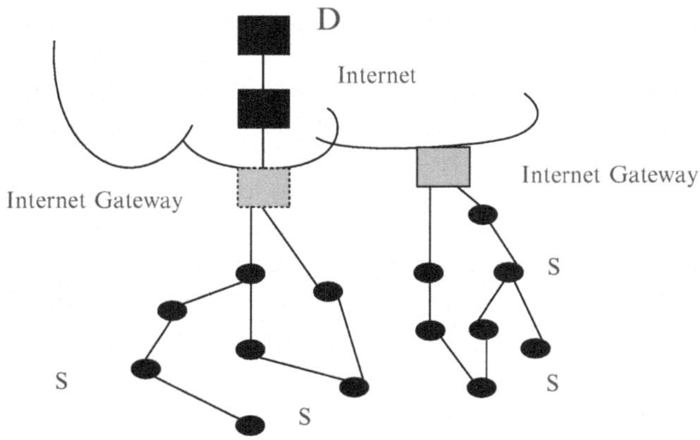

Fig. 28.3 MANET connected to Internet scenario

est path to a previously used gateway. The TTL of the periodic gateway message is used as a parameter to adjust the network conditions. A heuristic algorithm called "Minimal Benefit Average" [25] decides the next TTL to be used for the periodic gateway advertisement messages.

The goal of the adaptive advertisement scheme is to send advertisement packets only when the gateway detects the movement of nodes, which would result in the paths used by the source mobile nodes communicating with the gateway to be changed. Adaptive advertisement is performed when needed, regardless of the time interval used for periodic advertisement. [26]. In this approach there is reduction in overhead messages since the periodic advertisements are sent at a long time interval and perform adaptive advertisement only if there is mobility in the network.

The various parameters that affect the control overhead created by interoperating the ad hoc routing protocols and IP based mobility management protocols is addressed in [27]. Mobile IP is used as the baseline mobility management protocol and AODV is chosen as the ad hoc routing protocol. IP tunneling is used to separate the ad hoc network from the fixed network. In mobile IP, a mobile node can tell which router is available by listening to router advertisements, which are periodically broadcasted by the routers.

A fixed access router is assigned the role of mobility agent and has connection to at least one of the MANET nodes. Such router is referred to as Ad hoc Internet Access Router (AIAR) and it maintains a list called ad hoc list, which keeps a list of IP address of the mobile nodes that wish to have Internet connectivity. In an integrated network the control overhead comprises of AIAR registration packets, routing protocol control packets, mobile IP registration packets and mobile IP router advertisement.

In mobile IP majority of the overhead is due to the route advertisement packets that are being repeatedly and periodically forwarded among the mobile nodes. Also, the router advertisement used by the mobility management protocol to carry network information is the major source of unnecessary control overhead within MANET. Varying TTL value is an effective mechanism to control the amount of advertisement packets [27].

A multihop router is developed in [28] for non-uniform route connectivity with low routing overhead. To achieve efficient route discovery and route maintenance a new routing scheme called hopcount based routing (HBR) protocol is developed. HBR is an on demand routing protocol. When a source node needs to discover a route to a node on the wired network, it initiates a route discovery to the nearest access point by broadcasting a route request packet. By utilizing the hop counts referring to access points, the route discovery region is localized to a small area. On receiving the route request packet, the access point responds by sending a route reply packet to the source node. Once a route is found, the source node begins to forward data to the access point. After the access point receives these data packets from the source node, it then forwards these packets through the wired line to the destination node on the wired network using the routing protocol in the wired network. By using the hop count information an attempt is made to reduce the number of nodes to whom the route request is propagated. Thus in HBR routing protocol to construct routes from mobile nodes to access points hop count information is utilized to localize the route discovery within a limited area in order to reduce the routing overhead.

In this chapter a description of several techniques that are proposed for minimizing the routing overhead in ad hoc routing protocols is discussed. Different algorithms are classified into several categories such as clustering, hierarchical, header compression and Internet connectivity to mobile ad hoc networks based on main objective of minimizing the routing overheads. Clearly, the selection of this area in this paper is highly subjective. Besides, the routing overhead minimizing schemes discussed in this chapter, there are dozen of research schemes that are currently the focus of the research community.

With this tutorial, readers can have a comprehensive understanding of different schemes that are employed to reduce the routing overhead. We hope that this discussion can facilitate researchers to move in new direction and devise new methods to reduce the control overheads that are inherently associated with the routing protocols.

References

1. R. C. Timo and L. W. Hanlen, “MANETs: Routing Overhead and Reliability”, IEEE, 2006.
2. Z.-H. Tao, “An Analytical Study on Routing Overhead of Two level Cluster Based Routing Protocols for Mobile Ad hoc Networks”, IEEE, 2006.
3. J. Broch, D. A. Maltz, D. B. Johnson, Y.-C. Hu, and J. Jetcheva, “A Performance Comparison of Multi-hop Wireless Ad Hoc Network Routing Protocols,” inProceedings of ACM MobiCom, Oct. 1998.
4. S. R. Das, R. Castaneda, and J. Yan, “Simulation Based Performance Evaluation of Routing Protocols for Mobile Ad Hoc Networks,” Mobile Networks and Applications, vol. 5, no. 3, pp. 179–189, Sept. 2000.

5. A. Boukerche, "Performance Evaluation of Routing Protocols for Ad Hoc Wireless Networks," Mobile Networks and Applications, vol. 9, no. 4, pp. 333–342, Aug. 2004.
6. "The network simulator – NS-2," http://www.isi.edu/nsnam/ns
7. X. Zeng, R. Bagrodia, and M. Gerla, "GloMoSim: A Library for Parallel Simulation of Large-Scale Wireless Networks," Proceedings of 12th Workshop on Parallel and Distributed Simulations, May 1998.
8. T. H. Clausen et al., "The Optimized Link State Routing Protocol, Evaluation Through Experiments and Simulation", Proceedings of IEEE Symposium on Wireless Personal Mobile Communications, 2001.
9. "OPNET Modeler," http://www.opnet.com/
10. R. S. Gray, D. Kotz, C. Newport, N. Dubrovsky, A. Fiske, J. Liu, C. Masone, S. McGrath, and Y. Yuan, "Outdoor Experimental Comparison of Four Ad Hoc Routing Algorithms," inProceedings of ACM International Symposium on Modeling, Analysis and Simulation of Wireless and Mobile Systems, Oct. 2004.
11. E. Borgia, "Experimental Evaluation of Ad Hoc Routing Protocols,"in Proceedings of IEEE International Conference on Pervasive Computing and Communications Workshops, May 2005.
12. N. Zhou, H. Wu, and A. A. Abouzeid, "Reactive Routing Overhead in Networks with Unreliable Nodes," in Proceedings of ACM MobiCom, Sept. 2003.
13. H. Wu and A. A Abouzeid, "Cluster-Based Routing Overhead in Networks with Unreliable Nodes," in Proceedings of IEEE WCNC, Mar. 2004.
14. N. Zhou and A. A. Abouzeid, "Routing in Ad Hoc Networks: A Theoretical Framework with Practical Implications," in Proceedings of IEEE INFOCOM, Mar. 2005.
15. J. Sucec and I. Marsic, "Hierarchical Routing Overhead in Mobile Ad Hoc Networks", IEEE Transactions on Mobile Computing, vol. 3, no. 1, pp. 46–56, Jan.–Mar. 2004.
16. T. C. Chiang and Y. M. Huang, "A Limited Flooding Scheme for Mobile Ad Hoc Networks", IEEE, 2005.
17. C. E. Perkins and E. M. Royer, "Ad Hoc on Demand Distance Vector Routing", Proceedings of IEEE Workshop on Mobile Computing Systems and Applications , pp. 90–100, Feb. 1999.
18. C. E. Perkins and P. Bhagwat, "Highly Dynamic Destinatio-Sequenced Distance Vector Routing (DSDV) for Mobile Computers", Proceedings of ACM SIGCOMM, 1994.
19. J. Sucec and I. Marsic, "Hierarchical Routing Overhead in Mobile Ad Hoc Networks," Clustering Overhead for Hierarchical Routing in Mobile Ad Hoc Networks", IEEE INFOCOM, 2002.
20. Y. Qin and J. He, "The Impact of Throughput of Hierarchical Routing in Ad Hoc Wireless Networks", IEEE, 2005.
21. X. Mingqiang, E. R. Inn-Inn and K. G. S. Winston, "Analysis of Clustering and Routing Overhead for Clustered Mobile Ad Hoc Networks", IEEE ICDCS, 2006.
22. Sprint. "IP Monitoring Project". Feb. 6, 2004. http://ipmon.sprint.com/packstat/packetoverview.php
23. A. Jesus, A. Syeed, and H. Daniel, "Header Compression for Ad Hoc Networks", IEEE, 2004
24. H. Wang et al., "A Robust Header Compression Method for Ad hoc Network", IEEE, 2006.
25. P. M. Ruiz and A. F. Gomez Skarmet, "Enhanced Internet Connectivity Through Adaptive Gateway Discovery", 29th Annual IEEE International Conference on Local Computer Networks, pp. 370–377, Nov. 2004.
26. R. Vanzara and M. Misra, "An Efficient Mechanism for Connecting MANET and Internet Through Complete Adaptive Gateway Discovery", IEEE, 2006.
27. K. A. Chew and R. Tafazolli, "Overhead Control in Interworking of MANET and Infrastructure Backed Mobile Networks", International Conference on Wireless Ad Hoc Networks, IEEE, 2004.
28. R. Teng, H. Morikawa, et al., "A Low Overhead Routing Protocol for Ad Hoc Networks with Global Connectivity", IEEE, 2005.
29. X. Li and L. Cuthbert," Stable Node-Disjoint Multipath Routing with Low Overhead in Mobile Ad Hoc Networks", Proceedings of the 12th Annual International Symposium on Modeling, Analysis, and Simulation of Computer and Telecommunications System, IEEE, 2004.

30. A. M. Abbas and B. N. Jain, "Mitigating Overheads and Path Correlation in Node-Disjoing Multipath Routing for Mobile Ad Hoc Networks", IEEE, 2006.
31. R. C. Timo and L. W. Hanlen," Routing Overhead and Reliability", IEEE, 2006.
32. P. M. Ruiz and A. F. Gomez Skarmet, "Reducing Data Overhead of Mesh Based Ad Hoc Multicast Routing Protocols by Steiner Trees", IEEE, 2004.
33. S. Bansal, R. Shorey, and A. Misra, "Comparing the Routing Energy Overheads of Ad Hoc Routing Protocols", IEEE, 2003.
34. R. Teng, H. Morikawa, and T. Aoyama, "A Low Overhead Routing Protocol for Ad Hoc Networks with Global Connectivity", IEEE, 2005.
35. E. R. Inn Inn and K. G. S. Winston, "Clustering Overhead and Convergence Time Analysis of the Mobility Based Multi hop Clustering Algorithm for Mobile Ad Hoc Networks", Proceedings of the 11th International Conference on Parallel and Distributed Systems, IEEE, 2005.
36. S.-H. Wang, M.-C. Chan, and T.-C. Hou, "Zone ased Controlled Flooding in Mobile Ad Hoc Networks", International Conference on Wireless Networks, Communications and Mobile Computing, IEEE, 2005.
37. www.scalable-networks.com

Chapter 29
Context Aware In-Vehicle Infotainment Interfaces

H. Sharma and A.K. Ramani

Abstract The current state-of-practice for developing automotive software for Infotainment and Telematics systems offers little flexibility to accommodate such heterogeneity and variation. This chapter presents a framework that enables development of context aware interaction interfaces for in-vehicle infotainment applications. Essential framework concepts are presented with development and discussion of a lightweight component model. Finally, the design of an example application is presented and evaluated.

Keywords In-Vehicle Infotainment · Context Aware · Interface · Context Management · Interaction Architecture

Today's cars provide an increasing number of new functionalities that enhance the safety and driving performance of drivers or raise their level of comfort. In addition, infotainment systems are innovating themselves, providing luxury facilities or enabling modern communication mechanisms. With every new generation of cars, there are more built-in software functions.

The current state-of-practice for developing automotive software for Infotainment and Telematics systems offers little flexibility to accommodate such heterogeneity and variation. Currently, application developers have to decide at design time what possible uses their applications will have and the applications do not change or adapt once they are deployed on an infotainment platform. In fact, In-vehicle infotainment applications are currently developed with monolithic architectures, which are more suitable for a fixed execution context.

This chapter presents a framework that enables development of context aware interaction interfaces for in-vehicle infotainment applications. Essential framework concepts are presented with development and discussion of a lightweight component model. Finally, the design of an example application is presented and evaluated.

H. Sharma (✉)
Software Engineer at Delphi Delco Electronics Europe GmbH, Bad Salzdetfurth, Germany,
E-mail: hemant.sharma@delphi.com

S.I. Ao and L. Gelman (Eds.), *Advances in Electrical Engineering and Computational Science*, Lecture Notes in Electrical Engineering.

29.1 In-Vehicle Infotainment

With increased commuting times, vehicles are becoming a temporary mobile workplace and home place for both drivers and passengers, implying significant additional content and tools related to depiction and interaction with information about one's own identity, entertainment, relationship-building/maintenance (e.g., communication with others), self-enhancement (including education and health monitoring), access to information resources (e.g., travel information, references, and other databases), and commerce. Vehicles as communication, entertainment, and information environments may involve distributed displays that depict virtual presence in other spaces, games played with other vehicles' drivers and passengers, and status or instruction concerning driving/parking maneuvers, repairs, parts reordering, etc.

In-vehicle infotainment systems offer a new generation of ultra-compact embedded computing and communication platforms, providing occupants of a vehicle with the same degree of connectivity and the same access to digital media and data that they currently enjoy in their home or office. These systems are also designed with an all-important difference in mind: they are designed to provide an integrated, upgradeable control point that serves not only the driver, but the individual needs of all vehicle occupants.

Infotainment system manufacturers seeking to serve worldwide markets face a complex matrix of 15 major OEMs (car companies) with 78 major marks (brands) and perhaps as many as 600 vehicle models. Models may each have a number of trim levels; serve multiple languages and regulatory regimes (countries). To compete, the suppliers will also need to deliver new functionality; features and a fresh new look on an annual basis. Building Human Machine Interface (HMI) for these complex applications is difficult and code-intensive, consuming disproportionate amount of resources and adding considerably to project risk.

29.1.1 Infotainment System Interaction

An In-vehicle Infotainment device is usually connected to the vehicle network as well as to GPS network. Further it not only, may have WiFi or cellular connectivity, but also interface to various ad hoc networks using the infrared or Bluetooth interfaces. The potential for interaction with its environment is great. However, the system only provides limited HMI primitives for this. The result is that such devices are still seen as stand-alone and independent system, which interacts mainly to offer static services -interaction with their environment and peers is either not considered or is very limited. Thus, although physically mobile, they are logically static systems.

This HMI interaction model in current infotainment systems has various disadvantages: There is little code sharing between applications running on the same device. There is no framework providing higher level interoperability and communication primitives for HMI service applications running on different devices. HMI

Applications are monolithic, composed of a single static interaction interface, which makes it impossible to update part of HMI structure. The procedure needed to host third party dynamic service applications is difficult.

29.2 Context Aware Interaction

Unlike fixed distributed systems, infotainment systems execute in an extremely dynamic context. By context, it is meant everything that can influence the behavior of an infotainment application, from resources within the physical device, such as memory, power cycle, screen size, and processing power, to resources outside the physical device, such as location, remote service available, network bandwidth and latency. Infotainment applications need to be aware of changes occurring in their execution context, and adapt to these changes, to continue to deliver a high quality-of-service to their users.

29.2.1 Context Aware Design

In order to ease the development of context-aware applications, middleware layered between the underlying software system and the HMI applications have to provide application developers with powerful abstractions and mechanisms that relieve them from dealing with low-level details. For example, applications must be able to specify, in a uniform way, which resources they are interested in and which behaviors to adopt in particular contexts. The middleware, then, maintains, on behalf of the applications, updated context information, detects changes of interest to them and reacts accordingly.

29.2.1.1 Context Awareness by Reflection

It is argued that reflection is a powerful means to build mobile computing frameworks that supports the development of context-aware applications. The key to the approach is to make some aspects of the internal representation of the framework explicit and, hence, accessible from the application, through a process called *reification*. Applications are then allowed to dynamically inspect middleware behavior (*introspection*) and also to dynamically change it (*adaptation*) by means of a meta interface that enables runtime modification of the internal representation previously made explicit. The process where some aspects of the system are altered or overridden is called *absorption*.

By definition, reflection allows a program to access, reason about and alter its own interpretation. The key to the approach is to offer a meta-interface supporting the inspection and adaptation of the underlying virtual machine (the meta-level).

29.2.1.2 Context Awareness by Logical Mobility

As a highly dynamic system, an infotainment system similar to mobile system encounters, by definition, changes to its requirements to which it needs to adapt.

The acceptance of logical mobility (LM) techniques, or the ability to ship part of an application or even a complete process from one host to another, has recently been witnessed in automotive domain with upcoming infotainment applications. It has been shown that providing the flexible use of LM primitives to mobile computing applications through a middleware system, allow for a greater degree of application dynamicity, and will provide new ways to tackle interoperability and heterogeneity as well as ease deployment.

Infotainment system can associate the context information to the applications and adopt a suitable policy, reflection or logical mobility or both, to achieve context awareness. The infotainment applications need suitable mechanisms for management of context information and application of context awareness policy.

29.2.2 Context Management

The evolving nature of automotive services makes complete formalization of all context knowledge a challenging mission. Nevertheless, it is observed that certain contextual entities (e.g., location, user, vehicle and computational entity, etc.) are most fundamental for capturing the general condition about the context, as they not only form the skeleton of overall situation, but also act as indices into associated information.

There are several types of service components that are involved in the sensing, interpreting, managing and utilizing context information in an intelligent automotive environment. All these components are packaged in form of pluggable services such that they can be remotely deployed and managed.

29.3 Interaction Architecture

29.3.1 Layered Model

The framework is designed according to layered architecture shown in the figure below (Fig. 29.1). The architecture consists of components organized into four layers.

The Application layer defines a standard framework to create and run context-aware HMI applications on top of the framework's application model.

Application management layer contains components that define context and QoS policies for the HMI application. The layer is divided into *Context Management* and

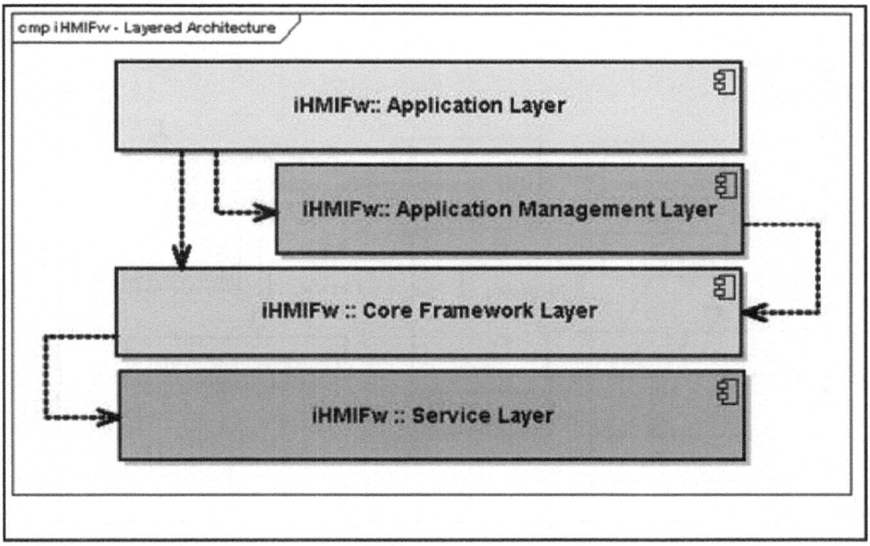

Fig. 29.1 Layered architecture of HMI framework

QoS Management components. The context management components are responsible for specification and management of HMI specific contexts.

The purpose of *core layer* of the framework, in general, is to provide higher level interaction primitives than those provided by the vehicle infotainment network and infotainment service system as a layer, upon which HMI applications are then constructed.

The components of *service layer* offers application service patterns, design-to-implementation mappings, platform profiles, and extensibility mechanisms for interoperating with other infotainment applications.

29.3.2 Component Model

The term component model refers to a description of components and a component infrastructure that abstracts from the details of a concrete implementation, such as the exact format of the component executable. The goal in doing so is to understand the choices that can be made in the design of component architecture, without getting distracted by machine-specific or platform specific details. This is especially important in infotainment systems, as it is believed that the diversity of process and architectures will mean that many different implementations of any given component model will be needed.

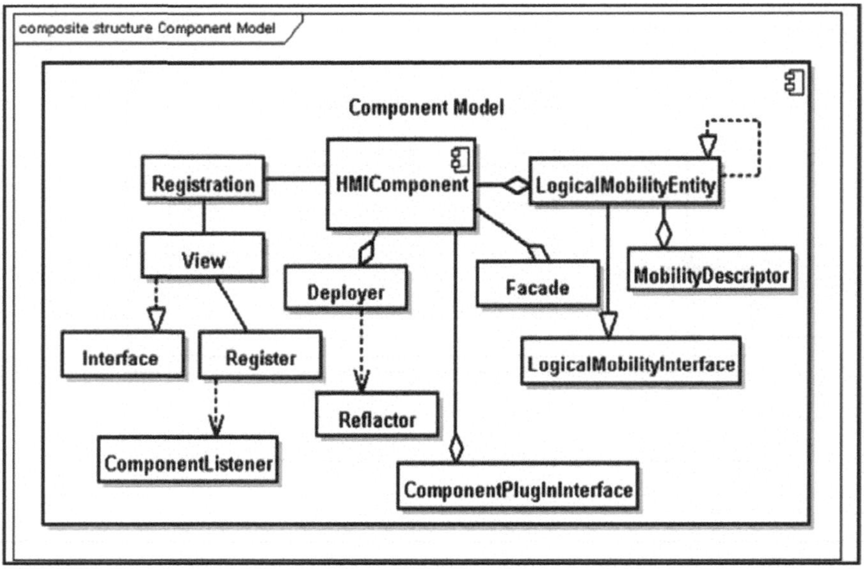

Fig. 29.2 Component meta model overview

29.3.2.1 Component Meta-Model Overview

The component metamodel, as shown in Fig. 29.2, is a Meta Object Facility (MOF) compliant extension of the UML metamodel.

It builds upon and extends the UML concepts of Classifier, Node, Class, Interface, Data Type, and Instance. The most novel aspect of the component model is the way in which it offers distribution services to local components, allowing instances to dynamically send and receive components at runtime.

The component metamodel is a local, or in process, reflective component metamodel for HMI applications hosted on infotainment platforms. The model uses logical mobility primitives to provide distribution services and offers the flexible use of those primitives to the applications; instead of relying on the invocation of remote infotainment services via the vehicle network. The HMI framework components are collocated on the same address space. The model supports the remote cloning of components between hosts, providing for system autonomy when application service connectivity is missing or is unreliable. As such, an instance of HMI framework, as part of HMI application, is represented as a collection of local components, interconnected using local references and well-defined interfaces, deployed on a single host. The model also offers support for structural reflection so that applications can introspect which components are available locally, choose components to perform a particular task, and dynamically change the system configuration by adding or removing components.

29.3.2.2 Component Model Elements

This section describes the primitive elements of the component model.

Components

The framework components encapsulate particular functionality, such as, for instance, a user interface, a service advertisement protocol, a service, a graphics framework, or a widget library. The components separate interfaces and implementations. A component is implemented by one or several HMI framework classes. It can implement one or more interfaces, called *facades*, with each facade offering any number of operations. A metamodel for components that are going to be deployed across autonomous domain boundaries needs to ensure that interfaces that have once been defined cannot be changed.

Each framework component implements at least one facade, the Component façade. The purpose of this facade is to allow an application to reason about the component and its attributes. This permits access to the properties of the component by retrieving, adding, removing, and modifying attributes. The component facade also contains a constructor, which is used to initialize the component, and a destructor, which is used when removing the component from the system.

Containers

The central component of every HMI application is the container component. A container is a component specialization that acts as a registry of components installed on the system. As such, a reference to each component is available via the container. The container component implements a specialization of the component facade that exports functionality for searching components that match a given set of attributes.

An adaptive system must also be able to react to changes in component availability. For example, a *media player* interface for *iPOD* must be able to reason about which streams it can decode. Hence, the container permits the registration of listeners (represented by components that implement the *ComponentListener* facade) to be notified when components matching a set of attributes given by the listener are added or removed.

To allow for dynamic adaptation, the container can dynamically add or drop components to and from the system. Registration and removal of components is delegated to one or more registrars. A registrar is a component that implements a facade that defines primitives for loading and removing components, validating dependencies, executing component constructors, and adding components to the registry.

29.3.2.3 Component Life Cycle

The HMI framework supports a very simple and lightweight component life cycle. When a component is passed on to the container for registration by loading it from persistent storage, using a *Deployer*, etc., the container delegates registration to a registrar component. The registrar is responsible for checking that the dependencies of the component are satisfied, instantiating the component using its constructor, and adding it to the registry. Note that the component facade prescribes a single constructor. An instantiated component can use the container facade to get references to any other components that it may require. A component deployed and instantiated in the container may be either enabled or disabled. The semantics of those and the initial state of the component depend on the component implementation. The functionality needed to manipulate the state of the component is exported by the component facade.

29.3.3 Framework Core

The aim of HMI Framework core in general is to provide higher level interaction primitives than those provided by the vehicle infotainment network and infotainment service system as a layer upon which HMI applications are then constructed. In doing so, the framework hides the complexities of addressing distribution, heterogeneity, and failures (Fig. 29.3).

The framework core along with underlying middleware provides a number of services to applications. The services themselves are seen as regular components built on top of the container. As such, they can be dynamically added and removed.

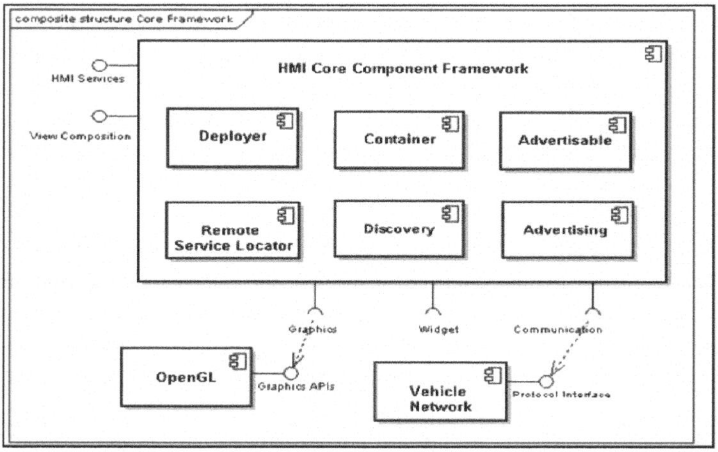

Fig. 29.3 Overview of framework core

29.3.4 Distribution and Logical Mobility

An HMI application built using the framework can reconfigure itself by using logical mobility primitives. As different paradigms can be applied to different scenarios, our metamodel does not build distribution into the components themselves, but it provides it as a service; implementations of the framework metamodel can, in fact, dynamically send and receive components and employ any of the above logical mobility paradigms.

The framework considers four aspects of Logical Mobility: Components, Classes, Instances, and Data Types; the last is defined as a bit stream that is not directly executable by the underlying architecture. One such, the Logical Mobility Entity (LME), is defined as an abstract generalization of a Class, Instance, or Data Type. In the framework component metamodel, an LMU is always deployed in a Reflective component. A Reflective component is a component specialization that can be adapted at runtime by receiving LMUs from the framework migration services. By definition, the container is always a reflective component, as it can receive and host new components at runtime.

29.3.5 Application Model

The Application Model (Fig. 29.4) of HMI framework provides support for building and running context-aware infotainment HMI applications on top of the framework context model.

HMI applications access the framework functionality through an *Application-Interface*: each time an application is started, an *ApplicationSkeleton* is created to allow interactions between the framework and the application itself. In particular, application interfaces allow applications to request services from the underlying framework, and to access their own application context profile through a well-defined reflective meta-interface. The HMI application is realized as composition of components based on the component model of *the HMIFramework.*

The application model shall guide modeling of infotainment HMI application components with the use of interfaces of component of the framework in an efficient manner. The application model is also expected to ease the integration of context awareness to applications by proper use of the framework interfaces.

29.4 Application Development

This section represents an implementation of a simple *Traffic Message Notification* application using the HMI framework core. Components that implement Traffic Message decoding and text presentation inherit message format from the radio tuner running on the same platform. As such, the Traffic HMI application uses the

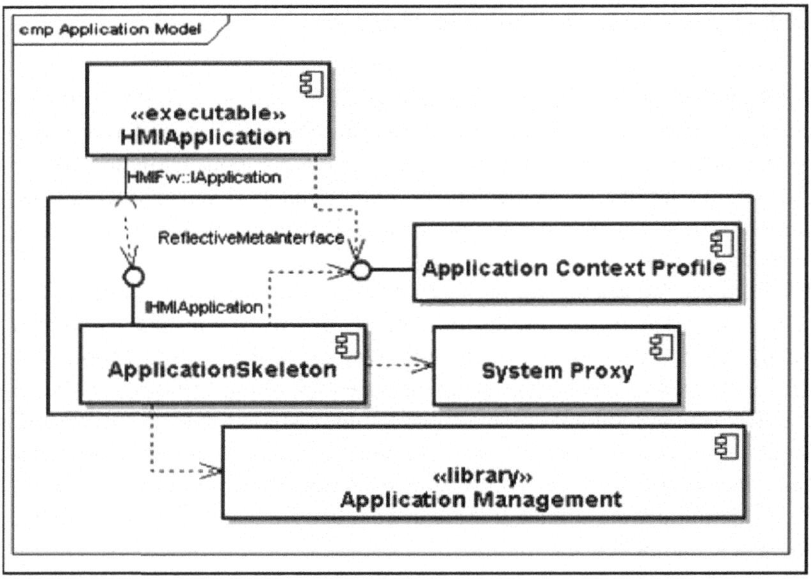

Fig. 29.4 Application model for the framework

notification service to be notified whenever the tuner façade component that has Traffic Message attribute implemented is registered. Moreover, it uses the *deployer* and the discovery components to premium Traffic message service that are found remotely.

29.4.1 Application Model

Figure 29.5 presents the platform independent application level model for TMC HMI application.

The Traffic Message HMI demonstrates an application that uses the container to listen to the arrival of new components, adapting its interface and functionality upon new TMC service component arrival. It also demonstrates reaction to context changes, as the application monitors the discovery services for new service components and schedules them for use as soon as they appear.

29.4.2 Evaluation

The TMC HMI application demonstrates an application that uses the container to listen to the arrival of new components and then adapts its interface and functionality to reflect the arrival of a new component. It also demonstrates reaction to context changes as the application monitors the TMC services for new messages

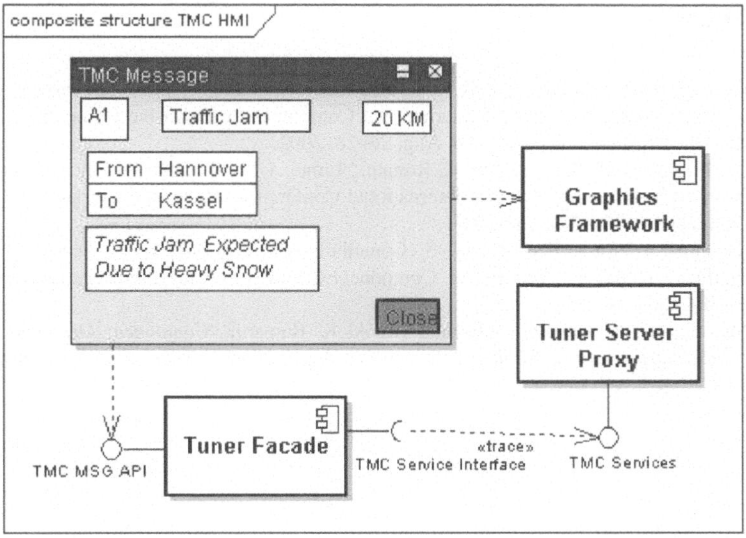

Fig. 29.5 Traffic message application overview

and schedules them for navigation reroute as soon as they appear. The operation is transparent to the end user.

Finally, it illustrates the ease with which existing code can be adapted to run under the framework environment. The adapted code gains the ability to be deployed dynamically at runtime and to be used as part of a component-based application, such as the TMC HMI application. Hence, it can express dependencies on other components or platforms, it can be required by other components, and a component-based application can be built with it.

29.5 Summary

This chapter presented an approach for building context aware HMIs for In-vehicle infotainment software applications by means of components. The HMI framework enable this by using a logical mobility based component model. The lightweight component metamodel is instantiated as a framework system for adaptable HMI application and systems. The framework offers logical mobility primitives as first-class citizens to achieve context awareness.

References

1. K. Henricksen, J. Indulska, and A. Rakotonirainy, "Modeling Context Information in Pervasive Computing Systems", 1st International Conference on Pervasive Computing (Zurich, Switzerland), Springer, pp. 167–180, Aug. 26–28, 2002.
2. A. L. Murphy, G. P. Picco, and G.-C. Roman, "Lime: A Middleware for Physical and Logical Mobility," Proceedings of 21st International Conference Distributed Computing Systems (ICDCS 21), pp. 368–377, May 2001.
3. F. Bachman, L. Bass, S. Buhman, S. Comella-Dorda, F. Long, R. C. Seacord, and K. C. Wallnau, Technical Concepts of Component-Based Software Engineering. Tech. Rep. CMU/SEI-2000-TR-008.
4. H. Sharma, Dr. R. Kuvedu-Libla, and Dr. A. K. Ramani, "*Component Oriented Human Machine Interface for In-Vehicle Infotainment Applications*", WEC 2008 at London, pp. 192–197, 2–4 July 2008.

Chapter 30
Facial Expression Analysis Using PCA

V. Praseeda Lekshmi, M. Sasikumar, Divya S. Vidyadharan, and S. Naveen

Abstract Face recognition has drawn considerable interest and attention from many researchers. Generally pattern recognition problem rely upon the features inherent in the pattern for efficient solution. Conversations are usually dominated by facial expressions. A baby can communicate with its mother through the expressions on its face. But there are several problems in analyzing communication between human beings through non-verbal communication such as facial expressions by a computer. In this chapter, video frames are extracted from image sequences. Using skin color detection techniques, face regions are identified. PCA is used to recognize faces. The feature points are located and their coordinates are extracted. Gabor wavelets are applied to these coordinates and to the images as a whole.

Keywords Facial Expression · PCA · Face recognition · video frame · skin color detection · Gabor wavelet

30.1 Introduction

Face recognition has drawn considerable interest and attention from many researchers for the last 2 decades because of its potential applications, such as in the areas of surveillance, secure trading terminals, control and user authentication. The problem of face recognition is either to identify facial image from a picture or facial region from an image and recognize a person from set of images. A robust face recognition system has the capability of distinguishing among hundreds, may be even thousands of faces. For automated access control, most common and accepted method is face recognition. A number of face recognition methods have been proposed.

V. P. Lekshmi (✉)
College of Engineering, Kidangoor, Kottayam, Kerala, India
E-mail: vplekshmi@yahoo.com

S.I. Ao and L. Gelman (Eds.), *Advances in Electrical Engineering and Computational Science*, Lecture Notes in Electrical Engineering.

Generally pattern recognition problem rely upon the features inherent in the pattern for efficient solution. The challenges associated with face detection and recognition problem are pose, occlusion, expression, varying lighting conditions, etc. Facial expression analysis has wide range of applications in areas such as Human Computer Interaction, Image retrieval, Psychological area, Image understanding, Face animation etc. Humans interact with each other both verbally and non-verbally.

Conversations are usually dominated by facial expressions. A baby can communicate with its mother through the expressions on its face. But there are several problems in analyzing communication between human beings through non-verbal communication such as facial expressions by a computer because expressions are not always universal. It varies with ethnicity. Further facial expressions can be ambiguous. They have several possible interpretations. To analyze the facial expression, face regions have to be detected first. Next step is to extract and represent the facial changes caused by facial expressions. In facial feature extraction for expression analysis, there are two types of approaches, namely Geometric based methods and Appearance based methods. In Geometric based method, the shape and location of facial features are extracted as feature vectors. In Appearance based method, image filters are applied either to whole face or to specific regions of facial image to extract facial features.

In this paper, video frames are extracted from image sequences. Using skin color detection techniques, face regions are identified. PCA is used to recognize faces. The feature points are located and their coordinates are extracted. Gabor wavelets are applied to these coordinates and to the images as a whole.

Rest of the paper is organized as follows. Section 30.2 gives background and related works. Section 30.3 discusses the proposed method. Results are given in Section 30.4. Conclusion and future work are given in Section 30.5.

30.2 Background and Related Work

Most face recognition methods fall into two categories: Feature based and Holistic [1]. In feature-based method, face recognition relies on localization and detection of facial features such as eyes, nose, mouth and their geometrical relationships [2]. In holistic approach, entire facial image is encoded into a point on high dimensional space. Images are represented as Eigen images. A method based on Eigen faces is given in [3]. PCA [4] and Active Appearance Model (AAM) for recognizing faces are based on holistic approaches. In another approach, fast and accurate face detection is performed by skin color learning by neural network and segmentation technique [5]. Facial asymmetry information can be used for face recognition [7]. In [8], ICA was performed on face images under two different conditions: In one condition, image is treated as a random variable and pixels are treated as outcomes and in the second condition which treated pixels as random variables and image as outcome. Facial expressions are extracted from the detailed analysis of eye region images is given in [9]. In the appearance based approaches given in [10], facial images are recognized by warping of face images. Warping is obtained by automatic AAM processing.

Another method of classification of facial expression is explained in [11] in which the geometry and texture information are projected in to separate PCA spaces and then recombined in a classifier which is capable of recognizing faces with different expressions. Kernel Eigen Space method based on class features for expression analysis is explained in [12]. In [13] facial expressions are classified using LDA, in which the gabor features extracted using Gabor filter banks are compressed by two stage PCA method.

30.3 Method

Our proposed method consists of training stage, face recognition and expression classification stages.

30.3.1 Training stage

The face image sequences with certain degree of orientation, wearing glasses and large variations in the facial expressions are considered. As a first step, frames with peak expression called key frames are identified. Face regions from these key frames are extracted using skin color detection method. Skin regions are identified and connected component labeling is done to classify the sub regions in the image.

Faces are detected from these detected skin regions. Figure 30.1 shows detected face regions from skin areas. Frontal looking faces with neutral expressions called normal faces and faces with set of non-neutral expressions form the database. There were 'K' frames with 'N' expressions for each face so that 'K x N' face images were used as the database. Normalization is done to make the frames with uniform scale. The normalized faces are shown in Fig. 30.2.

Facial expressions are highly reflected in eyes and mouth regions. Fourteen markers as mentioned in [11] are used to automatically select for registering important facial features. A triangulation method is applied to fit the mask on the faces. The marking points represented as white dots are shown in Fig. 30.3. The coordinates are used for further verification.

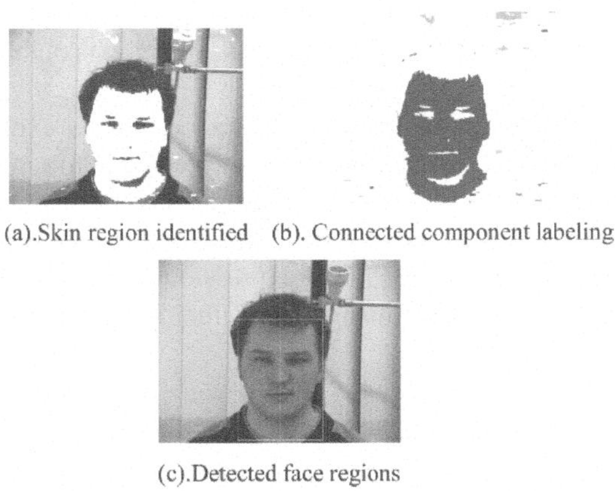

Fig. 30.1 (**a**) Skin region identified, (**b**) connected component labeling, (**c**) detected face regions

Fig. 30.2 Normalized faces

Fig. 30.3 Marking points

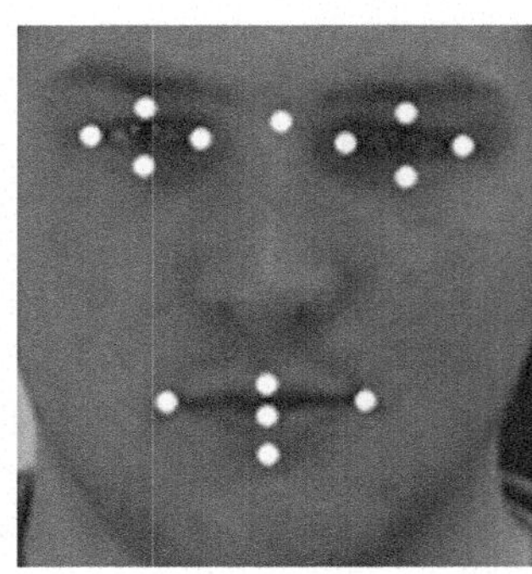

30.3.2 Face Recognition and Expression Classification

Face recognition has long been a primary goal of computer vision, and it has turned out to be a challenging task. The primary difficulty in attempting to recognize faces from image gallery comes from immense variability of face appearance due to several factors including illumination, facial expressions, and view points of camera, poses, and occlusions. This method treated face recognition as a 2-D recognition problem.

PCA is a useful statistical technique that has found applications in the fields such as face recognition; image compression, etc. It is a common technique for finding patterns in the data of high dimensions. In PCA, face images are projected into feature space or face space. Weight vector comparison was done to get the best match.

Let the training set of face images be X_1, X_2, X_3,...,Xn, then the average set or mean of faces be defined as

$$m = \frac{\sum_{i=1}^{n} X_i}{n} \tag{30.1}$$

Notice that the symbol m to indicate the mean of set X. The average distance of each face from the mean of the data set is given by

$$Q_1 = X_1 - m; Q_2 = X_2 - m \ldots Qn = Xn - m \tag{30.2}$$

which is the standard deviation.

The covariance matrix is given by

$$\mathrm{C} = \mathrm{A}^* \mathrm{A}' \tag{30.3}$$

where A = $[Q_1\, Q_2\, Q_3 \ldots\ldots\ldots\ldots Qn]$.

In order to reduce the dimensionality, co-variance can be calculated as

$$\mathrm{C} = \mathrm{A}'^* \mathrm{A}.$$

Eigen values and Eigen vectors are calculated for the covariance matrix. All the face images in the database are projected in to Eigen space and weight for each image is calculated.

Then image vectors for each face image is obtained as

$$\text{Image Vector} = \sum\nolimits_{i=1}^{10} weight(i)^* Eigevector(i) \tag{30.4}$$

Usually while using PCA, normal images are used as reference faces. To overcome the large variations in transformations, mean image is used as the reference face.

Ten face image sequences are used here, each with five facial expressions as the database. Mean faces of five key frames with peak expressions and Eigen faces are shown in Figs. 30.4 and 30.5.

The probe image is also subjected to preprocessing steps before projecting it into feature space. The weight vector is calculated to identify the image from the database with closest weighting vector.

So far a reference face for each testing face is identified. After recognizing the face, the coordinates of the facial features are extracted as explained in Section 30.1. The coordinates of each testing face is compared with its reference face by calculating the mean square error between the testing face and all the prototypes of same individual. This mean square error tells us how far the expression on the testing face is from each type of expressions and thus can be used to classify expressions.

Fig. 30.4 Mean faces

Fig. 30.5 Eigen faces

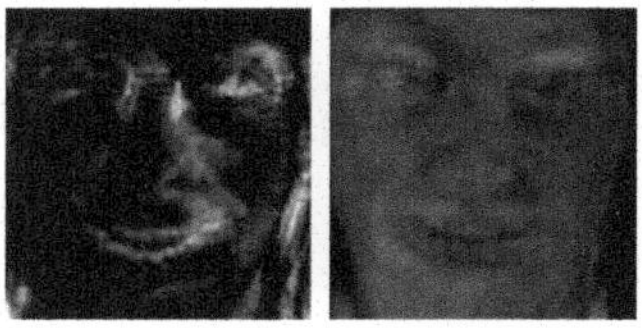

As a second method, Gabor wavelet transformation is applied to the extracted coordinates and to the image itself. The main characteristics of wavelet are the possibility to provide a multi-resolution analysis of the image in the form of coefficient matrices. Strong arguments for the use of multi-resolution decomposition can be found in psycho visual research, which offers evidences that the human visual system processes the image in a multi scale way. Wavelets provide a time and frequency decomposition at the same time. Computational complexity of wavelets is linear with the number of computed coefficients. Thus the complexity is less compared to other fast transformations.

The Gabor wavelet representation of images allows description of spatial frequency structure in the image while preserving information about spatial relations. The response image of Gabor filter can be written as a correlation of input image $I(x)$ with Gabor kernel $P_k(x)$.

$$ak(x0) = \int\int I(x)Pk(x - x0)dx \tag{30.5}$$

$$p_k(x) = \frac{k^2}{\sigma^2}\exp\left(-\frac{k^2}{2\sigma^2}x^2\right)\left(\exp(ikx) - \exp\left(-\frac{\sigma^2}{2}\right)\right) \tag{30.6}$$

where k is the characteristic vector.

$$k = (k_x, k_y) = (k_v \cos\theta_w, k_v \sin\theta_w) \text{ and } k_v = 2^{-\frac{v+2}{2}}\pi; \quad \theta_w = w\frac{\pi}{8}$$

Here v is the discrete set of different frequencies and w is the orientation.

The Gabor coefficients are projected to feature space or face space. The feature vectors of each testing face is compared with its reference face by calculating the mean square error between the testing face and all the prototypes of same individual.

30.4 Results

Ten face image sequences from FG-NET consortium with variations in lighting conditions, small orientations, wearing glasses, heavy expressions, etc. selected. The expressions used were 'happy', 'sad', 'normal', 'surprise', and 'anger'. Frames with peak expressions from color face image sequences were extracted. Face regions were identified using skin color. Eigen faces were used for recognizing faces. These faces were converted to gray scale and normalized to a size 192X192. Fourteen points were marked at the highly expression reflected face regions. The face images in the database and the test image were projected to face space for face recognition and their coordinates were verified to identify which expression belongs to the test face. The performance ratios are 100% for expression recognition from extracted faces, 88% for expression recognition from frames and 88% for the combined recognition. As a comparison two experiments were conducted on theses face images. Gabor wavelet transformation is applied to all these coordinates and the resultant Gabor coefficients were projected to feature space or PCA space.

As the second part, the whole face images were subjected to Gabor wavelet transformation. Figure 30.6 shows Gabor faces. The high dimensional Gabor coefficients were projected to PCA space to reduce he dimension. Mean square error

Fig. 30.6 Gabor faces

Table 30.1 Expression recognition from extracted face

Expressions	I1	I2	I3	I4	I5	I6	I7	I8	I9	I10
Anger	Y	Y	Y	Y	Y	Y	Y	Y	Y	Y
Happy	Y	Y	Y	Y	Y	Y	Y	Y	Y	Y
Neutral	Y	Y	Y	Y	Y	Y	Y	Y	Y	Y
Sad	Y	Y	Y	Y	Y	Y	Y	Y	Y	Y
Surprise	Y	Y	Y	Y	Y	Y	Y	Y	Y	Y

Table 30.2 Expression recognition from frames

Expressions	I1	I2	I3	I4	I5	I6	I7	I8	I9	I10
Anger	N	Y	Y	Y	Y	Y	Y	Y	Y	Y
Happy	Y	N	Y	Y	N	Y	Y	Y	Y	Y
Neutral	Y	Y	Y	Y	Y	Y	Y	Y	Y	Y
Sad	Y	N	N	Y	Y	Y	Y	Y	Y	Y
Surprise	N	Y	Y	Y	Y	Y	Y	Y	Y	Y

Table 30.3 Combined performance

Expressions	I1	I2	I3	I4	I5	I6	I7	I8	I9	I10
Anger	N	Y	Y	Y	Y	Y	Y	Y	Y	Y
Happy	Y	N	Y	Y	N	Y	Y	Y	Y	Y
Neutral	Y	Y	Y	Y	Y	Y	Y	Y	Y	Y
Sad	Y	N	N	Y	Y	Y	Y	Y	Y	Y
Surprise	N	Y	Y	Y	Y	Y	Y	Y	Y	Y

Table 30.4 Expression recognition by applying Gabor wavelet on fudicial points

Expressions	I1	I2	I3	I4	I5	I6	I7	I8	I9	I10
Anger	Y	Y	N	Y	Y	Y	Y	Y	Y	Y
Happy	N	Y	Y	Y	Y	Y	Y	N	Y	Y
Neutral	Y	N	Y	Y	N	Y	Y	Y	Y	Y
Sad	Y	Y	N	Y	Y	Y	Y	Y	Y	Y
Surprise	Y	Y	Y	Y	Y	Y	Y	N	Y	Y

is calculated between feature vectors of test image and all the reference faces. The performance ratio for the application of Gabor wavelets on extracted features is 86% and to the whole image is 84%.

A comparison was made between the approaches for different databases are given in the Tables 30.1, 30.2 and 30.3. Table 30.4 and 30.5 shows the expression recognition performance by applying Gabor wavelets to the extracted fudicial points and to the whole image.

Table 30.5 Expression recognition by applying Gabor wavelet to the whole image

Expressions	I1	I2	I3	I4	I5	I6	I7	I8	I9	I10
Anger	Y	Y	Y	N	Y	Y	N	Y	Y	Y
Happy	N	Y	Y	Y	Y	Y	N	Y	Y	Y
Neutral	Y	Y	Y	Y	Y	Y	Y	N	Y	Y
Sad	Y	Y	Y	N	Y	Y	Y	Y	Y	Y
Surprise	Y	Y	Y	Y	Y	N	Y	Y	Y	N

30.5 Conclusion

In this paper, face recognition and expression classification from video image sequences are explained. Frames are extracted from image sequences. Skin color detection method is applied to detect face regions. A holistic based approach in which whole face is considered for the construction of Eigen space. As a first step, images are projected to PCA space for recognizing face regions. After recognizing the face, our system could efficiently identify the expression from the face.

To compare the performance, two experiments are also conducted. Gabor wavelet transformation is applied to the extracted fudicial points and to the images as a whole. The Gabor coefficients are projected to feature space or face space for comparison with that of test image.

This logic performs well for recognition of expressions from face sequences. The computational time and complexity was also very small.

Acknowledgements We thank FG-NET consortium [13] for providing us the database for the experiment.

References

1. Rama Chellappa, Charles L. Wilson and Saad Sirohey, "Human and Machine Recognition of Faces: A Survey", In Proceedings of IEEE, Vol. 83(5), 705–740, 1995
2. Stefano Arca, Paola Campadelli and Raffaella Anzarotti, "An Automatic Feature Based Face Recognition System", In Proceedings of the 5th International Workshop on Image Analysis for Multimedia Interactive Services (WIAMIS 2004)
3. Matthew Turk and Alex Pentland, "Face Recognition Using Eigen Faces", In Proceedings of IEEE Conference on Computer Vision and Pattern Recognition, pp. 586–591, 1991
4. Sami Romdhani, Shaogang Gong and Alexandra Psarrou, A Multi-view Nonlinear Active Shape Model Using Kernel PCA, BMVC99
5. Hichem Sahbi, Nozha, Boueimma, Tistarelli, J. Bigun and A.K. Jain (Eds), "Biometric Authentication" LNCS2539, Springer, Berlin/Heidelberg, 2002, pp. 112–1206
6. Sinjini Mitra, Nicola A. Lazar and Yanxi Liu, "Understanding the Role of Facial Asymmetry in Human Face Identification", Statistics and Computing, Vol. 17, pp. 57–70, 2007. DOI 10.1007/s 1222-006-9004-9

7. Marian Stewrt Bartlett, R. Javier, Movellan and Terrence J. Seinowski, "Face Recognition by Independent Component Analysis", IEEE Transactions on Neural Networks, Vol. 113, No. 6, Nov. 2002
8. Tsuyoshi Moriyama, Takeo Kanade, Jing Xiao and Jeffrey F. Cohn, "Meticulously Detailed Eye Region Model and It's Application to Analysis of Facial Images", IEEE Transactions on Pattern Analysis and Machine Intelligence, Vol. 28, No. 5, May 2006
9. Hui Li, Hui Lin and Guang Yang, "A New Facial Expression Analysis System Based on Warp Images", Proceedings of Sixth World Congress on Intelligent Control and Automation, Dalian, China, 2006
10. Xiaoxing Li, Greg Mori and Hao Zhang, "Expression Invariant Face Recognition with Expression Classification", In Proceedings of Canadian Conference on Computer and Robot Vision (CRV), pp. 77–83, 2006
11. Y. Kosaka and K. Kotani, "Facial Expression Analysis by Kernal Eigen Space Method Based on Class Features (KEMC) Using Non-linear Basis for Separation of Expression Classes." International Conference on Image Processing (ICIP), 2004
12. Hong-Bo Deng, Lian-Wen Jin, Li-Xin Zhen and Ian-Cheng Huang, "A New Facial Expression Recognition Method Based on Local Gabor Filter Bank and PCA Plus LDA", International Journal of Information Technology, Vol. 11, No. 11, 2005
13. Frank Wallhoff, Facial Expressions and Emotion Database http://www.mmk.ei.tum.de/∼waf/fgnet/feedtum.html,Technische Universität München 2006

Chapter 31
The Design of a USART IP Core

A. H. El-Mousa, N. Anssari, A. Al-Suyyagh, and H. Al-Zubi

Abstract Embedded systems have drastically grown in importance and complexity in recent years. Many systems are now designed using Field Programmable Gate Array (FPGA) due to its size, flexibility, and resources. This chapter presents the design and implementation of a flexible and user reconfigurable Universal Synchronous Asynchronous Receive Transmit (USART) IP core suitable for use in embedded systems and Systems on Chip (SoC). The design scheme employed, allows the USART to be used in various modes of operation such as standalone and 9-bit addressable mode for multi-drop network of serial devices. It also supports high speed data rates of up to 3 Mb/s. The design utilizes Hardware Description Language (HDL) to describe the operation, ease implementation and allow cross platform utilization. The chapter shows through a comprehensive testing methodology that the proposed design functions properly while consuming minimum resources from the target FPGA.

Keywords Embedded systems · addressable USART · IP core · FPGA · multi-drop networks · HDL

31.1 Introduction

Interest in embedded systems has grown drastically in recent years; the world market for embedded software will grow from about \$1.6 billion in 2004 to \$3.5 billion by 2009, at an average annual growth rate (AAGR) of 16%. On the other hand, embedded hardware growth will be at the aggregate rate of 14.2% to reach \$78.7 billion in 2009, while embedded board revenues will increase by an aggregate rate

A. H. El-Mousa (✉)
University of Jordan Computer Engineering Department,
Faculty of Engineering & Technology, University of Jordan, Amman, Jordan
E-mail: elmousa@ju.edu.jo

S.I. Ao and L. Gelman (Eds.), *Advances in Electrical Engineering and Computational Science*, Lecture Notes in Electrical Engineering.

of 10% [1]. At the same time, embedded systems are increasing in complexity and more frequently they are also networked. As designs become more complex, embedded systems based on FPGA are required to interact with software running on stock commercial processors [2]. This interaction more often than not makes use of a serial communications transmission link. Design techniques based on hardware-software co-design are generally implemented on platforms that utilize FPGAs as accelerators together with embedded CPU cores for control and operational procedure definition [3]. Frequently these platforms also require high speed serial data communication blocks.

USARTs have been around for years and they have become established for easy and simple serial transmission. However, most of these take the form of hardwired specialized ICs which make them unattractive and unsuitable for use in recent embedded systems; especially those utilizing FPGA technology, since they cannot be incorporated within the HDL design. Also, most are designed with limited features and capabilities; for example: limited speed and no capability for use in multi-drop networks. Attempts at the design of an HDL-based USART have been reported in the literature. Many are just HDL implementation of the well known industry standard 16550 UART without additional features [4–6].

This chapter presents the design, implementation and verification of a high speed configurable USART suitable to be used on platforms that utilize FPGAs. The architectural design allows serial communications in multi-drop networks using 9-bit operational mode using master-slave operation. It also features configurable high speed transmission rates and transmission error detection and recovery. User configuration is accomplished through specialized internal registers. The design is suitable to be used for inter-chip, inter-processor, and inter-system communications among others. The design implements both modes of operation synchronous and asynchronous.

The rest of the chapter is organized as follows: Section 31.2 presents a general description of USART theory of operation and describes the modes of operation. Section 31.3 provides a detailed description of the specifications of the developed USART. Section 31.4 discusses the design methodology followed and provides a description and operational features of the various blocks used in the design. Section 31.5 is concerned with the testing and verification procedures followed.

31.2 USART Theory of Operation

USARTs operate as parallel to serial converters at the transmitter's side where data is transmitted as individual bits in a sequential fashion whereas at the receiver's side, USARTs assemble the bits into complete data and thus act as serial to parallel converters.

USARTs mandate preliminary configuration of data format, transfer rates and other options specific to the design model. The user can send individual or a block of data, whose size is determined by the design model, to the transmitter section of the USART. Data is stored in the internal buffer and framed according to the transfer mode and user defined options. Finally, the frame is sent to its destination one bit at a time. On the receiver side, the data frame bits are received and sampled. The extracted data from the received frame resides in the internal receiver buffer waiting to be read. The receiver monitors the reception for possible errors and informs the recipient of their existence should the proper control bit(s) be set. Most USARTs offer a status monitoring mechanism via dedicated status register(s).

31.2.1 Modes of Operation

There are two major modes of operation in USARTs: synchronous and asynchronous modes. The latter prevails in most applications.

31.2.1.1 Synchronous Mode

In synchronous mode, both the transmitter and receiver are synchronized to the same clock signal which is usually generated by the transmitter and sent along with the data on a separate link to the receiver. The receiver in turn utilizes the received clock in extracting the timing sequence and determining the beginning and end of the received bit interval. Therefore, the receiver knows when to read the bit's value and when the next bit in the sequence begins. However, when the line is idle (i.e. no data is exchanged), the transmitter should send a fill character.

A synchronous communication session begins by sending one or more synchronizing frames to indicate the beginning of transmission then the sequence of data frames follow (a parity bit is appended to the end of the data frame if single error detection is required), transmission is terminated by sending a stop frame after which the line returns to the idle state.

31.2.1.2 Asynchronous Mode

An idle line remains at a predetermined level. The frame consists of a *start bit* which differs in polarity to that of the line's idle state, followed by the data bits and a parity bit – if single error detection is used – and ends with at least one *stop bit* which has the same polarity as that of an idle line. A stop bit might be followed by another frame – back to back transmission – or an idle state of transmission. Both the transmitter and receiver are preconfigured to run at the same fixed clock rate which is an exact multiple of the required baud rate. Once the receiver recognizes the transition from the idle state polarity to the opposing polarity, it waits a half bit interval duration and verifies the presence of a start bit, if start bit arrival is confirmed, the receiver reads the values of the bits every full bit-width interval until the reception of a stop bit is confirmed denoting end of frame.

31.3 Specifications of the Developed USART

Performance oriented features include an interrupt driven approach and universal eight bit addressability which make the USART ideal for industrial and control applications. Eight-level buffering allows for data block transfers that is beneficial considering the high speeds the USART can handle in data transfer which can reach 3 MHz.

Reliability oriented specifications include programmable odd/even parity bit with the ability to detect parity, framing and overrun errors.

Table 31.1 lists the detailed specifications of the proposed USART.

31.4 USART System Design Methodology

The methodology adopted in the USART system design was based on systems and software engineering approaches. It used a variation of both the waterfall and incremental approaches suited to the design environment and constraints. The design steps that the system went through are [7]:

1. System Requirements Definition. The requirements definition phase specified the functionality as well as essential and desirable system properties. This involved the process of understanding and defining services required from the system and identifying constraints on its operation.
2. System/Subsystem Design. This phase was concerned with how system functionality was to be provided by the components and how system specification was converted into an executable system specification.
3. Subsystems Implementation and Testing. Here subsystems were implemented and mapped into hardware code using HDL. In this stage, individual modules were extensively tested for correct functional operation.
4. System Integration. During system integration, the independently developed subsystems were merged together to build the overall USART system in an incremental approach.
5. System Testing. The integrated system was subjected to an extensive set of tests to assure correct functionality, reliability and performance.

Figure 31.1 shows the major parts of the USART system:

The input/output signals involved are:

Sys. clock: the oscillator clock	Reset: master reset of the system
Serial out: transmitted data	Serial in: received data
TXInt: transmitter interrupt	RXInt: receiver interrupt
ErrInt: error interrupt	

Table 31.1 Functional specifications of the USART

Specification	Justification
Support for the following transmission modes (programmable): Asynchronous (full/half duplex modes) Synchronous (full/half duplex modes)	Full duplex mode is employed in modern systems while half duplex mode is retained for compatibility with legacy systems
Supports a wide range of transmission/reception rates (from 50 Hz to 3 MHz)	High frequencies are essential for high speed communication. Lower speeds are needed to communicate with older USARTs. Moreover, lower speeds can be used to minimize cross talk if similar links are adjacent
Eight-level transmitter/receiver buffer	To account for the high speeds of communication that the USART can reach, blocks of data can be received and buffered until read by the user. Also, this allows the user to issue the transmission of a block of eight-frame size in a single operation. This will also reduce the load on the module that controls the USART operation in the system
Parity supported (programmable – enable/disable parity and odd/even parity)	Single error detection techniques might prove beneficial in noisy operation environments
Variable data lengths supported (programmable – 5–8 bits)	Byte communication is the modern norm. Five to seven bits data length is to retain compatibility with legacy systems
Variable stop bits supported (asynchronous mode) (programmable – 1 or 2 stop bits)	This is to comply with the RS232 standard where two stop bits mode is used to accommodate slightly different clocks in the transmitter and receiver sides when USARTs from different vendors are connected together
Error detection of the following errors: Parity error Overrun error Framing error (asynchronous mode)	Parity error detection provides a measure of the reliability of communication. Framing error detection indicates the necessity of reconfiguring the internal clocking sources at both ends correctly. Finally, overrun error informs that data has been lost and the need to frequently read the received data
Supports interrupts (programmable – with ability of global masking)	Most modern systems are interrupt-driven for the reason that interrupt techniques save processing clock cycles in comparison with polling techniques and are essential in real time applications
Supports addressability (8-bit universal – addresses up to 256 devices) while sending 9-bits	Used in industrial and applications in multi-drop networks where a master can communicate with a certain other slave

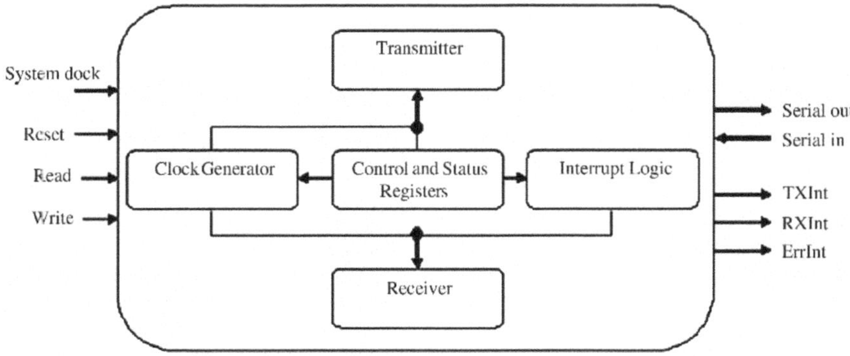

Fig. 31.1 The USART module

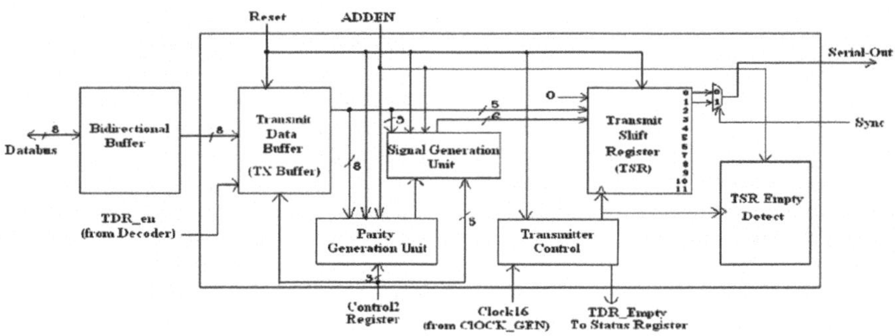

Fig. 31.2 The transmitter module

31.4.1 Transmitter Module

Figure 31.2 shows the functional block diagram of the transmitter module.

It consists of the following sub modules:

- Transmitter buffer
- Bypass logic
- Transmit shift register (TSR)
- Parity generation logic
- Shift logic
- TSR empty detection logic

In the following, due to space constraints, we only show the structure of the transmitter buffer.

31.4.1.1 The Transmitter Buffer

The transmitting buffer is the memory where data to be sent are stored waiting for the transmit shift register (TSR) to be empty. It becomes an essential component when the inter arrival time of transmitted frames becomes very small. Moreover, when the system is accessed using a processor/DSP that operates at a higher

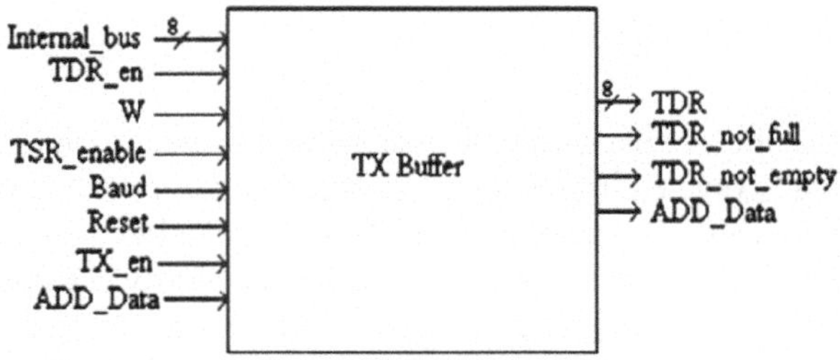

Fig. 31.3 The input/output signals associated with the TX buffer

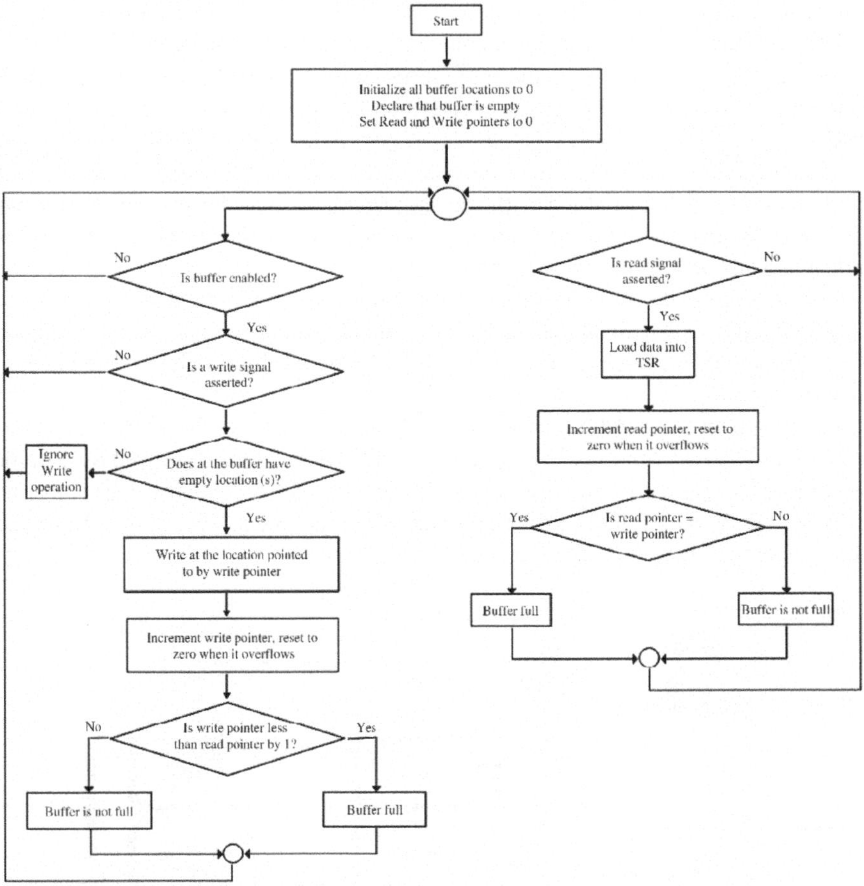

Fig. 31.4 Transmitting buffer dataflow diagram

frequency than the transmission baud clock, this buffer will reduce the number of times the processor is interrupted to request for new data. The signal TDR_empty is generated to indicate that the buffer has at least one empty slot. Figure 31.3 shows the input/output signals associated with the transmitting buffer while Fig. 31.4 shows the data flow diagram for it.

31.4.2 Receiver Module

In the asyn. mode, the receiver waits for a transition from a mark to space after an idle line state or an expected stop bit to initiate the data reception logic provided that the transition is not caused by a noise notch. This is ensured by sampling each of the received bits at three different times and then using a majority detection circuit. The end of asynchronous reception is detected by waiting for a stop bit at the end of the frame.

In the synch. mode, instead of waiting for logic transition, the receiver waits for a synchronizing character which if received after an idle state line, a start of reception is detected. An end of reception is signaled if a certain stop character is received.

When data reception is detected at the serial_in input, the internal receiver logic is enabled and the received data bits are serially shifted into the Receiver Shift Register (RSR). Meanwhile, the parity is calculated per each received bit for the received *word* and finally compared to the received parity value. Parity and framing errors are evaluated and stored along with the data in the receiver buffer. The buffer can hold up to seven received words, if data are received while the buffer is full, the data is dropped and an overrun error is indicated.

If 9-bit address detection mode is enabled, transmission is fixed at eight bits and the ADD-Data bit is substituted for parity. The address of the receiving node must be received with ADD-Data bit is set to "1" in order for the frame to be considered an address frame.

In the synchronous addressable mode of operation, a synchronizing character with ADD-Data bit value set to zero must be initially received, followed by a frame containing the address of the receiving node but with ADD-Data bit value set to one, followed by the data frames with ADD-Data bit reset again. Figure 31.5 shows the functional block diagram of the receiver.

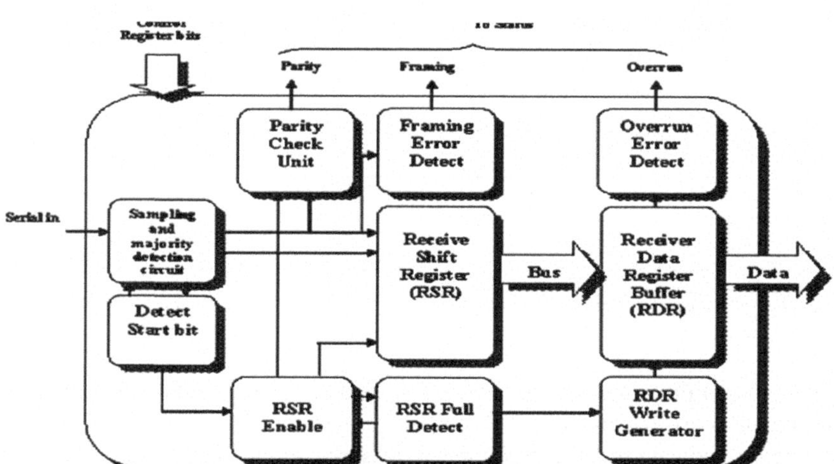

Fig. 31.5 Receiver module

The receiver module consists of the following sub-modules:

- Sampling and majority detection logic
- Detect start bit logic
- Rx shift register enable logic
- Rx shift register (RSR)
- Rx shift register full detection logic
- Parity error detection logic
- Framing error and stop bit detection logic
- Rx buffer write with synch. and stop character detect logic
- Rx buffer with overrun detection logic

Due to space constraints, we only show the structure of the receiver buffer write with synchronous and stop character detect logic sub-module as an example of the methodology of design.

31.4.2.1 Rx Buffer Write with Synch and Stop Character Detect Logic

This sub-module is responsible for generating an internal signal to write the received frames in RSR, together with their corresponding parity and framing information, into RDR buffer only if certain conditions are met. In synchronous mode of operation, this involves checking for the reception of valid sync and stop characters that delimit a block of consecutive frames. If 9th bit mode is used, all received frames are dropped if the USART is not the targeted node, which is indicated by receiving a valid address frame prior to receiving data. Figure 31.6 shows the data flow diagram for this sub-module.

31.5 Testing and Verification Procedures

In general, the system testing process has two distinct goals:

1. To demonstrate to the developer and the customer that the system meets its requirements
2. To discover faults or defects in the system where the behavior of the system is incorrect, undesirable or does not conform to its specification [7]

The first goal leads to validation testing, where the system is expected to perform correctly using a given set of test cases that reflect the system's expected use. The second goal leads to defect testing, where the test cases are designed to expose defects. The test cases can be deliberately obscure and need not reflect how the system is normally used. For validation testing, a successful test is one where the system performs correctly. For defect testing, a successful test is one that exposes a defect that causes the system to perform incorrectly.

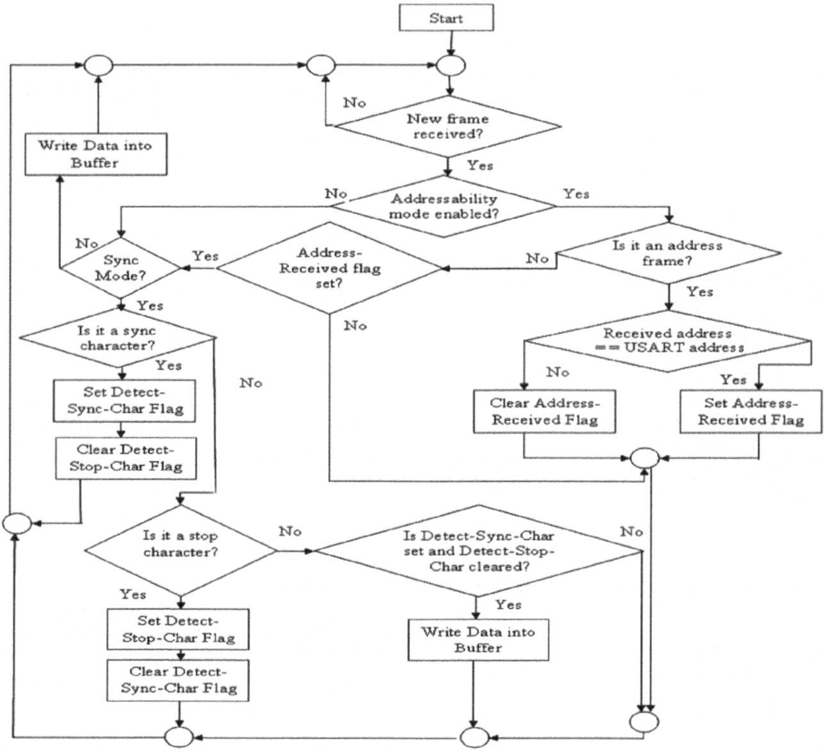

Fig. 31.6 Receiver buffer write logic dataflow diagram

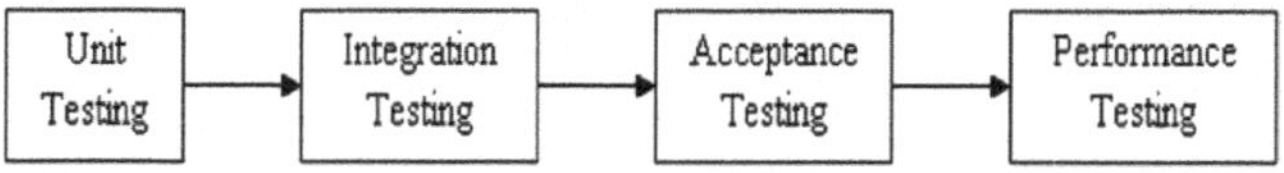

Fig. 31.7 Phases of the testing process

The proposed USART design was implemented and tested using firmware from Xilinx Corporation. For software, the free ISE Webpack version 8.2i was used [8]. As for hardware, different development kits were used. These include: Digilab 2 XL (D2XL) Development Board [9], Digilent Spartan-3 System Board [10], and Spartan-3E Starter Kit Board [11]. Since the design was entirely implemented using a universal hardware description language, Verilog HDL, it is expected to be directly interoperable with any environment provided by other vendors.

In general, the project went through several phases during the testing process as illustrated in Fig. 31.7.

A unit test is a test of the functionality of a system module in isolation, and the unit test should be traceable to the detailed design. A unit test consists of a set of test cases used to establish that a subsystem performs a single unit of functionality to some specification. Test cases should be written with the express intent of uncovering undiscovered defects [12].

After the units of a system have been constructed and tested, they are integrated into larger subcomponents leading eventually to the construction of the entire system. The intermediate subsystems must be tested to make sure that components operate correctly together. The purpose of integration testing is to identify errors in the interactions of subsystems [12].

The primary goal of an acceptance test is to verify that a system meets its requirement specification. To demonstrate that the system meets its requirements, it must be shown that it delivers the specified functionality, performance and dependability, and that it does not fail during normal use. Ideally, the acceptance test plan is developed with the engineering requirements and is traceable to them. Acceptance testing is usually a black-box testing process where the tests are derived from the system specification. The system is treated as a black box whose behavior can only be determined by studying its inputs and the related outputs [7].

Before connecting the USART to an external system, it was tested at first by connecting the serial output of its transmitter section to the serial input of its receiver section to locate any potential errors within the system itself. Similarly, the clock output of the transmitter section was connected to the clock input of the receiver section in the synchronous mode.

The entire system was implemented on two Xilinx development boards. Both were programmed with the same USART design. However, one was used to transmit data, while the other was used to receive the sent data. Special temporary modifications to the internal design were implemented to allow certain internal signals to be observed with a digital storage scope. Initially, data was exchanged with the PC using the EPP mode of the parallel port. The PC was used to configure the USARTs with the different communication options by sending the appropriate control words to the respective registers and also to supply the data to be transmitted serially.

From this test, more indications were obtained that the system complies with its requirements specification. Error conditions reflected the true state of the received data when the two USARTs were deliberately configured with conflicting communications options. Moreover, the USARTs functioned as expected in the 9-bit mode of operation.

Figure 31.8 illustrates a snapshot of a communication session that was established between the two USARTs.

In performance testing, an effective way to discover defects is to design tests around the limits of the system; that is, stressing the system (hence the name stress testing) by making demands that are outside the design limits of the system until the system fails.

During the test cases described previously, the USART design was subjected to some extreme conditions to explore various aspects of the limits of its operation. For example, the output of the baud rate generator was examined at the highest baud rate possible, and the buffers were tested with increasing read and write speeds. Moreover, the operation of the entire USART was checked while operating at the highest baud rate possible when two systems on separate boards were connected together,

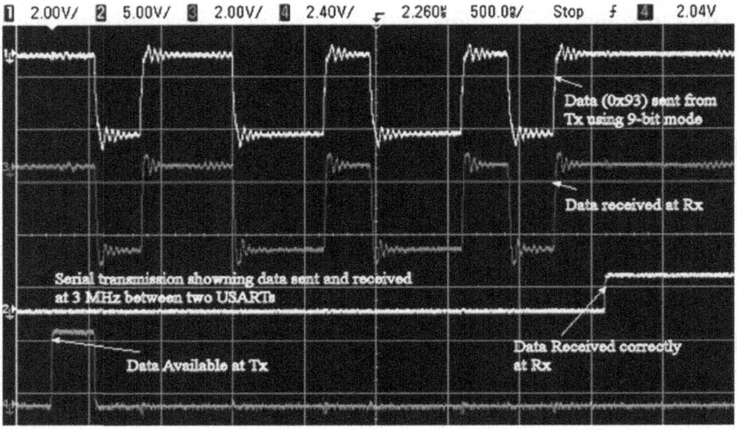

Fig. 31.8 Asynchronous 9-bit address transmission and detection between two USARTs

as well as when the transmitter section was used to drive the receiver section of the same system.

References

1. Ravi Krishnan, Future of Embedded Systems Technology, BCC Research Report, ID: IFT016B, June 2005.
2. John A. Stankovic, Insup Lee, Aloysius Mok and Raj Rajkumar, Opportunities and Obligations for Physical Computing Systems, *IEEE Computer Magazine*, **11**, pp. 23–31 (2005).
3. Wayne Wolf, *High Performance Embedded Computing* (Elsevier, Amsterdam, Netherlands 2007), pp. 383–387.
4. Mohd Yamani Idna Idris, Mashkuri Yaacob and Zaidi Razak, A VHDL Implentation of UART Design with BIST Capabilty, *Malaysian Journal of Computer Science*, **19 (1)**, pp. 73–86 (2006).
5. Azita Mofidian, DesignWare Foundation DW_16550: A fine work of UART, Designware Technical bulletin, Technical Forum for Design Automation Information, 4 3 (1999); http://www.synopsys.com/news/pubs/dwtb/q499/dwtb_art1.html
6. Shouqian Yu, Lili Yi, Weihai Chen and Zhaojin Wen, Implementation of a Multi-channel UART Controller Based on FIFO Technique and FPGA, Proceedings of 2nd IEEE Conference on Industrial Electronics and Applications ICIEA 2007, pp. 2633–2638 (2007).
7. Ian Sommerville, *Software Engineering*, 8th Edition (Addison-Wesley, England, 2007).
8. http://www.xilinx.com/ise/logic_design_prod/webpack.htm
9. Digilent D2XL System Board Reference Manual,Revision: (June 9, 2004); www.digilentinc.com
10. Spartan-3 Starter Kit Board User Guide, version 1.1: (May 13, 2005); www.xilinx.com
11. Spartan-3E Starter Kit Board User Guide, version 1.0: (March 9, 2006); www.xilinx.com
12. Ralph M. Ford and Chris Coulston, *Design for Electrical and Computer Engineers* (McGraw-Hill, New York, 2005).

Chapter 32
Multilayer Perceptron Training Optimization for High Speed Impacts Classification*

Angel Garcia-Crespo, Belen Ruiz-Mezcua, Israel Gonzalez-Carrasco, and Jose Luis Lopez-Cuadrado

Abstract The construction of structures subjected to impact was traditionally carried out empirically, relying on real impact tests. The need for design tools to simulate this process triggered the development in recent years of a large number of models of different types. Taking into account the difficulties of these methods, poor precision and high computational cost, a neural network for the classification of the result of impacts on steel armours was designed. Furthermore, the numerical simulation method was used to obtain a set of input patterns to probe the capacity of themodel development. In the problem tackled with, the available data for the network designed include, the geometrical parameters of the solids involved - radius and length of the projectile, thickness of the steel armour - and the impact velocity, while the response of the system is the prediction about the plate perforation.

Keywords Impacts Classification · Multilayer Perceptron · Optimization · neural network · simulation

32.1 Introduction

There are a wide number of systems which, during their service life, can suffer the impact of objects moving at high speed (over 500 m/s). This phenomenon is named impact ballistic and the most characteristic examples are found in the military field. However over the last decades, this kind of problems has become of interest in civil

A. Garcia-Crespo, B. Ruiz-Mezcua, I. Gonzalez-Carrasco, and J. L. Lopez-Cuadrado
Department of Computer Science, Universidad Carlos III, Av. Universidad 30 - 28911, Leganes (Madrid) Spain;
E-mails: acrespo@ia.uc3m.es,{bruiz,igcarras,jllopez}@inf.uc3m.es.

* This research was done with the financial support of the Comunidad Autonoma de Madrid under Project GR/MAT/0507/2004.

S.I. Ao and L. Gelman (Eds.), *Advances in Electrical Engineering and Computational Science*, Lecture Notes in Electrical Engineering.

applications. In them, the structural elements are required to absorb the projectile energy so that it does not damage critical parts of the global system.

Due to this, there are new possibilities in similar fields, among which passive safety in vehicles stands out. In this field, it is especially relevant the design of structures whose mission is to absorb energy in crashes of Crashworthiness type (200 m/s $\leq$ speed $\leq$ 500 m/s), as well as those that can appear in road or railway accidents, helicopters' emergency landings, etc. Therefore, what is being sought is to design structures capable of absorbing energy to avoid or lessen the damages to the passengers of the concerned vehicles.

The construction of structures subjected to impact was traditionally carried out empirically, relying on real impact tests, each one using a given configuration of projectile and target. The mathematical complexity of solving the equations that rule the impact phenomenon, and the relative ignorance of the mechanical behaviour of the materials at high strain rates, discouraged any simulation of the problem.

The need for design tools to simulate this process triggered the development in recent years of a large number of models of different types; all of them belong to two families: analytical modelling and numerical simulation. Thus, the use of expensive experimental tests has been postponed to the final stage of the design. All the previous stages can be covered by the use of this kind of simulation tools.

Taking into account the difficulties of these methods, poor precision and high computational cost, a neural network for the classification of the result of impacts on steel armours was designed. Furthermore, the numerical simulation method was used to obtain a set of input patterns to probe the capacity of the model development. In the problem tackled with, the available data for the network designed include, the geometrical parameters of the solids involved – radius and length of the projectile, thickness of the steel armour – and the impact velocity, while the response of the system is the prediction about the plate perforation.

32.2 Ballistic Impact

The ballistic terminology is defined as the scientific study of everything related to the movement of the projectile. Within this science, the Terminal Ballistic or Effect Ballistic discipline stands out. It is in charge of studying the results produced in the body or object that suffers the impact of the bullet or projectile. Moreover it analyses the way the projectile behaves when it reaches its target, how the projectile ends up, how the transference of kinetic energy is carried out and what effects it has on the objective, etc.

The conditions in which this projectile impacts on the objective depend most of the times on different parameters, namely the materials, both of the projectile and the target, the speed, the impact angle, the thickness of the target, the diameter of the projectile and finally the shape and weight. The combination of these factors leads to the existence of a vast variety of possible behaviours once the contact between them is produced.

The model developed was simplified to reduce the computational complexity of the numeric simulations needed to recreate impact tests of this type. For this purpose the obliquity of the projectile was not taking into account. Therefore, a right angle between the axis of the projectile and the surface of the plate where it impacts against, has been considered.

From a purely physique point of view, the impact of a projectile on a protection can originate three different results: arrest, perforation or ricochet. Meanwhile the process of the projectile's entrance inside the protection during an impact is called penetration. A perforation is produced when the projectile goes through the protection completely. On the other hand, an arrest takes place when the projectile goes inside the protection but does not go through it. Finally if the angle of attack in the impact comes into play, the projectile can rebound, what is known as ricochet (Fig. 32.1).

The investigation carried out in this study analyses the behaviour of metallic projectiles when the impact against metallic plates, both recreated with the same SAE 1006 steel material. This material has been chosen because it is isotope, its parameters are known and it allows carrying out further tests in the laboratory at a low cost.

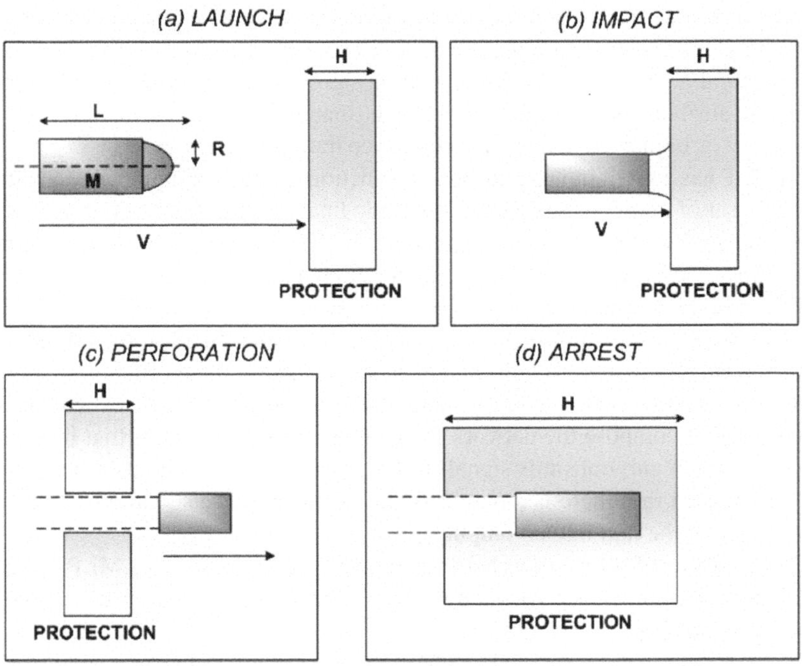

Fig. 32.1 Scheme of the projectile's impact on steel armours

32.3 Use of ANN in Impacts Situations

Artificial Neural Networks (ANNs) are statistical models of real world systems which are built by tunning a set of parameters. These parameters, known as weights, describe a model which forms a mapping from a set of given values, the inputs, to an associated set of values, the outputs.

The mathematical model that the neural network builds is actually made up of a set of simple functions linked together by the weights. The weights describe the effect that each simple function (known as unit or neuron) will have on the overall model [1].

Within the field of research of this article, the ANNs have been applied successfully within the limits of the fracture mechanic [2] to estimate material breakage parameters such as concrete [3], and in no-destructive tests to detect breaks in more fragile materials [4]. In all these applications, the input data needed for the training has been obtained by means of experimentation and numeric simulation.

Nevertheless there are not enough studies dedicated to the application of ANN to solve problems of ballistic impacts. At the present time the investigations made have focused on low speed impacts or situations where it is necessary an energy absorption (Crashworthiness) [5–7].

Therefore, and due to the mentioned experience on the mechanical problems and on the good results shown on the previous researches [8], a MultiLayer Perceptron (MLP) neural network with backpropagation algorithm has been defined. The most important attribute of this kind of ANN is that it can learn a mapping of any complexity or implement decision surfaces separating pattern classes.

An MLP has a set of inputs units whose function it is to take input values from the outside, a set of outputs units which report the final answer, and a set of processing hidden units which link the inputs to the outputs. The function of the hidden neurons is to extract useful features from the input data which are, in turn, used to predict the values of the output units.

MLPs are layered neural network, see Fig. 32.2, that means they are based on several layers of adaptive weights and neurons. There are three different layers: the input layer, at least one hidden layer and finally the output layer. Between the units or neurons that compose the network there are a set of connections that link one to each other. Each unit transmits signals to the ones connected with its output. Associated with each unit there is a output signal or transference signal, that transform current state of the unit into an output signal.

The feedforward connections between the units of the layers in a MLP represent the adaptive weights that permit the mapping between the input variables and the output variables [9].

Different authors have showed that MLP networks with as few as a single hidden layer are universal approximators, in other words ANN are capable to approximate, with accurate, arbitrary regions, if enough hidden units are available [10, 11].

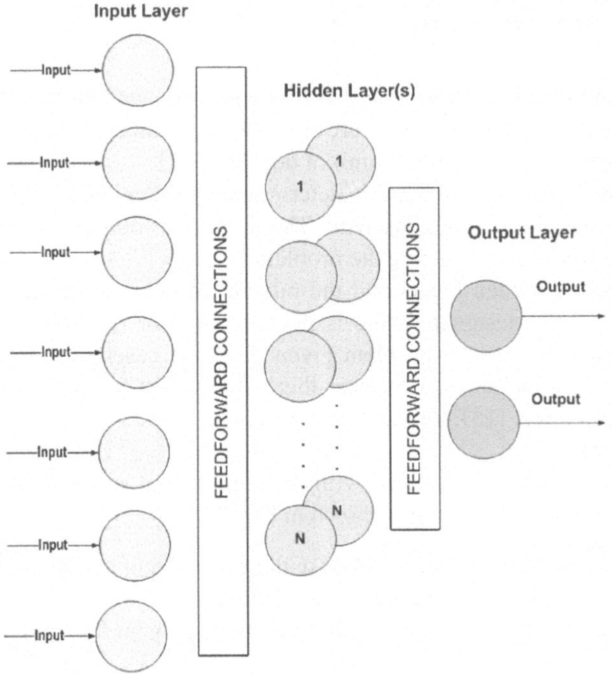

Fig. 32.2 Typical multilayer perceptron architecture

32.3.1 Impact Scenario Parameters

Within the limits this article deals with, there are different variables or parameters that characterize the behaviour of the projectile when it impacts on an steel armour. Therefore, there are various parameters to shape the input data to the network, and these define the number of neurons of the input layer.

The available variables are: kinetic energy (K), velocity (V), radius (R), length (L), mass (M) and quotient L/R, being all of them related to the projectile, and on the other hand the thickness of the objective (H).

However, the use of all the available variables is not always necessary to carry out the training. In some cases, an information overload or the existing connections between the variables can saturate the prediction model that adjusts the output of the network, therefore complicating the learning and reducing the generalization rates. In this domain, K is correlated with V and M. On the other hand, the ratio L/R and M are correlated with R and L because density is constant. So for the neural network performance it is better to remove K and ratio L/R from the list of available variables.

Moreover, in order to support the latter assertion, a series of studies were designed to measure the influence that each variable has, separately and in connection with the remainder, on the learning and the network generalization ability.

32.3.2 Network Structure

For the network design, a three level MLP architecture was selected. According to Lippman's studies, this type of structure allows to solve most of the problems, being able to form any complex random limit of decision [12].

The number of inputs in a MLP is determined by the number of available input parameters in the problem dealing with. The number of outputs is the same as the number of classes needed to solve the problem.

The number of hidden layers and the number of neurons of these layers have to be chosen by the designer. There is not a method or rule that determines the optimum number to solve a problem given. In some cases these parameters are determined by test and error. However, there are current techniques for obtaining automatic estimations [13], although this research follows the steps described by Tarassenko [14].

The neural network was created with the commercial software Neurosolutions for Excel v4 and its architecture is shown in Fig. 32.3.

- Five neurons in the input layer linked with the five input variables (R,L,H,M,V). The chosen transference function is the identity.
- Four neurons in the hidden layer with hyperbolic tangent transference function.

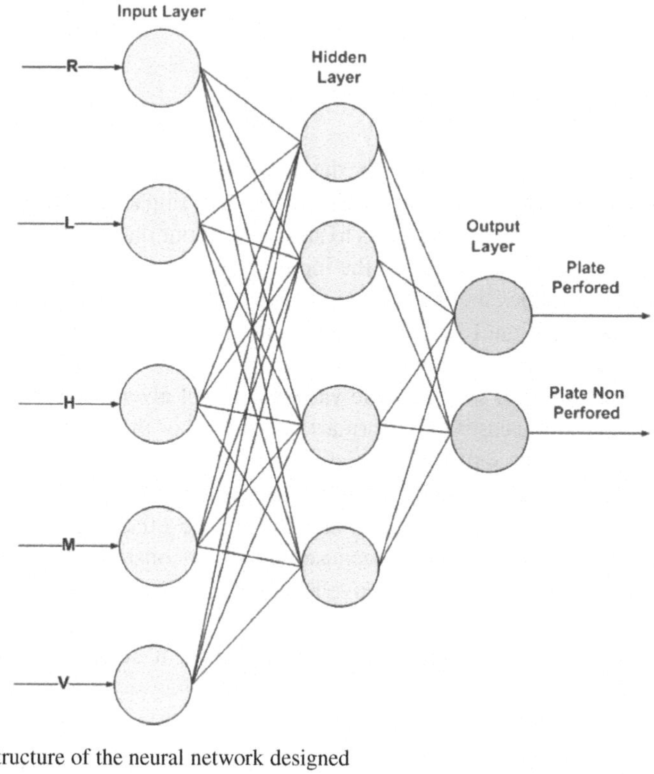

Fig. 32.3 Structure of the neural network designed

- Two neurons in the output layer associated to the output variable plate perforation (both outputs are complementary as it allows to improve the learning process). The chosen transference function in this case is the hyperbolic tangent.

32.3.3 Input Data

It is important to make an estimation of the necessary information to train the network in an appropriate way. For this purpose, it is essential to have a number of data large enough to ensure that. While the network is trained, it is essential to recognise and answer to the wide set of conditions that may appear during the testing process. If the number of data is too small, the complete range of connections that the neural network should learn will not be covered. In this sense, there are different approximations to calculate the optimum number of input data, like [9] or [15]. The designed system has followed the guidelines set by Tarassenko to calculate the number of examples needed [14].

Fifty impact examples were generated by numerical simulation. Some of them were used for the training and testing process and the other for the validation of the network after learning. The modelling and simulation were done with the commercial software ABAQUS/Explicit v6.4.1 [16], a finite element explicit code widely used in highly non-linear dynamic simulation.

The stopping criterion selected was to perform a specific number of epochs to train the network, in this case, 10,000 epochs. Consequently the neural network does not use validation data in the training process because the stopping criterion is set by the epochs and not by the cross validation error. Therefore the ANN only uses a train subset for the training process and a test set to validate the accurate of the network.

The set of training patterns is repeatedly presented in random order to the network and the weight update equations are applied after the presentation of each pattern. When all the patterns have been presented once to the network, one epoch of training is made. After the training process, the unseen patterns of the test set are presented to the network. This demonstrates the generalization performance of the trained network.

32.4 Solution Proposed

In the light of the results obtained by Garcia et al. in 2006 [8], the neural networks present themselves as an alternative to be borne in mind to recreate impact problems. The network results reliable to predict the projectile arrest with the advantage, opposing to the use of simulation tools, that it boasts a computational cost very inferior once the training has been carried out.

However, these conclusions lead to another question. Could there be a more efficient ANN model? That is to say a model that would need less input variables and input data to be able to approximate the output function correctly.

The research carried out to solve this question established two fundamental goals, first, to minimize the number of training data without affecting the generalization ability of the model; and second, to analyse the influence of the different input variables on the network learning ability.

These input variables form the independent variables of the function that allows to approximate the perforation function. The result of this answer function is a dichotomic variable that depends on the input variables, and admits "Yes" or "No" values.

Hence it is possible to infer that the easier the function that describes the network's behaviour is, the less costly will be generate the set of training data through numeric simulation.

The type of heuristic selection of input variables and training set size selected is well documented in the early literature of Neural Networks within other domains, and the results obtained certifies the suitability of this election [17].

32.4.1 Randomness Elimination

The results obtained in the former research [8], present the uncertainty if they depend on a possible randomness of the data intended for training and testing. In other words, if the assignment of the available data influence on the predictions of the network.

To carry out a research with conclusive results, it was established a number of trials by far superior to the previous work. The study carried out is broken out in a series of experiments, in which the number of patterns intended for training and test varies in each series. This way, the minimum number of data needed to ensure an acceptable percentage of correct answers will be found.

It has been accomplished 100 different trials in each experiment, exchanging in each of them and in a random way the data destined to train and test the network. Moreover, in two different trials of a same experiment the same training and test data will never coincide, therefore increasing the reliability on the results obtained. Thanks to this, the possible random data that are provided to the network, which could lead to obtain not very reliable results, are eliminated.

The catalogue of experiments to be done in the study is the following: the first one includes 100 trials with 50 training data and 10 testing data. From this point onwards, in each of them the number of data intended to training is reduced to 5. The last experiment consists of 100 trials with 15 training data and 10 testing data.

The result obtained in each one of these 100 trails is processed and the average error porcentage is showed for each experiment. In this way, the results obtained in each of the eight experiments can be compared. Thanks to this, it can be determined how many training data are necessary to achieve good generalization rates;

and besides, it ensures that the possible random data of the input patterns does not influence on these results.

32.4.2 Determination of the Influence of the Independent Variables

The second part of this study is focused on the analysis of the influence of each variable within the training process and, therefore, of the mathematical model that regulates its output.

So the main question is to find the explanatory variables, in other words, to include in the model just those variables that can influence on the answer, rejecting those ones that do not provide information. It is also necessary look for the possible interactions between the independent variables that affect the answer variable. For example, this work will analyse the results obtained by including the thickness and the velocity separately, as well as when they are both included.

32.5 Evaluation

The data shown in Table 32.1 are the average error percentages that the network makes for each experiment. The different experiments are grouped in columns, where it is indicated the parameters or variables used, e.g the column "All-H" points out that the experiment uses all the variables except the parameter H. One result example in Table 32.1 is the following one: in the model with all the variables, the value 3,2 is the average error of the network for 100 trials with 50 data to train and 10 to testing.

Table 32.1 Error percentage obtained by varying the input variables and the number of training data

Train. data	All	All -H	All -L	All -M	All -R	All -V	All -H&M	All -H&V	All -H,M&V
50	3,2	7,7	3,3	4	5,6	26,3	10,2	30,1	37,9
45	5,3	12	6,5	5,9	6,2	34,1	10	33	45,6
40	5,3	15,3	4	5,7	6,5	30,1	10,6	36,5	38,4
35	5,6	10	6,9	5,5	4,8	27,2	11,3	36,3	39,4
30	11	14,1	6,6	7,3	9,7	32,9	14,5	42,6	43,1
25	11,2	11,2	9,5	8,8	9,4	36,6	16,5	39	46,6
20	13,4	14,2	14,6	11,8	14,2	35,7	17,3	39,1	44,1
15	13,9	11,9	13,6	10,6	13,8	40,7	12,9	42,8	45,2
Average	8,61	12,05	8,13	7,45	8,78	32,95	12,91	37,43	42,54

Regarding the objective sought, to find the network's architecture that makes the smallest error, when the data is analysed it could be established that the more input variables the network has, the better it learns.

However, it can occur that the networks saturation, that is to say to introduce all the available parameters, is not the best option. In this problem, this network architecture has the lowest error probability only for 50 and 45 train data. In addition to this, there are two designs that have lower average probability of error, the ones without mass and without length. Specifically the design that omits the mass as a network learning variable is the one that has the smallest average error.

On the other hand, for the average of the error obtained, it can be observed that the velocity is a very relevant magnitude for the network learning. Its omission leads the error probability to increase 342%, or what is the same, 4.4 times in relation to the network's design considered as the most efficient (architecture of network without mass).

Finally as expected, the network with the worst results is the one with less information, that is, with less input variables. However, it is advisable to highlight that the network without the thickness and mass variables has quite acceptable results (12.91%).

Taking the best configuration for the most architectures, 50 train and 10 testing, the Figure 32.4 depicts the average error percentage made for each network architecture in which some variable is removed with regard to the network that holds all of them.

Table 32.2 portrays the differences of the error percentages for each of the network designs with regard to the one with the minimum error. As it has been

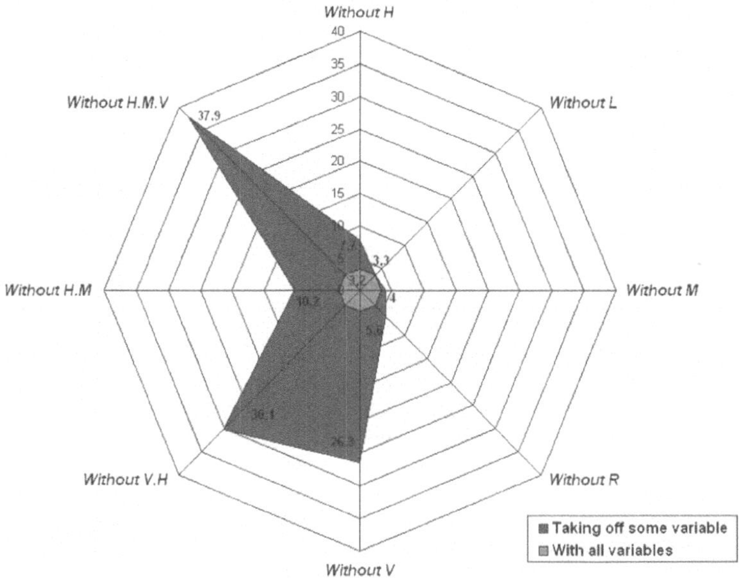

Fig. 32.4 Error made in each architecture for the configuration 50 train and 10 test

Table 32.2 Deviation percentage made in each simulation in relation to the one that makes the minimum error

Train. data	All	All -H	All -L	All -M	All -R	All -V	All -H&M	All -H&V	All -H,M&V
50	0	4,5	0,1	0,8	2,4	23,1	7	26,9	34,7
45	0	6,7	1,2	0,6	0,9	28,8	4,7	27,7	40,3
40	1,3	11,3	0	1,7	2,5	26,1	6,6	32,5	34,4
35	0,8	5,2	2,1	0,7	0	22,4	6,5	31,5	34,6
30	4,4	7,5	0	0,7	3,1	26,3	7,9	36	36,5
25	2,4	2,4	0,7	0	0,6	27,8	7,7	30,2	37,8
20	1,6	2,4	2,8	0	2,4	23,9	5,5	27,3	32,3
15	3,3	1,3	3	0	3,2	30,1	2,3	32,2	34,6
Total	13,8	41,3	9,9	4,5	15,1	208,5	48,2	244,3	285,2

mentioned, the network that includes all the available variables in the learning is the one that makes the smaller error for 50 and 45 data; whilst the network without mass is the one that boasts the smallest error for a smaller number of trials. Based on these results, it can be concluded that the influence of the mass on learning is not very relevant, and therefore the network that does not include it, it is considered the most efficient.

32.6 Conclusions

In the light of the results and taking into account that perceptron is one of the simplest topologies, this work shows clearly the possibilities of the neural networks in the prediction of the material behaviour at high deformation speeds. Furthermore, the architecture chosen presents a high reliability when predicting the result of a projectile impact. Moreover, its computational cost once the training has started, is smaller than the one of the simulations carried out with tools of finite elements.

It is crucial to highlight the small number of training data needed for the network to learn with a relative small error in its predictions. With only 40 numeric simulations of finite elements, the network obtains an error below 6% in both designs with the best results (with all the variables and without the mass variable). In spite of the small number of input data, the network does not present overlearning problems.

The experiments developed help to better understand the ballistic impact, analysing the influence of the parameters that appear in the penetration problem. The results obtained verify that the variable with the most influence is the velocity. Furthermore, any network architecture where the velocity variable is removed obtains error percentages quite high, what confirms this variable importance.

On the other hand, the research determines that the influence of the mass on learning is not very relevant. Therefore and taking into account its numeric simulation cost, the network without this variable is considered the most efficient.

The network with the worst results is the one with less information, that is to say, with less input variables. However, it is advisable to highlight that the network without thickness and mass variables has quite acceptable results taking into account the little information it receives in comparison with the rest.

Finally, the knowledge acquired as a result of this research, can be spread out to other fields of great practical interest. Among these, the design of structural components of energy absorption stands out, being of great relevance in the field of passive safety in vehicles.

References

1. K. Swingler. *Applying Neural Networks: A Practical Guide*. Morgan Kaufmann, San Francisco, CA 1996.
2. Z. Waszczyszyn and L. Ziemianski. Neural networks in mechanics of structures and materials. New results and prospects of applications. *Computers & Structures*, 79 IS - 22-25:2261 EP–2276, 2001.
3. R. Ince. Prediction of fracture parameters of concrete by artificial neural networks. *Engineering Fracture Mechanics*, 71(15):2143–2159, 2004.
4. S.W. Liu, J.H. Huang, J.C. Sung, and C.C. Lee. Detection of cracks using neural networks and computational mechanics. *Computer methods in applied mechanics and engineering (Comput. methods appl. mech. eng.)*, 191(25-26):2831–2845, 2002.
5. W. Carpenter and J. Barthelemy. A comparison of polynomial approximations and artificial neural nets as response surfaces. *Structural and multidisciplinary optimization*, 5(3):166, 1993.
6. P. Hajela and E. Lee. Topological optimization of rotorcraft subfloor structures for crash worthiness consideration. *Computers and Structures*, 64:65–76, 1997.
7. L. Lanzi, C. Bisagni, and S. Ricci. Neural network systems to reproduce crash behavior of structural components. *Computers structures*, 82(1):93, 2004.
8. A. Garcia, B. Ruiz, D. Fernandez, and R. Zaera. Prediction of the response under impact of steel armours using a multilayer perceptron. *Neural Computing & Applications*, 2006.
9. C. Bishop. *Neural Networks for Pattern Recognition*. Oxford University Press, USA, 1996.
10. G. Cybenko. Approximation by superpositions of a sigmoidal function. *Mathematics of Control, Signals, and Systems*, 2:303–314, 1989.
11. K. Hornik and M. Stinchcombe. Multilayer feedforward networks are universal approximators. *Neural Networks*, 2(5):359–366, 1989.
12. R. Lippmann. An introduction to computing with neural nets. *ASSP Magazine, IEEE [see also IEEE Signal Processing Magazine]*, 4(2):4–22, 1987.
13. P. Isasi and I. Galvan. *Redes de neuronas artificiales: un enfoque practico*. Pearson Prentice Hall, Madrid, 2004.
14. L. Tarassenko. *A guide to neural computing applications*. Arnol/NCAF, 1998.
15. B. Widrow. 30 years of adaptive neural networks: perceptron, madaline, and backpropagation. *Proceedings of the IEEE 78*, 9:1415–1442, 1990.
16. USA. ABAQUS Inc., Richmond. Abaqus/explicit v6.4 users manual, 2003.
17. M. Gevrey, I. Dimopoulos, and S. Lek. Review and comparison of methods to study the contribution of variables in artificial neural network models. *Ecological Modelling*, 160(16):249–264, 2003.

Chapter 33
Developing Emotion-Based Pet Robots

W.-P. Lee, J.-W. Kuo, and P.-C. Lai

Abstract Designing robots for home entertainment has become an important application of intelligent autonomous robot. Yet, robot design takes considerable amount of time and the short life cycle of toy-type robots with fixed prototypes and repetitive behaviors is in fact disadvantageous. Therefore, it is important to develop a framework of robot configuration so that the user can always change the characteristics of his pet robot easily. In this paper, we present a user-oriented interactive framework to construct emotion-based pet robots. Experimental results show the efficiency of the proposed framework.

Keywords Pet robot · robot emotion · human-robot interaction · behavior coordination · neural network

33.1 Introduction

In recent years, designing robots for home entertainment has become an important application of intelligent autonomous robot. One special application of robot entertainment is the development of pet-type robots and they have been considered the main trend of the next-generation electronic toys [1, 2]. This is a practical step of expanding robot market from traditional industrial environments toward homes and offices.

There have been many toy-type pet robots available on the market, such as Tiger Electronics' Furby, SONY's AIBO, Tomy's i-SOBOT and so on. In most cases, the robots have fixed prototypes and features. With these limitations, their life cycle is thus short, as the owner of pet robots may soon feel bored and no longer interested in their robots. Sony's robot dog AIBO and humanoid robot QRIO are sophisticated

W.-P. Lee (✉)
Department of Information Management, National Sun Yat-sen University, Kaohsiung, Taiwan,
E-mail: wplee@mail.nsysu.edu.tw

S.I. Ao and L. Gelman (Eds.), *Advances in Electrical Engineering and Computational Science*, Lecture Notes in Electrical Engineering.

pet robots with remarkable motion ability generated from many flexible joints [2,3]. But these robots are too expensive to be popular. Also the owners are not allowed to reconfigure the original design. Therefore, it would be a great progress to have a framework for robot configuration so that the user can always change the characteristics of his robot according to his personal preferences to create a new and unique one.

Regarding the design of pet robots, there are three major issues to be considered. The first issue is to construct an appropriate control architecture by which the robots can perform coherent behaviors. The second issue is to deal with human-robot interactions in which natural ways for interacting with pet robots must be developed [4, 5]. And the third issue to be considered is emotion, an important drive for a pet to present certain level of intelligence [6, 7]. In fact, Damasio has suggested that efficient decision-making largely depends on the underlying mechanism of emotions [8]. By including an emotional model, the pet robot can explicitly express its internal conditions through the external behaviors, as the real living creature does. On the other hand, the owner can understand the need and the status of his pet robot to then make appropriate interactions with it.

To tackle the above problems, in this paper we present an interactive framework by which the user can conveniently design (and re-design) his personal pet robot according to his preferences. In our framework, we adopt the behavior-based architecture ([9, 10]) to implement control system for a pet robot to ensure the robot functioning properly in real time. A mechanism for creating behavior primitives and behavior arbitrators is developed in which an emotion model is built for behavior coordination. Different interfaces are also constructed to support various human-robot interactions. To evaluate our framework, we use it to construct a control system for the popular LEGO Mindstorms robot. Experimental results show that the proposed framework can efficiently and rapidly configure a control system for a pet robot. In addition, experiments are conducted in which a neural network is used to learn a user-specified emotion model for behavior coordination. The results and analyses show that an emotion-based pet robot can be designed and implemented successfully by the proposed approach.

33.2 Building Emotion-Based Pet Robots

33.2.1 A User-Oriented Framework

Our aim is to develop a user-oriented approach that can assist a user to rapidly design (and re-design) a special and personal robot. Robot design involves the configuration of hardware and software. Expecting an ordinary user to build a robot from a set of electronic components is in fact not practical. Therefore, instead of constructing a personal pet robot from the electronic components, in this work we concentrate on how to imbue a robot with a personalized control system.

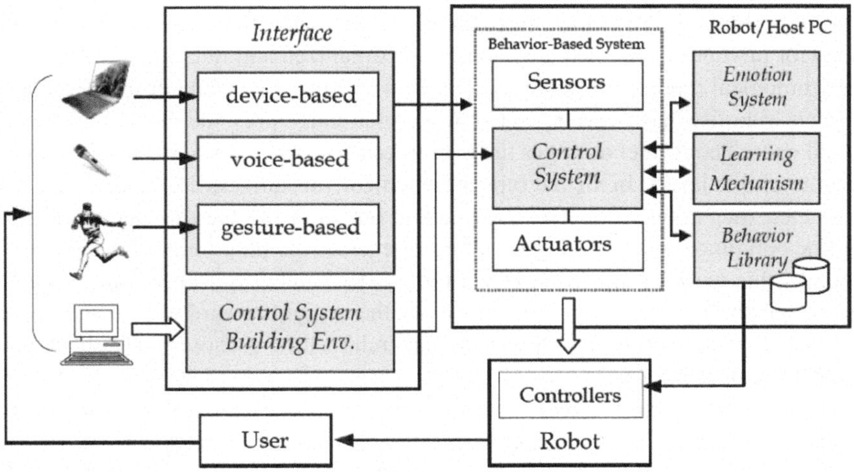

Fig. 33.1 The overall architecture of our system framework

Figure 33.1 illustrates the system framework of the proposed approach. As can be seen, our system mainly includes three modules to deal with the problems in building a control system for a pet robot. The first module is to design an efficient control architecture that is responsible for organizing and operating the internal control flow of the robot. Because behavior-based control architecture has been used to construct many robots acting in the real world successfully, our work adopts this kind of architecture to design control systems for robots. The second module is about human-robot interaction. As a pet robot is designed to accompany and entertain its human partner in everyday life, interactions between the owner and his pet are essential. Here we develop three ways for human-robot interaction and communication, including device-based (keyboard and mouse), voice-based and gesture-based methods. The third module is an emotion system that works as the arbitration mechanism to resolve the behavior selection problem within the behavior-based control architecture. With the emotion system, a pet robot can act autonomously. It can choose whether to follow the owner's instructions, according to its internal emotions and body states. Our system is designed to be user-oriented and has a modular structure. With the assist of the presented system, a user can build his robot according to his preferences. If he is not satisfied with what the robot behaves, he can change any part of the control software for further correction. Details of each module are described in the following subsections.

33.2.2 Control Architecture

As in the behavior-based control paradigm [9, 10],the behavior system here takes the structure of parallel task-achieving computational modules to resolve the control problem. In order to achieve more complex tasks in a coherent manner, the behavior modules developed have to be well organized and integrated. Inspired by the ethological models originally proposed to interpret the behavior motivations of animals, robotists have developed two types of architectures: the flat and the hierarchical ones. The former arranges the overall control system into two levels; and the latter, multiple levels. In the flat type arrangement, all subtasks are independent and have their own controllers. The independent outputs from the separate controllers will be combined appropriately in order to generate the final output for the overall task. As this work means to provide an interactive and easy-to-use framework for the design and implementation of pet robots, the straightforward way (that involves less task decomposition techniques), the flat architecture, is chosen to be the default control structure for a pet.

Using behavior-based control architecture, one needs to deal with the corresponding coordination (or action selection) problem. It is to specify a way to combine the various outputs from the involved controllers. Our work here takes the method of *command switching* that operates in a winner-take-all fashion. That is, only one of the output commands from the involved behavior controllers is chosen to take over the control at any given moment, according to certain sensory stimuli.

As mentioned above, we intend to build low cost toy-type robots as embodied digital pets. Yet, since a toy-type robot only has limited computational power and storage resources, it is generally not possible to perform all the above computation on such a robot. Therefore, to implement the above architecture, an external computing environment is needed to build robots. The experimental section will present how we construct an environment for the LEGO robots.

33.2.3 User–Robot Interaction

To communicate with a pet robot, our framework provides two natural ways for interacting with the robot by using oral or body languages. For the oral communication, we implement the popular speech-control method, command-and-control, to communicate with the robot, and adopt the Microsoft Speech API to implement the speech recognition mechanism in our Windows application. Though more sophisticated language interface can be developed, here we simply parse a sentence into individual words, distinguish whether it is an affirmative or negative sentence, and then recognize the user commands from the words.

Gesture recognition is another way used in this work to support natural communication between human and robots. Gesture recognition is the process by which the user's gestures are made known to the system via appropriate interpretation. To infer the aspects of gesture needs to sense and collect data of user position, configuration, and movement. This can be done by directly using sensing devices attached to the user (e.g., magnetic field trackers), or by indirectly using cameras and computer vision techniques. Here we take a dynamic gesture recognition approach, and use digital cameras with image processing techniques for gesture recognition.

For simplicity, in our current implementation, we use light spots to represent the hand positions of a pet robot owner, and extract the light track with the cubic spline interpolation method to obtain a behavior curve. Then we take a curve-fitting approach to match the curve produced by the continuous hand movement to the ones recorded previously as user's instructions in the database. Figure 33.2 shows a typical example. A successful match means that the gesture has been recognized and its corresponding instruction is then identified. As the traditional approach of dynamic programming for curve fitting has inherent drawback caused by the curve discretization-based distance measurement, in this work we employ the approach reported in [11] that takes the underlying differential equation into account and finds a continuous solution for curve fitting. The fitting result is then used to recognize gestures.

Figure 33.3 illustrates our speech/gesture recognition mechanism that mainly includes three parts. The first part is to develop and maintain a behavior library within which different behavior modules are pre-constructed and recorded. The second part is to capture and record an owner's spoken-sentences/images, and to extract the words/curves for further interpretation. The third part is a matching procedure that tries to match the extracted words/curves to the ones recorded in the mapping table in which each command-word/curve represents a behavior controller. If a match is found, the corresponding behavior module is then retrieved and activated from the behavior library. The behavior controller is executed and the control

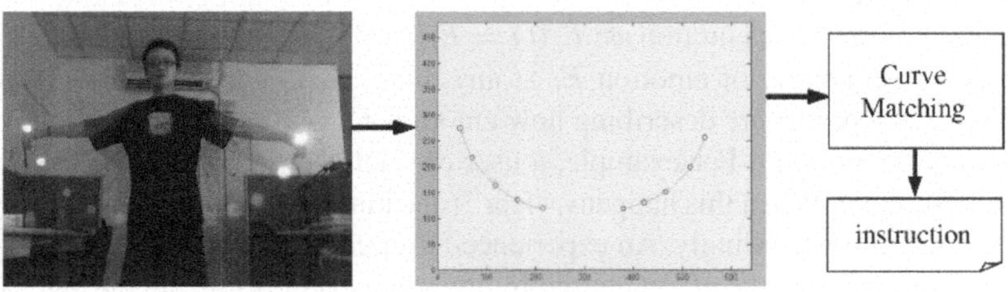

Fig. 33.2 An example of extracting trajectory curve

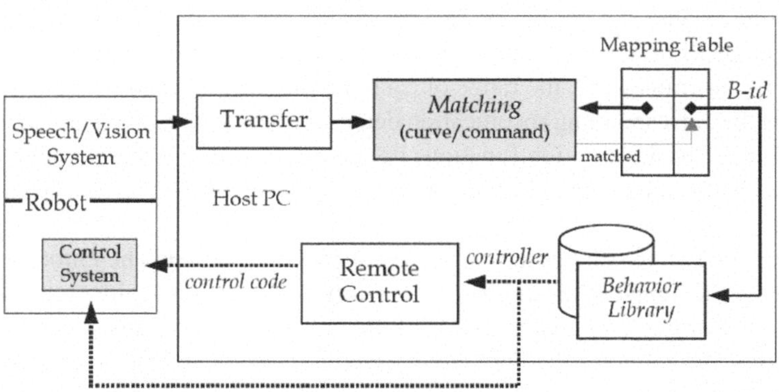

Fig. 33.3 Speech/gesture-based control

command is used to drive the robot. Alternatively, if the behavior controllers can be pre-uploaded to the on-board computer of the robot, the mapping result will send an identification signal to trigger the corresponding controller on the robot. Here the user is allowed to construct the mapping table to decide how to interact with his robot.

33.2.4 Emotion Model

To coordinate different behavior controllers, our framework uses a special mechanism that exploits emotions for selection of behavior. Emotions are often categorized into two kinds: basic emotions and higher cognitive emotions. Basic emotions (such as joy, anger, fear, etc.) are considered as universal and innate; they are encoded in the genome. Higher cognitive emotions (such as love, shame, pride, etc.) are universal too, but they exhibit more cultural variations. Basic emotions tend to be reflex-like responses over which animals have little conscious control. They involve less cognitive processes and are thus much faster in controlling motion than those culturally determined experiences residing in the cognitive system (long term memory). As our goal here is to establish a framework for constructing personal pet robots, rather than to investigate the interactions and relationships between cognitive and emotion systems, our work only models basic emotions to coordinate the behave controllers the user has pre-chosen for his pet. To quantify the emotions, we define each of the basic emotion as: $E_i(t) = E_i(t-1) + \alpha \times Event_i$, in which $E_i(t)$ represents the quantity of emotion E_i at any time t, α is a user-specified weight, and $Event_i$ is a procedure describing how emotion E_i is affected by a set of events pre-defined by the user. For example, a user can define an event to be the appearance of a stranger. When this happens, "fear" (one kind of emotion of the robot) will increase one unit accordingly. An experienced user can define a more sophisticated procedure to alleviate the effect caused by the same event occurring during a short period of time.

In addition to emotions, another set of internal variables is defined to describe a robot's body states (e.g., hungry). These variables represent the basic requirements the robot has to be satisfied, and they must be regulated and maintained in certain ranges. In our framework, the range of each internal variable can be defined by the user, and the corresponding specification determines the innate characteristics of his robot. Similar to the emotion functions defined above, an internal variable here is described by a mathematical formula with a set of involved events, also specified by the user.

Figure 33.4 shows the aspect of the emotion system used. The emotion model determines the innate characteristics of a robot, which are highly related to its abilities of learning and adaptation. With the above model, at any given time, the emotions of the robot can be derived and used to trigger the behavior controller to generate a specific action. After the action is performed, the external environment conditions will thus change and that can further affect the emotion of the robot. The

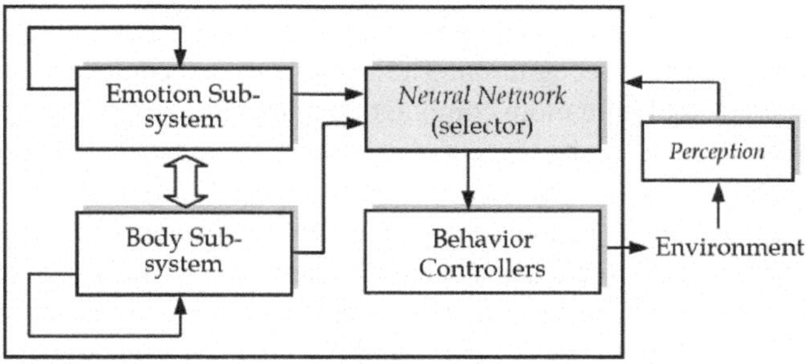

Fig. 33.4 Emotion-based behavior coordination

pet robot then makes new decision for behavior selection, based on the modified quantities of emotions. For example, a pet dog in a hungry state may be angry, may keep looking for food, and would eat anything as soon as the pet dog finds it. After that, the dog may not be hungry any more (the body state has been changed). Then it is happy (the emotion has been changed too) and may want to sleep or fool around (now new behavior is selected). The above procedure is repeated and the emotion model continuously works as a decision-making mechanism for behavior selection.

As can be observed, in the above operating model, the most important part is the one for selecting appropriate behavior at any time. In this work, we use a feedforward neural network to implement such a mechanism that maps the emotions and body states of a robot into a set of desired behaviors. To allow the user to determine the characteristics of his pet, our work also provides an interface by which the user can define examples to train the neural network to achieve the specified mapping of emotions and behaviors. Once the neural network is obtained, it works as a behavior selector to choose appropriate controllers for the robot.

33.3 Experiments and Results

33.3.1 Implementation

To evaluate our approach, two robots have been built and a distributed and networked computing environment has been developed for the robots. The robot used in the experiments is LEGO Mindstorms NXT 9797. It has light sensors for light detection, ultrasonic sensors for distance measurement, and micro-switches for touch detection. In addition, to enhance its vision ability, we equip an extra mini-camera on the head of the robot so that it can capture images for further interpretation.

Because the LEGO NXT only has limited computational power and memory devices, a distributed and networked computing environment is thus constructed for the robot, in which the individual pre-programmed behavior controllers are installed in the NXT on-board memory, and two PCs are connected through the Internet for other computation. Here, one PC is responsible for the computation of emotion modeling, speech recognition, and communication management, and the other PC is used for image/gesture recognition. The network-based computing environment can be extended to deal with more computation when more robots are involved.

The emotion system is also executed in one of the host PCs. It interprets the received signals to activate different events to change emotions and body states accordingly, as described in Section 33.2.4. The newly derived values of emotions and states are used to choose a behavior controller, according to the mapping strategy that can be pre-wired manually or learnt automatically. Once the behavior controller is selected, the host PC then sends an identification signal through the Bluetooth channel to evoke the corresponding controller from the set of controllers pre-stored on the robot.

After the above computing environment was established, experiment has been conducted to show how our LEGO robots can achieve a user-specified task. As soon as the owner asked one of the robots to push a red box, it started moving around to look for the box. Once the robot found the box, it tried to push the box. As the box was too heavy and the robot could not push it away alone, it then sent a message to another robot through the built-in Bluetooth communication channel to ask for help. The second robot came over, found the box and tried to push it too. As shown in Fig. 33.5, after the two robots pushed the box together, the box was moved away successfully.

33.3.2 Building Emotion-Based Behavior Selector

To verify our approach of emotion-based control, this section describes how the emotion model can be built, trained and modified. In the experiments, the simple types of emotions are modeled. They are so-called "basic emotions", including "happy", "angry", "fear", "bored", "shock", and "sad". Also three variables, "hungry", "tired", and "familiar" are defined to indicate the internal body states of the robot. As mentioned in Section 33.2.4, the user is allowed to define event procedures and the relevant weight parameters for the above emotions and body states to describe how the quantities of different emotions vary over time for their own

Fig. 33.5 Two LEGO NXT robots push the red box together

robots. For naïve users, our framework also provides three default emotion models (aggressive, gentle, and shy) and events as choices to represent different characteristics of a robot. For example, a robot with an aggressive model will change its emotions rapidly than others. Users can choose a default model from the interface without extra settings. Experienced users can develop more sophisticated models to guide the variations of emotions.

Currently, ten basic behaviors are built, including target seeking, barking, wandering, shaking, sniffing, sleeping, escaping, scratching, wailing, and wagging. As mentioned above, with the emotion model developed, a user can build a behavior selector manually or automatically and use it to map the emotions and body states into appropriate behavior controllers. At each time step, the internal emotions and states of the robot change and the newly obtained values are used as the input of an emotion model to select behavior controller at that moment. Figure 33.6 shows the interface that presents the numerical values of the internal and emotion variables over time during a trial. These values are illustrated to provide information about the body state and emotion of the robot, so that the user can inspect the detailed information related to his personal robot accordingly.

As can be seen in Fig. 33.6, the interface also includes a set of event buttons on the right hand side to simulate the happening of different events. Each button here is associated with an event procedure that describes how emotions and body states are changed by this event. It corresponds to a situated event in reality. That is, when the pre-conditions of an event procedure are satisfied in the real world, the same effect will be given to change the emotions and body states of a robot. With the assistance

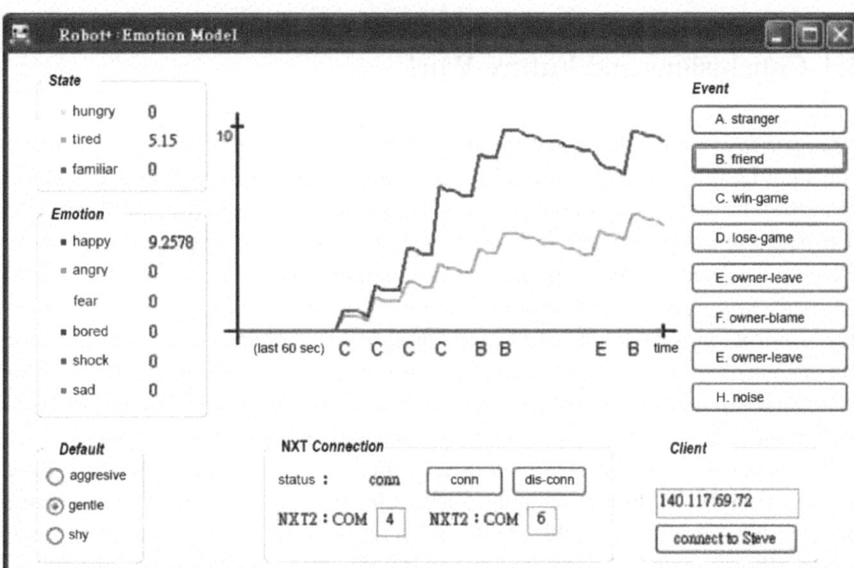

Fig. 33.6 The interface for showing the typical results

of event buttons, users can examine the correctness of the effect caused by the event procedure he defined in an efficient way.

In addition, our framework offers a learning mechanism to train a feedforward neural network from examples as a behavior selector. The above variables (emotions and body states) are arranged as the input of the network, and it output is used to determine which behavior to perform at a certain time. In the training phase, the user is allowed to give a set of training examples and each specifies which behavior the robot is expected to perform when the set of emotions and states reaches the values he has assigned. The back-propagation algorithm is then used to derive a model for the set of data examples. Figure 33.7 presents the interface through which a user can edit the training set. Based on the training set provided by the user, the system then tries to learn a mapping strategy with best approximation.

It should be noted that it is possible the examples provided by the user are inconsistent and consequently a perfect strategy cannot be obtained. In the latter case, the user can use the interface shown in Fig. 33.7 to correct the training examples to re-build the behavior selector (i.e., the neural network) again. If the model has been derived successfully but the behavior of the robot did not satisfy the owner's expectation, he can still correct the robot behavior for any specific time step by editing the output produced by the mapping strategy learnt previously through the interfaces shown in Fig. 33.7. Then the modified outputs can be used as new training examples to derive a new strategy of behavior arbitration. In this way, the user can easily and conveniently design (and re-design) the characteristics of his personal robot.

33.4 Conclusions and Future Work

In this paper, we have described the importance of developing toy-type pet robots as an intelligent robot application. We have also proposed to integrate knowledge from different domains to build low-cost pet robots. To realize the development of pet robot, a user-oriented interactive framework has been constructed with which the user can conveniently configure and re-configure his personal pet robot according to his preferences. Our system framework mainly investigates three issues: robot control, human-robot interaction, and robot emotion. Different interfaces have also been constructed to support various human-robot interactions. Most importantly, an emotion-based mechanism has been developed in which different emotions and internal states have been modeled and used to derive a behavior selector. The behavior selector is a neural network and the user is allowed to define training examples to infer a behavior selector for his robot. To evaluate our framework, we have used it to build LEGO NXT robots to achieve a cooperation task successfully.

Based on the presented framework, we are currently trying different toy-type robots with more sensors and actuators to evaluate our approach extensively. Also

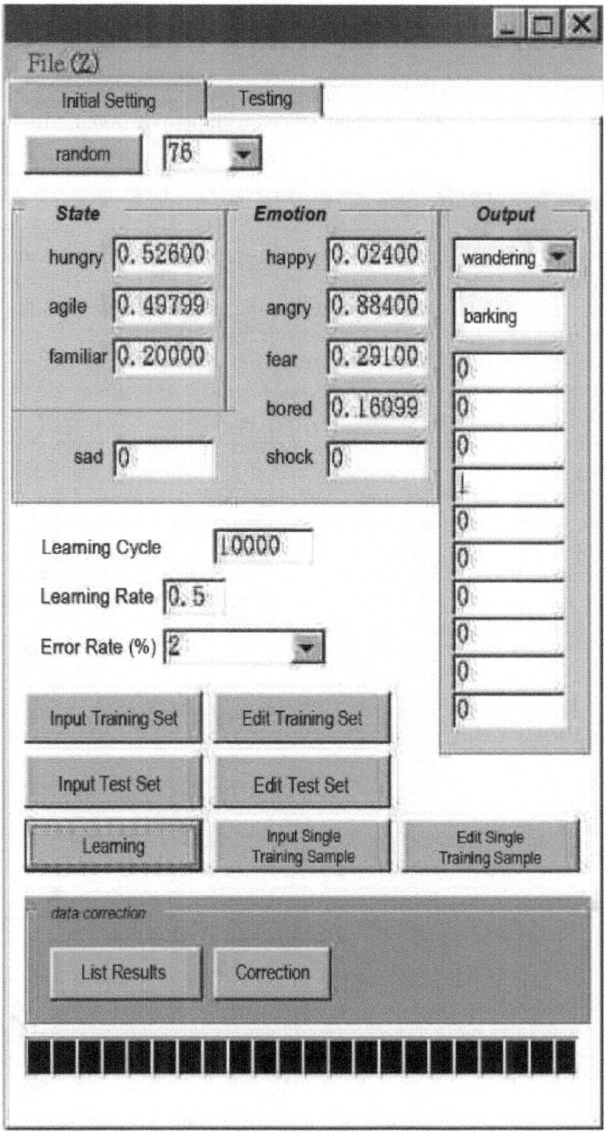

Fig. 33.7 The interface for preparing training data

we are implementing a new vision module for the robot so that it can recognize human facial expressions and interact with people accordingly. In addition, we plan to define a specific language and construct a message-passing channel through which different types of robots can communicate with each other.

References

1. M. M. Veloso, Entertainment robotics, *Communication of the ACM*, 45(3), 59–63, 2002.
2. M. Fujita, On activating human communications with pet-type robot AIBO, *Proceedings of the IEEE*, 92(11), 1804–1813, 2004.
3. T. Ishida, Y. Kuroki, and J. Yamaguchi, Mechanical system of a small biped entertainment robot, *Proceedings of IEEE/RSJ International Conference on Intelligent Robots and Systems*, pp. 1129–1134, 2003.
4. S. Thrun, Toward a framework for human-robot interaction, *Human-Computer Interaction*, 19(1–2), 9–24, 2004.
5. D. Perzarcowski, A. Schultz, W. Adams, E. Marsh, and M. Bugajska, Building a multimodal human-robot interface, *IEEE Intelligent Systems*, 16(1), 16–21, 2001.
6. J. E. LeDoux, *The Emotional Brain*, New York: Simon & Schuster, 1996.
7. J.-M. Fellous and M. Arbib (eds.), *Who Needs Emotions? The Brain Meets the Robot*, New York: Oxford University Press, 2005.
8. A. R. Damasio, *Descarte's Error: Emotion, Reason, and Human Brain,* New York: Grosset/ Putnam, 1994.
9. R. C. Arkin, *Behavior-Based Robotics*, Cambridge, MA: MIT Press, 1998.
10. R. A. Brooks, A robust layered control system for a mobile robot, *IEEE Journal of Robotics and Automation*, 2(1), 14–23, 1986.
11. M. Frenkel and R. Basri, Curve matching using the fast marching method, *Proceedings of International Workshop on Energy Minimization Methods in Computer Vision and Pattern Recognition*, LNCS-2683, pp. 35–51, 2003.

Chapter 34
Designing Short Term Trading Systems with Artificial Neural Networks

Bruce Vanstone, Gavin Finnie, and Tobias Hahn

Abstract There is a long established history of applying Artificial Neural Networks (ANNs) to financial data sets. In this paper, the authors demonstrate the use of this methodology to develop a financially viable, short-term trading system. When developing short-term systems, the authors typically site the neural network within an already existing non-neural trading system. This paper briefly reviews an existing medium-term long-only trading system, and then works through the Vanstone and Finnie methodology to create a short-term focused ANN which will enhance this trading strategy. The initial trading strategy and the ANN enhanced trading strategy are comprehensively benchmarked both in-sample and out-of-sample, and the superiority of the resulting ANN enhanced system is demonstrated.

Keywords Trading System · Short-Term · Artificial Neural Network · trading strategy · Vanstone and Finnie methodology

34.1 Introduction

There is a long established history of applying Artificial Neural Networks (ANNs) to financial data sets, with the expectation of discovering financially viable trading rules. Despite the large amount of published work in this area, it is still difficult to answer the simple question, "Can ANNs be used to develop financially viable stockmarket trading systems?" Vanstone and Finnie [1] have provided an empirical methodology which demonstrates the steps required to create ANN-based trading systems which allow us to answer this question.

B. Vanstone (✉)
Faculty of Business, Technology and Sustainable Development Bond University Gold Coast, Queensland, 4229 Australia
E-mail: bvanston@bond.edu.au

S.I. Ao and L. Gelman (Eds.), *Advances in Electrical Engineering and Computational Science*, Lecture Notes in Electrical Engineering.

In this paper, the authors demonstrate the use of this methodology to develop a financially viable, short-term trading system. When developing short-term systems, the authors typically site the neural network within an already existing non-neural trading system. This paper briefly reviews an existing medium-term long-only trading system, and then works through the Vanstone and Finnie methodology to create a short-term focused ANN which will enhance this trading strategy.

The initial trading strategy and the ANN enhanced trading strategy are comprehensively benchmarked both in-sample and out-of-sample, and the superiority of the resulting ANN enhanced system is demonstrated. To prevent excessive duplication of effort, only the key points of the methodology outlined are repeated in this paper. The overall methodology is described in detail in Vanstone and Finnie [1], and this methodology is referred to in throughout this paper as 'the empirical methodology'.

34.2 Review of Literature

There are two primary styles of stockmarket trader, namely Systems traders, and Discretionary traders. Systems traders use clearly defined rules to enter and exit positions, and to determine the amount of capital risked. The strategies created by systems traders can be rigorously tested, and clearly understood. The alternative, discretionary trading, is usually the eventual outcome of an individual's own experiences in trading. The rules used by discretionary traders are often difficult to describe precisely, and there is usually a large degree of intuition involved. In many cases, some of the rules are contradictory – in these cases, the discretionary trader uses experience to select the appropriate rules. Despite these obvious drawbacks, however, it is commonly accepted that discretionary traders produce better financial results [2].

For the purposes of this paper, it is appropriate to have a simple, clearly defined mathematical signal which allows us to enter or exit positions. This allows us to accurately benchmark and analyze systems.

This paper uses the GMMA as the signal generator. The GMMA is the Guppy Multiple Moving Average [3], as created and described by Daryl Guppy [4], a leading Australian trader. Readers should note that Guppy does not advocate the use of the GMMA indicator in isolation (as it is used in this study), rather it is appropriate as a guide. However, the GMMA is useful for this paper, as it is possible to be implemented mechanically. In essence, any well defined signal generator could be used as the starting point for this paper.

The GMMA is defined as:

$$GMMA = \left(\begin{pmatrix} ema(3) + ema(5) \\ +ema(8) + ema(10) \\ +ema(12) + ema(15) \end{pmatrix} - \begin{pmatrix} ema(30) + ema(35) \\ +ema(40) + ema(45) \\ +ema(50) + ema(60) \end{pmatrix} \right) \tag{34.1}$$

Creation of the ANNs to enhance this strategy involves the selection of ANN inputs, outputs, and various architecture choices. The ANN inputs and outputs are a cut-down version of those originally described in Vanstone [5]. The original list contained 13 inputs, and this paper uses only five. These five variables, discussed later in this paper, were selected as they were the most commonly discussed in the main practitioners' journal, 'The Technical Analysis of Stocks and Commodities'. Similarly, the choices of output and architecture are described in the empirical methodology paper. Again, these are only briefly dealt with here.

For each of the strategies created, an extensive in-sample and out-of-sample benchmarking process is used, which is also further described in the methodology paper.

34.3 Methodology

This study uses data for the ASX200 constituents of the Australian stockmarket. Data for this study was sourced from Norgate Investor Services [6]. For the in-sample data (start of trading 1994 to end of trading 2003), delisted stocks were included. For the out-of-sample data (start of trading 2004 to end of trading 2007) delisted stocks were not included. The ASX200 constituents were chosen primarily for the following reasons:

1. The ASX200 represents the most important component of the Australian equity market due to its high liquidity – a major issue with some previously published work is that it may tend to focus too heavily on micro-cap stocks, many of which do not have enough trading volume to allow positions to be taken, and many of which have excessive bid-ask spreads.
2. This data is representative of the data which a trader will use to develop his/her own systems in practice, and is typical of the kind of data the system will be used in for out-of-sample trading.

Software tools used in this paper include Wealth-Lab Developer, and Neuro-Lab, both products of Wealth-Lab Inc (now owned by Fidelity) [7]. For the neural network part of this study, the data is divided into two portions: data from 1994 up to and including 2003 (in-sample) is used to predict known results for the out-of-sample period (from 2004 up to the end of 2007). In this study, only ordinary shares are considered.

The development of an ANN to enhance the selected strategy is based on simple observation of the GMMA signals. One of the major problems of using the GMMA in isolation is that it frequently whipsaws around the zero line, generating spurious buy/sell signals in quick succession.

One possible way of dealing with this problem is to introduce a threshold which the signal must exceed, rather than acquiring positions as the zero line is crossed. The method used in this paper, however, is to forecast which of the signals is most likely to result in a sustained price move. This approach has a major advantage over the threshold approach; namely, in a profitable position, the trader has entered earlier, and therefore, has an expectation of greater profit. By waiting for the threshold to be exceeded, the trader is late in entering the position, with subsequent decrease in profitability.

However, for the approach to work, the trader must have a good forecast of whether a position will be profitable or not. This is the ideal job for a neural network.

In Fig. 34.1, there are a cluster of trades taken between June 2006 and September 2006, each open for a very short period of time as the GMMA whipsaws around the zero line. Eventually, the security breaks out into a sustained up trend. What is required is an ANN which can provide a good quality short-term forecast of the return potential each time the zero line is crossed, to allow the trader to discard the signals which are more likely to become whipsaws, thus concentrating capital on those which are more likely to deliver quality returns.

The neural networks built in this study were designed to produce an output signal, whose strength was proportional to expected returns in the 5 day timeframe. In essence, the stronger the signal from the neural network, the greater the expectation of return. Signal strength was normalized between 0 and 100.

The ANNs contained five data inputs. These are the technical variables deemed as significant from the review of both academic and practitioner publications, and details of their function profiles are provided in Vanstone [5]. The formulas used to compute these variables are standard within technical analysis. The actual variables used as inputs, and their basic statistical characteristics are provided in Table 34.1.

For completeness, the characteristics of the output target to be predicted, the 5 day return variable, are shown in Table 34.2. This target is the maximum percentage

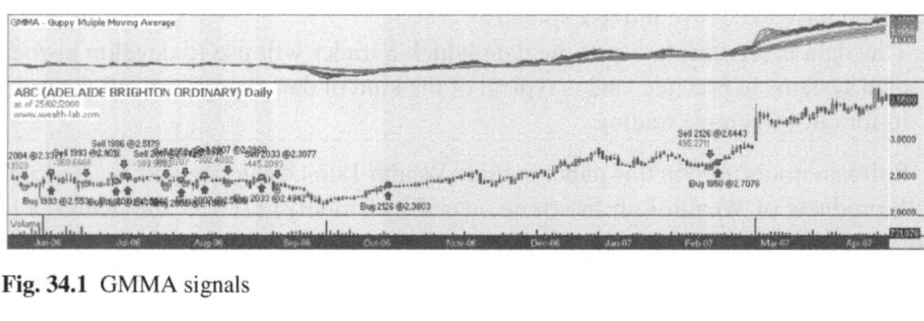

Fig. 34.1 GMMA signals

Table 34.1 Technical variables: statistical properties

Variable	Min	Max	Mean	Std dev
ATR(3)/ATR(15)	0.00	3.71	1.00	0.30
STOCHK(3)	0.00	100.00	54.52	36.63
STOCHK(15)	0.00	100.00	64.98	27.75
RSI(3)	0.12	100.00	58.07	25.00
RSI(15)	32.70	98.03	58.64	8.48

Table 34.2 Target variable: statistical properties

Variable	Min	Max	Mean	Std dev
Target	0.00	100.00	5.02	21.83

change in price over the next 5 days, computed for every element i in the input series as:

$$\left(\frac{\left(highest\ \left(close_{i+5\ldots j+1}\right) - close_i\right)}{closei}\right) \times 100 \qquad (34.2)$$

Effectively, this target allows the neural network to focus on the relationship between the input technical variables, and the expected forward price change.

The calculation of the return variable allows the ANN to focus on the highest amount of change that occurs in the next 5 days, which may or may not be the 5-day forward return. Perhaps a better description of the output variable is that it is measuring the maximum amount of price change that occurs within the next 5 days. No adjustment for risk is made, since traders focus on returns and use other means, such as stop orders, to control risk.

As explained in the empirical methodology, a number of hidden node architectures need to be created, and each one benchmarked against the in-sample data.

The method used to determine the hidden number of nodes is described in the empirical methodology. After the initial number of hidden nodes is determined, the first ANN is created and benchmarked. The number of hidden nodes is increased by one for each new architecture then created, until in-sample testing reveals which architecture has the most suitable in-sample metrics. A number of metrics are available for this purpose, in this study, the architectures are benchmarked using the Average Profit/Loss per Trade expressed as a percentage. This method assumes unlimited capital, takes every trade signaled, and includes transaction costs, and measures how much average profit is added by each trade over its lifetime. The empirical methodology uses the filter selectivity metric for longer-term systems, and Tharp's expectancy [8] for shorter term systems. This paper also introduces the idea of using overall system net profit to benchmark, as this figure takes into account both the number of trades (opportunity), and the expected return of each trade on average (reward).

34.4 Results

A total of 362 securities had trading data during the test period (the ASX200 including delisted stocks), from which 11,897 input rows were used for training. These were selected by sampling the available datasets, and selecting every 25th row as an input row.

Table 34.3 In-sample characteristics

Strategy (in-sample data)	Overall net system profit	Profit/loss per trade (%)	Holding period (days)
Buy-and-hold naïve approach	1,722,869.39	94.81	2,096.03
GMMA alone	632,441.43	1.09	35.30
ANN – 3 hidden nodes + GMMA	878,221.68	2.32	46.15
ANN – 4 hidden nodes + GMMA	1,117,520.33	3.69	59.20
ANN – 5 hidden nodes + GMMA	353,223.61	3.00	42.64

Table 34.3 reports the Overall Net System Profit, Average Profit/Loss per Trade (as a percentage), and Holding Period (days) for the buy-and-hold naïve approach (first row), the initial GMMA method (second row), and each of the in-sample ANN architectures created (subsequent rows). These figures include transaction costs of $20 each way and 0.5% slippage, and orders are implemented as day +1 market orders. There are no stops implemented in in-sample testing, as the objective is not to produce a trading system (yet), but to measure the quality of the ANN produced. Later, when an architecture has been selected, stops can be determined using ATR or Sweeney's [9] MAE technique.

The most important parameter to be chosen for in-sample testing is the signal threshold, that is, what level of forecast strength is enough to encourage the trader to open a position. This is a figure which needs to be chosen with respect to the individuals own risk appetite, and trading requirements. A low threshold will generate many signals, whilst a higher threshold will generate fewer. Setting the threshold too high will mean that trades will be signalled only rarely, too low and the trader's capital will be quickly invested, removing the opportunity to take higher forecast positions as and when they occur.

For this benchmarking, an in-sample threshold of 5 is used. This figure is chosen by visual inspection of the in-sample graph in Fig. 34.2, which shows a breakdown of the output values of a neural network architecture (scaled from 0 to 100) versus the average percentage returns for each network output value. The percentage returns are related to the number of days that the security is held, and these are shown as the lines on the graph. Put simply, this graph visualizes the returns expected from each output value of the network and shows how these returns per output value vary with respect to the holding period. At the forecast value of 5 (circled), the return expectation rises above zero, so this value is chosen.

As described in the empirical methodology, it is necessary to choose which ANN is the 'best', and this ANN will be taken forward to out-of-sample testing. It is for this reason that the trader must choose the in-sample benchmarking metrics with care. If the ANN is properly trained, then it should continue to exhibit similar qualities out-of-sample which it already displays in-sample.

From the above table, it is clear that ANN – four hidden nodes should be selected. It displays a number of desirable characteristics – it shows the highest level of Profit/Loss per Trade. Note that this will not necessarily make it the best ANN for

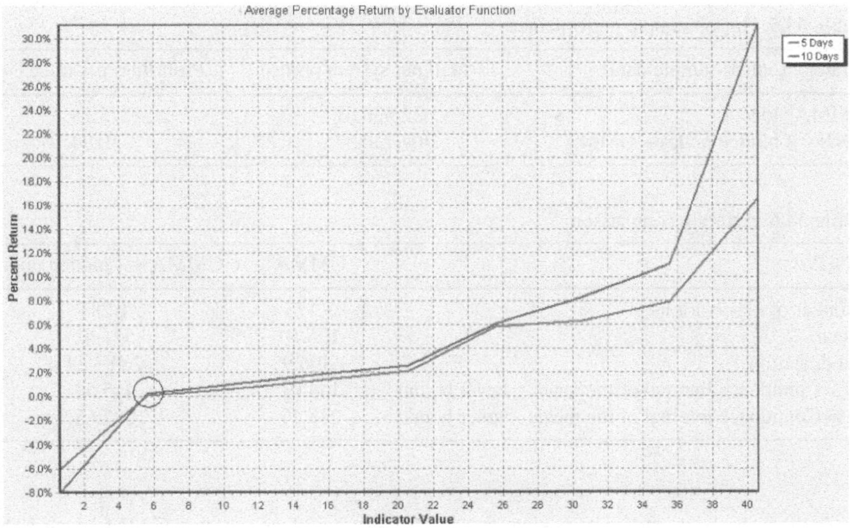

Fig. 34.2 In-sample ANN function profile

Table 34.4 Number of trades signalled

Strategy (in-sample data)	Number of trades signaled
Buy-and-hold naïve approach	362
GMMA alone	11,690
ANN – 3 hidden nodes + GMMA	7,516
ANN – 4 hidden nodes + GMMA	6,020
ANN – 5 hidden nodes + GMMA	2,312

a trading system. Extracting good profits in a short time period is only a desirable trait if there are enough opportunities being presented to ensure the traders capital is working efficiently.

Therefore, it is also important to review the number of opportunities signalled over the 10-year in-sample period. This information is shown in Table 34.4.

Here the trader must decide whether the number of trades signalled meets the required trading frequency. In this case, there are likely to be enough trades to keep an end-of-day trader fully invested.

This testing so far covered data already seen by the ANN, and is a valid indication of how the ANN should be expected to perform in the future. In effect, the in-sample metrics provide a framework of the trading model this ANN should produce.

Table 34.5 shows the effect of testing on the out-of-sample ASX200 data, which covers the period from the start of trading in 2004 to the end of trading in 2007. These figures also include transaction costs and slippage, and orders are implemented as next day market orders.

This was a particularly strong bull market period in the ASX200.

Table 34.5 Out-of-sample performance

Strategy (out-of-sample data)	Overall net system profit	Profit/loss per trade (%)
GMMA alone	622,630.01	3.88
ANN – 4 hidden nodes + GMMA	707,730.57	10.94

Table 34.6 ANOVA comparison

ANOVA	GMMA	GMMA + 4 hidden nodes
Number of observations	3,175	1,283
Mean	196.10	551.62
Std deviation	1,496.99	2,483.64
95% Confidence internal of the mean – lower bound	144.01	415.59
95% Confidence internal of the mean – upper bound	248.19	687.65

Although there appears a significant difference between the GMMA, and the ANN enhanced GMMA, it is important to quantify the differences statistically. The appropriate test to compare two distributions of this type is the ANOVA test (see supporting work in Vanstone [5]). The results for the ANOVA test are shown in Table 34.6 below.

The figures above equate to an F-statistic of 34.26 (specifically, $F(14,456) = 34.261$, $p = 0.00$ ($p < 0.05$)), which gives an extremely high level of significant difference between the two systems.

34.5 Conclusions

The ANN out-of-sample performance is suitably close to the ANN in-sample performance, leading to the conclusion that the ANN is not curve fit, that is, it should continue to perform well into the future. The level of significance reported by the ANOVA test leads to the conclusion that the ANN filter is making a statistically significant improvement to the quality of the initial GMMA signals.

The trader now needs to make a decision as to whether this ANN should be implemented in real-life.

One of the main reasons for starting with an existing successful trading strategy is that it makes this decision much easier. If the trader is already using the signals from a system, and the ANN is used to filter these signals, then the trader is still only taking trades that would have been taken by the original system. The only difference in using the ANN enhanced system is that trades with low expected profitability should be skipped.

Often in trading, it is the psychological and behavioural issues which undermine a traders success. By training ANNs to support existing systems, the trader can have additional confidence in the expected performance of the ANN.

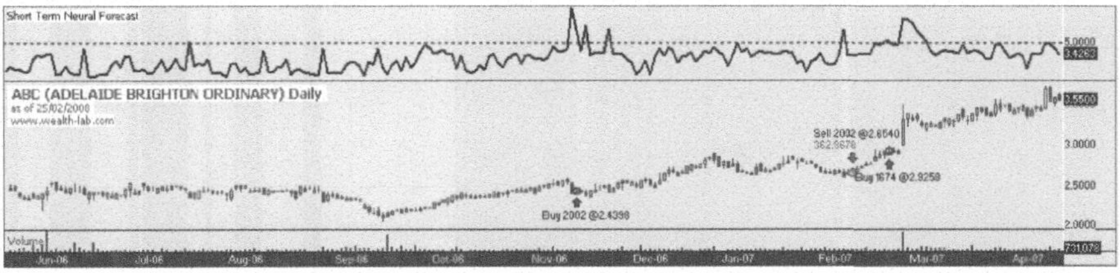

Fig. 34.3 GMMA signals filtered with an ANN

Finally, Fig. 34.3 shows the same security as Fig. 34.1. The ANN has clearly met its purpose of reducing whipsaws considerably, which has resulted in the significant performance improvement shown in Tables 34.3 and 34.5.

Of course the result will not always be that all whipsaws are removed. Rather, only whipsaws which are predictable using the ANN inputs will be removed.

References

1. Vanstone, B. and Finnie, G. (2007). "An Empirical Methodology for developing Stockmarket Trading Systems using Artificial Neural Networks." *Expert Systems with Applications*. In-Press (DOI: http://dx.doi.org/10.1016/j.eswa.2008.08.019).
2. Elder, A. (2006). *Entries & Exits: Visits to Sixteen Trading Rooms*. Hoboken, NJ: Wiley.
3. guppytraders.com. "Guppy Multiple Moving Average." Retrieved 04-05-2007, from www.guppytraders.com/gup329.shtml
4. Guppy, D. (2004). *Trend Trading*. Milton, QLD: Wrightbooks.
5. Vanstone, B. (2006). Trading in the Australian stockmarket using artificial neural networks, Bond University. Ph.D.
6. "Norgate Premium Data." (2004). Retrieved 01-01-2004, from www.premiumdata.net
7. "Wealth-Lab." (2005) from www.wealth-lab.com
8. Tharp, V. K. (1998). *Trade Your Way to Financial Freedom*. New York: McGraw-Hill.
9. Sweeney, J. (1996). *Maximum Adverse Excursion: Analyzing Price Fluctuations for Trading Management*. New York, Wiley.

Chapter 35
Reorganising Artificial Neural Network Topologies
Complexifying Neural Networks by Reorganisation

Thomas D. Jorgensen, Barry Haynes, and Charlotte Norlund

Abstract This chapter describes a novel way of complexifying artificial neural networks through topological reorganisation. The neural networks are reorganised to optimise their neural complexity, which is a measure of the information-theoretic complexity of the network. Complexification of neural networks here happens through rearranging connections, i.e. removing one or more connections and placing them elsewhere. The results verify that a structural reorganisation can help to increase the probability of discovering a neural network capable of adequately solving complex tasks. The networks and the methodology proposed are tested in a simulation of a mobile robot racing around a track.

Keywords Artificial Neural Network · topological reorganization · information-theoretic complexity · mobile robot racing · Complexification

35.1 Introduction

Artificial Neural Networks (ANNs) have been used in many different applications, with varying success. The success of a neural network, in a given application, depends on a series of different factors, such as topology, learning algorithm and learning epochs. Furthermore all of these factors can be dependent or independent of each other. Network topology is the focus of this research, in that finding the optimum network topology can be a difficult process. Ideally all network topologies should be able to learn every given task to competency, but in reality a given topology can be a bottleneck and constraint on a system. Selecting the wrong topology can result in a network that cannot learn the task at hand [1–3]. It is commonly known that a too small or too large network does not generalise well, i.e. learn a

T. D. Jorgensen (✉)
Department of Electronic and Computer Engineering,University of Portsmouth, UK
E-mail: Thomas.Jorgensen@port.ac.uk

S.I. Ao and L. Gelman (Eds.), *Advances in Electrical Engineering and Computational Science*, Lecture Notes in Electrical Engineering.

given task to an adequate level. This is due to either too few or too many parameters used to represent a proper and adequate mapping between inputs and outputs.

This chapter proposes a methodology that can help find an adequate network topology by reorganising existing networks by rearranging one or more connections, whilst trying to increase a measure of the neural complexity of the network. Assuming complex task solving requires complex neural controllers, a reorganisation that increases the controller complexity can increase the probability of finding an adequate network topology. Reorganising an existing network into a more complex one yields an increased chance of better performance and thus a higher fitness.

There are generally four ways to construct the topology of an ANN [3–5]. (1) Trial and Error is the simplest method. This essentially consists of choosing a topology at random and testing it, if the network performs in an acceptable way, the network topology is suitable. If the network does not perform satisfactory, select another topology and try it. (2) Expert selection; the network designer decides the topology based on a calculation or experience [3, 6]. (3) Evolving connections weights and topology through complexification. Extra connections and neurons can be added as the evolutionary process proceeds or existing networks can be reorganised [7–12]. (4) Simplifying and pruning overly large neural networks, by removing redundant elements [3–5]. This chapter seeks to add another one to this list, as reorganisation of existing networks can potentially help discover the appropriate topology of a network.

One advantage of the proposed methodology, compared to complexification through adding components, is that the computational overhead is constant because no extra components and parameters are added. The time it takes to compute the output of the network is effected, as well as the time it takes for the genetic algorithm to find appropriate values for the connection weights. More parameters yield a wider and slower search of the search space and additionally it yields more dimensions to the search space.

35.2 Background

Most research in complexification has so far focused on increasing the structural complexity, i.e. increasing the number of network components, of a neural network, this is done to mimic natural evolution [13]. Different routes and techniques have been proposed to continuously complexify neural network for a continuous increase in fitness, most prominently is the NEAT framework [10]. In the NEAT model, mechanisms are introduced to evolve network structure, either by adding neurons or connections, in parallel with the normal evolution of weights. The results of these experiments with complexification achieve, in some cases, faster learning as well as a neural network structure capable of solving more complex tasks than produced by normally evolved controllers. Other approaches do not cross breed networks of different topology, but use mutation as the evolutionary operator that evolves the network. Angeline [8] proposes networks that are gradually evolved by adding connections or neurons and new components are frozen, so that fitness is not reduced. This is similar to the first method of topological complexification proposed by Fahlman [7], which increased network size by adding neurons.

Research into the use of neural complexity in complexification to produce biologically plausible structures is limited, this is due to the lack of proper calculation tools and the variety of definitions and focus. Neural complexity is a measure of how a neural network is both connected and differentiated [11]. It is measure of the structural complexity as well as the connectivity of the network. The measure was developed to measure the neural complexity of human and animal brains by estimating the integration of functionally segregated modules. This measure reflects the properties that fully connected networks and functionally segregated networks have low complexity, whilst networks that are highly specialised and also well integrated are more functionally complex. Yaeger [12] has shown, that when optimising an artificial neural network with a fixed number of neurons for neural complexity, the fitness increases proportionally, suggesting a link between neural and functional complexity. The more complex a network, the greater the likelihood that it will be capable of solving complex tasks and surviving in complex environments [9–12].

35.3 Neuroscientific Foundations

Complexification in artificial neural networks can prove to be as important, as it is in the development of natural neural systems. It is important in artificial development to unleash fitness potential otherwise left untouched and constrained by a fixed neural topology. Complexification in neural networks is a vital process in the development of the brain in any natural system [14]. Complexification in human brains happens in several different ways, by growth, by pruning and by reorganisation. The first form of complexification happens from before birth and goes on up to adulthood, as the brain is formed. During this period neurons and interconnections grow and hence complexifies the network. The second form of complexification happens through continuous pruning. Connections between neurons have to be used for them not to fade away and eventually possibly disappear. This concept is called Neural Darwinism, as it is similar to normal evolution, where the fittest, in this case connections, survive [15]. The third form of complexification happens through reorganisation. In some cases, for yet unknown reasons, connections detach themselves from a neuron and reconnects to another. Mostly, reorganisations in natural systems have a detrimental effect, but some might have unexpected positive effects. The effects of reorganisation in artificial systems are investigated in this chapter.

35.4 Reorganising Neural Networks

Artificial neural network can be reorganised in several different ways and to different extents. The methodology proposed herein operates with two degrees of reorganisation. Reorganising one connection is defined as a minor reorganisation, whereas reorganising more connections is defined as a major reorganisation. Networks in this chapter are only reorganised once, where the objective is to increase the neural complexity of the network.

35.4.1 Neural Complexity

The neural complexity measure is an information-theoretic measure of the complexity of the neural network and not a measure of the magnitude of weights or of the number of elements in the network [11]. The neural complexity measure uses the correlation between neuron output signals to quantify the integration and the specialisation of neural groups in a neural network. Complex systems are characterised highly specialised clusters, which are highly integrated with each other. Systems that have very highly independent functional components or have very highly integrated clusters will have a low complexity. X is a neural system with n neurons, represented by a connection matrix. The standard information entropy H(X) is used to calculate the integration between components [16]. The integration between individual neurons can be expressed by:

$$I(X) = \sum_{i=1}^{n} H(x_i) - H(X) \tag{35.1}$$

The integration I(X) of segregated neural elements equals the difference between the sum of entropies of all of the individual components x_i of the neural network and the entropy of the network as a whole. In order to be able to give an estimate of the neural complexity of a neural network, not only the integration between the individual elements is needed, but also the integration of any neural clusters in the network. It is very likely that neurons in an artificial neural network cluster together form some sort of functional cluster. The average integration between functionally segregated neural groups with k (out of n) elements is expressed with <I(X)> j is an index indicating that all possible combinations of subsets with k components are used. The average integration for all subsets with k components is used to calculate the neural complexity:

$$C_N(X) = \sum_{k=1}^{n} [(k/n) \cdot I(X) - <I(X_j^k)>] \tag{35.2}$$

The neural complexity C_N of a neural system X is the sum of differences between the values of the average integration <I(X)> expected from a linear increase for increasing subset size k and the actual discrete values observed. This neural complexity measure yields an estimate of the information-theoretic complexity of a neural network by measuring the integration between individual components and possible combinations of subsets.

35.4.2 Using the Complexity Measure

The neural complexity measure is used to optimise the complexity of the neural network. A reorganisation only takes place if this complexity increases. The reorganisation methodology proposed is summarised by the following algorithm:

1. Determine a starting topology of sufficient size and complexity. This network can be chosen randomly or based on the experience of the designer.
2. The starting network is trained to proficiency given some predefined measure.
3. The network is now reorganised. The number of connections to be reorganised decides the degree of reorganisation. A connection is chosen at random, removed and reinserted elsewhere in the network.
4. If this reorganisation increases the neural complexity of the network, the reorganisation is deemed valid and the network is retrained. If the reorganisation does not increase the neural complexity the reorganisation has been unsuccessful and it is undone. Another reorganisation can be attempted or the process stopped. In the experiments conducted here, five reorganisation attempts are made before the process stops.
5. If it is desired and previous reorganisations have been successful further reorganisations can take place.

Ideally it would be preferable to remove and reinsert the connection in the place that yields the largest possible increase in the complexity out of all possible reorganisations. This requires knowledge of all possible topologies given this number of connections and neurons, which is not computationally efficacious. Only one connection is reorganised at any time, this could be increased to several connections if desired.

35.4.3 The Simulated Track and Robot

The controllers evolved here are tested in a simulated environment with a robot. In this environment a robot has to drive around a track, which consists of 32 sections. The objective of this task is to complete 3 laps in the shortest amount of time. If a robot fails to complete three laps, the distance covered is the measure of its performance. The robot has to drive around the track covering all of the sections of the track, it is not allowed to skip any sections. In total the robot has to complete three laps, with 32 sections in each lap, all visited in the correct order. If the robot is too slow at driving between two sections the simulation is terminated. The following Fig. 35.1, illustrates the task to be completed and the robot and its perception of the track:

Figure 35.1-left illustrates the track and the robot driving around it. The robot is not limited in its movement, i.e. it can drive off the track, reverse around the track or adapt to any driving patterns desired, as long at its drives over the right counter clockwise sequence of sections. Figure 35.1-right illustrates the robot driving on the

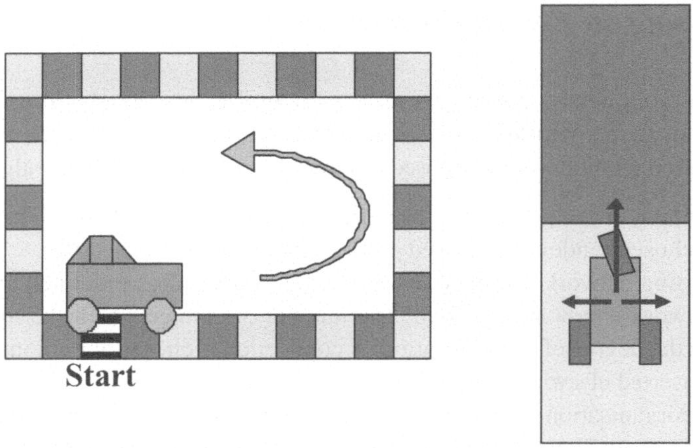

Fig. 35.1 Left is the track and the robot driving around it; right is a top-view of the robot and its sensory perception of the track

track seen from above. The track sections have alternating colours to mark a clear distinction between sections. The arrows illustrate the three sensors of the robot. The front sensor measures the distance to the next turn and the two side sensor measures the distance to the edge of the track. As illustrated, the simulated robot has three wheels and not four, to increase the difficulty of evolving a successful controller. The risk when driving this three wheeled robot, in contrast to a four wheeled vehicle, is that it will roll over if driven to abruptly. A robot that has rolled over is unlikely to be able to continue. The front wheel controls the speed as well as the direction.

35.5 Experiments and Results

A total of three sets of experiments have been conducted. One set of experiments, with a randomly selected neural network, acts as a benchmark for further comparisons. This network only has its connection weights evolved, whereas the topology is fixed. The second set of experiments starts with the benchmark network selected in the first experiment, which is then reorganised. This reorganisation is only minor, in that only one connection is removed and replaced in each network. The new network resulting from the reorganisation is tested. The third and final set of experiments also uses the network from experiment one as a starting condition. The network is then reorganised more extensively, by reshuffling several connections to create a new network based on the old network. The results from these experiments are used to compare the different strategies of evolution.

35.5.1 The Simulation Environment

The evolved neural network controllers are tested in a physics simulator to mimic a real world robot subject to real world forces. The genetic algorithm has in all tests a population size of 50 and the number of tests per method is 25. Uniformly distributed noise has been added on the input and output values to simulate sensor drift, actuator response, wheel skid and other real world error parameters. To give the simulations similar attributes and effects as on real racing track, the track has been given edges. Whenever the robot drives off the track it falls off this edge onto another slower surface. This means, that if the robot cuts corners, it could potentially have wheels lifting off the ground thus affecting stability and speed of the robot, due to the edge coming back onto the track.

35.5.2 The Fitness Function

Fitness is rewarded according to normal motorsport rules and practice. Three laps of the track have to be completed and the controller that finishes in the fastest time wins the race, i.e. it is the fittest controller. If a controller fails to finish three laps, the controller with the most laps or longest distance travelled wins. In the case that two controllers have reached the same distance the racing time determines the fittest controller. The following fitness function states that the longest distance covered in the shortest amount of time yields the best fitness. Time is the time it takes to complete the track. If a controller fails to finish this Time is set to 480 s, which is the absolute longest time a controller is allowed to be in existence, before a simulation is stopped. In the likely event that two controllers have covered the same distance, the controller with the fastest time will be favoured for further evolving. The precise version of the fitness function is:

$$\text{Fitness} = \frac{\text{Sections} = (\text{Laps} * \text{Track Length})}{\text{Time}} \tag{35.3}$$

The fitness is equal to the distance divided by the time. The distance is equal to the number of track sections covered in the current lap, plus the number of sections covered in previous laps. Track length is the total number of sections, which is 32. The minimum fitness obtainable is $1/480 \approx 0.002$.

35.5.3 The Benchmark and the Reorganised Networks

The first set of experiments was conducted with a fixed structure network, where the connection weights were evolved, is used as the benchmark for the other experiments. The network used is a feed-forward connected network with three input

Table 35.1 Results form experiments

Method	Minimum fitness	Average fitness	Maximum fitness	Standard deviation	Complexity
Benchmark network	0.857	0.964	1.042	0.062	14.71
Minor reorganisation	0.866	0.998	1.138	0.112	15.03
Major reorganisation	1.002	1.058	1.120	0.047	15.40

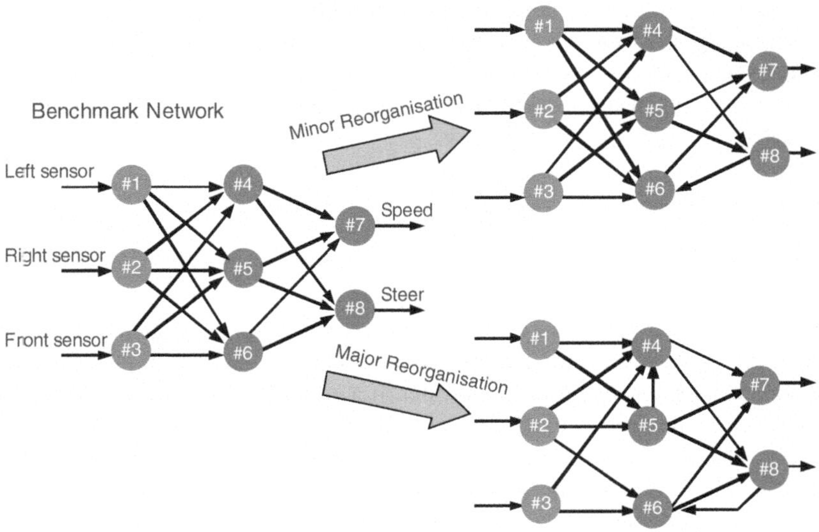

Fig. 35.2 The benchmark neural network and the reorganised networks

neurons, three hidden layer neurons and two output neurons. The inputs are the sensor values as described previously and the outputs are the direction and speed of the front wheel. This network was trained to competency and the results are shown in Table 35.1. The neural complexity of this network is 14.71, calculated with Eq. (35.2). The network is shown to left in Fig. 35.2.

The second set of experiments was conducted with a network that has been reorganised. The benchmark network has undergone a minor reorganisation, which is shown in Fig. 35.2 upper right. The connection between neuron 6 and neuron 8 has been rearranged and is now a recursive connection. The new network has increased its neural complexity by the reorganisation to 15.03. Immediately after the reorganisation the network loses some of its fitness, this fitness is regained by retraining. The reorganised network was retrained with the same weights as before the reorganisation and in all cases the network, as a minimum, regained all of its previous fitness and behaviour. Additionally, all of the connection weights were re-evolved in another experiment to see if the results and tendencies were the same, and as expected the results were the same.

The final set of experiments conducted used a network, which has had a major reorganisation. The benchmark network was changed by removing a connection between neuron 3 and 5 and between neuron 1 and 6. These connections are moved to between neuron 5 and 4 and between neuron 8 and 6. As the benchmark network is feed-forward connected only recursive connections are possible for this particular network. The new network is shown in Fig. 35.2 lower right. The neural complexity of the new network has risen to 15.40, which is a 5% increase. Similar to the previously reorganised network, this network was, after a reorganisation, subject to a fitness loss, but it was retrained to competency. The controller increased its fitness over the original network. Even re-evolving all of the connection weights yields a better overall performance.

35.5.4 Evaluation and Comparison of the Proposed Methods

The results from all of the experiments show that all networks learn the task proficiently, however some networks seem to perform better than others. Figure 35.3 shows the route that controllers choose to drive around the track. The route reflects the results which are summarised in Table 35.1. The race car starts in (0, 0) and drives to (20, 0) where it turns. Hereafter it continues to (20, 11) where it turns and continues to (−2.5, 11) and from here it continues to (−2.5, 0) and on to (0, 0). The controller tries to align the car on the straight line between the points. Figure 35.3 illustrates the difference between an average lap of the benchmark networks and of the reorganised networks, and it clearly illustrates the route that the robot takes around the track.

Figure 35.3 illustrates the degree of overshoot when turning and recovering to drive straight ahead on another leg of the track. Figure 35.3 clearly shows that the controllers that have been reorganised overshot less than the benchmark networks. Less overshot, ultimately means that the racing car is able to move faster, which means it has a better fitness. Table 35.1 shows the fitness from the benchmark network experiments, and the fitness regained by the new networks after a reorganisation and retraining. To put these results into context, the best human driver on the same race track has a record of 1.036, which is worse than the average major reorganised controller.

The hypothesis that artificial neural networks that have undergone a minor reorganisation, where the neural complexity is optimised, are statistically better than the fixed structure network it originates from does not hold true for these experiments. A t-test, with a 5% significance level, indicates that there is no statistical difference between the two methods, despite the higher minimum, average and maximum values. The second hypothesis tested in this chapter, states that artificial neural networks that have undergone a major reorganisation, where the neural complexity is optimised, are better than the networks they originate from, holds true. A t-test, with a 5% significance level, indicates that there is a statistical difference between the two methods. This can be due to the increased neural complexity of the new

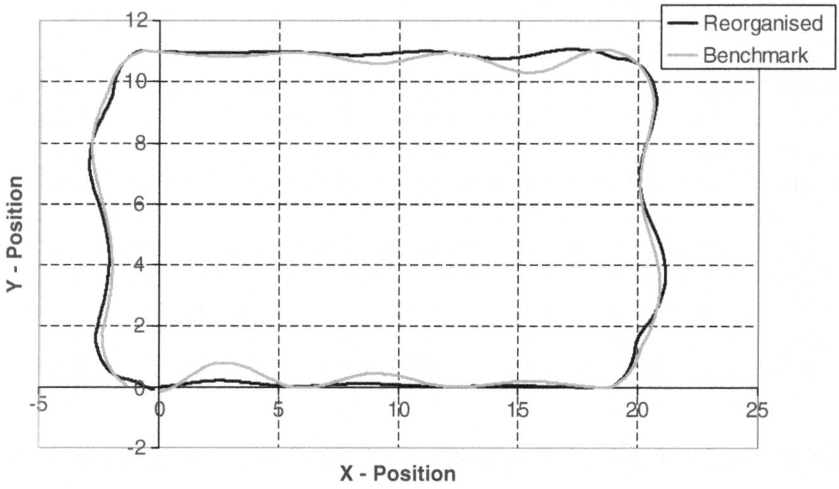

Fig. 35.3 The route of an average benchmark network and a reorganised network

network created by the reorganisation. Some of this increased performance can possibly be accredited to the fact that one of the networks has a recursive connection, which is a common way to increase the neural complexity and performance of a network, but the experiments indicate that this is only part of the explanation. The experiment clearly indicates that increased neural complexity yield a higher probability of finding suitable and well performing neural networks, which is in line with other research in the field [9].

The results from the experiments don't indicate any significant difference in the speed of learning produced by either methodology. This means that it takes about the same number of iterations to learn a given task for any network topology used in the experiments, this was expected as they all have the number of parameters.

35.6 Conclusion

This chapter has presented a new methodology for complexifying artificial neural networks through structural reorganisation. Connections were removed and reinserted whilst trying to increase the neural complexity of the network. The evolved neural networks learned to control the vehicle around the track and the results indicate the viability of the newly reorganised networks. The results also indicate that it might be necessary to rearrange more than one connection in order to achieve significantly better results. This chapter indicates that neural complexity in conjunction with reorganisation can help unleash potential and increase the probability of finding neural network controllers of sufficient complexity to adequately solve complex tasks. Furthermore, the results indicate that a reorganisation can substitute structural elaboration as a method for improving network potential, whilst keeping the computational overhead constant. These results are in line with previous research done in the field and they reconfirm the importance of high neural complexity and structural change.

References

1. A.S. Weigend, D.E. Rumelhart, and B.A. Huberman, Back-propagation, weight-elimination and time series prediction, *Proceedings of the 1990 Summer School on Connectionist Models* (1990).
2. J. Denker, D. Schwartz, B. Wittner, S. Solla, R. Howard, L. Jackel, and J. Hopfield, Large Automatic Learning, Rule Extraction and Generalization, *Complex Systems* **1**(5): 877–922 (1987).
3. S. Nolfi, and D. Floreano, *Evolutionary Robotics; The Biology, Intelligence, and Technology of Self-Organizing Machines* (Cambridge, MA: MIT Press 2000).
4. X. Yao, and Y. Liu, A New Evolutionary System for Evolving Artificial Neural Networks, *IEEE Transactions on Neural Networks* **8**(3): 694–713 (1997).
5. X. Yao, Evolving Artificial Neural Networks, *Proceedings of the IEEE* **87**(9) (1999).
6. R. Jacobs, and M. Jordan, Adaptive Mixtures of Local Experts, *Neural Computation* **3**: 79–87 (1991).
7. S.E. Fahlman, and C. Lebiere, The Cascade-Correlation Learning Architecture, *Advances in Neural Information Processing Systems* **2**: 524–532 (1990).
8. P. Angeline, and J. Pollack, Evolutionary Module Acquisition, *Proceedings of the Second Annual Conference on Evolutionary Programming* (1993).
9. O. Sporns, and M. Lungarella, Evolving Coordinated Behaviours by Maximizing Informational Structure, *Proceedings of the Tenth International Conference on Artificial Life* (2006).
10. K.O. Stanley, and R. Miikkulainen, Continual Coevolution Through Complexification, *Proceedings of the Genetic and Evolutionary Conference* (2002).
11. G. Tononi, O. Sporns, and G.M. Edelman, A Measure for Brain Complexity: Relating Functional Segregation and Integration in the Nervous System, *Proceedings of the National Academy of Science of USA* (1994).
12. L.S. Yaeger, and O. Sporns, Evolution of Neural Structure and Complexity in a Computational Ecology, *Proceedings of the Tenth International Conference on Artificial Life* (2006).
13. R. Dawkins, *Climbing Mount Improbable* (Reprint by Penguin Books, London, England, 2006).
14. G.N. Martin, *Human Neuropsychology* (London: Prentice Hall, 1998, Reprinted 1999).
15. G. Edelman, *Neural Darwinism – The Theory of Neuronal Group Selection* (New York: Basic Books, 1989, Print by Oxford Press 1990).
16. C.E. Shannon, A Mathematical Theory of Communication, *The Bell System Technical Journal* **27**: 379–423/623–656 (1948).

Chapter 36

Design and Implementation of an E-Learning Model by Considering Learner's Personality and Emotions

S. Fatahi, M. Kazemifard, and N. Ghasem-Aghaee

Abstract Emotion, personality and individual differences are those effective parameters on human's activities such as learning. People with different personalities show different emotions in facing an event. In the case of teaching and learning, personality difference between learners plays an important role. In virtual learning projects this point should consider that the learners' personalities are various and the teaching method used for each learner should be different from the other learners. In this chapter, a new model presented according to the learning model based on emotion and personality and the model of virtual classmate. Based on their knowledge base, the virtual teacher and classmate express suitable behaviors to improve the process of learning according to the learner's emotional status.

Keywords E-Learning · Learners Personality · Emotion · Implementation · virtual classmate

36.1 Introduction

E-learning is usually defined as a type of learning supported by information and communication technology (ICT) that improves quality of teaching and learning. E-learning system is a powerful tool for achieving strategic objectives of the university (teaching, research and serving the society) [1]. E-Learning like all other tools offers advantages such as: access to differentiated online resources, Self-directed learning, and Learning matches learners' lifestyles, etc. Despite of all the advantages this kind of learning lacks the necessary attractiveness most of the time. It seems that regarding the human characteristics and inserting them in virtual learning environments, it would be possible to show these environments more real.

S. Fatahi (✉)
Department of Computer Engineering, University of Isfahan, Isfahan, Iran
E-mail: fatahi_somayeh@yahoo.com

S.I. Ao and L. Gelman (Eds.), *Advances in Electrical Engineering and Computational Science*, Lecture Notes in Electrical Engineering.

Emotion, personality and individual differences are those effective parameters on human's activities such as learning. People with different personalities show different emotions in facing an event. Difference in the characteristics of the individuals is reflected in their daily activities and their works. In the case of teaching and learning, personality difference between learners plays an important role. The learner's personality will be effective in his learning style [2]. In virtual learning projects this point should consider that the learners' personalities are various and the teaching method used for each learner should be different from the other learners.

36.2 Previous Works

In virtual learning systems created up to now, the learner's emotions received much more attention and the emotional agents were more employed. In a few of these systems personality drew our attention as an independent parameter that some of them are mentioned here:

In ERPA architecture by using ID3 algorithm, the learner's emotional reaction towards an event is predicted (for example, taking an exam score) [3]. Chaffar and his colleagues used the Naïve Bayes Classifier method to predict the learner's emotions [4]. In ESTEL architecture, the Naïve Bayes Classifier method is used to predict the optimized emotional status. In this architecture, in addition to emotion, the learner's personality is also considered. In this system, a module tries to create and induce an optimized emotional state. For instance, when the learner enters the system, after the identification of learner's personality, for example extrovert, and recognition of optimal emotional state, such as happiness, an emotion is induced to that learner by showing various interfaces (e.g. music, picture, and etc.) to him [5]. In Passenger software designed by German researchers, cooperative learning methods are used. This software examines a series of emotions for the virtual teacher that is present in the system based on OCC model, and invites the learners to group work. The virtual teacher traces the learners' activities and helps the learners who are not able to do cooperative activities [6]. Abrahamian and his colleagues designed an interface for computer learners appropriate for the type of their personality using MBTI test and concluded that learning through this interface as a result of using personality characters leads into developments in learning process [7]. In implementation performed by Maldonado and his colleagues, a virtual classmate agent is used. This agent is placed beside the learners, and mostly plays the role of a co-learner and a support. In this project each of the teacher, learner, and classmate has own emotions and the learner's emotions affected his/her classmate [8].

36.3 Psychological Principles

Emotion, personality and individual differences are those effective parameters on human's activities. Everybody needs special learning style according to his/her personality characteristics. Some tools are used to evaluate the different learning style to determine the learner's learning style. MBTI is the well-known questionnaire used for personality and learning style determination [9, 10]. According to MBTI classification, every individual has a set of instinctive preferences that determine how he or she behaves in different situations [9]. This questionnaire helps to identify personality characteristics and learning preferences of the individuals and to elicit the suitable learning styles from these characteristics [11].

MBTI uses four two-dimensional functions according to the Jung's theory. According to the theory that Jung proposed the four functions of mind are thought, emotion, comprehension, and exploration. These functions are the main ways for understanding, and they explain the truth. These functions are related to each other and they simulate one another. Nevertheless, one of the functions is often dominant and that dominant function inclines the person to a state. The Jung theory distinguished three dimensions of Extroversion/Introversion (E/I), Sensing/Intuition (S/N) and Thinking/Feeling (T/F), but in MBTI another dimension of, Judging/ Perceiving (J/P) also was added [7, 11]. Irrational mental functions, Sensing (S) or Intuition (N), relate to how an individual perceives information, while rational mental functions, Thinking (T) or Feeling (F), provide insight into how one makes judgments or decisions based upon their perceptions. Mental functions of Extrovert/Introvert and Judgment/Perception are related to how individuals interact with the external environment and the around world. Sixteen personality types are resulted from mixing these 4 two-dimensional functions that each learner would be placed in one group [11, 12].

36.4 Proposed Model

In this paper, a new model presented according to the learning model based on emotion and personality [13] and the model of virtual classmate [14] in previous our studies. The outline of the improved model is shown in Fig. 36.1. The model contains six major modules:

Personality identification module: In first step, learner comes across MBTI questionnaire and his personality will be identified (for example ISFJ, ESTP, INTJ, etc.).

Module of choosing a learning style commensurate with learner's personality: Generally, there are three kinds of learning environment: individual, competitive and collaborative [15]. System based on the identified personality of learner, put him in one of three groups of independence, contribution with virtual classmate or competition with virtual classmate [16].

Module of choosing virtual classmate agent: If the learner is put in the independence group the process of learning and education will be started, otherwise the system at first chooses a virtual classmate that matches the type of learner's personality, then the process of learning and education will get started. This module will explain at the next section with more details.

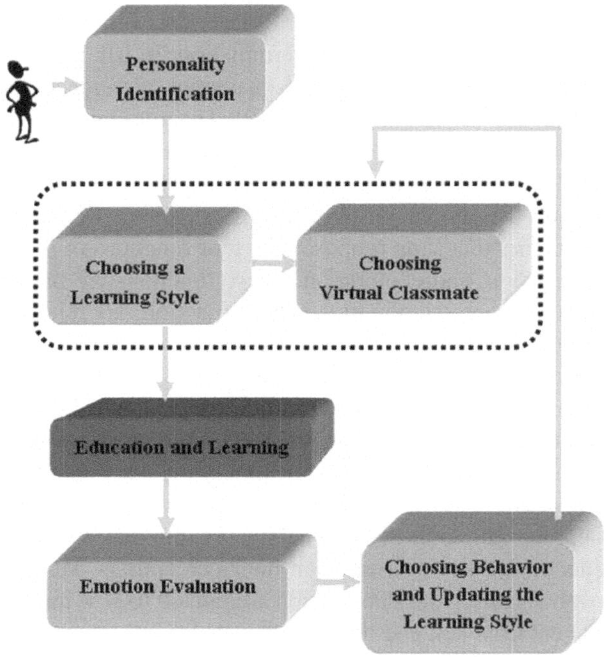

Fig. 36.1 Proposed model

Education module: In this module, lesson's points are presented to learner as exercises.

Module of emotion evaluation: When doing exercises and evaluating the extend of learning, there are some emotion expressed in learner which are relevant to level of learner's learning and the events happen in the environment (as have liking for virtual classmate, be disappointed in doing exercises, etc.). According to the performed studies, we found out only special emotions are effective in the process of learning [17, 18]. Accordingly, the first and the third branch of emotions in OCC model are used. The first branch of emotions in OCC model are the effective emotions in process of learning and the third branch are those emotions that a person shows them when facing the others (for example virtual classmate agent).

Module of choosing behavior and updating the learning style: The module changes the style of education according to the events happen in the environment that cause changes to the learner's emotion and also the learner's personality characteristics.

Based on their knowledge base, the virtual teacher and classmate express suitable behaviors to improve the process of learning according to the learner's emotional status.

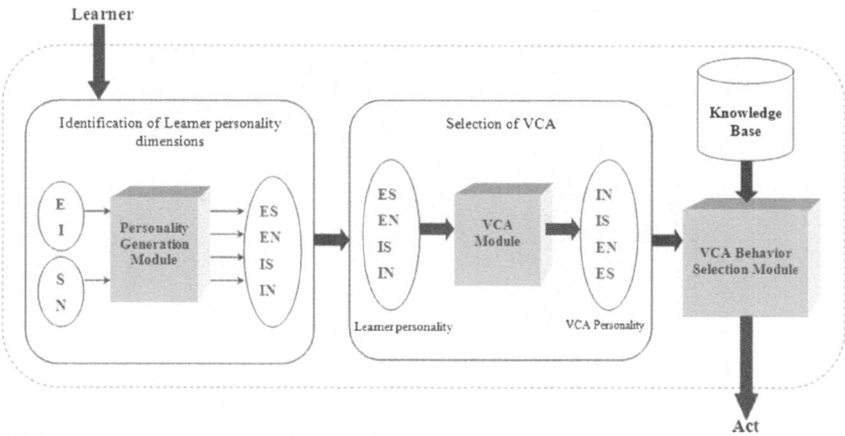

Fig. 36.2 Module of choosing virtual classmate agent

36.5 Virtual Classmate Model

In this section we explained module of choosing virtual classmate agent with more details. This module is displayed in Fig. 36.2. The module includes three main parts, each of them described below:

The part of personality generation: In this part, using the MBTI questionnaire, the personality of the learner is recognized. In this paper we only considered two dimensions of E/I and S/N which are important in learning process [7]. Considering two dimensions, four types of personality that are EI, EN, IS, and IN would be resulted.
The part of classmate selection: In this part, a VCA appropriate for the learner's personality is selected. Selected VCA is completely opposite in his MBTI dominant with learner. Based on research, the opposite personality displays a higher performance than the similar personality [19–21]. The personality selection for the VCA is so that it would result in improvements in learning process.
The part of classmate behavior selection: During the learning process, regarding the events that happen in the environment and the learning situation of the individual, the VCA exhibits appropriate behaviors. Tactics knowledgebase is used to interact with the learner.

For two personality dimensions considered in this paper four parameters are elicited. Independence and replying speed parameters for the E/I dimension and detail-oriented attitude and attention for the S/N dimension, now based on the extent of these parameters in each type of personality, the VCA exhibits a certain behavior. These behaviors are shown separately for each dimension in Table 36.1 and Table 36.2. Mixing the two dimensions' functions, four personality types would be resulted, that are IN, IS, EN, and ES. A sample of these tactics is presented in Table 36.3.

Table 36.1 The VCA behavior with E and I personality dimensions

Learner's personality	VCA's	Characteristics	Independence parameter	Characteristic	Replying speed parameter	Events and VCA tactics (solving problem)
I	E	The introvert person mostly acts independently and is inclined to do the exercises alone. The extrovert person is interested in group work and rarely acts lonely	The E VCA decreases the independence of I	The introvert person takes a lot of time solving the problem, the extrovert person mostly acts without thinking and replies	The E VCA tries to increase the speed of I	1. The E VCA tries to cooperate with I 2. The E VCA tries to activate I by announcing the remaining time so that I answer the questions sooner
E	I		The I VCA increases the independence of E		The I VCA tries to decrease the speed of E	1. The I VCA do not cooperate with E so that he act independently 2. The E VCA tries to make I more relaxed so that he thinks more on problems

Table 36.2 The VCA behavior with N and S personality dimensions

Learner's personality	VCA's	Characteristics	detail-oriented attitude parameter	Characteristic	Attention parameter	Events and VCA tactics (solving problem)
S	N	The S person mostly pays attention to details, while N pays attention to the relation in the problems, and the result is important to him	The N VCA tries to explain the relation between the problems for S	The N person never pays attention to your words and predicts the words you want to say. The S person if there was any need to ask for help asks N and do not react negatively to his mistaken views	The N VCA pays attention to S and helps him if necessary	1. The N VCA tries to shed some light on the relation between problems for S, clarify the problem to S and explains the general points to the learner. 2. The N VCA helps and cooperates with the learner if he asks for that
N	S	.	The S VCA tries to decrease the amount of attention N pays to the details		The S VCA tries to attract the N's attention to himself	1. The S VCA reminds the details to N 2. The S VCA tries to convey his views to N

Table 36.3 VCA tactics

Learner's personality	VCA's personality	Events and VCA tactics (solving problem)
IS	EN	1. The E VCA tries to cooperate with I 2. The E VCA tries to activate I by announcing the remaining time so that I answer the questions sooner 3. The N VCA tries to shed some light on the relation between problems and explains the general points of the exercise for S 4. If the learner asks for help, the N VCA helps and cooperates with him

36.6 Simulation of Proposed Model

In our educational environment two agents – VTA and VCA – are used. These two agents with identifying the learner's emotions after each event, choose suitable tactics when are face to face with the learner. In this environment the way of choosing the tactics to face the learner is related to the learner's type of group. Depending on the point that which groups the learner belongs to, special tactic is chosen.

Knowledge base of this system contains 65 rules (Table 36.4): 16 rules are to identify the learner's group, 10 rules for independent learning group, 20 rules for collaborative learning group and 19 rules for competitive group. Four examples of these rules are in the following:

The first rule is an example of learner's classifying into learning groups. According to the first rule, system groups the learner with ISFJ personality that is "Introvert", "Sensing", "Feeling" and "Judging" groups in independent group. The second rule is an example of the rules of teacher's dealing in a situation that the learner is in independent group. The third and fourth rules are examples of situations that the learner is in collaborative and competitive groups respectively. As the rules shows in these two situations relevant to the learner's emotions the virtual teacher and classmate use special tactics in interaction with learner.

36.7 Implementation

For implementing the e-learning environment, a series of English language exercises were used. In this environment it was assumed that the learner has already learnt the subject and refers to this software for solving the exercises. The exercises are categorized in five levels by difficulties.

The learner begins to solve the exercises together with the VCA, and regarding the events that happen in the environment interacts with the VCA. An image of this environment is presented in Fig. 36.3. Visual C#.Net and SQL Server database were

Table 36.4 Rules

Rule 1:				
If student S1 has personality ISFJ				
Then his/her group is independent				
Rule 2:				
IF		Student Group	IS	Independent
	AND	Satisfaction	IS	High
	OR	Satisfaction	IS	Medium
	AND	Event	IS	Correct answer
THEN		Teacher_tactic1	IS	Congratulate student
	AND	Teacher_tactic2	IS	Change student group to competitive
Rule 3:				
IF		Student Group	IS	Collaborative
	AND	Like	IS	High
	OR	Like	IS	Medium
	AND	Disappointment	IS	High
	OR	Disappointment	IS	Medium
	AND	Event	IS	Wrong answer
THEN		Classmate_tactic1	IS	Increase student self ability
	AND	Classmate_tactic2	IS	Increase student effort
	AND	Classmate_tactic3	IS	Persuade student to think more for problem
Rule 4:				
IF		Student group	IS	Competitive
	AND	Like	IS	High
	OR	Like	IS	Medium
	AND	Fear	IS	High
	OR	Fear	IS	Medium
	AND	Virtual classmate's personality	IS	EN
	OR	Virtual classmate's personality	IS	ES
	AND	Event	IS	While student is accomplishing to a task
	AND	Student's response speed	IS	Lower than threshold
THEN		Classmate tactic1	IS	Increase student effort
	AND	Classmate tactic2	IS	Notify student for deadline
	AND	Teacher tactic1	IS	Increase student self ability
	AND	Teacher tactic2	IS	Change student group to collaborative
	AND	Classmate tactic2	IS	Notify student for deadline

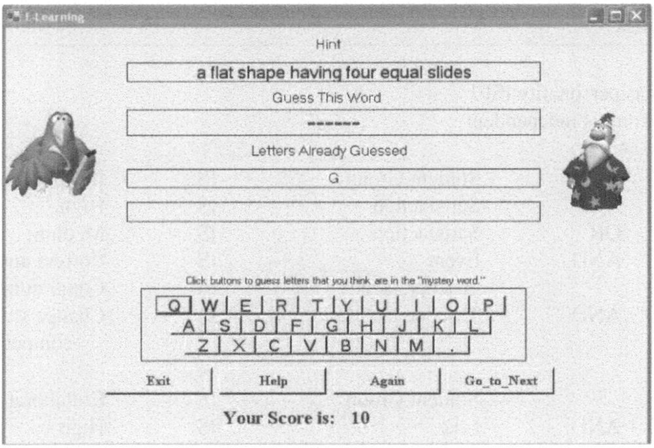

Fig. 36.3 Educational environment with classmate and teacher agent

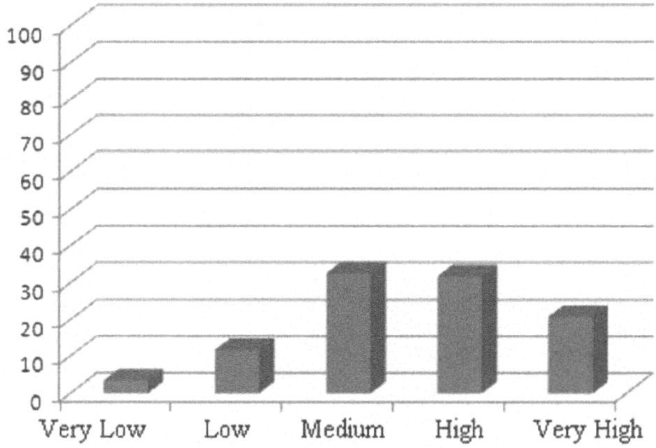

Fig. 36.4 Learner satisfaction of learning environment

used for implementing. Also, to show the agents – VTA and VCA – we used two present agents in Microsoft that are Merlin and Peedy respectively.

36.8 Results

We tested our model in real environment with 30 students. Students work with our system, then they answered ten questions for evaluating our learning environment. We got rate of learner satisfaction Based on four question of our questioner (Fig. 36.4).

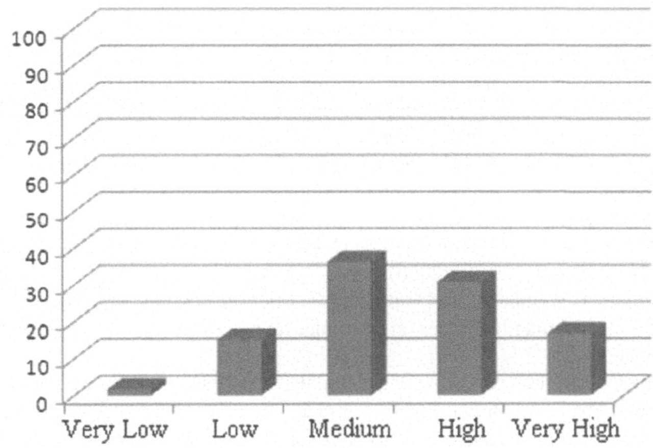

Fig. 36.5 Presence effect of VCA in learning environment

Results show that learners are satisfied of learning environment based on learner's emotion and personality.

We got rate of presence effect of VCA Based on six question of our questioner (Fig. 36.5).

The results show that the presence of the VCA leads advancements in the learning process and attractiveness of e-learning environment.

36.9 Conclusion and Future Works

In this paper a model for using in e-learning was presented. In this model some modules for personality recognition and selecting an appropriate VCA for the learner's personality, were considered to develop the interaction with the learner. The Behavior of VCA saved in knowledgebase of system. The results show that placing the learner beside an appropriate VCA, lead to improvement in learning and makes the virtual learning environment more enjoyable.

In the future we will try to improve the system with considering the parameters of culture, case based reasoning and agent's learning and also makes the virtual teacher and classmate agents more credible for the user.

References

1. N. Begičević, B. Divjak, and T. Hunjak, Decision Making Model for Strategic Planning of e-Learning Implementation (17th International Conference on Information and Intelligent Systems IIS, Varaždin, Croatia, 2006).
2. J. Du, Q. Zheng, H. Li, and W. Yuan, The Research of Mining Association Rules Between Personality and Behavior of Learner Under Web-Based Learning Environment (ICWL, 2005), pp. 406–417.

3. P. Chalfoun, S. Chaffar, and C. Frasson, Predicting the Emotional Reaction of the Learner with a Machine Learning Technique (Workshop on Motivaional and Affective Issues in ITS, International Conference on Intelligent Tutoring Systems, Jhongli, Taiwan, 2006).
4. S. Chaffar, G. Cepeda, and C. Frasson, Predicting the Learner's Emotional Reaction Towards the Tutor's Intervention (7th IEEE International Conference, Japan, 2007), pp. 639–641.
5. S. Chaffar and C. Frasson, Inducing Optimal Emotional State for Learning in Intelligent Tutoring Systems (lecture notes in computer science, 2004), pp. 45–54.
6. B.F. Marin, A. Hunger, and S. Werner, Corroborating Emotion Theory with Role Theory and Agent Technology: A Framework for Designing Emotional Agents as Tutoring Entities, Journal of Networks (1), 29–40 (2006).
7. E. Abrahamian, J. Weinberg, M. Grady, and C. Michael Stanton, The Effect of Personality-Aware Computer-Human Interfaces on Learning, Journal of Universal Computer Science (10), 27–37 (2004).
8. H. Maldonado, J.R. Lee, S. Brave, C. Nass, H. Nakajima, R. Yamada, K. Iwamura, and Y. Morishima, We Learn Better Together: Enhancing e-Learning with Emotional Characters (In Proceedings, Computer Supported Collaborative Learning, Taipei, Taiwan, 2005).
9. D.J. Pittenger, Measuring the MBTI ... and Coming Up short, Journal of Career Planning and Employment (54), 48–53 (1993).
10. M.D. Shermis, and D. Lombard, Effects of computer-based Test Administration on Test Anxiety and Performance, Journal of Computers in Human Behavior (14), 111–123 (1998).
11. S. Rushton, J. Morgan, and M. Richard, Teacher's. Myers-Briggs Personality Profiles: Identifying Effective Teacher Personality Traits, Journal of -Teaching and Teacher Education (23), 432–441 (2007).
12. S.A. Jessee, P.N. Neill, and R.O. Dosch, Matching Student Personality Types and Learning Preferences to Teaching Methodologies, Journal of Dental Education (70), 644–651 (2006).
13. N. Ghasem-Aghaee, S. Fatahi, and T.I. Ören, Agents with Personality and Emotional Filters for an e-Learning Environment (Proceedings of Spring Agent Directed Simulation Conference, Ottawa, Canada, 2008).
14. S. Fatahi, N. Ghasem-Aghaee, and M. Kazemifard, Design an Expert System for Virtual Classmate Agent (VCA) (Proceedings of World Congress Engineering, UK, London, 2008), pp. 102–106.
15. S. Ellis, and S. Whalen, Cooperative Learning: Getting Started, Scholastic, New York (1996).
16. http://www.murraystate.edu
17. B. Kort, R. Reilly, and R.W. Picard, An Affective Model of Interplay between Emotions and Learning: Reengineering Educational Pedagogy Building a Learning Companion (Proceedings IEEE International Conference on Advanced Learning Technology, Madison, 2001), pp. 43–48.
18. S.M. Al Masum, and M. Ishizuka, An Affective Role Model of Software Agent for Effective Agent-Based e-Learning by Interplaying between Emotions and Learning (WEBIST, USA, 2005), pp. 449–456.
19. K.S. Choi, F.P. Deek, and I. Im, Exploring the Underlying of Pair Programming: The Impact of Personality, Journal of Information and Software Technology, doi: 10.1016/j.infsof.2007.11.002 (2007).
20. L.F. Capretz, Implications of MBTI in Software Engineering Education, ACM SIGCSE Bulletin – Inroads, ACM Press, New York, vol. 34, pp. 134–137 (2002).
21. A.R. Peslak, The impact of Personality on Information Technology Team Projects (Proceedings of the 2006 ACM SIGMIS CPR Conference on Computer Personnel Research: Forty Four Years of Computer Personnel Research: Achievements, Challenges & the Future, Claremont, California, USA, 2006).

Chapter 37
A Self Progressing Fuzzy Rule-Based System for Optimizing and Predicting Machining Process

Asif Iqbal and Naeem U. Dar

Abstract Researchers have put forward variety of knowledge acquisition methods for automatic learning of the rule-based systems. It has been suggested that two basic anomalies of rule-based system are incompleteness and incorrectness. In the chapter, a fuzzy rule-based system has been presented that not only self-learns and self-corrects but also self-expands. The chapter moves ahead with description of the configuration of the system, followed by explanation of the methodology that consists of algorithms for different modules. At the end, the operation of the self-progressing fuzzy rule-based system is explained with the help of examples related to optimization of machining process.

Keywords Fuzzy Rule-Based System · Self Progressing · Machining Process · Optimizing

37.1 Introduction

Rule-based systems represent the earliest and most established type of AI systems that tend to embody the knowledge of a human expert in a computer program [1]. Though, rule-based systems have been quite effectively utilized in approximating complicated systems – for which analytical and mathematical models are not available – with considerable degree of accuracy, they still pose a very static picture. The foremost shortcoming, in their design, is the lack of dynamism. The price paid by this inadequacy is their failure in coping with fast changing environments. This describes the main reason of inability of rule-based system technology to find its full fledge application at industrial levels.

A. Iqbal (✉)
Department of Mechanical Engineering, University of Engineering & Technology, Taxila, Pakistan
E-mail: asif.asifiqbal@gmail.com

S.I. Ao and L. Gelman (Eds.), *Advances in Electrical Engineering and Computational Science*, Lecture Notes in Electrical Engineering.

It is predominantly important to keep the rule-base (or knowledge-base in general) updated and corrected all the time in order to maintain the efficacy and viability of the system. Obviously, it is not prudent to engage the services of a knowledge engineer for this purpose, as the efficacy calls for rapid upgrading of the rule-base. The dire need is to find the means for automatic accomplishment of this requirement.

Most of the time required for the development of a rule-based system can be attributed to the knowledge acquisition phase [2]. Researchers have put forward variety of knowledge acquisition methods for automatic learning of the rule-based systems. It has been suggested that two basic anomalies of rule-based system are incompleteness and incorrectness [3]. It was further proposed that by integrating machine learning (ML) techniques in the validation step of the rule-based systems' evolutionary life cycle model, the knowledge-base can be refined and corrected throughout a refinement cycle [3]. In other article [4] the authors presented an adaptive fuzzy learning algorithm, used for nonlinear system identification that provided a new way for representing the consequent part of the production rule. In other work [5], a knowledge factory has been proposed that allows the domain expert to collaborate directly with ML system without needing assistance of a knowledge engineer. In other paper [2] the authors have presented an inductive learning algorithm that generates a minimal set of fuzzy IF-THEN rules from a set of examples. In another work [6], the authors have presented a self-testing and self-learning expert system, which is based on fuzzy certainty factor and it checks the rule-base for correctness and completeness.

Few papers can be found that describe the application of machine learning in manufacturing domain. In a paper [7] researchers have presented a fuzzy expert system for design of machining conditions dedicated to the turning process. The learning module contained by the system corrects the empirical relationships by changing the fuzzy membership functions. Another paper [8] has presented an approach for building knowledge-base from numerical data, which has proved to be useful for classification purposes. In other article [9] the authors have utilized Support Vector Regression, a statistical learning technique, to diagnose the condition of tool during a milling process. Finally, in another paper [10], the authors have presented the comparison of three ML algorithms for the purpose of selection of the right dispatching rule for each particular situation arising in a flexible manufacturing system.

In the current research work a fuzzy rule-based system has been presented that not only self-learns and self-corrects but also self-expands. Following are the distinguishing features of the self-progressing fuzzy rule-based system:

1. For any physical process, it predicts the values of response (output) variables based on the values of predictor (input) variables
2. Suggests the best values of predictor variables that maximize and/or minimize the values of selected set of response variables
3. Automatically adjusts newly entered variable at any stage of development
4. Learns and corrects itself according to new set of data provided

5. Automatically generates fuzzy sets for newly entered variables and regenerates sets for other variables according to newly added data
6. Automatically generates rules for optimization and prediction rule-bases
7. Provides conflict resolution facility among contradictory rules
8. Updates interface of the system according to newly entered variables

The first two points depict the main objectives of the rule-based system, in connection with any manufacturing process, while the others portray a picture of high level automation required for the system to self-progress. The chapter moves ahead with description of the configuration of the system, followed by explanation of the methodology that consists of algorithms for different modules. At the end, the operation of the self-progressing fuzzy rule-based system is explained with the help of examples related to optimization of machining process.

37.2 System Configuration

Figure 37.1 describes the configuration of the self-progressing rule-based system in graphical form. The main purpose of the system is to self-generate the rule-bases, from the experimental data, that could be used to: (1) optimize the settings of predictor variables (of any physical process) for the purpose of maximization and/or minimization of set of selected response variables; (2) predict the values of response variables according to the finalized settings of the predictor variables.

The pattern of the system consists of four parts: the knowledge-base, the shell, the inference engine, and the self-development mode. The knowledge-base is the

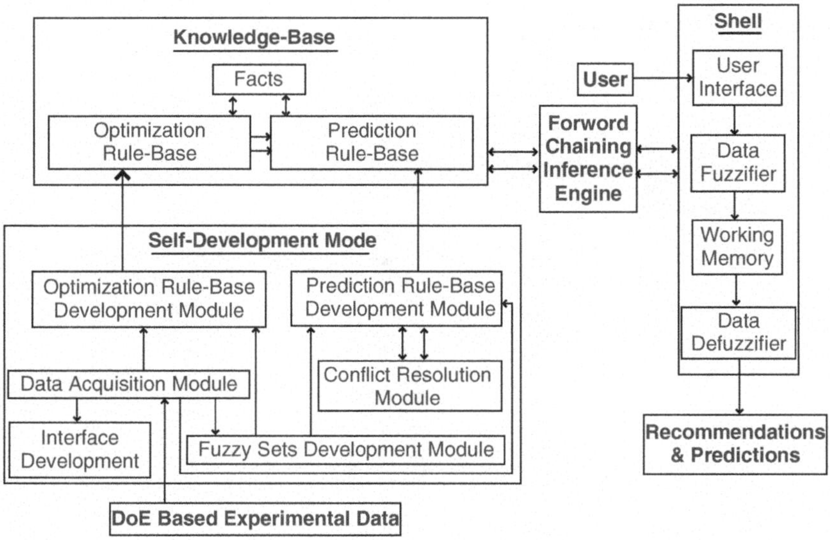

Fig. 37.1 The configuration of the self-progressing rule-based system

combination of facts, the optimization rule-base, and the prediction rule-base. The functional details of optimization and prediction rule-bases are also available in reference [11]. The shell consists of a user interface through which the input is taken from the user. The data fuzzifier fuzzifies the values of numeric predictor variables according to the relevant fuzzy sets. The user-interface and the fuzzy sets are also auto-developed by the system itself. For development of the knowledge-base consisting of two rule-bases, the inference mechanism of forward chaining shell, Fuzzy CLIPS (C Language Integrated Production Systems) was used [12].

37.3 The Self-Development Mode

The most important and distinguishing constituent of the system is the self-development mode. This mode consists of following four modules: data acquisition module; fuzzy sets development module; optimization rule-base development module; and prediction rule-base development module integrated with the conflict resolution sub-module. The detail is as follows.

37.3.1 Data Acquisition

This module facilitates the automation of intake, storage, and retrieval of data.

The data could be the specifications of a new variable or the values of input and output variables obtained from experiments. Data related to specifications of a new variable are stored in file Variable.dat, while that related to the new values/records are stored in Data.dat on hard disk. Figure 37.2 shows the flow chart of data acquisition algorithm.

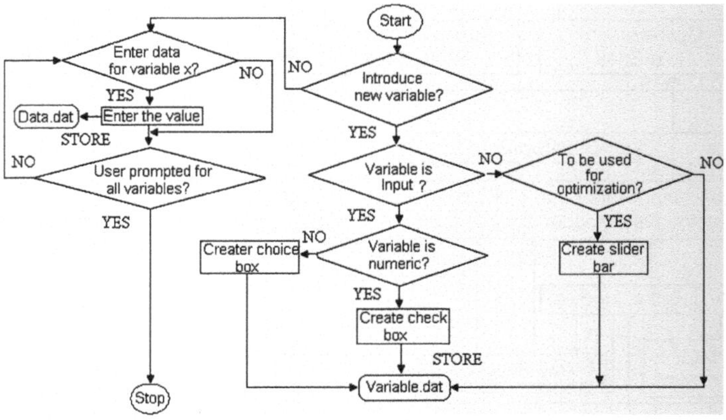

Fig. 37.2 The flow chart of the data acquisition module

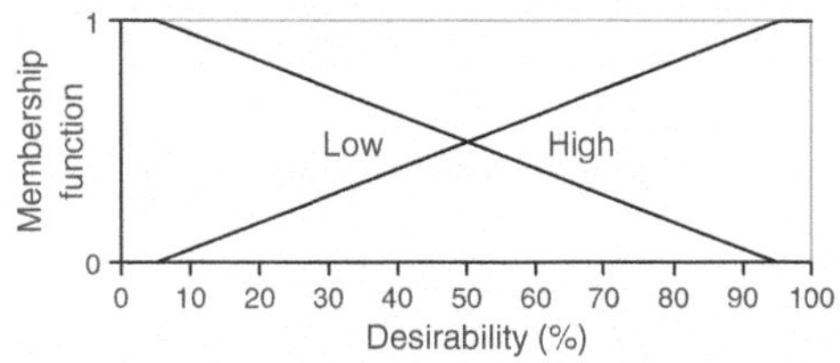

Fig. 37.3 Fuzzy sets for maximizing/minimizing output variable

37.3.2 Self-Development of Fuzzy Sets

This module covers three areas: (1) Rearranging the fuzzy sets for already entered variables according to the newly entered data records; (2) development of fuzzy sets for newly entered numeric variables; and (3) development of two fuzzy sets (*low* & *high*) for each output variable that is included for optimization purpose. The set *low* represents the minimization requirement and the other one represents maximization. The design of sets for category (3) is fixed and is shown in Fig. 37.3, while design of first two categories is dynamic and based upon the data values of respective variables.

The desirability values shown in Fig. 37.3 are set by the user using the slider bar available on interface of the rule-based system. Any value below 5% means desirability is of totally minimizing the output variable (performance measure), and total desirability of maximization is meant if the value is above 95%. The desirability of 50% means optimization of that output variable makes no difference.

Figure 37.4 shows the customized flow chart for the methodology used for self-development of fuzzy sets. The user has to decide the maximum allowable number of fuzzy sets for input as well as for output variables.

The logic involved in methodology is that a value of input variable, which has higher frequency of appearance in the data records, has more right to be picked up for allocation of a fuzzy set, while for output variable any value having greater difference from its previous and next values in the list – termed as Neighbor Distance (*Neighb_dist* in Fig. 37.4) – possesses more right for allocation of a fuzzy set. Neighbor Distance can mathematically be represented as follows:

$$Neighbor\_Distance = \begin{cases} Value[i+1] - Value[i]; & if(i = first) \\ Value[i] - Value[i-1]; & if(i = last) \\ {}^{1}\!/_{2}\,(Value[i+1] - Value[i-1])\,; & otherwise \end{cases} \tag{37.1}$$

37.3.3 Self-Development of Prediction Rule-Base

This step consists of following two parts: (1) automatic development of rules, for prediction of the manufacturing process's performance measures, based on the data

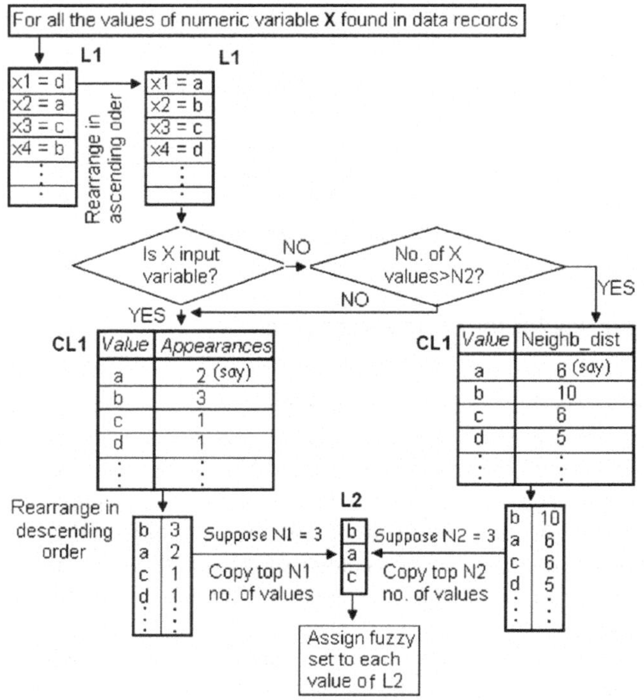

Fig. 37.4 Customized flow chart for self-development of fuzzy sets

records provided by the users; and (2) conflict resolution among self-developed contradictory rules.

Figure 37.5 provides the graphical description of the first part. The objective is to convert each node of 2-D linked list *Data_output* (including list of related values of input variables *Data_input)* into a rule. 2-D linked list is a list that expands in two directions as shown in Fig. 37.5. The objective is achieved by finding and assigning the most suitable fuzzy sets for all of the values involved per node of *Data_output*. The list *Data_output* is navigated from first node to last and for all of its values the closest values in fuzzy sets of respective variables are matched. If the match is perfect then certainty factor (*CF*) of 1 is assigned to the match of the data value and the fuzzy set. If the suitable match of any fuzzy set for a given data value is not found then the data value is assigned the intersection of two closest fuzzy sets. All the rules are stored in a 2-D linked list, named *Rule_Consequent*, each node of whose represents a rule. Each node contains the assigned fuzzy set of the output variable and also a linked list (*Rule_antecedent*) containing assigned fuzzy sets of all the relevant input variables.

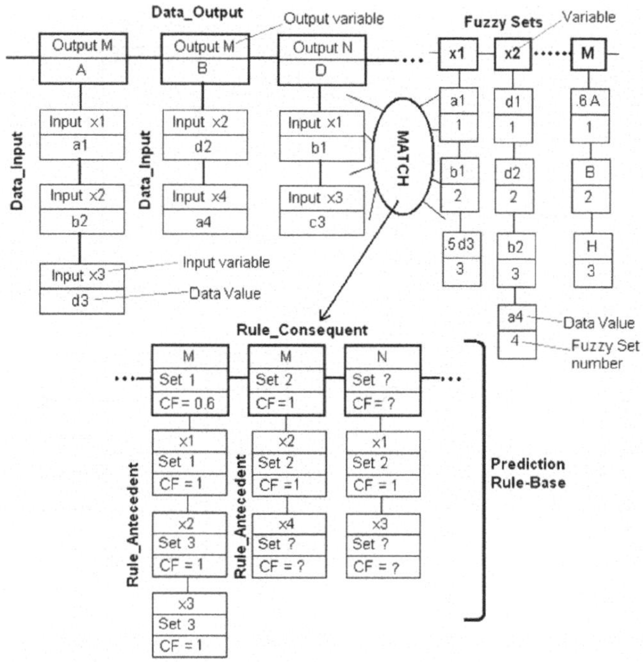

Fig. 37.5 Framework for self-development of prediction rule-base

37.3.3.1 Conflict Resolution among Contradictory Rules

There is always a possibility that some anomalous data might be entered by the user that could lead to self-development of some opposing rules. So it is necessary to develop a mechanism that would detect such possible conflict and provide a way for its resolution.

The mechanism of conflict resolution can be described as follows: Compare each and every rule of the prediction rule-base to all the other rules of the same rule-base. If, in the consequent parts of any two rules, following two conditions satisfy: (1) output variables are same; and (2) assigned fuzzy sets are different, then check whether the antecedent parts of both the rules are same (i.e., same input variables with same fuzzy sets assigned). If yes, then these two rules form a pair of contradictory rules. Next the user is inquired regarding which one of the two contradictory rules needs to be abandoned. The *CF* value of the rule to be abandoned is set to zero. The same procedure is continued for whole of the rule-base. At the completion of the process, all the rules possessing the *CF* values greater than zero are printed to the CLIPS file.

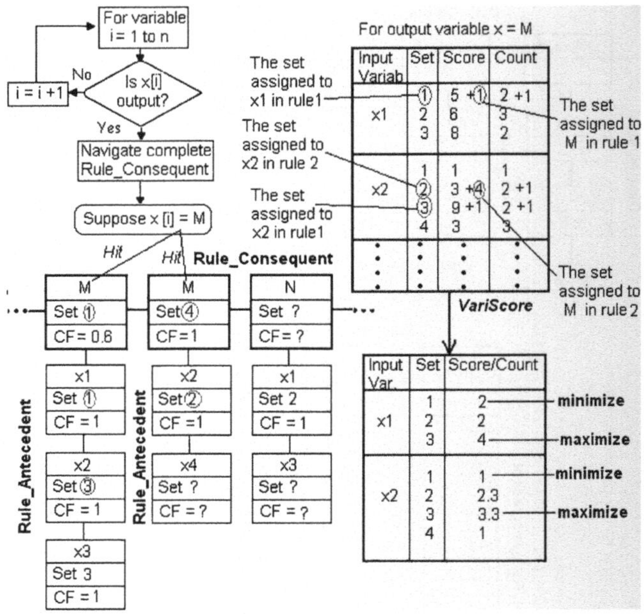

Fig. 37.6 Framework for self-development of optimization rule-base

37.3.4 Self-Development of Optimization Rule-Base

This module generates a set of rules that is responsible for providing the optimal settings of input variables that would best satisfy the maximization and/or minimization of the selected output variables. Figure 37.6 describes the framework.

The idea exploited in development of this module is that for maximization of any output variable select an ideal fuzzy set for each numeric input variable, which, on average, would generate the maximum value of that output variable. For the minimization purpose, select those fuzzy sets for respective input variables that would result in the least possible value of the output variable, available in the data records.

37.4 Optimization and Prediction of Machining Process

In order to demonstrate the versatility and self-progressing capabilities of the presented system, two application examples are presented.

Machining is the most wide-spread of all the manufacturing processes and amount of investment being done in it is an indication of wealth of the nations [13]. For demonstration of applicability of the self-progressing rule-based system in optimization and prediction of the processes included in the machining domain, the process of milling (a machining process in which tool rotates and workpiece remains stationary) has been chosen. Table 37.1 presents a set of limited data related

Table 37.1 Limited experimental data for rookie knowledge-base

No.	Predictor variables		Response variables		
	Speed (m/min)	Rake (°)	Orientation	Tool life (mm²)	Cutting force (N)
1	150	−8	Up	3,601	910
2	150	−8	Down	5,500	643
3	150	5	Up	3,583	885
4	150	5	Down	4,912	532
5	250	−8	Up	2,991	1,008
6	250	−8	Down	4,004	769
7	250	5	Up	2,672	986
8	250	5	Down	3,609	704

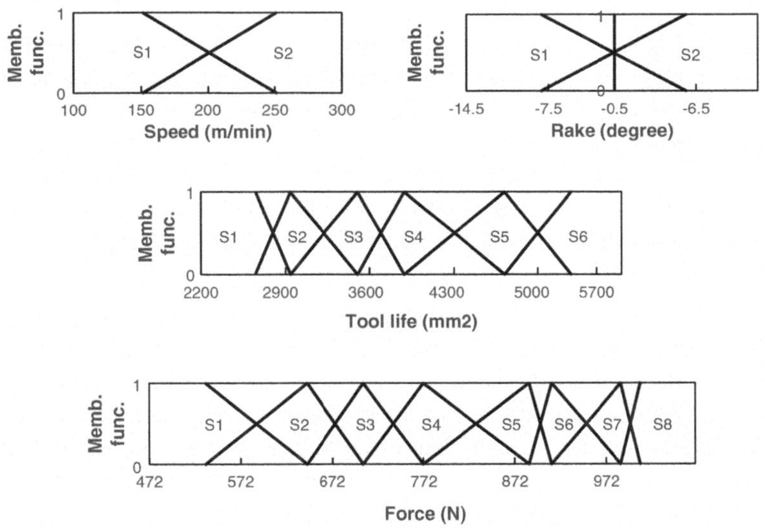

Fig. 37.7 Self-developed fuzzy sets for numeric variables

to milling process. The first three variables, namely: speed, rake, and orientation (milling-orientation) are predictor (input) ones, while the other two, tool life and cutting force are response (output) variables. If the knowledge-base is developed based entirely on the data presented in the table, it is very likely that the system may provide anomalous results because of the fact that the other influential milling parameters have not been taken into account, and thus, the self-progressed knowledge-base can be termed as "Rookie Knowledge-Base".

Suppose the system is asked to develop its rule-bases and update its interface based on the data provided and it is also asked to include tool life, but not the cutting force, as output variable for optimization. Figure 37.7 shows the detail of the triangular fuzzy sets of numeric variables, developed itself by the rule-based system, in addition to the sets of the objective, as shown in Fig. 37.3. Following is the detail of the six rules, self-generated by the self-development mode and to be operated by the optimization module of the rule-based system:

Table 37.2 Self-generated prediction rule-base

Rule No.	Antecedents			Consequents			
	Speed	Rake	Orientation	Tool life	CF	Force	CF
1	S1	S1	Up	S3	0.88	S6	1
2	S1	S1	Down	S6	1	S2	1
3	S1	S2	Up	S3	1	S5	1
4	S1	S2	Down	S5	1	S1	1
5	S2	S1	Up	S2	1	S8	1
6	S2	S1	Down	S4	1	S4	1
7	S2	S2	Up	S1	1	S7	1
8	S2	S2	Down	S3	0.82	S3	1

Rule 1: IF *Objective Tool_life* is High AND *Speed* is not fixed THEN *Speed* is S1.
Rule 2: IF *Objective Tool_life* is High AND *Rake* is not fixed THEN *Rake* is S1.
Rule 3: IF *Objective Tool_life* is High AND *Orientation* is not fixed THEN *Orientation* is Down.
Rule 4: IF *Objective Tool_life* is Low AND *Speed* is not fixed THEN *Speed* is S2.
Rule 5: IF *Objective Tool_life* is Low AND *Rake* is not fixed THEN *Rake* is S2.
Rule 6: IF *Objective Tool_life* is Low AND *Orientation* is not fixed THEN *Orientation* is Up.

Out of these six rules the first three perform the maximization operation, while the others perform minimization. Table 37.2 presents the detail of eight rules, self-generated by the rule-based system and to be operated by its prediction module.

Figure 37.8 shows the interface of the system related to the rookie knowledge-base. The slider bar, shown at middle of the figure, prompts the user whether to maximize or minimize the selected output variable and by how much desirability. Suppose the rule-based system is provided with following input:

- Objective: maximize tool life with desirability of 98%.
- Rake angle of tool prefixed to 0°.
- Cutting speed and milling-orientation: *open* for optimization.

Pressing the *Process* button starts the processing and following results are displayed in the information pane:

- The recommended orientation is down-milling cutting speed is 154.2 m/min.
- The predicted tool life is $4,526.4\,\mathrm{mm}^2$ and cutting force is 649.68 N.

Suppose the same rule-based system is provided with more experimental data, covering effects of all the influential parameters of milling process. When the system is asked to generate knowledge-base from that set of data, the resulting knowledge-base would be a veteran knowledge-base. As more and more data will be provided to the rule-based system it will keep improving its accuracy of optimization and prediction processes.

Figure 37.9 presents the interface of the rule-based system from the experimental data provided in papers [14, 15] in addition to that provided in Table 37.1.

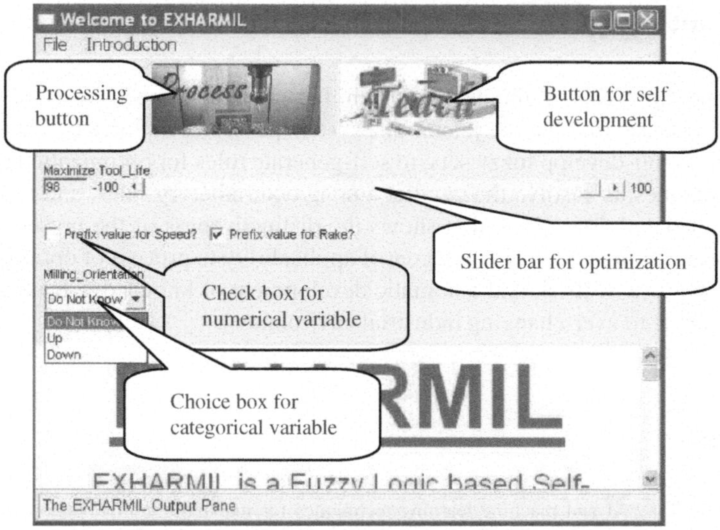

Fig. 37.8 Self-generated interface of rule-based system utilizing rookie knowledge-base

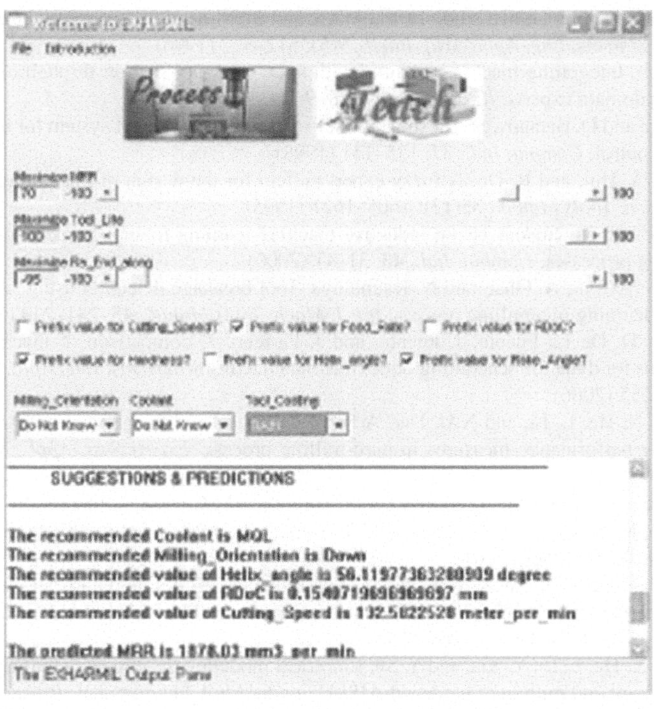

Fig. 37.9 Self-generated interface of rule-based system utilizing veteran knowledge-base

37.5 Conclusion

The chapter presents a unique approach for designing mechanism of a self-progressing fuzzy rule-based system. The system possesses abilities to manage new variables, to self-develop fuzzy sets, to self-generate rules for optimization and prediction modules, to resolve the conflict among contradictory rules, and to keep its interface updated. The discussion shows the distinctiveness of the presented rule-based system along with its high degree of applicability to process of optimizing the machining process. Brisk and automatic development of knowledge-base makes it well adaptable to ever-changing industrial environment.

References

1. L. Monostori, AI and machine learning techniques for managing complexity, changes, and uncertainty in manufacturing, *Eng. Appl. Artif. Intell.*, **16**: 277–291 (2003)
2. J.L. Castro, J.J. Castro-Schez, and J.M. Zurita, Use of a fuzzy machine learning technique in the knowledge-acquisition process, *Fuzzy Sets Syst.*, **123**: 307–320 (2001)
3. H. Lounis, Knowledge-based systems verification: A machine-learning based approach, *Expert Syst. Appl.*, **8**(3) 381–389 (1995)
4. K.C. Chan, A comparative study of the MAX and SUM machine-learning algorithms using virtual fuzzy sets, *Eng. Appl. Artif. Intell.*, **9**(5): 512–522 (1996)
5. G.I. Webb, Integrating machine learning with knowledge-acquisition through direct interaction with domain experts, *Knowl-Based Syst.*, **9**: 253–266 (1996)
6. A. Lekova and D. Batanov, Self-testing and self-learning fuzzy expert system for technological process control, *Comput. Ind.*, **37**: 135–141 (1998)
7. Y. Chen, A. Hui, and R. Du, A fuzzy expert system for the design of machining operations, *Int. J. Mach. Tools Manuf.*, **35**(12): 1605–1621 (1995)
8. B. Filipic and M. Junkar, Using inductive machine learning to support decision making in machining processes, *Comput. Ind.*, **43**: 31–41 (2000)
9. S. Cho, S. Asfour, A. Onar, and N. Kaundinya, Tool breakage detection using support vector machine learning in a milling process, *Int. J. Mach. Tools Manuf.*, **45**: 241–249 (2005)
10. P. Priore, D. De La Fuente, J. Puente, and J. Parreno, A comparison of machine learning algorithms for dynamic scheduling of flexible manufacturing systems, *Eng. Appl. Artif. Intell.*, **19**: 247–255 (2006)
11. A. Iqbal, N. He, L. Li, and N.U. Dar, A fuzzy expert system for optimizing parameters and predicting performance measures in hard-milling process, *Expert Syst. Appl.*, **32**(4): 1020–1027 (2007)
12. R.A. Orchard, *Fuzzy CLIPS, V6.04A Users' Guide*, NRC, Canada (1998)
13. T. Childs, K. Maekawa, T. Obikawa, and Y. Yamane, *Metal Machining: Theory and Applications*, Arnold, London (2000)
14. A. Iqbal, N. He, L. Li, W.Z. Wen, and Y. Xia, Influence of tooling parameters in high-speed milling of hardened steels, *Key Eng. Mater.* (*Advances Machining & Manufacturing and Technology* 8), **315–316**: 676–680 (2006)
15. A. Iqbal, N. He, L. Li, Y. Xia, and Y. Su, Empirical modeling the effects of cutting parameters in high-speed end milling of hardened AISI D2 under MQL environment. *Proceedings of 2nd CIRP Conference on High Performance Cutting*, Vancouver, Canada, 2006

Chapter 38
Selection of Ambient Light for Laser Digitizing of Quasi-Lambertian Surfaces

D. Blanco, P. Fernández, E. Cuesta, C. M. Suárez, and N. Beltrán

Abstract Present work deals with the influence of ambient light on the quality of surfaces digitized with a laser triangulation probe. Laser triangulation systems project a laser beam onto the surface of the workpiece, so that an image of this projection is captured in a photo-sensor. The system is then able to establish vertical position for each point on the projection, processing the information contained in the mentioned image. As the sensor does not only capture the light projected by the laser, but also captures the ambient light emitted in the same wavelength of the laser beam, ambient light becomes a potential error source. A methodology for testing different light sources under the same digitizing conditions has been developed and applied to the digitizing of a 99% quasi-lambertian reflectance standard. Tests have been carried out for six different types of ambient light sources and the resulting point clouds have been set for comparison. Three different criteria have been used to analyze the quality of the point cloud: the number of captured points, the average dispersion of the test point cloud with respect to a reference point cloud, and the distribution of such geometric dispersion across the whole surface. Results show that best quality is obtained for low pressure sodium lamps and mercury vapor lamps.

Keywords Laser triangulation · digitizing · influence of ambient light

38.1 Introduction

In laser triangulation (LT), the projection of a laser beam onto a surface is captured as an image in a charged coupled device (CCD). Applying image processing techniques and the triangulation principle, 3D coordinates of the surface points are

D. Blanco (✉)
Department of Manufacturing Engineering, University of Oviedo,
Campus de Gijón, 33203 Gijón, Spain
E-mail: dbf@uniovi.es

S.I. Ao and L. Gelman (Eds.), *Advances in Electrical Engineering and Computational Science*, Lecture Notes in Electrical Engineering.

acquired (Fig. 38.1). If the distance between a particular point P and the CCD matches exactly the value of the reference distance (stand-off), its image in the CCD will be placed exactly in a reference point P'. Otherwise, if the point onto the surface were further away a distance H in the direction of the laser beam, its image on the CCD will be placed a distance h from the reference point. This way, it is possible to determine the spatial position of every single point from its image position on the sensor [1]. When digitizing a part, the laser beam projection sweeps the target surface, capturing a set of digitized points (point cloud) from its surface.

Accurate calculation of the spatial position for each point of the laser stripe depends on the accurate calculation of the centroid of its light distribution in the sensor [2]. If the intensity of the light distribution captured in the sensor is too weak, the system can not properly calculate the position of the points. Otherwise, if laser intensity is too high, the sensor will turn into saturation, so that the system could not calculate the position of points. For intermediate situations, the light distribution is analysed to determine its centroid position, which corresponds to distance h measured from reference point. Consequently, the light distribution affects the accuracy of distance H calculation (Fig. 38.1).

The result of the scanning process depends on the LT system characteristics, the geometry and quality of the surface and the environmental conditions [3]. These elements determine the shape and contrast of the laser stripe onto the surface and the image captured by the sensor. Since surface quality is an important influence factor, most LT systems allow for adjusting laser intensity according to surface colour and roughness requirements, achieving an improvement in the sharpness of the laser beam projection.

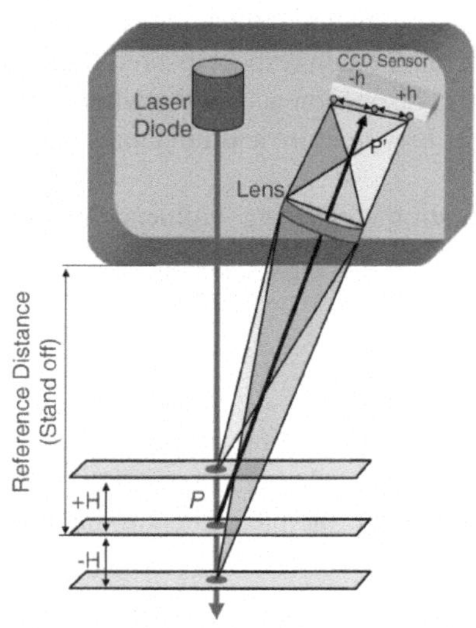

Fig. 38.1 Scheme of the laser triangulation principle

The ambient light present at the scanning process is one of the possible environmental influences [4]. Usually, LT systems incorporate optical filters to reduce or eliminate the influence of the ambient light. These filters accept only those wavelengths in the laser emission band. Commercial light sources emit light in a wide spectrum of frequencies. Since ambient light emitted in the laser emission band will not be filtered, it will become part of the information captured by the sensor and will be used in the calculation of point position.

38.2 Objectives

In this work, ambient light influence on the quality of digitized point clouds is evaluated. The tests have been set for the digitizing of a quasi-lambertian surface. In this type of surfaces, the reflexion is ideally diffused, as the whole incident energy is reflected in a uniform way in all spatial directions. The quasi-lambertian surface has been chosen as its behaviour is accepted to provide the best results for LT digitizing.

Digitizing tests have been carried out to compare results under different ambient light conditions. Although there is a wide range of commercial light sources, the present work deals with the most commonly used lamps. Tests have been carried out under laboratory conditions, where illumination for each experiment is reduced to a single light source. For each test configuration, nature and importance of the uncertainty introduced by ambient light have been established. Results for each light source have been afterwards compared to elaborate usage suggestions.

38.3 Configuration of the Tests

Tests have been carried out using a LT stripe commercial system from *Metris* (model *Metris LC*50) which has been mounted on a *Brown & Sharpe Global* CMM (model *Image*). The *LC*50 uses a laser beam emitting in the red visible spectrum with a wavelength emission band between 635 and 650 nm. The maximum peak power is 1 mW.

A reflectance standard from *Labsphere* has been used as the surface to be digitized [5]. This surface (aka *reference surface*) is 99% reflectance certified. This means that its surface is quasi-lambertian, so 99% of the received energy is reflected. Orientation of the light source with respect to the reference surface and the LT system has been selected so that the light direction theoretically makes an angle of 45° ($\varphi = 45°$) with the standard surface.

Moreover, theoretical direction of the incident light is orthogonal to the sweep direction. For every point, the system calculates the z coordinate value taking into account distance H (Fig. 38.1) and sensor position and orientation. Therefore, influence of ambient light affects the calculated vertical position of the points.

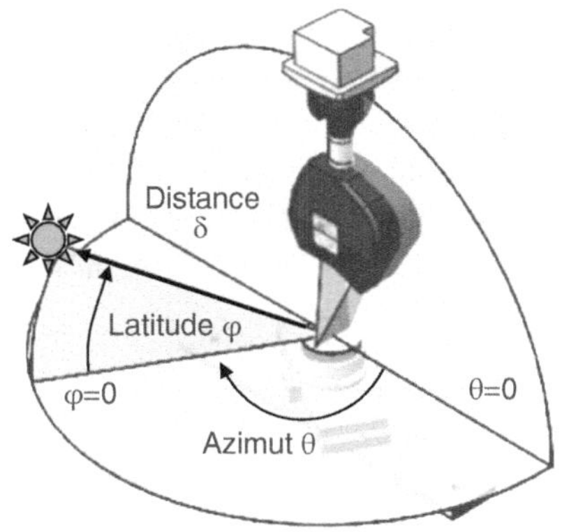

Fig. 38.2 Spherical coordinate system used to orientate the light source

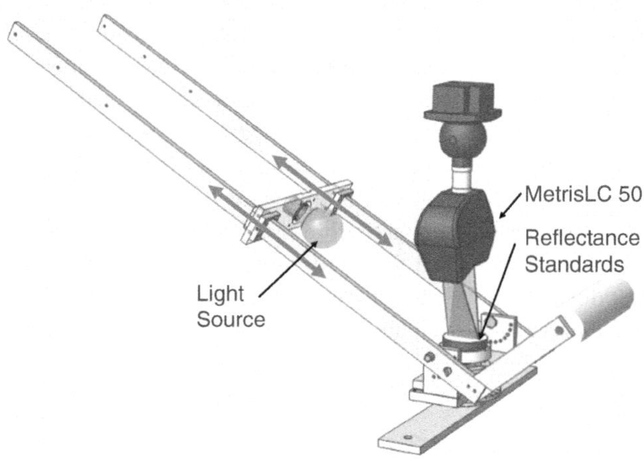

Fig. 38.3 Scheme of the main elements used in the test

Figure 38.2 shows the spherical coordinate system used for the light sources orientation. The origin of this coordinate system is the centre of the *reference surface*. In order to incorporate the influence of light source intensity to this work, tests have been carried out with two different positions for the light sources: 200 mm ($\delta 1$) or 400 mm ($\delta 2$) from the origin of the coordinate system.

Light sources and *reference surface* are mounted in a test-bench designed *ad hoc*. This test-bench provides a proper position and orientation of the light source according to the spherical coordinate system (Fig. 38.3). This mounting allows for comparing point clouds obtained under different test conditions. The set formed by the test-bench, the light source and the *reference surface* has been installed on the CMM table.

The light sources used for this work are among the most usual types on a metrological laboratory or a workshop. This way, three types of incandescent lamps (clear,

Table 38.1 Different lamps used in the experiments

K	Model	Manufacturer	Lamp type	Lumens
1	CLAS A CL	OSRAM	Clear incandescent	1,360
2	DECOR A	OSRAM	Blue incandescent	–
3	64476 BT	OSRAM	Halogen	1,600
4	PLE-T	PHILIPS	Fluorescent	1,500
5	HQL 50 SDL	OSRAM	Mercury vapour	1,600
6	SOX 18	OSRAM	Low pressure sodium	1,800

tinted in blue and halogen), one fluorescent lamp, one low pressure sodium lamp and one mercury vapour lamp constitute the final selection.

Although there is a wide variety of light sources for each class that can be tested (attending to power or shape), the selected lamps have similar values for the luminous flux (lumens). This selection criterium is based on finding alternatives that offer the operator a similar visual comfort, when performing long-time running digitizing processes.

Commercial references for the light sources used in this work are in Table 38.1.

38.4 Experimental Procedure

Test where the only illumination comes from the laser itself will not be altered by any external energy. Assuming this, a point cloud that has been digitized in the absence of light will suffer no distortions. Hence, digitizing in the dark appears to be the most appropriated way for scanning surfaces, although working in the absence of light is an unapproachable situation for human operators. Nevertheless, a point cloud obtained in the absence of light can be used as a reference when evaluating the quality of point clouds digitized under normal ambient lighting.

Experimental procedure used in this work allows for comparing the results obtained when digitizing under particular light sources with the results obtained in the absence of ambient light.

Although a single reference point cloud for all test may seem a suitable option, in practice, this approach is not recommended. Sensitivity of the internal geometry of the sensor to thermal variations (related to the time the laser remains switched on) must be taken into account. The fact that the manufacturer of the sensor recommends a minimum 40 min warm-up period since switching on the laser until a proper stability is reached, confirms the importance of this effect. Therefore, instead of using a single reference cloud for all tests, specific reference clouds have been used in each test comparison. These reference clouds must be digitized immediately after the capture of each single test cloud. This procedure will minimize the possible alteration of the sensor internal geometry due to thermal drift.

Thus, the first cloud $N^{k\delta}$ is obtained by digitizing the reference surface under a particular type of light source (k), placed at a given distance from the origin of the coordinate system (δ). Immediately after the first one, a second point cloud, known as reference point cloud $P^{k\delta}$, is obtained by digitizing the same surface in the absence of ambient light.

Comparison between these two clouds requires each point in the test cloud to have its equivalent in the reference cloud. To ensure this relationship, a computer application has been implemented, capable of selecting and classifying a group of 400 points (20×20) in the central area of the *reference surface*. Matrix constructed in this way allows for the comparison between each point and its equivalent.

LT system test parameters (as laser light intensity) have been adjusted to avoid loos of points due to saturation in the reference cloud. Distance between digitized points has been set to be 1.5 mm in both X and Y directions of the CMM coordinate system.

38.5 Analysis Criteria

Three criteria have been used for comparing the quality of point clouds obtained under different sources of light.

The first criterium evaluates the influence of lighting on the number of points captured in the test cloud. As discussed previously, an excessive input of light energy turns the sensor to saturation. Therefore, it becomes impossible to calculate a proper value for the z coordinate of the saturated points.

The saturated points are not included in the cloud $N^{k\delta}$ as the system rejects them. The parameter used to characterize this criterium is the number of valid points ($n^{k\delta}$) on the cloud.

The second criterium evaluates the influence of lighting in the proper calculation of z coordinate value for each point. Improper values will cause the points of the test cloud to appear in a higher or lower place than they really are.

The absolute difference between the z values for each equivalent pair of valid points in both the test cloud $N^{k\delta}$ and its reference cloud $P^{k\delta}$ is calculated (38.1).

$$\left|d_i^{k\delta}\right| = \left|z_i^{Nk\delta} - z_i^{Pk\delta}\right| \tag{38.1}$$

The standard deviation $\sigma^{k\delta}$ of the calculated differences $d_i^{k\delta}$ has been used as the characteristic parameter for this second criterium (38.3).

$$\mu^{k\delta} = \frac{1}{n}\sum_{i=1}^{n} d_i^{k\delta} \tag{38.2}$$

$$\sigma^{k\delta} = \sqrt{\frac{1}{n}\sum_{i=1}^{n}\left(d_i^{k\delta} - \mu^{k\delta}\right)^2} \tag{38.3}$$

The last criterium used in this work is qualitative. It consists on a graphical representation of deviations $d_i^{k\delta}$ for each digitized point cloud. This representation shows how the ambient light modifies the position of each single point. It allows for determining whether light influence is equal across the whole surface or not.

38.6 Results Discussion

Attending to first criterium ($n^{k\delta}$), the results of the tests in Table 38.2, illustrate how certain types of light sources cause a high percentage of points to become saturated.

Thus, in three of the tests (N^{11}, N^{12}y N^{31}), no point has been captured due to saturation caused by the great amount of energy in the laser wavelength band. The sensor can not obtain properly the z coordinate value of these points, therefore the point cloud is empty.

A partial loose of points has only occurred in one case (test N^{21}). However, the same light source placed on a further position (test N^{22}) provides a complete point cloud. The rest of the tests provide complete point clouds, so that proper information from surface geometry can be easily obtained.

An order of preference between different light sources can be established by using the second criterium (explained in previous section) referring to the standard deviation ($\sigma^{k\delta}$).

From these tests it can be concluded that the best results for both testing positions are obtained for the low pressure sodium lamp. This light source causes a lower distortion for the test cloud over the reference cloud. Comparison between pairs of clouds, all obtained in the absence of light, provides a medium value of 1.2 μm as the systematic error attributable to the system itself. Then, the result obtained for

Table 38.2 Results of the tests for the number of valid points and his standard deviation with respect to the reference cloud

K	δ	$N^{K\delta}$	$n^{K\delta}$	$\sigma^{K\delta}$ (μm)	$d_{i\ min}$(μm)	$d_{i\ max}$ (μm)
1	1	N^{11}	0	–	–	–
	2	N^{12}	0	–	–	–
2	1	N^{21}	44	8.36	−29.24	17.33
	2	N^{22}	400	4.64	−21.49	21.12
3	1	N^{31}	0	–	–	–
	2	N^{32}	400	4.83	−19.4	25.15
4	1	N^{41}	400	4.33	−7.33	21.97
	2	N^{42}	400	2.14	−8.97	9.89
5	1	N^{51}	400	2.15	−10.75	9.21
	2	N^{52}	400	1.88	−7.75	9.89
6	1	N^{61}	400	1.99	−18.19	13.79
	2	N^{62}	400	1.13	−5.92	5.07

the sodium lamp and the furthest position ($\sigma^{k\delta} = 1.13\,\mu$m) indicates a moderate influence of this light source on points positions.

The behaviour of the fluorescent lamp is clearly worse than the mercury vapour one for the closest position, while for the furthest one the difference is not so evident.

For the tinted blue incandescent lamp and the furthest position, the standard deviation is approximately 4.1 times greater than the value calculated for the sodium lamp. For the closest position, its standard deviation is extremely high, but it must be remarked that this value has been calculated considering a small number of valid points, as most of the theoretical points in the cloud have not been captured due to saturation.

In the case of the halogen lamp, the results for deviation $\sigma^{k\delta}$ are the worst of all. At the furthest position, the deviation is the maximal observed, approximately 4.3 times greater than the sodium lamp. This result was predictable as this lamp causes all the points to become saturated for the distance of 200 mm.

Finally, the third criterium (graphic representation of the parameter $d_i^{k\delta}$ along the whole area) shows how the distribution of these values depends on the type of light source and is clearly non-uniform (Fig. 38.4 and 38.5).

Thus, when testing the blue incandescent lamp at a distance of 200 mm from the origin of the coordinate system (Fig. 38.4), the graph shows how the lack of points due to saturation is produced in points of the *reference surface* that are close to the light source.

On the other hand, valid points are only registered in a narrow strip placed in the area where the points of the surface are further from the light source. In certain areas of each test cloud, some of the points are located in an upper vertical position from their equivalent ones in the reference cloud, whereas points in other areas are located in a lower position.

However, this distortion in the test clouds does not seem to be related to the proximity of points to the light source, even when such a relationship can be established for the saturation of points.

By contrast, the distribution of peaks and valleys in Fig. 38.4 and 38.5 shows a parallel orientation to the laser stripe projection onto the surface.

This result does not fit with any of the previous assumptions. The effect may be related to local irregularities on the reference surface properties. This should be confirmed by later work.

The appearance of the differences plotted in Fig. 38.4 confirms the conclusions obtained for the second criterium. Since sodium lamp generates less distortion in the cloud, it provides the best performance among the tested lamps.

This distortion is increased for mercury lamp and for fluorescent lamp. On the other hand, the blue incandescent lamp causes a huge distortion effect.

When testing the lights on the furthest position ($\delta = 400$ mm), five completely-full clouds have been obtained (Fig. 38.5). As it was set for the closest distance, distribution of $d_i^{k\delta}$ shows a non-uniform behaviour. Furthermore, the parallelism between the preferential direction of peaks and valleys in the graphs and the orientation of the laser stripe can be noticed as in previous Fig. 38.4.

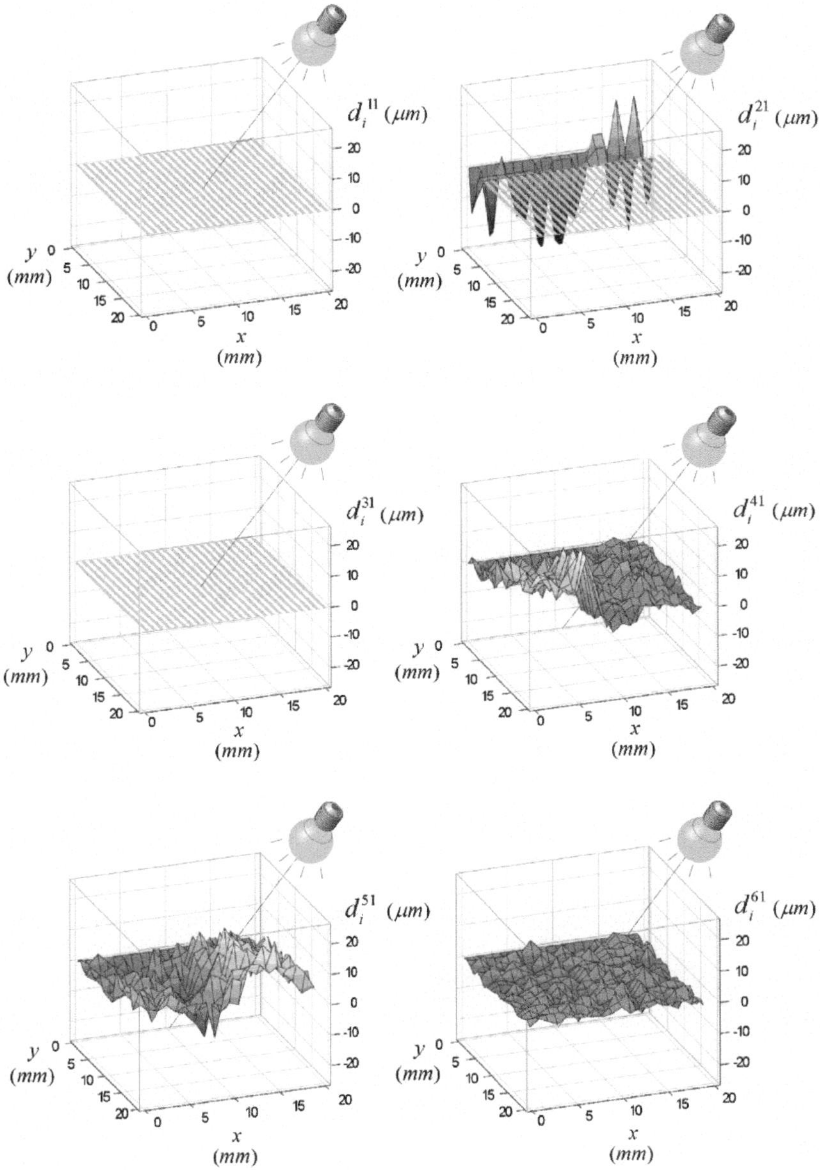

Fig. 38.4 $d_i^{k\delta}$ for a distance $\delta = 200$ mm

Moreover, the result in terms of the level of distortion for the test clouds follows the same previously established pattern according to the type of light source.

The sodium lamp is again the source that has a lower influence in the distortion introduced in the test cloud. For this position, the distortion introduced by mercury

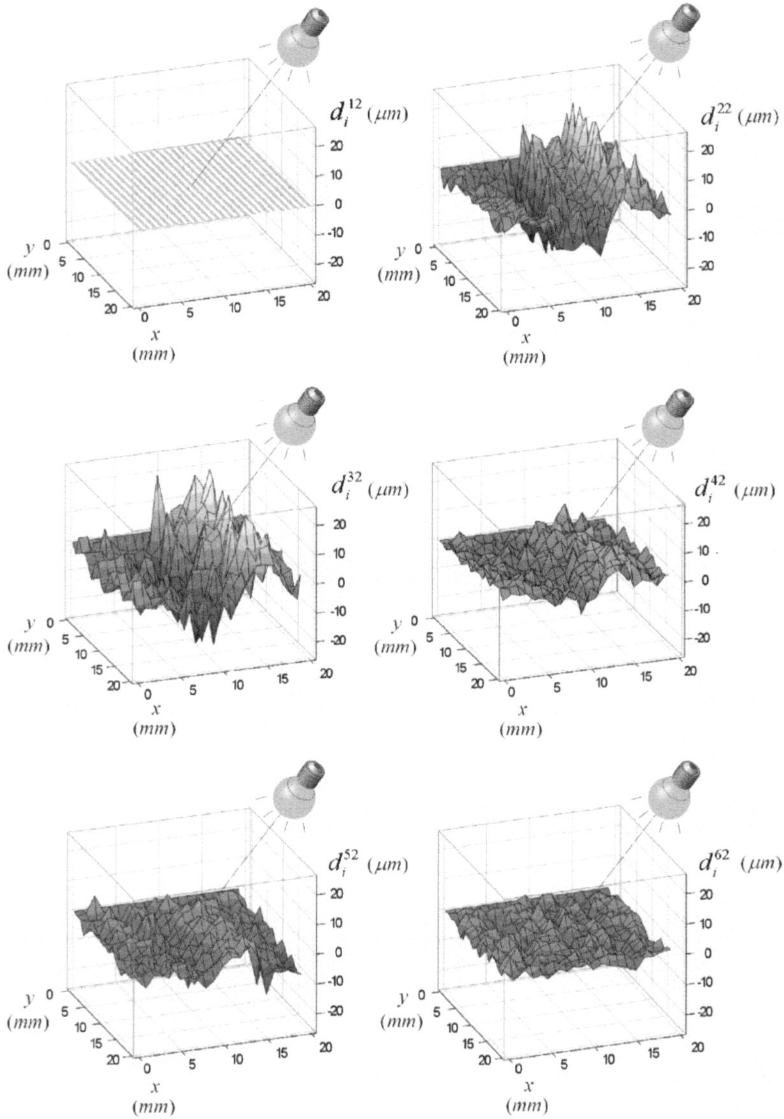

Fig. 38.5 $d_i^{k\delta}$ for a distance $\delta = 400$ mm

and fluorescent lamps is very similar, when a better behaviour of the mercury one has been established for the closest distance.

The results for the incandescent lamps (both the blue one and the halogen) are the worst among the tested lamps. The differences (both positive and negative) are significantly higher than for the rest of lamps.

38.7 Conclusions

This work has established the influence of ambient light on the quality of laser triangulation digitized surfaces. Results of the tests show how the influence of the light sources affects the digitizing in different ways. Sources that introduce a huge amount of energy on the laser wavelength band will cause some of the points to saturate. In severe cases this affects the whole point cloud and no information will be obtained. It has been also demonstrated that different types of sources cause different results when calculating vertical position for every point of the cloud.

The experimentation carried out confirms that laser digitizing of surfaces in complete absence of external sources of light provides the best results. In the usual case that this requirement can not be satisfied, results lead to recommend using those sources of light that cause less distortion of the point cloud: low pressure sodium lamps and the mercury vapour lamps. Sodium lamps emit in the orange range (589 nm) of the visible spectrum, which is especially annoying when working for long time periods, as it disables the operator for distinguishing different colors. This leads to recommend mercury vapour lamps as the most appropriate election.

Acknowledgements The authors wish to thank the Spanish Education and Science Ministry and the FEDER program for supporting this work as part of the research project MEC-04-DPI2004-03517.

References

1. Hüser-Teuchert, D., Trapet, E., Garces, A., Torres-Leza, F., Pfeifer, T., Scharsich, P., 1995. Performance test procedures for optical coordinate measuring probes final project report. European Communities.
2. Hüsser, D., Rothe, H., 1998. Robust averaging of signals for triangulation sensors. Measurement Science and Technology. Vol. 9, pp.1017–1023.
3. Boehler, W., Bordas Vicent, M. Marbs, A., 2003. Investigating Laser Accuracy. In: CIPA XIXth. International Symposium, 30 Sept.–4 Oct., Antalya, Turkey, pp. 696–701.
4. Blais, F. 2003. A Review of 20 Years of Ranges Sensor Development, SPIE Proceedings, Electronic Imaging, Videometrics VII. Santa Clara, CA, USA. Vol. 5013, pp. 62–76.
5. Forest, J., Salvi, J., Cabruja, E., Pous, C., 2004. Laser Stripe Peak Detector for 3D Scanners. A FIR Filter Approach. Proceedings of the 17th International Conference on Pattern Recognition. Vol. 3, pp. 646–649.

Chapter 39
Ray-Tracing Techniques Applied to the Accessibility Analysis for the Automatic Contact and Non Contact Inspection

B. J. Álvarez, P. Fernández, J. C. Rico, and G.Valiño

Abstract Accessibility analysis represents one of the most critical tasks in inspection planning. The aim of this analysis is to determine the valid orientations of the inspection devices used in probe operations and non-contact scanning operations on a Coordinate Measuring Machine (CMM). A STL model has been used for discretizing the inspection part in a set of triangles, which permits the application of the developed system to any type of part, regardless of its shape and complexity. The methodology is based on the application of computer graphics techniques such as ray-tracing, spatial subdivision and back-face culling. This analysis will take into account the real shape and geometry of the inspection device and the constraints imposed by the CMM on which it is mounted. A simplified model has been developed for each inspection device component using different basic geometrical shapes. Finally, collision-free orientations are clustered for minimizing the orientation changes during the inspection process.

Keywords Accessibility · CMM · inspection · probe · laser stripe · scanning

39.1 Introduction

Among the activities of an automatic process planning system for inspection on a Coordinate Measuring Machine (CMM), the determination of inspection device orientation with regard to the part stands out. These inspection device orientations are obtained from a methodology based on the accessibility analysis [1] and the application of ray-tracing algorithms [2]. Moreover, different computer graphics techniques

B. J. Álvarez (✉)
Department of Manufacturing Engineering, University of Oviedo, Campus de Gijón, 33203 Gijón, Spain
E-mail: braulio@uniovi.es

S.I. Ao and L. Gelman (Eds.), *Advances in Electrical Engineering and Computational Science*, Lecture Notes in Electrical Engineering.

like space partitioning and back-face culling have been applied in order to speed up the searching of valid orientations.

The methodology has been applied to inspection processes based on a touch-trigger probe [3] and to non-contact scanning processes based on a laser stripe [4]. In both cases, different constraints have been considered: real shape and dimensions of the inspection devices, process parameters and possible orientations of the motorized head of the CMM where the inspection device has been mounted. This motorized head (PH10MQ) provides 720 feasible orientation of the inspection device by rotating at a resolution of 7.5° about both horizontal and vertical axes (A, B).

39.2 Analysis Methodology

The accessibility analysis for a touch-trigger probe deals with determining all the feasible probe orientations that allow for performing the part inspection avoiding collisions with the part or any other obstacle in the environment of the inspection process. Moreover, the valid orientations of the non-contact scanning device will be obtained to guarantee the visibility of the surface to be scanned. The methodology applied in both cases will be exactly the same. In both cases several simplifications have to be made. In a first stage, the inspection devices have been abstracted by means of infinite half-lines. In a second stage, the real shape and dimensions of the inspection devices are taken into account.

Another simplification is related to the model of the part being inspected. The CAD model of the part, that may include complex surfaces, is converted to a STL model, where each surface is discretized by a set of simple triangular facets. For solving the accessibility analysis, the naïve approach consists on testing the intersection between the part triangles and any feasible orientation of the inspection device. In order to reduce the calculation time, different computer graphics techniques like space partitioning based on kd-tree, ray traversal algorithm, back-face culling and ray-triangle intersection tests have been also applied [5].

39.3 Analysis Considering Inspection Devices as Infinite Half-Lines

Figure 39.1 shows the abstractions made to contact and non contact inspection devices. In the case of touch-trigger probe, each probe orientation is replaced by an infinite half-line with the same direction of the orientation being analyzed. In the case of the laser triangulation system, two infinite half-lines have been used. One represents the laser beam while the second represents the vision space of the CCD sensor. Then, the analysis is divided in two phases: *local* and *global* analysis.

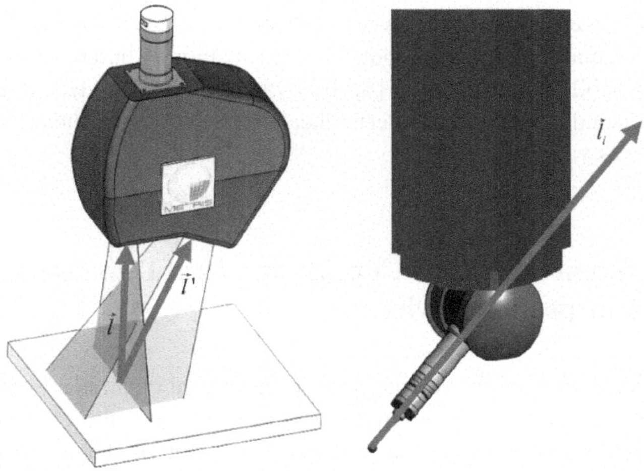

Fig. 39.1 Inspection devices as infinite half-lines

First, the *local* analysis only takes into account the possible interference between the infinite half-line and the part surface being inspected (probed or digitized) whereas the possible interferences with the rest of part surfaces or any other obstacle are ignored. Hence, valid orientations **l** will be those which make an angle between 0 and $\pi/2$ with the surface normal vector **n** .

In a second phase, the orientations obtained in the *local* analysis will be checked in order to determine if they have or not collision with the rest of part surfaces or any other obstacle. This analysis is called *global* analysis and it is complex and expensive from a computational point of view because it involves the calculation of multiple interferences tests. To make this calculation easier, a STL model of the part is used, where each surface is discretized by a set of triangles. Thus the global accessibility analysis is reduced to determine if there exist interferences between orientations **l** obtained in the local analysis and the triangles that compose the STL model of either the part or the obstacle.

The use of space partitioning structures like kd-trees allows for reducing the number of the triangles to test in the global analysis because it involves checking intersection exclusively with triangles which can potentially be traversed by each inspection device orientation. The part is partitioned in regions bounded by planes (*bounding boxes*) and each part triangle is assigned to the region within which it is located. Then, regions traversed by each inspection device orientation are selected by means of the ray-traversal algorithm that was first developed and applied by Kaplan [2].

Before checking intersection between each inspection device orientation and all triangles included in the traversed regions previously determined, the number of intersection tests can be reduced even more by applying a *back-face culling* algorithm [5]. Thus, from the initial set of triangles included in the traversed regions, a subset is extracted that do not include those triangles whose visibility according to an analyzed inspection device orientation is completely blocked by other triangles.

Finally, the intersection test between each undiscarded triangle and the inspection device orientation **l** is carried out. This verification is based on the algorithm developed by Möller and Trumbore [6]. If any intersection is detected then the orientation is rejected. If there is no interference with any triangle, the orientation will be considered as valid.

39.4 Interference Analysis Considering Real Dimensions of the Inspection Device

The orientations obtained in previous sections are based on an ideal representation of the inspection device as an infinite half-line. To check if these orientations are really valid will be necessary to take into account the real shape and dimensions of the inspection device. Therefore, intersection between triangles of the STL part model and each of the inspection device components must be checked. Figure 39.2 shows the components for each of the inspection devices that have been considered in this work:

1. Touch-trigger probe: column, head, touch probe, stylus and tip (Fig. 39.2a)
2. Laser stripe system: column, machine head, adapter and laser head (Fig. 39.2b)

The intersection analysis is speeded up by using a simplified model of each inspection device component (Fig. 39.2). The column and laser head have been modeled by straight prisms, the machine head by a sphere and the laser adapter, touch probe and stylus by a capsule.

Similarly to Section 39.2, a kd-tree algorithm has been implemented in order to test exclusively the interference between each of the inspection device components and the triangles included in the part regions that they traverse. To carry out this task effectively, each inspection device component is enclosed in a bounding volume and only the part regions that overlap with that volume are analyzed. Then, intersections between part triangles included in these regions and the component are checked.

Since several geometrical shapes have been used to model the components of each inspection device, different algorithms for checking intersections are applied:

1. Sphere-triangle intersection algorithm to analyze interferences with the machine head [7]. This algorithm simply calculates the minimum distance between a point (sphere centre) and a triangle. If this distance is smaller than the radius of the sphere, intersection will occur.
2. Prism-triangle intersection algorithm to analyze interferences with the column. This algorithm derives from the *separating axis theorem* which was initially developed to determine if two convex polyhedral are disjoint [8]. The theorem states that two convex polyhedra, A and B, are disjoint if they can be separated along either an axis parallel to a normal of a face of either A or B, or along an axis formed from the cross product of an edge from A with an edge from B. The application of this theorem to a prism-triangle test involves checking their relative position with regard to different potential *separation axes* [9].

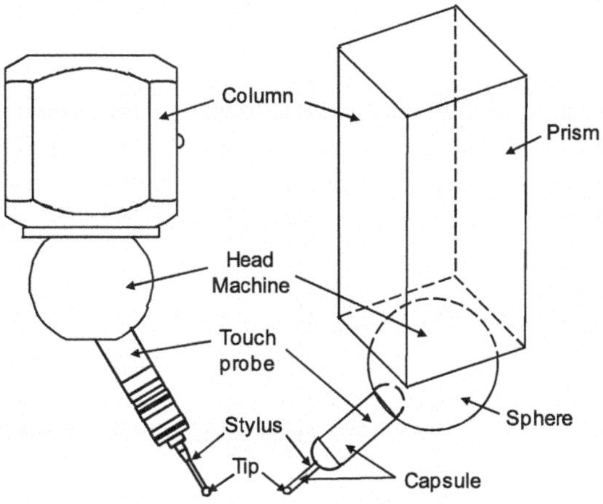

(a) Touch-trigger probe

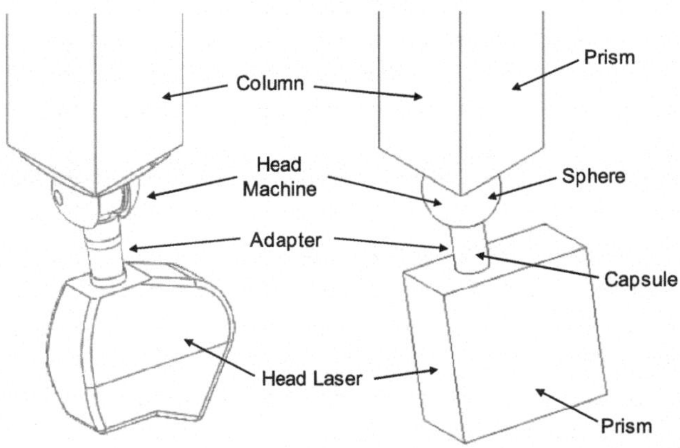

(b) Laser stripe system

Fig. 39.2 Components of the inspection device and their simplified models

3. Capsule-triangle intersection algorithms to analyze interferences with the touch probe and the stylus-tip [10]. In this case, intersection analysis is based on finding the minimum distance between each edge of the triangle and the capsule line segment, as well as the minimum distance between each extreme point of the capsule segment and the triangle. If any of these distances is smaller than the radius of the capsule, then intersection will occur.

39.5 Clustering

From the previous analysis, the inspection device orientations that are valid for probing each point or scanning each part triangle have been determined. These orientations are mathematically represented by means of a binary matrix $A_i(q, r)$ where each element corresponds to a combination of discrete values of A and B angles:

$$A_i(q,r) = \left\{ \begin{array}{ll} 1 & \text{if } (\mathrm{A} = a_q, \mathrm{B} = b_r) \\ & \quad \text{is a valid orientation for point } \mathrm{P}_i \\ 0 & \text{if } (\mathrm{A} = a_q, \mathrm{B} = b_r) \\ & \quad \text{is a not valid orientation for point } \mathrm{P}_i \end{array} \right\} \tag{39.1}$$

To reduce the process operation time related to device orientation changes, orientations (a_q, b_r) common to the greatest number of points to probe (clusters of points) or triangles to scan (clusters of triangles) must be found. The classification of points or triangles continues until no intersection can be found between the final clusters.

The algorithm used is similar to that developed by Vafaeesefat and ElMaraghy [11]. Next, the algorithm is explained for an inspection process using a touch-trigger probe.

Each point $P_i(i = 1, 2, \ldots, n)$ to be probed is associated to a binary matrix of valid orientations $A_i^k(i = 1, 2, \ldots, n)$. Initially $(k = 1)$, the clusters will be the same as each of the points to be probed: $C_i^k = P_i$.

Starting from the clusters C_i^k and from the binary matrices A_i^k, a new matrix $CI^k(i, j) = A_i^k \cap A_j^k$ is built showing the common probe orientations to the clusters two against two.

With the purpose of creating clusters whose points are associated with the greatest number of valid orientations, the algorithm searches the indices (s, t) of CI^k that correspond to the maximum number of common valid orientations. After that, clusters C_s^k and C_t^k associated to points P_s and P_t respectively will be regenerated as follows:

$$C_s^k = C_s^k \cup C_t^k \text{ and } C_t^k = \emptyset \tag{39.2}$$

and the binary matrix of valid orientation associated to the new cluster C_s^k will be $A_s^k = A_s^k \cap A_t^k$.

With these new clusters, matrix CI^k is updated to:

$$CI^k(i,j) = \left\{ \begin{array}{ll} A_i^k \cap A_j^k & \text{for } i = t \text{ and } j = s \\ \emptyset & \text{for } i = t \text{ and } j = t \end{array} \right\} \tag{39.3}$$

The clustering process finishes when the number of common orientations corresponding to all the elements above the main diagonal of matrix CI^k have become zero. A similar process is used to determine the clusters of triangles T_i $(i = 1, 2, \ldots, n)$ for a part to be scanned.

39.6 Application Results

39.6.1 Inspection Process by Means of a Touch-Trigger Probe

In order to analyze the application results of accessibility and clustering algorithms to the inspection process, the part shown in Fig. 39.3 has been considered. For this part, the STL model contains 2,438 triangles and 38 inspection points located on three different surfaces (s7, s19 and s39). As an example, Fig. 39.3 shows the accessibility maps for the inspection point P4s39 (point P_4 on surface s39) considering the different geometrical abstractions of the probe. As it can be seen, when real dimensions and different probe components are taken into account, the accessibility map is substantially reduced. For the rest of inspection points, similar accessibility maps can be obtained.

Furthermore, the application of the clustering algorithm allows for obtaining the clusters shown in Table 39.1. In this case seven clusters have been found.

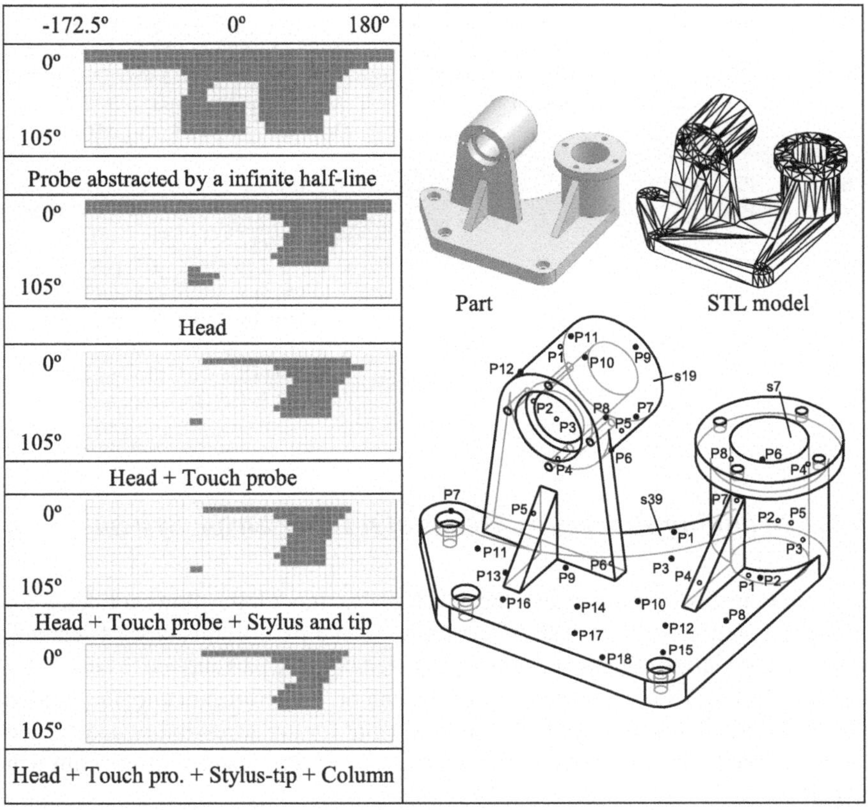

Fig. 39.3 Accessibility map for the point P4 on the surface s39

Table 39.1 Cluster of points and common orientations in these clusters

Cluster	Points/surface	Orientations (A, B)
1	P1s19, P2s19, P12s19, P4s7, P4s39, P7s39, P11s39, P13s39	(37.5, 67.5) (45, 67.5)
2	P3s19, P1s39, P5s39, P6s39	(67.5, −60)
3	P4s19, P5s19	(97.5, −90) (97.5, −82.5) (97.5, −75) (97.5, −67.5) (97.5, −60)
4	P6s19, P8s7	(30, −165) (30, 180) (37.5, −172.5) (37.5, −165) (37.5, −157.5) (37.5, −150) (37.5, −142.5) (37.5, −135) (37.5, 157.5) (37.5, 165) (37.5, −172.5) (37.5, 180) (45, −172.5) (45, −165) (45, −157.5) (45, −150) (45, −142.5) (45, −135) (45, −127.5) (45, −120) (45, 180) (52.5, −172.5) (52.5, −157.5) (52.5, −150) (52.5, −142.5) (52.5, −135) (52.5, −127.5) (52.5, −120) (52.5, −112.5)
5	P7s19, P8s19, P9s19, P10s19, P11s19, P6s7, P2s39, P3s39, P8s39, P9s39, P10s39, P12s39, P14s39, P15s39, P16s39, P17s39, P18s39	(15, 120) (15, 127.5) (22.5, 112.5) (22.5, 120) (22.5, 127.5) (22.5, 135) (22.5, 142.5) (22.5, 150) (22.5, 165) (30, 112.5) (30, 120) (30, 127.5) (30, 135) (30, 142.5) (30, 150) (30, 165)
6	P2s7, P3s7	(7.5, −15)
7	P7s7	(7.5, −165) (7.5, −157.5) (7.5, −150) (7.5, 165) (7.5, 172.5) (7.5, 180) (15, −172.5) (15, −165) (15, −157.5) (15, −150) (15, −135) (15, −142.5) (15, 150) (15, 157.5) (15, 165) (15, 172.5) (15, 180) (22.5, −172.5) (15, −165) (15, −157.5) (15, 172.5) (15, 180) (30, −172.5)

39.6.2 Scanning Process by Means of a Laser Stripe System

Apart from the inspection process, the developed methodology allows for determining the orientations of a laser stripe system to scan a part. In this type of scanning systems a laser stripe of known width is projected onto the part surface to be scanned and the reflected beam is detected by a CCD camera. Therefore, not only the incident laser beam orientation has been taken into account for the accessibility analysis but also the possible occlusion due to the interference of the reflected laser beam with the part.

Figure 39.4 shows the laser head orientation map associated to the local and global accessibility analysis for triangle 558 of the example part. The darkest color

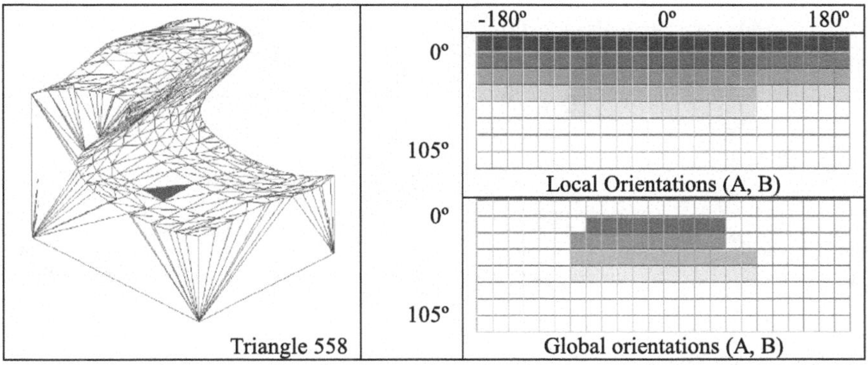

Fig. 39.4 Local and global accessibility maps for triangle 558

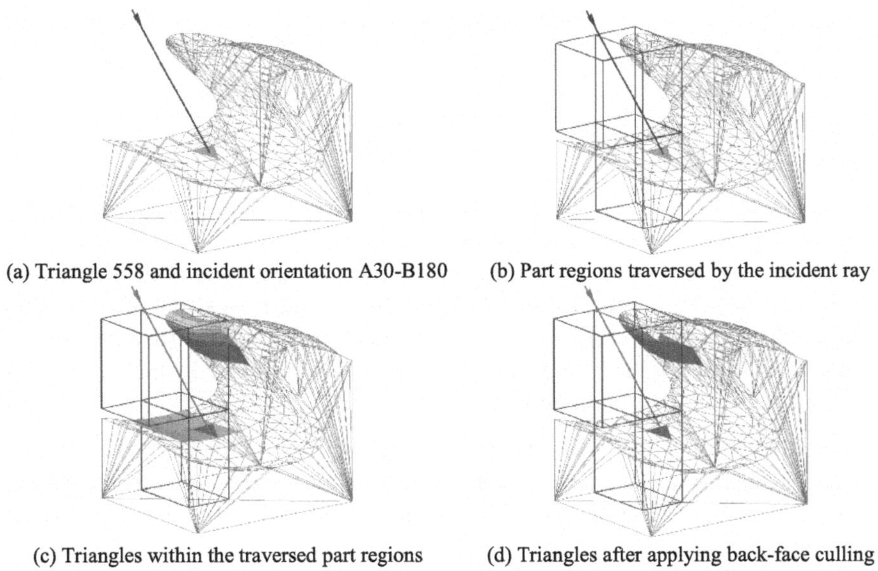

Fig. 39.5 Different stages to determine the global visibility map for a triangle and an incident laser beam orientation

represents the head orientations (A, B) that coincide or are closest to the normal direction of the triangle analyzed. These are considered as optimal orientations. Grey colors represent head orientations far from the optimal value which lead to worse scanning quality. White color represents head orientations that do not enable to scan the considered triangles. For a better visualization of the orientation map, increments of 15° have been considered for angles A and B.

For triangle 558 an incident laser beam orientation (A = 30°, B = 180°) has been selected in order to show the part regions (bounding boxes) that it traverses (Fig. 39.5). Triangles partially or totally enclosed in these bounding boxes are shown in the figure before and after applying *back-face culling*.

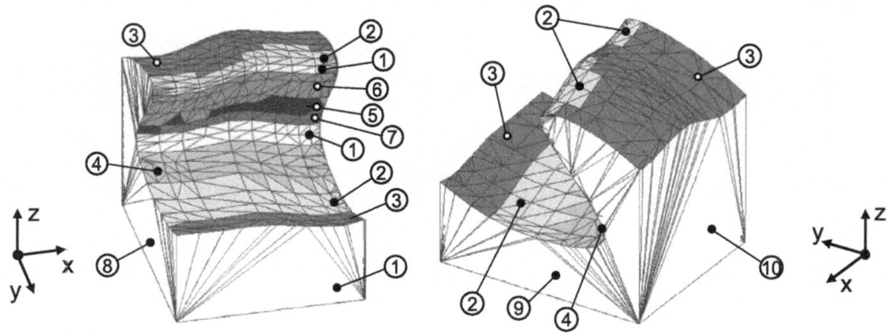

Fig. 39.6 Clusters associated to the example part

Figure 39.6 shows in a grey scale the ten triangle clusters obtained for the example part.

39.7 Conclusions

Most of the accessibility analysis presented in other works only deal with a limited number of the inspection device orientations, simple parts with only planar surfaces or specific geometrical shapes, simplified device representations or a short number of points in the case of inspection by touch-trigger probe. However, in this paper, a new methodology for accessibility analysis is presented which allows for overcoming the previous limitations:

1. The methodology has been extended to the inspection process by means of a touch-trigger probe and the scanning process by means of a laser stripe system.
2. All the possible orientations (720) of the inspection device are taken into consideration.
3. The use of the STL model permits the application of the developed system to any type of part, regardless of its shape and complexity.
4. The real shape and dimensions of the inspection device are considered for the analysis.
5. The implemented algorithms based on Computer Graphics reduce computation time and consequently can deal with a high number of inspection points and complex surfaces.
6. Moreover, a clustering algorithm is applied that efficiently groups the inspection points and triangles of the STL of the part to be scanned in order to reduce the number of probe orientation changes.

The developed system has been applied to different parts with satisfactory results demonstrating the application in practice. Future research will concentrate on developing new algorithms that further reduce computation time, and on generating the inspection paths from the orientations obtained in the clustering process.

Acknowledgements This paper is part of the result of a Spanish State Commission for Science and Technology Research Project (DPI2000-0605) and was supported by this Commission and FEDER.

References

1. Spyridi, A. J. and Requicha, A. A. G., 1990, Accessibility analysis for the automatic inspection of mechanical parts by coordinate measuring machines, *Proceedings of the IEEE International Conference on Robotics and Automation*. Vol. **2**, 1284–1289.
2. Kaplan, M., 1985, Space-tracing: A constant time ray-tracer, *Proceedings of the SIGGRAPH'85*, 149–158.
3. Limaiem, A. and ElMaraghy, H. A., 1999, CATIP: A computer-aided tactile inspection planning system, *Int. J. Prod. Res.* **37**(3): 447–465.
4. Lee, K. H. and Park, H. -P., 2000, Automated inspection planning of free-form shape parts by laser scanning, *Robot. Comput. Integr. Manuf.* **16**(4): 201–210.
5. Foley, J. D., Van Dam, A., Feiner, S. K., and Hughes, J. F., 1996, *Computer Graphics, Principles and Practice*, 2nd ed., Addison-Wesley, Reading, MA, pp. 663–664.
6. Möller, T. A. and Trumbore, B., 1997, Fast, minimum storage ray/triangle intersection, *J. Graph. Tools* **2**(1): 21–28.
7. Schneider, P. J. and Eberly, D. H., 2003, *Geometrical Tools for Computer Graphics*, Morgan Kaufmann, San Francisco, CA, pp. 376–382.
8. Gottschalk, S., Lin, M. C., and Manocha, D., 1996, OBBTree: A hierarchical structure for rapid interference detection, *Proceedings of the SIGGRAPH'96*, 171–180.
9. Möller, T. A., 2001, Fast 3D triangle-box overlap testing, *J. Graph. Tools* **6**(1): 29–33.
10. Eberly, D. H., 2000, *3D Game Engine Design. A Practical Approach to Real-Time Computer Graphics*, Morgan Kaufmann, San Diego, CA, pp. 53–57.
11. Vafaeesefat, A. and ElMaraghy, H. A., 2000, Automated accessibility analysis and measurement clustering for CMMs, *Int. J. Prod. Res.* **38**(10): 2215–2231.

Chapter 40
Detecting Session Boundaries to Personalize Search Using a Conceptual User Context

Mariam Daoud, Mohand Boughanem, and Lynda Tamine-Lechani

Abstract Most popular Web search engines are carachterized by "one size fits all" approaches. Involved retrieval models are based on the query-document matching without considering the user context, interests ang goals during the search. Personalized Web search tackles this problem by considering the user interests in the search process. In this chapter, we present a personalized search approach which adresses two key challenges. The first one is to model a conceptual user context across related queries using a session boundary detection. The second one is to personalize the search results using the user context. Our experimental evaluation was carried out using the TREC collection and shows that our approach is effective.

Keywords Personalized search · Session Boundaries · Conceptual User Context · search engine

40.1 Introduction

Most popular web search engines accept keyword queries and return results that are relevant to these queries. Involved retrieval models use the content of the document and their link structure to assess the relevance of the document to the user query. The major limitation of such systems is that information about the user is completely ignored. They return the same set of results for the same keyword query even though these latter are submitted by different users with different intentions. For example, the query *python* may refer to *python* as a snake as well as the programming language.

Personalized Web search aims at tailoring the search results to a particular user by considering his profile, which refers to his interests, preferences and goals during

M. Daoud (✉)
IRIT, University of Paul Sabatier, 118, Route de Narbonne, Toulouse, France
E-mail: daoud@irit.fr

S.I. Ao and L. Gelman (Eds.), *Advances in Electrical Engineering and Computational Science*, Lecture Notes in Electrical Engineering.

the search. One of the key challenges in personalized web search are how to model accurately the user profile and how to use it for an effective personalized search.

User profile could be inferred from the whole search history to model long term user interests [1] or from the recent search history [2] to model short term ones. According to several studies [2], mining short term user interests is more effective for disambiguating the Web search than long term ones. User profile representation model has also an impact on the personalized retrieval effectiveness. Involved models could be arranged from a very simple representation based on bags of words to complex representation based on concept hierarchy, namely the ODP [3, 4].

This chapter focuses on learning short term user interests to personalize search. A short term user interest is represented by the user context in a particular search session as a set of weighted concepts. It is built and updated across related queries using a session boundary identification method. Search personalization is achieved by re-ranking the search results of a given query using the short term user context.

The chapter is organized as follows. Section 40.2 presents some related works of search personalization approaches and session boundary identification propositions. Section 40.3 presents our contribution of search personalization and session boundary detection. In Section 40.4, we present an experimental evaluation, discussion and results obtained. In the last section, we present our conclusion and plan for future work.

40.2 Related Works

Personalized Web search consists of two main components: the user profile modeling and the search personalization processes.

User profile is usually represented as a set of keyword or class of vectors [5, 6] or by concept hierarchy issued from the user search history [7, 8]. As we build the user context using the ODP ontology, we review some related works based on the same essence. The ODP ontology is used in [3] to learn a general user profile applicable to all users represented by a set of concepts of the first three levels of the ontology. Instead of using a set of concepts, an ontological user profile in [9] is represented over the entire ontology by classifying the web pages browsed by the user into its concepts. Similar to this last work, an ontological user profile is described in [4] where the concept weights are accumulated using a spreading activation algorithm that activates concepts through the hierarchical component of the ontology.

The user profile is then exploited in a personalized document ranking by means of query reformulation [10], query-document matching [5] or result re-ranking [9]. Mining short term user interests in a session based search requires a session boundary mechanism that identifies the most suitable user interests to the user query. Few studies have addressed this issue in a personalized retrieval task. We cite the UCAIR system [2] that defines a session boundary detection based on a semantic similarity measure between successive queries using mutual information.

Unlike previously related works, our approach has several new features. First, we build a semantic user context as a set of concepts of the ODP ontology and not as an instance of the entire ontology [9]. The main assumption behind this representation is to prune non relevant concepts in the search session. Second, we build the user profile across related queries, which allow using the most suitable user interest to alleviate an ambiguous web search.

40.3 Detecting Session Boundaries to Personalize Search

Our general approach for search personalization consists of modeling a conceptual user context built across a search session to improve the web search accuracy of related queries. We define a session boundary recognition method that allows grouping a sequence of related queries in the same search session. Involved approach is detailed in next sections by: (a) modeling the user context and use it to personalize search, (b) describing the session boundary recognition mechanism.

40.3.1 A Conceptual User Context Modelling for Personalizing Search

We present in this section how to build and update the user context in a search session and how to exploit it in a personalized document re-ranking.

40.3.1.1 Building the User Context Using the ODP Metadata

We build a short-term user context that refers generally to the user's topics of interests during a search session. It is built for each submitted query using the relevant documents viewed by the user and the ODP ontology. The first step consists of extracting the keyword user context K^s for a submitted query q^s. Let D_r^s be the set of documents returned with respect to query q^s and judged as relevant by the user, each represented as a single term vector using the tf*idf weighting scheme. The keyword user context K^s is a single term vector where the weight of term t is computed as follows:

$$K^s(t) = \frac{1}{|D_r^s|} \sum_{d \in D_r^s} w_{td} \tag{40.1}$$

w_{td} is the weight of term t in document d. The concept-based user context is built by first mapping it on the ODP ontology, then disambiguating the obtained concept set using a sub-concept aggregation scheme.

(A) Mapping the Keyword User Context on the Ontology

The user context K^s is matched with concepts of the ODP ontology using the cosine similarity measure. Each concept of the ODP is related to sub-concepts with "*is-a*" relations and is associated to a set of web pages classified under that concept. We represent each concept by a single term vector $\mathbf{c_j}$ where terms are extracted from all individual web pages classified under that concept as detailed in a previous work [11]. Given a concept c_j, its similarity weight $sw(c_j)$ with $\mathbf{K^s}$ is computed as follows:

$$sw(c_j) = cos(\mathbf{c_j}, \mathbf{K^s}) \tag{40.2}$$

We obtain a set of concepts that contain relevant concepts at different levels and having different weights. In the next section, we proceed by disambiguating the concept set in order to rank the most relevant concepts of general level in the top of the user context representation.

(B) Disambiguating the Concept Set

We aim at disambiguating the obtained concept set using a sub-concept aggregation scheme. We retain the level three of the ontology to represent the user context. Indeed, the level two of the ontology is too general to represent the user interests, and leaf nodes are too specific to improve the web search of related queries.

Then, we recomputed the weight of each weighted concept by summing the weights of its descendants. We are based on the assumption that the most relevant concepts are those having a greater number of descendant concepts weighted according to the ontology. As shown in Fig. 40.1a, we identify a cluster of weighted concepts having a common general depth-three concept; we assign to this latter a relevance score computed by adding the weights of its descendant concepts as shown in Fig. 40.1b. The weight of a general concept c_j, having a set of n related descendant concepts $S(c_j)$, is computed as follows:

$$sw(c_j) = \frac{1}{n} \cdot \sum_{1 \leq k \leq n \wedge c_k \in S(c_j)} sw(c_k) \tag{40.3}$$

We finally represent the user context C^s performed at time s as a set of depth-three weighted concepts, noted $< c_j, p(c_j) >$.

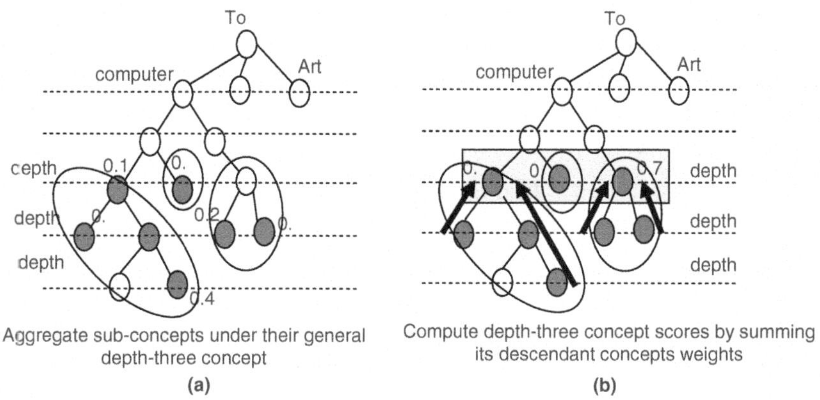

Fig. 40.1 Disambiguating the user context

40.3.1.2 A Session-Based Context Updating

We update the user context across queries identified as related using a session boundary recognition mechanism described in the next section. Let C^{s-1} and C^s be respectively the user contexts for successive and related queries. Updating method is based on the following principles: (1) enhance the weight of possible common concepts that can appear in two successive user contexts, (2) alter the weight of non-common concepts using a decay factor β. This allows taking into account the most recent concepts of interests to the user in the search session. The new weight of a concept c_j in the user context C^s is computed as follows:

$$sw_{C^s}(c_j) = \begin{cases} \beta * sw_{C^{s-1}}(c_j) + (1-\beta) * sw_{C^s}(c_j) \; if \; c_j \in C^{s-1} \\ \beta * sw_{C^s}(c_j) \; otherwise \end{cases} \tag{40.4}$$

where $sw_{C^{s-1}}(c_j)$ is the weight of concept c_j in context C^{s-1}, $sw_{C^s}(c_j)$ is the weight of concept c_j in context C^s.

40.3.1.3 Personalizing Search Using the User Context

We personalize search results of query q^{s+1} related to the user context C^s represented as an ordered set of weighted concepts $<c_j, sw(c_j)>$ by combining for each retrieved document d_k, its initial score S_i and its contextual score S_c as follows:

$$S_f(d_k) = \gamma * S_i(q, d_k) + (1-\gamma) * S_c(d_k, C^s)$$
$$0 < \gamma < 1 \tag{40.5}$$

Contextual score S_c of document d_k is computed using the cosine similarity measure between d_k and the concepts of the user context C^s as follows:

$$S_c(d_k, C^s) = \frac{1}{h} * \sum_{j=1..h} sw(c_j) * cos(\mathbf{d_k}, \mathbf{c_j}) \tag{40.6}$$

Where c_j is a concept in the user context, $score(c_j)$ is the weight of concept c_j in the user context C^s.

40.3.2 How to Detect Session Boundaries

We propose a session boundary recognition method using the *Kendall* rank correlation measure that quantifies the conceptual correlation ΔI between the user context C^s and the query q^{s+1}. We choose a threshold σ and believe the queries are from the same session if the correlation is above the threshold.

Here, the term-based query vector $\mathbf{q_t^{s+1}}$ (where terms are weighted according to their frequency in the query) is matched with concepts of the ontology in order to represent the concept vector $\mathbf{q_c^{s+1}}$. We adopt a context-sensitive query weighting scheme by introducing the query frequency (QF) in the current search session, in order to rank concepts in the top of $\mathbf{q_c^{s+1}}$ when they are close to the user context. Indeed, query vector $\mathbf{q_c^{s+1}} =< w_1, w_2, ..., w_i, ... >$ is computed as follows:

$$w_i = CW(q^{s+1}, c_i) * QF(c_i) \tag{40.7}$$

where the query frequency (QF) and the concept weight (CW) are formally defined as:

$$QF(c_i) = \frac{|\mathbf{q}|^S}{< |\mathbf{q}|^S, c_i >}, CW(q^{s+1}, c_i) = cos(\mathbf{q_t^{s+1}}, \mathbf{c_i}) \tag{40.8}$$

$|\mathbf{q}|^S$ is the total number of related queries submitted in search session S, $< |\mathbf{q}|^S, c_i >$ is the number of user contexts built in the current search session S and containing concept c_i.

Thus, the similarity ΔI gauge the changes in the concept ranks between the query and the user context as follows.

$$\Delta I = (\mathbf{q} o \mathbf{C^s}) = \frac{\sum_c \sum_{c'} S_{cc'}(\mathbf{q}) \times S_{cc'}(\mathbf{C^s})}{\sqrt{\sum_c \sum_{c'} S_{cc'}^2(\mathbf{q}) \times \sum_c \sum_{c'} S_{cc'}^2(\mathbf{C^s})}} \tag{40.9}$$

$$S_{c_i c_j}(\mathbf{v}) = sign(\mathbf{v}(c_i) - \mathbf{v}(c_j)) = \frac{\mathbf{v}(c_i) - \mathbf{v}(c_j)}{|\mathbf{v}(c_i) - \mathbf{v}(c_j)|}$$

Where, c_i and c_j are two concepts issued from both the query and the user context, $\mathbf{q_c^{s+1}}(c_i)$ (resp. $\mathbf{C^s}(c_i)$) is the weight of the concept c_i in $\mathbf{q_c^{s+1}}$ (resp. $\mathbf{C^s}$).

40.4 Experimental Evaluation

Our experimental evaluation is designed to evaluate empirically the accuracy of the session boundary recognition measure and the effectiveness of the search personalization approach.

40.4.1 Experimental Data Sets

The experiments were based on the test data provided by TREC collection especially from disks 1&2 of the ad hoc task that contains 741670 documents. We particularly tested topics from $q_{51} - q_{100}$ presented in Table 40.1. The choice of this test collection is due to the availability of a manually annotated domain for each query. This allows us, on one hand, to enhance the data set with simulated user interests associated for each TREC domain. On the other hand, we can define a search session as a set of related queries annotated in the same domain of TREC.

40.4.2 Experimental Design and Results

Our experimental methodology is designed to evaluate three particular topics:

- The accuracy of our proposed session boundary recognition mechanism
- The effectiveness of our personalized search approach comparatively to typical search

40.4.2.1 Evaluating the Session Boundary Recognition Method

The goals of the session boundary recognition experiments are: (A) analyzing query – context correlation values (ΔI) according to the *Kendall* coefficient measure, (B) computing the accuracy of the session identification measure and identifying the best threshold value (σ).

For this purpose, we apply a real evaluation scenario that consists of choosing a query sequence holding six successive sessions related to six domains of TREC listed in Table 40.1. For testing purpose, we build the user context for each query using 30 of its relevant documents listed in the TREC assessment file and update it using formula (8) across related queries of the same domain using $\beta = 0.2$.

(A) Analyzing Query-Context Correlations

In this experiment, we computed the query-context correlations values between a particular query and the user context built across previous and queries related to the same TREC domain with respect to the query sequence. Figure 40.2 shows a query sequence holding six search sessions presented on the X-axis and the query–context correlation values on the Y-axis. A fall of the correlation curve means a decrease of the correlation degree with the previous query and possible session boundary identification. A correct session boundary is marked by a vertical line according to

Table 40.1 TREC domains with annotated queries

Domains	Queries
Environment	59 77 78 83
Military	62 71 91 92
Law and government	70 76 85 87
International relations	64 67 69 79 100
US economics	57 72 84
International politics	61 74 80 93 99

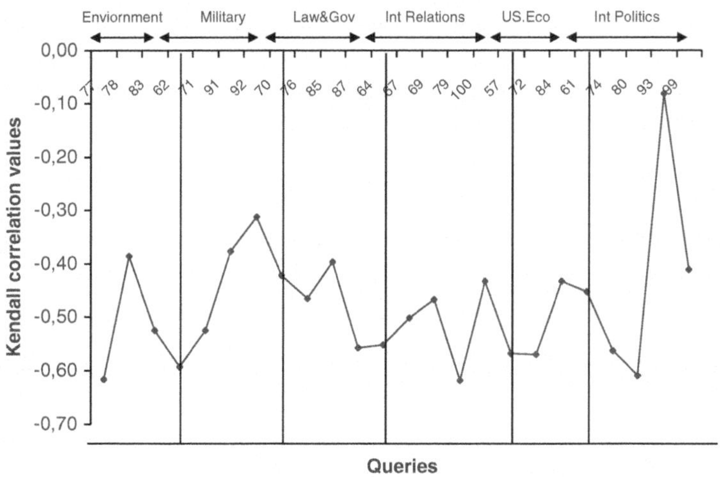

Fig. 40.2 Kendall correlation values computed across a query sequence

the annotated queries in each TREC domain. We can notice that correlation values vary between queries in the same domain. Indeed, in the domain "Environment" of TREC, some queries are related to the environmental concepts of the ODP, while a specific query (q_{59}) is related to the "Weather" topic that has no match with the set of the environmental concepts.

Based on the range of correlation values $[-0.61, -0.08]$, we identify in the next paragraph the best threshold cut-off value (σ).

(B) Measuring the Session Boundary Measure Accuracy

The goal of this experiment is to evaluate the accuracy of the session boundary detection measure. It is computed for each threshold value σ as follows:

$$P(\sigma) = \frac{|CQ|}{|Q|} \tag{40.10}$$

$|CQ|$ is the number of queries identified as correctly correlated to the current user context along the query sequence, and $|Q|$ is the total number of correlated queries in the query sequence.

We show in Fig. 40.3 the accuracy of the Kendall correlation measure with varying the threshold in the range of ($[-0.61, -0.08]$. The optimal threshold value is identified at -0.58 achieving the optimal accuracy of 70%. We can conclude that the Kendall measure achieves significant session identification accuracy. Indeed, it takes into account the concept ranks in both the query and the user context representations, which makes it tolerant for errors of allocating related queries in different search sessions.

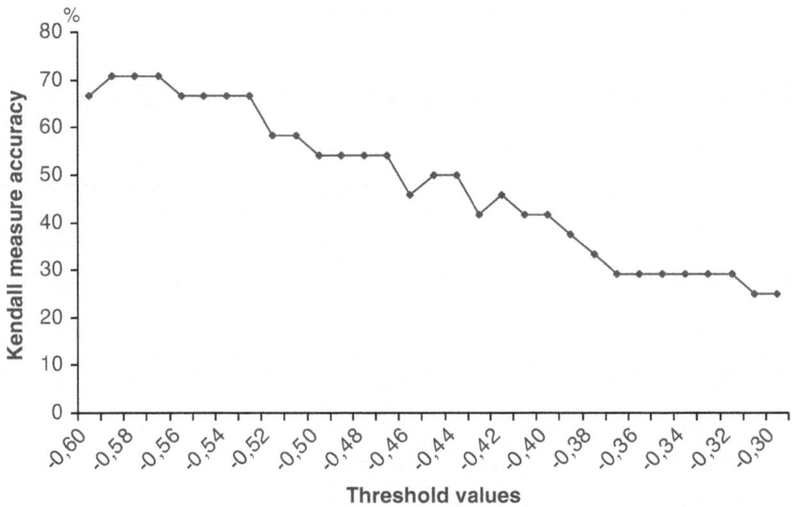

Fig. 40.3 Kendall correlation accuracy with varying the threshold value

40.4.2.2 Retrieval Effectiveness

Our experimental design for evaluating the retrieval effectiveness consists of comparing the personalized search performed using the query and the suitable user context to the standard search performed using only the query ignoring any user context.

We conducted two sets of controlled experiments: (A) study the effect of the re-ranking parameter γ in the re-ranking formula (9) on the personalized precision improvement, (B) evaluate the effectiveness of the personalized search comparatively to the standard search.

We used "*Mercure*" as a typical search engine where the standard search is based on the BM25 scoring formula retrieval model. We measure the effectiveness of re-ranking search results in terms of Top-n precision (P5, P10) and Mean average precision (MAP) metrics.

The evaluation scenario is based on the k-fold cross validation explained as follows:

- For each simulated TREC domain in Table 40.1, divide the query set into k equally-sized subsets, and using $k - 1$ training subsets for learning the user context and the remaining subset as a test set.
- For each query in the training set, an automatic process generates the associated keyword user context based on its top n relevant documents listed in the TREC assessment file provided by TREC using formula (1), and then maps it on the ODP ontology to extract the semantic user context.
- Update the user context concept weights across an arbitrary order of the queries in the training set using $\beta = 0.2$ in formula (8) and use it for re-ranking the search results of the queries in the test set using $h = 3$ in formula (9).

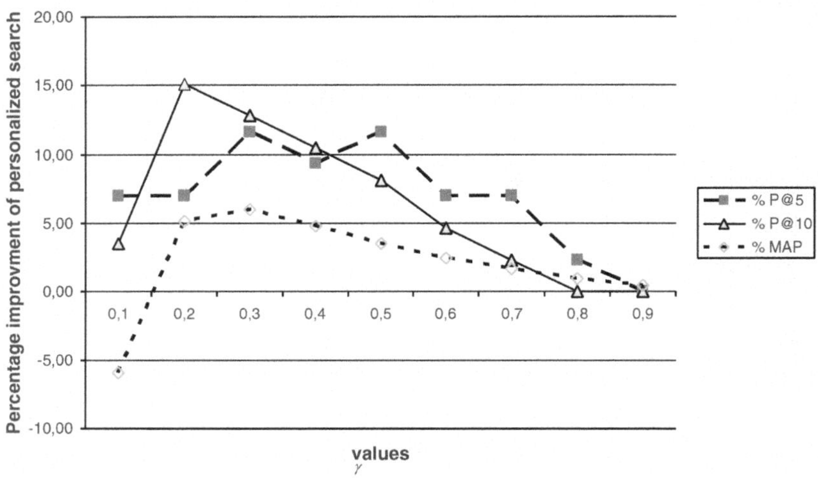

Fig. 40.4 Percentage of improvement achieved by personalized search as a result of varying γ in the re-ranking formula (9)

(A) Effects of Re-ranking Parameter γ on the Retrieval Effectiveness
We present in Fig. 40.4 the precision improvement graph obtained for the personalized search compared to the standard search at each cutoff of P5, P10 and MAP averaged over the queries belonging to the same domain. In this experiment, we fix the number of relevant documents per query used to represent the user context randomly at 30 for testing purpose. We see that the setting ($\gamma = 0.3$) produces the best improvement in personalized search since it produces higher precision improvement at P5 (11.63%). This proves that favoring contextual score returned by the system in the result ranking using small values of $\gamma (\gamma < 0.5)$ against the original score gives better performance for improving the web search ranking.
(B) Personalized Retrieval Effectiveness
In this experiment, we evaluated the effectiveness of the personalized search over various simulated domains. We have computed the percentage of improvement of personalized search comparatively to the standard search computed at P5, P10 and MAP and averaged over the queries belonging to the same domain. Precision improvement (*Impro*) is computed as follows: $Impro = (P_{personalized} - P_{baseiine})/P_{baseline} * 100$. We fixed γ at the best value occurring at 0.3 in the re-ranking formula (9). Then, we ran experiment to identify the best number of the relevant document used to represent the user context per query and get 20 at the best value. Results are presented in Table 40.2.

We see that personalized search improves the retrieval precision of almost the queries in the six simulated domains. However, the precision improvement varies between domains. This is probably due on one hand, to the accuracy level of the user context representation, and on the other hand to the correlation degrees between queries of the same domain. Indeed some queries annotated in the same domain of TREC may not share concepts of the ontology, then re-ranking search results with

Table 40.2 Result effectiveness of the personalized search

	The baseline model			Our personalized retrieval model					
Domain	P@5	P@10	MAP	P@5	Impro	P@10	Impro	MAP	Impro
Environment	0.25	0.32	0.18	0.35	40%	0.37	15.38%	0.19	1.73%
Military	0.25	0.27	0.05	0.35	40%	0.32	18.18%	0.07	46.46%
Law & Gov.	0.40	0.42	0.12	0.50	25%	0.45	5.88%	0.14	12.33%
Inter. Rel.	0.16	0.12	0.01	0.16	0%	0.16	33.33%	0.02	36.59%
US Eco.	0.26	0.30	0.09	0.33	25%	0.36	22.22%	0.10	8.35%
Int. Pol.	0.16	0.10	0.05	0.20	25%	0.16	60%	0.07	42.26%

not related concepts influences the precision improvement and probably reduce the retrieval performance especially for *Law & Gov* TREC domain.

40.5 Conclusion and Outlook

In this chapter, we described our approach for a session-based personalized search. It consists of learning short term user interests by aggregating concept-based user contexts identified within related queries. We proposed a session boundary identification based on the Kendall rank correlation mesure, which tracking changes in the dominant concepts between a new query and the current user context. Our experimental results show that the session boundary identification based on the Kendall measure achieves significant accuracy. Moreover, our experimental evaluation shows a significant improvement of personalized retrieval effectiveness compared to the typical search.

In future work, we plan to enhance the user context representation with semantically related concepts according to the ontology. We plan also to evaluate the accuracy of the user context representation as well as its impact on the retrieval effectiveness. Moreover, we intend to evaluate the session boundary recognition measure using real user data provided by web search engine log file, which can reveal comparison with time-based session boundary recognition approaches.

References

1. Bin Tan, Xuehua Shen, and ChengXiang Zhai. Mining long-term search history to improve search accuracy. In *KDD '06: Proceedings of the 12th ACM SIGKDD international conference on Knowledge discovery and data mining*, pages 718–723, New York, NY, USA, 2006. ACM.
2. Smitha Sriram, Xuehua Shen, and Chengxiang Zhai. A session-based search engine. In *SIGIR'04: Proceedings of the International ACM SIGIR Conference*, 2004.
3. Fang Liu, Clement Yu, and Weiyi Meng. Personalized web search for improving retrieval effectiveness. *IEEE Transactions on Knowledge and Data Engineering*, 16(1):28–40, 2004.

4. Ahu Sieg, Bamshad Mobasher, and Robin Burke. Web search personalization with ontological user profiles. In *Proceedings of the CIKM'07 conference*, pages 525–534, New York, NY, USA, 2007. ACM.
5. Lynda Tamine-Lechani, Mohand Boughanem , Nesrine Zemirli. Personalized document ranking: exploiting evidence from multiple user interests for profiling and retrieval. to appear. *In Journal of Digital Information Management*, vol. 6, issue 5, 2008, pp. 354–366.
6. John Paul Mc Gowan. *A multiple model approach to personalised information access*. Master thesis in computer science, Faculty of science, Universit de College Dublin, February 2003.
7. Alessandro Micarelli and Filippo Sciarrone. Anatomy and empirical evaluation of an adaptive web-based information filtering system. *User Modeling and User-Adapted Interaction*, 14(2–3):159–200, 2004.
8. Hyoung R. Kim and Philip K. Chan. Learning implicit user interest hierarchy for context in personalization. In *Proceedings of IUI '03*, pages 101–108, New York, NY, USA, 2003. ACM.
9. Susan Gauch, Jason Chaffee, and Alaxander Pretschner. Ontology-based personalized search and browsing. *Web Intelli. and Agent Sys.*, 1(3–4):219–234, 2003.
10. Ahu Sieg, Bamshad Mobasher, Steve Lytinen, Robin Burke. Using concept hierarchies to enhance user queries in web-based information retrieval. In *The IASTED International Conference on Artificial Intelligence and Applications. Innsbruck, Austria*, 2004.
11. Mariam Daoud, Lynda Tamine-Lechani, and Mohand Boughanem. Using a concept-based user context for search personalization. to appear. In *Proceedings of the 2008 International Conference of Data Mining and Knowledge Engineering (ICDMKE'08)*, pages 293–298. IAENG, 2008.

Chapter 41
Mining Weather Information in Dengue Outbreak: Predicting Future Cases Based on Wavelet, SVM and GA

Yan Wu, Gary Lee, Xiuju Fu, Harold Soh, and Terence Hung

Abstract Dengue Fever has existed throughout the contemporary history of mankind and poses an endemic threat to most tropical regions. Dengue virus is transmitted to humans mainly by the *Aedes aegypti* mosquito. It has been observed that there are significantly more *Aedes aegypti* mosquitoes present in tropical areas than in other climate regions. As such, it is commonly believed that the tropical climate suits the life-cycle of the mosquito. Thus, studying the correlation between the climatic factors and trend of dengue cases is helpful in conceptualising a more effective pre-emptive control measure towards dengue outbreaks. In this chapter, a novel methodology for forecasting the number of dengue cases based on climactic factors is presented. We proposed to use Wavelet transformation for data pre-processing before employing a Support Vector Machines (SVM)-based Genetic Algorithm to select the most important features. After which, regression based on SVM was used to perform forecasting of the model. The results drawn from this model based on dengue data in Singapore showed improvement in prediction performance of dengue cases ahead. It has also been demonstrated that in this model, prior climatic knowledge of 5 years is sufficient to produce satisfactory prediction results for up to 2 years. This model can help the health control agency to improve its strategic planning for disease control to combat dengue outbreak. The experimental result arising from this model also suggests strong correlation between the monsoon seasonality and dengue virus transmission. It also confirms previous work that showed mean temperature and monthly seasonality contribute minimally to outbreaks.

Keywords Dengue fever modelling · wavelet · SVM · infectious disease · data mining

---

Y. Wu (✉)
Agency for Science, Technology and Research
20 Biopolis Way, #07–01 Centros (A*GA), Singapore 138668
E-mail: yan_wu@scholars.a-star.edu.sg

S.I. Ao and L. Gelman (Eds.), *Advances in Electrical Engineering and Computational Science*, Lecture Notes in Electrical Engineering.

41.1 Introduction

In 2006, the World Health Organization reported that "dengue is the most rapidly spreading vector-borne disease". Between 1985 and 2006, in Southeast Asia alone, an estimated 1,30,000 people were infected annually with dengue virus, which is transmitted to humans mainly by the *Aedes aegypti* mosquito [1]. The Dengue virus is transmitted from a female *Aedes* mosquito to a human or vice versa during the feeding process, generally referred to as horizontal transmission. The Dengue virus may also be passed from parent mosquitoes to offspring via vertical transmission.

However, mosquitoes can only survive in certain climate conditions. For example, the *Aedes aegypti* mosquito cannot survive below the freezing temperature of 0°C [2]. However, climatic conditions worldwide have displayed enormous short-term abnormalities, which increases the likelihood of mosquitoes surviving longer and penetrating temperate regions that were previously free from mosquito-borne diseases [2].

In addition to climate conditions, other factors contribute to the time-series transmission profile of the dengue virus, e.g. geographical expansion of *Aedes* mosquito population and its density, the demographic distribution, mobility and susceptibility of the human population. This dynamic relationship is illustrated in the Fig. 41.1 below.

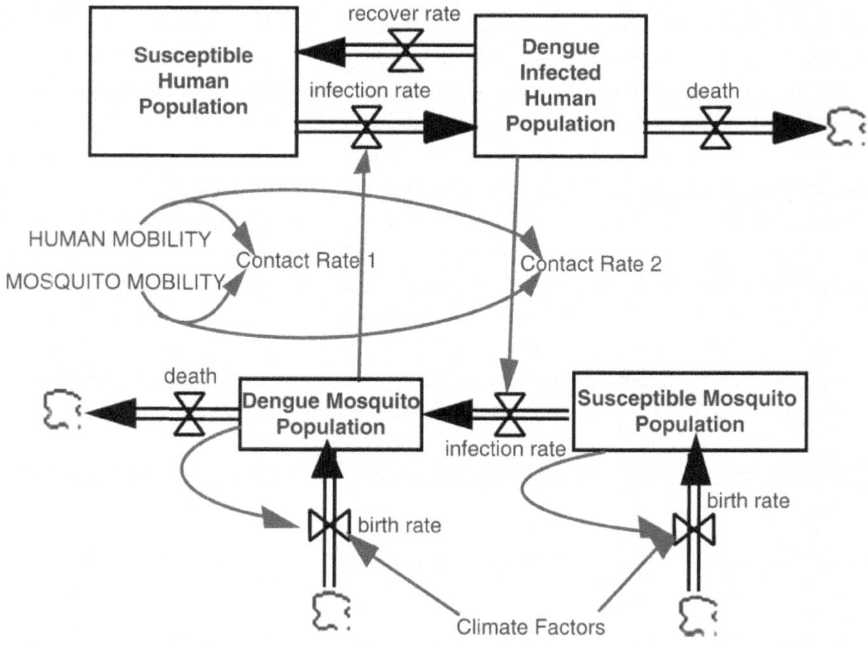

Fig. 41.1 The dynamics of dengue virus spread

The ideal method of modelling the dynamics of the dengue outbreak is to precisely capture each and every contributing element in the chain. However, associating factors such as mobility of human population to an exact function in the model remains a challenging problem. In this chapter, we proposed a method to simplify the problem in a small relatively well-defined area.

Previous work by Fu et al. [3] incorporated time-lags of weather features into their model and selected contributing features with a genetic algorithm. However, the work assumed the usefulness of all levels of detail, in terms of time resolutions/time lags, including noise that was likely present in every feature. Although the method was able to detect seasonal factors that contributed to the outbreak of dengue fever, the resulting model possessed a large number of features, which raised the curse of dimensionality [4]. The work also ignored the use of regression learning for testing the generality of its feature selection model for case forecasting.

The following sections of this chapter describe how a simple yet effective model can be constructed to accommodate the above limitations. Briefly, the proposed method pre-processes the data using Wavelet Transformation (WT) to separate information at different time resolutions superposed in the time-series. Genetic Algorithms (GA) with Support Vector Machines (SVM) are then applied to select features. Following this, regression based on SVM is used to perform forecasting using the model. Conclusions drawn from the empirical modelling using the dengue cases in Singapore are studied, discussed and compared against [3].

41.2 Model Construction

41.2.1 Data Representation

Data selected for the model should represent the many variables affecting the spread of the dengue virus. Prior studies have shown that the population size of the mosquito depends largely on the presence of suitable breeding sites and climate conditions [5]. As such, we can make the following assumptions of the input data:

- Variations in the weather data represent the breeding patterns of the *Aedes* mosquitoes and
- The time-series of dengue cases in a relatively confined area represents the ensemble of dengue spread information such as susceptibility of human population to the virus

The coding pattern of the above two pieces of information is unknown but likely non-linear. Moreover, not all information at different time resolutions superposed in the time-series of an input feature, such as temperature, contributes to the fluctuation of dengue cases. Perhaps only one or a few seasonal levels of detail in a particular input feature contribute to the outbreak. Hence, the use of an input feature without eliminating irrelevant levels of details induces unnecessary noise into the model.

In our model, the input features consisted of the past weekly dengue cases and eight daily weather features, namely rainfall intensity, cloudiness (maximum, mean and minimum), temperature and humidity. As the dengue cases were only recorded when a patient was diagnosed and the incubation period of dengue virus is known to differ across individuals, daily dengue cases do not accurately represent the outbreak. Thus, summing the daily cases into weekly buckets reduces such inaccuracy. As such, all daily weather factors were averaged into weekly means after performing wavelet decomposition as described in Section 41.2.2 below.

41.2.2 Wavelet Decomposition

In order to separate the different levels of details encoded in the input elements of the system, wavelet decomposition of the time-series inputs was chosen. WT are similar to Fourier transforms. The fundamental difference is that instead of deriving the spectral density of a give signal, WT projects the signal onto functions called *wavelets* with dilation parameter $\boldsymbol{a}$ and translation parameter $\boldsymbol{b}$ of their mother wavelet:

$$\psi_{a,b}(t) = \frac{1}{\sqrt{a}}\psi\left(\frac{t-b}{a}\right) \tag{41.1}$$

subject to the condition of:

$$\Psi_o = \int_0^{+\infty} \frac{|\Psi(\omega)|^2}{|\omega|} d\omega \text{ is bounded} \tag{41.2}$$

where $\Psi(\omega)$ is the Fourier transform of $\psi(t)$

The continuous WT is defined as follows:

$$T_{a,b}(f) = \int_{-\infty}^{+\infty} f(t)\psi^*(a,b)dt \text{ or}$$

$$T_{a,b}(f) = \langle f(t), \psi_{a,b}(t)\rangle \tag{41.3}$$

where $\langle *, *\rangle$ denotes the scalar product in the space of $f(t)$

Hence the reconstructed signal from the WT is:

$$f(t) = \frac{1}{\Psi_{\emptyset}} \int_{-\infty}^{+\infty}\int_0^{+\infty} T_{a,b}(f)\psi(a,b)da/a^2dc \tag{41.4}$$

If a finite number of terms are taken to represent the original signal, Eq. (41.4) can be rewritten as a multi-resolution approximation [4]:

$$\hat{f}(t) = \sum_{i=1}^{N} \Psi_i \psi(a_i, b_i) \tag{41.5}$$

This can be interpreted as an ensemble of N levels of details $\{D_1 \ldots D_N\}$ and a level of approximation $\{A_N\}$ which is illustrated in Eq. (41.6).

$$\hat{f}(t) = A_0 = A_N + \sum_{n=1}^{N} D_n, \text{ where } A_n = A_{n+1} + D_{n+1} \tag{41.6}$$

While the decomposed signals resemble the fundamental ideas of the use of time-lags, it separates signals at different levels which in turn separate potential noise from signals.

41.2.3 Genetic Algorithm

After Wavelet decomposition, each signal input into the model generates a series of daughter Wavelets inputs. As mentioned earlier, not all levels of detail of every input signal may be relevant. Thus, it is important to filter sub-signals that are irrelevant to the output.

For this purpose, we implemented a biologically-inspired, robust, general-purpose optimisation algorithm: the genetic algorithm (GA). In the GA, each input feature is represented as an allele in a chromosome [6]. The evolution of chromosomes begins from the original (randomly initialized) population and propagates down generationally. During each generation, the fitness of all chromosomes in the population is evaluated. Then, competitive selection and modification are performed to generate a new population. This process is repeated until a predefined limit or convergence is achieved. In order to confirm and validate the feature selection model and to generate a better representative set of contributing factors, a 10 cross-validation technique was introduced into the methodology. The converged solution of weights in zeros and ones represented the non-selection/selection of a particular sub-signal by the GA.

41.2.4 Support Vector Machines

The core component of GA is to construct the learning classifier. Among the supervised learning techniques used to correlate input and output data, the Support Vector Machine (SVM) stands firmly as one of the most successful methods. The SVM maps input vectors to a higher dimensional space where a maximum hyperplane is drawn to separate classes [7]. Given a set of input $\{(\mathbf{x}_i, y_i)\}$, where y_i is the corresponding output $\{-1, 1\}$ to the ith input feature vector $\mathbf{x}_i\{\mathbf{x}_i \in R^n\}$. $\Phi(\mathbf{x}_i)$

maps $\mathbf{x}_i$ into a higher dimensional space where a possible linear estimate function can be defined:

$$f(\mathbf{x}) = \mathbf{w}^T \Phi(\mathbf{x}) + b = \mathrm{sgn}\left(\sum_{i \varepsilon SVs} \alpha_i y_i K(\mathbf{x}_i, \mathbf{x}) + b\right) \tag{41.7}$$

where $K(\mathbf{x}_i, \mathbf{x}) = \Phi(\mathbf{x})^T \Phi(\mathbf{x}_i)$ is the kernel function, *SVs* are the Support Vectors, α is the Lagrange multipliers and b is the bias. The SVM problem can be formulated as a quadratic programming problem by optimising α as:

The SVM problem can be formulated as a quadratic programming problem by optimising:

$$\min_{\alpha_i} \sum_{i=1}^{1} \alpha_i - \frac{1}{2} \sum_{i=1}^{1} \sum_{j=1}^{1} \alpha_i \alpha_j y_i y_j K(\mathbf{x}_i, \mathbf{x}_j) \tag{41.8}$$

subject to the condition of:

$$0 \le a_i \le C \quad \forall_i \text{ and } \sum_i a_i y_i = 0$$

where C is the capacity control penalty term.

41.2.5 Regression Learning

After the SVM-based GA selects the most relevant features, the input is fed into a regression learning algorithm which generates a prediction model. An SVM regression (SVR) method based on the material discussed in Section 41.2.4 above can be used for non-linear learning by introducing a threshold ε analogous to the function of C in SVM Classification [8].

41.2.6 Model Assessment Criteria

To assess the quality of the final model, we compared predicted results to the actual data using two simple metrics: the Mean Squared Error (MSE) and the Coefficient of Determination (R^2). A small MSE and large $R^2 (0 <= R^2 <= 1)$ suggest a better modelling of the data.

41.3 Results and Discussions

Because Dengue fever occurs throughout the year in Singapore, we gathered the dengue case and weather data in Singapore from 2001 to 2007. As shown in Fig. 41.2 below, this set of dengue data includes three significant peaks at Weeks 190, 245 and 339.

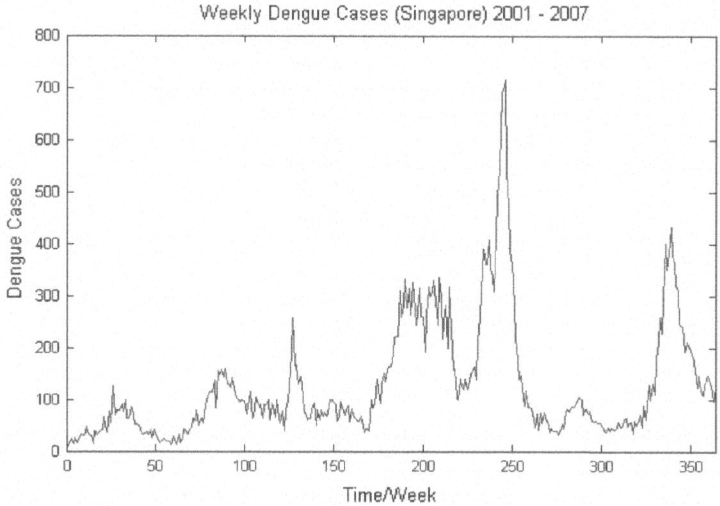

Fig. 41.2 Weekly dengue cases in Singapore from 2001 to 2007

Dengue cases over these years in Singapore are reported weekly while climate data from weather stations were collected daily. We corrected the resolution difference by first performing Wavelet transform of the weather data before sorting the samples into weekly bins. Additionally, we discarded the finest three levels of detail generated by Wavelet transforms because the down-sampling by seven makes $\log_2(7) \approx 3$ levels of detail redundant. The mother Wavelet employed here is the Daubechies [9] 6 Wavelet which was selected because reconstruction of the signal after suppression of the finest three levels of detail showed the best approximation of the original signal compared to Daubechies 2 and 4.

Before the GA and SVM were applied, all the input and output vectors were normalised. The normalisation process ensured that none of the variables overwhelmed the others in terms of the distance measure, which was critical for proper feature selection. In our experiment, a zero-mean unit-variance normalisation was employed to give equal distance measures to each feature and levels of detail.

The current dengue cases and decomposed current weather series were used as input features with future dengue cases as output into GA with 10 cross-validations to generate a general model of the data represented. A feature was finally selected if and only if it was selected in at least seven of the ten validation cycles. The method employed in GA was a tournament with the aim of minimising the root mean squared error. A recursive method was used to tune the parameter C for best performance ($C = 1$ in this case). There are only 13 terms selected by GA out of the 63 input signals, tabulated in Table 41.1 below. As compared to the 50 features selected by the model in [3], the dimensionality required to describe the number of dengue cases was significantly smaller in this model. In the Table, the first term of a feature denotes its most general trend (A_N) while the second to seventh term,

Table 41.1 GA selected features

Features	Terms selected
Current dengue cases	Entire series
Cloudiness	Third, fourth, seventh
Maximum temperature	Fourth
Mean temperature	None
Minimum temperature	Fifth
Maximum humidity	Fifth, seventh
Mean humidity	Fifth
Minimum humidity	Second
Rainfall intensity	Second, fifth

D_N to D_1 respectively, denotes different levels of detail from the coarsest to the finest.

From Table 41.1, we observed that either the fourth or fifth term was selected in almost all the input features. These levels of detail corresponded to the averaged details between 2–4 months. In Singapore, the climate can be divided into two main seasons, the Northeast Monsoon and the Southwest Monsoon season, separated by two relatively short inter-monsoon periods [10]. Each of these periods has a span of 2–4 months, which correlates well with our finding and increases the likelihood of a close relationship between dengue outbreak and seasonality in Singapore, i.e. the seasonal change of temperature, humidity and rainfall. From the literature reviewed in [2], it had been concluded that the rise in temperature does accelerate the life-cycle of the mosquito. This agrees with the findings from our model, i.e. the first terms of all temperature features were not selected.

In [3], Fu suggested that time-lags of (0, 2, 6, 8, 12) weeks are important. In contrast, our model selected a time-lag of 4 weeks which corresponded to an averaged detail of 4 weeks, i.e. sixth term in our feature vectors. From Table 41.1, none of the sixth terms were selected in this model. Taken together, both models suggest that the contribution of monthly fluctuations to dengue virus transmission may not be significant. Moreover, both models also suggest that mean temperature does not contribute to the prediction models at any level. The general trend (seventh term) present in Cloudiness and Maximum Humidity suggests that some positive correlation exists between the growth of dengue cases and these two features.

In our previous work [11], we demonstrated that for future dengue outbreak prediction, SVR was likely a better model than linear regression due to its robustness even in the event of over-fitting. In [11], to investigate our claim that the coding pattern of weather data into dengue cases was non-linear, we performed regression with a Radial Basis Function (RBF) SVR with 5 years of training data and 1 year of testing input. The width γ used in RBF was set to be the inverse of the sample size (312). The range of data used for training and testing were taken from 2001–2006.

To confirm the validity of the model, we carried out a similar experiment with a different range of training and testing data (2002–2007) while keeping the model variables unchanged. Instances of the results from these two sets of data are tabu-

Table 41.2 Performance of an instance of RBF SVR approach

RBF SVR	With GA	MSE	R^2	Figure
Training data: 01–05	No	0.27	0.75	3a
Testing data: 06	Yes	0.091	0.92	3b
Training data: 02–06	No	0.37	0.71	3c
Testing data: 07	Yes	0.085	0.91	3d

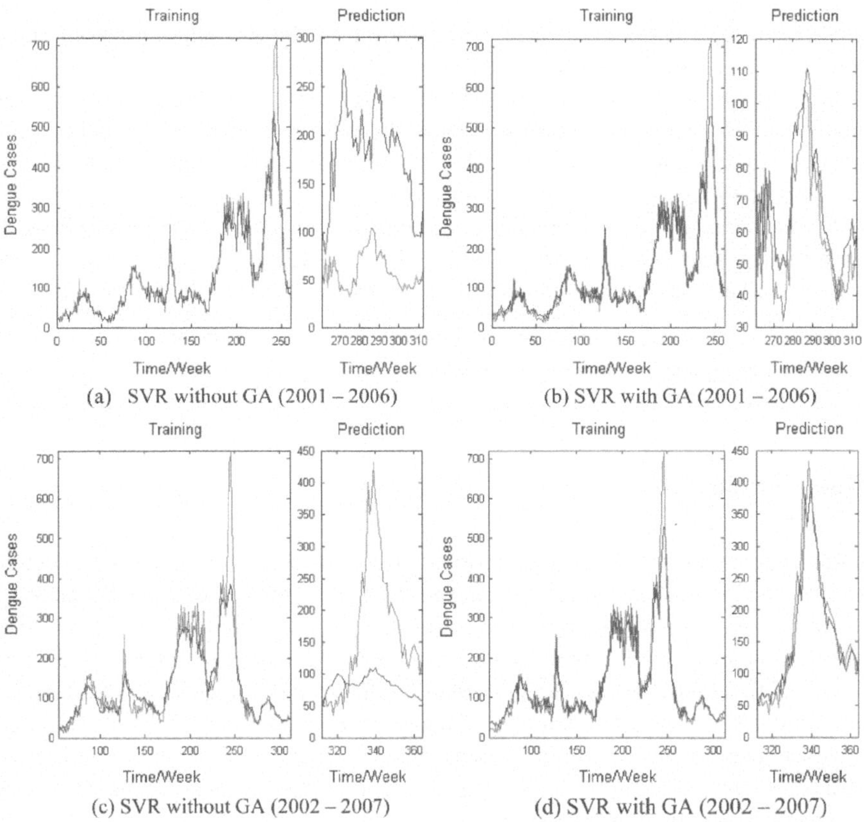

Fig. 41.3 Prediction results of SVR with and without GA (**a**) SVR without GA (2001–2006), (**b**) SVR with GA (2001–2006), (**c**) SVR without GA (2002–2007), (**d**) SVR with GA (2002–2007)

lated in Table 41.2 and plotted in Fig. 41.3 (red and blue lines denote actual and predicted No. of cases respectively). The performance statistics show that mapping the weather data into the RBF space (which is non-linear) produced a high correlation between the input data and the actual dengue cases, reinforcing our claim of non-linearity.

Statistics in Table 41.2 show that on average, SVR with GA yields a 70% reduction in MSE and a 25% increase in correlation with the actual data as compared to that of SVR without GA. From this observation, we can infer that a significant

Table 41.3 Performance of different sample combinations

RBF SVR with GA	MSE	R^2
Training: 01–05 Testing: 06	0.091	0.92
Training: 01–05 Testing: 06 & 07	0.094	0.90
Training: 02–06 Testing: 07	0.085	0.91

number of irrelevant input features were removed by the GA, supporting our claim that not all levels of detail influenced the spread of dengue.

In order to test the stability of our proposed model, a further experiment was carried out. We extended the testing samples to two consecutive years (2006–2007) using the same training data (2001–2005). The result is tabulated in Table 41.3 in comparison with the previous two experiments.

The performance statistics in Table 41.3 clearly show that the difference in MSE is minimal (< 0.01) and the correlation coefficient is within 2% of each other. These empirical results suggest that the proposed model is capable of the followings:

1. Producing relatively reliable predictions for up to 2 years ahead
2. Constructing a stable model instance using only 5 years of data

41.4 Conclusion

In this chapter, a novel model for predicting future dengue outbreak was presented. We proposed the use of Wavelet decomposition as a mean of data pre-processing. This technique, together with SVM-based GA feature selection and SVR learning, proved to be effective for analysing a special case of infectious diseases – dengue virus infection in Singapore. Our empirical results strongly suggest that weather factors influence the spread of dengue in Singapore. However, the results also showed that mean temperature and monthly seasonality contribute minimally to outbreaks. Analysis of the results in this chapter gives rise to further epidemiological research questions on dengue transmissions:

- As low frequency (higher order) terms denote general trends of the signal, the results tabulated in Table 41.1 indicate that dengue incidence seems to be closely related to the rainfall and humidity trends. Can this be proven epidemiologically?
- Although the rise in average temperature is thought to accelerate mosquito's breeding cycle, why was there no low frequency terms of temperature represented in the model? Moreover, why are some moderate frequencies (between 2 and –4 months) of both maximum and minimum temperatures selected in the model?
- As Singapore is an area surrounded by sea, does this research finding fit well into other regions with dengue prevalence?

To provide answers to these questions, further work has to be performed, particularly in the collection of more precise data over a longer period of time. Predictions of longer durations ahead, such as 2 weeks onwards, can also be constructed to measure the performance of the model.

Because outbreaks are stochastic events, it is not possible to precisely predict the specific realization of one particular outbreak. However, models, such as the one proposed in our work, are useful for providing quantitative risk assessments of disease spread in a well-defined area, which is essential information for constructing effective public-health strategies.

Acknowledgements Our heartfelt gratitude is expressed to William C. Tjhi from Institute of High Performance Computing, Singapore for reviewing the Chapter. We also wish to acknowledge NOAA of the USA for releasing daily weather data free-of-charge for research and education.

References

1. World Health Organisation: Dengue Reported Cases. 28 July 2008. WHO <http://www.searo.who.int/en/Section10/Section332_1101.htm>
2. Andrick, B., Clark, B., Nygaard, K., Logar, A., Penaloza, M., and Welch, R., "Infectious disease and climate change: detecting contributing factors and predicting future outbreaks", *IGARSS '97: 1997 IEEE International Geoscience and Remote Sensing Symposium*, Vol. 4, pp. 1947–1949, Aug. 1997.
3. Fu, X., Liew, C., Soh, H., Lee, G., Hung, T., and Ng, L.C. " Time-series infectious disease data analysis using SVM and genetic algorithm", *IEEE Congress on Evolutionary Computation (CEC) 2007*, pp. 1276–1280, Sept. 2007.
4. Mallat, S.G., "Multiresolution approximations and wavelet orthonormal bases of $L^2(R)$", *Transactions of the American Mathematical Society*, Vol. 315, No. 1, pp. 69–87, Sept. 1989.
5. Favier, C., Degallier, N., Vilarinhos, P.T.R., Carvalho, M.S.L., Yoshizawa, M.A.C., and Knox, M.B., "Effects of climate and different management strategies on Aedes aegypti breeding sites: a longitudinal survey in Brasflia (DF, Brazil)", *Tropical Medicine and International Health 2006*, Vol. 11, No. 7, pp. 1104–1118, July 2006.
6. Grefenstette, J.J., *Genetic algorithms for machine learning*, Kluwer, Dordrecht, 1993.
7. Burges, C.J.C., "A tutorial on support vector machines for pattern recognition", *Data Mining and Knowledge Discovery*, Vol. 2, No. 2, pp. 955–974, 1998.
8. Drucker, H., Burges, C.J.C., Kaufman, L., Smola, A., and Vapnik, V., "Support Vector Regression Machines", *Advances in Neural Info Processing Systems 9*, MIT Press, Cambridge, pp. 155–161, 1996.
9. Daubechies, I. "Orthonormal Bases of Compactly Supported Wavelets." *Communications on Pure and Applied Mathematics,* Vol. 41, pp. 909–996, 1988.
10. National Environment Agency, Singapore: Climatology of Singapore. 20 Aug. 2007. NEA, Singapore. <http://app.nea.gov.sg/cms/htdocs/article.asp?pid=1088>
11. Wu, Y., Lee, G., Fu, X., and Hung, T., "Detect Climatic Factors Contributing to Dengue Outbreak based on Wavelet, Support Vector Machines and Genetic Algorithm", *World Congress on Engineering 2008*, Vol. 1, pp. 303–307, July 2008.
12. Bartley, L.M., Donnelly, C.A., and Garnett, G.P., "Seasonal pattern of dengue in endemic areas: math models of mechanisms", *Transactions of the Royal Society of Tropical Medicine and Hygiene*, pp. 387–397, July 2002.
13. Shon, T., Kim, Y., Lee, C., and Moon, J., "A machine learning framework for network anomaly detection using SVM and GA", *Proceedings from the Sixth Annual IEEE SMC Information Assurance Workshop 2005*, pp. 176–183, June 2005.

14. Nakhapakorn, K. and Tripathi, N. K., "An information value based analysis of physical and climatic factors affecting dengue fever and dengue haemorrhagic fever incidence", *International Journal of Health Geographics*, Vol. 4, No. 13, 2005.
15. Ooi, E., Hart, T., Tan, H., and Chan, S., "Dengue seroepidemiology in Singapore", *The Lancet*, Vol. 357, No. 9257, pp. 685–686, Mar 2001.
16. Ministry of Health, Singapore: Weekly Infectious Diseases Bulletin. 28 July 2008. M.O.H. Singapore. <http://www.moh.gov.sg/mohcorp/statisticsweeklybulletins.aspx>
17. Gubler, D.J., "Dengue and dengue hemorrhagic fever", *Clinical Microbiology Reviews*, Vol. 11, No. 3, pp. 480–496, July 1998.

Chapter 42
PC_Tree: Prime-Based and Compressed Tree for Maximal Frequent Patterns Mining

Mohammad Nadimi-Shahraki, Norwati Mustapha, Md Nasir B Sulaiman, and Ali B Mamat

Abstract Knowledge discovery or extracting knowledge from large amount of data is a desirable task in competitive businesses. Data mining is an essential step in knowledge discovery process. Frequent patterns play an important role in data mining tasks such as clustering, classification, and prediction and association analysis. However, the mining of all frequent patterns will lead to a massive number of patterns. A reasonable solution is identifying maximal frequent patterns which form the smallest representative set of patterns to generate all frequent patterns. This research proposes a new method for mining maximal frequent patterns. The method includes an efficient database encoding technique, a novel tree structure called PC_Tree and PC_Miner algorithm. Experiment results verify the compactness and performance.

Keywords Maximal frequent pattern · frequent pattern · prime number · database encoding

42.1 Problem of Maximal Frequent Patterns Mining

Since the introduction of the Apriori algorithms [1], frequent patterns mining plays an important role in data mining research for over a decade. Frequent patterns are itemsets or substructures that exist in a dataset with frequency no less than a user specified threshold.

Let L = $\{i_1, i_2 \ldots i_n\}$ be a set of items. Let D be a set of database transactions where each transaction T is a set of items and |D| be the number of transactions in D. Given P = $\{i_j \ldots i_k\}$ be a subset of L ($j \leq k$ and $1 \leq j, k \leq n$) is called a

M. Nadimi-Shahraki (✉)
Department of Computer Engineering, Islamic Azad University, Najafabad branch, Iran and PhD student of Computer Science, Faculty of Computer Science and Information Technology, Putra University of Malaysia, 43400 UPM, Selangor, Malaysia
E-mail: admin1@iaun.ac.ir

S.I. Ao and L. Gelman (Eds.), *Advances in Electrical Engineering and Computational Science*, Lecture Notes in Electrical Engineering.

pattern. The support of a pattern P or S (P) in D is the number of transactions in D that contains P. Pattern P will be called frequent if its support is no less than a user specified support threshold min_sup $\sigma(0 \leq \sigma \leq |D|)$.

In many real applications especially in dense data with long frequent patterns enumerating all possible 2^L – two subsets of an L length pattern is infeasible [2]. A reasonable solution is identifying a smaller representative set of patterns from which all other frequent patterns can be derived [3]. Maximal frequent patterns (MFP) form the smallest representative set of patterns to generate all frequent patterns [4]. In particular, the MFP are those patterns that are frequent but none of their supersets are frequent. The problem of maximal frequent patterns mining is finding all MFP in D with respect to σ.

This research introduces a new method to find all MFP using only one database scan. The method includes an efficient database encoding technique, a novel tree structure called Prime-based encoded and Compressed Tree or PC_Tree and also PC_Miner algorithm. The database encoding technique utilizes prime number theory and transforms all items from each transaction into only a positive integer. The PC_Tree is a novel and simple tree structure but yet efficient to capture whole of transactions information. The PC_Miner algorithm traverses the PC_Tree efficiently using pruning techniques to find the MFP.

42.2 Related Work

Many efficient algorithms have been introduced to solve the problem of maximal frequent pattern mining more efficiently. They are almost based on three fundamental frequent patterns mining methodologies: Apriori, FP-growth and Eclat [5]. Mostly, they traverse the search space to find MFP. The key to an efficient traversing is the pruning techniques which can remove some branches in the search space. The pruning techniques can be categorized into two groups:

SUBSET FREQUENCY PRUNING: THE ALL SUBSETS OF ANY FREQUENT PATTERN ARE PRUNED BECAUSE THEY CAN NOT BE MAXIMAL FREQUENT PATTERN.
SUPERSET INFREQUENCY PRUNING: THE ALL SUPERSETS OF ANY INFREQUENT PATTERN ARE PRUNED BECAUSE THEY CAN NOT BE FREQUENT PATTERN.

The Pincer-Search algorithm [6] uses horizontal data layout. It combines a bottom-up and a top down techniques to mine the MFP. However search space is traversed without an efficient pruning technique. The MaxMiner algorithm [4] uses a breadth-first technique to traverse of the search space and mine the MFP. It makes use of a look ahead pruning strategy to reduce database scanning. It prunes the search space by both subsets frequency and supersets infrequency pruning. The Depth Project [7] finds MFP using a depth first search of a lexicographic tree of patterns, and uses a counting method based on transaction projections. The DepthProject demonstrated an efficient improvement over previous algorithms for mining MFP. The Mafia [2] extends the idea in DepthProject. It uses a search strategy has been improved by an effective pruning mechanisms. The MaxMiner, DepthProject and Mafia use Rymon's set enumeration [8] to enumerate all the patterns. Thus these algorithms avoid having to compute the support of all the frequent patterns. The Flex [9] is a lexicographic tree designed in vertical layout to store pattern P and list of transaction identifier where pattern P appears. Its structure is restricted test-and-generation instead of Apriori-like is restricted generation-and-test. Thus nodes generated are certainly frequent. The Flex tree is constructed in depth-first fashion. Recently a new two-way-hybrid algorithm [10] for mining MFP uses a flexible two-way-hybrid search method. The two-way-hybrid search begins the mining procedure in both the top-down and bottom-up directions at the same time. Moreover, information gathered in the bottom-up can be used to prune the search space in the other top down direction.

42.3 Proposed Method

This research proposes a new method to mine all MFP in only one database scan efficiently. The method introduces an efficient database encoding technique, a novel tree structure called PC_Tree to capture transactions information and PC_Miner algorithm to mine MFP.

42.3.1 Database Encoding Technique

The presentation and encoding of database is an essential consideration in almost all algorithms. The most commonly database layout is the horizontal and vertical layout [11]. In both layouts, the size of database is very large. The database encoding is a useful technique which can reduce the size of database. Obviously, reducing of the size of database can enhance performance of mining algorithms. Our method uses a prime-based database encoding technique to reduce the size of transaction database. It transforms each transaction into a positive integer called Transaction Value (TV) during of the PC_Tree construction as follows: Given transaction T = (tid, X) where tid is the transaction-id and X = $\{i_j \ldots i_k\}$ is the transaction-items or pattern X. While the PC_Tree algorithm reads transaction T, the encoding procedure considers a prime number p_r for each item i_r in pattern X, and then TV_{tid} is computed by Eq. (42.1). Therefore, all transactions can be represented in new layout using this encoding technique.

$$TV_{tid} = \prod_{j}^{k} p_r \quad \mathrm{T} = (\mathrm{tid}, \mathrm{X}) \tag{42.1}$$

$$\mathrm{X}* = \{\mathrm{i}_j \ldots \mathrm{i}_k\} \text{ and } \mathrm{i}_r \text{ is presented by } p_r)$$

Table 42.1 The transaction database DB and its Transaction Values

TID	Items	Encoded	TV
1	A, B, C, D, E	2, 3, 5, 7, 11	2,310
2	A, B, C, D, F	2, 3, 5, 7, 13	2,730
3	A, B, E	2, 3, 11	66
4	A, C, D, E	2, 5, 7, 11	770
5	C, D, F	5, 7, 13	455
6	A, C, D, F	2, 5, 7, 13	910
7	A, C, D	2, 5, 7	70
8	C, D, F	5, 7, 13	455

The encoding technique utilizes Eq. (42.1) based on simple following definitions.

A positive integer N can be expressed by unique product
$N = p_1^{m_1} p_2^{m_2} \dots p_r^{m_r}$ where p_i is prime number,
$p_1 \prec p_2 \prec \cdots p_r$ and m_i is a positive integer, called the multiplicity of p_i [12].

For example, $N = 1,800 = 2^3 * 3^2 * 5^2$. Here we restrict the multiplicity to $m_i = 1$ because there is no duplicated item in transaction T.

To facilitate the process of the database encoding technique used in our method, let's examine it through an example. Let item set L = {A, B, C, D, E, F} and the transaction database, DB, be the first two columns of Table 42.1 with eight transactions. The fourth column of Table 42.1 shows TV_{tid} computed for all transactions.

42.3.2 PC_Tree Construction

Using tree structure in mining algorithms makes two possibilities to enhance the performance of mining. Firstly, data compressing by well-organized tree structures like FP-tree. Secondly, reducing search space by using pruning techniques. Thus the tree structures have been considered as a basic structure in previous data mining research [5, 9, 13]. This research introduces a novel tree structure called PC_Tree (Prime-based encoded and Compressed Tree). The PC_Tree makes use of both possibilities data compressing and pruning techniques to enhance efficiency.

A PC_Tree includes of a root and some nodes that formed sub trees as children of the root or descendants. The node structure consisted mainly of several different fields: value, local-count, global-count, status and link. The value field stores TV to records which transaction represented by this node. The local-count field set by 1 during inserting current TV and it is increased by 1 if its TV and current TV are equal. The global-count field registers support of pattern P which presented by its TV.

In fact during of insertion procedure the support of all frequent and infrequent patterns is registered in the global-count field. It can be used for interactive mining where min_sup is changed by user frequently [13]. The status field is to keep track-

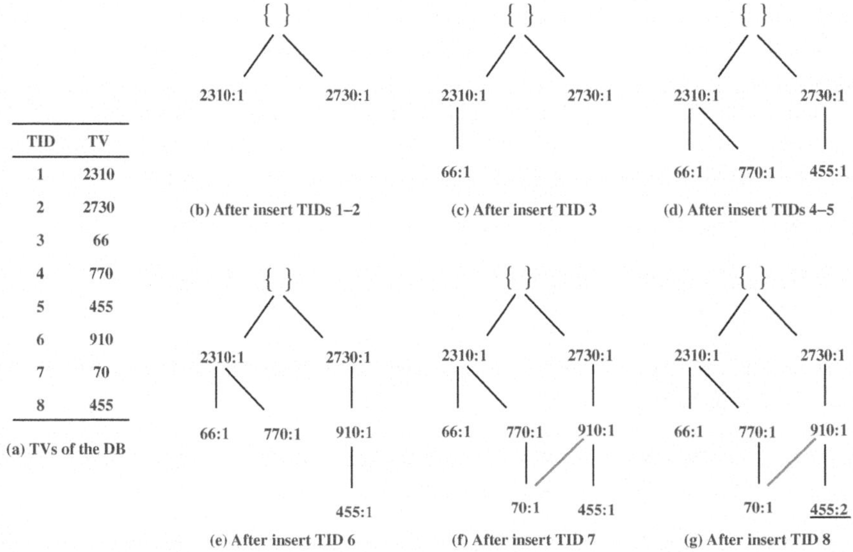

Fig. 42.1 Construction of PC_Tree

ing of traversing. When a node visited in the traversing procedure the status field is changed from 0 to 1. The link field is to form sub trees or descendants of the root.

Figure 42.1 shows step by step construction of PC_Tree for transactions shown in Table 42.1 which summarized in Fig. 42.1a.

The construction operation mainly consists of insertion procedure that inserts TV(s) into PC_Tree based on definitions below:

Definition 42.1. IF TV OF NODE n_r AND n_s IS EQUAL THEN $r = s$. INSERTION PROCEDURE INCREASES LOCAL – COUNT FIELD OF NODE n_r BY 1 IF THE CURRENT TV IS EQUAL WITH TV OF n_r.

Definition 42.2. $R = (n_i, n_{i-1} \ldots n_j, root)$ IS A DESCENDANT IFF TV OF NODE $n_r \in R(i \leq r \leq j)$ CAN DIVIDE ALL TVS KEPT IN NODES DESCENDANT $R_r = (n_{r+1}, n_{r+2}, \ldots n_j, root)$.

Definition 42.3. TV OF THE ROOT IS ASSUMED NULL AND CAN BE DIVIDED BY ALL TVS.

Definition 42.4. NODEn_rCAN BE BELONGED TO MORE THAN ONE DESCENDANT.

Figure 42.1b shows insertion of the first and second transactions. The second TV with value 2,730 can not be divided by the first TV with value 2,310 and it creates a new descendant using definition 2 and 3. Transactions 3–6 are inserted into their descendant based on definition 2 shown in Fig. 42.1c–e. Insertion of the seventh transaction applies definition 4 when TV 70 is belonged to two descendants (second descendant shown in the red and bold line) shown in Fig. 42.1f. The TV of eighth transaction with value 455 is equal with the fourth TV, then the local-count field of forth TV is increased by 1 using definition 1 shown in Fig. 42.1g (shown in underline).

Each TV in PC_Tree represents a pattern P and the support of pattern P or S (P) is registered in the global-count field. Given pattern P and Q have been presented by TVP and TVQ respectively, the PC_Tree has some nice below properties:

Property 42.1. The S (P) is computed by traversing only descendants of TVP.

Property 42.2. P and Q belong to descendant R and S (P) < S (Q) iff TVP can be divided by TVQ.

Property 42.3. S (P) $\geq \sigma$ and S (P's fathers in its descendants) $\prec \sigma$ IFF P is a Maximal Frequent Pattern.

Property 42.4. Nodes are arranged according to TV order, which is a fixed global ordering. In fact the PC_Tree is an independent-frequency tree structure.

Property 42.5. important procedures are almost done only by two simple mathematic operations product and division. Obviously using mathematic operation enhances the performance instead of string operation.

42.3.2.1 PC_Miner Algorithm

As explained in previous section, during of insertion each TV in the PC_Tree, the following procedures are done.

(a) Item-frequency counting
(b) Local-counting
(c) Global-counting
(d) Descendant constructing

The PC_Miner algorithm traverses the completed PC_Tree to discover the MFP in top-down direction. There is no need to database scan again, because all information about items and patterns are stored in the PC_Tree. However Fig. 42.1g as a completed PC_Tree didn't show some information like global-count stored in the tree. The miner algorithm makes use of a combined pruning strategy including both superset infrequency and subset frequency pruning. As a result the search space is reduced, which dramatically reduces the computation and memory requirement and enhances the efficiency. The superset infrequency pruning is assisted by the frequency of items computed during the PC-Tree construction. Table 42.2 shows the item frequency and considered prime number for transaction database shown in Table 42.1.

To illustrate the pruning strategy used in the PC_Miner algorithm let's examine it by some examples. Figure 42.2 shows the completed PC_Tree using TVs and patterns respectively. Given $\sigma = 5$, according to Table 42.2, infrequent item set (IFI) consists of {B, E, F} then all superset of IFI like node (A, B, C, D, E) can be

Table 42.2 Frequency of items and considered prime numbers

Item	Prime number	Item frequency
A	2	6
B	3	3
C	5	7
D	7	7
E	11	3
F	13	4

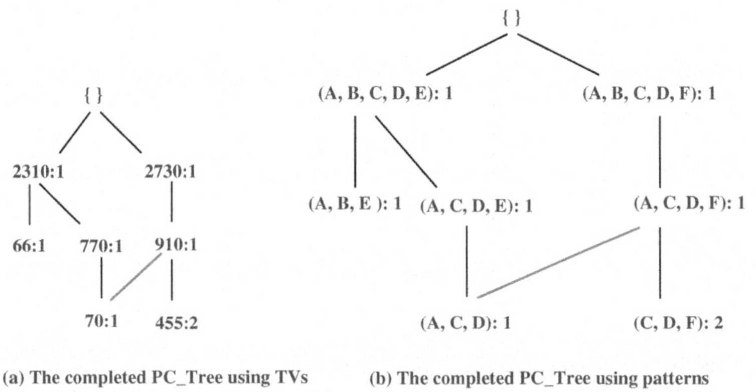

Fig. 42.2 The completed PC_Tree

pruned – the superset infrequency pruning. Given σ has been changed from 5 to 2 then all items are frequent. While the PC_Miner traverses left sub tree or descendant of node (A, B, C, D, E), node (A, C, D, E) is found as a maximal frequent – the property 3. Then all its subsets (here is only node (A, C, D)) are pruned – the subset frequency pruning.

42.4 Experimental Results

In this section, the accuracy and performance of the method is evaluated. All the experiments are performed on PC with CPU Intel P4 2.8 GHz, 2 GB main memory, and running Microsoft Windows XP. All the algorithms are implemented using Microsoft Visual C++6.0.

In first experiment, the Synthetic data used in our experiments are generated using IBM data generator which has been used in most studies on frequent patterns mining. We generate five datasets with number of items 1,000, average transaction length ten and number of transaction 1,000–10,000 that called D1, D2, D4, D8, and D10 respectively. In this experiment, the compactness of our database encoding technique is verified separately. In fact all TVs transformed are stored in a flat file

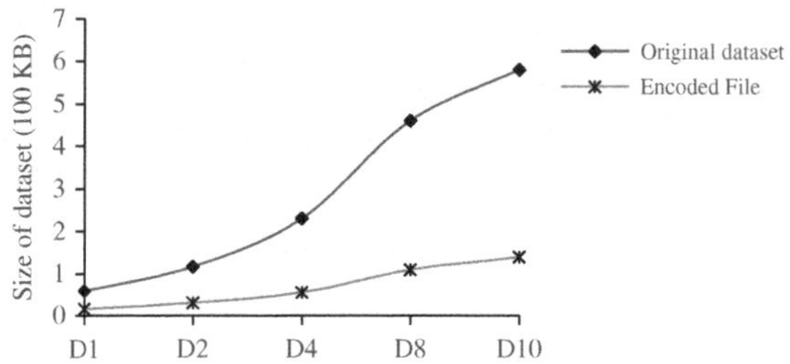

Fig. 42.3 The compactness of the database encoding technique

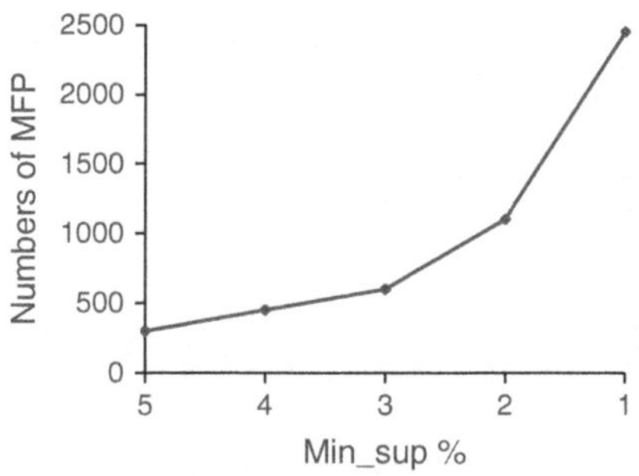

Fig. 42.4 The number of MFP discovered for T10.I6.D10k

called encoded file then size of this file is compared with the size of original dataset. It shows good results and reducing the size of these datasets more than half as shown in Fig. 42.3.

In second experiment, we show accuracy and correctness of the method. The test dataset T10.I6.D10K is also generated synthetically by the IBM data generator. Figure 42.4 shows the numbers of maximal frequent patterns discovered for the tests at varying min_sup on this dataset.

In third experiment, in order to evaluate the scalability of our new algorithm, we applied it as well as Apriori to four IBM dataset generated in experiment 1. Figure 42.5 shows the performance of two algorithms as a function of the numbers of transactions for min_sup 2%. When number of transaction is less than 5,000 Apriori slightly outperforms the PC_Miner in execution time. When the numbers of transactions are increased, the execution time of Apriori degraded as compared to the PC_Miner.

Fourth experiment is to compare the performance of the PC-Miner with the Apriori and Flex algorithms on the dataset mushroom, which is a real and dense dataset contains characteristics of various species of mushrooms [14]. In dataset mushroom, there are 8,124 transactions, the number of items is 119 and the average transaction length is 23. Figure 42.6 shows the PC_Miner algorithm outperforms both Apriori and Flex algorithms.

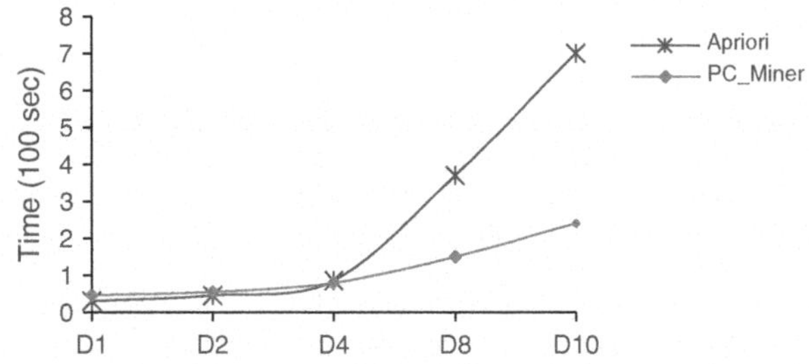

Fig. 42.5 The scalability of PC_Miner vs. Apriori on datasets D1-D10

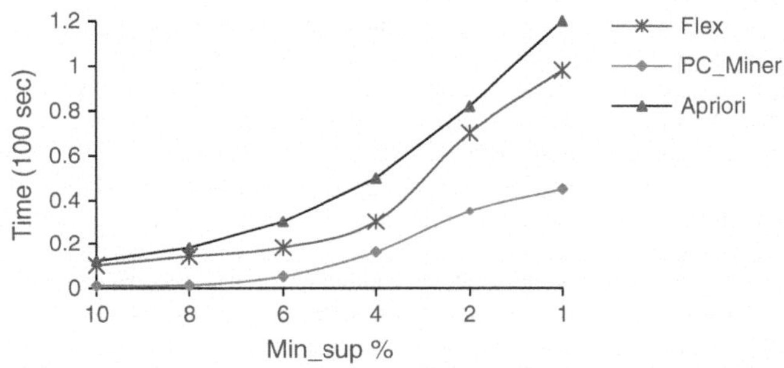

Fig. 42.6 The performance of PC-miner vs. Flex on dataset mushroom

42.5 Conclusion and Future Works

In this paper, we proposed a new method to discover maximal frequent patterns efficiently. Our method introduced an efficient database encoding technique, a novel tree structure called Prime-based encoded and Compressed Tree or PC_Tree and also PC_Miner algorithm. The experiments verified the compactness of the database encoding technique. The PC_Tree presented well-organized tree structure with nice properties to capture transaction information. The PC_Miner reduced the search space using a combined pruning strategy to traverses the PC_Tree efficiently. The experimental results showed the PC_Miner algorithm outperforms the Apriori and Flex algorithms. For interactive mining where min_sup can be changed frequently, the information kept in the PC_Tree can be used and tree restructuring is no needed.

In fact this research introduced a number-theoretic method to discover MFP that made use of prime number theory and simple computation based on division operation and a combined pruning strategy. There are some directions for future improvement optimal data structures, better memory management and pruning method to enhance the efficiency. This method also can be extended for incremental frequent patterns mining where transaction database is updated or minimum support threshold can be changed [13].

References

1. R. Agrawal and R. Srikant, Fast algorithms for mining association rules, 20th International Conference. Very Large Data Bases, VLDB, 1215: 487–499 (1994).
2. D. Burdick, M. Calimlim, and J. Gehrke, Mafia: A maximal frequent itemset algorithm for transactional databases, 17th International Conference on Data Engineering: pp. 443–452 (2001).
3. S. Bashir and A.R. Baig, HybridMiner: Mining Maximal Frequent Itemsets Using Hybrid Database Representation Approach, 9th International Multitopic Conference, IEEE INMIC: pp. 1–7 (2005).
4. R.J. Bayardo Jr, Efficiently mining long patterns from databases, ACM SIGMOD International Conference on Management of Data: pp. 85–93 (1998).
5. J. Han, H. Cheng, D. Xin, and X. Yan, Frequent pattern mining: Current status and future directions, Data Mining and Knowledge Discovery, 15(1): 55–86 (2007).
6. D.I. Lin and Z.M. Kedem, Pincer-Search: A New Algorithm for Discovering the Maximum Frequent Set, Advances in Database Technology–EDBT'98: 6th International Conference on Extending Database Technology, Valencia, Spain (1998).
7. R.C. Agarwal, C.C. Aggarwal, and V.V.V. Prasad, Depth first generation of long patterns, Sixth ACM SIGKDD International Conference on Knowledge Discovery and Data Mining: pp. 108–118 (2000).
8. R. Rymon, Search through systematic set enumeration, Third International Conference on Principles of Knowledge Representation and Reasoning: pp. 539–550 (1992).
9. N. Mustapha, M.N. Sulaiman, M. Othman, and M.H. Selamat, FAST DISCOVERY OF LONG PATTERNS FOR ASSOCIATION RULES, International Journal of Computer Mathematics, 80(8): 967–976 (2003).
10. F. Chen, and M. Li, A Two-Way Hybrid Algorithm for Maximal Frequent Itemsets Mining, Fourth International Conference on Fuzzy Systems and Knowledge Discovery, FSKD 2007.
11. M.J. Zaki, Scalable algorithms for association mining, IEEE Transactions on Knowledge and Data Engineering, 12(3): 372–390 (2000).
12. T.T. Cormen, C.E. Leiserson, and R.L. Rivest, Introduction to algorithms: MIT Press, Cambridge, MA (1990).
13. M. Nadimi-Shahraki, N. Mustapha, M.N. Sulaiman, and A. Mamat, Incremental updating of frequent pattern: basic algorithms, Second International Conference on Information Systems Technology and Management (ICISTM 08) (1): 145–148 (2008).
14. C.B.a.C. Merz, UCI Repository of Machine Learning, 1998.

Chapter 43
Towards a New Generation of Conversational Agents Based on Sentence Similarity

Karen O'Shea, Dr. Zuhair Bandar, and Dr. Keeley Crockett

Abstract The Conversational Agent (CA) is a computer program that can engage in conversation using natural language dialogue with a human participant. Most CAs employ a pattern-matching technique to map user input onto structural patterns of sentences. However, every combination of utterances that a user may send as input must be taken into account when constructing such a script. This chapter was concerned with constructing a novel CA using sentence similarity measures. Examining word meaning rather than structural patterns of sentences meant that scripting was reduced to a couple of natural language sentences per rule as opposed to potentially 100s of patterns. Furthermore, initial results indicate good sentence similarity matching with 13 out of 18 domain-specific user utterances as opposed to that of the traditional pattern matching approach.

Keywords Conversational Agent · Sentence Similarity · word meaning

43.1 Introduction

The concept of 'intelligent' machines was first conceived by the British mathematician Alan Turing [1]. The imitation game, known as the 'Turing Test', was devised to determine whether or not a computer program was 'intelligent'. This led to the development of the Conversational Agent (CA) [1] – a computer program that can engage in conversation using natural language dialogue with a human participant.

CAs can exist in two forms: 'Embodied' agents [2] possess an animated humanoid body and exhibit attributes such as facial expressions and movement of eye gaze. 'Linguistic' agents [3, 4] consist of spoken and/or written language without

K. O'Shea (✉)
The Intelligent Systems Group, Department of Computing and Mathematics, Manchester Metropolitan University, Chester Street, Manchester M1 5GD
E-mail: k.oshea@mmu.ac.uk

S.I. Ao and L. Gelman (Eds.), *Advances in Electrical Engineering and Computational Science*, Lecture Notes in Electrical Engineering.

embodied communication. One of the earliest text-based CAs developed was ELIZA [3]. ELIZA was capable of creating the illusion that the system was actually listening to the user simply by answering questions with questions. This was performed using a simple pattern matching technique, mapping key terms of user input onto a suitable response. Further advancements on CA design led to PARRY [4], capable of exhibiting personality, character, and paranoid behavior by tracking its own internal emotional state during a conversation. Unlike ELIZA, PARRY possessed a large collection of tricks, including: admitting ignorance by using expressions such as "I don't know" in response to a question; changing the subject of the conversation or rigidly continuing the previous topic by including small stories about the theme [4]. CAs can also engage in social chat and are capable of forming relationships with a user. ALICE [5], an online chatterbot and InfoChat [6] are just two such examples. By conversing in natural language these CAs are able to extract data from a user, which may then be used throughout the conversation.

Considerable research has been carried out on the design and evaluation of embodied CAs [2,7]; however, little work appears to have been focused on the actual dialogue. This paper will concentrate on text-based CAs and the development and evaluation of high-quality dialogue.

Most text-based CA's scripts are organized into contexts consisting of a number of hierarchically organized rules. Each rule possesses a list of structural patterns of sentences and an associated response. User input is then matched against the patterns and the pre-determined response is sent as output. InfoChat [6] is one such CA capable of interpreting structural patterns of sentences. However, every combination of utterances must be taken into account when constructing a script – an evidently time-consuming, high maintenance task, which undoubtedly suggests scope for alternative approaches. It is, therefore, envisaged that the employment of sentence similarity measures could reduce and simplify CA scripting by using a few prototype natural language sentences per rule.

Two successful approaches to the measurement of sentence similarity are: Latent Semantic Analysis (LSA) [8] and Sentence Similarity based on Semantic Nets and Corpus Statistics [9]. LSA is a theory and method for extracting and representing the contextual-usage meaning of words by statistical computations applied to a large corpus of text [8]. A word by context matrix is formed based on the number of times a given word appears in a given set of contexts. The matrix is decomposed by Singular Value Decomposition (SVD) into the product of three other matrices, including the diagonal matrix of singular values [10]. This dimension reduction step collapses the component matrices so that words that occurred or did not occur in some contexts now appear with a greater or lesser frequency [8]. Reconstruction of the original matrix enables LSA to acquire word knowledge among large numbers of contexts. Although LSA makes no use of syntactic relations, it does, however, offer close enough approximations of people's knowledge to underwrite and test theories of cognition. Sentence Similarity based on Semantic Nets and Corpus Statistics will be employed as the measure in this research and will be described in further detail in Section 43.2.

This chapter is organized as follows: Section 43.2 will describe and illustrate the sentence similarity measure; Section 43.3 will describe a traditional CA scripting approach; Section 43.4 will describe CA scripting using sentence similarity; Section 43.5 will present an experimental analysis of the two approaches; Section 43.6 will evaluate the results and Section 43.7 will conclude and highlight areas for further work.

43.2 Sentence Similarity Measure

Sentence Similarity based on Semantic Nets and Corpus Statistics [9] is a measure that focuses directly on computing the similarity between very short texts of sentence length. Through the use of a lexical/semantic knowledge-base such as WordNet [11], the length of separation between two words can be measured, which in turn, can be used to determine word similarity. The synset – a collection of synonyms – at the meeting point of the two paths is called the subsumer. The depth of the subsumer is similarly measured by counting the levels from the subsumer to the top of the hierarchy. Li et al. [9, 12] proposed that the similarity between two words be a function of the attributes: path length and depth. The algorithm initiates by combining the two candidate sentences (T1 and T2) to form a joint word set using only distinct words. For example:

T1 = Mars is a small red planet
T2 = Mars and Earth orbit the sun
A joint word set 'T' is formed where:
T = Mars is a small red planet and earth orbit the sun

As a result, each sentence is represented by the use of the joint word set with no surplus information. Raw semantic vectors are then derived for each sentence using the hierarchical knowledge-base WordNet [11], in order to determine the separation between words. Taking a non-linear transfer function as an appropriate measure, the following formula denotes a monotonically decreasing function of l, where l = path length between words and α is a constant.

$$f(l) = e^{-\alpha l} \tag{43.1}$$

As for the depth of the subsumer, the relationship of words at varying levels of the hierarchy must be taken into consideration. For example, words at the upper layers are far more general and less semantically similar than words at lower layers [9]. Therefore, subsuming words at upper layers must be scaled down whereas words at lower layers must be scaled up, resulting in a monotonically increasing function of h, where h = depth of subsumer and β is a constant.

$$f(h) = (e^{\beta l} - e^{-\beta h})/(e^{\beta l} + e^{-\beta h}) \tag{43.2}$$

As such, the raw similarity *s(w1, w2)* between two words is calculated as:

$$s(w1, w2) = e^{-\alpha l}.(e^{\beta l} - e^{-\beta h})/(e^{\beta l} + e^{-\beta h}) \quad (43.3)$$

where $\alpha = 0.2$ and $\beta = 0.45$.

Each word is then weighted, that is, assigned an information content value, based on its significance and contribution to contextual information. By combining the raw semantic vector *s(w1, w2)* with the information content of each word, $I(w1)$ and $I(w2)$, semantic vectors are created:

$$s_i = s(w1, w2).I(w1).I(w2) \quad (43.4)$$

Finally, the semantic similarity *Ss* between two sentences, $s1$ and $s2$, is calculated as:

$$Ss = si1.si2/\sqrt{si1}/\sqrt{si2} \quad (43.5)$$

where *si*1 is the resultant semantic vector of sentence 1 and *si*2 is the resultant semantic vector of sentence 2.

Word order also plays an active role in sentence similarity. Each word is assigned a unique index number which simply represents the order in which the word appears in the sentence. For example, take the following sentences denoted T1 and T2:

T1 = The cat ran after the mouse
T2 = The mouse ran after the cat
A joint word set 'T' is formed where:
T = The cat ran after the mouse

Each sentence is than compared to that of the joint word set. If the same word is present – or if not, the next most similar word – then the corresponding index number from T1 will be placed in the vector, r1. As such, the word order vectors r1 and r2 for the example sentence pair T1 and T2 would be formed as follows:

$$\text{r1} = \{123456\}$$
$$\text{r2} = \{163452\}$$

Therefore, word order similarity *Sr* is calculated as:

$$Sr = 1 - \sqrt{(r1 - r2)}/\sqrt{(r1 - r2)} \quad (43.6)$$

Finally, the sentence similarity is derived by combining both semantic similarity and word order similarity. The overall sentence similarity between two sentences $S(T1, T2)$ is calculated as:

$$S(T1, T2) = \delta Ss + (1 - \delta)Sr \quad (43.7)$$

where δ takes into account that word order plays rather a less significant role when determining sentence similarity.

43.3 Traditional CA Scripting

Traditional approaches [6] interpret structural patterns of sentences by using scripts consisting of rules organized into contexts. A context may be described as a collection of rules relating to a particular topic. Each context contains a number of hierarchically organized rules each possessing a list of structural patterns of sentences and an associated response. A user's utterance is then matched against the patterns and the associated response is 'fired' (selected) and sent as output. The following steps 1–3 illustrate the procedure.

1. Natural language dialogue from the user is received as input and is matched to a pattern contained in a rule.
2. Match-strength is calculated based on various parameters, including the activation level of each rule.
3. The pattern with the highest strength is thus 'fired' and sent as output.

Scripts are constructed by first assigning each rule a base activation level, a number between 0 and 1. The purpose of the activation level is to resolve conflicts when two or more rules have patterns that match the user's input [13]. The scripter must then decide which patterns a user may send in response to output. Each pattern is assigned a pattern-strength value, typically ranging between 10 and 50. For example, a rule may be constructed as follows:

```
<Rule_01>
a:0.5
p:50 *help*
p:50 I do not *<understand-0>*
r: How can I help you
```

where a = activation level, p = pattern strength/pattern, r = response.

Patterns can also contain wildcard elements * which will match with one or more consecutive characters. In addition, the macro < understand-0 > enables the scripter to incorporate stock patterns into a rule [6]. Writing such scripts is a time-consuming and highly skilled craft [14]. For example, a script typically consists of a number of contexts each denoting a particular topic of conversation. Each context contains a hierarchically organized list of rules each possessing a collection of structural patterns of sentences. However, modifying one rule or introducing a new rule into the script invariably has an impact on the remaining rules. As such, a reassessment of the entire script would be warranted, without which would render the CA futile. The scripter is, therefore, required to remember the rankings of the rules and predict how the introduction of new rules will interact with existing rules [13]. The huge overhead and maintenance of this type of scripting undoubtedly suggests scope for an alternative approach.

43.4 CA Scripting Using Sentence Similarity

The new proposed approach using sentence similarity will maintain the same script as that of the traditional approach; however, all patterns will be replaced with natural language sentences. This considerably reduces the burden and skill required to produce CA scripts. Through the use of a sentence similarity measure [9], a match is determined between the user's utterance and the natural language sentences. The highest ranked sentence is fired and sent as output. Figure 43.1. and the following steps 1–3 illustrate the procedure.

1. Natural language dialogue is received as input, which forms a joint word set with each rule from the script using only distinct words in the pair of sentences. The script is comprised of rules consisting of natural language sentences.
2. The joint word set forms a semantic vector using a hierarchical semantic/lexical knowledge-base [11]. Each word is weighted based on its significance by using information content derived from a corpus.
3. Combining word order similarity with semantic similarity the overall sentence similarity is determined. The highest ranked sentence is 'fired' and sent as output.

The proposed scripts are simply constructed by assigning a number of prototype natural language sentences per rule. For example, one such rule may be constructed as follows:

<Rule_01>
I need help
I do not understand
r: How can I help you

where s = sentence and r = response.

The precise number of sentences per rule will start at one and increase to 'n' where 'n' is determined by experimental analysis. However, it is expected that the value of 'n' will be small and significantly less than the number of patterns used in traditional scripting methodologies.

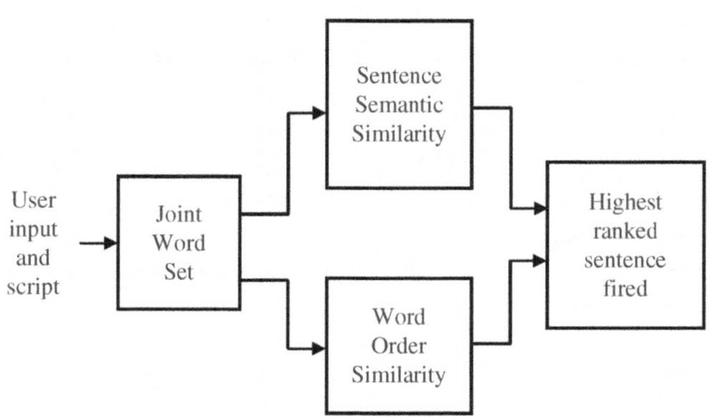

Fig. 43.1 New proposed CA algorithm

43.5 Experimental Methodology

43.5.1 Domain

The real world domain is concerned with advising students at University on debt management and the payment of tuition fees. For the purpose of experimentation, one script, which consists of 18 rules, was taken from a substantially extensive script developed by Convagent Ltd. [6]. These rules were representative of the domain and its complexity.

43.5.2 Experiments

Two sets of experiments were undertaken to compare the traditional scripted CA and the sentence similarity based CA. The first experiment examined the traditional approach using structural pattern of sentences [6]. The rules consisted of patterns, which were in some cases consolidated with macros. This accumulated the count of patterns into the 100s. In comparison, the second experiment examined the new proposed approach, which used a sentence similarity measure and was scripted using natural language sentences. Through the use of a sentence similarity measure, the level of scripting was reduced to a couple of generic prototype sentences. Table 43.1 illustrates the scripting by the two approaches for the same rule.

Approach one consists of structural patterns of sentences consolidated with macros. The macro <confused-0> contains 16 patterns. Similarly, the macros <confusing-0>, <sure-neg-0>, <sure-neg-1>, and <understand-0> contain a further 8, 21, 10 and 13 additional patterns respectively. This accumulates the final number of patterns, including the patterns *help* and *not* to 70. Approach two,

Table 43.1 Example scripting by two approaches to CA design

Approach one Traditional pattern scripting	Approach two New proposed scripting
<Rule_01>	<Rule_01>
a: 0.5	s: I need help
p: 50 *<confused-0>	s: I do not understand
p: 50 *<confused-0>*	s: This is confusing
p: 50 *<sure-neg-0>*	r: How can I help you
p: 50 *<sure-neg-1>*	
p: 50 *help*	
p: 50 *not *<understand-0>*	
r: How can I help you	

however, replaces the above patterns for three generic natural language sentences: "I need help", "I do not understand" and "This is confusing".

43.6 Results and Discussion

The first experiment examined the traditional approach using structural patterns of sentences [6], while the second approach examined the new proposed approach using natural language sentences. The experiments entailed sending as input 18 domain-specific user utterances. The 18 selected user utterances were deemed to be a good sample of the domain. The resulting output, that is the fired pattern/sentence, for the 18 cases are displayed in Table 43.2.

Table 43.2 Results of user input for two approaches to CA design

Utterance User input	Approach one Traditional pattern scripting Fired pattern	Approach two New proposed scripting Fired sentence
1. I am having trouble with mybenefactor	*	I have a problem with my sponsor
2. I want assistance	* Will pay *	I need help
3. I have not quit my course	*I* not *quit* course	I have not received my funding
4. Could I pay a tiny quantityof the cost	Could I *	I would like to pay a small amount of the fee
5. I have no finance	* No *	I have no funding
6. I have already paid the fee	* Have paid *	I could pay part of the fee
7. I have a different reason	* Have a *	It is none of those reasons
8. I have not sent any payment	* Not sent * payment *	Payment has not been sent
9. I am no longer studying atthe University	* No *	I am still attending my course
10. I have to wait for my career development loan draft	* Wait * loan	I am still waiting for my loan
11. I have not sent any payment however I have not quit	* However *	Payment has not been sent in the post
12. Could you repeat the choices	*	Please repeat the option
13. I have not yet obtained my student loan	* Student loan *	I have not received my student loan
14. My local education authority appraisal has been delayed	*	I have not received my local education authority assessment
15. My hardship finance has failed to arrive	* Hardship *	I have not received hardship funding
16. I am having trouble with my direct debit	* Direct debit *	I have direct debit problems
17. I am broke	*	I am not at the University
18. I sent you the cash weeks ago	* Sent *	Payment was sent in the post to the University last week

The results of the user utterances are: The outputs generated after the input of user utterances 3, 6, 8, 10, 13, 15, 16, and 18 indicate a correct firing by approach one. As a result, approach one appears to have found a structurally comparable match. The outputs generated after the input of user utterances 1, 2, 4, 5, 7, 8, 10, 12, 13, 14, 15, 16, and 18 indicate a correct firing by approach two. As a result, approach two appears to have found sufficient semantic similarity between the user utterances and the corresponding natural language sentences.

The outputs generated after the input of user utterances 1, 2, 4, 5, 7, 9, 11, 12, 14, and 17 indicate a miss-firing by approach one. As a result, approach one appears to have failed to find an identical or comparable match to that of the user utterance. The outputs generated after the input of user utterances 3, 6, 9, 11, and 17 indicate a miss-firing by approach two. As a result, approach two appears to have failed to identify sufficient semantic similarity between the user utterances and the natural language sentences.

In the cases where approach one miss-fired, this was due to the script not possessing an identical or comparable structural match. This, however, may be rectified by incorporating the missing patterns into the script. In the cases where approach two miss-fired, this was due to insufficient sentences representative of that specific rule. This, however, can be rectified by incorporating additional natural language sentences into the script. Furthermore, the sentence similarity measure could be adjusted so as to consider other parts-of-speech.

In totality, approach one correctly matched 8 out of 18 user utterances, whereas approach two correctly matched 13 out of 18 user utterances. Typically the number of patterns per rule for the traditional pattern script was between 50 and 200. In contrast, the average number of sentences per rule for the natural language script was three. Moreover, a considerable amount of time is required to write and maintain the scripts of approach one as opposed to that of approach two in which scripting is greatly simplified.

43.7 Conclusions and Further Work

Most CAs employ a pattern-matching technique to map user input onto structural patterns of sentences. However, every combination of utterances that a user may send as input must be taken into account when constructing such a script. This paper was concerned with constructing a novel CA using sentence similarity measures. Examining word meaning rather than structural patterns of sentences meant that scripting was reduced to a couple of natural language sentences per rule as opposed to potentially 100s of patterns. Furthermore, initial results indicate good sentence similarity matching with 13 out of 18 domain-specific user utterances as opposed to that of the traditional pattern matching approach.

Further work will entail considerable development of the new proposed approach. The aim will be to incorporate the use of context switching whereby each context defines a specific topic of conversation. The CA will be robust, capable of tolerating a variety of user input. The new approach will become easily adaptable via automatic or other means. It is intended that a user evaluation of the two approaches to CA design will be conducted. Firstly, each approach would be subjected to a set of domain-specific utterances. Each CA would then compute a match between the user utterance and the rules within the scripts, firing the highest strength pattern/sentence as output. A group of human subjects would evaluate the scripts and their corresponding outputs in order to judge whether the correct pattern/sentence had been fired. This would provide a means for evaluating the opposing approaches and their scripting methodologies.

References

1. A. Turing, "Computing Machinery and Intelligence" Mind, Vol. 54 (236), 1950, pp. 433–460.
2. D. W. Massaro, M. M. Cohen, J. Beskow, S. Daniel, and R. A. Cole, "Developing and Evaluating Conversational Agents", Santa Cruz, CA: University of California, 1998.
3. J. Weizenbaum, ELIZA – "A Computer Program for the Study of Natural Language Communication between Man and Machine", Communications of the Association for Computing Machinery, Vol. 9, 1966, pp. 36–45.
4. K. Colby, "Artificial Paranoia: A Computer Simulation of Paranoid Process". New York: Pergamon, 1975.
5. R. S. Wallace, ALICE: Artificial Intelligence Foundation Inc. [Online]. Available: http://www.alicebot.org (2008, February 01).
6. D. Michie and C. Sammut, Infochat^{TM} Scripter's Manual. Manchester: Convagent, 2001.
7. G. A. Sanders and J. Scholtz, "Measurement and Evaluation of Embodied Conversational Agents", in Embodied Conversational Agents, Chapter 12, J. Cassell, J. Sullivan, S. Prevost and E. Churchill, eds., Embodied Conversational Agents, MIT Press, 2000.
8. T. K. Landauer, P. W. Foltz, and D. Laham, "Introduction to Latent Semantic Analysis". Discourse Processes, Vol. 25 (2–3), 1998, pp. 259–284.
9. Y. Li, D. McLean, Z. A. Bandar, J. D. O'Shea, and K. Crockett, "Sentence Similarity Based on Semantic Nets and Corpus Statistics", IEEE Transactions on Knowledge and Data Engineering, Vol. 18 (8), 2006, pp. 1138–1149.
10. D. Landauer, D. Laham, and P. W. Foltz, "Learning Human-Like Knowledge by Singular Value Decomposition: A Progress Report", in Advances in Neural Information Processing Systems 10, M. I. Jordan, M. J. Kearns, and S. A. Solla, eds., Cambridge, MA: MIT Press, 1998, pp. 45–51.
11. G. A. Miller, "WordNet: A Lexical Database for English", Communications of the Association for Computing Machinery, Vol. 38 (11), 1995, pp. 39–41.
12. Y. Li, Z. A. Bandar, and D. Mclean, "An Approach for Measuring Semantic Similarity between Words using Multiple Information Sources", IEEE Transactions on Knowledge and Data Engineering, Vol. 15 (4), 2003, pp. 871–881.
13. C. Sammut, "Managing Context in a Conversational Agent", Electronic Transactions on Artificial Intelligence, Vol. 3 (7), 2001, pp. 1–7.
14. D. Michie, "Return of the Imitation Game", Electronic Transactions in Artificial Intelligence, Vol. 6 (2), 2001, pp. 203–221.

Chapter 44
Direction-of-Change Financial Time Series Forecasting Using Neural Networks: A Bayesian Approach

Andrew A. Skabar

Abstract Conventional neural network training methods find a single set of values for network weights by minimizing an error function using a gradient descent-based technique. In contrast, the Bayesian approach infers the posterior distribution of weights, and makes predictions by averaging the predictions over a sample of networks, weighted by the posterior probability of the network given the data. The integrative nature of the Bayesian approach allows it to avoid many of the difficulties inherent in conventional approaches. This paper reports on the application of Bayesian MLP techniques to the problem of predicting the direction in the movement of the daily close value of the Australian All Ordinaries financial index. Predictions made over a 13 year out-of-sample period were tested against the null hypothesis that the mean accuracy of the model is no greater than the mean accuracy of a coin-flip procedure biased to take into account non-stationarity in the data. Results show that the null hypothesis can be rejected at the 0.005 level, and that the t-test p-values obtained using the Bayesian approach are smaller than those obtained using conventional MLP methods.

Keywords Direction-of-change forecasting · financial time series · Markov Chain Monte Carlo · neural networks

44.1 Introduction

Financial time series forecasting has almost invariably been approached as a regression problem; that is, future values of a time series are predicted on the basis of past values. In many cases, however, correctly predicting the direction of the change (i.e., *up* or *down*) is a more important measure of success. For example, if a trader

A.A. Skabar
Department of Computer Science and Computer Engineering, La Trobe University, Australia,
E-mail: a.skabar@latrobe.edu.au

S.I. Ao and L. Gelman (Eds.), *Advances in Electrical Engineering and Computational Science*, Lecture Notes in Electrical Engineering.

is to make buy/sell decisions on the basis of forecast values, it is more important to be able to correctly predict the direction of change than it is to achieve, say, a small mean squared error. We call this *direction-of-change forecasting*, and its importance has been acknowledged in several recent papers [1–4].

Clearly, if one has made a prediction for the future value of a time-series, then that prediction can trivially be converted into a direction-of-change prediction by simply predicting up/down according to whether the forecast value is greater/smaller than the present value. There may, however, be advantages in predicting the direction of change directly (i.e., without explicitly predicting the future *value* of the series). For example, traders are likely to base their trading decisions primarily on their opinion of whether the price of a commodity will rise or fall, and to a lesser degree on the extent to which they believe that it will rise or fall. Speculatively, this may create in financial systems an underlying dynamic that allows the direction of change to be predicted more reliably than the actual value of the series. Therefore in this paper we conceptualize direction-of-change forecasting not as a regression problem, but as a binary classification problem, and use multilayer perceptrons (MLPs) to perform the classification.

Whereas conventional (i.e., gradient descent-based) MLP training methods attempt to find a single set of values for the network weights by minimizing an error function, Bayesian MLP methods infer the posterior distribution of the weights given the data. To perform a classification, samples are drawn from this distribution, each sample representing a distinct neural network. The final prediction is obtained by averaging the predictions of the individual sampled networks weighted by the posterior probability of the network given the training data. Thus, whereas the conventional approach optimizes over parameters, the Bayesian approach integrates over parameters, allowing it to avoid many of the difficulties associated with conventional approaches.

This paper reports on the application of Bayesian MLP methods [5–7] to the financial prediction problem and expands on preliminary results that have been presented in Skabar [8]. The paper is structured as follows. Section 44.2 outlines the operation of MLPs and highlights some of the difficulties inherent in using gradient descent-based MLP training methods on financial forecasting tasks. Section 44.3 then describes how Bayesian methods can be applied to MLPs. Section 44.4 provides empirical results from applying the technique to predicting the direction of change in the values of the Australian All Ordinaries Index over a 13 year out-of-sample period, and Section 44.5 concludes the paper.

44.2 MLPs for Financial Prediction

A *multilayer perceptron* (MLP) is a function of the following form:

$$f(x^n) = h(u) \text{ where } u = \sum_{j=0}^{N_1} w_{kj} g\left(\sum_{i=0}^{N_0} w_{ji} x_i^n\right) \tag{44.1}$$

where N_0 is the number of inputs, N_1 is the number of units in a hidden layer, w_{ji} is a numerical weight connecting input unit i with hidden unit j, and w_{kj} is the weight connecting hidden unit j with output unit k. The function g is either a sigmoid (i.e., $g(x) = (1 + \exp(-x))^{-1}$) or some other continuous, differentiable, nonlinear function. For regression problems h is the identity function (i.e., $h(u) = u$), and for classification problems h is a sigmoid.

Thus, an MLP with some given architecture, and weight vector **w**, provides a mapping from an input vector **x** to a predicted output y given by $y = f(\mathbf{x}, \mathbf{w})$. Given some data, D, consisting of n independent items $(\mathbf{x}^1, y^1), \ldots, (\mathbf{x}^N, y^N)$, the objective is to find a suitable **w**. In the case of financial prediction, the raw data usually consists of a series of values measured at regular time intervals; e.g. the daily close value of a financial index such as the Australian All Ordinaries. The input vector, $\mathbf{x}^N$, corresponds to the Nth value of the series, and usually consists of the current value, in addition to time-lagged values, or other quantities which can be derived from the series, such as moving averages.

The conventional approach to finding the weight vector **w** is to use a gradient descent method to find a weight vector that minimizes the error between the network output value, $f(\mathbf{x}, \mathbf{w})$, and the target value, y. For regression problems, it is usually the squared error that is minimized, whereas for binary classification problems it is usually more appropriate to minimise the cross-entropy error, which will result in output values that can be interpreted as probabilities (See Bishop [9] for a detailed discussion of error functions). This approach is generally referred to as the *maximum likelihood* (ML) approach because it attempts to find the most probable weight vector, given the training data. This weight vector, $\mathbf{w}_{\mathbf{ML}}$, is then used to predict the output value corresponding to some new input vector $\mathbf{x}^{n+1}$.

Because of the very high level of noise present in financial time series data, there is an overt danger that applying non-linear models such as MLPs may result in overfitting – a situation in which a model performs well on classifying examples on which it has been trained, but sub-optimally on out-of-sample data. While the risk of overfitting can be reduced through careful selection of the various parameters that control model complexity (input dimensionality, number of hidden units, weight regularization coefficients, etc.), this usually requires a computationally expensive cross-validation procedure. Bayesian methods provide an alternative approach which overcomes many of these difficulties.

44.3 Bayesian Methods for MLPs

Bayesian methods for MLPs infer the posterior distribution of the weights given the data, $p(\mathbf{w}|D)$. The predicted output corresponding to some input vector $\mathbf{x}^n$ is then obtained by performing a weighted sum of the predictions over all possible weight vectors, where the weighting coefficient for a particular weight vector depends on $p(\mathbf{w}|D)$. Thus,

$$\hat{y}^n = \int f(\mathbf{x}^n, \mathbf{w})\, p(\mathbf{w}|D)\, d\mathbf{w} \tag{44.2}$$

where $f(\mathbf{x}^n, \mathbf{w})$ is the MLP output. Because $p(\mathbf{w}|D)$ is a probability density function, the integral in Eq. (44.2) can be expressed as the expected value of $f(\mathbf{x}^n, \mathbf{w})$ over this density:

$$\int f(\mathbf{x}^n, \mathbf{w})p(w|D)\,d\mathbf{w} = E_{p(w|D)}\left[f(\mathbf{x}^n, \mathbf{w})\right] \simeq \frac{1}{N}\sum_{i=1}^{N} f(\mathbf{x}^n, \mathbf{w}) \tag{44.3}$$

Therefore the integral can be estimated by drawing N samples from the density $p(\mathbf{w}|D)$, and averaging the predictions due to these samples. This process is known as *Monte Carlo* integration.

In order to estimate the density $p(\mathbf{w}|D)$ we use the fact that $p(w|D) \propto p(D|w)p(w)$, where $p(D|w)$ is the likelihood, and $p(\mathbf{w})$ is the prior weight distribution. Assuming that the target values, t^n, are binary, then the likelihood can be expressed as

$$p(D|\mathbf{w}) = \exp\sum_{n}\{t^n \ln f(\mathbf{x}^n, \mathbf{w}) + (1-t)^n \ln(1 - f(\mathbf{x}^n, \mathbf{w}))\}. \tag{44.4}$$

The distribution $p(\mathbf{w})$ can be used to express our prior beliefs regarding the complexity of the MLP. Typically, $p(\mathbf{w})$ is modelled as a Gaussian with zero mean and inverse variance α:

$$p(\mathbf{w}) = \left(\frac{\alpha}{2\pi}\right)^{m/2} \exp\left(-\frac{\alpha}{2}\sum_{i=1}^{m} w_i^2\right) \tag{44.5}$$

where m is the number of weights in the network [7]. The parameter α therefore controls the smoothness of the function, with large values of α corresponding to smoother functions, and hence weights with smaller magnitudes. Because we do not know what variance to assume for the prior distribution, it is common to set a distribution of values. As α must be positive, a suitable form for its distribution is the gamma distribution [7]. Thus,

$$p(\alpha) = \frac{(a/2\mu)^{a/2}}{\Gamma(a/2)}\alpha^{a/2-1}\exp(-\alpha a/2\mu) \tag{44.6}$$

where the a and μ are respectively the shape and mean of the gamma distribution, and are set manually. Because the prior depends on α, Eq. (44.2) should be modified such that it includes the posterior distribution over the α parameters:

$$\hat{y}^n = \int f(\mathbf{x}^n, w)\, p(\mathbf{w}, \alpha|D)\, dw d\alpha \tag{44.7}$$

where

$$p(\mathbf{w}, \alpha|D) \propto p(D|\mathbf{w})p(\mathbf{w}, \alpha). \tag{44.8}$$

The objective in Monte Carlo integration is to sample preferentially from the region where $p(\mathbf{w}, \alpha|D)$ is large, and this can be done using the Metropolis algorithm [10]. This algorithm operates by generating a sequence of vectors in such a way that each successive vector is obtained by adding a small random component to the previous vector; i.e., $\mathbf{w}_{\text{new}} = \mathbf{w}_{\text{old}} + \varepsilon$, where ε is a small random vector. Preferential sampling is achieved using the criterion:

$$\text{if } p(\mathbf{w}_{\text{new}}|D) > p(\mathbf{w}_{\text{old}}|D) \text{ accept}$$

$$\text{if } p(\mathbf{w}_{\text{new}}|D) < p(\mathbf{w}_{\text{old}}|D) \text{ accept with probability } \frac{p(\mathbf{w}_{\text{new}}|D)}{p(\mathbf{w}_{\text{old}}|D)}$$

A problem with applying this sampling method to estimate integrals for MLPs is that the strong correlations in the posterior weight distribution will lead to many candidates in the sampling chain being rejected due to the fact that they lead to a decrease in $p(\mathbf{w}|D)$ [9]. However, this problem can be alleviated by using gradient information to reduce the random walk behaviour, and the resulting algorithm is known as the Hybrid Monte Carlo algorithm [11]. While the Hybrid Monte Carlo algorithm allows for the efficient sampling of parameters (i.e., weights and biases), the posterior distribution for α should also be determined. In this paper we use Neal's approach [6], and use Gibbs sampling [12] for the α s.

44.4 Empirical Results

The experimental results reported in this section were obtained by applying the technique described above to the daily close values of the Australian All Ordinaries Index (AORD) for the period from November 1990 to December 2004.

44.4.1 Input Pre-processing

Almost invariably, successful financial forecasting requires that some pre-processing be performed on the raw data. This usually involves some type of transformation of the data into new variables more amenable to learning from. Thus, rather than using, for example, the past prices of a stock, transformed variables might include the absolute change in the price $(p_t - p_{t-1})$, the rate of change of price $(p_t - p_{t-1})/p_{t-1}$, or the price or change in price relative to some index or other baseline such as a moving average. The advantage of using variables based on changes in price (either relative or absolute) is that they help to remove non-stationarity from the time series.

In this study, the input variables used are the relative change in close value from the previous day to the current day (r), and the 5, 10, 30 and 60 day moving averages $(ma_5, ma_{10}, ma_{30}, ma_{60})$. The moving averages are calculated by simply averaging the x previous closing values, where x is the period of the moving average. Thus, the input to the network is the vector

$$(r(t), ma_5(t), ma_{10}(t), ma_{30}(t), ma_{60}(t)) \text{ where}$$
$$r(t) = (p_t - p_{t-1}) / p_{t-1} \tag{44.9}$$

and

$$ma_n(t) = \frac{1}{n} \sum_{t-n}^{t} p_t \tag{44.10}$$

Note that there would almost certainly exist some other combination of inputs and pre-processing steps that might lead to better performance than that which can be achieved using this particular combination. However, in this paper we are not concerned with optimizing this choice of inputs; rather, we are concerned with comparing the performance of the Bayesian approach with that of the ML approach.

44.4.2 Hypothesis Testing

Assuming that a return series is stationary, then a coin-flip decision procedure for predicting the direction of change would be expected to result in 50% of the predictions being correct. We would like to know whether our model can produce predictions which are statistically better than 50%. However, a problem is that many financial return series are not stationary, as evidenced by the tendency for commodity prices to rise over the long term. Thus it may be possible to achieve an accuracy significantly better than 50% by simply biasing the model to always predict up.

A better approach is to compensate for this non-stationarity, and this can be done as follows. Let x_a represent the fraction of days in an out-of-sample test period for which the *actual* movement is up, and let x_p represent the fraction of days in the test period for which the *predicted* movement is up. Therefore, under a coin-flip model the expected fraction of days corresponding to a correct upward prediction is $(x_a \times x_p)$, and the expected fraction of days corresponding to a correct downward prediction is $(1 - x_a) \times (1 - x_p)$. Thus the expected fraction of correct predictions is

$$a_{exp} = (x_a \times x_p) + ((1 - x_a) \times (1 - x_p)) \tag{44.11}$$

We wish to test whether a_{mod} (the accuracy of the predictions of our model) is significantly greater than a_{exp} (the compensated coin-flip accuracy). Thus, our null hypothesis may be expressed as follows:

$$\textit{Null Hypothesis}: H_0 : a_{mod} \leq a_{exp} \quad H_1 : a_{mod} > a_{exp}$$

We test this hypothesis by performing a paired one-tailed t-test of accuracies obtained using a collection of out-of-sample test sets. Specifically, we take the period from January 1992 to December 2004, and divide this into 120 30-day test periods. Predictions for each 30-day test period are based on a model constructed using the 200 trading days immediately preceding this test period. The training and prediction windows are each then advanced by 30 days. For each 30-day prediction period we calculate a_{mod} and a_{exp}. We then use a paired t-test to determine whether the means of these values differ statistically.

44.4.3 Setting the Priors

The Bayesian approach requires that we specify the prior weight distribution, $p(\mathbf{w})$. Recall that $p(\mathbf{w})$ is assumed to be Gaussian, with inverse variance α, and that α is assumed to be distributed according to a Gamma distribution with shape a and mean μ., which remain to be specified. In order to gain some insight into the range of values may be suitable for these parameters, we conducted a number of trials using the ML approach, with weight optimization performed using the scaled conjugate gradients algorithm [13].

Table 44.1 shows the training and test accuracies corresponding to various values of α. Accuracies are averaged over the 120 30-day prediction windows. The values in the fourth column of the table represent the p-values obtained through performing the t-test. Italics show best performance.

Figure 44.1 plots training and test set accuracies against the α value. Low values for α, such as 0.01, impose a small penalty for large weights, and result in overfitting; i.e., high accuracy on training data, but low accuracy on test data. In this case, the null hypothesis cannot be rejected at the 0.05 level. In contrast, when α is very high (e.g., 10.0), large weights will penalised more heavily, leading to weights with small magnitudes. In this case the MLP will be operating in its linear region and the MLP will display a strong bias towards predictions that are in the same direction as the direction of the majority of changes on the training data. Thus, if the number of upward movements on the training data is greater than the number of negative movements, the MLP will be biased towards making upwards predictions on the test data; however, this is not likely to lead to a rejection of the null hypothesis because the null hypothesis takes the non-stationarity of the data into account. It can be seen

Table 44.1 Train accuracy, test accuracy and p-value for various α values

α value	Train. acc.	Test. acc.	p-value (Ho)
0.010	0.725	0.490	0.402
0.100	0.701	0.501	0.459
0.500	0.608	0.516	0.169
0.750	0.593	0.520	0.070
1.000	***0.585***	***0.524***	***0.007***
1.500	0.570	0.521	0.038
2.000	0.562	0.518	0.587
5.000	0.549	0.526	0.528
10.00	0.542	0.525	0.479

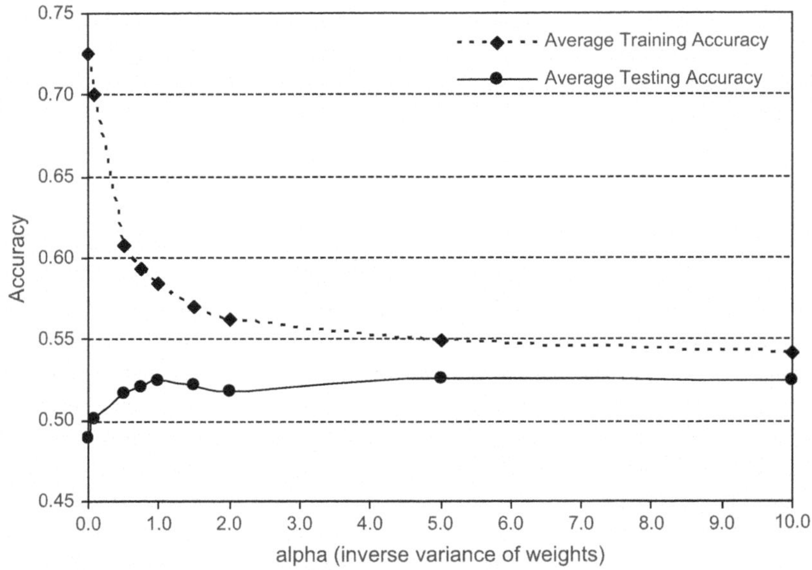

Fig. 44.1 Training and test set accuracy averaged over 120 training/test set pairs. A local maximum test accuracy corresponds to an α value of approximately 1.0

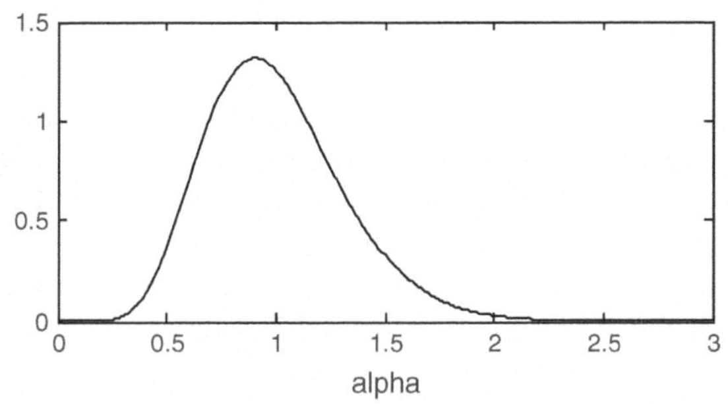

Fig. 44.2 Gamma distribution with mean a = 1.0 and shape parameter $\mu = \mathbf{10}$

from Fig. 44.1 that a local maximum for the test set accuracy occurs for an α value of approximately 1.0, and in this case the null hypothesis can clearly be rejected at the 0.01 level.

The range of α values for which the null hypothesis can be rejected is very narrow, and ranges from a lower α value in the vicinity of 0.5–0.75, to an upper α value in the vicinity of 1.52.0. After visualizing the pdf for Gamma distributions with mean 1.0 and various values for the shape parameter, a shape parameter of 10 was chosen. The pdf is shown in Fig. 44.2. Note that the pdf conforms roughly to the α value identified in the Table 44.1 as leading to a rejection of the null hypothesis.

Table 44.2 Train accuracy, test accuracy and p-value for Bayesian MLPs (MCMC) and conventionally trained (ML) MLPs ($a = 1.0, \mu = 10$)

Method	Train. acc.	Test. acc.	p-value (Ho)
MCMC	0.571	0.528	0.0011
ML	0.585	0.524	0.0068

44.4.4 MCMC Sampling

We now describe the application of the Bayesian approach, which relies on MCMC sampling to draw weight vectors from the posterior weight distribution.

Monte Carlo sampling must be allowed to proceed for some time before the sampling converges to the posterior distribution. This is called the *burn-in period*. In the results presented below, we allowed a burn-in period of 1,000 samples, following which we then saved every tenth sample until a set of 100 samples was obtained. Each of the 100 samples was then applied to predicting the probability of an upwards change in the value of the index on the test examples, and the probabilities were then averaged over the 100 samples. The resulting accuracies and p-values are shown in the first row of Table 44.2. The second row shows results obtained using the maximum-likelihood approach. Note that the p-values for MCMC are much smaller than those for the maximum likelihood approach, indicating increased confidence in the rejection of the null hypotheses. Also note that the average test accuracy for MCMC (52.8%) is greater than that for ML (52.4%), while the average training accuracy for the MCMC (57.1%) is less than that for ML (58.5%), thereby supporting the claim that the Bayesian approach is better at avoiding overfitting.

44.5 Conclusions

The superior performance of the Bayesian approach can be attributed to its integrative nature: each individual weight vector has its own bias, and by integrating over many weight vectors, this bias is decreased, thus reducing the likelihood of inferior generalization performance resulting from overfitting on the training data.

The most important decision to be made in using the Bayesian approach is the choice of prior. In this study, we used a relatively narrow distribution for α, the parameter controlling the degree of regularization present in the error function. This choice was made based on experimenting with different α values within the ML approach. The criticism could be made that this prior distribution was selected based on the same data that we had used for testing, and hence that the significance of our results may be overstated; however, this is unlikely to be the case, as the prior depends heavily on factors such as the degree of noise in the data, and this is relatively constant over different periods of the same time series. Moreover, the fact that we need only select parameters describing the distribution of α, rather than a specific value for α, further diminishes the possibility that our prior is biased towards the particular dataset that we have used.

One of the advantages of the Bayesian approach is its inherent ability to avoid overfitting, even when using very complex models. Thus, although the results presented in this paper were based on MLPs with six hidden units, the same performance could, in principle, have been achieved using a more complex network. It is not clear, however, whether we should expect any coupling between the number of hidden layer units and the prior distribution. For this reason, we would recommend preliminary analysis using the ML approach to ascertain the appropriate range for α and selection of priors based on values which lead to significant predictive ability.

References

1. Y. Kajitani, A.I. McLeod, K.W. Hipel, Forecasting Nonlinear Time Series with Feed-forward Neural Networks: A Case Study of Canadian Lynx data, *Journal of Forecasting,* **24**, 105–117 (2005).
2. J. Chung, Y. Hong, Model-Free Evaluation of Directional Predictability in Foreign Exchange Markets, *Journal of Applied Econometrics,* **22**, 855–889 (2007).
3. P.F. Christoffersen, F.X. Diebold, Financial Asset Returns, Direction-of-Change Forecasting, and Volatility Dynamics, Penn Institute for Economic Research PIER Working Paper Archive 04-009, 2003.
4. S. Walczak, An Empirical Analysis of Data Requirements for Financial Forecasting with Neural Networks, *Journal of Management Information Systems*, **17**(4), 203–222 (2001).
5. D.J.C. MacKay, A Practical Bayesian Framework for Back Propagation Networks, *Neural Computation*, **4**(3), 448–472 (1992).
6. R.M. Neal, Bayesian Training of Backpropagation Networks by the Hybrid Monte Carlo Method, Department of Computer Science, University of Toronto Technical Report CRG-TR-92-1, 1992.
7. R.M. Neal, *Bayesian Learning for Neural Networks* (Springer-Verlag, New York, 1996).
8. A. Skabar, Application of Bayesian Techniques for MLPs to Financial Time Series Forecasting, *Proceedings of 16th Australian Conference on Artificial Intelligence*, 888–891 (2005).
9. C. Bishop, *Neural Networks for Pattern Recognition* (Oxford University Press, Oxford, 1995).
10. N.A. Metropolis, A.W. Rosenbluth, M.N. Rosenbluth, A. Teller, and E. Teller, Equation of State Calculations by Fast Computing Machines, *Journal of Chemical Physics*, **21**(6), 1087–1092 (1953).
11. S. Duane, A.D. Kennedy, B.J. Pendleton, and D. Roweth, Hybrid Monte Carlo, *Physics Letters* B, **195**(2), 216–222 (1987).
12. S. Geman, G. Geman, Stochastic Relaxation, Gibbs Distributions and the Bayesian Restoration of Images, *IEEE Transactions on Pattern Analysis and Machine Intelligence*, **6**, 721–741 (1984).
13. M.F. Moller, A scaled conjugate gradient algorithm for fast supervised learning, *Neural Networks*, **6**, 525–533 (1993).

Chapter 45
A Dynamic Modeling of Stock Prices and Optimal Decision Making Using MVP Theory

Ramin Rajabioun and Ashkan Rahimi-Kian

Abstract In this paper, first a precise mathematical model is obtained for four competing or cooperating companies' stock prices and then the optimal buy/sell signals are ascertained for five different agents which are trading in a virtual market and are trying to maximize their wealth over 1 trading year period. The model is so that gives a good prediction of the next 30th day stock prices. The companies used in this modeling are all chosen from Boston Stock Market. Genetic Programming (GP) is used to produce the predictive mathematical model. The interaction among companies and the effect imposed by each of five agents on future stock prices are also considered in our modeling. Namely, we have chosen eight companies in order that there is some kind of interrelation among them. Comparison of the GP models with Artificial Neural Networks (ANN) and Neuro-Fuzzy Networks (trained by the LoLiMoT algorithm) shows the superior potential of GP in prediction. Using these models; five players, each with a specific strategy and all with one common goal (wealth maximization), start to trade in a virtual market. We have also relaxed the short-sales constraint in our work. Each of the agents has a different objective function and all are going to maximize themselves. We have used Particle Swarm Optimization (PSO) as an evolutionary optimization method for wealth maximization.

Keywords Stock market model · price prediction · Genetic Programming · mean-variance portfolio selection · Particle Swarm Optimization (PSO)

R. Rajabioun (✉)
School of Electrical and Computer Engineering, Control and Intelligent Processing Center of Excellence, University of Tehran, Tehran, Iran
E-mail: r.rajabioun@ece.ut.ac.ir

S.I. Ao and L. Gelman (Eds.), *Advances in Electrical Engineering and Computational Science*, Lecture Notes in Electrical Engineering.

45.1 Introduction

Forecasting the change in market prices and making correct decisions is one of the most principal needs of anyone who economical environments concerns him. Time series are the most common methods used in price prediction [1–3]. But the predominant defect of these methods is that they use only the history of a company's price to do a prediction. Recently, there has been growing attention to the models that concern the interaction among companies in modeling and the use of game theory [4–6] in decision making because of providing more realistic models. Because of complexity of the mutual effects of each company on the others, methods like Artificial Neural Networks (ANN), Neuro-Fuzzy Networks and State Space (SS) models are used more often for the stock price modeling. In references [7–10] Neural Network is used to model the stock market and make prediction. In reference [8], Genetic algorithm (GA) is incorporated to improve the learning and generalizability of ANNs for stock market prediction. In reference [11] the difference between the price and the moving average, highest and lowest prices is used as inputs for one-day-ahead price prediction. More over, volume of transactions, market indicators and macro economic data are also considered as input variables [12]. There are also some studies being performed on the fluctuations and the correlations in stock price changes in physics communities, using the concepts and methods in physics [13, 14]. In reference [15] a neuro-genetic stock prediction system is introduced, which uses genetic algorithm to select a set of informative input features for a recurrent neural network. In references [16, 17], the neuro-genetic hybrids for stock prediction are proposed. The genetic algorithm is used to optimize the weights of ANN.

Producing the right buy/sell signals are also important for those who trade in the stock markets. In reference [18], two simple and popular trading rules including moving average and trading range breakout are tested in the Chilean stock market. Their results were compared with the buy-and-hold strategy, and both trading rules produced extra returns compared to the buy-and-hold strategy.

Genetic Programming (GP) is a symbolic optimization technique, developed by Koza [19]. It is an evolutionary computational technique based on the so-called "tree representation". This representation is extremely flexible because trees can represent computer programs, mathematical equations, or complete models of process systems [20]. In reference [21] GP is used to produce a one-day-ahead model to predict stock prices. This model is tested for a fifty consecutive trading days of six stocks and has yielded relatively high returns on investment.

In this paper, we use the GP to find the best mathematical models for the four companies' stock prices under study. Our GP models are able to predict these stock prices for up to the next 30 days with acceptable prediction errors in the market. Because, the GP is a well known algorithm we will not present it in details. However, reference [22] provides a good review of the GP algorithm.

The modeling is done for four companies in Boston Stock Market [23]. Selected companies include: Advanced Micro Devices (AMD), Ericsson (ERIC), Sony (SNE), Philips (PHG), International Business Machines (IBM), Intel Corporation (INTC), Microsoft (MSFT) and Nokia (NOK). These companies are assumed to have a relationship like competition or cooperation and so their stock prices could affect on each other. Letters allocated in parentheses are the symbols using which one can access the price data of each company. We use the price history of these eight companies as inputs to predict our four objective companies' prices including: Ericsson (ERIC), International Business Machines (IBM), Sony (SNE) and Philips (PHG). Obtained four models precision is compared with two traditional methods: (1) Multi Layer Perceptron (MLP) and (2) Neuro-Fuzzy Network trained by Locally Linear Model Tree (LoLiMoT) method.

After modeling the four companies' stock prices, we create five agents who trade in a virtual market in order to maximize their wealth. These agents (players) will buy or sell their in hand stocks according to their uniquely defined strategies. Each player has a unique objective function. The Buy/Sell actions of each player are obtained so as to maximize its objective function in each trading period. The maximization is done using the Particle Swarm Optimization (PSO) method [24].

At rest of the paper, in Section 45.2, modeling and prediction is discussed. Section 45.3 demonstrates the virtual stock market and argues its constraints and presumptions. Then in Section 45.4 the results of our simulations are shown and finally the conclusion is done in Section 45.5.

45.2 Modeling and Prediction

As stated earlier, our primary goal is to obtain a predictive model that is able to predict the future stock prices precisely. The companies that we are going to predict their stocks include: Ericsson (ERIC), International Business Machines (IBM), Sony (SNE) and Philips (PHG). We presume that these companies have some kind of interrelations with four other companies including: Advanced Micro Devices (AMD), Intel Corporation (INTC), Microsoft (MSFT) and Nokia (NOK). So we downloaded these eight companies' price data from the Boston Stock Market [23]. The downloaded data encompasses some information like: daily opening price, daily closing price, daily highest price, daily lowest price and exchange volume. In this paper, we predict the average of daily opening and closing prices. Our data set contains sampled price data for the interval of 2001/07/08 to 2006/03/11. The criterion used to evaluate the models is the Normalized Mean Square Error (NMSE), which is defined as follows:

$$\mathrm{NMSE} = \frac{\sum_{i=1}^{n} (y_i - \hat{y}_i)^2}{\sum_{i=1}^{n} y_i^2} \tag{45.1}$$

where yi and ŷi are the original and predicted price values, respectively. Figure 45.1 shows the NMSE values for the train data set (the first 70%) using GP, ANN (MLP) and Neuro-Fuzzy networks (trained using LoLiMoT). Figure 45.2 depicts this comparison for the test data set (the last 30%).

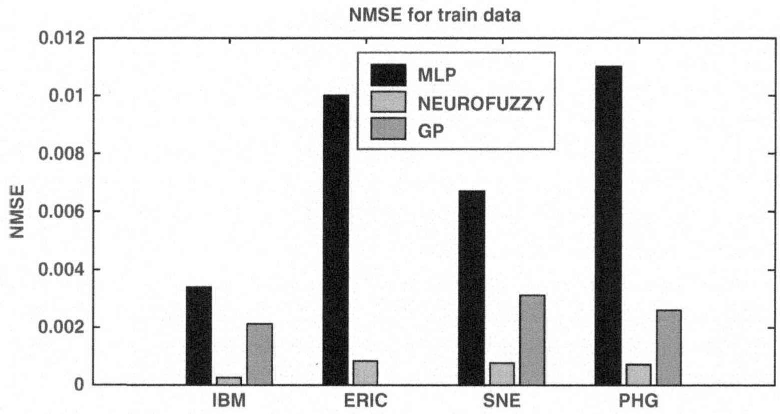

Fig. 45.1 The prediction error (NMSE) for all companies (using train data)

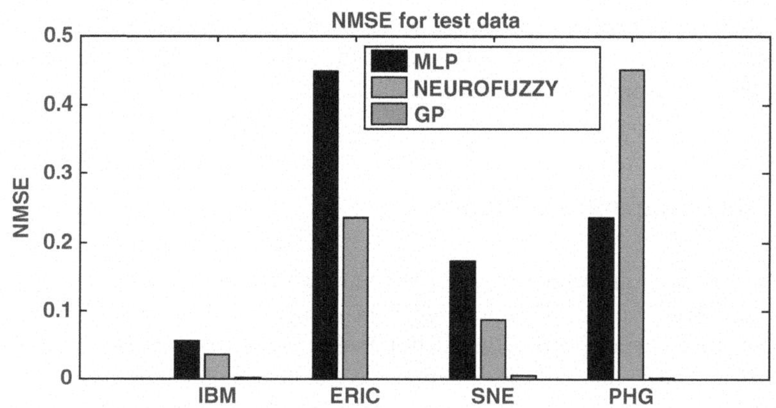

Fig. 45.2 The prediction error (NMSE) for all companies (using test data)

The GP-based stock price models were initialized with some functions and terminals. The terminals included random number generators together with integers from 1 to 16. The functions included: $\{+, -, \times, \div, \log x, e^x, x^y\}$. The population sizes were set to 600 except for ERIC, which was set to 400. Meanwhile the number of iterations was set to 80. As it can be seen from Figs. 45.1 and 45.2, the GP-based price model prediction errors are acceptable for the training data set and less than both of the MLP and Neuro-Fuzzy models for test data set. The only drawback of the GP algorithm is its time-consuming modeling characteristics, which is acceptable comparing to its precise modeling.

Until now we have modeled the interactions of eight different companies that affect the future price values. But due to the fact that buyers/sellers also affect future stock prices of the companies, it is essential to include such interactions in the modeling. Therefore, we augment a new term to our price models in order to include the effects of the market players' actions (buy/sell weights) into the future price

Table 45.1 The PSO model parameters

Parameter range	[−1, 1]
Maximum optimization iterations each day	200
Population size	180
Acceleration constant 1	2
Acceleration constant 2	2

changes. Since there are not much available data on how the buy/sell volumes of the market players affect the future prices, we decided to add a new term to show these effects in our price prediction models as follows:

$$augmented\ term = \gamma \times W \times a \times Price\_vector \tag{45.2}$$

where:

γ: is a weighting coefficient that regulates the impact of the augmented term on the models. When γ is large the augmented term makes the model deviate from the trend of the time-series market historical data.
W: is a weight vector that its elements show each company's stock trade impact on future prices. The elements of this vector are between 0 and 1.
a: is the action vector of all players. Its elements are between -1 and 1 that show the buy/sell rates of the stocks.
Price_vector: contains the current stock price values in the market.

The best value for the γ factor obtained to be 0.1. The W-vector was chosen as follows: **W** = [0.1 0.05 0.1 0.2 0.2 0.2 0.05 0.1]. The corresponding companies' symbol vector is: (AMD ERIC IBM INTC MSFT NOK PHG SNE).

The augmented term makes it possible for us to see the effect of each player's market decision on the stock prices and other players' wealth (similar to a non-cooperative game).

Our objective in the next section would be to find the best market actions (sell/buy of each stock) of each player so as to maximize its expected objective function (wealth) in the market. Our market simulation studies are done in a Virtual Stock Market and by means of an evolutionary optimization algorithm (PSO). In our simulations a common PSO with inertia was used. Table 45.1 shows the parameters used in the PSO optimization.

45.3 Virtual Stock Market

We assume five players (agents) in the stock market. We also assume that these players have no stocks at the beginning of the market. They just have US$5,000,000 and intend to buy stocks that would maximize their expected wealth in the market. The players are free to buy and sell stocks in each round of the market. There are

1,000,000 stocks assumed to be available from each company in our virtual stock market (4,000,000 stocks in total). The only limitation imposed by the market is the maximum number of stocks each player can buy or sell each day. This buy/sell volume is limited to 1,000 stocks trading per day for each company. This constraint is essential because if there is no limitation the whole stocks might be bought at the beginning of the trading period by one of the players. This way there will be no chance for other players to buy some of the stocks. Through the augmented term added to the stock price models we can see the effect of each agent's action (sell/buy stocks) on the future prices and other agents' wealth in the market.

We assume five players (agents) with different objective functions and different strategies in the market, but we assume that all the agents have access to the stock price models (developed in Section 45.2) symmetrically.

The players' strategies are as follows:

Strategy of player 1:

This player buys (sells) the maximum number of allowed stocks when the prediction shows an increase (decrease) in next 30 day prices compared to the average prices of the last 10 days.

Strategy of player 2:

This player uses the Mean-Variance Analysis (MVA). He chooses the standard deviation of the expected return (r_p) as a measure of risk (σ_p). He plots the opportunity set (efficient frontier) for a four-asset portfolio and takes an average risk and for an average return each day. A sample opportunity set for a four-asset portfolio is shown in Fig. 45.3.

Strategy of player 3:

This player believes in Random Walk Theory. He believes that the stock prices are unpredictable and therefore, he buys and sells stocks randomly.

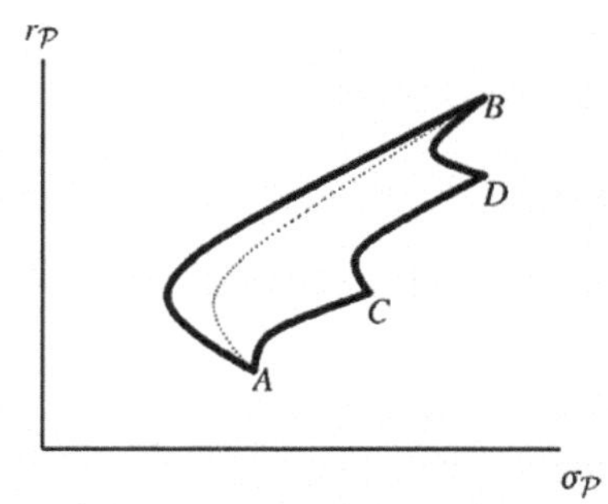

Fig. 45.3 A sample opportunity set for a four-asset portfolio (dotted line: the opportunity set with A and B assets only)

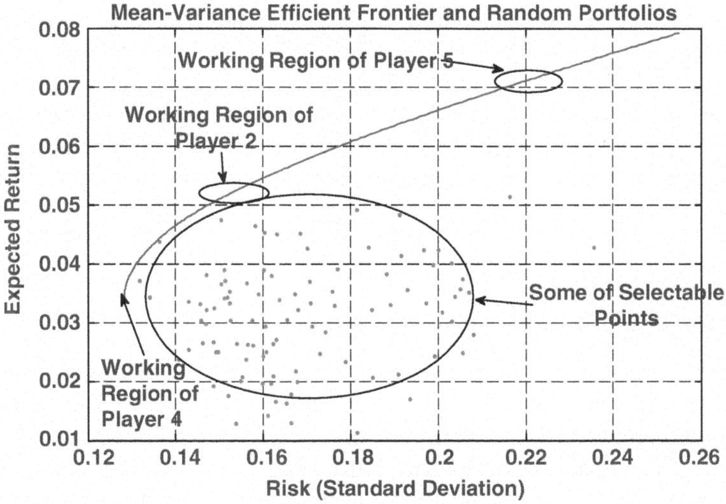

Fig. 45.4 The working regions of players 2, 4 and 5 on the risk-return efficient frontier (the red points can be selected in each trading day)

Strategy of player 4:

This player acts just like player 2. The only difference is in his risk averse behavior. To reach the minimum risk, in each stage, he selects the buy/sell weights on the knee of efficient frontier curve.

Strategy of player 5:

This player also acts like player 2 with the difference that this player is risk lover. Therefore, in each stage this player selects the buy/sell weights with the maximum risk and maximum expected return.

The working regions of players 2, 4 and 5 on the risk-return efficient frontier are shown in Fig. 45.4.

These five players buy and sell in the virtual stock market. In the related literature it is usually seen that short-sales are disregarded when optimizing the players' objective functions and the optimization is just done through stock purchases. However, in this paper we have relaxed this constraint and allowed the players to buy and sell their stocks when needed.

In the following, we define the objective functions for all players and demonstrate their optimization process. For players 2, 4 and 5 that the risk values are important in their decisions, we define their objective function as follows:

$$E_i = \lambda E(r_{p_i}) - (1 - \lambda)\sigma_{p_i}, \quad i = 2, 4, 5 \tag{45.3}$$

where E_i: is the Expected return of player i, λ: is a constant between 0 and 1. In fact, this is a weight that shows the relative importance of the expected return

($E(r_p)$) versus the risk (σ_p) of player-i, for $\lambda = 1$ the risk term disappears form the objective function of player-i: $E_i = E(r_{pi})$. In our market simulation studies, we chose $\lambda = \{0, 0.5, 1\}$ for players {2, 4, 5} respectively according to their defined risk behaviors in the market. The players' objective functions were optimized with respect to their decision variables (stock sell/buy actions) using the Particle Swarm Optimization method and the results are presented and analyzed in the following section.

45.4 The Market Simulation Results

The market simulation results for the five players are presented and analyzed in this section. Figures 45.5 and 45.6 show the optimal buy/sell actions for players 1 and 5 for each company's stock (ERIC, IBM, PHG and SNE). The optimal buy/sell actions for players 2, 3 and 4 are shown in the appendix. In these figures, the buy actions are positive, sell actions are negative and no-actions are zero.

If the buy action gets +1 (or −1), then the player, should buy (or sell) the maximum number of stocks allowed for that company. In Fig. 45.7, the wealth of each player is shown for 1 trading year period. The wealth is measured as the values of the in-hand stocks added to the cash amount in-hand for each player.

As can be seen from Fig. 45.7, players 1 and 5 have done better than the rest of them in terms of the wealth maximization for 1 year stock trading. Figure 45.8 shows the expected risk values in each trading day for each player. As we expected, player 1 has the minimum expected risk over the trading period and has also obtained the maximum return from the market.

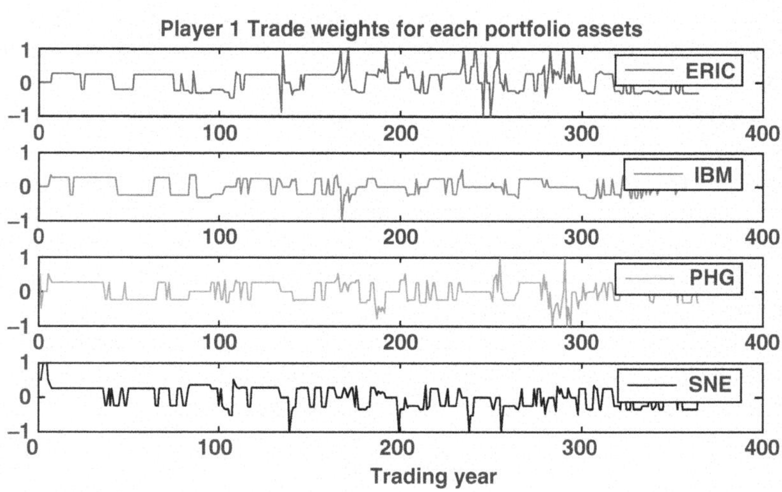

Fig. 45.5 The optimal trading actions of player 1 for all companies' stocks

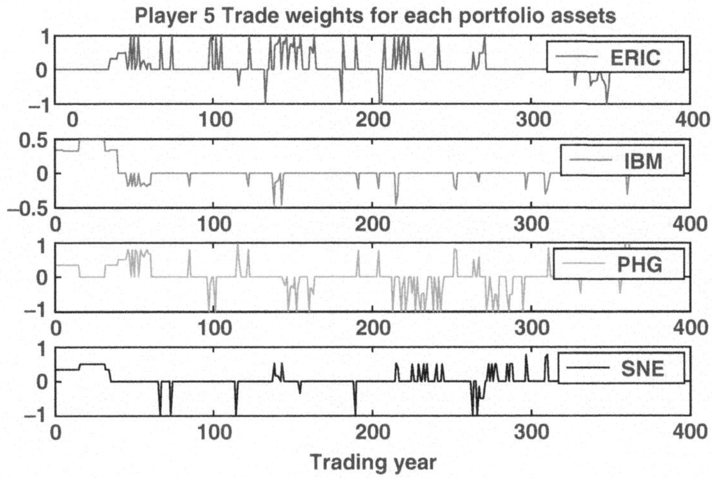

Fig. 45.6 The optimal trading actions of player 5 for all companies' stocks

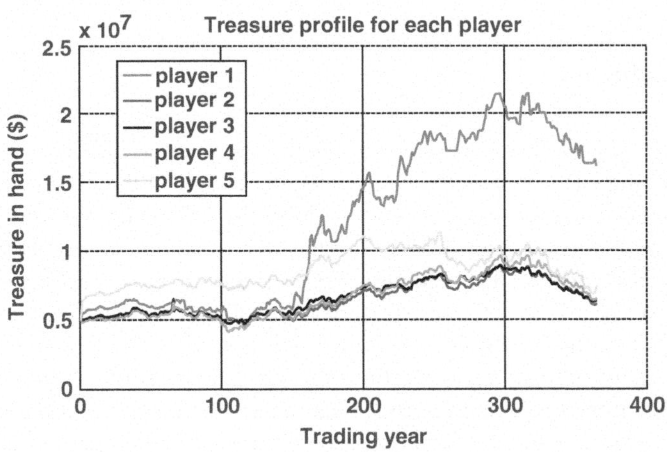

Fig. 45.7 The wealth of all players for all days of the trading year

Its strategy was to buy/sell maximum stocks with respect to the comparison of the predicted future-prices' trends with those of the moving average 10-day-before prices. Since the GP prediction models had small prediction errors for the test data, this player did well in the market by relying on the prediction results.

Among players 2, 4 and 5 (referring to Fig. 45.7), we can see that player 5 with the maximum risk level has made the most wealth (expected returns) and stands in the second rank (after player 1) in terms of market returns. Player 3's strategy was to buy and sell randomly; by referring to Figs. 45.7 and 45.8, one can see that his expected returns are similar to those for player 2 (see the Fig. 45.9, 45.10 and 45.11 in the Appendix) but, his expected risks values are more than other players.

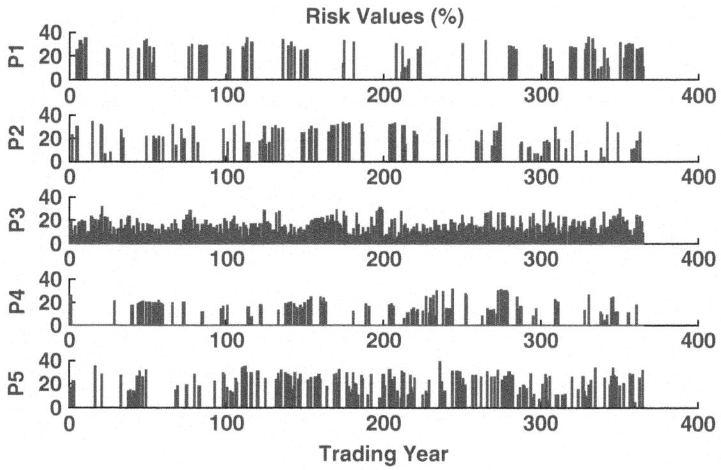

Fig. 45.8 The expected risk values for each player over the 1 trading year (P1 to P5 from top to bottom graphs)

45.5 Conclusion

In this paper, precise price predictive models were obtained for four companies' stocks using the GP. This model incorporated the effects of the players' actions on the stock prices and other players' wealth. After the GP model was verified (using the test data), it was used for making sell/buy decisions by five agents that traded in a virtual stock market. The trading period was considered 1 year for our market simulation studies. Five different strategies and objective functions were defined for the market trading agents (with different risk attitudes). The PSO algorithm was used to obtain the optimal buy/sell actions for each player in order to maximize their objective functions (expected returns). The players' strategies and their expected risk-returns were obtained and analyzed for the 1 year trading period. Our market simulation studies showed that the player (P1) was the most successful one in our virtual stock market.

Appendix

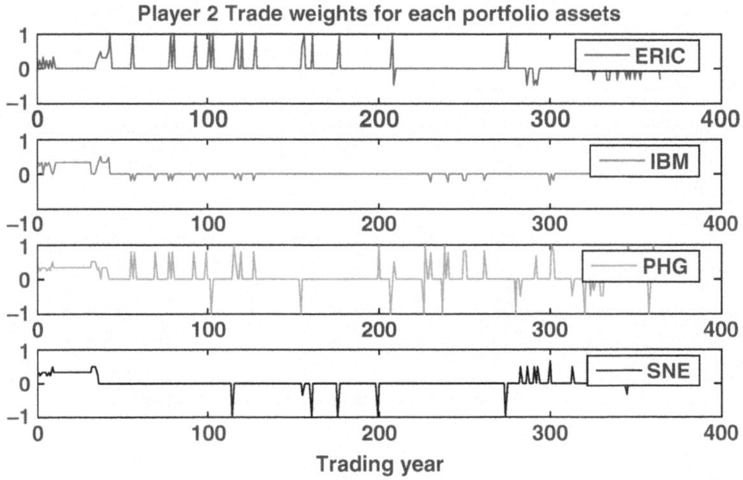

Fig. 45.9 The optimal trading actions of player 2 for all companies' stocks

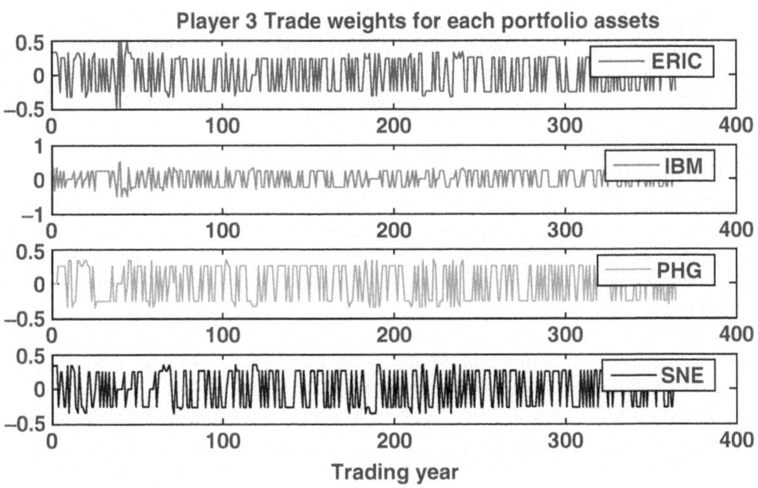

Fig. 45.10 The optimal trading actions of player 3 for all companies' stocks

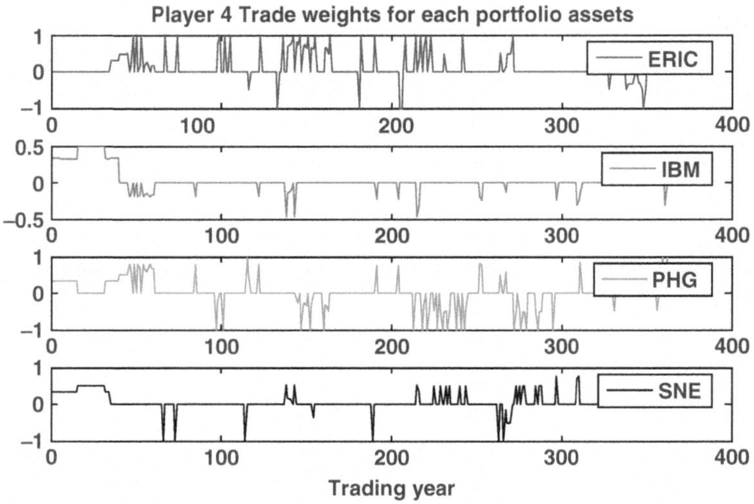

Fig. 45.11 The optimal trading actions of player 4 for all companies' stocks

References

1. R. Thalheimer and M. M. Ali, Time series analysis and portfolio selection: An application to mutual savings banks, *Southern Economic Journal*, **45**(3), 821–837 (1979).
2. M. Pojarliev and W. Polasek, Applying multivariate time series forecasts for active portfolio management, *Financial Markets and Portfolio Management*, **15**, 201–211 (2001).
3. D. S. Poskitt and A. R. Tremayne, Determining a portfolio of linear time series models, *Biometrika*, **74**(1), 125–137 (1987).
4. T. Basar and G. J. Olsder, *Dynamic Noncooperative Game Theory*. (San Diego, CA: Academic, 1995).
5. J. W. Weibull, *Evolutionary Game Theory*. (London: MIT Press, 1995).
6. D. Fudenberg and J. Tirole, *Game Theory*. (Cambridge, MA: MIT Press, 1991).
7. S. Lakshminarayanan, An Integrated stock market forecasting model using neural network, *M. Sc. dissertation*, College of Engineering and Technology of Ohio University (2005).
8. K. J. Kim and W. B. Lee, Stock market prediction using artificial neural networks with optimal feature transformation, *Journal of Neural Computing & Applications*, *Springer*, **13**(3), 255–260 (2004).
9. K. Schierholt and C. H. Dagli, Stock market prediction using different neural network classification architectures, *Computational Intelligence for Financial Engineering*, *Proceedings of the IEEEIAFE Conference*, pp. 72–78 (1996).
10. K. Nygren, Stock Prediction – A Neural Network Approach, *M.Sc. dissertation*, Royal Institute of Technology, KTH (2004).
11. E. P. K. Tsang, J. Li, and J. M. Butler, EDDIE beats the bookies, *International Journal of Software, Practice and Experience*, **28**(10), 1033–1043 (1998).
12. D. S. Barr and G. Mani, Using neural nets to manage investments, *AI EXPERT*, **34**(3), 16–22 (1994).
13. R. N. Mantegna and H. E. Stanley, *An Introduction to Econophysics: Correlation and Complexity in Finance* (Cambridge University Press, Cambridge, MA, 2000).
14. J. P. Bouchaud and M. Potters, *Theory of Financial Risks: From Statistical Physics to Risk management* (Cambridge University Press, Cambridge, MA, 2000).
15. Y. K. Kwon, S. S. Choi, and B. R. Moon, Stock Prediction Based on Financial Correlation, *In Proceedings of GECCO'2005*, pp. 2061–2066 (2005).

16. Y. K. Kwon and B. R. Moon, Daily stock prediction using neuro-genetic hybrids, *Proceedings of the Genetic and Evolutionary Computation Conference*, pp. 2203–2214 (2003).
17. Y. K. Kwon and B. R. Moon, Evolutionary ensemble for stock prediction, *Proceedings of the Genetic and Evolutionary Computation Conference*, pp. 1102–1113 (2004).
18. F. Parisi and A. Vasquez, Simple technical trading rules of stock returns: evidence from 1987 to 1998 in Chile, *Emerging Market Review*, **1**, 152–64 (2000).
19. J. Koza, *Genetic Programming: On the Programming of Computers by Means of Natural Evolution*. (Cambridge, MA: MIT Press, 1992).
20. J. Madar, J. Abonyi, and F. Szeifert, Genetic Programming for the identification of nonlinear input-output models, *Industrial Engineering Chemistry Research*, **44**, 3178–3186 (2005).
21. M. A. Kaboudan, Genetic Programming prediction of stock prices, *Computational Economics*, **16**(3), 207–236 (2000).
22. S. Sette and L. Boullart, Genetic Programming: principles and applications, *Engineering Applications of Artificial Intelligence*, **14**, 727–736 (2001).
23. Boston Stock Group web page (August 1, 2007); http://boston.stockgroup.com
24. J. Kennedy and R. Eberhart, Particle Swarm Optimization, *Proceedings of the IEEE International Conference on Neural Networks*, Perth, Australia, pp. 1942–1945 (1995).

Chapter 46
A Regularized Unconstrained Optimization in the Bond Portfolio Valuation and Hedging

Yury Gryazin* **and Michael Landrigan**

Abstract In this chapter, a numerical approach to the valuation and hedging of a portfolio of bond and options is presented in the case of strong dependency of bond principal on the market interest rate metric. Collateralized Mortgage Obligations (CMO) represents one of the important classes of such bond type. To improve the CMO valuation methodology and to develop a robust procedure for the construction of the optimal hedge for CMO, we introduce an optimization approach to minimize the dispersion of the portfolio spread distribution by using available on market options. In doing so, we design an optimal hedge with respect to the set of available benchmarks and we obtain the two new valuation metrics that represent the quality of hedge with respect to this set. Our two main outputs are the mean and the standard deviation of the individual spreads for the optimal portfolio. These metrics can be used in comparison analysis in addition to the standard OAS valuation.

Keywords Regularized Unconstrained Optimization · Bond Portfolio · Valuation · Hedging · Collateralized Mortgage Obligation

46.1 Introduction

This paper presents a numerical approach to the valuation and hedging of a portfolio of bond and options in the case of strong dependency of bond principal on the market interest rate metric. Collateralized Mortgage Obligations (CMO) represent one of the important classes of such bond type. CMOs can have a high degree of variability in cash flows. Because of this, it is generally recognized that a yield to maturity of

Y. Gryazin (✉)
Department of Mathematics, Idaho State University, Pocatello, ID 83209, USA,
E-mail: gryazin@isu.edu

*The work of this author was supported in part by Mexican Consejo Nacional de Ciencia y Tecnologia (CONACYT) under Grant # CB-2005-C01-49854-F.

static spread calculation is not a suitable valuation methodology. Since the 1980s Option-Adjusted Spread (OAS) has become a ubiquitous valuation metric in the CMO market. There have been many criticisms of OAS methodology, and some interesting modifications have focused on the prepayment side of the analysis, e.g., [1,2]. One of the problems with using OAS analysis is the lack of information about the distribution of the individual spreads, which in turn leads to the difficulties in the construction of the hedging portfolio for CMO.

To improve the CMO valuation methodology and to develop a robust procedure for the construction of the optimal hedge for CMO, we introduce an optimization approach to minimize the dispersion of the portfolio spread distribution by using available on market options. In doing so, we design an optimal hedge with respect to the set of available benchmarks and we obtain the two new valuation metrics that represent the quality of hedge with respect to this set. Our two main outputs are the mean and the standard deviation of the individual spreads for the optimal portfolio. These metrics can be used in comparison analysis in addition to the standard OAS valuation.

This new methodology can lead to quite different conclusions about CMO than OAS does. In particular, in comparing two bonds, the more negatively convex bond may look cheaper on an OAS basis, but be more attractive according to our analysis since we provide a way to estimate the quality of available hedge.

The main difficulty in implementing our new methodology is in the minimization of the spread-variance functional. The difficulty is partly because the optimization problem is ill-conditioned, and in many situations, requires introduction of some type of regularization. Our approach is to use the standard Tikhonov regularization [3], which has the strong intuitive appeal of limiting the sizes of benchmark weights used in hedging.

A word about static versus dynamic hedging may be in order. Our methodology is to set up a static hedge and conduct bond valuation based on the static optimal portfolio. It may be argued that an essentially perfect hedge may be created dynamically. However, a dynamically-hedged portfolio will not have the OAS as a spread. Moreover, ones hedging costs will be related to the amount of volatility in the future so can be quite uncertain. Our methodology greatly reduces dynamic hedging costs by setting up an optimized static hedge, and thus reduces the uncertainty in dynamic hedging costs.

The rest of the paper is organized as follows. In the second section we briefly review OAS and point out in more detail the problems we see with it. In the third section we explicitly define our hedging methodology. In the forth section we give a brief description of the regularized numerical method. In the fifth section we present some details on the term-structure model used; on our prepayment assumptions and summarize numerical results from our analysis.

46.2 Option-Adjusted Spread Analysis

Standard OAS analysis is based on finding a spread at which the expected value of the discounted cash flows will be equal to the market value of a CMO. This is encapsulated in Eq. (46.1):

$$MV_{CMO} = \widetilde{E} \sum_{i=1}^{L} d(t_i, s) \cdot c(t_i). \tag{46.1}$$

Here MV_{CMO} is the market value of CMO, $\widetilde{E}$ denotes expectation with respect to the risk neutral measure, $d(t_i, s)$ is the discount factor to time t_i with a spread of s, and $c(t_i)$ is the cash flow at time t_i, for $i = 1, 2, \dots, L$. Note that in the OAS framework, a single spread term is added to the discounted factors to make this formula a true equality, this spread, s, is referred to as *the OAS*.

The goal of OAS analysis is to value a CMO relative to liquid benchmark interest rate derivatives, and, thus, the risk-neutral measure is derived to price those benchmarks accurately. To calculate expected values in practice one uses Monte-Carlo simulation of a stochastic term-structure model. We use the two-factor Gaussian short-rate model G2++ as in [4] calibrated to U.S. swaption prices, but other choices may be suitable. In terms of numerical approximation, it will give us Eq. (46.2):

$$MV_{bench}^{k} = \frac{1}{N} \sum_{n=1}^{N} \left\{ \sum_{i=1}^{L} cf_{bench}^{k}(n,t) \prod_{t=1}^{t} \frac{1}{1 + \Delta t_i \cdot r(n, t_i)} \right\} + Err^{k}. \tag{46.2}$$

Here $\Delta t_i = t_i - t_{i-1}$, $i = 1, \dots, L$, $t_0 = 0$, MV_{bench}^{k}, $k = 1, \dots, M$ are the market values of the benchmarks, $cf_{bench}^{k}(n, t_i)$, $n = 1, \dots, N$, $i = 1, \dots, L$, $k = 1, \dots, M$ are the future cash flows of the benchmarks, N is the number of generated trajectories, L is the number of time intervals until expiration of all benchmarks and CMO, and M is the number of benchmarks in the consideration. The last term Err^{k} represents the error term. Though, the detailed consideration of calibration procedure is outside the scope of this presentation, it is worth to mention that the absolute value of the Err^{k} term is bounded in most of our experiments by five basis points.

The second step in the OAS analysis of CMOs is to find the spread term from the Eq. (46.3):

$$MV_{CMO} = \frac{1}{N} \sum_{n=1}^{N} \left\{ \sum_{i=1}^{i} cf(n, t_i) \prod_{j=1}^{i} \frac{1}{1 + \Delta t_j \cdot (r(n, t_j) + s)} \right\}. \tag{46.3}$$

Here $cf(n, t_i)$, $n = 1, \dots, N$, $i = 1, \dots, L$, are cash flows of the CMO. These cash flows come from the structure of the CMO, a perfectly known quantity, and a prepayment model, a more subjectively known quantity. The parameter s, the OAS, is an indicator as to whether the CMO is underpriced or overpriced: If the OAS is positive then the CMO is underpriced, if it is negative then the CMO is overpriced. Not only its sign, but also the magnitude of the OAS commonly quoted as a measure of cheapness of a CMO.

An important issue is managing a portfolio of CMOs is how to hedge the portfolio by using actively traded benchmarks. We return to this question little bit later but for now let's assume that we somehow have found the list and corresponding weights of benchmarks that would provide a sufficient hedge for our portfolio. To evaluate this portfolio, we will extend the OAS analysis to include the calculation of spread for the portfolio of CMO and the benchmarks. This straightforward extension of OAS approach will result in Eq. (46.4):

$$MV_{CMO} + \sum_{k=1}^{M} w^k MV_{bench}^k = \frac{1}{N}\sum_{n=1}^{N}\left\{\sum_{i=1}^{L}\left[cf(n,t_i) + \sum_{k=1}^{M} w_k \cdot cf_{bench}^k(n,t_i)\right] \times \prod_{j=1}^{i}\frac{1}{1+\Delta t_j \cdot (r(n,t_j)+s)}\right\}. \tag{46.4}$$

We will see in more detail later how consideration of the OAS of various portfolios makes some serious drawbacks of OAS analysis apparent. Some of the drawbacks are:

1. It matches a mean to market value, but provides no information on the distribution of the individual discounted values.
2. One can use it to calculate some standard risk metrics, but it gives no way to design a refined hedge.
3. It is sensitive to positions in your benchmarks.

In our approach an affective hedging strategy is proposed and new valuation metrics are considered.

46.3 The Spread Variance Minimization Approach

In our proposed approach, instead of using the same spread for all paths, we are looking at the individual spreads for every path of the portfolio of CMO and benchmarks. The dispersion of the distribution can be considered a measure of risk or alternately as a measure of the quality of a hedge. In a classical mean-variance portfolio optimization problem (see, e.g., [5]) there are two parts of the formulation usually considered: (1) the task of the maximizing the mean of the portfolio return under a given upper bound for the variance; (2) the task of minimizing the variance of the portfolio return given a lower bound on the expected portfolio return. We approach the problem of hedging CMOs with swaptions from a similar point of view, but focus on the distribution of spreads. The goal is to apply an optimization algorithm to find weights so that the variance of the $s(n, w_1, \dots, w_M)$ is a minimum.

The spread for path n is the value $s(n, w_1, \dots, w_M)$ such that Eq. (46.5) satisfied:

$$M V_{CMO} + \sum_{k=1}^{M} w_k M V_{bench}^{k} = \sum_{i=1}^{L} \left[cf(n, t_i) + \sum_{k=1}^{M} w_k \cdot cf_{bench}^{k}(n, t_i) \right] \prod_{j=1}^{i} \frac{1}{1 + \Delta t_i \cdot (r(n, t_j) + s(n, w_1, \dots, w_M))}. \quad (46.5)$$

Here $w_k, k = 1, \dots, M$ are the weights of the individual benchmarks (a negative indicates a short position).

The goal of our analysis is to minimize the variance of the individual spreads. In this form the problem is not well defined; the weights may exhibit some instability and tend to drift to infinity. A common approach to the solution of this type of problem is to add regularization term to the target functional. Instead of minimizing only variance, we will use Tikhonov regularization [3]. We introduce this in more detail in the next section.

46.4 Numerical Method

As mentioned before, we are considering that for every interest rate trajectory the individual spread depends on the weights for benchmarks in the portfolio. Then the target functional is defined in Eq. (46.6):

$$f(w_1, \dots, w_n) = \frac{1}{N} \sum_{n=1}^{N} s(n, w_1, \dots, w_n)^2 - \mu^2. \quad (46.6)$$

Here $\mu = \frac{1}{N} \sum_{n=1}^{N} s(n, w_1, \dots, w_M)$. The Jacobian and Hessian of the functional are given by Eq. (46.7):

$$\frac{\partial f}{\partial w_i} = \frac{2}{N} \sum_{n=1}^{N} (s(n, w_1, \dots, w_n) - \mu) \frac{\partial s(n, w_1, \dots, w_n)}{\partial w_i}, i = 1, \dots, M, \quad (46.7)$$

$$\frac{\partial^2 f}{\partial w_i \partial w_l} = \frac{2}{N} \sum_{n=1}^{N} \left\{ \frac{\partial s}{\partial w_i} \cdot \frac{\partial s}{\partial w_l} + (s - \mu) \frac{\partial^2 s}{\partial w_i \partial w_l} \right\} -2 \cdot \left\{ \frac{1}{N} \sum_{n=1}^{N} \frac{\partial s}{\partial w_i} \right\} \cdot \left\{ \frac{1}{N} \sum_{n=1}^{N} \frac{\partial s}{\partial w_l} \right\}, i, l = 1, \dots, M. \quad (46.8)$$

Using implicit differentiation one can find $\partial s/\partial w_i, \partial^2 s/(\partial w_i \partial w_l), i, l = 1, ..., M$. But this functional in general is ill-conditioned. To ensure the convergence of the optimization method, we introduce standard Tikhonov regularization. The modified target functional could be presented as in Eq. (46.9):

$$\tilde{f}(w_1, \ldots, w_M) = f(w_1, \ldots, w_M) + \alpha||w||_2^2. \tag{46.9}$$

In most situations this guaranties the convergence of the numerical method to the unique solution. In our approach we are using the optimization technique based on the combination of the Broyden-Fletcher-Goldfarb-Shanno (BFGS) (see, e.g., [6]) and Newton methods. The BFGS approach is applied on the early iterations to ensure the convergence of Newton's method to a minimum. When the $l_2 - norm$ of the gradient of the target function becomes less then 10^{-10} we apply the Newton method assuming that the approximation is already close enough to the solution and the quadratic convergence rate of the Newton method can be achieved.

As we already mentioned, the regularization keeps the value of the target functional small and stabilizes the optimization method by preventing the weights of the benchmarks in the portfolio $w_i, i = 1, \ldots, M$ from becoming too large. To keep the condition number of the Hessian bounded, one has to use fairly large regularization parameter α. On the other hand, in order to keep the regularized problem reasonably close the original one, we expect that α needs to be small. In addition to these pure mathematical requirements, in the case of managing a portfolio, one has to take into consideration the cost of the hedging. Since regularization term prevents the optimal weights from drifting to infinity, the regularization parameter becomes a desirable tool in keeping hedging cost under control. In our experiments we found that the parameter $\alpha = 10^{-9}$ represents a good choice.

46.5 Numerical Results

To illustrate our proposed methodology we consider hedging an unstructured trust IO (interest only strip) with a basket of European swaptions of different strikes and expiration dates, all with 10-year tenor. To generate the trajectories of the short rate, we use the two-additive-factor Gaussian model G2++ [4]. The dynamics of the instantaneous-short-rate process under the risk-neutral measure are given by $r(t) = x(t) + y(t) + \varphi(t)$, $r(0) = r_0$, where the processes $\{x(t) : t \geq 0\}$ and $\{y(t) : t \geq 0\}$ satisfy Eq. (46.10):

$$\begin{aligned} dx(t) &= -ax(t)dt + \sigma dW_1(t),\ x(0) = 0, \\ dy(t) &= -by(t)dt + \eta dW_2(t),\ y(0) = 0. \end{aligned} \tag{46.10}$$

Here (W_1, W_2) is a two-dimensional Brownian motion with instantaneous correlation $\rho : dW_1(t)dW_2(t) = \rho dt$. The parameters a, b, σ, η, ρ are defined in the calibration procedure to match the prices of a set of actively traded European swaptions. The l_2-norm of the difference vector of the model swaption prices and the market prices in our experiments is less than five basis points.

A key component in valuing mortgage bonds is to calculate the principal balance. This calculation is based on normal amortization and prepayment. In order to calculate amortization rates, we assume that the loans backing the bond have some specific parameters. Those parameters can take a wide range of values, but here we assume the following. The loan age (WALA) is 60 months, the interest rate on the loans is (WAC) is 6%. We also assume that the coupon on the bond is 5.5%, and it has a price of \$22.519. The bond under consideration has the following parameters: WALA = 60 months, WAC = 6%, coupon = 5.5%, and a price of \$22.519. We are using very simple prepayment model defined by linear interpolation of the data from Table 46.1.

For optimization we use a basket of 64 payer and receiver European swaptions with yearly expiration dates from years 1 to 16, including at the money (ATM) payers and receivers, +50 basis points out of the money payers and −50 basis points out of the money receivers. The regularization parameter used is 10^{-9}, and the number of trajectories in all experiments was 500. The SVM numerical method was implemented in Matlab with using the BFGS standard implementation available in the Matlab optimization toolbox. The computer time for the Matlab optimization routine was 42 s on the standard PC with 2 Hz processor frequency.

Figure 46.1 represents the convergence history of the iterations in the optimization method, showing the log of the norm of the gradient.

Figure 46.2 shows the square root of the target functional, which is approximately the standard deviation of the spread distribution. One can see that as a result of the application of the new methodology, the standard deviation of the spread distribu-

Table 46.1 Prepayment rates

Interest rates (%)	3.0	3.5	4.0	4.5	5.0	5.5	6.0	6.5	7.0
CPR (%)	70	50	30	15	10	8	4	3	3

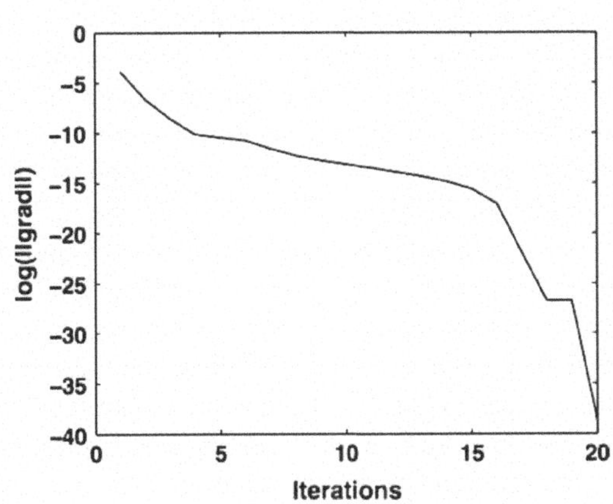

Fig. 46.1 Convergence history

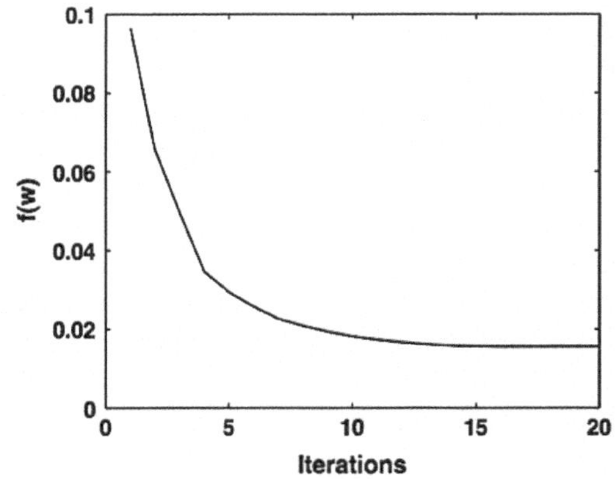

Fig. 46.2 Convergence of target functional

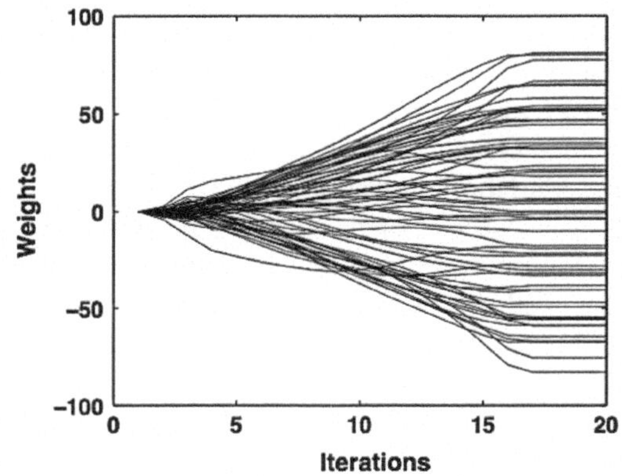

Fig. 46.3 Evolution of weights

tion is reduced significantly. This can be considered as reduction in riskiness of the portfolio of CMO and benchmarks.

Figure 46.3 presents the evolution of weights in our portfolio.

Figure 46.4 presents the distribution of the spreads for the original portfolio on the left and the optimal portfolio on the right.

In the next series of experiments, we illustrate the quality of the hedge constructed using different baskets of swaptions. Table 46.2 presents the results of these experiments. In our experiments we used the IO strip in combination with different numbers of swaptions. For reference we include results with no hedge at all; this is the common OAS analysis (but we also include the mean spread). For our SVM approach, we first use just two ATM swaptions, one payer and one receiver, both with expiration date of 1 year. Then we use eight swaptions with expiration dates of 1 and 2 years. For each of these expirations we include an ATM payer, an ATM receiver, an ATM +50 bps payer, and an ATM-50 bps receiver.

The test with twenty options uses expiration dates 1–5 years, again with four swaptions per expiration date. In the last experiment we use swaptions with

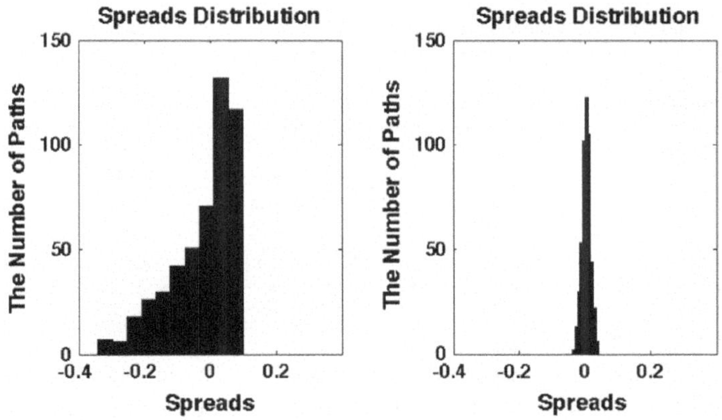

Fig. 46.4 Spread distributions

Table 46.2 Hedging results

Number of swaptions	0	2	8	20	64
Cost of hedge($)	0	–0.002	5.628	60.14	11.361
σ(bsp)	900	614	407	224	113
Mean spread (bsp)	–71	–18	37	49	33
OAS (bsp)	127	101	95	68	40

expiration dates 1–16. In all of these experiments we used one unit of the trust IO with the cost of $22.519.

As we can see from the first line of the table, the cost of the hedge is an increasing function of the number of options. But the most importantly, we manage to significantly decrease the standard deviation of the spread distribution by using spread variance minimization methodology.

Notice that, when using just two options, the cost of our hedging portfolio is essentially zero, we are buying the ATM receiver swaption and selling an equal amount of the ATM payer swaption. This is effectively entering into a forward rate agreement, and so this case could be considered as a hedging strategy of offsetting the duration of the CMO. The experiments with the larger sets of swaptions refine this strategy and take into account more detail about the structure of the cash flows of the bond. It proves to be a very successful approach to the hedging of the path-dependent bond.

The last two rows of the table present two different metrics: The mean spread and the OAS of the portfolio of bond and hedges. We can see that they become close as standard deviation decreases. In fact, they would be the same if the standard deviation becomes zero. With no swaptions, or with very few of them, these metrics can result in drastically different conclusions about the cheapness of the bond. Contrary to a common interpretation of OAS, it does not represent the mean spread of an unhedged bond. In fact, it can be expected to be close to the mean spread only when

Table 46.3 Results for short receiver, long payer

Number of payer/ receiver pairs	0	1	10	100	500
OAS (bps)	122	123	125	149	267
σ (bps)	893	907	968	1,636	5,363
OAS_{opt}	44	44	44	44	45
$\bar{s}_{opt}$	37	37	38	38	38
σ_{opt}	133	133	133	133	138

considering a refined hedging portfolio of swaptions. Though the discussion of the advantages of the SVM analysis is outside of the scope of this paper, it is worth mentioning that it caries significant information about the path-dependent bond and is worth taking into account alongside the standard OAS parameter.

In the last set of experiments, we see that we can increase the OAS of a portfolio, even without changing its cost. We include in our portfolio the IO, a long position in a payer, and a short position in a receiver, with the swaptions being ATM and having an expiration of 1 year. Table 46.3 presents the results of experiments with the portfolio of an IO strip and different amounts of the swaptions. We can see that we can actually increase the OAS (second row in Table 46.3) by increasing the number of fairly priced payer/receiver combinations, but it comes with the price of increasing the standard deviation (third row in the Table 46.3) of the spreads for this portfolio. Usually this information is missing in OAS analysis, and it could give a skewed impression about the cheapness of a portfolio. It should be expected that a more risky portfolio will have a better return, but in normal OAS analysis there is no assessment of the tradeoff between risk and return. So the idea is to try first decrease the standard deviation of the spread distribution as much as possible by using in the hedge the same type of swaptions as in original portfolio, and only then use the OAS or/and mean spread constant as a metric for the cheapness of the portfolio.

As indicated in the second row of Table 46.3, the OAS of the portfolio may increase dramatically, even though we are not changing the cost of setting up or portfolio, i.e., strictly speaking there is no leveraging. On the other hand, rows three, four and five show that the optimized portfolio of swaptions, from SVM analysis remains stable. We can see that OAS and the mean of spreads stay the same in all experiments and serves as a better metric for the cheapness or richness of the portfolio.

46.6 Conclusions

In this paper we presented a regularization approach to the construction of an optimal portfolio of CMO and swaptions. The standard Tikhonov regularization term serves as an important tool for preventing the weights of the benchmarks in the optimal portfolio drift to the infinity and so keeps the hedging procedure practical.

The optimization method demonstrates excellent convergent properties and could be used in practical applications for hedging a portfolio of CMO. The numerical results demonstrate the effectiveness of the proposed methodology. The future development may include the construction of new target functional based on the combination of the spread variance and cost function. This modification might improve the efficiency of the developed hedging strategy.

There appears to be a wide variety of application of our SVM approach, and in general of systematic path-by-path analysis of securities with stochastic cash flows. We plan to apply our analysis to other CMO structures directly. In addition, our analysis lends itself to comparing structured CMO with liquid strip interest only and principal only bonds.

References

1. Cohler, G., Feldman, M. & Lancaster, B., Price of Risk Constant (PORC): Going Beyond OAS. The Journal of Fixed Income, 6(4), 1997, 6–15.
2. Levin, A. & Davidson, A., Prepayment Risk and Option-Adjusted Valuation of MBS. The Journal of Portfolio Management, 31(4), 2005.
3. Tikhonov, A. N., Regularization of incorrectly posed problems. Soviet Mathematics Doklady, 4, 1963, 1624–1627.
4. Brigo, D. & Mercurio, F., Interest Rate Models: Theory and Practice. Berlin Heidelberg, Springer-Verlag, 2001, pp. 132–165.
5. Korn, R. & Korn, E., Option Pricing and Portfolio Optimization. AMS, Providence, Rhode Island, V. 31, 2000.
6. Vogel, C. R., Computational methods for inverse problems. SIAM, Philadelphia, PA, 2002.

Chapter 47
Approximation of Pareto Set in Multi Objective Portfolio Optimization

I. Radziukyniene and A. Zilinskas

Abstract In this chapter we experimentally investigate several evolutionary multi objective optimization methods and compare their efficiency in problems of portfolio selection with the efficiency of specially tailored method of adjustable weights. Test problems were based on standard portfolio quality criteria, and data on stocks of ten Lithuanian companies. We do not concern here in mach between analytical properties of the criteria functions and such properties favorable for the considered methods; we believe, however that general (global) structure of multi-criteria portfolio selection problem will be invariant with respect to switching from criteria defined by simple analytical formula to criteria defined by complicated numerical methods.

Keywords Multi Objective Portfolio Optimization · Pareto Set · Approximation · portfolio selection · evolutionary optimization

47.1 Introduction

Many computational finance problems ranging from asset allocation to risk management, from option pricing to model calibration can be solved efficiently using modern optimization techniques. The question of optimal portfolio allocation has been of long-standing interest for academics and practitioners in finance. In 1950s Harry Markowitz published his pioneering work where he has proposed a simple quadratic program for selecting a diversified portfolio of securities [1]. His model for portfolio selection can be formulated mathematically either as a problem of maximization of expected return where risk, defined as variance of return, is (upper)

I. Radziukyniene (✉)
Department of Informatics, Vytautas Magnus University, 8 Vileikos str.,
Kaunas LT 44404, Lithuania,
E-mail: i.radziukyniene@if.vdu.lt

S.I. Ao and L. Gelman (Eds.), *Advances in Electrical Engineering and Computational Science*, Lecture Notes in Electrical Engineering.

bounded or as a problem of minimization of risk where expected return is (lower) bounded. The classical Markowitz approach to portfolio selection reduces the problem of two criteria optimization to a one criterion optimization where the second criterion is converted to a constraint. Reduction of a multi-criteria problem to one criterion problem not always is the best method to solve multi-criteria problems especially in the case of vague a priory comparability of criteria. In such problems some evaluation of a whole Pareto set is of interest. It is the case for a portfolio selection problem formulated as a multi-criteria optimization problem where compromise between criteria crucially depends on the subjective priorities of a decision maker. Therefore Pareto set approximation is an important method facilitating rational portfolio selection. We have implemented a method of adjustable weights for Pareto set approximation where the classical scalarization idea of weighted summation of criteria is enhanced using branch and bound type procedure. To attack the problem of Pareto set approximation evolutionary optimization methods are claimed very promising; see e.g. [2–7]. In this paper we experimentally investigate several evolutionary multi objective optimization methods and compare their efficiency in problems of portfolio selection with the efficiency of specially tailored method of adjustable weights. Test problems were based on standard portfolio quality criteria, and data on stocks of ten Lithuanian companies. We do not concern here in mach between analytical properties of the criteria functions and such properties favorable for the considered methods; we believe, however that general (global) structure of multi-criteria portfolio selection problem will be invariant with respect to switching from criteria defined by simple analytical formula to criteria defined by complicated numerical methods. The paper is organized as follows. In Section 47.2 the multi-objective portfolio optimization problem is outlined, Sections 47.3 and 47.4 describe the considered multi objective optimization methods and their characteristics. In Sections 47.5 and 47.6 we discuss the used performance metrics and the experimental results. The paper is completed with conclusions on properties of the considered methods relevant to their applicability to multi objective portfolio selection.

47.2 Multi-Objective Portfolio Optimization Problem

Risk plays an important role in modern finance, including risk management, capital asset pricing and portfolio optimization. The problem of portfolio selection can be formulated as the problem to find an optimal strategy for allocating wealth among a number of securities (investment) and to obtain an optimal risk-return trade-off. The portfolio optimization problem may be formulated in various ways depending on the selection of the objective functions, the definition of the decision variables, and the particular constraints underlying the specific situation [7–10]. Beyond the expected return and variance of return, like in Markowitz portfolio model [1], the additional objective function can include number of securities in a portfolio, turnover, amount of short selling, dividend, liquidity, excess return over of a benchmark random variable and other [7]. In the bank portfolio management, the additional criteria such as the prime rate, processing cost, expected default rate, probability of unexpected losses, quantity of the long-term and short-term can be considered [8]. For example, the multi-objective portfolio selection problem can include the following objectives [9]: (to be maximized) portfolio return, dividend, growth in sales, liquidity, portfolio return over that of a benchmark, and (to be minimized) deviations from asset allocation percentages, number of securities in portfolio, turnover (i.e., costs of adjustment), maximum investment proportion weight, amount of short selling. We considered two multi-objective portfolio problems. The first problem was based on a simple two objectives portfolio model including the standard deviation of the returns and mean of the returns, where the return R_i is 1 month return of stock i; return means percentage change in value. The second problem included three objectives, where annual dividend yield is added to two above mentioned objectives. For the experiment we used a data set of ten Lithuanian companies' stock data from Lithuanian market.

Financial portfolio selection is one of real world decision making problems with several, often conflicting, objectives which can be reduced to multi objective optimization. There are many methods to attack multi objective optimization problems. In some cases a solution to a multi objective problem can be defined as a special point of Pareto set by means of scalarization, i.e. by converting the initial multi objective problem to a single criterion problem. However, in many cases decision can not be made without some information about the whole Pareto set. The theoretical problem to find the whole Pareto set normally (e.g. in case of continuum cardinality of Pareto set) can not be solved algorithmically, thus the theoretical problem of finding of whole Pareto set should reformulated to algorithmic construction of an appropriate its approximation. In the present paper we consider several algorithms of Pareto set approximation including the newly proposed method of adjustable weights and some evolutionary algorithms well assessed in recent publications. Over the past decades evolutionary algorithms have received much attention owing to its intrinsic ability to handle optimization problems with both single and multiple objectives including problems of financial optimization [2–7]. Let us note that financial optimization problems were attacked also by means of other heuristics, e.g. by simulated annealing in [3], and by Tabu search in [8]. Comparisons of the performance of different heuristic techniques applied to solve one criterion portfolio choice problems are given by Chang et al. [3]. However, to the best knowledge of the authors' similar comparative analysis of performance of recently proposed evolutionary multi-criteria algorithms has not been yet reported.

47.3 Multi-Objective Optimization by the Method of Adjustable Weights

In this paper we investigated approximation of Pareto set by means of a finite set of points uniformly distributed in close vicinity of the Pareto set; the terms "uniformly" and "close vicinity" are defined more precisely in the section on experimental results.

First method that attracted our attention was the widely used one scalarization method of weighted criteria summation. Besides of the status of "classics" of multi objective optimization, an important argument to consider the method of weighted criteria summation is simplicity of its implementation. Since a multi criteria problem is converted into a single criterion optimization problem where the objective function is a weighted sum of the criteria functions, weights can be used to express the relative significance of different criteria. For some multi objective optimization problems it is convenient to take into account in this way the preferences of decision makers. However, we do not assume availability of information about importance of criteria, and for us weights play the role of parameters defining points in the Pareto set corresponding to the solutions of parametric single criteria problems. The mathematical model of the weighted sum method takes the form of:

$$\min f(x), \; f(x) = \sum_{i=1}^{m} \omega_i f_i(x), \tag{47.1}$$

where ω_i is the weight of ith criterion, $0 \leq \omega_i \leq 1, i = 1, ..., m$, and $\sum_{i=1}^{m} \omega_i$.

Assume that all $f_i(x)$ are convex functions; then for every point of Pareto set there exist weights $\omega_i, i = 1, ..., m$ such that this Pareto point is a minimizer of $f(x)$ defined by Eq. (47.1). The violation of the convexity assumption can imply not existence of weights implying location of the minimizer of the corresponding composite objective function $f(x)$ in some subsets of Pareto set. However, in the problems of portfolio selection objectives $f_i(x)$ normally are defined by rather simple analytical expressions, and checking of the validity of convexity assumption is easy. The fact of existence of the correspondence between points in Pareto set and minimizers of the parametric minimization problem (Eq. (47.1)) theoretically justifies further investigation of properties of this correspondence. In many cases it is not an easy task to choose the set of weights defining solutions of Eq. (47.1) that would be well (in some sense uniformly) distributed over the Pareto set. To generate such a subset of the Pareto set by means of repeatedly solution of Eq. (47.1) with different weights the latter should be chosen in a special but a priory unknown way.

We propose a branch and bound type method for iterative composing of a set of weights implying the desirable distribution of solutions of Eq. (47.1). The feasible region for weights

$$\Omega = \left\{ (\omega_1, ..., \omega_m) : 0 \leq \omega_i \leq 1, \; i = 1, ..., m, \; \sum_{i=1}^{m} \omega_i = 1 \right\}, \tag{47.2}$$

is a standard simplex. Our idea is to partition Ω into sub simplices whose vertices are mapped to the Pareto set via Eq. (47.1). The sequential partition is controlled aiming to generate such new sub simplices that mapping of their vertices would generate points uniformly distributed in Pareto set. The partition procedure is arranged as a branching of a tree of nods corresponding to sub simpices. The original standard simplex is accepted as the root of the tree. Branching means partition of a simplex into two sub simplices where the new vertex is the midpoint of the favorable edge defined later. Simplicial partitioning procedure has been applied to construct algorithms for single criterion global optimization in [11, 12]; this procedure can be applied also in multi objective optimization if only preconditions for selection of a simplex for partition would be reconsidered.

We aim to subdivide the original simplex into sub simplices in such a way that solutions of Eq. (47.1) corresponding to the vertices of these sub simplices would be well distributed over the Pareto set. Let us consider the current set of sub simplices in the space of weights, and points in the criteria space corresponding to the vertices of these simplices. The eligibility of a simplex to partition is defined by the longest distance between the points in the criteria space corresponding to the vertices of the considered simplex. Branching strategy is based on depth first search, and the selected simplices are partitioned until the longest distance between the corresponding points in criteria space is reduced up to predefined tolerance.

The method of adjustable weights (AdjW) was implemented for two and three criteria cases in MATLAB using fminconstr for solving minimization problem (Eq. (47.1)). To illustrate the advantage of the proposed method of adjustable weights over the standard weighted sum method both methods were applied for the construction of Pareto set of two criteria portfolio optimization with the data mentioned above. Figure 47.1 shows the distributions of the Pareto points found by both methods. In the standard method weights have been changed with the step for 0.05. Because of space limitation we do not describe in detail the tolerance for branching of the method of adjustable weights; it have been set aiming at generation of similar number of points as generated by weighting method. In this experiment the number of points generated by the weighting method was equal to 21, and that generated by our method was equal to 26.

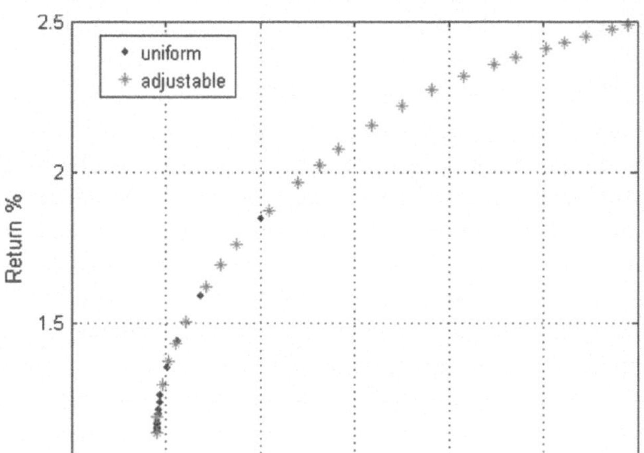

Fig. 47.1 Points in Pareto set generated by the standard weighted sum method and by the method of adjustable weights

47.4 Evolutionary Methods for Multi-Objective Optimization

Three methods: Fast Pareto genetic algorithm (FastPGA) [2], Multi-Objective Cellular genetic algorithm (MOCeLL) [3], and Archive-based hybrid Scatter Search algorithm [4] were proposed during the last 2 years. Their efficiency for various problems has been shown in original papers, but their application to portfolio optimization problem was not yet explored. NSGA-II [5], the state-of-the-art evolutionary method, was chosen following many authors who use it as a standard for comparisons.

FastPGA. Eskandari and Geiger [2] have proposed framework named fast Pareto genetic algorithm that incorporates a new fitness assignment and solution ranking strategy for multi-objective optimization problems where each solution evaluation is relatively computationally expensive. The new ranking strategy is based on the classification of solution into two different categories according to dominance. The fitness of non-dominated solutions in the first rank is calculated by comparing each non-dominated solution with one another and assigning a fitness value computed using crowding distance. Each dominated solution in the second rank is assigned a fitness value taking into account the number of both dominating and dominated solutions. New search operators are introduced to improve the proposed method's convergence behavior and to reduce the required computational effort. A population regulation operator is introduced to dynamically adapt the population size as needed up to a user-specified maximum population size, which is the size of the set of non-dominated solutions. FastPGA is capable of saving a significant number of solution evaluations early in the search and utilizes exploitation in a more efficient manner at later generations.

Characteristics of FastPGA: the regulation operator employed in FastPGA improves its performance for fast convergence; proximity to the Pareto optimal set, and solution diversity maintenance.

MOCeLL. Nebro et al. [3] presented MOCeLL, a multi-objective method based on cellular model of GAs, where the concept of small neighborhood is intensively used, i.e., population member may only interact with its nearby neighbors in the breeding loop. MOCell uses an external archive to store the non-dominated solutions found during the execution of the method, however, the main feature characterizing MOCell is that a number of solutions are moved back into the population from the archive, replacing randomly selected existing population members. This is carried out with the hope of taking advantage of the search experience in order to find a Pareto set with good convergence and spread.

MOCell starts by creating an empty Pareto set. The Pareto set is just an additional population (the external archive) composed of a number of the non-dominated solutions found. Population members are arranged in a two-dimensional toroidal grid, and the genetic operators are successively applied to them until the termination condition is met. Hence, for each population member, the method consists of selecting two parents from its neighbourhood for producing an offspring.

An offspring is obtained applying operators of recombination and mutation. After evaluating of the offspring as the new population member it is inserted in both the auxiliary population (if it is not dominated by the current population member) and the Pareto set. Finally, after each generation, the old population is replaced by the auxiliary one, and a feedback procedure is invoked to replace a fixed number of randomly chosen population members of the population by solutions from the archive. In order to manage the insertion of solutions in the Pareto set with the goal to obtain a diverse set, a density estimator based on the crowding distance has been used. This measure is also used to remove solutions from the archive when this becomes full.

Characteristics of MOCeLL: the method uses an external archive to store the non-dominated population members found during the search; the most salient feature of MOCeLL with respect to the other cellular approaches for multi-objective optimization is the feedback of members from archive to population.

AbYSS. This method was introduced by Nebro et al. [4]. It is based on the scatter search using a small population, known as the reference set, whose population members are combined to construct new solutions. Furthermore, these new population members can be improved by applying a local search method. For local search the authors proposed to use a simple (1 + 1) Evolution Strategy which is based on a mutation operator and a Pareto dominance test. The reference set is initialized from an initial population composed of disperse solutions, and it is updated by taking into account the solutions resulting from the local search improvement. AbYSS combines ideas of three state-of-the-art evolutionary methods for multi criteria optimization. On the one hand, an external archive is used to store the non-dominated solutions found during the search, following the scheme applied by PAES [5], but using the crowding distance of NSGA-II [6] as a niching measure instead of the adaptive grid used by PAES; on the other hand, the selection of solutions from the initial set to build the reference set applies the density estimation used by SPEA2 [4].

Characteristics of AbYSS: it uses an external archive to store the non-dominated population members found during the search; salient features of AbYSS are the feedback of population members from the archive to the initial set in the restart phase of the scatter search, as well as the combination of two different density estimators in different parts of the search.

NSGA-II. The evolutionary method for multi-criteria optimization NGSA-II contains three main operators: a non-dominated sorting, density estimation, and a crowded comparison [6]. Starting from a random population the mentioned operators govern evolution whose aim is uniform covering of Pareto set.

Non-dominated sorting maintains a population of non dominated members: if a descendant is dominated, it immediately dies, otherwise it becomes a member of population; all members of parent generation who are dominated by descendants die. The density at the particular point is measured as the average distance between the considered point and two points representing the neighbour (left and right) population members. The crowded comparison operator defines selection for crossover oriented to increase the spread of current approximation of Pareto front. Population members are ranked taking into account "seniority" (generation number) and local crowding distance.

The worst-case complexity of NSGA-II algorithm is $O(mN^2)$, where N is the population size and m is the number of objectives [6].

Characteristics of NSGA-II: this method is of the lower computational complexity than that of its predecessor NSGA; elitism is maintained, no sharing parameter needs to be chosen because sharing is replaced by crowded-comparison to reduce computations.

47.5 Description of Experimental Investigation

Theoretical assessment of available multi objective optimization methods with respect to their applicability to portfolio optimization does not seem possible. Therefore such an assessment has been performed experimentally. There are some publications on experimental comparison of different performance aspects of evolutionary methods. For example, in the study presented in [13] some general conclusions are drawn, however they could not be directly applied to a specific problem. For assessing of the considered methods several different performance measures can be taken into account: the distance between the approximated Pareto set generated by the considered method and the true Pareto set, the spread of the solutions, and computational time. To determine the first measure the true Pareto set should be known. In this experimental investigation we didn't know true Pareto sets. Therefore, the best approximation found by means of combining results of all considered methods was used instead of true Pareto set. Although, computational time is one of the most important performance measures in comparison of optimization methods, it has not been included in our study. This can be justified only by the severe differences in implementations of the algorithms (e.g. in C and in MATLAB) making direct comparison of running times unfair. We compared methods according to three performance measures:

1. *Generational distance* (GD) shows how far the approximation is from true Pareto set [14].
2. *Inverted generational distance* (IGD) [15]. This quality indicator is used to measure how far the elements are in the Pareto optimal set from those in the set of non-dominated vectors found.
3. *Hypervolume* (HV) [14]. This quality indicator calculates the volume (in the objective space) covered by members of a non-dominated set of solutions. Methods with larger values of HV are desirable.

Before evaluating the fitness function in FastPGA, MOCeLL, AbYSS, and NSGAII the proportions of stocks in the portfolio were normalized as in reference [7]. These methods were run with different parameters that were recommended by the authors [14, 16–18].

They were:

1. **FastPGA** Maximum population = 100, initial population = 100, crossover probability = 1.0

2. **MOCeLL** Population = 100, archive = 100, crossover probability = 0.9
3. **AbYSS** Population = 20, archive = 100, crossover probability = 1.0, setref1 = 10, setref2 = 10
4. **NSGAII** Population = 100, crossover probability = 0.9

To evaluate each of these methods, while solving two objectives portfolio optimization problem we performed three series of experiments. First, we ran all the methods for 15,000 function evaluations, and then repeated them with the execution of 25,000 and 35,000 function evaluations as the stopping condition. In the case of three objectives portfolio, two series of experiments with 25,000 and 50,000 function evaluations have been performed. For each problem we have executed 50 independent runs.

47.6 Discussion on Experimental Results

Experiments were performed to compare performance of the selected evolutionary methods and the method of adjustable weights (AdjW). The evolutionary methods are randomized, but the method of adjustable weights is deterministic. Because of this difference the experimental results in tables below are presented correspondingly. The averages (Avg.) and standard deviations (Std.) of all performance measures for each evolutionary method were calculated from the samples of 50 runs. The best results in tables are printed in boldface.

It follows from the Table 47.1 that MOCeLL outperforms the other methods in all cases except of comparison with respect to Inverted general distance where number of functions evaluations was equal to 25,000; in the latter case NSGAII was considerably better. The differences in performance weaken with increase of number of evaluations as it is well illustrated by the results of the last block of results in the Table 47.1.

The upper part of the Pareto set shown in Fig. 47.1 is most difficult to reveal for all evolutionary methods. Approximation of the Pareto set with a curve of changing curvature shows that the curve is flattening at its upper part. Similar dependence between flatness of the Pareto front and decrease of quality of its approximation using evolutionary methods is mentioned also by the other authors. For illustration of this phenomenon the points generated by all methods in the mentioned part of Pareto set are presented in Fig. 47.2; maximal number of function evaluations was fixed equal to 25,000. It can be noticed that the method of adjustable weights does not suffer from flattening of the Pareto front.

The experimental results for three criteria problem are given in Table 47.2. These results show that the best of the evolutionary methods according GD and HV measures is MOCeLL. NSGAII and FastPGA are the best methods with respect to IGD. It can be noticed that GD value of the method of adjustable weights is equal to zero; this means that all solutions lie precisely in Pareto set. The approximations of Pareto sets obtained by evolutionary methods with 50,000 function evaluations and the adjustable weights method were visualized using color stereo display. Since such

Table 47.1 Performance metrics for two objective problem; the performance measures for AdjW were GD = 0.0, IGD = 2.97e-5, HV = 0.886; **F** denotes the number of function evaluations by the evolutionary methods

Method	GD		IGD		HV	
	Avg.	Std.	Avg.	Std.	Avg.	Std.
F = 15,000						
AbYSS	5.47e-4	4.09e-4	1.86e-3	2.14e-3	8.55e-1	3.31e-2
FastPGA	4.12e-4	2.81e-4	2.28e-3	1.64e-3	8.56e-1	2.38e-2
MOCeLL	**3.10e-4**	2.31e-4	**1.23e-3**	1.65e-3	**8.69e-1**	2.12e-2
NSGAII	4.17e-4	2.75e-4	1.40e-3	1.24e-3	8.69e-1	1.42e-2
F = 25,000						
AbYSS	2.06e-4	5.95e-5	1.60e-4	5.59e-4	8.82e-1	5.50e-3
FastPGA	2.26e-4	7.04e-5	4.88e-4	9.77e-4	8.79e-1	1.26e-3
MOCeLL	**9.29e-5**	1.88e-5	1.11e-4	2.92e-4	**8.84e-1**	1.90e-3
NSGAII	2.36e-4	2.87e-5	**9.90e-5**	1.28e-5	8.82e-1	3.00e-4
F = 35,000						
AbYSS	1.63e-4	3.57e-5	7.20e-5	2.40e-6	8.83e-1	3.20e-4
FastPGA	2.11e-4	2.72e-5	1.04e-4	1.15e-4	8.83e-1	4.70e-4
MOCeLL	**6.34e-5**	8.90e-6	**6.70e-5**	9.00e-7	**8.84e-1**	8.50e-5
NSGAII	2.41e-4	3.02e-5	9.70e-4	4.70e-6	8.82e-1	1.80e-4

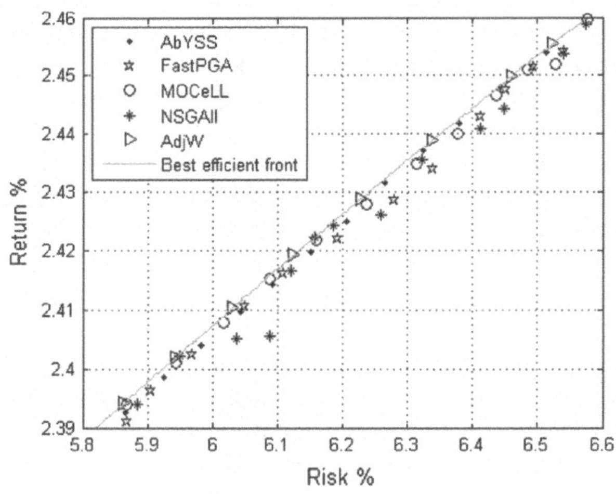

Fig. 47.2 Pareto sets of AbYSS, FastPGA, MOCeLL, NSGAII, and AdjW

a visualization can not be included in a paper we will present only heuristically obvious conclusion that the quality of approximation of Pareto set by adjustable weights method is better than that by evolutionary methods. However the latter are less computing intensive. Therefore development of a hybrid method seems promising.

Table 47.2 Performance metrics for two objective problem; the performance measures for AdjW were GD = **0.0**, IGD = **1.48e-4**, HV = **0.735**; **F** denotes the number of function evaluations by the evolutionary methods

Method	GD		IGD		HV	
	Avg.	Std.	Avg.	Std.	Avg.	Std.
F = 25,000						
AbYSS	1.44e-3	4.68e-4	2.16e-4	1.16e-4	7.15e-1	6.30e-3
FastPGA	1.42e-3	3.85e-4	**2.08e-4**	1.90e-5	7.16e-1	2.10e-3
MOCeLL	**1.16e-3**	3.62e-4	2.12e-4	1.90e-5	**7.18e-1**	1.30e-3
NSGAII	1.33e-3	3.02e-4	2.10e-4	2.20e-5	7.15e-1	1.90e-3
F = 50,000						
AbYSS	1.22e-3	4.04e-4	2.13e-4	1.90e-5	7.16e-1	1.50e-3
FastPGA	1.24e-3	3.59e-4	**2.06e-4**	1.50e-5	7.18e-1	1.40e-3
MOCeLL	**1.14e-3**	2.77e-4	2.12e-4	1.80e-5	**7.19e-1**	1.20e-3
NSGAII	1.65e-3	5.05e-4	2.12e-4	2.20e-5	7.16e-1	1.50e-3

47.7 Conclusions

From the results of the experiments for two criteria portfolio optimization it follows that MOCeLL is the best of four considered evolutionary methods with respect to all three performance criteria. The results of the experiments with these methods for three criteria portfolio optimization reveal that MOCeLL provides the best results in terms of Hypervolume, and Generational distance, but is slightly outperformed by FastPga with respect to the Inverted generational distance. The evaluated performance criteria of evolutionary methods are only slightly worse than those of method of adjustable weights who is advantageous in the considered cases. Summarizing the results it seems promising to develop a hybrid method trying to combine the advantages of evolutionary methods with those of the method of adjustable weights.

Acknowledgements The second author acknowledges the support by the Lithuanian State Science and Studies Foundation.

References

1. Markowitz, H.: Portfolio selection. Journal of Finance, **7**, 77–91 (1952).
2. Li, J., Taiwo, S.: Enhancing Financial Decision Making Using Multi-Objective Financial Genetic Programming. Proceedings of IEEE Congress on Evolutionary Computation 2006, pp. 2171–2178 (2006).
3. Chang, T.-J., Meade, N., Beasley, J. E., Sharaiha, Y. M.: Heuristics for cardinality constrained portfolio optimisation. Computers and Operations Research, **27**, 1271–1302 (2000).
4. Wang, S.-M., Chen, J.-C., Wee, H. M., Wang, K. J.: Non-linear Stochastic Optimization Using Genetic Algorithm for Portfolio Selection. International Journal of Operations Research, **3**, No. 1, 16–22 (2006).
5. Ehrgott, M., Klamroth, K., Schwehm, C.: Decision aiding an MCDM approach to portfolio optimization. European Journal of Operational Research, **155**, 752–770 (2004).

6. Xia, Y., Wang, S., Deng, X.: Theory and methodology: a compromise solution to mutual funds portfolio selection with transaction costs. European Journal of Operation Research, **134**, 564–581 (2001).
7. Mukerjee, A., Biswas, R., Deb, K., Mathur, A. P.: Multi-objective evolutionary algorithm for the risk-return trade-off in bank loan management. International Transactions in Operational Research, **9**, 583–597 (2002).
8. Stummer, C., Sun, M.: New Multiobjective Metaheuristic Solution Procedures for Capital Investment Planning. Journal of Heuristics, **11**, 183–199 (2005).
9. Ehrgott, M., Waters, C., Gasimov, R. N., Ustun, O.: Multiobjective Programming and Multiattribute Utility Functions in Portfolio Optimization. 2006. Available via http://www.esc.auckland.ac.nz/research/tech/esc-tr-639.pdf, cited 20 June 2008.
10. Mukerjee, A., Biswas, R., Deb, K., Mathur, A. P.: Multi-objective evolutionary algorithm for the risk-return trade-off in bank loan management. International Transactions in Operational Research, **9**, 583–597 (2002).
11. Steuer, R. E., Qi, Y., Hirschberger, M.: Portfolio Selection in the Presence of Multiple Criteria. In Zopounidis, C., Doumpos, M., Pardalos, P. M. (eds.) Handbook of Financial Engineering, Springer, New York (2008).
12. Clausen, J., Zilinskas, A.: Global Optimization by Means of Branch and Bound with simplex Based Covering, Computers and Mathematics with Applications, **44**, 943–955 (2002).
13. Zilinskas, A., Zilinskas, J.: Global Optimization Based on a Statistical Model and Simplicial Partitioning, Computers and Mathematics with Applications, **44**, 957–967 (2002).
14. Nebro, A. J., Durillo, J. J., Luna, F., Dorronsoro, B., Alba, E.: A Cellular Genetic Algorithm for Multiobjective Optimization. Proceedings of NICSO 2006, pp. 25–36.
15. Deb, K., Pratap, A., Agarwal, S., Meyarivan, T.: A Fast and Elitist Multiobjective Genetic Algorithm: NSGA-II. IEEE Transactions on Evolutionary Computation, **6**, No. 2, 182–197 (2002).
16. Zizler, E., Deb, K., Thiele, L.: Comparison of Multi Objective Evolutionary Algorithms: Empirical Results, Evolutionary Computation, **8**, No. 2, 173–195 (2000).
17. Eskandari, H., Geiger, C. D.: A Fast Pareto Genetic Algorithm Approach for Solving Expensive Multiobjective Optimization Problems, Journal of Heuristics, **14**, No. 3, 203–241 (2008).
18. Nebro, A. J., Luna, F., Alba, E., Beham, A., Dorronsoro, B.: AbYSS Adapting Scatter Search for Multiobjective Optimization. Technical Report ITI-2006–2, Departamento de Lenguajes y Ciencias de la Computaci6n, University of Malaga (2006).

Chapter 48
The Scaling Approach for Credit Limit Management and Its Application

M.Y. Konovalikhhin, D.O. Sergienko, and O.V. Balaeva

Abstract When bank decides to provide the loan to a client, decision has to be made on the credit amount and the term. Some studies looked into possibility of taking into consideration the probability of default while choosing the optimal credit policy by introducing synthetic coefficient or VaR technology. This paper suggests the scaling methodology for calculation of coefficients, which are required for optimal credit limit estimation. Scaling rationale is based on the goal-oriented parameters. The method is built on comparison of two competing forces: (i) potential increase of the limit caused by client's positive characteristics such as income, good credit history, ownership of movable and immovable properties etc. and (ii) potential limit decrease, which is a result of the probability of default calculated from the client score. Such model can be related to the quantitative cause-effect model type, which is focused on the final result without consideration of the process' dynamics.

Keywords Credit Limit Management · Scaling Approach · goal-oriented parameters · cause-effect model

48.1 Introduction

One of the consequences of impressive growth of Russian retail market is a very strong competition between the banks. This makes Basel Committee on Banking Supervision (Basel II) [1] requirement related to the necessity of development of the modern flexible bank systems for client reliability assessment [2] becomes particularly important. Increased competition and growing pressures for revenue generation have made financial institutions to look for more effective ways to attract new

M.Y. Konovalikhhin (✉)
The head of analytical division of risk analysis department, Ph.D., Bank VTB 24 (Closed joint-stock company), Moscow, Russia,
E-mail: Konovalihin.MY@vtb24.ru

S.I. Ao and L. Gelman (Eds.), *Advances in Electrical Engineering and Computational Science*, Lecture Notes in Electrical Engineering.

creditworthy customers and at the same time to control the losses. Aggressive marketing efforts have resulted in deeper penetration of the risk pool amongst potential customers. The need to process them rapidly and effectively has initiated growing automation of the credit and insurance application and adjudication processes.

Banks need rapid credit decision making process in order to maintain high borrowing power, but on the other hand it should not result in deterioration of portfolio quality. The risk manager is now challenged to provide solutions that not only assess the creditworthiness, but also keep the per-unit processing cost low, while reducing turnaround time for the customers. In addition, customer service quality requires this automated process to be able to minimize the denial of credit to creditworthy customers, while keeping out as many potentially delinquent customers as possible.

In the recent years a particular attention has been paid to risk scoring, which along with other predictive models is a tool evaluating the risk level associated with applicants or customers. While it does not identify "good" (no negative behavior expected) or "bad" (negative behavior expected) applications on individual basis, it provides the statistical odds, or probability, that an applicant with any given score will be either "good" or "bad". These probabilities or scores along with other business considerations, such as expected approval rates, profits and losses are used as a basis for decision making.

When bank decides to provide the loan to a client, decision has to be made on the credit amount and the term. Credit limit is the approved level of a loan amount, which theoretically must represent client's financial rating. In other words, correct calculation of the optimal credit limit is a powerful tool for minimizing the credit losses.

At the same time it should be noted that most of the existing methodologies for credit limit calculation do not take into consideration the explicit assessment of the probability of default. Traditional formula for credit limit calculation does not deal with the credit risk. It is postulated that the limit, calculated by this formula, provides zero credit risk, or, in other words, the probability of default reduces to zero [3].

Some studies looked into possibility of taking into consideration the probability of default while choosing the optimal credit policy by introducing synthetic coefficient [4] or VaR technology [5]. These studies are based on the estimation of a hypothecation value for inter-bank credits, which is not always applicable to retail business.

This paper suggests the scaling methodology for calculation of coefficients, which are required for optimal credit limit estimation. While developing such scaling methodology, three principles have been considered:

First, scaling rationale is based on the goal-oriented parameters. In this case, scaling parameters are determined from credit limit quality perspective.

Second, while seeking the simplicity of the scaling equation, it is required to reduce the (potential) effects of characteristics, involved in calculations, on component phenomena.

Third, equations and correlations that were used in development of the scaling methodology are transparent and can be separately validated.

The method is built on comparison of two competing forces: (i) potential increase of the limit caused by client's positive characteristics such as income, good credit history, ownership of movable and immovable properties etc. and (ii) potential limit decrease, which is a result of the probability of default calculated from the client score. Such model can be related to the quantitative cause-effect model type, which is focused on the final result without consideration of the process' dynamics.

48.2 Analysis

48.2.1 Model Description

Existing credit limit calculation methodology, which is currently used at some banks is presented below. Here, the total client income (D) is defined as a sum of documented and undocumented incomes:

$$D = DCI + B3 \times UCI, \tag{48.1}$$

where $B3$ is a coefficient showing the presence of some attributes (movable and immovable properties for instance), that could indicate additional income.

Then the D value is corrected according to the client's score:

$$BP = D \times B2 - OL, \tag{48.2}$$

where $B2$ is a coefficient, which represents client's rating, OL is a total amount of promissory notes. Client's expected income is influenced by a number of dependencies, which is reflected by $B1$ coefficient.

$$SD = BP \times B1 \tag{48.3}$$

The credit limit is calculated as following:

$$L = \frac{SD \times T}{1 + \frac{R \times T}{12}}, \tag{48.4}$$

where T is the loan term and R is the bank interest rate (%)/100.

The weakness of this method is in the fact, that coefficients $B1$, $B2$ and $B3$ are defined by the expert judgment. Calculation of the limits, which is based on the expert judgement may result in ungrounded over- or understatement of credit limits. In the case of understatement of the credit limits the bank cannot use all the credit resources, that in turn leads to reduced profit. One of the main consequences of credit limit overstatement is the increase in credit risks, which leads to additional losses.

On the other hand, applying precise methods to the credit limit calculation allows to reach full bank credit potential and to maximize bank's profit. Also, one should take into account client's rating (scoring) in order to minimize the risk of default.

48.2.2 B3 Coefficient Calculation

$B3$ coefficient (Eq. (48.1)) can be estimated by examining the dependency between the ratio $\frac{UCI}{DCI+UCI}$, which indicates the share of client's undocumented income in the total income, and the default probability $P(X) = \frac{1}{1+e^{-\frac{1}{X}}}$, where X is the client score. We assume, that this value can grow only with the improvement of client's rating or with the score increase and, respectively, with the decrease of the probability of default:

$$\frac{UCI}{DCI + UCI} = f(P(X)) = C(X) \times \frac{1}{P(X)}, \tag{48.5}$$

where $C(X)$ – a proportional coefficient, which may be independent of the score X. In order to assess this dependency a pool of 37,890 clients (for one product within certain timeframe) was analyzed. Then a number of clients were selected that had declared *UCI* incomes. From this data two groups were selected – 9,681 clients without delinquency and 1,334 with delinquency over 30 days.

The first problem, which had to be investigated, was to check for a potential correlation between $\frac{UCI}{DCI+UCI}$ and the presence of delinquency. The comparison of average values in both cases (with and without delinquency) showed that the difference between these two values is about 6%, which can be explained by statistical uncertainty (Figs. 48.1 and 48.2). This fact confirms the random nature of $\frac{UCI}{DCI+UCI}$.

With the knowledge of relation (48.5) it is possible to find the level of undocumented income *UCI*, which statistically corresponds to the documented income *DCI*:

$$\begin{aligned}(UCI)_{calc} &= DCI \times C(X) \times \frac{1}{P(X)} / \left(1 - C(X) \times \frac{1}{P(X)}\right) \\ &= \frac{DCI \times C(X)}{P(X) - C(X)}\end{aligned} \tag{48.6}$$

First, it was necessary to estimate parameter $C(X)$ by using statistical data. For this purpose two regions of possible risk levels (high and low) were selected. The line "good clients average ratio" was chosen as upper conservative boundary because the region, which is located above this line and correspondingly above the line "bad clients average ratio" was considered a high risk zone. The lower boundary was estimated by averaging (dashed line "low risk region", Figs. 48.1 and 48.2) bad client data, which occurs under the line "good clients average ratio". The region under the line "low risk region" was defined as a low risk region, which contains the values $(UCI)_{calc}$.

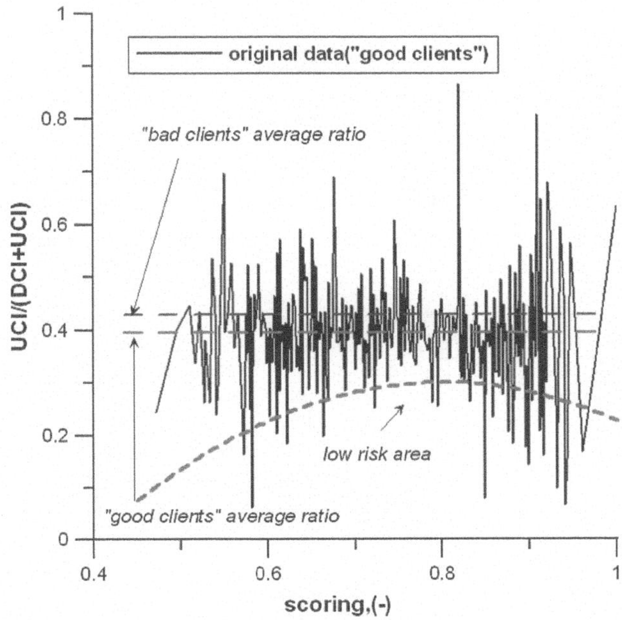

Fig. 48.1 Dependency of average $\frac{UCI}{DCI+UCI}$ value on the score of the "good" clients

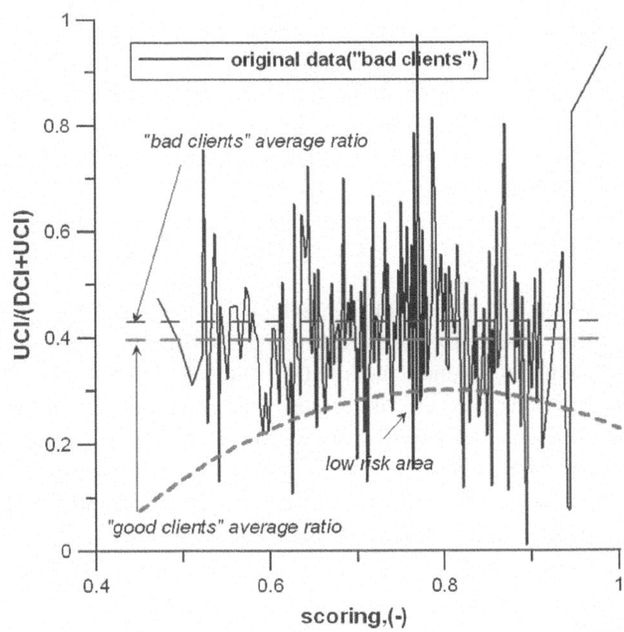

Fig. 48.2 Dependency of average $\frac{UCI}{DCI+UCI}$ value on the score of the "bad" clients

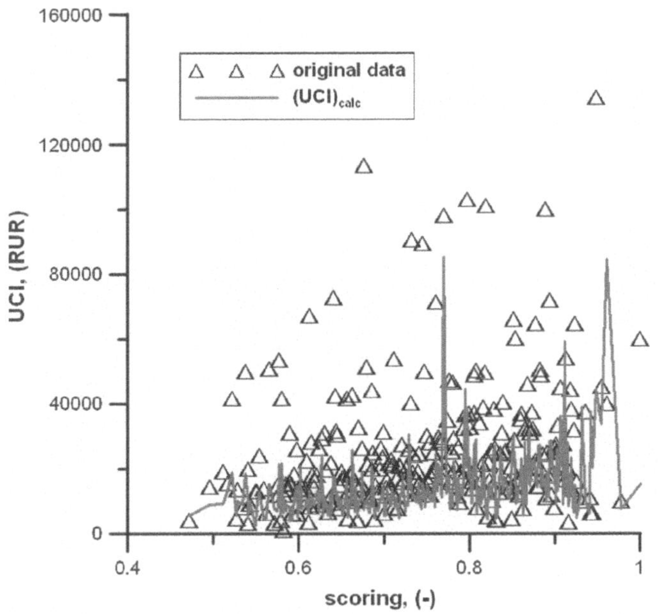

Fig. 48.3 Comparison of calculated UCI values and original data

Then two alternatives were examined, when Eq. (48.1) applied undocumented income $(UCI)_{appl}$ is less than $(UCI)_{calc}$ and Eq. (48.2) when it is greater than $(UCI)_{calc}$. In the first case the whole UCI amount can be taken into account and be employed for client income calculation or, in other words, $B3 = 1$ in Eq. (48.1). In the second case the *UCI* value can be calculated as a sum of two items: $(UCI)_{calc}$ and some part of the value $[(UCI)_{appl} - (UCI)_{calc}]$:

$$UCI = (UCI)_{calc} + C_3 \times [(UCI)_{appl} - (UCI)_{calc}] \tag{48.7}$$

During the next step the coefficient C_3, which demonstrates the optimal degree of confidence in the client, should be estimated. Figure 48.3 shows the difference between real client undocumented income *UCI* and calculated values $(UCI)_{calc}$. One of the possible solutions is to use the cause-effect relations, which have been successfully applied to quantitative analysis of various technical problems [6]. Such equations provide solution to the problem without describing process' dynamics. The result in this case is the ratio between positive and negative characteristics, that are represented by client's financial condition, such as movable and immovable properties, etc. in the numerator and the default probability $P(X)$ in the denominator:

$$C_3 = C_{3,corr} \times \frac{\sum_i A_i Z_i}{P(X)}, \tag{48.8}$$

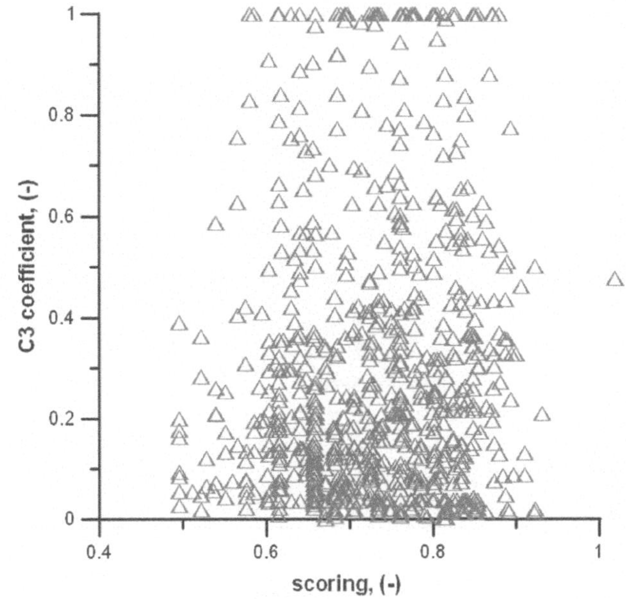

Fig. 48.4 Distribution of C3 coefficient

where $C_{3,\,corr}$ is the coefficient, which correlates Eq. (48.8) to statistical data, Z_i – parameters of client financial condition, A_i – regressive coefficients, that are calculated by the *Interactive Grouping* methodology. In order to estimate C_3 the following parameters were used: cost of movable properties (car, yacht etc.), country house, apartment and business shareholding. Results of the calculation are presented in Fig. 48.4.

48.2.3 B2 Coefficient Calculation

Similar approach could be applied for calculation of $B2$ coefficient in Eq. (48.2). Here again we compare two competing parameters through their ratio. The upper parameter, which is responsible for potential limit increase, is represented by a regressive sum of characteristics such as documented income, occupation, employment position etc. The lower parameter, which is pushing down the credit limit, is the probability of default $P(X)$.

$$B2 = C_2, corr \times \frac{E'}{P(X)}, \qquad (48.9)$$

where

$$E' = \sum_i D_i E_i, \qquad (48.10)$$

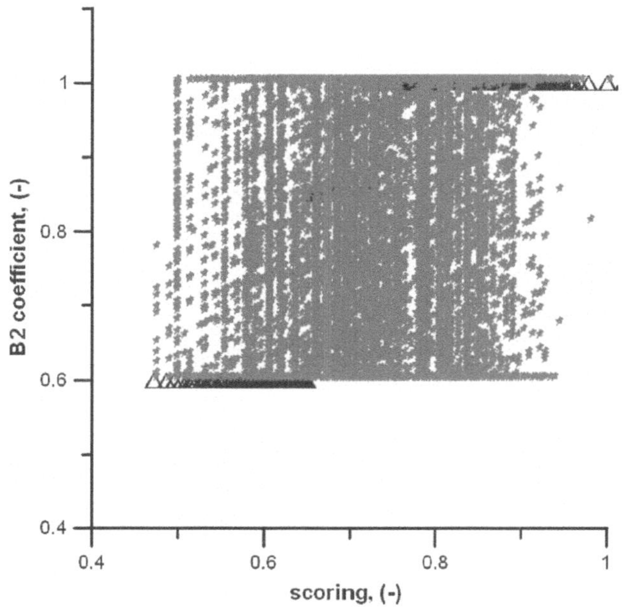

Fig. 48.5 Calculated B2 coefficient distribution

where E_i are values of the positive characteristics, D_i are regressive coefficients, and $C_2, corr$ is a correlation coefficient between calculated values and statistical data range.

In order to estimate $B2$ coefficient two samples were chosen – 33,664 clients with good credit histories and 4,226 clients with delinquency over 30 days. The $B2$ coefficient distribution, which is based on an expert judgment, is shown on the Fig. 48.5 (black points). Then specific weights of "good" and "bad" clients in each score range were determined. The comparison of these values indicates that they are almost equal and this fact attests to independency of $B2$ distribution on the client quality. Also Fig. 48.5 shows that this expert distribution covers a small part of possible $B2$ values calculated for clients from the "good" sample. Each score corresponds to a number of $B2$ values, which are determined by the personal characteristics of each client.

48.2.4 B1 Coefficient Calculation

In order to find $B1$ coefficient in Eq. (48.3) it would be natural to use $P(X)$ value, which reflects client's reliability:

$$B1 = C_1, corr \times \frac{BP}{FM \times P(X)}, \tag{48.11}$$

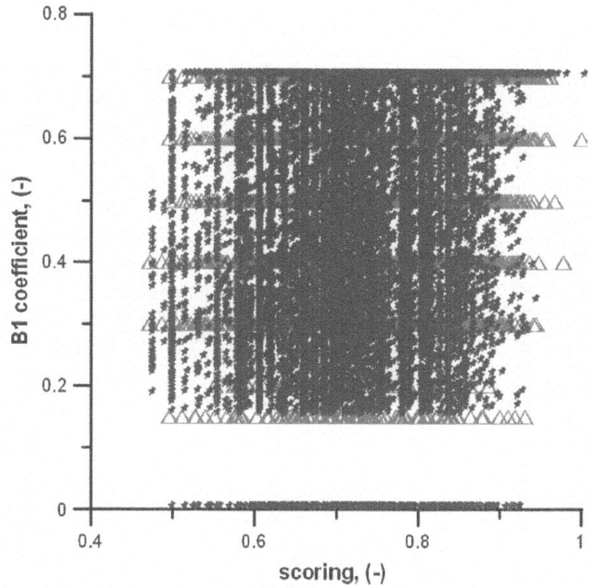

Fig. 48.6 Calculated B1 coefficient distribution

where FM is a number of dependencies, BP is the income, calculated from Eq. (48.2), C1, corr is the coefficient, which represents the correlation between this ratio and statistical data. As shown on Figs. 48.7 and 48.8 B1 calculation through Eq. (48.11) provides mapping of discrete values to the value field (Fig. 48.6), demonstrating individual approach to each client.

48.3 Analysis Results

Based on the described approach the credit limits for all clients from the samples were recalculated. The results are summarized in Table 48.1.

From the analysis results the following observations can be made:

- Total increase of the sample portfolio was +14.4% while maintaining the initial risk level. Calculations were performed in relative values.
- Number of clients, that had undocumented income *(UCI)* in the low risk region $((UCI)_{calc})$, was 32.3% of the total number of clients, that had undocumented income.
- Table 48.1 shows that the maximum average increase was obtained in the low score range. This can be explained by the fact that clients in the range were undervalued.
- It should be noted that all B coefficients can be adjusted by the corresponding coefficients C, that depend on the product statistical data.

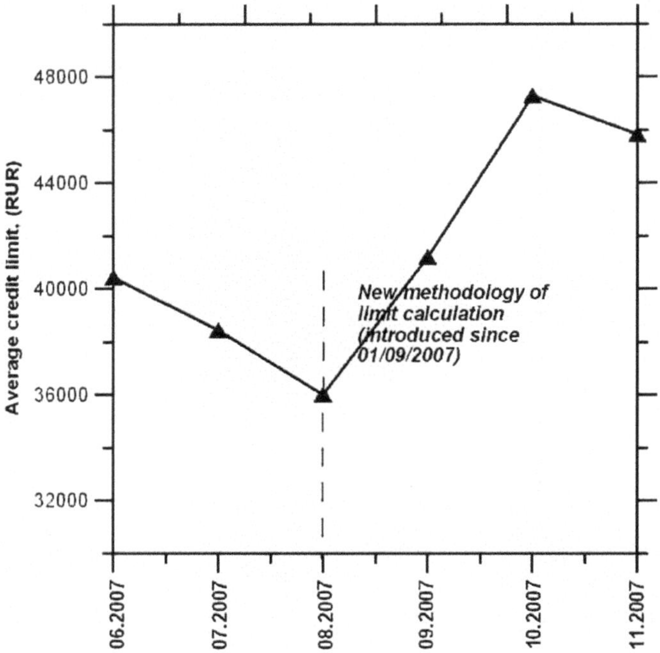

Fig. 48.7 The average credit limit dynamic of one product (the results are calculated at the end of each month)

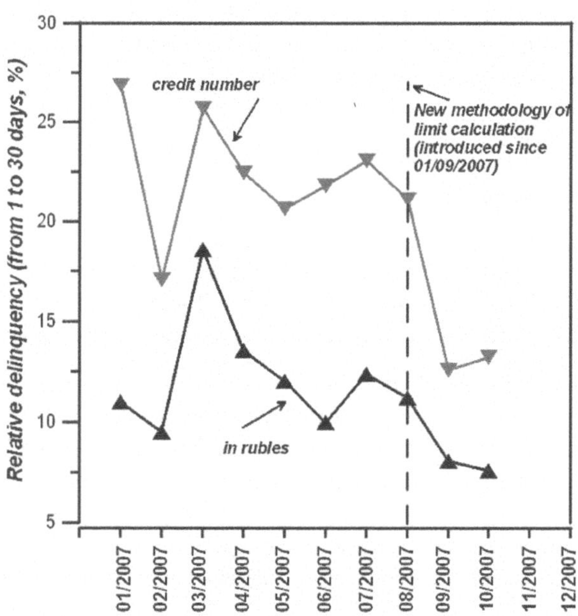

Fig. 48.8 The relative delinquency (from 1 to 30 days) level dynamic of one product (the results are calculated at the end of each month)

Table 48.1 The credit limits for all clients from the samples

The score, X	X < 0.6	0.6 ≤ X < 0.8	0.8 ≤ X ≤ 1
Average limit change (%)	+38.6	+14.34	+7.9

- Because of the normal score distribution most clients have scores in the range $0.6 \leq X < 0.8$. This explains the fact that the average changes in this range and in the total portfolio are very close.
- The model enabled to substantially increase (Fig. 48.7) the credit portfolio (about +17%) while maintaining or improvement (Fig. 48.8) its quality.

48.4 Conclusions

This paper presented the scaling model that can be applied to the calculation of the optimal credit limit. Particular attention has been paid to this problem due to growing competition between the banks in the retail market, that leads to the increase in credit risk. Correct credit limit calculation plays one of the key roles in risk reduction.

The model described in the paper is based on comparison of two opposite client characteristics, one leading to potential credit limit increase, and the second leading to decrease of the credit limit. Credit limit calculation involved client 's personal data, which allows to approach each client on the individual basis during credit amount allocation. The scaling method was applied in order to analyze the data obtained from the real client database. The scaling ratio provides reasonable predictive capability from the risk point of view and therefore has been proposed to serve as a model for credit limit allocation. The model's flexibility allows coefficients' adjustments according to new statistical data. Although the present work is quite preliminary, it does indicate that presented solution allows to substantially increase the credit portfolio while maintaining its quality.

References

1. Bank for International Settlements, 2004. International Convergence of capital Measurement and Capital Standards. A Revised Framework, Basel, 2004.
2. R. Jankowitsch, S. Pichler, W.S.A. Schwaiger, "Modeling of Economic Value of Credit Rating System", Journal of Banking & Finance 31, 181–198, 2007.
3. M. Leippold, S. Ebnoether, P. Vanini, "Optimal Credit Limit Management", National Centre of Competence in Research Financial Valuation and Risk Management, Working Paper No. 72, September 2003.
4. I.T. Farrahov, "Credit Limit Calculation. Non-traditional approach", The Analytical Bank Journal, 04(83), 2002.

5. I.V. Voloshin, "Time Factor Consideration in the Credit Limit Calculation with the VaR Technology", Thesis report, Bank analyst club, http://www.bankclub.ru/library.htm
6. M.Y. Konovalikhin, T.N. Dinh, R.R. Nourgaliev, B.R. Sehgal, M. Fischer, "The Scaling Model of Core Melt Spreading: Validation, Refinement and Reactor Applications", Organization of Economical Cooperation and Development (OECD) Workshop on Ex-Vessel Debris Coolability, Karlsruhe, Germany, 15–18 November 1999.

Chapter 49
Expected Tail Loss Efficient Frontiers for CDOS of Bespoke Portfolios Under One-Factor Copula Marginal Distributions

Diresh Jewan, Renkuan Guo, and Gareth Witten

Abstract The global structured credit landscape has been irrevocably changed with the innovation of Collateralized Debt Obligations (abbreviated as CDOs). As of 2006, the volume of synthetic CDO structures outstanding grew to over $1.9 trillion. Understanding the risk/return trade-off dynamics underlying the bespoke collateral portfolios is crucial when optimising the utility provided by these instruments. In this paper, we study the behaviour of the efficient frontier generated for a collateral portfolio under heavy-tailed distribution assumptions. The convex and coherent credit risk measures, ETL and Copula Marginal ETL (abbreviated as CMETL), are used as our portfolio optimisation criterion. iTraxx Europe IG S5 index constituents are used as an illustrative example.

Keywords Collateralized Debt Obligations · Expected Tail Loss · Efficient Frontiers · Bespoke Portfolio · One-Factor Copula Marginal Distribution

49.1 Introduction

The global structured credit landscape has been irrevocably changed with the innovation of Collateralized Debt Obligations (abbreviated as CDOs). As of 2006, the volume of synthetic CDO structures outstanding grew to over $1.9 trillion, making it the fastest growing investment vehicle in the financial markets. Bespoke deals made up 21% of the total volume [1]. Understanding the risk/return trade-off dynamics underlying the bespoke collateral portfolios is crucial when optimising the utility provided by these instruments.

R. Guo (✉)
Department of Statistical Sciences, University of Cape Town, Private Bag, Rhodes' Gift, Rondebosch 7701, Cape Town, South Africa
E-mail: Renkuan.Guo@uct.ac.za

S.I. Ao and L. Gelman (Eds.), *Advances in Electrical Engineering and Computational Science*, Lecture Notes in Electrical Engineering.

The loss distribution, a key element in credit portfolio management, is generated through one-factor copula models. This approach has been widely used in credit portfolio optimisation [2–4].

The main feature of these models is that default events, conditional on some latent state variable are independent, and simplifies the computation of the aggregate loss distribution [5]. For a comparative analysis, both the Gaussian and Clayton copula models shall be used to model the default dependence structure inherent in credit portfolios.

According to [6], there exist two types of risk measures: relevant and tractable. Relevant measures capture key properties of a credit loss distribution, while tractable measures can be optimized using computationally efficient methods. Efficient frontiers are defined by a collection of optimal risk/return portfolios.

Expected Tail Loss (abbreviated as ETL) was initially introduced by [7], in the portfolio optimisation context. ETL has proved a more consistent risk measure, since it is sub-additive and convex [8]. Reference [7] has shown that the portfolio optimisation under this risk measure results in a linear programming problem. ETL is a superior measure to derive empirical efficient frontiers that act as a useful approximation to the true efficient frontier [9, 10].

In this paper, we study the behaviour of the efficient frontier generated for a collateral portfolio under heavy-tailed distribution assumptions. The structure of the paper is as follow: Section 49.2 introduces bespoke CDOs and its underlying mechanism. Section 49.3 introduces the Stable Paretian distribution and the copulas for heavy-tail default dependence modelling. Section 49.4 then introduces convex and coherent credit risk measures, particularly, ETL and Copula Marginal ETL (abbreviated as CMETL), which is used as our portfolio optimisation criterion. Section 49.5 provides the results of the numerical analysis conducted on a collateral test portfolio. iTraxx Europe IG S5 index constituents are used as an illustrative example. Concluding remarks on the analysis are presented in the final section.

49.2 Bespoke CDO Mechanics

A bespoke CDO is a popular second-generation credit product. This standalone single-tranche transaction is referred to as a bespoke because it allows the investor to customise the various deal characteristics such as the collateral composition, level of subordination, tranche thickness, and rating. Other features, such as substitution rights, may also play an important role [11].

In a typical bespoke CDO transaction, the main decision step for potential investors is to select the portfolio of credits, to which they want exposure. A short position can also be held on a subset of names to have overweight views on certain credits or sectors. The analysis will focus on the portfolio optimisation problem that exists within this step of the structuring process.

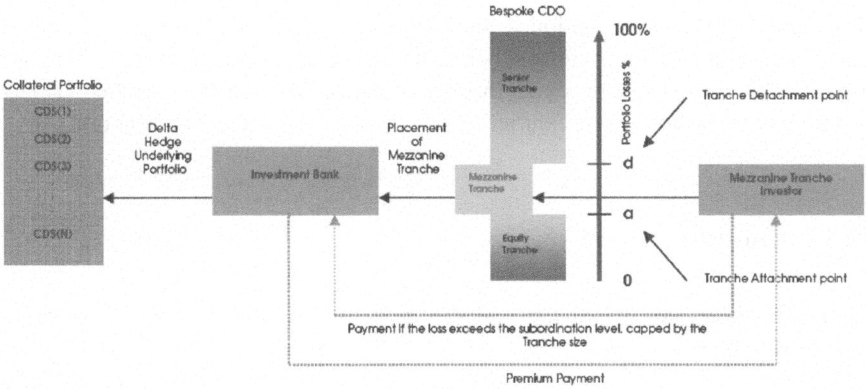

Fig. 49.1 Placement of a mezzanine tranche in a bespoke CDO transaction

The next step is to determine the level of subordination, and tranche thickness corresponding to the risk appetite of the investor. These deal parameters determine the degree of leverage and the required premium payments [11]. This step of the process adds a further two constraints to the transaction optimisation problem. The first is related to the tranche thickness, and the second is the credit rating assigned to the tranche. At this level of the problem, from an investor's perspective, the return generated by the CDO tranche is maximised.

A typical placement of a bespoke CDO is outlined in the following Fig. 49.1. The cash flow payments that occur between the investor and the bank is indicated by the dashed lines.

The challenge of issuing bespoke CDOs is the ongoing need and expense of the risk management of the tranche position [12].

49.3 Heavy-Tail Modelling and Copulas

The use of heavy-tailed distributions for the margins of copula models is becoming an industry standard for credit risk applications.

49.3.1 Stable Paretian Distributions

The Stable Paretian distribution can be expressed by its characteristic function, $\varphi(t)$, is given by [13]:

$$\mathrm{In}\,\phi(t) = \begin{cases} -\sigma|t|^{\alpha}\left[1 - i\beta\,\mathrm{sign}(t)\tan\left[\frac{\pi\alpha}{2}\right]\right] + i\mu t, & \text{if } \alpha \neq 1 \\ -\sigma|t|\left[1 - i\beta\,\mathrm{sign}(t)\frac{2}{\pi}\mathrm{in}t\right] + i\mu t, & \text{if } \alpha = 1 \end{cases} \quad (49.1)$$

Maximum likelihood estimation techniques cannot be used directly. The density function must first be approximated before it can be used in the parameter estimation process. This is usually done either through the direct numerical integration [13], or the application of the inverse Fourier transform to the characteristic function [14].

49.3.2 Copulas

Copula functions are useful tools in the modelling of the dependence between random variables.

Definition 49.1. A copula is defined as a multivariate cumulative distribution function (abbreviated cdf) with uniform marginal distributions:

$$C(u_1, u_2, \ldots, u_k), \quad u_i \quad \epsilon[0,1] \text{ for all } \dot{s} \;=\; 1, 2, \ldots, n,$$

where,

$$C(u_i) = u_i \text{ for all } \dot{s} = 1, 2, \ldots, n$$

The ideas behind multivariate distribution theory carries through to copula functions. The above definition indicates that copulas represent the joint cdf of uniform marginal distributions. The probability integral transformation of continuous non-uniform random variables allows copulas to define the joint multivariate distribution H with margins $F_1, F_2, \ldots,, F_n$. By the strict monotocity of the cdf,

$$\begin{aligned} H(F_1, F_2, \ldots, F_k) &= \Pr(X_1 \leq \omega_1, \ldots, X_k \leq \omega_k) \\ &= C(F_1(\omega_1), F_2(\omega_2), \ldots, F_k(\omega_k)). \end{aligned} \tag{49.2}$$

An alternative representation of copulas is used for credit portfolio modelling. This approach introduced by [15], is analogous to the Capital Asset Pricing Model (abbreviated as CAPM) in Modern Portfolio Theory.

De Finetti's theorem for exchangeable sequence of binary random variables provides a theoretical background for the use of factor models in credit risk applications [5].

Reference [16] relates the factor and copula approaches. These one-factor copula models are presented in Section 49.5.

49.4 Credit Risk Measures

Recently, growing literature has been devoted to an axiomatic theory of risk measures [8], in which authors presented an axiomatic foundation of coherent risk measures to quantify and compare uncertain future cashflows. [17] extended the notion of coherent risk measures to convex risk measures.

49.4.1 Convex and Coherent Risk Measures

Let X denote the set of random variables defined on the probability space $(\Omega, \Im, P)$. We define Ω as a set of finitely many possible scenarios for the portfolio. Financial risks are represented by a convex cone $M \subseteq (\Omega, \Im, P)$ of random variables. Any random variable X in this set, will be interpreted as a possible loss of some credit portfolio over a given time horizon. The following provides a definition of a convex cone.

Definition 49.2. M is a convex cone if $\forall X_1, X_2 \in M$ then $X_1 + X_2 \in M$ and $\lambda X_1 \in M$ for every $\lambda \geq 0$.

Definition 49.3. Given some convex cone, M of random variables, a measure of risk Θ with domain M is a mapping: $\Theta : M \rightarrow \mathbb{R}$.

From an economic perspective Θ, can be regarded as the capital buffer that should be set aside for adverse market movements. In order to measure and control the associated risk [8] introduced an axiomatic framework of coherent risk measures which were recently 'generalized' by [17] to convex risk measures.

Definition 49.4. A mapping Θ is called a convex risk measure, if and only if it is,

- **Convex:** for every $X_1 \in M$ and $X_2 \in M$, we have $\Theta(\lambda X_1 + (1 - \lambda)X_2) \leq \lambda\Theta(X_1) + (1 - \lambda)\Theta(X_2)$, for some $\lambda \in \mathbb{R}$,
- **Monotone::** for every $X_1 \in M$ and $X_2 \in M$ with $X_1 \leq X_2$ we have $\Theta(X_1) \leq \Theta(X_2)$, and
- **Translation Invariant:** if a is a constant then $\Theta(X + aI) \leq \Theta(X) - a$ where I denotes the unit vector.

Adding to these properties, positive homogeneity one obtains:

Definition 49.5. A risk measure Θ is called coherent, if in addition it is, **Positive homogeneous:** for every $\kappa \geq 0$ we have $\Theta(\kappa X) = \kappa\Theta(X)$.

49.4.2 Expected-Tail-Loss

In the CDO portfolio optimisation problem, we represent the portfolio risk objective function by the Expected-Tail-Loss measure.

Definition 49.6. Expected-Tail-Loss is defined by the following:

$$ETL_p(\beta) = E[L_p | L_p > UL_p(\beta)],$$

where L_p denotes the loss for portfolio p and UL_p, the unexpected losses at the confidence level β.

Portfolio optimisation with *ETL* as object function will result in a smooth, convex problem with a unique solution [7].

49.4.3 Portfolio Optimisation Under Expected-Tail-Loss

A Copula Marginal Expected Tail Loss (abbreviated as CMETL) which is similar to that of Stable (Distribution) Expected Total Loss (abbreviated as SETL) optimal portfolio discussed in [9], is one that minimizes credit portfolio ETL subject to a constraint of achieving expected credit portfolio returns at least as large as an investor defined level, along with other typical constraints on weights, where both quantities are evaluated in the CMETL framework.

In order to define the above CMETL precisely we use the following quantities:

R_p: the random return of portfolio p,
$CMER_p$: the expected return of portfolio p with respect to the copula marginal distribution, and
β: tail loss probability.

The following assumptions are imposed by [9] for the *CMETL* investor:

- The universe of assets is $\underline{Q}$(the set of admissible credit portfolios)
- The investor may borrow or deposit at the risk-free rate r_f without restriction
- The investor seeks an expected return of at least μ

Definition 49.7. The CMETL investor's optimal portfolio is then given by the following:

$$X_0(\mu \mid \beta) = \arg\min_{\underline{q} \in Q} \left\{ CMETL_p(\beta) \right\},$$

$$\text{s, t.}$$

$$\sum_{\tau} x_\tau = 1,$$

$$\ell \leq x_\tau \leq u, \; \forall\, r,$$

$$CMER_p \geq \mu. \tag{49.3}$$

Where $X_{\epsilon}\cdot$ denotes the resulting portfolio's weights. The subscript c indicates that the problem is defined under the copula framework.

The constraints, in the order written above, require that (a) the sum of the portfolio weights should equal to one; (b) the position weights should lie within the trading limits l and u to avoid unrealistic long and short positions; (c) the expected return on the portfolio in the absence of rating transitions should be at least equal to some investor defined level μ.

Together, these optimal risk/return trade-offs define the efficient frontier.

49.5 Copula Marginal ETL Efficient Frontier for the CDO Collateral

The analysis is performed in a one-period, default-only mode structural framework. Consider a portfolio composed of n credit instruments. The exposure in each instrument is denoted by N_i. Under the binomial setting, the loss of counterparty i for scenario k is given by the following:

$$L_i^k = N_i - V_i^k, \quad for\ i = 1, \dots, n, \quad k = 1, \dots, s, \tag{49.4}$$

where V_i^k is the value of exposure i at the given horizon in scenario k. In particular:

$$V_i^k = \begin{cases} N_i, R_i, & default\ state, \\ N_i & otherwise, \end{cases} \tag{49.5}$$

where R_i is the recovery rate for reference entity i.

Definition 49.8. The portfolio loss function in scenario k, $L^k : \mathbb{R}^{n+1} \to \mathbb{R}$, over the chosen time horizon is given by the following:

$$L_k(\underline{x}) = \sum_{i=1}^{n} L_i^k x_i, \tag{49.6}$$

where $\underline{x} = (x_1, x_2, \cdots, x_n)^T$ is the vector of positions held in the portfolio.

In order to generate scenarios for the credits states of each reference entity, we follow the factor copula approach.

Definition 49.9. The 'default-like' stochastic process under the Gaussian copula assumption is defined by the following:

$$A_i(x_1, x_2, \dots, x_n, \epsilon_i) := \rho_i M + \sqrt{1 - \rho_i \epsilon_i}, \tag{49.7}$$

where M and ε_i are independent standard normal variants in the case of a Gaussian copula model, and $\text{cov}(\varepsilon_k, \varepsilon_l) \neq 0$, for all $k \neq l$.

The default dependence comes from the factor M. Unconditionally, the stochastic processes are correlated but conditionally they are independent. The default probability of an entity i, denoted by F_i, can be observed from market prices of credit default swaps, and defined by the following:

$$F_i(t) = 1 - \exp\left[- \int_0^t h_i(u) du \right], \tag{49.8}$$

where $h_i(u)$ represents the hazard rate for reference entity i. For simplicity, we assume that the CDS spread term structure is flat, and calibration of the hazard rates for all reference entities is straightforward.

The default barrier $B_i(t)$ is defined as: $B_i(t) = G^{-1}(F_i(t))$, where G defines the inverse distribution function. In the case of a Gaussian copula, this would be the inverse cumulative Gaussian distribution function.

A second type of copula model considered comes from the Archimedean family. In this family we consider the Clayton copula.

Definition 49.10. The 'default-like' stochastic process A_i under the one-factor Clayton copula assumption is defined by the following:

$$A_i := \varphi_{\frac{1}{\theta}}\left[-\frac{\text{In}(\varepsilon_i)}{M}\right] = \left[-\frac{\text{In}(\varepsilon_i)}{M} + 1\right]^{\frac{1}{\theta}}, \tag{49.9}$$

where $\phi(\cdot)$ is the Laplace transform of the $\texttt{Gamma}(1/\theta)$ distribution, ε_i is a uniform random variable and M is a $\texttt{Gamma}(1/\theta)$ distributed random variable.

Using the credit models presented above, the loss distribution for the test portfolio described in [18] can easily be derived. This is shown in Fig. 49.2. Reference [18] also provides empirical evidence for the stationarity of credit spreads. Stationarity is a crucial issue, since credit risk models implicitly assume that spreads follow stationary processes.

The following Table 49.1 displays the parameters of the two models for few of the iTraxx Europe IG index constituents.

The coefficients in the regression analysis for all constituents in the bespoke test portfolio were significant at a 99% confidence level.

The following Table 49.2 summarizes the expected loss, unexpected loss, and CMETL at the 99.9% confidence level.

Although the Gaussian copula predicts a slightly higher expected loss than the Clayton copula, there is a 1.758% absolute difference in the CMETL, and a 3% difference in the maximum loss between the copula models.

We optimise all positions and solve the linear programming problem represented by Eq. (49.3). Three scenarios are considered:

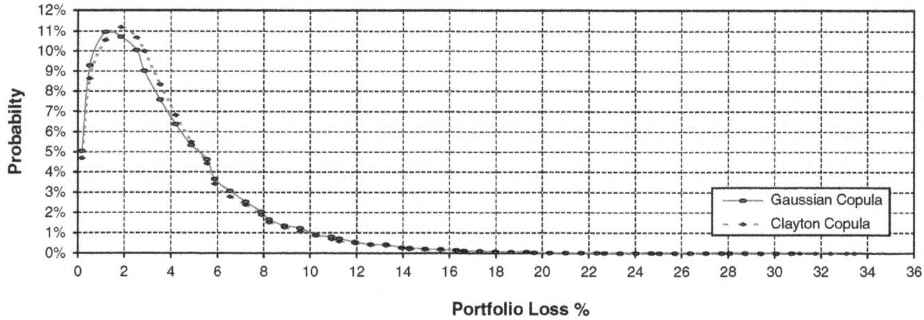

Fig. 49.2 Credit loss distributions for homogenous test portfolio at 5-year horizon under different default dependence assumptions

Table 49.1 Calibrated copula parameters

Corporate entity	Copula parameter	
	ρ (Gaussian)	θ (Clayton)
BAE Systems PLC	0.249616	0.043221
Unilever NV	0.239667	0.035097
Continental AG	0.264289	0.062691
Peugeot SA	0.243152	0.041494
Commerzbank AG	0.220350	0.025312

Table 49.2 Risk measure differences for initial homogenous test portfolio under both Gaussian & Clayton copulas assumptions

Risk measure	Gaussian copula	Clayton copula
Expected loss	3.7248%	3.6952%
Unexpected loss @ 99.9%	20.349%	20.328%
CMETL @ 99.9%	25.534%	27.292%
Maximum loss	30.447%	33.434%

1. An examination of the efficient frontiers for a well-diversified portfolio under both Gaussian and Clayton copula assumptions. Only long positions in the underlying credit default swaps are firstly allowed. The upper trading limit is set to 5% in this case.
2. An investigation of the behaviour of the efficient frontiers under the Clayton copula assumption when the upper trading limit is increased, is then studied. This consequently increases concentration risk.
3. The efficient frontiers under the Clayton copula assumption when both long and short positions are permitted, is lastly investigated.

In the first case lower and upper trading limit of 0.5% and 5% are set respectively. This is to ensure that no reference asset is excluded from the portfolio. This also maintains the well-diversified features of the iTraxx Europe IG portfolio, in the bespoke structure. The following Fig. 49.3 presents the efficient frontiers under the two default dependence assumptions.

The above Fig. 49.3 shows the difference in efficient frontiers between the Gaussian and Clayton copula assumptions. For a given level of risk, a higher return should be expected if the portfolio composition is based on the Clayton copula assumption rather then the Gaussian copula. The above Fig. 49.3 also indicates the inefficient risk/return levels for the original portfolio under both the Gaussian and Clayton copula assumptions. Under the Clayton copula assumption, the same level of return is achieved with less than one-fifth of the original risk.

In the second case, the upper trading limit is increased from 10% to 100%. The effect of concentration risk on the efficient frontiers is then investigated. The following Figs. 49.4 and 49.5 presents the efficient frontiers under these circumstances.

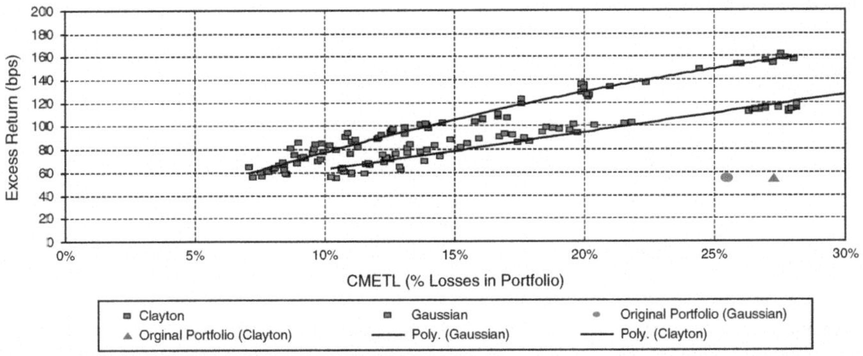

Fig. 49.3 Comparison of efficient frontiers under Gaussian and Clayton copula assumptions

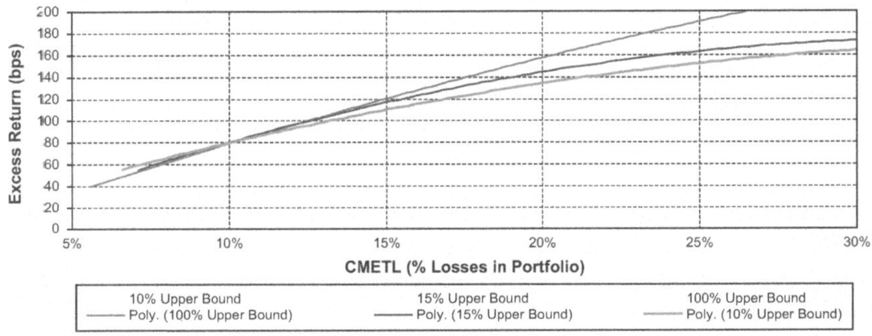

Fig. 49.4 Behaviour of second order polynomial fit of efficient frontiers, under increasing upper trading limit

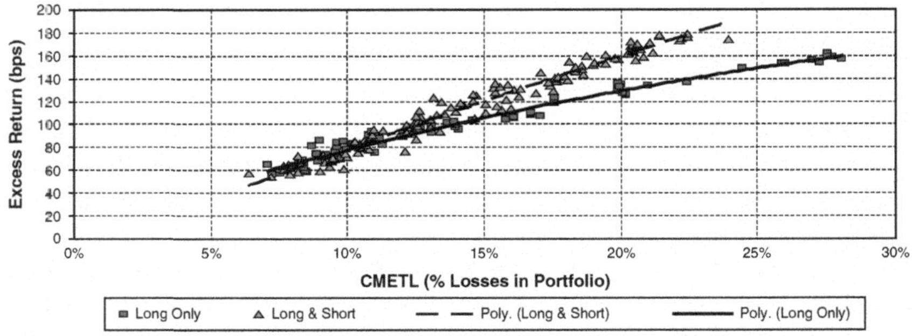

Fig. 49.5 Effect of short positions on the efficient frontier

The figure displays the expected variation in efficient frontiers when the upper trading limit is slowly relaxed. Under these circumstances, concentration risk is introduced into the portfolio. Investors demand a higher premium for taking on this risk. For a 20% level of risk, investors demand an extra 25 bps premium for holding a more concentrated portfolio. At these levels, the concentrated portfolio only holds positions in 49 of the 114 names, with the largest position size being 19.5%.

In the final case, the effect on the efficient frontier for allowing short positions is examined. Under this scenario, only the well-diversified portfolio case is considered. The lower and upper trading limits are set at −5% and 5% respectively.

The figure displays an important result: allowing short positions in credits, provides the investor with superior returns to those in the long-only case. At a 20% level of risk, investors can earn an extra 30 bps premium for taking on overweight views on certain credits in the portfolio. This indicates why hedge fund strategies involving bespoke CDO structures have become increasingly popular.

The results for all efficient frontier second order polynomial regressions were significant at a 95% confidence level. The resulting R^2 coefficients were all above 90%.

49.6 Concluding Remarks

In this paper, we propose a new framework, the CMETL model, for the investigations on CDO market patterns and the efficient frontier. The Gaussian copula asset allocation is proved sub-optimal. Clayton copula efficient frontiers provided a higher return for a given level of credit risk. A closer examination of the original risk profile shows that the credit risk can be reduced to one-fifth of the original amount, under the Clayton asset allocation method.

The permission of short positions in the bespoke CDO portfolio allows investors to increase returns beyond a specific risk tolerance level. In the case study considered, a maximum increase of 37.5% in investor-defined return is achieved by allowing overweight positions in certain credits.

References

1. E. Beinstein, "A review of themes from 2006 and outlook for 2007," JP Morgan, in *Corporate Quantitative Research*, 2007.
2. C. Finger, "Conditional approaches for credit metrics portfolio distributions," *Credit Metrics Monitor*, 2(1), 1999, 14–33.
3. P. Schonbucher, "Taken to the limit: Simple and not so simple loan loss distributions," Working paper, Department of Statistics, Bonn University.
4. O. Vasicek, "Probability of loss on a loan portfolio," Technical report, KMV Corporation, 1987, Available: www.moodys.com.
5. J. P. Laurent, & J. Gregory, *"Basket Default Swaps, CDO's and Factor Copulas,"* Working paper, 2002.
6. H. Mausser, & D. Rosen, "Efficient risk/return frontiers for credit risk," in *Algo Research Quarterly*, 2(4), 1999, 9–22.
7. R. T. Rockafellar, & S. Uryasev, "Optimisation of conditional value-at-risk," in *Journal of Risk*, 3, 2000, 21–41.

8. P. Artzner, F. Delbaen, J. M. Eber, & D. Heath, "Coherent measures of risk," in *Mathematical Finance*, 9, 1999, 203–28.
9. S. T. Rachev, R. D. Martin, B. Racheva, & S. Stoyanov, "Stable ETL optimal portfolios & extreme risk management," FinAnalytica, 2006, Available: www.statistik.uni-karlsruhe.de/download/Stable_Distributions_White_Paper.pdf
10. M. Kalkbrener, H. Lotter, & L. Overbeck, "Sensible and efficient capital allocation for credit portfolios," Deutsche Bank AG, 2003 Available: http://www.gloriamundi.org/detailpopup.asp?ID = 453057574
11. A. Rajan, G. McDermotte, & R. Roy, *The Structured Credit Handbook*, Wiley Finance Series, New Jersey, 2006, pp. 153–155.
12. G. Tejwani, A. Yoganandan, & A. Jha, *"How do dealers risk manage synthetic CDOs,"* Lehman Brothers, in Structured Credit Research, 2007.
13. J. P. Nolan, "Numerical Calculation of Stable densities and distribution functions," in *Communications in Statistics – Stochastic Models*, 13, 1997, 759–774.
14. S. Mittnik, T. Doganoglu, & D. Chenyao, "Computing the probability density function of Stable Paretian distributions," in *Mathematical & Computer Modelling*, 29, 1999, 235–240.
15. O. Vasicek, "Probability of loss on a loan portfolio," Technical report, KMV Corporation, 1987, Available: www.moodys.com
16. F. R. Frey, A. McNiel, & M. Nyfeler, "Copulas and credit models," in *Risk*, October, 2001, pp. 111–113.
17. H. Föllmer, & A. Schied, "Robust preferences and convex measures of risk," In S. Klaus, P. J. Schonbucher, & D. Sondermann, editors, Advances in Finance and Stochastics, Springer, New York, pp. 39–56, 2002.
18. D. Jewan, R. Guo, & G. Witten, "Copula marginal expected tail loss efficient frontiers for CDOs of bespoke portfolios," in *Proceedings of the World Congress in Financial Engineering Conference,* 2, 1124–1129, London, July 1–7, 2008.

Chapter 50
Data Mining for Retail Inventory Management

Pradip Kumar Bala

Abstract As part of the customer relationship management (CRM) strategy, many researchers have been analyzing 'why' customers decide to switch. However, despite its practical relevance, few studies have investigated how companies can react to defection prone customers by offering the right set of products. Consumer insight has been captured in the work, but not used for inventory replenishment. In the present research paper, a data mining model has been proposed which can be used for multi-item inventory management in retail sale stores. The model has been illustrated with an example database.

Keywords Retail Inventory Management · Data Mining · multi-item · Decision tree · Association rule

50.1 Introduction

Customer relationship management (CRM) aims at stronger loyalty of customers with greater market share. With competition for shelf space intensifying, there is a pressing need to provide shoppers with a highly differentiated value proposition through "right product mix in right amount at right time". Mining or extracting customer insight from structured and unstructured data, call center records, customer complaints, and other sources will be of tremendous importance for inventory management in retail stores. Purchase dependency is one type of consumer behavior which has not been addressed or used properly for managing inventory in retail sale. Purchase or non-purchase of item or items by a customer may depend on the purchase or non-purchase of some other item or items made by him or her. This is referred as purchase dependency in this paper.

P. K. Bala
Xavier Institute of Management, Bhubaneswar, PIN-751013, India
E-mail: p_k_bala@rediffmail.com

S.I. Ao and L. Gelman (Eds.), *Advances in Electrical Engineering and Computational Science*, Lecture Notes in Electrical Engineering.

In most of the real situations, we do not find a one-item inventory, but multi-item inventory with different periods of replenishment. These inventories are usually managed in aggregate because of the complexity of handling each individual item. Intuition and commonsense rules with the aid of historical data have been useful to some extent, but they are not often efficient or cost effective. Some examples of the multi-item inventory are retail sale store, spare parts for maintenance, medicine store etc.

Out of thousands of items held in an inventory of a typical organization, only a small percentage of them deserve management's closest attention and tightest control. It is not economical to exercise same degree of inventory control on all the items. Using selective inventory control, varying degree of control are exercised on different items. Most widely-used technique used to classify items for the purpose of selective inventory control is ABC classification. Inventory replenishment policy deals with 'how-much-to-order' and 'when-to-order' aspects. Purchase pattern is an important consumer insight. It is obvious that the knowledge of purchase pattern will be an important input for designing inventory replenishment policy.

Data mining is used to find new, hidden or unexpected patterns from a very large volume of historical data, typically stored in a data warehouse. Knowledge or insight discovered using data mining helps in more effective individual and group decision making. Irrespective of the specific technique, data mining methods may be classified by the function they perform or by their class of application. Using this approach, some major categories of data mining tools, techniques and methods can be identified as given below.

(i) Association Rule: Association rule is a type of data mining that correlates one set of items or events with another set of items or events. Strength of association rule is measured in the framework of support, confidence and lift [1].
(ii) Classification: Classification techniques include mining processes intended to discover rules that define whether an item or event belongs to a particular predefined subset or class of data. This category of techniques is probably the most broadly applicable to different types of business problems. Methods based on 'Decision Tree' and 'Neural Network' are used for classification.
(iii) Clustering: In some cases, it is difficult or impossible to define the parameters of a class of data to be analyzed. When parameters are elusive, clustering methods can be used to create partitions so that all members of each set are similar according to a specified set of metrics. Various algorithms like k-means, CLARA, CLARANS are used for clustering. Kohonen's map is also used for clustering.
(iv) Sequence Mining: Sequencing pattern analysis or time series analysis methods relate events in time, such as prediction of interest rate fluctuations or stock performance, based on a series of preceding events. Through this analysis, various hidden trends, often highly predictive of future events, can be discovered. GSP algorithm is used to mine sequence rules [2].
(v) Summarization: Summarization describes a set of data in compact form.

(vi) Regression: Regression techniques are used to predict a continuous value. The regression can be linear or non-linear with one predictor variable or more than one predictor variables, known as 'multiple regression'.

Numerous techniques are available to assist in mining the data, along with numerous technologies for building the mining models. Data mining is used by business intelligence organizations to learn consumer behavior for the purpose of customer relationship management, fraud detection, etc.

50.2 Literature Review

Inventory items have been *classified* or *clustered* in groups for the purpose of joint replenishment policy in [3–6]. Generic inventory stock control policies are derived using multi-item classification [3]. Clustering of items has also been done in production and inventory systems [4]. Multi-criteria inventory classification has been done by parameter optimization using genetic algorithm [5]. Artificial neural networks (ANNs) have been used for ABC classification of stock keeping units (SKUs). ANN has been used in a pharmaceutical company [6].

The interacting items showing the mutual increase in the demand of one commodity due to the presence of the other has been considered in the model given in [7]. Correlation between the demands of two items has been considered. The article gives a new approach towards a two-item inventory model for deteriorating items with a linear stock-dependent demand rate. It has also assumed a linear demand rate, that is, more is the inventory, more is the demand. The model has made an attempt to capture demand interdependency to some extent.

Market segmentation has been done using clustering through neural network and genetic algorithm [8]. Clustering of customers on the basis of their demand attributes, rather than the static geographic property has been done in [9]. However, the findings have not been used for the purpose of inventory management.

As part of the customer relationship management (CRM) strategy, many researchers have been analyzing 'why' customers decide to switch. However, despite its practical relevance, few studies have investigated how companies can react to defection prone customers by offering the right set of products [10]. Consumer insight has been captured in the work, but not used for inventory replenishment. For cross-selling consideration, a method to select inventory items from the association rules has been developed in [11] which gives a methodology to choose a subset of items which can give the maximal profit with the consideration of cross-selling effect. However, this does not talk about any inventory replenishment policy.

In the present research paper, a data mining model has been proposed which can be used for multi-item inventory management in retail sale stores. The model has been illustrated with an example database.

50.3 Purchase Dependencies in Retail Sale in the Context of Inventory Management

In the context of data mining, various aspects of purchase dependencies which may be useful for the purpose of retail inventory management have been discussed as given below.
Demand Interdependency: The problem of multi-item inventory is more challenging when there is association in the demand or usage pattern amongst the items or item-sets (Item-set refers to a set of one or more items, hence if we say item-set, it may refer to one item or a number of items.). The correlation in the demand amongst the items can be one to one, one to many, many to one or many to many. In many situations, a customer buys an item or item-set only when another item or item-set is also in stock.

To explain the above situation further, say, in a store, item B is in stock and item A is out of stock. One customer is interested in purchasing B, provided A is also available in the store, so that he can purchase both A and B. As the demand for B depends on demand for A, he will not purchase B, if A is not available. Under this situation, we can say that stock of B is as good as a stock-out situation for that customer. Hence, if A is not in stock, there will be no sale of B also in many cases. This example depicts a case of interdependency in demand. Interdependency in the demand of two items can be captured in various ways and can be used for inventory management. Association rules can be used to mine demand interdependencies.
Customer Profile and Demand Pattern: Customer profile is the detailed features of a customer. The profile may contain income, age, marital status, gender, education, number of cars, family size etc. The profile may also contain frequency of shopping, credit rating, loyalty index etc. In retail sale, demand pattern depends a lot on the customer profile along with the other factors. Hence, customer profile is an important parameter which may be used for learning the purchase pattern of a customer and this may be useful for inventory modeling. Classification and clustering can be used for learning the impact of customer profile on demand pattern. Customer profile will be a useful input for forecasting the demand of the items.
Sequence of Purchase: Many a times, a sequence of purchase gets repeated most of the times with a time gap between two purchases. A sequence may be of with two or more events of purchases. In each event of purchase, certain itemset is purchased. Once a repetitive sequence rule is identified, it can be used as an input for inventory modeling. Sequence rules are mined using GSP algorithm.
Time-Dependent Purchase Pattern: On different days of the week, there may be different purchase pattern of the customers. Purchase patterns on the weekdays and the weekends are generally different. Similarly, in different months or seasons, different patterns may be observed. Time-dependent purchase pattern may also be observed at different hours of the day. Purchase pattern in the evening hours happens to be different from the day hours. Segregated data with respect to time can be used to learn time-dependent purchase patterns. For example, data of weekend and weekdays can be segregated for this purpose.

Location-Dependent Purchase Pattern: There may be an impact of culture, climatic condition and other factors on the purchase pattern. Segregated data with respect to location or space can be used to learn location-dependent purchase patterns.
Negative Dependency: Sometimes, purchase of one item or itemset results in non-purchase of another item or itemset. Similarly, non-purchase of one item or itemset may result in purchase of another item or itemset. This kind of purchase dependency may be observed within a transaction made by a customer or in successive transactions made by the customer with a time gap.

As discussed, various purchase patterns can be mined using appropriate data mining tool.

50.4 Model Development

The model proposed has been described in Fig. 50.1. In the present research, it has been proposed to discover purchase patterns using data mining. For this purpose, sale transaction data of the inventories contained in the 'Materials' module can be

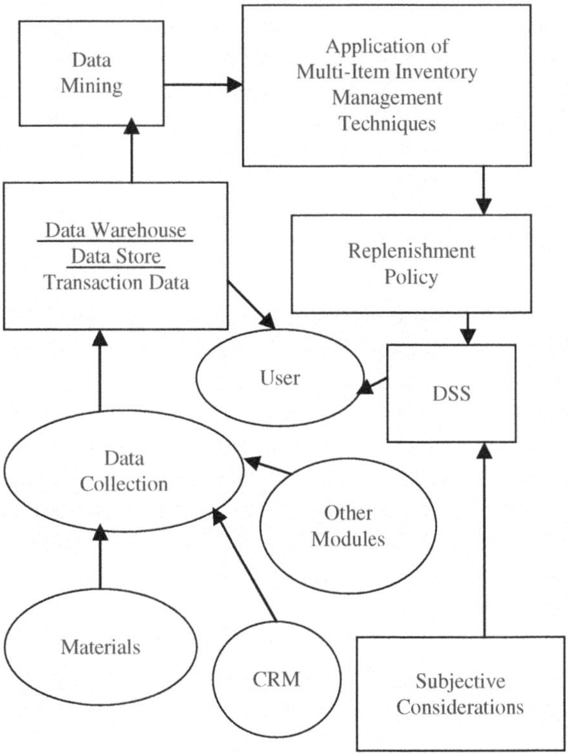

Fig. 50.1 Block diagram of the model

used for mining association rules describing demand interdependencies. Further, time-dependent and location or space-dependent association rules can be mined by proper segregation of the past sale transaction data from the "Materials' module. Using the modules of 'Materials' and 'CRM (Customer Relationship Management)' containing the demographical and other profiles of the customers, 'classification' technique of data mining can be applied to learn the impact of customer profiles on purchase pattern. Clustering of customers can also be done. The model can integrate various transactions based on the decision and action taken in other domains of finance, maintenance, human resource management, marketing, supply chain etc. In such cases, we find association rules with a mixture of real items with 'virtual items'. Virtual items are basically events, decisions and attributes etc. representing fields in a database or a data warehouse.

Transaction data is stored in database which is also known as data store and useful for operational decisions. These can not address the requirements of decision support system (DSS) for inventory management. Data Mining, facilitated by data warehousing, addresses the need of strategic decision-making. Knowledge discovered from data mining is used in Decision Support System (DSS). In this model, various types of consumer insights discovered using data mining can be used in multi-item inventory modeling. The consumer insights mined act as additional input in designing the inventory replenishment policies.

The model provides scope for subjective considerations also as an input to DSS along with the objective findings obtained using data mining. The focus in this model has been to reduce the subjectivity by objective decision-making.

One example in Section 50.5 has been illustrated where association rules have been mined and classification has been done by building decision tree.

50.5 Illustration with an Example

Eleven (11) items have been considered in a retail sale store with one thousand (1,000) past sale transaction data along with the profiles of the customers. Eleven items are named as 'a', 'b', 'c', 'd', 'e', 'f', 'g', 'h', 'i', 'j' and 'k'. A transaction contains the items purchased by a customer along with a small profile of the customers. The profile contains 'customer id', 'value' (as rated by the retailer), 'pmethod' (payment method by cheque/cash/card etc.), 'sex' (Male/Female expressed in M/F), 'hometown' (yes/no), 'income' and 'age'. Customer id is an identification code for the customer and it finds use for capturing the sequence of purchases made by the same customer and hence in mining sequence rules. Otherwise, customer id is not of any use for mining other patterns.

To understand the database, as it is not possible to show 1,000 records of the database in this paper, five records have been shown in Table 50.1 from which one can visualize various fields in the database.

Table 50.1 Sale transaction data with customer profile showing five transactions (part of the database with 1,000 rows)

Customer profile						Items										
Value	Payment method	Sex	Hometown (Y/N)	Income (in $ thousands)	Age	a	b	c	d	e	f	g	h	i	J	k
43	Ch	M	N	27	46	F	T	T	F	F	F	F	F	F	F	T
25	Cs	F	N	30	28	F	T	F	F	F	F	F	F	F	F	T
21	cs	M	N	13	36	F	F	F	T	F	T	T	F	F	T	F
24	cd	F	N	12	26	F	F	T	F	F	F	F	T	F	F	F
19	cd	M	Y	11	24	F	F	F	F	F	F	F	F	F	F	F

Table 50.2 Association Rules amongst d, f and g

Antecedent	Consequent	Association rule	Support (%)	Confidence (%)	Rule support (%)	Lift
d and g	F	(d, g) =>(f)	17.8	87.4	15.5	2.7
f and g	d	(f, g) =>(d)	18.1	85.9	15.5	2.7
d and f	g	(d, f) =>(g)	18.4	84.4	15.5	2.7

In Table 50.1, 'T' (under 'Items') implies purchase of the corresponding item in the corresponding transaction and 'F' (under 'Items') implies that the corresponding item has not been purchased in the corresponding transaction. Association rules have been mined from the database of 1,000 transactions with the threshold values of support and confidence to be 10% and 80% respectively. Association rules mined have been shown in Table 50.2.

Only three rules have qualified for the chosen threshold values of support and confidence. In each of these three rules, we observe that same three items (i.e., d, f and g) are involved.

With respect to the simultaneous purchase of all three items (i.e., d, f and g) in the same transaction, classification of the customers has been done based on their profiles. Using data mining, decision tree has been built up from the database for the purpose of classification which is given in Fig. 50.2. Considering threshold value of 80% confidence for the decision rules, only one rule (at node 3) qualifies. The rule (with 84.242% confidence) is – IF "income is less than or equal to 16,950 and sex is F (Female)", THEN "the customer purchases all three items, i.e., d, f, and g". The rule has been observed in 84.242% cases.

The association rules and the decision rule mined from the database can be used as input for designing inventory replenishment policy. Inventory managers must consider these rules for placing the orders for replenishment.

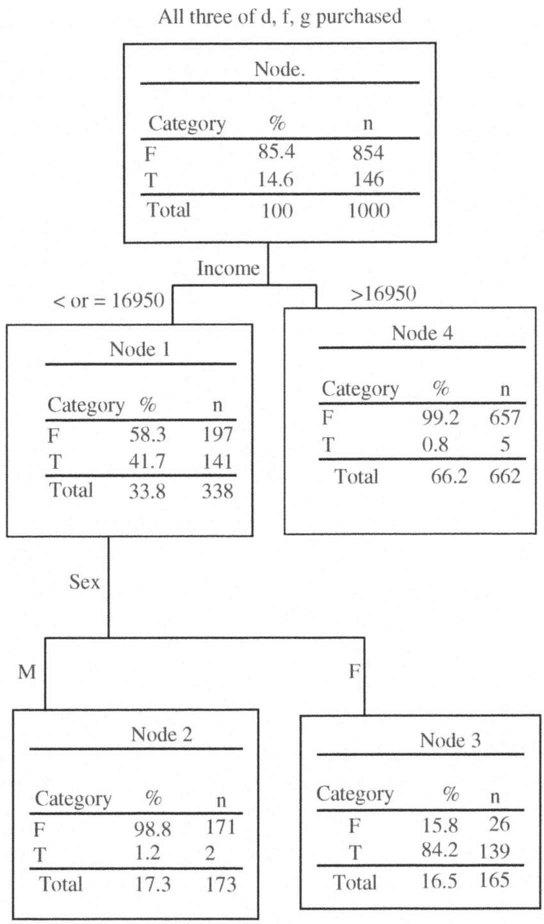

Fig. 50.2 Decision tree

50.6 Case Discussion

Based on focussed interview with sales people and customers in various grocery retail outlets, 45 items were selected for the research. The objective was to select a small number of frequently purchased items where some purchase dependencies can be anticipated and the list will also contain items where no purchase dependency is expected. In this way, for the purpose of research, the small subset of items can be representative of the superset of items with large number of items in retail store. Hence, the intention of the interview was to foresee a few purchase dependencies from the experience of sales people and customers and to include the related items for the purpose of the research. In this way, 45 items were selected where one may expect purchase dependency for some items and no purchase dependency for some other items. The list of 45 items with description is as given in Table 50.3. Names

Table 50.3 Names of the grocery tems

Item No.	Item-name	Description
1	All Out liquid (35 ml)	Mosquito repellent
2	Atta – Ashirvad (5 kg)	Flour
3	Basmati (1 kg)	Special rice
4	Bournvita – Cadbury (500 g)	Beverage
5	Cashew nuts (100 g)	Cashew nuts
6	Chilli sauce – Kissan (200 g)	Chilli sauce
7	Coffee – Nescafe (50 g)	Beverage
8	Hair oil – Parachute Jasmine (100 ml)	Hair oil
9	Hair oil – Shalimar Coconut (100 g)	Hair oil
10	Horlicks (500 g)	Beverage
11	Kismis (100 g)	Dry grapes
12	Lux – International soap (100 g)	Bathing soap
13	Maida – Rishta (1 kg)	Refined flour
14	MDH chicken masala	Packaged chicken spices
15	MDH meat masala	Packaged meat spices
16	Noodles – Maggi (100 g)	Snacks item
17	Noodles – Maggi (200 g)	Snacks item
18	OK powder (500 g)	Detergent powder
19	Parle – Hide & Seek (100 g)	Biscuit
20	Parle – Hide & Seek – orange cream (100 g)	Biscuit with orange cream
21	Pears soap – Green (100 g)	Bathing soap
22	Pepsodent G (80 g)	Toothpaste
23	Poha (Chura) (1 kg)	Foodstuff
24	Pond's Dreamflower powder (100 g)	Body powder
25	Priya pickles – Mango (200 g)	Pickles
26	Refined oil – Safflower – fora (1 l)	Refined oil
27	Rin Supreme (100 g)	Washing soap
28	Ruchi chilli powder (100 g)	Chilli powder
29	Ruchi chilli powder (50 g)	Chilli powder
30	Ruchi curry powder (50 g)	Curry powder
31	Ruchi haldi powder (100 g)	Turmeric powder
32	Salt – Tata (1 kg)	Salt
33	Shampoo – Clinic All Clear (50 ml)	Shampoo
34	Shampoo – Clinic Plus (50 ml)	Shampoo
35	Shaving cream – Old Spice (100 g)	Shaving cream
36	Sooji (1 kg)	Foodstuff
37	Sugar (1 kg)	Sugar
38	Sundrop refined oil (1 l)	Refined oil
39	Surf Excel (500 g)	Detergent powder
40	Tea – Tata Leaf (100 g)	Beverage
41	Tide detergent (200 g)	Detergent powder
42	Tide detergent (500 g)	Detergent powder
43	Tomato sauce (Lalls) (200 g)	Tomato sauce
44	Tomato sauce (Maggi) (200 g)	Tomato sauce
45	Vim bar (100 g)	Utensil cleaning soap

Table 50.4 Association rules amongst the grocery Items

Rule #	Antecedent	Consequent
1	Noodles – Maggi (200 g). and Tomato sauce (Maggi) (200 g)	MDH chicken masala
2	MDH chicken masala	Noodles – Maggi (200 g). and Tomato Sauce
3	MDH chicken masala. and Noodles – Maggi (200 g)	Tomato sauce (Maggi) (200 g)
4	Tomato sauce (Maggi) (200 g).	MDH chicken masala. and Noodles – Maggi (200 g)
5	Kismis (100 g)	Basmati (1 kg)
6	Tomato sauce (Maggi) (200 g)	MDH chicken masala
7	MDH chicken masala	Tomato sauce (Maggi) (200 g)
8	Coffee – Nescafe (50 g)	Bournvita – Cadbury (500 g)
9	MDH chicken masala. and Tomato sauce (Maggi) (200 g)	Noodles – Maggi (200 g)
10	Poha (chura) (1 kg). and Sooji (1 kg).	Noodles – Maggi (200 g)
11	Poha (chura) (1 kg)	Noodles – Maggi (200 g)
12	Sooji (1 kg)	Noodles – Maggi (200 g)
13	MDH chicken masala	Noodles – Maggi (200 g)
14	Tomato sauce (Maggi) (200 g)	Noodles – Maggi (200 g)

of the items have been written as usually told by the Indian shoppers. The list of 45 items contains 29 food items and 16 miscellaneous grocery items. Miscellaneous grocery items constitute of cosmetics, detergents, toothpaste and mosquito repellent. In view of identifying 'purchase dependencies within a group of items' and 'inter-group purchase dependencies' the variety of items were chosen.

A data volume of 8,418 sales transactions (at least one of the selected 45 items is present in each transaction) with quantity of the items was collected from various grocery retail outlets of different sizes (small and large) in two different cities in India.

For threshold of 20% support and 65% confidence, 14 association rules were obtained as given in Table 50.4. In this table, it is observed that rules # 1, 2, 3, 4 and 9 are amongst the items, Noodles – Maggi (200 g), Tomato Sauce (Maggi) (200 g) and MDH Chicken Masala. Rules 6,7,13 and 14 are between any two items of Noodles – Maggi (200 g), Tomato Sauce (Maggi) (200 g) and MDH Chicken Masala. This is like the example illustrated in Section 50.5 which involves three items d, f and g. The rules which involve these three items have been shown separately in Table 50.5.

50.7 Conclusions

Consumer insight extracted using data mining tools has been used for business decision making in marketing, shelf design, product management etc. There has been limited use in inventory management. Purchase dependency as a type of consumer

Table 50.5 Association rules amongst three selected items

Rule #	Antecedent	Consequent
1	Noodles – Maggi (200 g). and Tomato sauce (Maggi) (200 g)	MDH chicken masala
2	MDH chicken masala	Noodles – Maggi (200 g). and Tomato sauce (Maggi) (200 g)
3	MDH chicken masala. and Noodles – Maggi (200 g)	Tomato sauce (Maggi) (200 g)
4	Tomato sauce (Maggi) (200 g)	MDH chicken masala. and Noodles – Maggi (200 g)
6	Tomato sauce (Maggi) (200 g)	MDH chicken masala.
7	MDH chicken masala.	Tomato sauce (Maggi) (200 g)
9	MDH chicken masala. and Tomato sauce (Maggi) (200 g)	Noodles – Maggi (200 g)
13	MDH chicken masala	Noodles – Maggi (200 g)
14	Tomato sauce (Maggi) (200 g)	Noodles – Maggi (200 g)

insight which can be applied in inventory modeling has been discussed for the first time in this paper. The analysis and findings in this research throws some light on various aspects of purchase dependencies which may be useful for inventory management in retail stores. The model, illustrative example and the case discussion in this research paper will motivate researchers and inventory managers to develop methodologies for using the findings of data mining in multi-item inventory management.

There is scope for further work for identifying the guiding profile of the customers which causes in simultaneous purchase of three items, Noodles – Maggi (200 g), Tomato Sauce (Maggi) (200 g) and MDH Chicken Masala. This knowledge will of tremendous importance for designing inventory replenishment policy in retail sale.

References

1. Han, J., & Kamber, N., "Data Mining: Concepts and Techniques", Morgan Kaufmann, San Francisco, CA, 2006, pp. 227–284.
2. Pujari, A.K., "Data Mining Techniques", Hyderabad, University Press, 2002, pp. 259–260.
3. Cohen, M.A., & Ernst, R., "Multi-item classification and generic inventory stock control policies", Production and Inventory Management Journal, 29(3), 1988, pp. 6–8.
4. Ernst, R., & Cohen, M.A., "A clustering procedure for production/inventory systems", Journal of Operations Management, 9(4), 1990, pp. 574–598.
5. Guvenir, H.A., & Erel, E., "Multicriteria inventory classification using a genetic algorithm", European Journal of Operations Research, 105(1), 1998, pp. 29–37.
6. Partovi, F.Y., & Anandarajan, M., "Classifying inventory using an artificial neural network approach. Computers & Industrial Engineering", 41, 2002, pp. 389–404.
7. Bhattacharya, D.K., "Production, manufacturing and logistics on multi-item inventory", European Journal of Operational Research, 162(3), 2005, pp. 786–791.

8. Kuo, R.J., An, A.L., Wang, H.S., & Chung, W.J., "Integration of self-organizing feature maps neural network and genetic K-means algorithm for market segmentation", Expert Systems with Applications, 30, 2006, pp. 313–324.
9. Hu, T. L., & Sheu, J.B., "A fuzzy-based customer classification method for demand-responsive logistical distribution operations", Fuzzy Sets and Systems, 139, 2003, pp. 431–450.
10. Larivie're, B., & Piel, D.V.D., "Investigating the role of product features in preventing customer churn, by using survival analysis and choice modeling: The case of financial services", Expert Systems with Applications, 27, 2004, pp. 277–285.
11. Wong, W., Fu, A.W., & Wang, K., "Data mining for inventory item selection with cross-selling considerations", Data Mining and Knowledge Discovery, 11(1), 2005, pp. 81–112.

Chapter 51
Economic Process Capability Index for Product Design and Process Planning Economic Process Capability Index

Angus Jeang

Abstract The process capability index (PCI) is a value which reflects real-time quality status. The PCI acts as the reference for real-time monitoring that enables process controllers to acquire a better grasp of the quality of their on site processes. The PCI value is typically defined as the ability to carry out a task or achieve a goal. However, simply increasing the PCI value can easily create additional and unnecessary production costs that result from extra efforts and expensive devices for ensuring tolerance control. Hence, there is a need to balance customer demands for quality and production costs. In this regard, the off-line PCI value is introduced, in consideration of quality loss and production cost, simultaneously in this research. The quality loss is expressed by quality loss function, and the production cost is represented by tolerance cost function. Then, this new PCI expression can be used as linkage for concurrent product design and process planning, prior to actual production.

Keywords Process capability index · real-time quality status · production cost · off-line · tolerance cost function, concurrent product design

51.1 Introduction

In recent years, as the concept of concurrent engineering has become widely accepted, design engineers hope to achieve simultaneous product design and process planning, side by side, at an early stage of product development [1]. The goals are: to shorten the time span required for introducing the new product onto the market and to attain the lowest production cost and premium product quality. Hence,

A. Jeang
Department of Industrial Engineering and Systems Management, Feng Chia University, P.O. Box 25-150, Taichung, Taiwan, R.O.C.
E-mail: akjeang@fcu.edu.tw

S.I. Ao and L. Gelman (Eds.), *Advances in Electrical Engineering and Computational Science*, Lecture Notes in Electrical Engineering.

what is needed is a way to measure the degree of the producer's process capability, in satisfying the customer's quality requirement. More importantly, a growing number of producers include this measurement value in their purchase contracts with customers, as a documentation requirement [2]. One such measurement is the process capability index (PCI).

The process capability index (PCI) is a value which reflects real-time quality status. The PCI acts as the reference for real-time monitoring that enables process controllers to acquire a better grasp of the quality of their on site processes [3, 4]. Although the PCI is considered as one of the quality measurements employed during on-line quality management, several authors have pointed out that the PCI should be addressed at the beginning of the design stage rather than at the production stage, where process capability analysis is typically done [Shina, 1995]. For the sake of convenience, let us call the PCI for the former one, off-line PCI, and the latter one, on-line PCI. The on-line PCI has realized process mean and process variance that are obtained from the existing process. Conversely, the off-line PCI has the process mean and process variance as two unknown variables, which the product designer and process planner would have to determine. When cost is not considered as a factor for off-line PCI analysis; normally the process planners would do their best to set the process mean close to the design target, and minimize the process variance to the process limit. Because the additional cost incurred for tightening the variance is not considered, obviously, the establishment of mean and variance values will result in a high PCI scale [6]. Thus, a PCI expression which contains cost factors for an Off-line application is developed.

The PCI value is typically defined as the ability to carry out a task or achieve a goal. The controllable factors are the process mean and process variance [7]. The deviation between process mean and design target can be reduced by locating the process mean close to the design target without additional cost being incurred. The process variance can be lowered by tightening the process tolerance, with extra cost incurred. In case the conventional on-line PCI is used for process capability analysis during the product and process designs, designer engineers naturally intend to raise the PCI value by locating the process mean near the target value, and by reducing the tolerance value to ensure a better product quality. However, simply increasing the PCI value can easily create additional and unnecessary production costs that result from extra efforts and expensive devices for ensuring tolerance control. Hence, there is a need to balance customer demands for quality and production costs. In this regard, the off-line PCI value is introduced, in consideration of quality loss and production cost, simultaneously in this research. The quality loss is expressed by quality loss function, and the production cost is represented by tolerance cost function. Then, this new PCI expression can be used as linkage for concurrent product design and process planning, prior to actual production. The rationale will be discussed in the latter sections.

51.2 Process Capability Indices (PCI)

The frequently seen PCI includes C_p, C_{pk}, and C_{pm} expressions. C_p can be defined as follows [3, 4, 8–10]:

$$C_p = \frac{USL - LSL}{6\sigma} \tag{51.1}$$

The expression ($USL - LSL$) refers to the difference between the upper and lower limits which are specified by the customer's quality requirement; σ is the standard deviation which is actually incurred in the production process. However, during the production process, the process means U can be located at positions other than design target. If the process variance σ^2 did not change, the above C_p value would also remain unchanged; this was the major defect owing to the facts that only the spread of the process is reflected, and the deviation of process mean can not be reflected in the measurement. These are the main reasons why C_{pk} was developed; the C_{pk} expression is defined as below:

$$C_{pk} = Min\left(\frac{USL - U}{3\sigma}, \frac{U - LSL}{3\sigma}\right) \tag{51.2}$$

There is still a deficiency for C_{pk} expression: the same C_{pk} values may be constituted with different process means and variances. This situation has created a great deal of confusion, and uncertainty as to which would be the best process capability among the alternatives. To cope with the above arguments, another form of PCI, C_{pm}, was developed. C_{pm} is defined as follows:

$$C_{pm} = \frac{USL - LSL}{6\sqrt{\sigma^2 + (U - T)^2}} \tag{51.3}$$

When the process mean is equal to design target, the C_{pm} can be simplified as C_p. For the purpose of comparison, three processes: A, B, and C are depicted in Fig. 51.1. The C_p, C_{pk}, and C_{pm} values from processes A, B, and C are shown in Table 51.1. Because process C has the greatest C_p value, it is might be mistakenly concluded that process C had the best process capability among processes A, B, and C, when C_p is considered as a reference value. However, this erroneous conclusion originates from the fact that the C_p value is solely based on the magnitude of variance, and disregards the negative impact from the process deviation. Similarly, when C_{pk} was used in representing the levels of process capability, the process C_{pk} values for processes A, B, and C, are all just equal to one. Thus again, quite obviously, there is difficulty in ordering the superiority of process capability of the three processes. To overcome the defects appearing in C_p and C_{pk} expressions, another PCI expression, C_{pm} is introduced. Unlike the previous two expressions, C_{pm} can simultaneously reflect the impact from process deviation and process variance. This feature is particularly important because process mean and process variance are

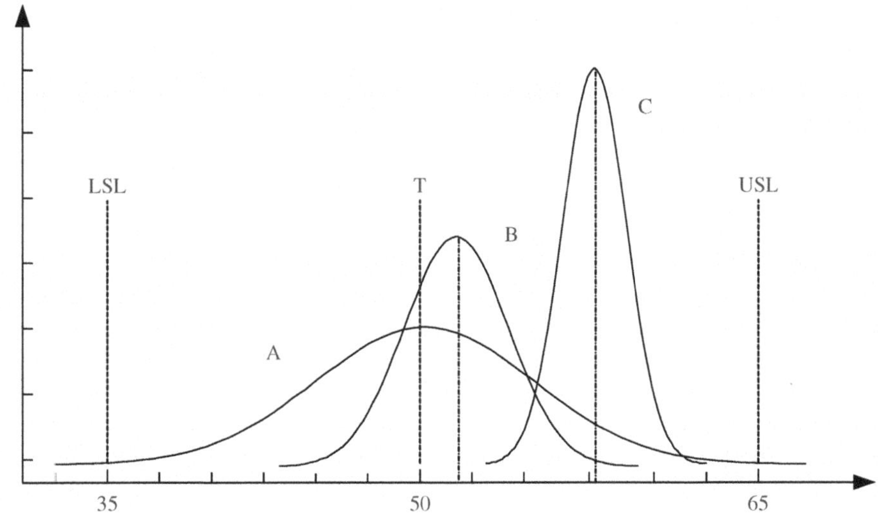

Fig. 51.1 The Distribution for process A, B and C

Table 51.1 PCI values for processes A, B, and C

Process	C_p	C_{pk}	C_{pm}	C_{pmc}
A	1	1	1	0.051298
B	1.25	1	1	0.047673
C	2	1	0.632456	0.031782

generally changed at the same time in most production process. Unfortunately, with C_{pm}, processes A and B are the best two choices. The non-different outcomes between processes A and B result from the fact that the contribution of C_{pm} value, from process mean deviation and process variance magnitude, is identical. Hence, there must be a way of measurement being provided to make mean deviation and process variance magnitude distinguishable in PCI expression.

Note: Process A: $U_A = 50.0, \sigma_A = 5.00$, $CM(t) = \$2{,}000$,
Process B: $U_B = 53.0, \sigma_B = 3.83$, $CM(t) = \$3{,}500$,
Process C: $U_C = 57.5, \sigma_C = 2.50$, $CM(t) = \$6{,}000$

The conventional PCI values are mainly applied to measure the on-line spread resulting from actual process deviation and process variance. The reality of this spread is a consequence of the production process, in which there are no more controllable elements. Hence, the conventional PCI expression is valid only in the on-line application. In this regard, there is an attempt to explore the process capability analysis, before production process, to enable designers to integrate the aspects of product design and process planning at an early time. According to the preceding discussion, unlike the other two PCI expressions, C_{pm} is able to simultaneously reflect the influences of process deviation and process variance. However, this is only legitimate after production process when realized U and t are no more controllable variables for a design. However, when C_{pm} is used as a measurement scale before production process, U and t become controllable variables. Then, it is possible that various combinations of U and t will result in the same C_{pm} value. Thus, it is difficult to make a distinction among alternatives, in order to make a correct choice among them. Additionally, the designers would most likely establish the process mean U as close as possible to the design target T, within the process feasible range, and attempt to decrease the process variance as much as possible within the process capability limits in order to attain a higher C_{pm} value. In other words, with the exclusive use of the process mean and process tolerance as the determinants of conventional C_{pm} expression, regardless of the cost impact on customer and production, there is a tendency for designers to position the process mean as close to the target value as possible, and solely cut down the process tolerance to lower capability limit in order to increase the C_{pm} value. Apparently, the found PCI value is erroneous.

The degree of proximity reflects the different quality loss according to the customer's view. Reducing the process variance is normally completed by tightening the tolerance value through tolerance design which usually involves additional cost. Therefore, in addition to the constraints from feasible ranges and capability limits, the influence exerted by the relevant costs representing the selected process mean and process tolerance, should be considered as well. This brings us to the next section, a discussion on the requirement that costs related to process mean and process tolerance must be contained in PCI expression, when referred to as off-line process capability analysis, during product design and process planning.

51.3 Proposed Process Capability Index

51.3.1 Single Quality Characteristic

Because various combinations of U and t will result in the same C_{pm} value, this unhelpful facet will prevent the conventional C_{pm} from being a suitable index for differentiating possible alternatives during product design or process planning. To overcome the above weakness, the lack of consideration of the cost influence from various U and t values, should be resolved. As is known, all costs incurred within a product life cycle, include the material and production costs which are incurred before the product reaches the consumer, and quality loss, which occurs after a sale. In these regards, let the denominator of C_{pm} be replaced with total cost, *TC*, which is the sum of quality and production related cost, which includes quality loss $K[\sigma^2 + (U - T)^2]$ and tolerance cost $C_M(t_i)$. To evaluate the tolerance cost, this paper adopts the tolerance cost function as developed in the literature [11]. $C_M(t) = a + b\ exp\ (-ct)$, where a, b, and c are the coefficients for the tolerance cost function, and t is the process tolerance.

Table 51.2 Various CPMC when CPM is 1.2

(U,t)	(50.00, 12.50)	(50.05, 12.49)	(51.00, 12.13)	(51.05, 12.09)	(52.00, 10.96)	(52.05, 10.88)
C_{pm}	1.2000000	1.2000000	1.2000000	1.2000000	1.2000000	1.2000000
C_{pmc}	0.0345994	0.0346019	0.0356385	0.0357504	0.0394258	0.039727

Have both U and t be controllable variables so that a maximum C_{pmc} can be achieved. This developed cost effectiveness PCI expression C_{pmc} is shown as follows.

$$C_{pmc} = \frac{USL - LSL}{6\sqrt{K[\sigma^2 + (U - T)^2] + C_M(t)}} \tag{51.4}$$

$$t_L \leq t \leq t_U \tag{51.5}$$

$$U_L \leq U \leq U_U \tag{51.6}$$

The values of t_L, t_U, U_U, U_L, S, and T are known in advance. σ is t/P, P is given [12]. There are infinite combinations of U and t which have the same C_{pm} value. Table 51.2 shows that different C_{pmc} values can be obtained under various combinations of U and t, when C_{pm} is fixed as 1.2. Apparently, the conventional C_{pm} is not capable of representing all possible alternatives for off-line application during design and planning stage. Table 51.1 shows that processes A and B are identical choices based on C_{pm}; however, process A is the only selection based on C_{pmc} value. Different selections are being made because C_{pm} lacks consideration of the combined influence from quality loss for customers and tolerance costs for production.

Example 1 – Let design target value $T = 30$ mm, design tolerance $S = 0.05$ mm, quality loss coefficient $K = 1{,}200$, coefficients of tolerance function $a = 50.11345$, $b = 119.3737$, $c = 31.5877$, $P = 3$, $t_L = 0.024$ mm, $t_U = 0.086$ mm. Substitute these values into Eqs. (51.3–51.6), to proceed the following discussion.

The optimal C^*_{pmc} is 0.0019. The optimal process mean U^* is 30.0000 and the process tolerance t^* is 0.05 mm. When the process mean is located at the target with fixed tolerance value, the maximum C_{pm} and C_{pmc} can be reached. The explanation is discernable by looking into the common expression, $(U - T)^2$, in the denominator of Eqs. (51.3 and 51.4). On the other hand, with fixed process mean at the target value, the maximum value C_{pm}, which is infinite, is reached when t is near to zero and the maximum C_{pmc}, which is finite, arrives when t is 0.04. The fact behind different optimal t values being found when the PCI is at its maximum, is comprehensible because the variance in C_{pmc} is cost related, while the variance in C_{pm} is cost unrelated. The t value in C_{pm} can be any small value regardless of the cost impact resulting from tolerance cost.

51.3.2 Multiple Quality Characteristics

As discussed in the preceding section, a lower quality loss (better quality) implies a higher tolerance cost, and a higher quality loss (poorer quality) indicates lower tolerance cost. Hence, the design parameters must be adjusted to reach an economic balance between reduction in quality loss and tolerance cost, so that the cost effectiveness PCI expression, C_{pmc}, is maximized. The model developed in the following illustration attempts to achieve this objective in the case of multiple quality characteristics.

Before model development, a functional relationship between the dependent variable Y and independent variable X should be identified and thoroughly understood. Based on this relationship, the resultant overall quality characteristic such as σ_Y and U_Y can be estimated from the set of individual quality characteristics in both product and process. The proposed process capability index, C_{pmc}, for the multiple quality characteristics is:

$$C_{pmc} = \frac{USL - LSL}{6\sqrt{K[(U_Y - T_Y)^2 + \sigma_Y^2] + \sum_{i=1}^{M_0} C_M(t_i)}} \tag{51.7}$$

where M_0 is the total number of quality characteristics in a product or process.

Example 2 – A DC circuit has four resistances: R_1, R_2, R_3, and R_4, as shown in Fig. 51.2 [13]. The design function representing the functional relationship between dependent variable Y and independent variables R_1, R_2, R_3, and R_4, is $\frac{E_1 \cdot \left(1+\frac{R_2}{R_1}\right)}{1+\frac{R_3}{R_4}} - \frac{E_2 \cdot R_2}{R_1}$, where E_1 is 1.0 V and E_2 is −1.0 V. M_0 is 4, so i is 1, 2, 3, 4.

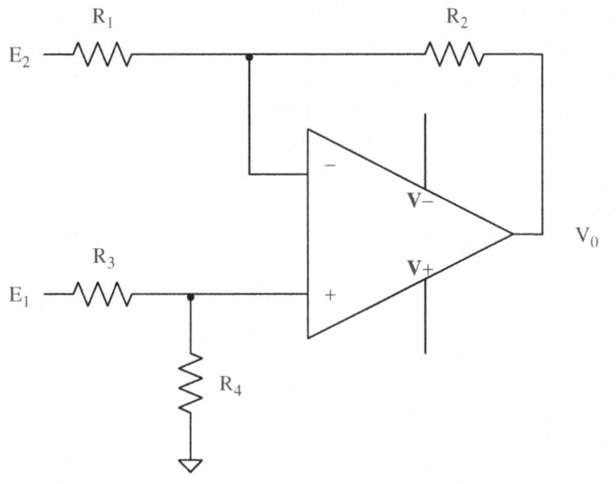

Fig. 51.2 A DC differential amplifier

Table 51.3 The numerical results of example 2

C/K	I		II	
	C_{pmc}	C_{pm}	C_{pmc}	C_{pm}
I	$U_1 = 11$	$U_1 = 11$	$U_1 = 11$	$U_1 = 11$
	$U_2 = 105.47$	$U_2 = 105$	$U_2 = 105.47$	$U_2 = 105$
	$U_3 = 8$	$U_3 = 8$	$U_3 = 8$	$U_3 = 8$
	$U_4 = 115$	$U_4 = 115$	$U_4 = 115$	$U_4 = 115$
	$t_1 = 0.6$	$t_1 = 0.6$	$t_1 = 0.6$	$t_1 = 0.6$
	$t_2 = 1.2$	$t_2 = 1.2$	$t_2 = 1.6769$	$t_2 = 1.2$
	$t_3 = 0.8$	$t_3 = 0.5$	$t_3 = 0.8$	$t_3 = 0.5$
	$t_4 = 6$	$t_4 = 4.9477$	$t_4 = 6$	$t_4 = 4.9477$
	$U_Y = 19.494$	$U_Y = 18.87$	$U_Y = 19.494$	$U_Y = 18.87$
	$t_Y = 1.0363$	$t_Y = 0.5704$	$t_Y = 1.0363$	$t_Y = 0.5704$
	$TC = 5.0377$	$TC = 26.895$	$TC = 4.2606$	$TC = 17.562$
	$C^{/}_{pm} = 1.1467$	$C_{pm} = 3.2544$	$C^{/}_{pm} = 1.1355$	$C_{pm} = 3.2544$
	$C_{pmc} = 0.1782$	$C_{pmc} =$ —	$C_{pmc} = 0.1937$	$C_{pmc} =$ —
II	$U_1 = 11$	$U_1 = 11$	$U_1 = 11$	$U_1 = 11$
	$U_2 = 105.47$	$U_2 = 105$	$U_2 = 105.47$	$U_2 = 105$
	$U_3 = 8$	$U_3 = 8$	$U_3 = 8$	$U_3 = 8$
	$U_4 = 115$	$U_4 = 115$	$U_4 = 115$	$U_4 = 115$
	$t_1 = 0.6$	$t_1 = 0.6$	$t_1 = 0.6$	$t_1 = 0.6$
	$t_2 = 1.2$	$t_2 = 1.2$	$t_2 = 1.2$	$t_2 = 1.2$
	$t_3 = 0.8$	$t_3 = 0.5$	$t_3 = 0.8$	$t_3 = 0.5$
	$t_4 = 6$	$t_4 = 4.9477$	$t_4 = 6$	$t_4 = 4.9477$
	$U_Y = 19.494$	$U_Y = 18.87$	$U_Y = 19.494$	$U_Y = 18.87$
	$t_Y = 1.0363$	$t_Y = 0.5704$	$t_Y = 1.0363$	$t_Y = 0.5704$
	$TC = 7.4205$	$TC = 31.215$	$TC = 6.6713$	$TC = 20.188$
	$C^{/}_{pm} = 1.1721$	$C_{pm} = 3.2544$	$C^{/}_{pm} = 1.1577$	$C_{pm} = 3.2544$
	$C_{pmc} = 0.1468$	$C_{pmc} =$ —	$C_{pmc} = 0.1548$	$C_{pmc} =$ —

The design target value T_Y is 20. Design specification S_Y is 2. Namely, $USL - LSL$ is 4.0. The quality coefficient K is 6,000. Assume P is 3. M_0 is 4. The coefficients of tolerance cost function, a, b, and c, are given in Table 51.3. The c_i value is further divided into coefficients I, II for the purpose of sensitivity analysis done in Table 51.2. The intention is to determine the process mean U and process tolerance t, so that the present PCI expression, C_{pmc}, is maximized. The formulation is given in the following expression.

First, let the expressions U_Y, σ_Y, t_Y be found by expansion design function through Taylor's series.

$$U_Y = \frac{U_2}{U_1} + 0.5\left[\frac{t_1^2\left(\frac{2U_2}{U_1^3} + \frac{2U_2}{U_1^3\left(1+\frac{U_3}{U_4}\right)}\right)}{P^2} + \frac{t_4^2\left(\frac{2\left(1+\frac{U_2}{U_1}\right)U_3^2}{\left(1+\frac{U_3}{U_4}\right)^3 U_4^4} - \frac{2\left(1+\frac{U_2}{U_1}\right)U_3}{\left(1+\frac{U_3}{U_4}\right)^2 U_4^3}\right)}{P^2} + \frac{2t_3^2\left(1+\frac{U_2}{U_1}\right)}{P^2\left(1+\frac{U_3}{U_4}\right)^3 U_4^2}\right] + \frac{1+\frac{U_2}{U_1}}{1+\frac{U_3}{U_4}} \quad (51.8)$$

$$\sigma_Y^2 = \frac{t_2^2\left(\frac{1}{U_1} + \frac{1}{U_1\left(1+\frac{U_3}{U_4}\right)}\right)^2}{P^2} + \frac{t_1^2\left(-\frac{U_2}{U_1^2} - \frac{U_2}{U_1^2\left(1+\frac{U_3}{U_4}\right)}\right)^2}{P^2} + \frac{t_4^2\left(1+\frac{U_2}{U_1}\right)^2 U_3^2}{P^2\left(1+\frac{U_3}{U_4}\right)^4 U_4^4} + \frac{t_3^2\left(1+\frac{U_2}{U_1}\right)^2}{P^2\left(1+\frac{U_3}{U_4}\right)^4 U_4^2} \tag{51.9}$$

$$t_Y^2 = t_2^2\left(\frac{1}{U_1} + \frac{1}{U_1\left(1+\frac{U_3}{U_4}\right)}\right)^2 + t_1^2\left(-\frac{U_2}{U_1^2} - \frac{U_2}{U_1^2\left(1+\frac{U_3}{U_4}\right)}\right)^2 + \frac{t_4^2\left(1+\frac{U_2}{U_1}\right)^2 U_3^2}{\left(1+\frac{U_3}{U_4}\right)^4 U_4^4} + \frac{t_3^2\left(1+\frac{U_2}{U_1}\right)^2}{\left(1+\frac{U_3}{U_4}\right)^4 U_4^2} \tag{51.10}$$

Objective function:

$$\text{Max } C_{pmc} = \frac{USL - LSL}{6\sqrt{K[(U_Y - T_Y)^2 + \sigma_Y^2] + \sum\limits_{i=1} C_M(t_i)}} \tag{51.11}$$

Subjective to:

$$|\ T_Y - U_Y\ | \leqq S_Y - t_Y \tag{51.12}$$

$$t_{Li} \leq t_i \leq t_{Ui} \tag{51.13}$$

$$\mathrm{U_{Li}} \leq \mathrm{U_i} \leq \mathrm{U_{Ui}} \tag{51.14}$$

decision variables are t_i and U_i, where i is 1, 2, 3, and 4.

U_Y, σ_Y^2, and t_Y^2 are obtained from Eqs. (51.8)–(51.10).

Then, the result is summarized in Table 51.2. For the purpose of comparison, let the above objective function C_{pmc}, Eq. (51.11), be replaced by C_{pm}, Eq. (51.3), with the same constraints. Then, relevant solutions are also listed in Table 51.2. The solutions from objective functions C_{pmc} and C_{pm} are arranged pair wisely for easy comparison. Clearly, the solutions found with C_{pm} as an objective function are insensitive to the changes of quality loss and tolerance cost; instead, the solutions based on C_{pmc} fully reveal the impact from various combinations of quality loss and tolerance cost. Let the optimal solutions, U^* and t^*, found from C_{pmc} as an objective function, be substituted into the C_{pm} expression to have $C_{pm}^{/}$, the values $C_{pm}^{/}$ still fall within acceptable range for general application. The reason for optimal value C_{pm} being greater than $C_{pm}^{/}$ is due to the irrational combination of U and t, when $C_{pm}$is used as an objective function. This irrational combination, resulting from disregarding the effect of quality loss and tolerance cost in C_{pm} expression, drives the C_{pm} value to become unusually high. Similar reasons can also explain why the total cost, TC, is always higher when C_{pm} is used as objective function than is the case for C_{pmc}. In other words, C_{pmc} is an appropriate expression for application in the process capability analysis during off-line application; particularly, for product design and process planning at the stage of blueprint.

51.4 Summary

Conventionally, production related issues are usually dealt with via process analysis after the product design or process planning has been completed. These approaches will likely result in poor product quality and high production cost, as a consequence of a lack of consideration concerning quality and production related costs. These are some of the reasons that concurrent engineers suggesting that possible issues occurring in the production stages should first be considered at the time that the new product is developed. That can reduce the time span for introducing a new product onto market, and increase the chance for obtaining a superior edge among competitors. Thus, the present research introduces a PCI measurement, C_{pmc}, for the process capability analysis, to ensure that lower production cost and high product quality can be achieved at the earlier time of the blue print stage. For an application, prior to production, process engineers can establish process mean and process tolerance based on the optimized mean and tolerance values via the above three approaches. As the expectation, the produced quality values after production process must be distributed with statistic values as the process mean and process tolerance established before production process. As a result, an effective PCI for product life cycle becomes actualized.

References

1. Carter, D. E. and Baker, B. S., 1992, Concurrent Engineering: The Product Development Environment for the 1990s, Addison-Wesley, New York.
2. Schneider, H. and Pruett, J., 1995, Use of Process Capability Indices in the Supplier Certification Process, Qual. Eng., 8: 225–235.
3. Kotz, S. and Johnson, N. L., 1993, Process Capability Indices, Chapman & Hall, London.
4. Kotz, S. and Lovelace, C. R., 1998, Process Capability Indices in Theory and Practice, Arnold, a Member of the Hodder Headline Group, New York.
5. Shina, S. G. and Saigal, A., 2000, Using C_{pk} as a Design Tool for New System Development, Qual. Eng., 12: 551–560.
6. Spiring, F. A., 2000, "Assessing Process Capability with Indices" in Statistical Process Monitoring and Optimization, edited by S. H. Park and G. G. Vining, Marcel-Dekker, New York.
7. Jeang, A., 2001, Combined Parameter and Tolerance Design Optimization with Quality and Cost Reduction, Int. J. Prod. Res., 39(5): 923–952.
8. Boyle, R. A., 1991, The Taguchi Capability Index, J. Qual. Technol., 23: 17–26.

9. Chan, L. K., Cheng, S. W., and Spring, F. A., 1989, A New Measurement of Process Capability: C_{pm}, J. Qual. Technol., 20: 162–175.
10. Kane, V. E., 1986, Process Capability Index, J. Qual. Technol., 18: 41–52.11.
11. Chase, K. W., Greenwood, W. H., Loosli, B. G., and Haugland, L. F., 1990, Least Cost Tolerance Allocation for Mechanical Assemblies with Automated Process Selection, Manuf. Rev., 3(1): 49–59.
12. Jeang, A., Liang, F., and Chung, C. P., 2008, Robust Product Development for Multiple Quality Characteristics Using Computer Experiments and an Optimization Technique, Int. J. Prod. Res., 46(12): 3415–3439.
13. Boyd, R. R., 1999, Tolerance Analysis of Electronic Circuits Using Matlab, CRC Press, New York.

Chapter 52
Comparing Different Approaches for Design of Experiments (DoE)

Martín Tanco, Elisabeth Viles, and Lourdes Pozueta

Abstract Design of Experiments (DoE) is a methodology for systematically applying statistics to experimentation. Since experimentation is a frequent activity at industries, most engineers (and scientists) end up using statistics to analyse their experiments, regardless of their background. OFAT (one-factor-at-a-time) is an old-fashioned strategy, usually taught at universities and still widely practiced by companies. The statistical approaches to DoE (Classical, Shainin and Taguchi) are far superior to OFAT. The aforementioned approaches have their proponents and opponents, and the debate between them is known to become heated at times. Therefore, the aim of this paper is to present each approach along with its limitations.

Keywords Design of Experiments · Classical · Shainin · Taguchi · statistical approach

52.1 Introduction

Lye [1] defined the Design of Experiments (DoE) as a methodology for systematically applying statistics to experimentation. More precisely, it can be defined as a series of tests in which purposeful changes are made to the input variables of a process or system so that one may observe and identify the reasons for these changes in the output response(s) [2].

Since experimentation is a frequent activity at industries, most engineers (and scientists) end up using statistics to analyse their experiments, regardless of their background [3]. Not all engineers are exposed to statistics at the undergraduate level, and this leads to problems when tools are required in practice; either they don't know

M. Tanco (✉)
Department of Industrial Management Engineering at TECNUN (University of Navarra), Paseo Manuel Lardizabal 13, 20018 San Sebastian, Spain
E-mail: mtanco@tecnun.es

S.I. Ao and L. Gelman (Eds.), *Advances in Electrical Engineering and Computational Science*, Lecture Notes in Electrical Engineering.

what type of experimental strategy is required for their problem and they select something inappropriate, or they select the correct strategy and apply it incorrectly, or they select the wrong strategy in which case it probably doesn't matter whether they use it correctly or not [4].

OFAT (one-factor-at-a-time) is an old-fashioned strategy, usually taught at universities and still widely practiced by companies. It consists of varying one variable at a time, with all other variables held constant. On the other hand, DoE is an efficient technique for experimentation which provides a quick and cost-effective method for solving complex problems with many variables. Numerous case studies in different areas of application prove the advantages and potential of DoE [5].

The statistical approach to Design of Experiments and the Analysis of Variance (ANOVA) technique was developed by R.A. Fisher in 1920. Since then, many have contributed to the development and expansion of this technique. Most techniques, defined throughout this article as "Classical", have adapted Fisher's ideas to various industries, including agriculture. However, engineers Shainin and Taguchi are especially influential due to their contribution of two new approaches to DoE. Both new approaches offer more than just Design of Experiments, as they can be considered quality improvement strategies [6].

All three approaches to DoE (Classical, Shainin and Taguchi) are far superior to OFAT. The aforementioned approaches have their proponents and opponents, and the debate between them is known to become heated at times. Dr. Deming once said, "Any technique is useful as long as the user understands its limitations." Therefore, the aim of this paper is to present each approach along with its limitations.

Most engineers have surely at least heard of Design of Experiments (DoE), Taguchi Methods or the Shainin System^{TM}. Yet how many of them can really say that they understand the differences among them? Or that they are capable of correctly deciding when to use which technique?

The answer to these questions is heavily influenced by the knowledge and experience one has of each approach to Design of Experiments. Since the majority of practitioners only have experience with a single approach, this article aims to open their minds and compare available approaches.

Each approach to Design of Experiments will be briefly described in chronological order in the following section. In Section 52.3, the main criticisms published in literature about each approach are highlighted. Finally, the conclusion and recommendations for engineers and managers working in the manufacturing industry are presented in Section 52.4.

52.2 Approaches to Design of Experiments

52.2.1 The Classical Approach

Although books on Design of Experiments did not begin to appear until the twentieth century, experimentation is certainly about as old as mankind itself [7]. The one-factor-at-a-time strategy (OFAT) was, and continues to be, used for many years.

However, these experimentation strategies became outdated in the early 1920s when Ronald Fisher discovered much more efficient methods of experimentation based on factorial designs [1]. Those designs study every possible combination of factor settings, and are especially useful when experimentation is cheap or when the number of factors under study is small (less than five). Fisher first applied factorial designs to solve an agricultural problem, where the effect of multiple variables was simultaneously (rain, water, fertilizer, etc.) studied to produce the best crop of potatoes. His experiences were published in 1935 in his book "Design of Experiments" [8].

Fractional Factorial designs were proposed in the 1930s and 1940s in response to the overwhelming number of experiments that are involved with full factorial designs. This design consists of a carefully selected fraction of the full factorial experimental design. They provide a cost-effective way of studying many factors in one experiment, at the expense of ignoring some high-order interactions. This is considered to be low risk, as high order interactions are usually insignificant and difficult to interpret anyway.

According to Montgomery [2], the second stage in the era of the classical approach to DoE began in the 1950s when Box & Wilson [9] developed what was later called Response Surface Methodology (RSM). Their methodology allowed DoE to be applied in the chemical industry and afterwards in other industries as well. They touted the advantages of industrial experiments compared to agricultural experiments in the two following areas: (a) *Immediacy:* response can be obtained quicker than with agricultural experiments, when results can sometimes take up to a year to be obtained; (b) *Sequentially:* the experimenter is able to carry out few experiments, analyse them and again plan new experiments based on findings obtained from the previous experiments. In this era Central Composite designs (CCD) and Box-Behnken designs (BBD) were created.

The third era of the classical approach started with the appearance of the Taguchi and Shainin approach in the US in the 1980s as a simple and efficient method of experimentation. Eventually, statisticians and academics began to acknowledge the value of certain engineering ideas of Taguchi and Shainin. This led to positive changes, adopting many ideas of the new approaches (for example, the reduction of variance became an important research area within classical design), and giving importance to constructing methodologies and guidelines to ease application.

In this last era, the democratization of statistics, thanks in part to software packages and the spread of Six Sigma thinking throughout industries [10], helped spread Design of Experiments to all types of industries. Moreover, an increasing interest in literature was advocated to Design of Experiments [11]. Furthermore, software packages have made the construction of graphs and calculus easier, further facilitating the application of DoE. Recent books by Funkenbusch (2005) [12] and Robinson (2000) [13] show how this approach can be understood by engineers.

Many scientists and statisticians have contributed to DoE development, making the classical approach a valid and robust methodology for Design of Experiments. Usual reference books are Box et al. [14] and Montgomery [2].

52.2.2 The Taguchi Approach

As a researcher at the Electronic Control Laboratory in Japan, engineer Genechi Taguchi carried out significant research on DoE techniques in the late 1940s. Although he published his first book in Japanese in the 1950s, the standardized version of DoE, popularly known as the Taguchi Method, wasn't introduced in the US until the early 1980s. His most influential books were "Introduction to Quality Engineering" [15] and "System of Experimental Design" [16]. The following decade was rife with heated debate mounted by two distinct camps of professionals, one unfailingly extolling the new-found virtues and power of Taguchi methods, while the other persistently exposed the flaws and limitations inherent to them [17].

Taguchi used and promoted statistical techniques for quality from an engineering perspective rather than from an statistical perspective [18]. Although Taguchi has played an important role in popularising DoE, it would be wrong to consider Taguchi Methods as just another way to perform DoE.

He developed a complete problem solving strategy [6], which he dubbed "Quality Engineering". However, there is general confusion in industry literature when dealing with systems aiming to reduce variation, as the terms robust design, Taguchi Methods, quality engineering and parameter design are used as synonyms [19].

The basic elements of Taguchi's quality philosophy can be summarized as follows [15, 18, 20]:

- A quality product is a product that causes a minimal loss to society during its entire life. The relation between this loss and the technical characteristics is expressed by the loss function, which is proportional to the square of the deviations of the performance characteristics from its target value.
- Taguchi breaks down his quality engineering strategies into three phases: System design, Parameter design and Tolerance design. System design deals with innovative research, looking for what factors and levels should be. Parameter design is what is commonly known as Taguchi Methods and is covered in this paper. This technique is intended to improve the performance of processes/products by adjusting levels of factors. Finally, Tolerance Design aims to determine the control characteristics for each factor level identified in earlier studies.
- Change experimentation objectives from "achieving conformance to specifications" to "reaching the target and minimising variability".

Since the core of Taguchi's parameter design is based on experimental methods [19], he went to great lengths to make DoE more user-friendly (easy to apply). Basically, Taguchi simplified the use of DoE by incorporating the following: a standard set of experimental design matrices (Orthogonal arrays), a graphical aid to assign the factors to the experimental matrix (linear graphs), clear guidelines for the interpretation of results (cookbook), special data transformation to achieve reduced variation (S/N Ratios) and a formal study of uncontrollable factors using the robust design technique, among others [20]. Finally, he simplified Tolerance Analysis through the use of DoE [21].

Taguchi's main contribution to experimental design was a strong emphasis on variation reduction. Quality is something that cannot be characterised solely by means of a defined quality characteristic such as yield. The variability of those characteristics must also be considered [22]. Therefore, he proposed a novel design, where factors (included in experimentation) are classified into two main groups: Control factors and Noise Factors. The first one includes parameters that can be easily controlled or manipulated, whereas noise factors are difficult or expensive to control. Therefore, the basic idea in parameter design is to identify, through exploiting interactions between control parameters and noise variables, the appropriate setting of control parameters at which the system's performance is capable of withstanding uncontrollable variation among noise factors [23]. Since the goal is to make the system resistant to variation of noise variables, the approach has also been called "Robust design".

A recent bibliography on Taguchi's approach to DoE may be found in Roy [20] and Taguchi et al.'s [24] Quality Engineering Handbook.

52.2.3 The Shainin Approach

The Shainin System™ is the name given to a problem solving system developed by Dorian Shainin. In 1975, he established his own consulting practice: Shainin LLC. His sons Peter and Richard later joined the family business. Shainin described his colourful method as the American approach to problem solving, with the same goals of the Taguchi approach [25].

Shainin viewed his ideas as private intellectual property, which he was known to sell to clients to help them gain a competitive advantage [26]. As Shainin Systems™ are legally protected trademarks and some of its methods are rarely discussed in literature, it is difficult to obtain a complete overview of the approach [27].

Keki R. Bhote was authorised to publish information in the first and only book about these methods. His company, Motorola, won the Malcolm Baldrige National Quality Award, which stipulates that its methods be shared with other US Companies [28]. Interest in Dorian Shainin's problem solving techniques rose with the 1991 publication of this book (a second edition was published in 2000 [28]).

Dorian Shainin included several techniques – both known and newly invented – in a coherent step-by-step strategy for process improvement in manufacturing environments [27]. Among those powerful tools, he considered Design of Experiments as the centrepiece. Moreover, he didn't believe that DoE was limited to the exclusive province of professionals, but could rather be extended so that the whole factory could be turned loose on problem-solving [28].

The foundation of Shainin's DoE strategy rests on:

- The Pareto Principle: Among many, even hundreds of candidate factors, a single one will be the root cause of variation of the response y. That root cause is called the Red X® and may be a single variable or the interaction of two more separate

variables [25]. There may be then a second or a third significant cause, called the Pink X® and Pale Pink X®, respectively.
- Shainin strongly objected to the use of the Fractional Factorial technique. He proposed instead to identify and diagnostically reduce most of the sources of variation down to a manageable number (three or four), at which time he allowed the use of full factorials [29].
- "Talk to the parts, they are smarter than engineers". First, talk to the parts. Then, talk to the workers on the firing line. Last, the least productive methods are to talk to the engineers [28].

The Shainin System presents many tools in a sequence of progressive problem solving. It can be divided into three main groups: Clue Generation tools, Formal DoE and Transition to SPC. The Shainin DoE technique considers as many variables as can be identified [30]. The first groups try to generate clues (like Sherlock Homes) with a number of tools (Multi Vary, Components Search™, Paired Comparison™, Process Search™ and Concentration Chart™) to reduce the number of variables involved in the problem through on-line experimentation. In the second stage, the Variable Search™ technique is used to reduce the number of variables by sequentially (not randomly) experimenting off-line, based on engineering judgement with binary search. Once a few factors are obtained, full factorials are used to analyse their effects and interactions. Afterwards, other techniques (B vs. C™, Response Surface, and ScaterPlots) are used to confirm the results and optimise them when necessary. Finally, in the last step, Positrol™, Process Certification and Pre-Control are recommended to guarantee that results will be obtained in the future.

52.3 Limitations of the Different Approaches

52.3.1 The Classical Approach

Up to the time when "Taguchi Methods" were propagated in the U.S., Design of Experiments (classical) and associated techniques were treated as mathematical tools, more like an adjunct to an engineer's technical resources for the study of product and process characteristics [17].

Taguchi and Shainin were the biggest critics of the Classical Approach. They believed that managers, engineers and workers found the use of DoE to be a complicated, ineffective and frustrating experience [28].

However, it is worth mentioning that after a decade of strong opposition to their new ideas, the "classical" proponents began to acknowledge the importance of some of the new ideas proposed by Taguchi and Shainin. For example, there have been many attempts to integrate Taguchi's parameter design principle with well-established statistical techniques [18]. As a consequence, the exploitation of response surface methodology for variance reduction has become a major area of research [31, 32]. Moreover, the simplicity demanded by critics was reached (or at least greatly improved) by the aid of software capable of designing and analysing experimental design. Furthermore, great emphasis was placed on presenting guidelines and including graphical tools to clarify every step in the planning stage. For example, the Pareto Charts of effects or the use of Multi-Vary charts to correctly define the problem were included.

52.3.2 Taguchi

There was great debate about these methods during the first decade after their appearance in the U.S. [21–23, 33, 34]. Taguchi's approach was criticised for being inefficient and often ineffective. Moreover, Shainin was highly critical of Taguchi, challenging the myth of the "secret super Japanese weapon" [25].

Nair [23] identified three general criticisms of Taguchi's work: (a) The use of the SNR as a measure of the basis of analysis, (b) his analysis methods and (c) his choice of experimental designs.

Many have criticized the use of SNR as a performance measurement. Better approaches to the parameter design problem have been addressed in recent years, such as Box's performance measurement [35] or the Response Surface approach [31].

There has also been much criticism of the new analysis techniques proposed by Taguchi. Most of them are more difficult and less effective than previous ones. An example is the use of accumulation analysis [21, 36].

Finally, and most importantly, is the criticism of experimental designs. Firstly, orthogonal arrays were criticised for underestimating interactions. In response to this criticism, Taguchi stated [29], "A man who does not believe in the existence of non-linear effects is a man removed from reality". He believes, however, in the ability of engineers to decide on the levels of factors (called sliding levels), in order to make some interactions for that particular experiment insignificant. Unfortunately, there is evidence that interaction should be avoided. This is accepted as a matter of faith among Taguchi's followers [17].

On the other hand, the designs proposed by Taguchi to simultaneously study both the mean and the variance (crossed arrays) were also criticized, since they require multiple runs and generally don't allow to study control factor interactions. Therefore, Welch [37], among others, proposed using a combined array to reduce the number of runs. There has been much debate in recent years on this topic and it is not yet clear what the best approach is. Pozueta et al. [38] and Kunert et al. [39] have demonstrated, for example, how classical designs are sometimes worse than Taguchi's designs.

It is worth mentioning that whereas some classical statisticians and academics have acknowledged the value of certain engineering ideas of Taguchi, the Taguchi camp has shown few signs of compromise, steadfastly vouching for the unchallengeable effectiveness of the Taguchi package of procedures in its original form [17].

For more details on the technical debate of Taguchi's approach, refer to Nair [23], Pignatello [21], Box [33] and Robinson [22], among others.

52.3.3 Shainin

As the Shainin System^{TM} is legally protected, one of the only ways to learn about his method is to go to his classes. The only other alternative is to read Bhote's book [28]. Unfortunately, this book is full of hyperbolic, over optimistic, extremely biased and sometimes even intellectually dishonest claims in its urge to prove that Shainin's techniques are superior to all other methods. For example, Bhote claims, [28] "We are so convinced of the power of the Shainin DoE that we throw out a challenge. Any problem that can be solved by classical or Taguchi Methods, we can solve better, faster, and less expensively with the Shainin methods". Moreover, he presents DoE as the panacea to all kinds of problems and criticises other alternatives such as Six Sigma and TQM without academic basis behind his claims.

Although there is little written in industry literature about Shainin, the existing material is enough to fuel criticism of his methods. The most criticised techniques are the Variable Search^{TM} [29, 30, 40, 41] and Pre-Control [26, 41]. Variable Search^{TM} has much in common with the costly and unreliable "one-factor-at-a-time" method [29]. Shainin's technique relies heavily on using engineering judgement. Its weakness lies in the skill and knowledge required to carry out two tasks: firstly, to correctly identify the variables and secondly, to allocate those variables to the experiment [30].

On the other hand, although Pre-Control is presented as an alternative to statistical process control (SPC), it is not an adequate substitution. In particular, it is not well-suited for poorly performing processes where its use will likely lead to unnecessary tampering [26].

A recent review and discussion by Steiner et al. [41, 42] provides an in-depth look at the Shainin System^{TM} and its criticisms.

52.4 Conclusions and Recommendations

Three different approaches to DoE have been presented throughout the last two sections. These approaches, compared with OFAT strategies, are more successful. However, this does not prove that each technique is necessarily the best. Therefore, we will give some recommendations and conclusions on each approach based on our experience and research.

Firstly, we must commended Shainin for stressing the importance of statistically designed experiments [29]. His methods are easy to learn and can be applied to ongoing processing during full production. Despite their obvious simplicity (which stems from the fact that they are, most of the time, simple versions of one-factor-at-a-time methods), they do not seem to offer a serious alternative to other well-established statistical methodologies. Classical and Taguchi approaches are still much more powerful, more statistically valid and more robust [29]. Shainin methods could perhaps have some applicability in medium-to-high volume processes for which a high level of quality has already been achieved [29]. They can also be used when dealing with binary response (for example, a machine works or not) and the reason behind the response may be due to a huge amount of variables.

On the other hand, the most important contribution of the Taguchi approach is in the area of quality philosophy and engineering methodology, which includes the loss function and robust design [18]. Taguchi Methods have the potential to bring about first-cut improvements in industrial applications. However, owing to their theoretical imperfections, success cannot be assured in every instance [17]. As a general rule, we don't recommend using Taguchi Methods unless you have to deal with two types of problems: Tolerance analysis and robustness to noise factors in products and processes. However, in those cases, the classical approach may also be suitable.

The two new approaches were a reaction to the existing complexity of the techniques used in industry. Both approaches are presented as an easy and effective technique for experimentation. However, since the appearance of both initiatives, the Classical approach has further developed the technique, including the engineering ideas of the new approaches. The introduction of software and graphical aids has made this approach much more user friendly than in the 1980s. Therefore, this approach is the most established and has the most statistically valid methods.

However, although the classical approach has outgrown its "competitors", there is still a need to shape an easy experimental design methodology to develop the information necessary for good planning, design and analysis [18,43,44]. Therefore, unless classical DoE is developed into easily understood strategies for solving specific problems and made widely available through teaching materials and software packages, we cannot expect it to become widely used on the factory floor.

Another strategy is to combine, with statistical rigor, the best DoE approaches in an easy, simple and integrated methodology from an engineering point of view. Instead of following one approach or another, we should use the most powerful tools available to gain the necessary knowledge of a process [34]. For example, the stress that Shainin placed on carrying out several on-line experiments in a diagnostic stage to analyse and characterise the problem before experimentation must be considered. Moreover, the awareness of the importance of variation in industrial experimentation carried out by Taguchi must force Quality Loss function and Crossed arrays to be integrated in this methodology. Finally, the importance of confirmatory runs, proposed by both of them, should also be stressed.

References

1. Lye, L.M., Tools and toys for teaching design of experiments methodology. In 33rd Annual General Conference of the Canadian Society for Civil Engineering. 2005 Toronto, Ontario, Canada.
2. Montgomery, D.C., Design and Analysis of Experiments. 2005, New York: Wiley.

3. Gunter, B.H., Improved Statistical Training for Engineers – Prerequisite to quality. Quality Progress, 1985. 18(11): pp. 37–40.
4. Montgomery, D., Applications of Design of Experiments in Engineering. Quality and Reliability Engineering International, 2008. 24(5): pp. 501–502.
5. Ilzarbe, L. et al., Practical Applications of Design of Experiments in the Field of Engineering. A Bibliographical Review. Quality and Reliability Engineering International, 2008. 24(4): pp. 417–428.
6. De Mast, J., A Methodological Comparison of Three Strategies for Quality Improvement. International Journal of Quality and Reliability Management, 2004. 21(2): pp. 198–212.
7. Ryan, T.P., Modern Experimental Design. 2007, Chichester: Wiley.
8. Fisher, R.A., The Design of Experiments. 1935, New York: Wiley.
9. Box, G.E.P. and K.B. Wilson, On the Experimental Attainment of Optimum Conditions. Journal of the Royal Statistical Society, 1951. Series B(13): pp. 1–45.
10. Montgomery, D.C., Changing Roles for the Industrial Statisticians. Quality and Reliability Engineering International, 2002. 18(5): pp. 3.
11. Booker, B.W. and D.M. Lyth, Quality Engineering from 1988 Through 2005: Lessons from the Past and Trends for the Future. Quality Engineering, 2006. 18(1): pp. 1–4.
12. Funkenbusch, P.D., Practical Guide to Designed Experiments. A Unified Modular Approach. 2005, New York: Marcel Dekker.
13. Robinson, G.K., Practical Strategies for Experimentation. 2000, Chichester: Wiley.
14. Box, G.E.P., J.S. Hunter, and W.G. Hunter, Statistics for Experimenters – Design, Innovation and Discovery. Second Edition. Wiley Series in Probability and Statistics, ed. 2005, New York: Wiley.
15. Taguchi, G., Introduction to Quality Engineering. 1986, White Plains, NY: UNIPUB/Kraus International.
16. Taguchi, G., System of Experimental Design: Engineering Methods to Optimize Quality and Minimize Cost. 1987, White Plains, NY: UNIPUB/Kraus International.
17. Goh, T.N., Taguchi Methods: Some Technical, Cultural and Pedagogical Perspectives. Quality and Reliability Engineering International, 1993. 9(3): pp. 185–202.
18. Tay, K.-M. and C. Butler, Methodologies for Experimental Design: A Survey, Comparison and Future Predictions. Quality Engineering, 1999. 11(3): pp. 343–356.
19. Arvidsson, M. and I. Gremyr, Principles of Robust Design Methodology. Quality and Reliability Engineering International, 2008. 24(1): pp. 23–35.
20. Roy, R.K., Design of Experiments Using the Taguchi Approach: 16 steps to Product and Process Improvement. 2001, New York: Wiley.
21. Pignatello, J. and J. Ramberg, Top Ten Triumphs and Tragedies of Genechi Taguchi. Quality Engineering, 1991. 4(2): pp. 211–225.
22. Robinson, T.J., C.M. Borror, and R.H. Myers, Robust Parameter Design: A Review. Quality and Reliability Engineering International, 2004. 20(1): pp. 81–101.
23. Nair, V.N., Taguchi's Parameter Design: A Panel Discussion. Technometrics, 1992. 31(2): pp. 127–161.
24. Taguchi, G., S. Chowdhury, and Y. Wu, Taguchi's Quality Engineering Handbook. First edition. 2004, New York: Wiley Interscience.
25. Shainin, D. and P. Shainin, Better than Taguchi Orthogonal Tables. Quality and Reliability Engineering International, 1988. 4(2): pp. 143–149.
26. Ledolter, J. and A. Swersey, An Evaluation of Pre-Control. Journal of Quality Technology, 1997. 29(2): pp. 163–171.
27. De Mast, J. et al., Steps and Strategies in Process Improvement. Quality and Reliability Engineering International, 2000. 16(4): pp. 301–311.
28. Bhote, K.R. and A.K. Bhote, Word Class Quality. Using Design of Experiments to Make it Happen. Second edition. 2000, New York: Amacom.
29. Logothetis, N., A perspective on Shainin's Approach to Experimental Design for Quality Improvement. Quality and Reliability Engineering International, 1990. 6(3): pp. 195–202.

30. Thomas, A.J. and J. Antony, A Comparative Analysis of the Taguchi and Shainin DoE Techniques in an Aerospace Enviroment. International Journal of Productivity and Performance Management, 2005. 54(8): pp. 658–678.
31. Vining, G.G. and R.H. Myers, Combining Taguchi and Response Surface Philosophies: A Dual Response Approach. Journal of Quality Technology, 1990. 22(1): pp. 38–45.
32. Quesada, G.M. and E. Del Castillo, A Dual Response Approach to the Multivariate Robust Parameter Design Problem. Technometrics, 2004. 46(2): pp. 176–187.
33. Box, G.E.P., S. Bisgaard, and C. Fung, An Explanation and Critique of Taguchi's Contribution to Quality Engineering. International Journal of Quality and Reliability Management, 1988. 4(2): pp. 123–131.
34. Schmidt, S.R. and R.G. Lausnby, Understanding Industrial Designed Experiments. Fourth Edition. 2005, Colorado Springs, CO: Air Academy Press.
35. Box, G.E.P., Signal to Noise Ratios, Performance Criteria, and Transformations. Technometrics, 1988. 30(1): pp. 1–17.
36. Box, G.E.P. and S. Jones, An Investigation of the Method of Accumulation Analysis. Total Quality Management & Business Excellence, 1990. 1(1): pp. 101–113.
37. Welch, W.J. et al., Computer Experiments for Quality Control by Parameter Design. Journal of Quality Technology, 1990. 22(1): pp. 15–22.
38. Pozueta, L., X. Tort-Martorell, and L. Marco, Identifying Dispersion Effects in Robust Design Experiments - Issues and Improvements. Journal of Applied Statistics, 2007. 34(6): pp. 683–701.
39. Kunert, J. et al., An Experiment to Compare Taguchi's Product Array and the Combined Array. Journal of Quality Technology, 2007. 39(1): pp. 17–34.
40. Ledolter, J. and A. Swersey, Dorian Shainin's Variables Search Procedure: A Critical Assessment. Journal of Quality Technology, 1997. 29(3): pp. 237–247.
41. De Mast, J. et al., Discussion: An Overview of the Shainin SystemTM for Quality Improvement. Quality Engineering, 2008. 20(1): pp. 20–45.
42. Steiner, S.H., J. MacKay, and J. Ramberg, An Overview of the Shainin SystemTM for Quality Improvement. Quality Engineering, 2008. 20(1): pp. 6–19.
43. Tanco, M. et al., Is Design of Experiments Really Used? A Survey of Basque Industries. Journal of Engineering Design, 2008. 19(5): pp. 447–460.
44. Viles, E. et al., Planning Experiments, the First Real Task in Reaching a Goal. Quality Engineering, 2009. 21(1): pp. 44–51.

Chapter 53
Prevention of Workpice Form Deviations in CNC Turning Based on Tolerance Specifications

Gonzalo Valiño Riestra, David Blanco Fernandez, Braulio Jose Álvarez, Sabino Mateos Diaz, and Natalia Beltrán Delgado

Abstract The simultaneous action of cutting forces and gradients of temperature due to different effects in turning, end result in deflections of the machine-workpiece-tool system and in a structural thermal drift. In turn, these factors lead to a deviation of the tool-point away from the theoretical path and consequently, to final dimensional and form errors on the workpiece manufactured. Traditionally, these non-desired deviations have been compensated by on-line corrections of the tool position, even although they were not strictly necessary according to the tolerance specifications on the workpiece drawings. These type of compensations require the use of machine-tools equipped with multiple sensors, already available in modern and precise machine-tools but not always in common ones, which are more reasonable from an economic point of view. On the other hand, on-line compensations modify the predicted working conditions and in turn, chatter or other undesirable effects can appear such as moving away from optimal cutting conditions. Taking this into account, a new approach is proposed in this work for prediction and compensation of deviations in turning by considering them at the planning stages prior to machining operations, mainly focused on development of CAPP systems for CNC turning. Tolerance specifications are considered as limiting values for the admissible workpiece deviations which are predicted by using a developed model based on the strain energy theory.

Keywords CNC turning · form deviation · machining error · tolerance · CAPP

G.V. Riestra (✉)
Department of Manufacturing Engineering, University of Oviedo, Campus of Gijón, Spain
E-mail: gvr@uniovi.es

S.I. Ao and L. Gelman (Eds.), *Advances in Electrical Engineering and Computational Science*, Lecture Notes in Electrical Engineering.

53.1 Introduction

The action of cutting forces in turning results in a deflection of the machine-workpiece-tool system. Moreover, thermal effects such as the increment of temperature at the cutting point, the temperature related to mechanical friction in movements and the environmental temperature also have influence on whole system deflection. In addition, dynamics of the cutting process also affects the accuracy of the manufactured workpiece. As a result, the actual workpiece obtained under certain cutting conditions is different from that expected initially. In this way, both dimensional and form deviations take place since the tool-point does not accurately follow the programmed path. A lack of flatness, cylindricity, coaxiality, perpendicularity, parallelism among other geometrical errors may occur as result, as well as distance and diameter deviations.

Most efforts for dealing with this scenery have focused on the prediction and compensation of workpiece dimensional deviations by means of different techniques. In this way, some authors [1] have implemented functions to modify the programmed depth of cut in the machine-tool numerical control. Others [2, 3] have made predictions of workpiece deflection based on FEM analysis. Only in a few cases [4–7] the influence of the whole machine-part-tool system stiffness has been considered. Taking apart the effect of cutting forces, some authors [8–10] have proposed models for compensation of deviations related to the machine-tool thermal drift.

Although numerous contributions have been made for error prediction and compensation, none of them analyzes whether error compensation enables to meet workpiece tolerances or even if it results really necessary to do any compensation when tight tolerances are not required.

Taking this into account, a new approach to deviations in CNC turning will be proposed in the next sections. The most common errors in turning are initially described and a model for the calculation of deviations is developed based on the strain energy theory. Workpieces with both external and internal features and variable section along the axis of rotation are supported by the model as well as the main clamping methods in turning (on a chuck, between-center and chuck-tailstock) are considered. Quantitative relationships between the different types of machining errors and the different types of tolerances are determined and proposed as constraints for setting up the optimal cutting conditions at the planning stages prior to machining.

53.2 Form Errors in Turning

Once finished machining operations, the deflection due to cutting forces disappears and the elastic recovery of material causes the workpiece axis to return to its original position, but this entails a defect of form in the workpiece surfaces equivalent to the deflection that the axis had undergone. Actually, *turning* of cylindrical or conical

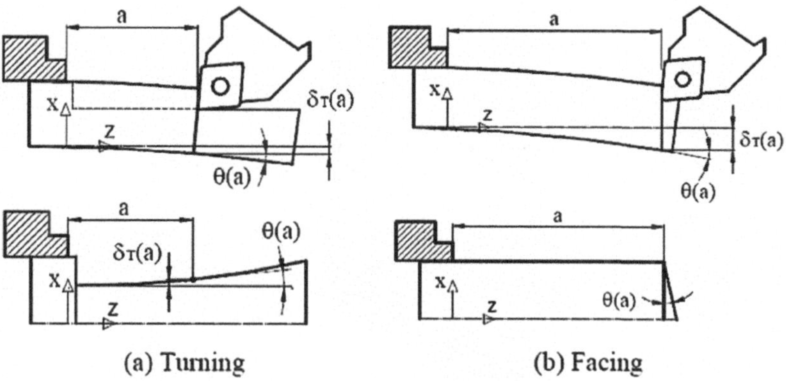

Fig. 53.1 Form errors in turning and facing due to deflections

surfaces leads to a third-order polynomial surface profile [4] (Fig. 53.1a). Similarly, *facing* of front surfaces leads to conical surfaces with a lack of perpendicularity with regard to the symmetry axis (Fig. 53.1b). The conicity of real front faces is directly related to the turn $\theta(z)$ of the section and to the total elastic deflection $\delta_T(z)$. For the section in which the cutting tool is applied ($z = a$) this relation will be:

$$\tan\theta(a) = \frac{d}{dz}\delta_T(a) \tag{53.1}$$

53.3 Calculation of Radial Deviations

The total deviation in the radial direction $\delta_T(z)$ of a turned component at a distance z from the chuck is composed by simultaneous factors related to deformations caused by cutting forces, such as the *spindle-chuck system* $\delta_{sc}(z)$, the *toolpost* $\delta_{tp}(z)$, the *workpiece deflection* $\delta_p(z)$ and the *thermal effects* [5, 11] $\delta_{th}(z)$. This can be expressed as:

$$\delta_T(z) = \delta_{sc}(z) + \delta_{tp}(z) + \delta_p(z) + \delta_{th}(z) \tag{53.2}$$

Deviations can be obtained for the tangential (Z) and radial (X) directions (Fig. 53.2). The contribution of the *spindle-chuck system* can be expressed for these directions [5] as:

$$\delta_{sc\,r}(z) = \left(\frac{F_r}{k_{2r}} + \frac{F_r}{k_{3r}}\right) z^2 + \frac{F_a}{2}\left(\frac{D}{k_{3r}} - \frac{D - 2\,F_r}{k_{2r}}\right) z + \left(\frac{F_r}{k_{1r}} - \frac{F_a\,D\,Lc}{2\cdot k_{2r}} + \frac{F_r\,Lc^2}{k_{2r}}\right) \tag{53.3}$$

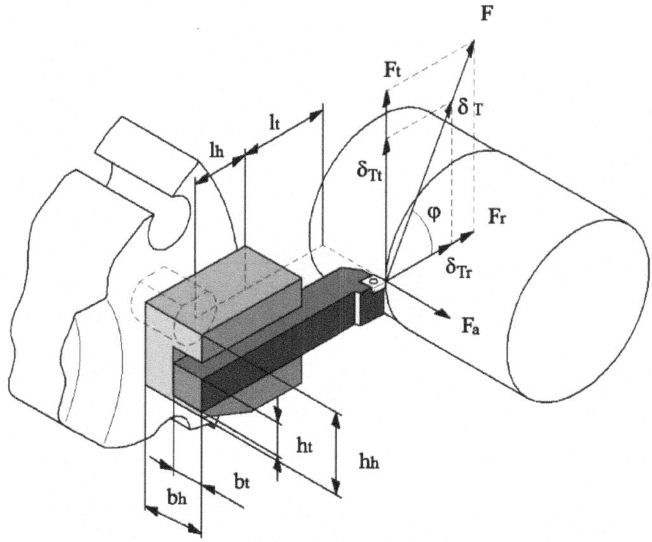

Fig. 53.2 Cutting force and deflection components

$$\delta_{sct}(z) = \left(\frac{F_t}{k_{2t}} + \frac{F_t}{k_{3t}}\right) z^2 + \frac{F_a}{2}\left(\frac{D}{k_{3t}} - \frac{D - 2\,F_t}{k_{2t}}\right) z + \left(\frac{F_t}{k_{1t}} - \frac{F_a\, D\, Lc}{2\, k_{2t}} + \frac{F_t\, Lc^2}{k_{2t}}\right) \tag{53.4}$$

where sub-indices r and t refer to radial and tangential directions respectively; F_r, F_t and F_a are the cutting force components in the radial, tangential and feed directions (Fig. 53.2); D is the machining diameter; k_1, k_2 and k_3 are the stiffness constants for the spring-model of the system [5, 6].

The contribution of the *toolpost* is mainly due to a shortening of the tool and the tool-holder in the radial direction. Therefore, it can be expressed as:

$$\delta_{tp}(z) = -F_r \cdot \left(\frac{l_t}{A_t \cdot E_t} + \frac{l_h}{A_h \cdot E_h}\right) \tag{53.5}$$

where F_r is the radial component of the cutting force, l_t, A_t and E_t the cantilever length, cross section and elastic modulus for the tool respectively, and l_h, A_h and E_h the length, cross section and elastic modulus for the tool-holder (Fig. 53.2).

Based on strain-energy, the contribution of the *workpiece deflection* for the *chuck* clamping method when a cutting force is at $z = a$ (distance between force and chuck) will be:

$$\delta_p(a) = \sum_i \left(\frac{4}{\pi} \cdot \frac{F}{E} \cdot \int_{Zi}^{Zi+1} \frac{(a-z)^2}{r_{i\,\text{ext}}^4 - r_{i\,\text{int}}^4} dz + \frac{\chi}{\pi} \cdot \frac{F}{G} \cdot \int_{Zi}^{Zi+1} \frac{1}{r_{i\,\text{ext}}^2 - r_{i\,\text{int}}^2} dz\right) \tag{53.6}$$

where i represents each zone with constant cross section, E is the elastic modulus, G the shear modulus, I the moment of inertia and χ is the shear factor. For the *between-centre* clamping method, the expression is:

$$\delta_p(a) = \frac{4\,F\,(L-a)^2}{\pi\,L^2\,E} \sum_i \left(\int_{Z_i}^{Z_i+1} \frac{z^2}{r_{i\,\text{ext}}^4 - r_{i\,\text{int}}^4} dz \right) + \frac{\chi\,F\,(L-a)^2}{\pi\,L^2\,G} \sum_i \left(\int_{Z_i}^{Z_i+1} \frac{1}{r_{i\,\text{ext}}^2 - r_{i\,\text{int}}^2} dz \right) \tag{53.7}$$

And in the case of the *chuck-tailstock* clamping, the expression becomes:

$$\delta_p(a) = \int_0^a \frac{\chi \cdot F\,(L-a)^2 \left[3L^2 - (L-a)^2\right]^2}{4L^6 A\,G} dz + \int_0^a \frac{F\,(L-a)^2 \left[\left(2L^2 - a\,(a-2L)\right) z + L \cdot a \cdot (a-2L)\right]^2}{4L^6 E\,I} dz \tag{53.8}$$

Finally, the contribution of the *thermal drift* takes place mainly in the radial direction and depends on factors such as spindle speed, feed rate, machine-tool operation time and ambient temperature. Some authors use these parameters as input data to a neural network, which output is the radial thermal drift. [10, 11]

53.4 Relationship Between Deviations and Tolerances

Although dimensional and geometrical specifications of a part are expressed on drawings by means of dimensional and geometrical tolerances, no relationships were found in literature between form deviations derived from machining process and value of tolerances. Nevertheless, it is essential to know these relationships for setting up the appropriate machining conditions. Therefore, in the next sections each type of tolerance will be analyzed according to ISO standard and geometrically related to the feasible deviation types in turned parts.

53.4.1 Length Tolerance

Due to the form error, the real distance between two front faces depends on the zone in which the measurement is taken. Actual measurements will be between the values L'_{min} and L'_{max} (Fig. 53.3a) and they will be admissible when the following expression is satisfied:

$$L'_{\text{max}} \le L_n + dL_u \text{ and } L'_{\text{min}} \ge L_n + dL_l \tag{53.9}$$

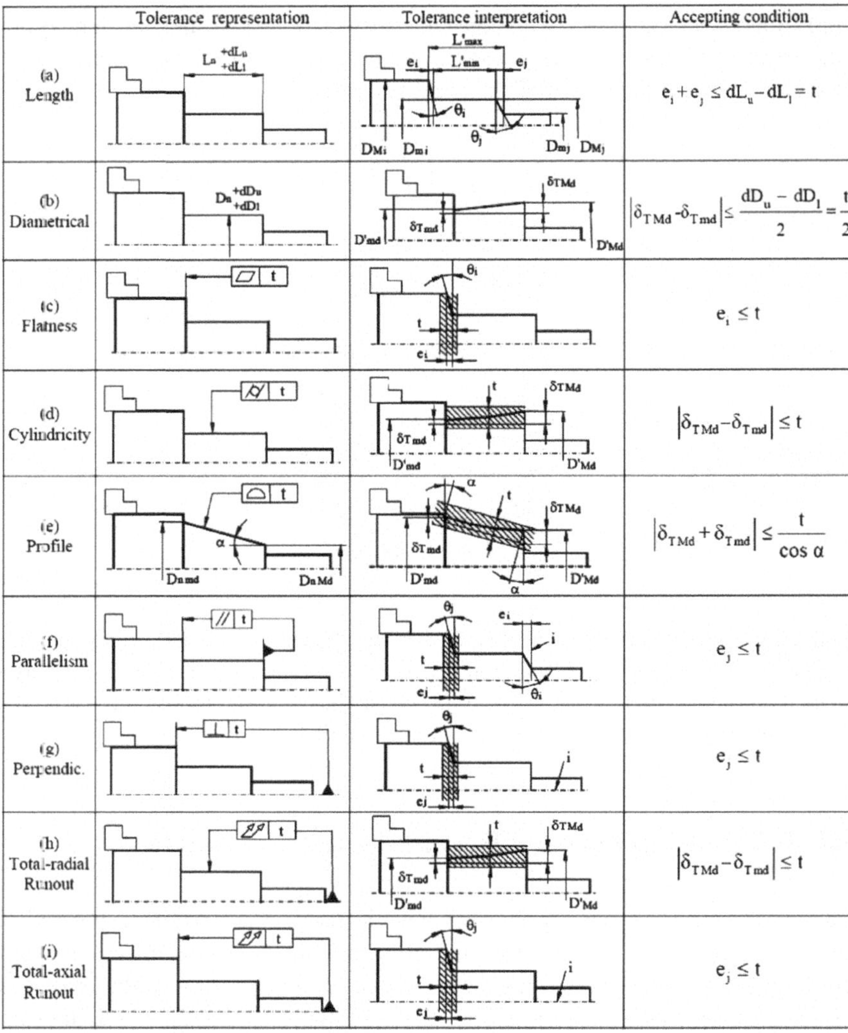

Fig. 53.3 Accepting conditions for deflections according to tolerance values

where L_n is the nominal distance and dL_u and dL_l are the upper and lower limits of longitudinal tolerance, respectively. Considering distances e_i and e_j in Fig. 53.3a, the following relation can be established:

$$L'_{max} - L'_{min} = e_i + e_j \tag{53.10}$$

Taking into account Eqs. (53.9)–(53.10) and that the tolerance zone t is the difference between the upper and lower limits, the following is the condition for the measurement to meet the tolerance:

$$e_i + e_j \leq dL_u - dL_l = t \tag{53.11}$$

where distances e_i and e_j can be expressed in terms of the turn of sections θ_i and θ_j:

$$e_i = \frac{D_{Mi} - D_{mi}}{2} \cdot \tan\theta_i \text{ and } e_j = \frac{D_{Mj} - D_{mj}}{2} \cdot \tan\theta_j \quad (53.12)$$

53.4.2 Diametrical Tolerance

Due to the form error of the cylindrical surfaces, the diameter of the workpiece depends on the location of the measurement point, varying between a maximum and a minimum value (D'_{Md} and D'_{md}) (Fig. 53.3b). The measurement is admissible when the following expression is satisfied:

$$D'_{Md} \leq D_n + dD_u \text{ and } D'_{md} \geq D_n + dD_l \quad (53.13)$$

where D_n is the nominal diameter and dD_u and dD_l are the upper and lower limits of the diametrical tolerance, respectively.

Once machined the cylindrical surface, the maximum error is given by the difference between the maximum δ_{TMd} and the minimum δ_{Tmd} deflections. The following relation can be deduced from geometry (Fig. 53.3b):

$$D'_{Md} - D'_{md} = 2\left|\delta_{TMd} - \delta_{Tmd}\right| \quad (53.14)$$

By considering Eqs. (53.13)–(53.14) and that the tolerance zone t is the difference between the upper and lower deviations, the following is the condition to meet the tolerance:

$$2\left|\delta_{TMd} - \delta_{Tmd}\right| \leq dD_u - dD_l = t \quad (53.15)$$

53.4.3 Geometrical Tolerance of Flatness

Definition of each type of geometrical tolerance has been considered based on ISO standard (ISO 1101:1983). According to that, the flatness tolerance establishes a zone of acceptance t with respect to the nominal orientation of the controlled face, within which the real machined surface must stand. In order to satisfy this condition, the form error must be lower than the width of the tolerance zone (Fig. 53.3c). Considering this and being distance e_i calculated as in Eq. (53.12), the condition will be:

$$e_i \leq t \quad (53.16)$$

53.4.4 Geometrical Tolerance of Cylindricity

The cylindricity tolerance establishes a volume between two coaxial cylinders whose difference of radii is the value of the tolerance zone t. The machined surface must lie between these two cylinders.

The maximum error obtained when machining the cylindrical surface is given by the difference between the maximum δ_{TMd} and minimum δ_{Tmd} deviations. Being D'_{Md} and D'_{md} the maximum and minimum surface diameters, the following relations can be deduced from geometry (Fig. 53.3d):

$$D'_{Md} - D'_{md} \leq 2 \cdot t \text{ and } D'_{Md} - D'_{md} = 2 \cdot |\delta_{TMd} - \delta_{Tmd}| \tag{53.17}$$

Therefore, it is obtained the condition:

$$|\delta_{TMd} - \delta_{Tmd}| \leq t \tag{53.18}$$

53.4.5 Geometrical Tolerance of Profile of a Surface

This tolerance specifies a zone between two surrounding spherical surfaces whose diameter difference is the value of the tolerance zone t and whose centres are located onto the theoretical surface. According to norm ISO 3040:1990, for the case of a conical surface, the tolerance zone becomes the space between two cones of the same angle than the datum and equidistant to that cone the half of the tolerance value.

All diameters of the real machined surface must lie within the tolerance zone, including the diameters in which deviation is maximum D'_{Md} and minimum D'_{md}, which will satisfy:

$$\frac{|D'_{Md} - D_{nMd}|}{2} \leq \frac{t}{2 \cdot \cos\alpha} \text{ and } \frac{|D'_{md} - D_{nmd}|}{2} \leq \frac{t}{2 \cdot \cos\alpha} \tag{53.19}$$

where D_{nMd} and D_{nmd} are the nominal diameters of the cone measured in the positions of maximum and minimum diametrical deviation (Fig. 53.3e).

On the other hand, workpiece deflections in the zones of maximum and minimum deviations (δ_{TMd} and δ_{Tmd}) can be expressed as:

$$\delta_{TMd} = \frac{|D'_{Md} - D_{nMd}|}{2} \text{ and } \delta_{Tmd} = \frac{|D'_{md} - D_{nmd}|}{2} \tag{53.20}$$

The final condition is derived from Eqs. (53.19–53.20):

$$\delta_{TMd} + \delta_{Tmd} \leq \frac{t}{\cos\alpha} \tag{53.21}$$

53.4.6 Geometrical Tolerance of Parallelism

Parallelism is applied between front faces in turned parts. Let be i the datum and j the related face. According to the definition, the related surface must lie within an space between two planes parallel to the datum and distanced one another the value of the tolerance t (Fig. 53.3f). Although even the datum has a form error, according to norm ISO 5459:1981, this surface will be considered perpendicular to the part axis and, consequently, also the planes which define the tolerance zone. Considering this, and calculating e_j as in Eq. (53.12), the geometrical condition to meet the tolerance will be:

$$e_j \leq t \tag{53.22}$$

53.4.7 Geometrical Tolerance of Perpendicularity

Perpendicularity is applied between a front face and a feature axis in turned parts. Two different situations must be considered depending on which of both elements is the datum and which the controlled one.

When *the axis is the datum and a face is controlled*, the tolerance zone is the space between two planes perpendicular to the datum axis and distanced one another the value of the tolerance t (Fig. 53.3g). Considering this, and calculating e_j as in Eq. (53.12), the condition will be same than in Eq. (53.22).

On the other hand, when *a face is the datum and the axis is controlled*, the tolerance zone is the space within a cylinder of diameter t, and axis perpendicular to the datum face. The elastic recovery of the part axis after machining implies that any deviation of this element does not depend directly on the cutting action but on other technological aspects such as a bad alignment of the workpiece in the lathe. Therefore, no relation can be established between this error and the deviation caused by cutting forces.

53.4.8 Geometrical Tolerance of Coaxiality

This tolerance limits the relative deviation between two zones of the part axis. As in the previous case, the elastic recovery of the workpiece axis after machining implies that the possible error is not due to deviations caused by cutting forces but to other causes.

53.4.9 Geometrical Tolerance of Circular Runout

For *circular-radial runout*, the tolerance zone is the area between two concentric circles located into a plane perpendicular to the axis, whose radii difference is the tolerance t and whose centre is located at datum axis. Since form errors derived from cutting forces are completely symmetrical with respect to the workpiece rotation axis, the error evaluated throughout this tolerance does not depend on these forces, but on the workpiece deflection caused by the clamping force or by a lack of workpiece alignment.

A similar situation takes place regarding *circular-axial runout*.

53.4.10 Geometrical Tolerance of Total Runout

The *total-radial runout* is used to control cumulative variations of circularity and cylindricity of surfaces constructed around a datum axis. As in the case of *circular-radial runout*, circularity does not depend on deviations from cutting forces but cylindricity does. Therefore, the condition to be satisfied in this case is the same than in Eq. (53.18) (Fig. 53.3h).

The *total-axial runout* controls cumulative variations of perpendicularity and flatness of front faces at a right angle to the datum axis. Therefore, the relation of this error with deviations coincides with expressions obtained for flatness and perpendicularity respectively (Fig. 53.3i).

53.5 Deviations as Optimization Conditions

The optimization of cutting conditions in turning is the final stage of process planning in which not only mathematical considerations about the objective function have to be done (e.g., time, cost or benefit) but also there are several constraints that restrict the best solution.

Common constraints are related to ranges of cutting parameters (cutting speed, feed-rate and depth of cut), ranges of tool-life and other operating limits such as surface finish, maximum power consumption and maximum force allowed. The cutting force constraint is imposed to limit the deflection of the workpiece or cutting tool, which result in dimensional error, and to prevent chatter. [12] Traditionally, the value of this constraint has not been clearly determined. Nevertheless, the relationships between workpiece deviations and tolerances described in the previous sections can also be considered as relationships between cutting forces and maximum deviations allowed and, therefore, all together can be used as optimization constraints.

53.6 Conclusions

This work provides a mathematical model based on strain energies for prediction of deviations in the turning process of workpieces with complex external and/or internal geometry and which takes into account the main clamping procedures in turning.

Likewise, an analysis of maximum deviations is developed according to each tolerance specification and an accepting criterion is proposed in each case so that compensation of the calculated errors must be carried out only when tolerances are not met. This becomes a different approach with regard to other works developed up to date which propose to make error compensations in all cases, even when they are not necessary according to tolerances. Moreover, the results of this analysis are utilized by a turning CAPP system as constraints for the optimization of cutting conditions and it also can be useful for deciding the most suitable clamping method for setting up the workpiece on a lathe.

Acknowledgements This work is part of a research project supported by the Spanish Education and Science Ministry (DPI2007-60430) and FEDER.

References

1. S. Yang, J. Yuan, and J. Ni, Real-time cutting force induced error compensation on a turning center, Int. J. Mach. Tools Manuf. **37**(11), 1597–1610 (1997).
2. A.-V. Phan, L. Baron, J.R. Mayer, and G. Cloutier, Finite element and experimental studies of diametrical errors in cantilever bar turning, J. App. Math. Model. **27**(3), 221–232 (2003).
3. L.Z. Qiang, Finite difference calculations of the deformations of multi-diameter workpieces during turning, J. Mat. Proc. Tech. **98**(3), 310–316 (2000).
4. L. Carrino, G. Giorleo, W. Polini, and U. Prisco, Dimensional errors in longitudinal turning based on the unified generalized mechanics of cutting approach. Part II: Machining process analysis and dimensional error estimate, Int. J. Mach. Tools Manuf. **42**(14), 1517–1525 (2002).
5. G. Valiño, S. Mateos, B.J. Álvarez, and D. Blanco, Relationship between deflections caused by cutting forces and tolerances in turned parts, Proceedings.of the 1st Manufacturing Engineering Society International Conference (MESIC), Spain (2005).
6. D. Mladenov, Assessment and compensation of errors in CNC turning, Ph.D. thesis, UMIST, UK (2002).
7. S. Hinduja, D. Mladenov, and M. Burdekin, Assessment of force-induced errors in CNC turning, Annals of the CIRP **52**(1), 329–332 (2003).
8. J. Yang, J. Yuan, and J. Ni, Thermal error mode analysis and robust modelling for error compensation on a CNC turning center, Int. J. Mach. Tools Manuf. **39**(9), 1367–1381 (1999).
9. V.A. Ostafiev and A. Djordjevich, Machining precision augmented by sensors, Int. J. Prod. Res. **37**(1), 1999, pp. 91–98.
10. X. Li, P.K. Venuvinod, A. Djorjevich, and Z. Liu, Predicting machining errors in turning using hybrid learning, Int. J. Adv. Manuf. Tech. **18**, 863–872 (2001).
11. X. Li and R. Du, Analysis and compensation of workpiece errors in turning, Int. J. of Prod. Res. **40**(7), 1647–1667 (2002).
12. Y.C. Shin and Y.S. Joo, Optimization of machining conditions with practical constraints, Int. J. Prod. Res. **30**(12), 2907–2919 (1992).

Chapter 54
Protein–Protein Interaction Prediction Using Homology and Inter-domain Linker Region Information

Nazar Zaki

Abstract One of the central problems in modern biology is to identify the complete set of interactions among proteins in a cell. The structural interaction of proteins and their domains in networks is one of the most basic molecular mechanisms for biological cells. Structural evidence indicates that, interacting pairs of close homologs usually interact in the same way. In this chapter, we make use of both homology and inter-domain linker region knowledge to predict interaction between protein pairs solely by amino acid sequence information. High quality core set of 150 yeast proteins obtained from the Database of Interacting Proteins (DIP) was considered to test the accuracy of the proposed method. The strongest prediction of the method reached over 70% accuracy. These results show great potential for the proposed method.

Keywords Protein–Protein Interaction · Prediction · Homology · Inter-domain Linker Region · yeast protein

54.1 Introduction

54.1.1 The Importance of Protein–Protein Interaction

The term protein–protein interaction (PPI) refers to the association of protein molecules and the study of these associations from the perspective of biochemistry, signal transduction and networks. PPIs occur at almost every level of cell function, in the structure of sub-cellular organelles, the transport machinery across

---

N. Zaki
Assistant Professor with the College of Information Technology, UAE University.
Al-Ain 17555, UAE,
E-mail: nzaki@uaeu.ac.ae

S.I. Ao and L. Gelman (Eds.), *Advances in Electrical Engineering and Computational Science*, Lecture Notes in Electrical Engineering.

the various biological membranes, the packaging of chromatin, the network of submembrane filaments, muscle contraction, signal transduction, and regulation of gene expression, to name a few [1]. Abnormal PPIs have implications in a number of neurological disorders; include Creutzfeld-Jacob and Alzheimer's diseases. Because of the importance of PPIs in cell development and disease, the topic has been studied extensively for many years. A large number of approaches to detect PPIs have been developed. Each of these approaches has strengths and weaknesses, especially with regard to the sensitivity and specificity of the method.

54.1.2 Current Methods to Predict PPI

One of the major goals in functional genomics is to determine protein interaction networks for whole organisms, and many of the experimental methods have been applied to study this problem. Co-immunoprecipitation is considered to be the gold standard assay for PPIs, especially when it is performed with endogenous proteins [2]. Yeast two-hybrid screen investigates the interaction between artificial fusion proteins inside the nucleus of yeast [3]. Tandem Affinity Purification (TAP) detects interactions within the correct cellular environment [4]. Quantitative immunoprecipitation combined with knock-down (QUICK) detects interactions among endogenous non-tagged proteins [5]. High-throughput methods have also contributed tremendously in the creation of databases containing large sets of protein interactions, such as Database of Interacting Proteins (DIP) [6], MIPS [7] and Human Protein Reference Database (HPRD) [8].

In addition, several *in silico* methods have been developed to predict PPI based on features such as gene context [9]. These include gene fusion [10], gene neighborhood [11] and phylogenetic profile [12]. However, most of the *in silico* methods seek to predict functional association, which often implies but is not restricted to physical binding.

Despite the availability of the mentioned methods of predicting PPI, the accuracy and coverage of these techniques have proven to be limited. Computational approaches remain essential both to assist in the design and validation of the experimental studies and for the prediction of interaction partners and detailed structures of protein complexes [13].

54.1.3 Computational Approaches to Predict PPI

Some of the earliest techniques predict interacting proteins through the similarity of expression profiles [14], description of similarity of phylogenetic profiles [12] or phylogenetic trees [15], and studying the patterns of domain fusion [16]. However, it has been noted that these methods predict PPI in a general sense, meaning joint involvement in a certain biological process, and not necessarily actual physical interaction [17].

Most of the recent works focus on employing protein domain knowledge [18–22]. The motivation for this choice is that molecular interactions are typically mediated by a great variety of interacting domains [23]. It is thus logical to assume that the patterns of domain occurrence in interacting proteins provide useful information for training PPI prediction methods [24].

An emerging new approach in the protein interactions field is to take advantage of structural information to predict physical binding [25, 26]. Although the total number of complexes of known structure is relatively small, it is possible to expand this set by considering evolutionary relationships between proteins. It has been shown that in most cases close homologs (>30% sequence identity) physically interact in the same way with each other. However, conservation of a particular interaction depends on the conservation of the interface between interacting partners [27].

In this chapter, we propose to predict PPI using only sequence information. The proposed method combines homology and structural relationships. Homology relationships will be incorporated by measuring the similarity between protein pair using Pairwise Alignment. Structural relationships will be incorporated in terms of protein domain and inter-domain linker region information. We are encouraged by the fact that compositions of contacting residues in protein sequence are unique, and that incorporating evolutionary and predicted structural information improves the prediction of PPI [28].

54.2 Method

In this work, we present a simple yet effective method to predict PPI solely by amino acid sequence information. The overview of the proposed method is illustrated in Fig. 54.1. It consists of three main steps: (a) extract the homology relationship by measuring regions of similarity that may reflect functional, structural or evolutionary relationships between protein sequences (b) downsize the protein sequences by predicting and eliminating inter-domain linker regions (c) scan and detect domain matches in all the protein sequences of interest.

54.2.1 Similarity Measures Between Protein Sequences

The proposed method starts by measuring the PPI sequence similarity, which reflects the evolutionary and homology relationships. Two protein sequences may interact by the mean of the amino acid similarities they contain [24]. This work is motivated by the observation that an algorithm such as Smith-Waterman (SW) [29], which measures the similarity score between two sequences by a local gapped alignment, provides relevant measure of similarity between protein sequences. This

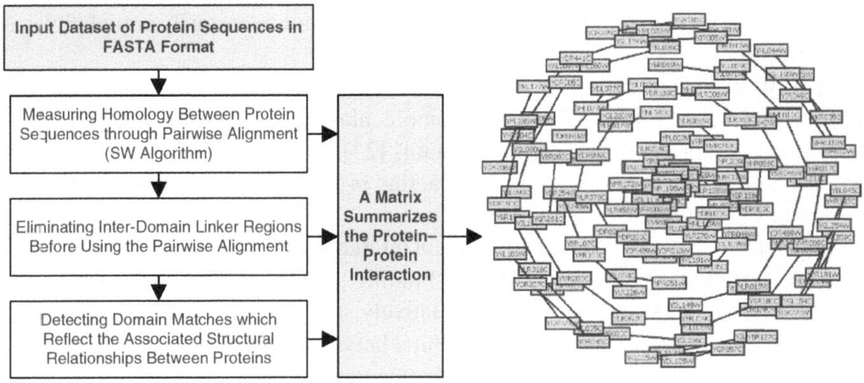

Fig. 54.1 Overview of the proposed method

similarity incorporates biological knowledge about protein evolutionary structural relationships [30].

The Smith-Waterman similarity score $SW(x_1, x_2)$ between two protein sequences x_1 and x_2 is the score of the best local alignment with gaps between the two protein sequences computed by the Smith-Waterman dynamic programming algorithm. Let us denote by μ a possible local alignment between x_1 and x_2, defined by a number n of aligned residues, and by the indices $1 \leq i_1 < ... < i_n \leq |x_1|$ and $1 \leq j_1 < ... < j_n \leq |x_2|$ of the aligned residues in x_1 and x_2 respectively. Let us also denote by $\prod(x_1, x_2)$ the set of all possible local alignments between x_1 and x_2, and by $p(x_1, x_2, \mu)$ the score of the local alignment $\mu \in \prod(x_1, x_2)$ between x_1 and x_2, the Smith-Waterman score $SW(x_1, x_2)$ between sequences x_1 and x_2 can be written as:

$$SW(x_1, x_2) = \max_{\mu \in \prod(x_1, x_2)} p(x_1, x_2, \mu) \tag{54.1}$$

The similarity matrix M can be calculated as follow:

$$M = \begin{bmatrix} SW(x_1, x_1) & SW(x_1, x_2) & ... & SW(x_1, x_m) \\ SW(x_2, x_1) & SW(x_2, x_2) & ... & SW(x_2, x_m) \\ \vdots & \vdots & \vdots & \vdots \\ SW(x_m, x_1) & SW(x_m, x_2) & ... & SW(x_m, x_m) \end{bmatrix} \tag{54.2}$$

where m is the number of the protein sequences. For example, suppose we have the following randomly selected PPI dataset: YDR190C, YPL235W, YDR441C, YML022W, YLL059C, YML011C, YGR281W and YPR021C represented by x_1, x_2, x_3, x_4, x_5, x_6, x_7 and x_8 respectively. The interaction between these eight proteins is shown in Fig. 54.2.

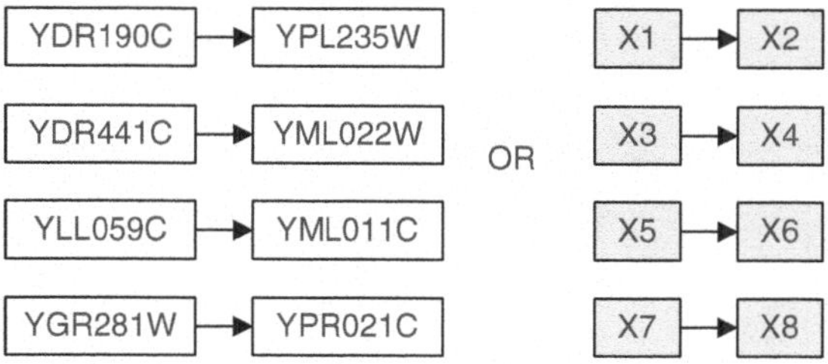

Fig. 54.2 The interaction between the randomly selected proteins

Then the SW similarity score matrix M will be calculated as:

	x_1	x_2	x_3	x_4	x_5	x_6	x_7	x_8
x_1	X	465	28	30	25	30	34	29
x_2	465	X	30	24	32	33	50	47
x_3	28	30	X	553	29	27	32	29
x_4	30	24	553	X	29	20	25	40
x_5	25	32	29	29	X	24	28	49
x_6	30	33	27	20	24	X	25	26
x_7	34	50	32	25	28	25	X	36
x_8	29	47	29	40	49	26	36	X

From M, higher score may reflect interaction between two proteins. $SW(x_1, x_2)$ and $SW(x_2, x_1)$ scores are equal to 465; $SW(x_3, x_4)$ and $SW(x_4, x_3)$ scores are equal to 553, which confirm the interaction possibilities. However, $SW(x_5, x_6)$ and $SW(x_6, x_5)$ scores are equal to 24; $SW(x_7, x_8)$ and $SW(x_8, x_7)$ scores are equal to 36, which are not the highest scores. To correct these errors more biological information is needed, which lead us to the second part of the proposed method.

54.2.2 Identify and Eliminate Inter-domain Linker Regions

The results could be further enhanced by incorporating inter-domain linker regions knowledge. The next step of our algorithm is to predict inter-domain linker regions solely by amino acid sequence information. Our intention here is to identify and eliminate all the inter-domain linker regions from the protein sequences of interest. By doing this step, we are actually downsizing the protein sequence to shorter ones with only domains information to yield better alignment scores. In this case, the prediction is made by using linker index deduced from a data set of domain/linker segments from SWISS-PROT database [31]. DomCut developed by Suyama et al. [32] is employed to predict linker regions among functional domains based on the difference in amino acid composition between domain and linker regions. Following

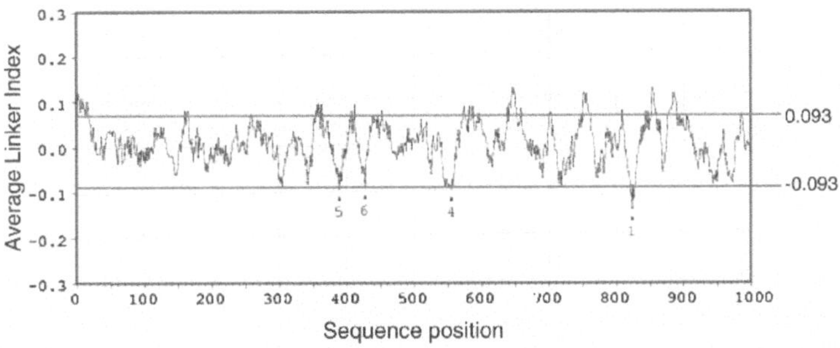

Fig. 54.3 An example of linker preference profile generated using Domcut. In this case, linker regions greater than the threshold value 0.093 are eliminated from the protein sequence

Suyama's method [32], we defined the linker index S_i for amino acid residue i and it is calculated as follows:

$$S_i = -\ln\left(\frac{f_i^{Linker}}{f_i^{Domain}}\right) \tag{54.3}$$

Where f_i^{Linker} is the frequency of amino acid residue i in the linker region and f_i^{Domain} is the frequency of amino acid residue$i$ in the domain region. The negative value of S_i means that the amino acid preferably exists in a linker region. A threshold value is needed to separate linker regions as shown in Fig. 54.3. Amino acids with linker score greater than the set threshold value will be eliminated from the protein sequence of interest.

When applying the second part of the method, the matrix M will be calculated as follows:

	x_1	x_2	x_3	x_4	x_5	x_6	x_7	x_8
x_1	X	504	30	30	25	32	34	27
x_2	504	X	30	21	32	32	50	36
x_3	30	30	X	775	29	24	38	29
x_4	30	21	775	X	19	21	53	37
x_5	25	32	29	19	X	28	28	24
x_6	32	32	24	21	28	X	23	27
x_7	34	50	38	53	28	23	X	339
x_8	27	36	29	37	24	27	339	X

From$M$, it's clearly noted that, more evidence is shown to confirm the interaction possibility between proteins x_7 and x_8, and therefore, the result is furthermore enhanced. In the following part of the method, protein domain knowledge will be incorporated in M for better accuracy.

54.2.3 Detecting Domain Matches and Associated Structural Relationships in Proteins

In this part of the method, protein domains knowledge will be incorporated in M. Protein domains are highly informative for predicting PPI as they reflect the potential structural relationships between them. In this implementation, we employed ps_scan [33] to scan one or several patterns, rules and profiles from PROSITE against our protein sequences of interest. Running ps_scan through the eight proteins identifies the following domains:

DR441C (x_3) → PS00103
YML022W (x_4) → PS00103
YGR281W (x_7) → PS00211, PS50893 and PS50929
YPR021C (x_8) → PS50929

Which reveals structural relationships between proteins x_3 and x_4; and proteins x_7 and x_8. Based on this relationship, $SW(x_3, x_4)$ and $SW(x_7, x_8)$ scores will be incremented by specified value (We have randomly selected a value equal to 300). Unfortunately, these results have not added more accuracy in this case, however, it confirmed the interacting possibilities between proteins x_3 and x_4; x_7 and x_8.

54.3 Experimental Work

To test the method, we obtained the PPI data from the Database of Interacting Proteins (DIP). The DIP database catalogs experimentally determined interactions between proteins. It combines information from a variety of sources to create a single, consistent set of PPIs in *Saccharomyces cerevisiae*. Knowledge about PPI networks is extracted from the most reliable, core subset of the DIP data [34]. The DIP version we used contains 4,749 proteins involved in 15,675 interactions for which there is domain information [6]. However, only high quality core set of 2,609 yeast proteins was considered in this experimental work. This core set is involved in 6,355 interactions, which have been determined by at least one small-scale experiment or two independent experiments [35]. Furthermore, we selected proteins interacts with only one protein and not involved in any other interactions. This process results in a dataset of 150 proteins with 75 positive interactions as shown in Fig. 54.4. The intention is to design a method capable of predicting protein interaction partner, which facilitate a way to construct PPI using only protein sequences information.

We started our experimental work by measuring the protein–protein sequence interaction similarity using SW algorithm as implemented in FASTA [36]. The default parameters are used: gap opening penalty and extension penalties of 13 and 1, respectively, and a substitution matrix BLOSUM62 matrix. The choice of which substitution matrix to use is not trivial because there is no correct scoring scheme for all circumstances. The BLOSUM matrix is another very common used

YBL045C	YPR191W	YLR456W	YPR172W	YGL057C	YJL135W	YLL059C	YML011C	YPL003W	YPR066W
YBR127C	YDL185W	YNL007C	YIR040C	YGL090W	YOR005C	YLR036C	YKR065C	YPL209C	YBR156C
YDR045C	YOR207C	YNL329C	YKL197C	YGL174W	YIR005W	YLR065C	YDL149W	YPR046W	YJR135C
YDR190C	YPL235W	YOR136W	YNL037C	YGL195W	YFR009W	YLR226W	YPR161C	YPR051W	YEL053C
YDR441C	YML022W	YPL195W	YJL024C	YGL254W	YGL154C	YLR240W	YBR097W	YBR107C	YDR254W
YEL041W	YJR049C	YPR029C	YLR170C	YGR057C	YKL015W	YLR317W	YNL140C	YDR080W	YDL077C
YER017C	YMR089C	YBR228W	YLR135W	YGR074W	YKL183W	YLR366W	YMR242C	YER069W	YJL071W
YGR180C	YJL026W	YDR001C	YLR270W	YGR208W	YKL177W	YLR417W	YPL002C	YER090W	YKL211C
YGR240C	YMR205C	YDR013W	YDR489W	YGR229C	YGR185C	YML119W	YLL032C	YGL008C	YCR024C-A
YGR261C	YBR288C	YDR086C	YLR378C	YHL044W	YKR035C	YMR052W	YFR008W	YGL236C	YMR023C
YHL027W	YJL056C	YDR098C	YGL220W	YHR193C	YDR252W	YMR228W	YFL036W	YGR075C	YBR152W
YHR024C	YLR163C	YDR139C	YLR306W	YJL006C	YML112W	YNL311C	YKL001C	YHR00[illegible] 0.093	[illegible]AL009W
YHR056C	YDR303C	YDR140W	YNR046W	YJL035C	YLR316C	YOL108C	YDR123C	YKL182W	YPL231W
YIL103W	YKL191W	YDR469W	YLR015W	YJL090C	YKL108W	YOL111C	YOR007C	YLR075W	YIR012W
YLR238W	YDR200C	YER159C	YDR397C	YKL160W	YKL036C	YOR269W	YLR254C	YNL259C	YDR270W

Fig. 54.4 Dataset of core interaction proteins used in the experimental work

amino acid substitution matrix that depends on data from actual substitutions. This procedure produces the matrix $M_{150X150}$. This matrix was then enhanced by incorporating inter-domain linker regions information. In this case, only well defined domains with sequence length ranging from 50 to 500 residues were considered. We skipped all the frequently matching (*unspecific*) domains. A threshold value of 0.093 is used to separate the linker regions. Any residue generates an index greater than the threshold value results in eliminating it. This procedure downsized the protein sequences without losing the biological information. In fact, running the SW algorithm on a sequence having pure domains, results in better accuracy. A linker preference profile is generated using the linker index values along an amino acid sequence using a siding window. A window of size $w = 15$ was used which gives the best performance (*different window sizes were tested*).

Further more, protein domains knowledge will be incorporated in $M_{150X150}$. In this implementation, ps_scan [33] is used to scan one or several patterns from PROSITE against the 150 protein sequences. All frequently matching (*unspecific*) patterns and profiles are skipped. The $M_{150X150}$ is then used to predict the PPI network.

54.4 Results and Discussion

The performance of the proposed method is measured by how well it can predict the PPI network. Prediction accuracy, whose value is the ratio of the number of correctly predicted interactions between protein pairs to the total number of interactions and non-interactions possibilities in network, is the best index for evaluating the performance of a predictor. However, approximately 20% of the data are truly interacting proteins, which leads to a rather unbalanced distribution of interacting and non-interacting cases. Suppose we have six proteins, then the interacting pairs are ^{1}®^{2}, ^{3}®^{4} and ^{5}®^{6}, which result in three interactions cases out of a total of 15 interaction possibilities (*12 non-interactions*).

To assess our method objectively, another two indices are introduced in this paper, namely specificity and sensitivity commonly used in the evaluation of

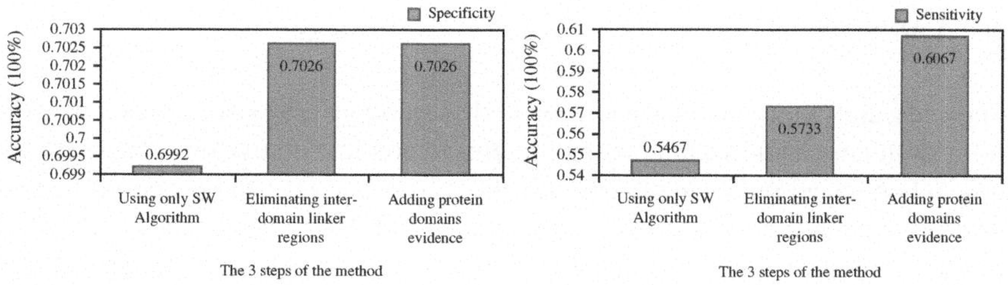

Fig. 54.5 Sensitivity and specificity results

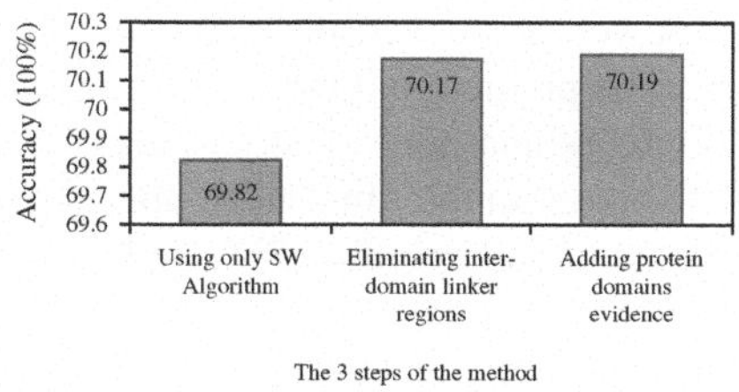

Fig. 54.6 Overall accuracy

Table 54.1 Overall performance evaluation

	tp	*fp*	*tn*	*fn*	Sens.	Spec.	Accuracy
Similarity using SW algorithm	82	68	15,523	6,677	0.5467	0.6992	60.82
Inter-domain linker regions	86	64	15,597	6,603	0.5733	0.7026	70.17
Structural domain evidence	91	59	15,597	6,603	0.6067	0.7026	70.19

information retrieval. A high sensitivity means that many of the interactions that occur in reality are detected by the method. A high specificity indicates that most of the interactions detected by the screen are also occurring in reality. Sensitivity and specificity are combined measures of true positive (*tp*), true negative (*tn*), false positive (*fp*) and false negative (*fn*) and can be expressed as:

Based on these performance measures, the method was able to achieve encouraging results. In Figs. 54.5 and 54.6, we summarized the sensitivity and specificity results based on the three stages of the method. The figures clearly show improvement in sensitivity but not much in specificity and that's because of the big number of non-interacting possibilities.

The overall performance evaluation results are summarized in Table 54.1.

54.5 Conclusion

In this research work we make use of both homology and structural similarities among domains of known interacting proteins to predict putative protein interaction pairs. When tested on a sample data obtained from the DIP, the proposed method shows great potential and a new vision to predict PPI. It proves that the combination of methods predicts domain boundaries or linker regions from different aspects and the evolutionary relationships would improve accuracy and reliability of the prediction as a whole. However, it is difficult to directly compare the accuracy of our proposed method because all of the other existing methods use different criteria for assessing the predictive power. Moreover, these existing methods use completely different characteristics in the prediction. One of the immediate future works is to consider the entire PPI network and not to restrict our work on binary interaction. Other future work will focus on employing more powerful domain linker region identifier such as profile domain linker index (PDLI) [37].

References

1. I. Donaldson, J. Martin, B. Bruijn, C. Wolting, V. Lay, B. Tuekam, S. Zhang, B. Baskin, G. D. Bader, K. Michalickova, T. Pawson, and C.W. Hogue, PreBIND and Textomy – mining the biomedical literature for protein-protein interactions using a support vector machine, *BMC Bioinformatics*, **4**(11) (2003).
2. E. Gharakhanian, J. Takahashi, J. Clever, and H. Kasamatsu, In vitro assay for protein–protein interaction: carboxyl-terminal 40 residues of simian virus 40 structural protein VP3 contain a determinant for interaction with VP1, *PNAS*, **85**(18), 6607–6611 (1998).
3. P. L. Bartel and S. Fields, *The yeast two-hybrid system.* In *Advances in Molecular Biology*, Oxford University Press, New York, 1997.
4. G. Rigaut, A. Shevchenko, B. Rutz, M. Wilm, M. Mann, and B. Seraphin, A generic protein purification method for protein complex characterization and proteome exploration, *Nature Biotechnology*, **17**, 1030–1032 (1999).
5. M. Selbach and M. Mann, Protein interaction screening by quantitative immunoprecipitation combined with knockdown (QUICK), *Nature Methods*, **3**, 981–983 (2006).
6. L. Salwinski, C. S. Miller, A. J. Smith, F. K. Pettit, J. U. Bowie, and D. Eisenberg, The Database of Interacting Proteins: 2004 update, *Nucleic Acids Research*, **1**(32), 449–51 (2004).
7. H. W. Mewes, MIPS: analysis and annotation of proteins from whole genomes, *Nucleic Acids Research*, **32**, 41–44 (2004).
8. S. Peri, Human protein reference database as a discovery resource for proteomics, *Nucleic Acids Research*, **32**, 497–501 (2004).
9. J. Espadaler, Detecting remotely related proteins by their interactions and sequence similarity, *Proceedings of the National Academy of Sciences USA*, **102**, 7151–7156 (2005).
10. E. Marcotte, Detecting protein function and protein–protein interactions from genome sequences, *Science*, **285**, 751–753 (1999).
11. T. Dandekar, Conservation of gene order: a fingerprint of proteins that physically interact, *Trends in Biochemical Sciences*, **23**, 324–328 (1998).
12. M. Pellegrini, E. M. Marcotte, M. J. Thompson, D. Eisenberg, and T. O. Yeates, Assigning protein functions by comparative genome analysis: protein phylogenetic profiles, *Proceedings of National Academy of Sciences USA*, **96**, 4285–4288 (1999).
13. A. Szilàgyi, V. Grimm, A. K. Arakaki, and J. Sholnick, Prediction of physical protein-protein interactions, *Physical Biology*, **2**, 1–16 (2005).

14. E. M. Marcotte, M. Pellegrini, M. J. Thompson, T. O. Yeates, and D. Eisenberg, A combined algorithm for genome-wide prediction of protein function, *Nature*, **402**, 83–86 (1999).
15. F. Pazos and A. Valencia, Similarity of phylogenetic trees as indicator of protein-protein interaction, *Protein Engineering*, **14**, 609–614 (2001).
16. J. Enright, I. N. Ilipoulos, C. Kyrpides, and C. A. Ouzounis, Protein interaction maps for complete genomes based on gene fusion events, *Nature*, **402**, 86–90 (1999).
17. D. Eisenberg, E. M. Marcotte, I. Xenarios, and T. O. Yeates, Protein function in the post-genomic era, *Nature*, **405**, 823–826 (2000).
18. J. Wojcik and V. Schachter, Protein-Protein interaction map inference using interacting domain profile pairs, *Bioinformatics*, **17**, 296–305 (2001).
19. W. K. Kim, J. Park, and J. K. Suh, Large scale statistical prediction of protein-protein interaction by potentially interacting domain (PID) pair, *Genome Informatics*, **13**, 42–50 (2002).
20. S. K. Ng, Z. Zhang, and S. H. Tan, integrative approach for computationally inferring protein domain interactions, *Bioinformatics*, **19**, 923–929 (2002).
21. S. M. Gomez, W. S. Noble, and A. Rzhetsky, Learning to predict protein-protein interactions from protein sequences, *Bioinformatics*, **19**, 1875–1881 (2003).
22. C. Huang, F. Morcos, S. P. Kanaan, S. Wuchty, A. Z. Chen, and J. A. Izaguirre, Predicting protein-protein interactions from protein domains using a set cover approach, *IEEE/ACM Transactions on Computational Biology and Bioinformatics*, **4**(1), 78–87 (2007).
23. T. Pawson and P. Nash, Assembly of cell regulatory systems through protein interaction domains, *Science*, **300**, 445–452 (2003).
24. N. M. Zaki, S. Deris, and H. Alashwal, Protein–protein interaction detection based on substring sensitivity measure, *International Journal of Biomedical Sciences*, **1**, 148–154 (2006).
25. P. Aloy and R. B. Russell, InterPreTS: protein interaction prediction through tertiary structure, *Bioinformatics*, **19**, 161–162 (2003).
26. L. Lu, Multiprospector: an algorithm for the prediction of protein–protein interactions by multimeric threading, *Proteins*, **49**, 350–364 (2002).
27. J. Espadaler, O. Romero-Isart, R. M. Jackson, and B. Oliva, Prediction of protein–protein interactions using distant conservation of sequence patterns and structure relationships, *Bioinformatics*, **21**, 3360–3368 (2005).
28. O. Keskin, A new, structurally nonredundant, diverse data set of protein–protein interfaces and its implications, *Protein Sciences*, **13**, 1043–1055 (2004).
29. T. Smith and M. Waterman, "Identification of common molecular subsequences", *Journal of Molecular Biology*, **147**, 195–197 (1981).
30. H. Saigo, J. Vert, N. Ueda, and T. Akutsu, Protein homology detection using string alignment kernels, *Bioinformatics*, **20**(11), 1682–1689 (2004).
31. A. Bairoch and R. Apweiler, The SWISS-PROT protein sequence database and its supplement TrEMBL in 2000, *Nucleic Acids Research*, **28**, 45–48 (2000).
32. M. Suyama and O. Ohara, DomCut: prediction of inter-domain linker regions in amino acid sequences, *Bioinformatics*, **19**, 673–674 (2003).
33. A. Gattiker, E. Gasteiger, and A. Bairoch, ScanProsite: a reference implementation of a PROSITE scanning tool, *Applied Bioinformatics*, **1**, 107–108 (2002).
34. I. Xenarios, L. Salwínski, X. J. Duan, P. Higney, S. Kim, and D. Eisenberg, DIP, the Database of Interacting Proteins: a research tool for studying cellular networks of protein interactions, *Nucleic Acids Research*, Oxford University Press, **30**, 303–305 (2002).
35. C. M. Deane, L. Salwinski, I. Xenarios, and D. Eisenberg, Protein interactions: two methods for assessment of the reliability of high throughput observations, *Molecular & Cellular Proteomics*, **1**, 349–56 (2002).
36. W. R. Pearson, Rapid and sensitive sequence comparisons with FASTAP and FASTA method, *Methods in Enzymology*, **183**, 63–93 (1985).
37. Q. Dong, X. Wang, L. Lin, and Z. Xu, Domain boundary prediction based on profile domain linker propensity index, *Computational Biology and Chemistry*, **30**, 127–133 (2006).

Chapter 55
Cellular Computational Model of Biology

Alex H. Leo and Gang Sun

Abstract One basic way that gene expression can be controlled by regulating transcription. In prokaryotes, our basic understanding of how transcriptional control of gene expression began with Jacob and Monod's 1961 model of the Lac operon. Recently, attempts have been made to model operon systems in order to understand the dynamics of these feedback systems. In the present study, it was independently attempted to model the Lac operon, based on the molecules and their activities on the lac operon of an *E. coli*. By various ways and means in engineering, a flow chart on the metabolic pathway of the lac operon activities of the *E. coli* was established.

Keywords Cellular Computational model · Lac operon · *E. coli* · metabolic pathway

55.1 Introduction

The control of gene expression has been a major issue in biology. One basic way that gene expression can be controlled by regulating transcription. In prokaryotes, our basic understanding of how transcriptional control of gene expression began with Jacob and Monod's 1961 model of the *Lac* operon (see Fig. 55.1), in which a regulator protein controls transcription by binding to a target region called the operator (see Fig. 55.2).

In the absence of this binding, RNA transcription is initiated when an RNA polymerase is able to bind to a region, called the promoter, just upstream from the operator. The RNA polymerase travels down stream and transcribes a series of structural genes coding for enzymes in a particular metabolic pathway (see Fig. 55.3). The operon system acts as a molecular switch involving positive or negative control [1].

G. Sun (✉)
Mechanics & Engineering Science Department, Fudan University, Shanghai 200433, China
E-mail: gang_sun@fudan.edu.cn

S.I. Ao and L. Gelman (Eds.), *Advances in Electrical Engineering and Computational Science*, Lecture Notes in Electrical Engineering.

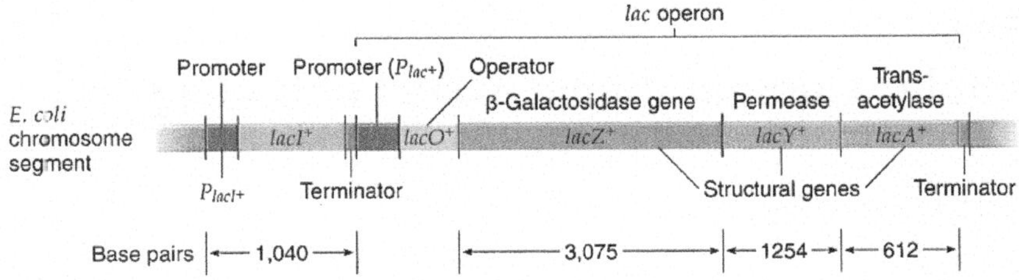

Fig. 55.1 Organization of the lac genes of *E. coli* and the associated regulatory elements, the operator, promoter, and regulatory gene [3]

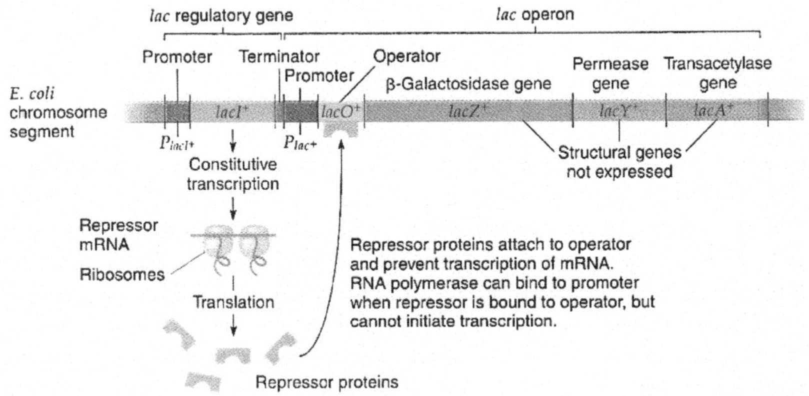

Fig. 55.2 Functional state of the lac operon in *E. coli* growing in the absence of lactose [3]

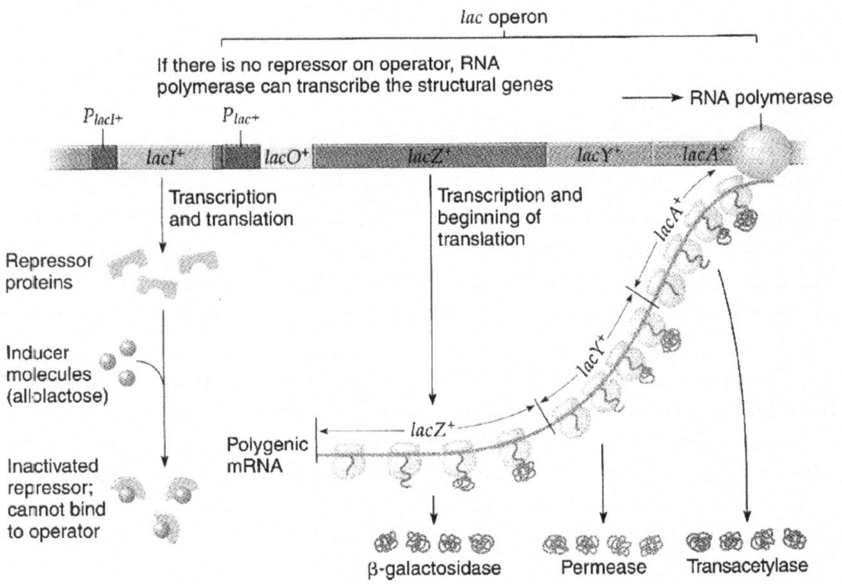

Fig. 55.3 Functional state of the lac operon in *E. coli* growing in the presence of lactose as the sole carbon source [3]

Recently, attempts have been made to model operon systems in order to understand the dynamics of these feedback systems. In 2003, Yidirim modeled the *Lac* operon as a series of coupled differential equations in terms of chemical kinetics [2]. Yidrim and Mackay note that their formulation assumes that they are dealing with a large number of cells and that the dynamics of small numbers of individual cells might be quite different. Meanwhile, it is hardly to obtain analytic solutions by using concentrations, average-values, as the variables of dynamics in differential equations, thus Laplace transform is necessary to solve the equations. Whereas, in the present study, it was independently attempted to model the *Lac* operon, based on the molecules and their activities on the *lac* operon of an *E. coli*. By various ways and means in engineering, a flow chart (Fig. 55.4) on the metabolic pathway of the *lac* operon activities of the *E. coli* was established. In terms of the amount of each relevant molecule present in the cell, a set of difference equations, mathematic model, comes on the basis of the flow chart. Since in form of difference equation, the numerical solution of molecular state-variable are available, by recurring the state-variables upon given a set of initial state or condition intuitionally.

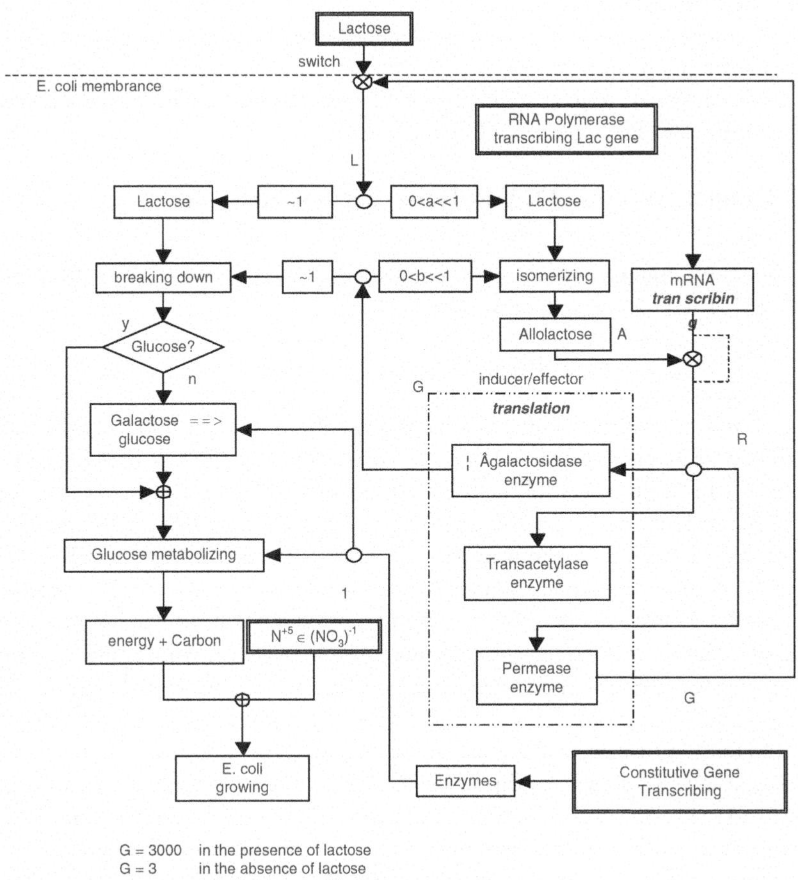

Fig. 55.4 Flow chart of the activity pathways of an *E. coli*

55.2 List of Variables

$mR(n)$_ amount of mRNA polymerases at step n
τ_R_ half-life of mRNA polymer
r_{TRSCRB}_ transcriptional rate of mRNA in the presence of allolactose
$G(n)$_ amount of β-galactosidase molecules at step n
τ_G_ half-life for β-galactosidase
r_{TRSLTN}_ translational rate of β-galactosidase, permease and transacetylase molecules per mRNA polymer
$L(n)$_ amount of lactose molecules transported by the permease at step n in the cell
$L_{ext}(n)$_ amount of lactose molecules outside of the cell at step n; $L_{ext}(0)$ is the amount of lactose molecules supplied at the beginning
$Maxr_{TRSPT}$_ the maximum rate of lactose transported in by each permease molecule
$A(n)$_ amount of ATP molecules, produced from the lactose at the step n
τ_A_ half-life of ATP
r_{meta}_ rate of a β-galactosidase molecule breaking down the lactose into the glucose, and thus into the ATP
Δ_ set step length, e.g. $\Delta = 1$(min), with regard to the half-life and the creating-rate for the substance presented by the variable

55.3 Determining mRNA Polymerases, mR(n + 1)

An *E. coli* cell contains about 3,000 copies of holoenzyme, a form of RNA polymerase [3]. The DNA-length of *lacZYA* is [1],

$$3{,}510 + 780 + 825 = 5{,}115\,\text{bp}$$

It was found that mRNA half-lives were globally similar in nutrient-rich media, determined from a least-squares linear fit. Approximately, 80% of half-lives ranged between 3 and 8 min in M9 + glucose medium, and 99% of the half-lives measured were between 1 and 18 min. In the LB, 99% of half-lives were between 1 and 15 min. The mean half-life was 5.7 min in M9 + glucose and 5.2 min in the LB [4]. The transcription of a polymerase approaches at a rate of average 30–50 nucleotides per second [3]. But in some defined media, the general time was approximately tripled [4]. The *lac* mRNA is extremely unstable, and decays with a half-life of only ~3 min [3]. Another article indicates the average half-life of an mRNA in *E. coli* is about 1.8 min. As lactose is no longer present, the repressor will be activated, and thus bind to the operator, making the transcription action stopped.

Considering the above conditions, $\mathrm{mR(n+1)}$, $\mathrm{r_{TRSCRB}}$, and τ_R are able to be determined in the presence of lactose. The variable, mR (n+1), equates, elapsing the period of Δ, the amount remained of mR (n) through the halving attenuation, plus the amount produced by transcription triggered in the presence of the allolactose from the lactose.

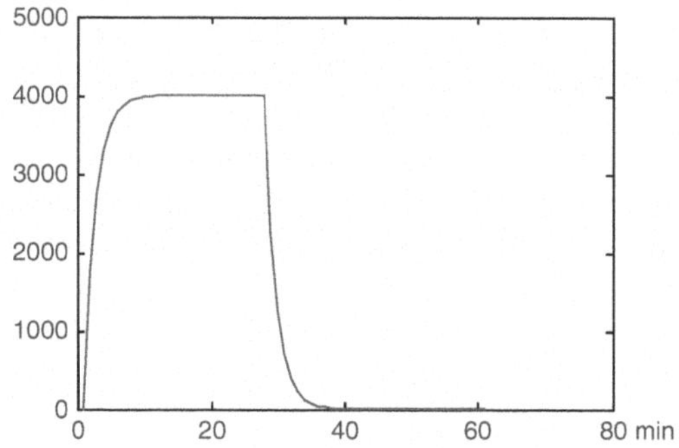

Fig. 55.5 Simulated amount of the mRNA molecules of the cell

Therefore, a polymerase needs 1.705–2.842 min for transcribing an mRNA, or 0.5865–0.3519 copies of mRNA come per minute per polymerase. So to speak, 1,759–1,055 copies of mRNA are maximally created per minute, with regard to 3,000 copies of the polymerases within an *E. coli*. Herein, $r_{TRSCRB} = 1{,}759 - 586$ mRNA/ min, $\tau_R = 5$ min. That is, the resulting formula is as follows:

$$mR(n+1) = e^{\frac{-0.69\Delta}{\tau_R}} \times mR(n) + r_{TRSCRB} \times \Delta \tag{55.1}$$

A simulated curve of *mR(n)* is shown as in Fig. 55.5.

55.4 Determining β-Galactosidases, $G(n+1)$

Given supplying *E. coli* cells lactose, β-galactosidase enzyme will appears in minutes, so do *lac* permease in membrane and transacetylase, a third protein. The level of β-galactosidase enzyme can accumulate to 10% of cytoplasmic protein [5]. Once a ribosome approaches away of the initiative site on an mRNA, another one will locate at the initiative site. Thus, many ribosomes may well simultaneously be translating each mRNA. An average mRNA has a cluster of 8–10 ribosomes, named as polysome, for synthesizing protein [3].

Where cells of *E. coli* are grown in the absence of lactose, there is no need of β-galactosidase. An *E. coli* cell contains 3–5 molecules of the enzyme [3]. That is, a dozen of the repressors bind and unbind rather than just bind and stay on the mRNA. In a fraction of a second, after one repressor unbinds and before another binds, an mRNA polymerase could initiate a transcription of the operon, even in the absence of lactose. [3]. In 2–3 min after adding lactose, there soon are 3,000–5,000 molecules of β-galactosidase per *E. coli* cell [1, 3]. The β-galactosidase in the *E. coli* is more stable than the mRNA, whose half-life is more than 20 h. The half-life of the permease is more than 16 h [1]. So that, the β-galactosidase activity remains at the induced level for longer. To add or change codons at the 5′ end of the gene to encode N-terminal amino acid within the DNA-*lac*ZYA will provide greater resistance to β-galactosidase production [6]. As a porter, the permease compound can be to concentrate lactose against the gradient across the cellular membrane in ad hoc manner for E. coli. The concentration of the substrate against a gradient can be achieved up to 10^3 to 10^6-fold. The existing β-galactosidase molecules will be diluted out by cell division [5]. Whereas, replicating the DNA of an *E. coli* needs 84 min [3], the half-life is 100 min for simplicity, or 30 min for conservation. This point of view could be used to derive a model in colony of the bacteria. The DNA-length of *lac*_ZYA [1]:

$$3{,}510 + 780 + 825 = 5{,}115\,\text{bp};$$

The mRNA-length of *lac*_ZYA:

$$5{,}115 \div 3 - 1 = 1{,}705 - 1 = 1{,}704 \text{ amino acids.}$$

Protein synthetic rate is 350–400 amino acids (1/4–1/5 β-galactosidase) per minute per ribosome [5]. There are approximate 8 ribosomes attached to an mRNA That is, for a ribosome to make a β-galactosidase needs 4–5 min (say 4.5 min) by running all the way of the mRNA, and there are 1.8 copies of the β-galactosidase translated per minute per mRNA (Fig. 55.6).

As a function of $mR(n)$, $G(n+1)$ is the amount of β-galactosidase, same as permease in count. $G(n+1)$, then r_{TRSLTN} and τ_G are able to be determined. $G(n+1)$ equates, elapsing the period of Δ, the β-galactosidase amount remained of $G(n)$ through the halving attenuation, plus the amount produced by translation of $mR(n)$, herein $\tau_G = 1{,}000$ min; $r_{TRSLTN} = 8 \div 4.5 = 1.8$; $G(0) = 3$. That is, the resulting formula is as follows:

$$G(n+1) = e^{\frac{-0.69\Delta}{\tau_G}} \times G(n) + r_{TRSLTN} \times \Delta \times mR(n) \tag{55.2}$$

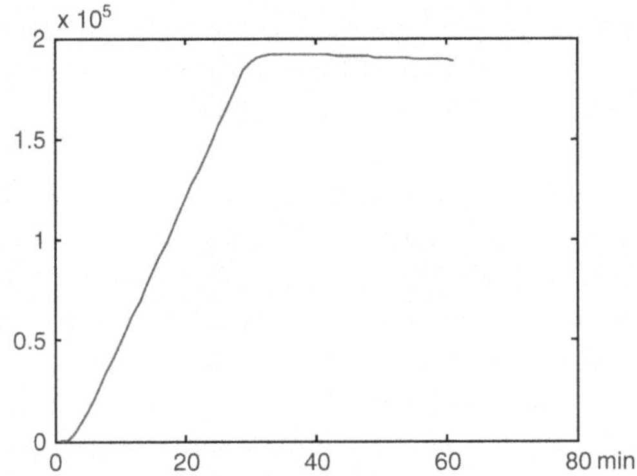

Fig. 55.6 The simulated curve of G(n), the permease or β-galactosidase of the cell

55.5 Determining Lactoses with in and out the Cell, Lext(n + 1) and L(n + 1)

Lac permease is a highly lipophilic protein with an apparent subunit molecular weight of about 30,000 [7]. The lactose transport-in rate of permease is 120 nmol per mg per min. That is, 2.4 copies of lactose are transported into the cell by a permease per minute.

r_{meta}, the rate of lactose breaking down into glucose by $G(n)$; 2 min is for to convert a lactose into 72 ATPs [3]. Therefore, r_{meta} is 0.5, and max r_{TRSPT} is 2.4.

r_L is the rate of transporting lactose into the cell by the permease, presented by G(n). $L(n+1)$ is the amount of the lactose at step n + 1 in the *E. coli.* For the period of Δ, $L(n+1)$ equates to the amount of the lactose brought-in by the permease, presented by $G(n)$, but subtracting the amount metabolized into glucose. Each glucose molecule, in turn, is immediately transformed into 72 ATPs (Figs. 55.7 and 55.8).

That is, the resulting formula is as follows:

$$L_{ext}(n+1) = L_{ext}(n) - \max r_{TRSPT} \times \Delta \times G(n) \times f(L_{ext}(n) - L(n)) \tag{55.3}$$

$$L(n+1) = L(n) + \max r_{TRSPT} \times \Delta \times G(n) \times f(L_{ext}(n) - L(n)) - r_{meta} \times \Delta \times L(n) \tag{55.4}$$

$$f(x) = \begin{cases} 2 - e^{\frac{-x}{\lambda}} & when\ x \geq 0 \\ e^{\frac{x}{\lambda}} & else \end{cases} \tag{55.5}$$

in which, λ is a constant, whose magnitude close to that of the lactose supplied.

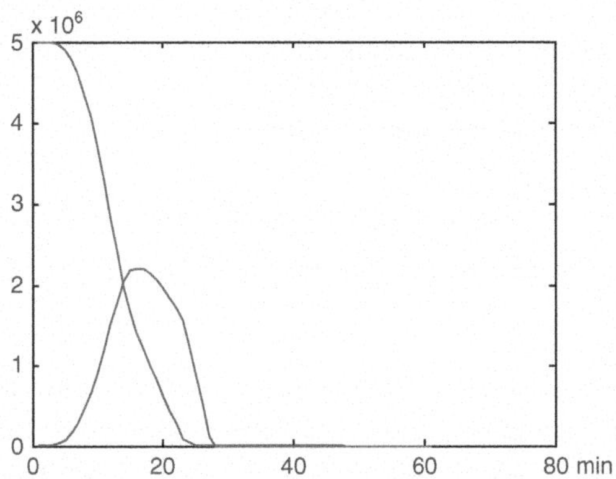

Fig. 55.7 The simulated curves of the L_{ext}(n) and $L(n)$, the external and internal lactose molecules of the cell

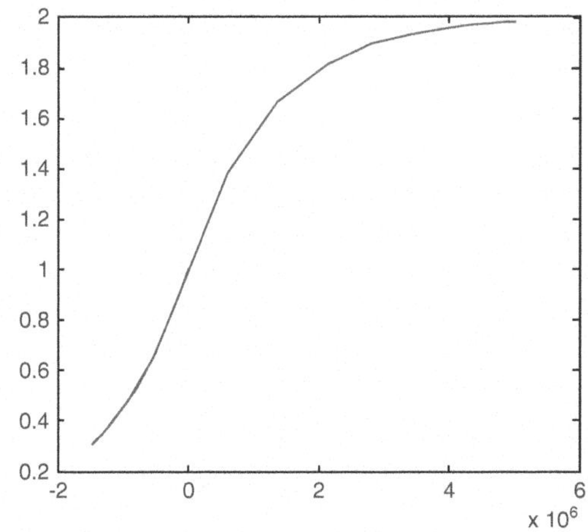

Fig. 55.8 The simulated function curve of the $f(x)$

55.6 Determining ATP Molecules, A(n)

ATP half-life, τ_A, is about 15 min in water at room temperature, estimating 2 min for to convert a copy of lactose into 72 ATPs. The energy resource for *E. coli* growth is from the lactose in form of ATP at 72 in scale. The number of ATP for the transportation-in of a lactose by a permease in energy equates about one. Therefore, the transcription of an mRNA needs 5,115 ATPs (in forms of ATP, UTP, CTP, or GTP), and the translation of an mRNA needs 1,700 ATPs.

Given the consumption of energy, comprising transcribing, translating, as well as transporting lactose into the cell, count 80% of total energy, as is the primary life activity, then a coefficient d for energy distribution set at 0.8. The remaining energy is for the other metabolic activities.

With regard to the r_{meta}, $G(n)$, Δ, steps n + 1 and n, the energy generated by the glucose is distributed among producing mRNA and β-Galactosidase, and bringing the lactose into *E. coli* cell (Fig. 55.9–55.11).

An energy-distributional equation is as follows:

$$A(n+1) = e^{\frac{-0.69\Delta}{\tau_A}} A(n) + 72\Delta \times r_{\text{meta}} \times G(n) - \Delta[5115 r_{\text{TRSCRB}} + 1700 r_{\text{TRSLTN}} \times mR(n) + \max r_{\text{TRSPT}} \times f(L_{ext}(n) - L(n)) \times G(n)] \quad (55.6)$$

In the above equation, first term is accumulated energy before step n; second term is the energy created from the lactose, and third term is the consumed for the three activities. If the third is greater or equal to the second, there will not be energy

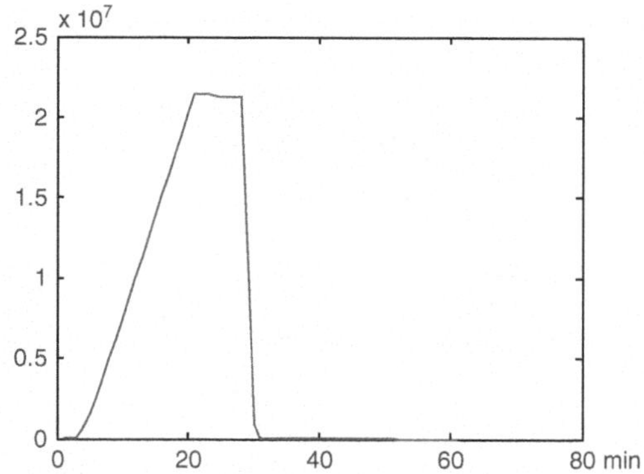

Fig. 55.9 Simulated curve of the ATP molecules consumed within the cell

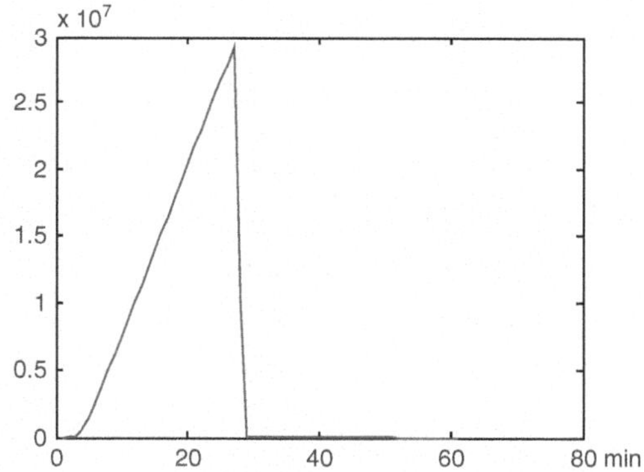

Fig. 55.10 Simulated function curve of the ATP molecules created within the cell

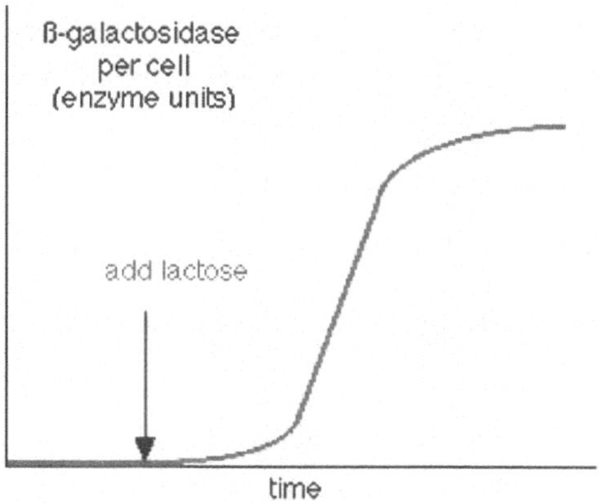

Fig. 55.11 A classic experimental data of *E. coli*

accumulation. An comprehensive energy distributive coefficient equation, c(n), can be determined as following,

$$c(n) = \begin{cases} e(n) & \textit{when } e(n) \le 1 \\ 1 & \textit{else} \end{cases} \tag{55.7}$$

$$\begin{aligned} e(n) = 72\Delta \times r_{\text{meta}} \times G(n) \times d \div [5115 r_{\text{TRSCRB}} + 1700 r_{\text{TRSLTN}} \times mR(n) \\ + r_{\text{TRSPT}} \times f(L_{ext}(n) - L(n)) \times G(n)] \end{aligned} \tag{55.8}$$

The domain of $c(n)$ is within [0, 1].

If the energy is inadequate, the energy-distributional coefficient equation comes, and thus the Eqs. (55.1)–(55.4) are modified into energy-constrained Eqs. (55.1)'–(55.4).

$$mR(n+1) = e^{\frac{-0.69\Delta}{\tau_R}} mR(n) + r_{\text{TRSCRB}}\Delta \times c(n), \tag{55.1}$$

$$G(n+1) = e^{\frac{-0.69\Delta}{\tau_G}} G(n) + r_{\text{TRSLTN}} \times \Delta \times m \times R(n) \times c(n), \tag{55.2}$$

$$\begin{aligned} L_{ext}(n+1) &= L_{ext}(n) - \max r_{\text{TRSPT}} \times \Delta \times G(n), \\ &\quad \times f(L_{ext}(n) - L(n)) \times c(n), \end{aligned} \tag{55.3}$$

$$\begin{aligned} L(n+1) &= L(n) + \max r_{\text{TRSPT}} \times \Delta \times G(n) \times f(L_{ext}(n) - L(n)), \\ &\quad \times c(n) - r_{\text{meta}} \times \Delta \times L(n). \end{aligned} \tag{55.4}$$

55.7 The System Structure

Eqs. (55.1)–(55.8), or Eqs. (55.1)'–(55.4)' in place if necessary, are integrated into a systematic function Eq. (55.9), with respect to the initial condition presented as Eq. (55.10).

$$\mathbf{X(n+1) = T(X(n)) + U} \tag{55.9}$$

$$\mathbf{X(n)} = [R(n), G(n), L_{ext}(n), L(n), A(n)]^T \tag{55.10}$$

They are solved numerically by MATLAB through setting the following initial condition:

$$\mathbf{X}(0) = [0 \quad 3 \quad \text{supply} \quad 0 \quad 0]^T$$
$$\mathbf{U} = [r_{TRSCRB}\Delta \quad 0 \quad 0 \quad 0 \quad 0]^T$$

55.8 In Comparison to Traditional Biological Experiment

A traditional experiment culturing E. coli in tubes offered the following graph [6].

In comparison of the outcomes of the presented equations and the traditional experimental shown as Fig. 55.2, the results are consistent in a mutual time scale; and so does mRNA in Fig. 55.1. In addition, more information on the cell, such as external and internal lactose, ATP consumption, etc. in a cell of *E. coli* are available by the model, but not in the other methods.

55.9 Conclusion

By computational simulation, the Figs. 55.2–55.7 were obtained. Supplied with mere lactose as nutrition, an *E. coli* cell of the Operon activities can be described by the eight difference equations, or a mathematical model. The Relevant outcomes of the model are highly consistent with traditional experimental results. The mathematic model can be developed into a general method to govern the growth of an *E. coli* colony by gene expression. The model is justified through the traditional biological experiment.

55.10 Discussion

This paper was based on the authors' background of computer science, Electrical Engineering, and Genetics. This paper discusses the growth of an *E. coli* cell supplied with mere lactose. As a core governing the state-variables of the *E. coli* cell's molecules on the *lac* operon, the model could be developed into another set of formula for colony, with regard to cellular divisions in compliance with normal distribution in mathematics. The set of formula may well model a colony of the bacteria by halving the values of state-variable in various periods, simulating cellular division. Finally, the formula could be transformed into corresponding differential equations. In addition, effects of parameters λ and d in the Eqs. (55.5) and (55.8) should be studied further respectively. The rate or behavior pattern of the permeases transporting lactose across the membrane of an *E. coli* cell is rather approximate. Finally, it is noticeable that the simulated numerical solutions of the model vary slightly with Δ value. The phenomenon probably is of the theoretical approximate hypothesis in the Section 55.2.

Since on individual cellular level, a colony model could be developed rather to describe and simulate more complex activities and regular patterns of the representative substances of the colony, by considering individual cell division, and could further be developed into a general quantitative analysis for cellular biology in high precise.

References

1. Lewis, Benjamin, 2004. *Genes VIII.* Prentice-Hall, Upper Saddle River, NJ
2. Yidirim, Necmettin et al., 2003. Feedback Regulation in the Lactose Operon: A Mathematical Modeling Study and Comparison with Experimental Data, *Biophys. J.* 84: 2841–2851.
3. Russell, Peter J., 2002. *iGenetics*, Benjamin Cumming, New York.
4. Bernstein, Jonathan, A. et al., 2002 July 23. Global Analysis of mRNA Decay and Abundance in *Escherichia coli* at Single-Gene Resolution Using Two-Color Fluorescent DNA Microarrays. *Proc. Natl. Acad. Sci. USA* 99(15): 9697–9702; http://www.pnas.org/cgi/content/full/99/15/9697
5. Terry, Thomas, M., 2002. *Regulation of Metabolism. Microbiology*, Chapter 12, pp. 237–250, 4th Edition; http://www.biologie.unihamburg.de/bonline/library/micro229/terry/229sp00/lectures/regulation.html.
6. Southern Illinois University, 2003. BIOTECHNOLOG.Y. *MICR 421*, chapter 6; http://www.science.siu.edu/microbiology/micr421/chapter6.html
7. Ullmann, Agnes, 2001. *Escherichia coli* Lactose Operon, *Encyclopedia of Lifesciences*; http://216.239.41.104/search?q = cache:mzCMeayarnMJ:www.cellcycle.bme.hu/oktatas/mikrofiz/extra/lac%2520operon.pdf + how + weight + lac + permease + is + in + e. + coli&hl = en&ie = UTF-8.
8. Weinglass, Adam, B. et al., 2001. Manipulating Conformational Equilibria in the Lactose Permease of Escherichia coli, *Encyclopedia of Life sciences*; http://darwin.bio.uci.edu/∼bardwell/weinglass02jmb.pdf.

Chapter 56
Patient Monitoring: Wearable Device for Patient Monitoring

Robert G. Lupu, Andrei Stan, and Florina Ungureanu

Abstract The increasing request of patients, suffering of chronic diseases, who wish to stay at home rather then in a hospital and also the increasing need of homecare monitoring for elderly people, have lead to a high demand of wearable medical devices. Also, extended patient monitoring during normal activity has become a very important target. Low power consumption is essential in continuously monitoring of vital-signs and can be achieved combining very high storage capacity, wireless communication, and ultra-low power circuits together with firmware management of power consumption. This approach allows the patient to move unconstraint around an area, city or country. In this paper the design of ultra low power wearable monitoring devices based on ultra low power circuits, high storage memory flash, bluetooth communication and the firmware for the management of the monitoring device are presented.

Keywords Patient Monitoring · Wearable Device · wearable medical Device · monitoring device

56.1 Introduction

Phenomena of ageing population observed in most developed countries [1] and prevalence of chronic diseases have increased the need for chronic and geriatric care at home [2]. The task of patient monitoring may be achieved by telemedicine (enabling medical information-exchange as the support to distant-decision-making) and telemonitoring (enabling simultaneous distant-monitoring of a patient and his vital functions) both having many advantages over traditional practice. Doctors can

R.G. Lupu (✉)
Gh. Asachi" Technical University of Iasi, Faculty of Automatic Control and Computer Engineering, 700050 Romania
E-mail: robert@cs.tuiasi.ro

S.I. Ao and L. Gelman (Eds.), *Advances in Electrical Engineering and Computational Science*, Lecture Notes in Electrical Engineering.

receive information that has a longer time span than a patient's normal stay in a hospital and this information has great long-term effects on home health care, including reduced expenses for healthcare.

Despite the increased interest in this area, a significant gap remains between existing sensor network designs and the requirements of medical monitoring. Most telemonitoring networks are intended for deployments of stationary devices that transmit acquired data at low data rates and rather high power consumption. By outfitting patients with wireless, wearable vital sign devices, collecting detailed real-time data on physiological status can be greatly simplified. But, the most important effect is the widely social integration of peoples with disabilities or health problems that need discreet and permanent monitoring.

It is well known that Bluetooth represents a reliable and easy solution for signal transmission from a portable, standalone unit to a nearby computer or PDA [3]. When a real time rudimentary analysis program detects pathological abnormality in the recorded data an alert is sent to the telemonitoring centre via internet or GSM/GPRS using a PDA or computer.

The needs of the society and market trend suggest that healthcare-related applications will be developed significantly. Particularly, PDA phones interfaced with wearable sensors have an enormous market as they will enlarge the goal of healthcare and mobile devices [4].

The development of ultra low power wearable monitoring device unit is propounded. The main functions of the wearable device consist in signal acquisition, rudimentary processing, local data storage and transmission to the remote care centre. In Fig. 56.1, the general architecture of the telemonitoring network is presented.

This paper presents the initial results and our experiences for a prototype medical wearable device for patient monitoring taking into account some hardware and software aspects regarding vital signs measurement, robustness, reliability, power consumption, data rate, security and integration in a telemonitoring network.

Fig. 56.1 The general structure of telemonitoring network

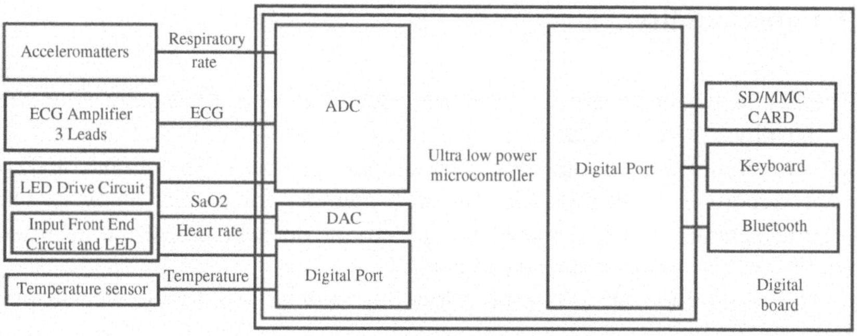

Fig. 56.2 Block diagram of the monitoring device

56.2 Wearable Monitoring Device

The monitoring device is build using custom developed hardware and application software. The block diagram is presented [5] in Fig. 56.2. Low power amplifiers and sensor are connected to the device [6], for vital parameters acquisition. The health parameters acquired are: heart rate, heart rhythm regularity, respiratory rate, oxygen saturation and body temperature. The monitoring device includes also a custom made three-lead ECG amplifier.

The digital board includes an ultra low power microcontroller and interface circuits for keyboard, SD/MMC card and Bluetooth. The low power sensors for the above mentioned parameters are connected to the digital board using digital port or specific converters.

The wireless communication via Bluetooth with a computer or PDA and the flash memory for raw data information storage are two of the most important facilities of the proposed wearable device.

The software application running on computer and PDA assures the communication via internet respectively GSM/GPRS [3, 7].

The monitoring device is built using an ultra low power microcontroller (MSP430 from Texas Instruments) that has a 16 bit RISC core with clock rates up to 8 MHz. An important advantage of this family of microcontrollers is the high integration of some circuits that offer the possibility to design devices with open architecture: digital ports, analog to digital converter (ADC) and digital to analog converter (DAC). MSP430 microcontroller also has a built in hardware multiplier which is a valuable resource for digital filter implementation. The custom made electronic boards used for data acquisition are all designed with low power circuits. The system is modular: every board is detachable.

The on board storage device (SD/MMC card) with FAT32 file system is used for raw data recording together with signals processing results. The radio module Bluetooth [8] designed for short-range communication is used for data transmission between monitoring device and PC or PDA. This module has low power consumption being used only when data transfer is performed.

56.3 Functionality

The ultra low power wearable monitoring device is able to acquire simultaneously the physiological parameters mentioned in the previous section and also to perform rudimentary digital signal processing on board. The signals are continuously recorded in separate files on flash memory for feature analysis. Once pathological abnormality is detected, the monitoring device requests a transmission trough PC or PDA to the remote care centre.

For the respiratory rate it is used a low power three axis low-g accelerometer MMA7260QT [9]. This is a low cost capacitive micromachined accelerometer that features signal conditioning, a one-pole low pass filter, temperature compensation and g-Select which allows for the selection among four sensitivities.

The accelerometer outputs are read and processed once at 10 ms (100 times per second).

Because this is a low power system, the accelerometer is disabled between two conversion cycles. Each conversion cycle has a 2 ms startup period for accelerometers to recover from sleep mode.

Due to the sensitivity of the accelerometer and the high resolution of analog to digital converter, the useful signal (the output voltages of accelerometer) has a large amount of superposed noise with possible sources in body moving or even heart beating. In Fig. 56.3, the acquired signal affected by noise is presented. There can be seen the initial rise of voltage corresponding to the start of chest movement. This is explained by the fact that the movable central mass of g cell moves in opposite direction of the movement. As the chest movement velocity becomes constant, the

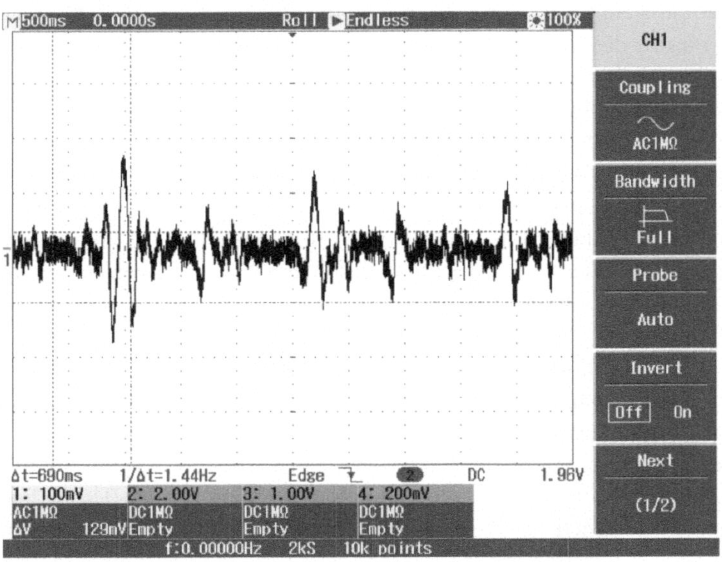

Fig. 56.3 The accelerometer time response

acceleration decreases and movable central mass comes back to initial position. This settling process implies acceleration opposite to the direction of movement.

Noise filtering is achieved by using a 128 point rolling average filter. Mathlab simulation was used with real datasets for filter calibration. Without filtering, the signal is almost unusable for further processing. After filtering, the noise is substantially reduced.

To easily distinguish between positive and negative accelerations (e.g. to determine the sense of movement) the accelerometer has an offset of half of supply voltage for 0 g acceleration. So, for positive or negative accelerations the corresponding output voltage is positive in all cases.

To make the difference between positive and negative accelerations, it is necessary to fix the reference value corresponding for 0 g. This is done at the application startup, when an average with 512 points is made.

In Fig. 56.4, the results of the algorithm are plotted. The first plot is the raw unfiltered data from accelerometers outputs in millivolts. This plot is a 2,500 points record of accelerometer output for a time period of 5 s. The second plot represents the filtered input acceleration.

The custom made ECG module uses three leads. The ECG signals are acquired using micro power instrumentation amplifiers and micro power operational amplifiers. Both are single power supply. A very important feature of the instrumentation amplifier is that it can be shutdown with a quiescent current of less than 1 μA (Fig. 56.5). Returning to normal operations within microseconds, the shutdown feature makes is optimal for low-power battery or multiplexing applications [10]. The propounded functioning is: on power on the instrumentation amplifier is shutdown. With 10 μs before a conversion cycle starts the instrumentation amplifier is returned from shutdown mode and back again when the conversion cycle finishes.

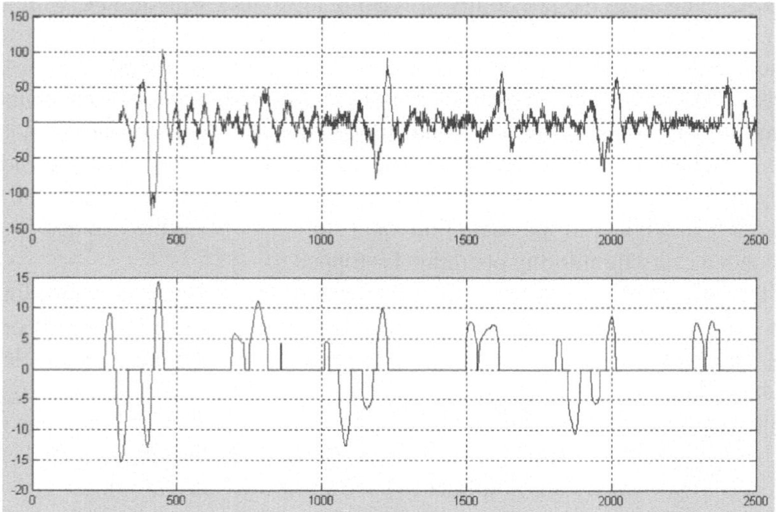

Fig. 56.4 The processed signals

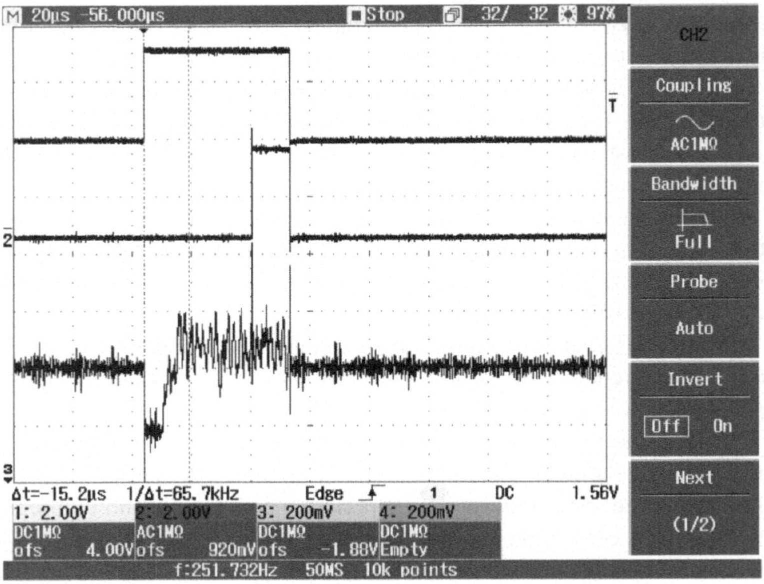

Fig. 56.5 The instrumentation amplifier output

The temperature is monitored every minute using an ultra low power temperature sensor.

Meanwhile, the sensor is shutdown saving maximum power by shutting down all device circuitry other then the serial interface, reducing current consumption to typically less than 0.5 µA [11].

Because the read/write operation with flash card can reach a current consumption of max 60 mA, these are done with blocks of data [12]. The file system on flash card is FAT32 witch gives the possibility of reading those files with any PC with a card reader.

The average current consumption of Bluetooth circuit is around of 35 mA. For this reason, the Bluetooth communication is used in two ways: always connected transmitting live or stored data (signal processing and data storing are not performed) and not connected, the circuit is in sleep mode (signal processing and data storing are performed). If a critical situation occurs the circuit is awaken, the communication starts, and in the same time data storage is performed. When files are downloaded every monitoring operation is stopped.

In the case of blood oxygenation measurement, for driving and controlling the LEDs a solution proposed by Texas Instruments was used [9]. It is based on two LEDs, one for the visible red wavelength and another for the infrared wavelength. The photo-diode generates a current from the received light. This current signal is amplified by a trans-impedance amplifier. OA0 (microcontroller built in), one of the three built in op-amps, is used to amplify this signal. Since the current signal is very small, it is important for this amplifier to have a low drift current [9].

Because of the high level of analog circuits' integration the external components involved in hardware development are very few. Furthermore, by keeping the LEDs ON for a short time and power cycling the two light sources, the power consumption is reduced [9].

56.4 Firmware Issue

The developed firmware consists of several tasks and each of them manages a particular resource as presented in Fig. 56.6. The communication between tasks is implemented with semaphores and waiting queues allowing a high level of parallelism between processes. Each process may be individually enabled or disabled. This feature is very important in increasing the flexibility of the application: if real time monitoring is desired, then SD Card Process may be disabled and Bluetooth Process is enabled, if only long term monitoring is desired then SD Card Process is enabled and Bluetooth Process may be disabled. This has a positive impact on power consumption because only the resources that are needed are enabled for use.

A logical cycle of the device operation is presented in Fig. 56.6. In the first step (thread), a buffer of data is acquired. The ADC module works in sequence mode and acquires data from several channels at a time. The ADC channels are associated with temperature, accelerometer, ECG signals and SaO2 inputs of electronic interface modules. One sequence of ADC conversion is made by one sample for each of these inputs. The data is formatted and then stored in a buffer. Only when the buffer is full the following thread may begin. Depending on application requirements the storage, transmission or analyze of the data may be performed.

Encryption of personal data is provided. Some files that store personal data related to the patient are encrypted and can be accessed only using the proper decryption algorithm. Only a part of the files are encrypted because there must be kept a balance between power consumption and computing power requirements. For space saving, a light compression algorithm may be activated as an extra feature of the device. The activation of this feature has a negative impact on the overall power consumption.

The communication process implements an application layer protocol for data exchange between the device and other Bluetooth enabled devices. The communication requirements must be limited to a minimum rate in order to save power. The Bluetooth module is powered on when a message has to be sent. The message transmission rate is kept low by using internal buffering and burst communication. The Bluetooth module can be also waked up by an incoming message that may embed commands to the device.

The analysis process implements some rudimentary signal processing tasks in order to detect anomalies in physiological activities. Only few vital parameters anomalies are locally detected (by the wearable device) and trigger an alarm event. More complex signal processing and dangerous condition detection are

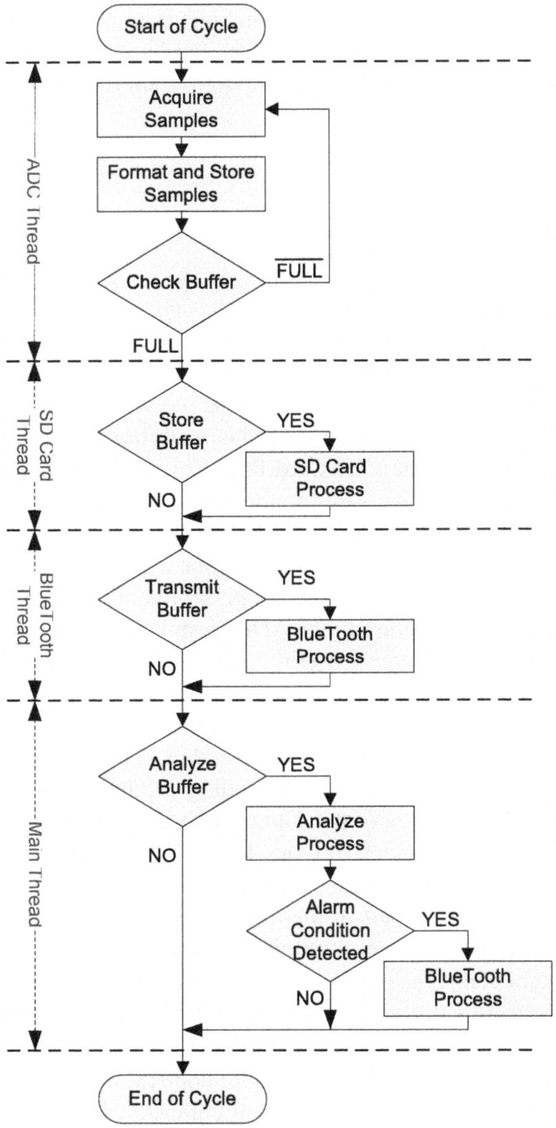

Fig. 56.6 The tasks cycle

implemented on a remote computer which has the required computing power. The remote computer receives raw data from the device and does the required analysis.

The physical processes that are monitored have a slow changing rate leading to a low sampling rate requirement. This allows the controller to be in sleep mode for an important percentage of operating time.

The acquisition of one full data buffer takes the most time of the operation cycle. The time needed for the execution of each remaining thread is shorter compared to the ADC process.

A double buffering scheme is used: meanwhile a buffer is filled up with new data samples, the second buffer may be processed (stored on SD card, transmitted via Bluetooth or locally analyzed in order to detect conditions that should trigger alarm events). When the new data buffer is acquired the buffers change (simple pointer assignment) occurs: the previous processed buffer becomes empty, the acquired data will be placed in it and the actual filled buffer will be processed.

The device must be reliable. For this purpose a power monitoring function is designed. To implement this feature the ADC module is used. One of its inputs monitors the voltage level across the batteries. The alarm events are designed to preserve as much power as possible. If a voltage threshold is reached then an alarm event is triggered and a resource could be disabled or its usage restricted only if configured so. For example, the messages content and its transmission period could be modified in order to save power but also to ensure a safe operation.

The computer software (see Fig. 56.7) is a MDI application (Multiple Document Interfaces). A connection can be established with more than one device (each one has a unique id). The software is always listening for connection with registered Bluetooth monitoring device. This means that before data transmission, the device id has to be known by the software. This is done by entering the code manually, or by answering yes, when the application prompts for a communication request. The ID is stored for further use. The communication session is established in two ways: by the device when a critical situation occurs or by the computer when a real time monitoring or file download is requested. This is done by sending a break condition to the device to wake it from the deep sleep state [8]. The software assures the connection to a server via Internet and also data transmission.

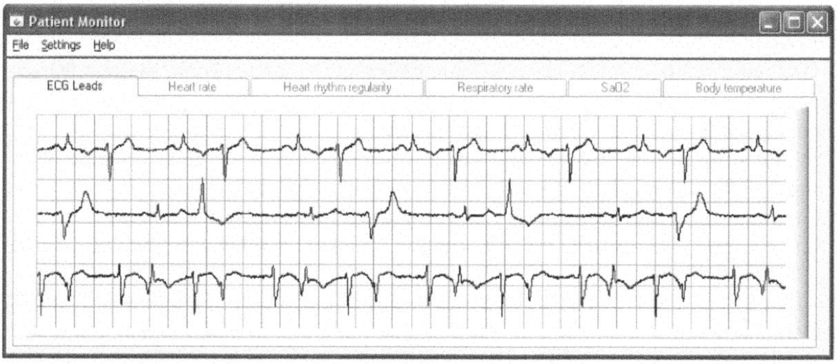

Fig. 56.7 Software user interface

56.5 Conclusion

The work of this paper focuses on design and implementation of an ultra low power wearable device able to acquire patient vital parameters, causing minimal discomfort and allowing high mobility. The proposed system could be used as a warning system for monitoring during normal activity or physical exercises. The active collaboration with The Faculty of Biomedical Engineering from University of Medicine and Pharmacy from Iasi (Romania) and with several hospitals offered the opportunity to test the prototype. Preliminary results were so far satisfactory. This wearable device will be integrated into SIMPA healthcare systems to provide real-time vital parameters monitoring and alarms.

In conclusion, this paper presents some challenges of hardware and software design for medical wearable device based on low-power medical sensors and microcontroller with a recent tremendous impact in many medical applications. Obviously, the results demonstrate that there is still significant work to be done if the wearable device is effectively integrated in a large network of medical sensors.

References

1. United Nations Population Division Department of Economic and Social Affairs. http://www.un.org/esa/population/publications/aging99/a99pwld.htm
2. United Nations Department of Economic and Social Affairs, Report on the World Social Situation 2007 http://www.un.org/esa/socdev/rwss/docs/rwss07_fullreport.pdf
3. K. Hung, *Wearable Medical Devices for Tele-Home Healthcare* Proceedings of the 26th Annual International Conference of the IEEE EMBS, CA, USA, September 1–5, 2004.
4. K. J. Liszka, *Keeping a beat on the heart* IEEE Pervasive Computing, vol. 3, no. 4, October–December 2004 p. 42.
5. Custom build using low power Texas Instruments technology http://www.ti.com
6. C. Rotariu, R. G. Lupu, *An ultra low power monitoring device for telemedicine*, 1st National Symposium on e-Health and Bioengineering, EHB 2007, November 16, 2007, IASI, ROMANIA.
7. R. G. Lupu, *Solution for home care monitoring via GSM network*, Master degree thesis, University of Naples "Federico II", Department of Electronic Engineering and Telecommunucations, Biomedical Engineering Unit, 2003.
8. F2M03GLA is a Low power embedded Bluetooth^{TM} v2.0 + EDR module with built-in high output antenna. Available: http://www.free2move.net/index.php?type = view&pk = 24&pk_language = 1&PHPSESSID = dc94542acd247be5916d98a1f5d088b9
9. A Single-Chip Pulsoximeter Design Using the MSP430. Available: http://focus.ti.com/lit/an/slaa274/slaa274.pdf
10. The INA322 family is a series of low cost, rail-to-rail output, micropower CMOS instrumentation amplifiers that offer wide range, single-supply, as well as bipolar-supply operation. Available: http://focus.ti.com/lit/ds/symlink/ina322.pdf
11. The TMP102 is a two-wire, serial output temperature sensor available in a tiny SOT563 package. http://focus.ti.com/lit/ds/symlink/tmp102.pdf
12. ATP MMC plus Specification. Available: http://flash.atpinc.com/articles/images/1031/MMC_Plus_spec.pdf

Chapter 57
Face Recognition and Expression Classification

V. Praseeda Lekshmi, Dr. M. Sasikumar, and Divya S Vidyadharan

Abstract In this chapter, faces are detected and facial features are located from video and still images. 'NRC-IIT Facial video database' is used as image sequences and 'face 94 color image database' is used for still images. Skin pixels and non-skin pixels are separated and skin region identification is done by RGB color space. From the extracted skin region, skin pixels are grouped to some meaningful groups to identify the face region. From the face region, facial features are located using segmentation technique. Orientation correction is done by using eyes. Parameters like inter eye distance, nose length, mouth position, and DCT coefficients are computed which is used for a RBF based neural network.

Keywords Face Recognition · Expression Classification · neural network · facial features

57.1 Introduction

The technological advancement in the area of digital processing and imaging has led to the development of different algorithms for various applications such as automated access control, surveillance, etc. For automated access control, most common and accepted method is based on face detection and recognition. Face recognition is one of the active research areas with wide range of applications. The problem is to identify facial image/region from a picture/image. Generally pattern recognition problems rely upon the features inherent in the pattern for efficient solution. Though face exhibits distinct features which can be recognized almost instantly by human eyes, it is very difficult to extract and use these features by a computer. Human can identify faces even from a caricature. The challenges associated with face detection

V.P. Lekshmi (✉)
College of Engineering, Kidangoor, Kottayam, Kerala, India
E-mail: vplekshmi@yahoo.com

S.I. Ao and L. Gelman (Eds.), *Advances in Electrical Engineering and Computational Science*, Lecture Notes in Electrical Engineering.

and recognition are pose, occlusion, skin color, expression, presence or absence of structural components, effects of light, orientation, scale, imaging conditions etc.

Most of the currently proposed methods use parameters extracted from facial images. For access control application, the objective is to authenticate a person based on the presence of a recognized face in the database.

Here skin and non-skin pixels are separated and the pixels in the identified skin region are grouped to obtain face region. From the detected face area, the facial features such as eyes, nose and mouth are located. Geometrical parameters of the facial features and the DCT coefficients are given to a neural network for face recognition. The expressions with in the face regions are analyzed using DCT and classified using a neural network.

57.2 Background and Related Work

A lot of research has been going on in the area of human face detection and recognition [1]. Most face detection and recognition methods fall into two categories: Feature based and Holistic. In feature-based method, face recognition relies on localization and detection of facial features such as eyes, nose, mouth and their geometrical relationships. In holistic approach, entire facial image is encoded into a point on high dimensional space. Principal Component Analysis (PCA) and Active Appearance Model (AAM) [2] for recognizing faces are based on holistic approaches. In another approach, fast and accurate face detection is performed by skin color learning by neural network and segmentation technique [3]. Independent Component Analysis (ICA) was performed on face images under two different conditions [4]. In one condition, image is treated as a random variable and pixels are treated as outcomes and in the second condition pixels are treated as random variables and image as outcome. Facial expressions are extracted from the detailed analysis of eye region images is given in [5]. Large range of human facial behavior is handled by recognizing facial muscle actions that produce expressions is given in [6]. Video based face recognition is explained in [7]. Another method of classification of facial expression using Linear discriminant analysis (LDA) is explained in [8] in which the Gabor features extracted using Gabor filter banks are compressed by two stage PCA method.

57.3 Discrete Cosine Transform

Discrete Cosine Transform of an $N \times N$ cosine transform matrix $C = \{c(k, n)\}$ is defined [6] as

$$
\begin{aligned}
C(k, n) &= 1/\sqrt{N}, k = 0, n = 0 \ldots N-1 \\
&= \sqrt{2/N} \cos(\Pi(2n+1)k)/2N, \\
k &= 1 \ldots N-1 \\
N &= 0 \ldots N-1
\end{aligned} \tag{57.1}
$$

The cosine transform is orthogonal, that is

$$C = C* => C - 1 = CT \tag{57.2}$$

On applying the DCT, the input signal will get decomposed into a set of basis images. For highly correlated data, cosine transforms show excellent energy compaction. Most of the energy will be represented by a few transform coefficients.

57.4 Radial Basis Function

Radial functions are a special class of function. Their characteristic feature is that their response decreases (or increases) monotonically with distance from a central point. The center, the distance scale, and the precise shape of the radial function are parameters of the model, all fixed if it is linear. A typical radial function is the Gaussian which, in the case of a scalar input, is

$$h(x) = \exp(-(x - c)^2/r^2). \tag{57.3}$$

Its parameters are its centre c and its radius r. Universal approximation theorems show that a feed forward network with a single hidden layer with non linear units can approximate any arbitrary function [Y]. No learning is involved in RBF networks. For pattern classification problems, the numbers of input nodes are equal to the number of elements in the feature vector and the numbers of output nodes are equal to the number of different clusters.A Radial Basis Function network is shown in Fig. 57.1.

57.5 Method

The method explains detection of faces from video frames and still images. This is followed by extraction of facial features, recognition of faces and analyzing facial expressions.

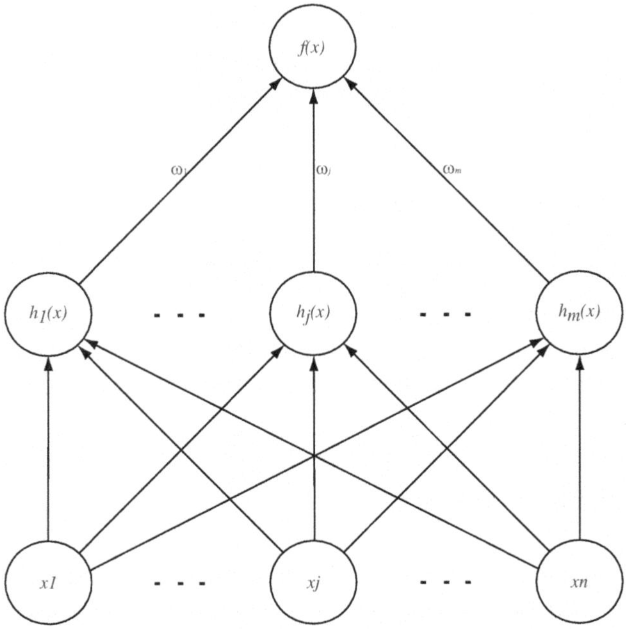

Fig. 57.1 Radial basis function network

57.5.1 Facial Region Identification

The first step in face detection problem is to extract facial area from the background. In our approach, both still and video images are used for face detection. Image frames from video are extracted first. The input images contain regions other than face such as hair, hat etc. Hence it is required to identify the face. Each pixel in the image is classified as skin pixel or non-skin pixel. Different skin regions are detected from the image. Face regions are identified from the detected skin region as in [9] which addressed the problem of face detection in still images. Some randomly chosen frames with different head pose, far away from the camera, and expressions from video face images are extracted as shown in Fig. 57.2. The difference image at various time instances is shown in Fig. 57.3.

The portion of the image that is moving is assumed as head. The face detection algorithm used RGB color space for the detection of skin pixels. The pixels corresponding to skin color of the input image is classified according to certain heuristic rules. The skin color is determined from RGB color space as explained in [10, 11]. A pixel is classified as skin pixel if it satisfies the following conditions.

$$\begin{aligned} &\mathrm{R} > 95 \text{ AND } \mathrm{G} > 40 \text{ AND } \mathrm{B} > 20 \text{ AND} \max\{\mathrm{R},\mathrm{G},\mathrm{B}\} - \min\{\mathrm{R},\mathrm{G},\mathrm{B}\} \\ &\quad > 15 \text{ AND } |\mathrm{R}-\mathrm{G}| > 15 \text{ AND} \mathrm{R} > \mathrm{G} \text{AND} \mathrm{R} > \mathrm{B} \end{aligned} \tag{57.4}$$

OR

Fig. 57.2 Some frames from face video database

Fig. 57.3 Difference image

Fig. 57.4 Edge detection and skin region identification from video frames

$$R > 220 \mathrm{AND} G > 210 \mathrm{AND} B > 170 \mathrm{AND} |R - G| \leq 15 \mathrm{AND} RB \mathrm{AND} G > B \quad (57.5)$$

Edge detection is performed in each frame. Edge detection and skin regions identified from the color images of video frames and still images are shown in Figs. 57.4 and 57.5.

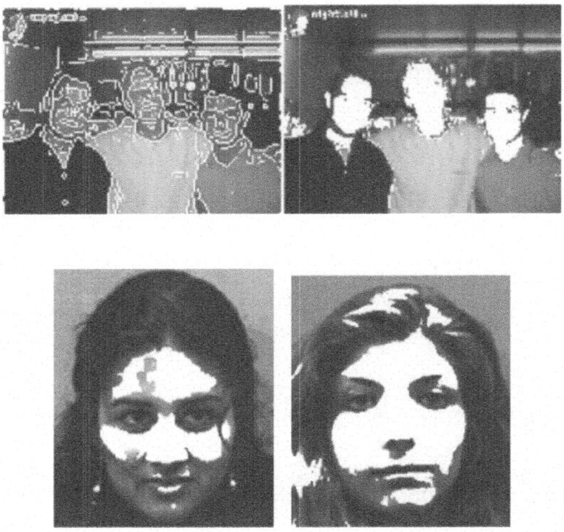

Fig. 57.5 Edge detection and skin region identification from still images

From these skin regions, it is possible to identify whether a pixel belongs to skin region or not. To find the face regions, it is necessary to categorize the skin pixels in to different groups so that it will represent some meaningful groups such as face, hand etc. Connected component labeling is performed to classify the pixels. In the connected component labeling operation, pixels are connected together geometrically. In this, 8-connected component labeling used so that each pixel is connected to its eight immediate neighbors. At this stage, different regions are identified and each region has to be classified as a face or not. This is done by finding the skin area of each region. If the height to width ratio of those skin region falls with in the range of golden ratio ($(1 + \sqrt{5})/2 \pm$ tolerance), then that region is considered as a face region.

57.5.2 Face Recognition

57.5.2.1 Segmentation

Image segmentation is a long standing problem in computer vision. There are different segmentation techniques which divides spatial area with in an image to different meaningful components. Segmentation of images is based on the discontinuity and similarity properties of intensity values. Cluster analysis is a method of grouping objects of similar kind in to respective categories. It is an exploratory data analysis tool which aims at sorting different objects in to groups in a way that degree of association between two objects is maximal if they belong to same group and minimal otherwise. K-Means is one of the unsupervised learning algorithms that solve the clustering problems. The procedure follows a simple and easy way to classify a given data set through a certain number of clusters (assume K clusters) fixed a priori. The idea is to define K centroids, one for each cluster. These centroids should be placed in a cunning way because of different location causes different results. So the better choice is to place them as much as possible far away from each other. The next step is to take each point belonging to a given dataset and associate it to the nearest centroid. When no point is pending, the first step is completed and an early groupage is done. At this point K new centroids need to be re-calculated as barycentres of the clusters resulting from previous step. After getting these K new centroids, a new binding has to be done between the same dataset points and the nearest new centroid. A loop has been generated. As a result of this loop it may be noticed that the K centroids change their location step by step until no more changes are done. If we know the number of meaningful groups/classes based on the range of pixel intensity, weighted K-means clustering can be used to cluster the spatial intensity values.

In facial images the skin color and useful components can be generally classified as two different classes. But if two class based clustering is used, it may result in components that may be still difficult to identify. So three classes are used and are able to cluster the data in useful manner. Initial cluster centers are calculated using histogram. Then K-means clustering algorithm computes distance between the different pixels and cluster centers and selects a minimum distance cluster for each pixel. This process continues until all pixels are classified properly.

The results of clustering algorithm are shown in Fig. 57.6. Class-I is selected since it is possible to separate the components properly compared to other classes. Then connectivity algorithm is applied to all the components in the clustered face.

57.5.2.2 Feature Extraction

Eye regions are located in the upper half of skin region and can be extracted using the area information of all the connected components. Using eye centers, orientation is corrected by rotation transformation. After calculating the inter eye distance, nose and mouth are identified as they generally appear along the middle of two eyes in the lower half. The inter eye distance, nose length and mouth area are computed.

To keep track of the overall information content in the face area, DCT is used. On applying DCT, most of the energy values can be represented by a few coefficients. First 64×64 coefficients are taken as part of the feature set. The face area calculated from the skin region is taken as another parameter. Parameters like inter eye

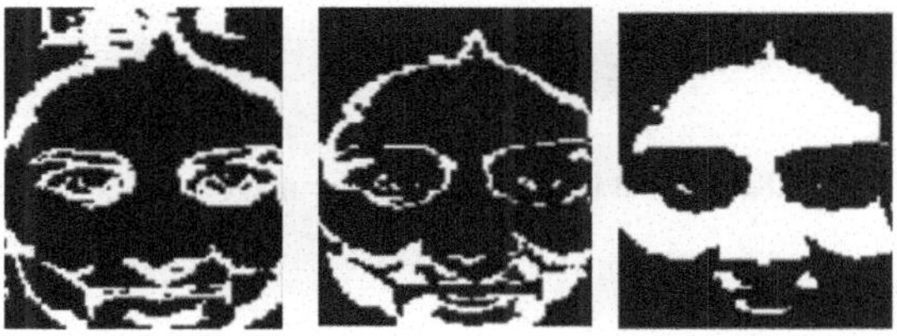

Fig. 57.6 Clustering

distance, nose length, mouth position, face area and DCT coefficients are computed and given to a RBF based neural network.

57.5.3 Facial Expression Analysis

For expression analysis, face images are considered from JAFFE face database [12]. Figure 57.7 shows some images in the database. There were 'K' images with 'N' expressions for each face so that 'K × N' face images are used as the database. Normalization is done to make the images with uniform scale. The facial expressions are analyzed by facial feature extraction. Facial expressions are dominated in eye and mouth regions. These regions are located first and then extracted by cropping the image. The face image with its features cropped is shown in Fig. 57.8.

DCT is applied to these cropped images. The DCT coefficients are given to the RBF based neural network which classifies the expressions. Four expressions namely 'Happy', 'Normal', 'Surprise' and 'Angry' are considered for the analysis.

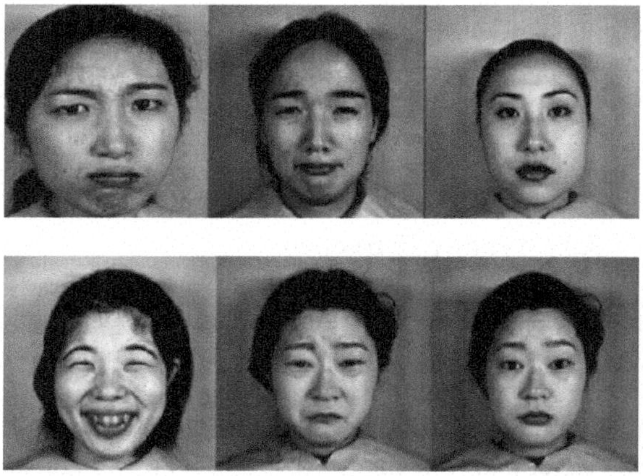

Fig. 57.7 Various facial expressions

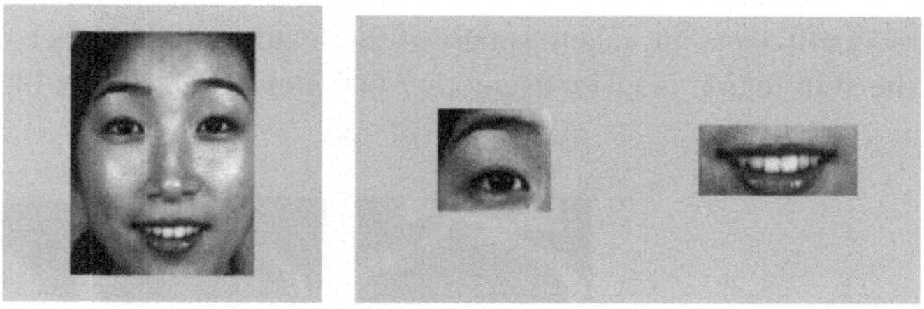

Fig. 57.8 Original image and cropped image

57.6 Results

In this experiment 200 frames of video images with considerable variations in head poses, expressions, camera viewing angle were used. The 'face 94' color image database was used for still images. One hundred face images were selected with considerable expression changes and minor variation in head turn, tilt and slant. The performance ratios were 90% for video image sequence and 92% for still images. The results of face detection for both still and video images are shown in Figs. 57.9 and 57.10. Figure 57.11 shows distinct connected components. Rotation transformation is done by eyes. Figure 57.12 shows the located features. Geometric parameters and DCT coefficients were given to a classification network for recognition.

Facial expressions are analyzed using JAFFE face database. Three sets of datasets from JAFFE database were used. Four expressions namely 'Happy', 'Normal', 'Surprise' and 'Anger' were analyzed. The DCT coefficients obtained from the cropped regions of faces which were very sensitive to facial expressions were given to the neural network which classifies the expressions with in the face. The average performance ratio is 89.11%. The efficiency plots for three sets of images are shown in Fig. 57.13.

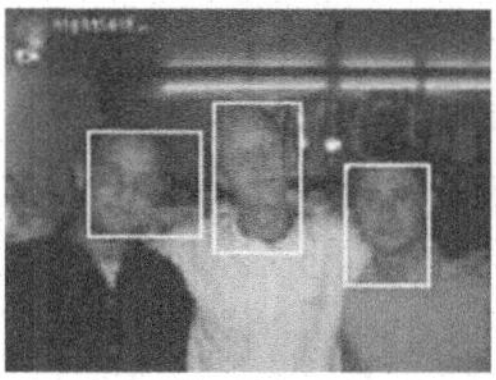

Fig. 57.9 Face detection from still images

Fig. 57.10 Face detection from video images

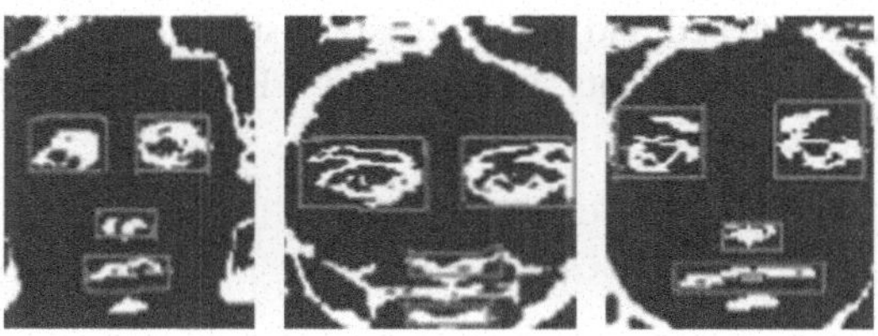

Fig. 57.11 Distinct connected components

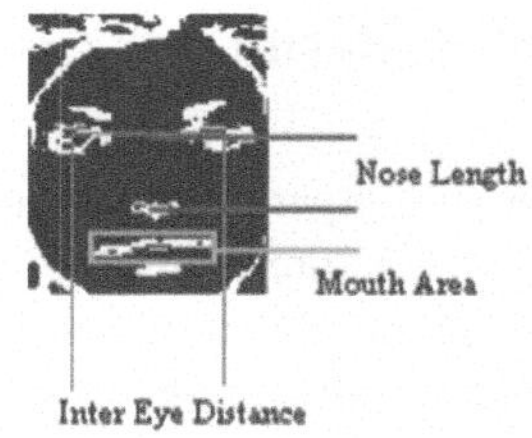

Fig. 57.12 Located features

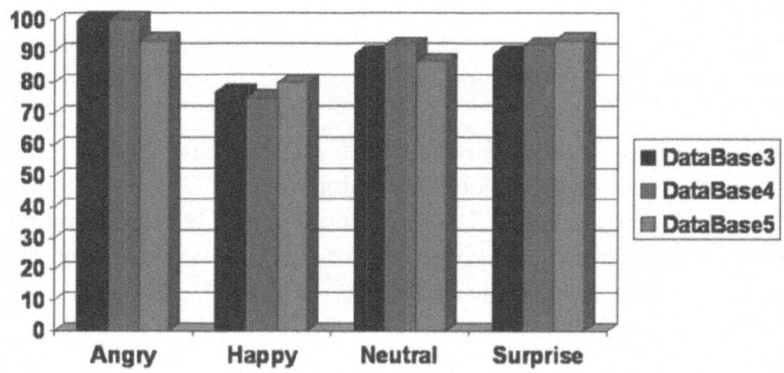

Fig. 57.13 Efficiency plot of facial expressions

57.7 Conclusion

In this paper, faces are detected and facial features are located from video and still images. 'NRC-IIT Facial video database' is used as image sequences and 'face 94 color image database' is used for still images. Skin pixels and non-skin pixels are separated and skin region identification is done by RGB color space. From the extracted skin region, skin pixels are grouped to some meaningful groups to identify the face region. From the face region, facial features are located using segmentation technique. Orientation correction is done by using eyes. Parameters like inter eye distance, nose length, mouth position, and DCT coefficients are computed which is used for a RBF based neural network. In this experiment only one image sequence is used for detection of faces.

Facial expressions namely 'Happy', 'Neutral', 'Surprise' and 'Anger' are analyzed using JAFFE face database. Facial features such as eyes and mouth regions are cropped and these areas are subjected to discrete cosine transformation. The DCT coefficients are given to a RBF neural network which classifies the facial expressions.

References

1. Rama Chellappa, Charles L. Wilson and Saad Sirohey, "Human and Machine Recognition of Faces: A Survey" In Proceedings of the IEEE, Vol. 83, No. 5, 1995, 705–740.
2. Nathan Faggian, Andrew Paplinski, and Tat-Jun Chin, "Face Recognition from Video Using Active Appearance Model Segmentation", 18th International Conference on Pattern Recognition, ICPR 2006, Hong Kong, pp. 287–290.
3. Hichem Sahbi, Nozha, Boueimma, Tistarelli, J. Bigun, A.K. Jain (Eds), and "Biometric Authentication" LNCS2539, Springer, Berlin/Heidelberg.
4. Marian Stewrt Bartlett, Javier R, Movellan and Terrence J Seinowski, "Face Recognition by Independent Component Analysis", IEEE Transactions on Neural Networks, Vol. 113, No. 6, November 2002.
5. Tsuyoshi Moriyama, Takeo Kanade, Jing Xiao and Jeffrey F. Cohn "Meticulously Detailed Eye region Model and Its Application to Analysis of Facial Images", IEEE Transactions in Pattern Analysis and Machine Intelligence, Vol. 28, No. 5, May 2006.
6. Yan Tong, Weinhui Lio and Qiang Ji, "Facial Action Unit Recognition by Exploiting Their Dynamic and Semantic Relationships", IEEE Transactions on Pattern Analysis and Machine Intelligence, Vol. 29, No. 10, October 2007.
7. Dmitry O. Gorodnichy "Video-based Framework for Face Recognition in Video", Second Workshop on Face Processing in Video. (FPiV'05) in Proceedings of Second Canadian Conference on Computer and Robot Vision (CRV'05), Victoria, BC, Canada, 9–11 May, 2005, pp. 330–338.
8. Hong-Bo Deng, Lian-Wen Jin, Li-Xin Zhen and Ian-Cheng Huang, "A New Facial Expression Recognition Method Based on Local Gabor Filter Bank and PCA plus LDA", International Journal of Information Technology, Vol. 11, No. 11, 2005 pp. 86–96.
9. K. Sandeep and A.N. Rajagopalan, "Human Face Detection in Cluttered Color Images Using Skin Color and Edge Information". Proc. Indian Conference on Computer Vision, Graphics and Image Processing, Dec. 2002.
10. P. Peer and F. Solina, "An Automatic Human Face Detection Method", Proceedings of 4th Computer Vision Winter Workshop (CVWW'99). Rastenfeld, Australia, 1999, pp. 122–130.
11. Franc Solina, Peter Peer, Borut Batagelj and Samo Juvan,"15 Seconds of Frame-An Interactive Computer Vision Based Art Installation", Proceedings of 7th International Conference on Control, Automation, Robotics and Vision (ICARCV 2002), Singapore, 2002, pp. 198–204.
12. Michael J. Lyons, Shigeru Akamatsu, Miyuki Kamachi and Jiro Gyoba,"Coding Facial Expressions with Gabor Wavelets", Proceedings, Third IEEE International Conference on Automatic Face and Gesture Recognition, April 14–16, 1998, Nara Japan, IEEE Computer Society, pp. 200–205.

Chapter 58
Spiking Neurons and Synaptic Stimuli: Neural Response Comparison Using Coincidence-Factor

Mayur Sarangdhar and Chandrasekhar Kambhampati

Abstract In this chapter, neural responses are generated by changing the Inter-Spike-Interval (ISI) of the stimulus. These responses are subsequently compared and a coincidence factor is obtained. Coincidence-factor, a measure of similarity, is expected to generate a high value for higher similarity and a low value for dissimilarity. It is observed that these coincidence-factors do not have a consistent trend over a simulation time window. Also, the lower-bound limit for faithful behaviour of coincidence factor shifts towards the right with the increase in the reference ISI of the stimulus. In principle, if two responses have a very high similarity, then their respective stimuli should be very similar and could possibly be considered the same.

However, as results show, two spike trains generated by highly-varying stimuli have a high coincidence-factor. This is due to limitations imposed by the one-dimensional comparison of coincidence-factor.

Keywords Spiking Neuron · Synaptic Stimuli · Neural Response · Coincidence-Factor · Inter-Spike-Interval

58.1 Introduction

The responses of a neuron to various types of stimuli have been studied extensively over the past years [1–9]. Stimulus-dependent behaviour of neurons has already been pursued to understand the spiking responses and it is thought that either the firing rate or firing time of individual spikes carries specific information of the neuronal response [3, 10–16]. The response of the neurons studied above has a constant magnitude whose variance is very low. In this paper, the neural responses fluctuate

---

M. Sarangdhar (✉)
Department of Computer Science, University of Hull, Cottingham Road, Hull, East-Yorkshire HU6 7RX
E-mail: M.Sarangdhar@2006.hull.ac.uk

S.I. Ao and L. Gelman (Eds.), *Advances in Electrical Engineering and Computational Science*, Lecture Notes in Electrical Engineering.

and a one-dimensional analysis based on firing times is shown to be insufficient for comparison.

A supra-threshold static current stimulus is sufficient to induce spiking behaviour in a neuron. The magnitude of these action potentials is considered to be almost the same and their variance is thus ignored. Such responses have been studied and models to depict their spiking behaviour have been proposed and implemented [17, 28]. On the other hand, a synaptic current is used to stimulate the same neuron [3]. This synaptic stimulus comprises of a static and a pulse component and is of particular interest as it induces fluctuations in the membrane voltage. These responses can be compared by their firing times [18,20,23–26] using a measure of comparison known as coincidence-factor. Here, the generality of this approach is investigated for a Hodgkin-Huxley (H-H) neuron [29] for which a synaptic current induces membrane fluctuations.

In this chapter, neural responses are generated by changing the Inter-Spike-Interval (ISI) of the stimulus. These responses are subsequently compared and a coincidence factor is obtained. Coincidence-factor, a measure of similarity, is expected to generate a high value for higher similarity and a low value for dissimilarity. It is observed that these coincidence-factors do not have a consistent trend over a simulation time window. Also, the lower-bound limit for faithful behaviour of coincidence factor shifts towards the right with the increase in the reference ISI of the stimulus. In principle, if two responses have a very high similarity, then their respective stimuli should be very similar and could possibly be considered the same. However, as results show, two spike trains generated by highly-varying stimuli have a high coincidence-factor. This is due to limitations imposed by the one-dimensional comparison of coincidence-factor. Elsewhere, [30,31] have worked on temporal patterns of neural responses but do not specifically address this issue. Thus, in order to differentiate spike trains with fluctuating membrane voltages, a two dimensional analysis is necessary taking both firing time and magnitude of the action potentials.

58.2 Neuronal Model and Synapse

58.2.1 The Neuron Model

The computational model and stimulus for an H-H neuron is replicated from [3]. The differential equations (Eqs. (58.1–58.3)) of the model are the result of non-linear interactions between the membrane voltage V and the gating variables m, h and n. for Na^+ and K^+.

$$\left.\begin{array}{l} C\frac{dv}{dt} = -g_{Na}m^3h(V - V_{Na}) - g_K n^4(V - V_K) \\ \qquad -g_L(V - V_L) + I_i \end{array}\right\} \tag{58.1}$$

$$\left.\begin{aligned}\frac{dm}{dt} &= -(\alpha_m + \beta_m)m + \alpha_m \\ \frac{dh}{dt} &= -(\alpha_h + \beta_h)h + \alpha_h \\ \frac{dn}{dt} &= -(\alpha_n + \beta_n)n + \alpha_n\end{aligned}\right\} \quad (58.2)$$

$$\left.\begin{aligned}\alpha_m &= 0.1(V + 40)/[1 - e^{-(V+40)/10}] \\ \alpha_h &= 0.07e^{-(V+65)/20} \\ \alpha_n &= 0.01(V + 55)/[1 - e^{-(V+55)/10}] \\ \beta_m &= 4e^{-(V+65)/18} \\ \beta_h &= 1/[1 + e^{-(V+35)/10}] \\ \beta_n &= 0.125e^{-(V+65)/80}\end{aligned}\right\} \quad (58.3)$$

The variable V is the resting potential where as V_{Na}, V_K, and V_L are the reversal potentials of the Na^+, K^+ channels and leakage. $V_{Na} = 50\,mV$, $V_K = -77\,mV$ and $V_L = -54.5\,mV$. The conductance for the channels are $g_{Na} = 120\,mS/cm^2$, $g_K = 36\,mS/cm^2$ and $g_L = 0.3\,mS/cm^2$. The capacitance of the membrane is $C = 1\,\mu F/cm^2$.

58.2.2 The Synaptic Current

An input spike train described in (Eq. (58.4)) is used to generate the pulse component of the external current.

$$U_i(t) = V_a \sum_n \delta(t - t_f) \quad (58.4)$$

where, t_f is the firing time and is defined as

$$t_{f_{(n+1)}} = t_{f_{(n)}} + T \quad (58.5)$$

$$t_{f_{(1)}} = 0 \quad (58.6)$$

T represents the ISI of the input spike train and can be varied to generate a different pulse component. The spike train is injected through a synapse to give the pulse current I_P.

$$I_P = g_{syn} \sum_n \alpha(t - t_f)(V_a - V_{syn}) \quad (58.7)$$

g_{syn}, V_{syn} are the conductance and reversal potential of the synapse. The α-function is defined in [32] as

$$\alpha(t) = (t/\tau)e^{-t/\tau}\Theta(t), \tag{58.8}$$

where, τ is the time constant of the synapse and Θ(t) is the Heaviside step function. $V = 30\,mV$, $\tau = 2\,ms$, $g_{syn} = 0.5\,mS/cm^2$ and $V_{syn} = -50\,mV$.

58.2.3 The Total External Current

The total external current applied to the neuron is a combination of static and pulse component

$$I_i = I_S + I_P + \varepsilon \tag{58.9}$$

where, I_S is the static and I_P is the pulse current, ε is the random Gaussian noise with zero mean and standard deviation $\sigma = 0.025$. [3] has ignored the noise in the external current and the current consists of only two terms. However, the presence of noise is necessary in the simulation of a biological activity and hence considered.

58.3 Comparison of Two Spike Trains

58.3.1 Responses of the Neuron

The static component I_S of the external current is set at 25 μ A. The H-H neuron is stimulated with a current $I_i = I_S + I_P$ and its response is recorded. The fluctuations in the membrane are due to the specific nature of synaptic stimulus. The amplitude of the action potential in Fig. 58.1 is not constant and the standard deviation is $\sigma_{Amp} = 3.0978$. Hence, the amplitude of the response is not ignored. This is one major difference between [3, 30, 31] and this work. The synaptic time constant of 2 ms defines the shape of the pulse current. As the refractory period of an H-H neuron is 1–2 ms, we choose a 2 ms bound for coincidence detection. The simulation activity is divided into three sets of ISIs. Each set has a corresponding reference ISI (T_{ref}). The first set compares responses generated using stimulus ISI between 14–16 ms while the second set compares responses of ISIs between 13–15 ms. The third set compares responses for ISIs varied between 15–17 ms. The responses for each set are compared with a fixed response known as the reference response. The reference response for each set is unique and is generated by varying the stimulus ISI. Reference ISIs for the sets are 15, 14 and 16 ms respectively. Neural responses are recorded for various ISIs within a set and compared with the reference response for that set. For set 1, the reference spike train is generated with T = 15 ms (T_{ref}) and compared with responses generated with T = 14–16 ms. Coincidence factors are calculated to estimate the similarity between these responses.

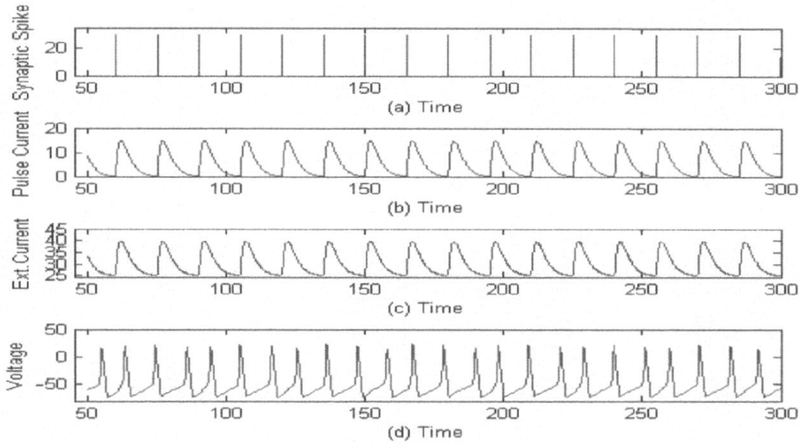

Fig. 58.1 Response of the H-H neuron to with T = 15 ms causing fluctuations in membrane voltage. (**a**) The synaptic spike train input that induces a pulse current. (**b**)The pulse current generated. (**c**) The total external current applied to the neuron. Note that there is a static offset. (**d**) The neuronal response to the current

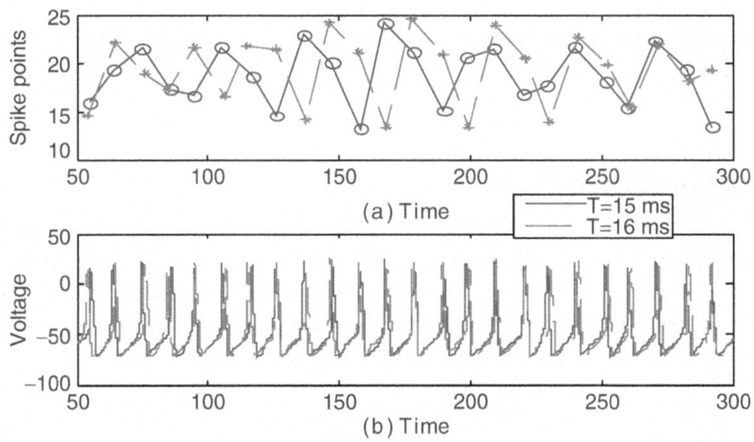

Fig. 58.2 Comparison of responses. (**a**) The corresponding magnitude of spikes for the responses at T = 16 ms and T = 15 ms. (**b**) The two spike trains not only differ in firing times but also in magnitudes

58.3.2 Comparison of Responses

The response of the neuron is specific to an input stimulus. In order to generate different stimuli, we varied the ISI of the synaptic input from T = 14–16 ms with T = 15 ms as the reference ISI. Figures 58.2 and 58.3 show that the response of the neuron differs with respect to both firing time and magnitude. The figures indicate that the variation in the input ISI causes the membrane voltage to fluctuate. They also

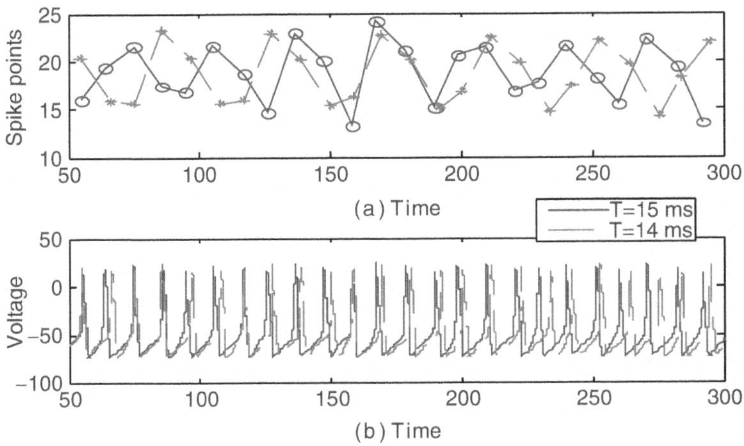

Fig. 58.3 Comparison of responses. (**a**) The corresponding magnitude of spikes for the responses at T = 14 ms and T = 15 ms. (**b**) The two spike trains not only differ in firing times but also in magnitudes

show the difference in responses generated with T = 14 ms & T_{ref} and $T = 16$ ms & T_{ref}.

58.3.3 Coincidence-Factor

The coincidence-factor, as described by [18, 20] is 1 only if the two spike trains are exactly the same and 0 if they are very dissimilar. Coincidence for an individual spike is established if its firing time is within 2 ms of the firing time of the corresponding spike in the reference spike train (in this case T = 15 ms).

$$\Gamma = \frac{N_{coinc} - \langle N_{coinc} \rangle}{1/2(N_1 + N_2)} \frac{1}{N} \tag{58.10}$$

where, N_1 is the number of spikes in the reference train, N_2 is the number of spikes in the train to be compared, N_{coinc} is the number of coincidences with a precision $\delta = 2ms$ between the spike trains. $\langle N_{coinc} \rangle = 2\nu\delta N_1$ is the expected number of coincidences generated by a homogeneous Poisson process with the same rate as the spike train to be compared. $N = 1 - 2\nu\delta$ is the normalising factor. For set 1, N_1 is the number of spikes in the reference spike train (T_{ref} = 15 ms) and N_2 is the number of spikes in the train to be compared (T = 14–16 ms). Figure 58.4 shows that the coincidence-factors for responses generated using T = 14–16 ms do not follow a fixed pattern. The coincidence-factor (Γ) is expectedly 1 when spike train generated with T = 15 ms is compared with the reference spike train T_{ref} (T = 15 ms). However, the coincidence factor for spike trains generated at T = 16 ms and T_{ref} is 1. This indicates that the two highly varying currents have an exactly similar

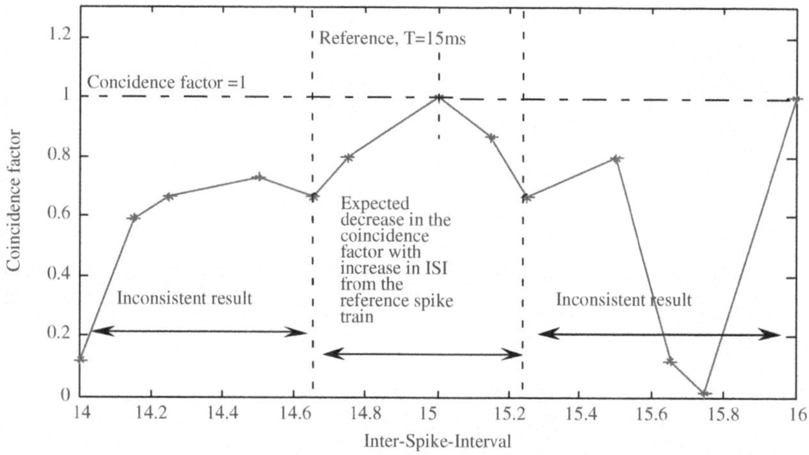

Fig. 58.4 Coincidence-factor versus ISI. The coincidence-factor decreases expectedly between T = 15–14.65 ms and T = 15–15.25 ms. At other times the result is inconsistent and does not have a fixed pattern

response or conversely as the responses are same; the two input stimuli are similar, which is an incorrect inference. The coincidence factor for the spike trains generated at T = 14 ms and T_{ref} is 0.1207 indicating very low similarity. From a mathematical and signal transmission standpoint, the coincidence-factor should decrease as the input stimulus increasingly varies from T_{ref}. However, this can only be observed between T = 14.65–15.25 ms (30% of the 2 ms time window). The coincidence-factor Γ increases from T = 14–14.5 ms but then drops till T = 14.65 ms. Γ steadily increases to 1 when T = 15 ms and drops for 0.25 ms. There is an upward rise from T = 15.25–15.5 ms, a sharp drop from T = 15.5–15.75 ms followed by a steep increase to $\Gamma = 1$ at T = 16 ms. Traversing from the reference the expected trajectory of the coincidence-factor breaks at T = 14.65 ms and T = 15.25 ms.

These are therefore taken as limits for faithful behaviour of the coincidence-factor approach. However, for set 2 reference spike train is chosen as T_{ref} = 14 ms, limits of faithful behaviour change (Fig. 58.5, left). The coincidence factor steadily rises to unity, stays there for 0.5 ms and drops gradually. Ideally, the coincidence-factor should be not 1 for T = 13.5, 13.65 and 13.75. While in set 3, Fig. 58.5, right, reference spike train chosen is at T_{ref} = 16 ms. The limits of faithful behaviour change with a change in the stimulus. There is a sharp rise in the coincidence factor from 15.75 to 16 ms where it reaches unity. From 16 to 17 ms the coincidence-factor executes a perfect curve as expected. From Figs. 58.4 to 58.6 it is conclusive that the lower-bound of faithful behaviour increases with the increase in the input reference ISI. The difference between the reference ISI (T_{ref}) and the lower-bound limit decreases with the increase in the reference ISI. It is also important to note that within each set of simulation, there are some false coincidences. The term false coincidence is used to identify comparisons whose coincidence factor is 1 – when it should not be. In Fig. 58.4, there is a false coincidence when ISI = 16 ms is

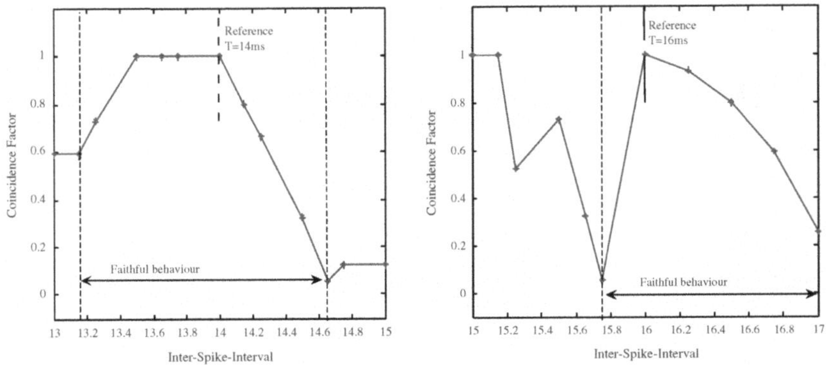

Fig. 58.5 Coincidence-factor versus ISI. *Left* – The coincidence-factor has a faithful behaviour between T = 13.15 ms and T = 14.65 ms. *Right* – The coincidence-factor has a faithful behaviour between T = 15.75–17 ms. It executes a perfect curve after 16 ms

compared with T_{ref} = 15 ms. In Fig. 58.5, left, false coincidences can be seen when ISI varied between 13.5–13.75 ms is compared with T_{ref} = 14 ms while in Fig. 58.5, right, false coincidences can be observed for ISI varied between 15 and 15.15 ms and compared with T_{ref} = 16 ms.

58.3.4 Two-Dimensional Analysis

The coincidence-factors over the 2 ms time window show an inconsistent trend. A one-dimensional approach of the coincidence-factor determination is thought to be the cause of this inconsistency. The coincidence-factor is highly accurate for spike trains with a constant amplitude response however; the coincidence-factor does not give a proper estimate of similarity between two spike trains with varying amplitudes. As a result, two visually distinct spike trains would still generate a high coincidence-factor (Figs. 58.2 and 58.3). A two-dimensional analysis of spike trains with fluctuating magnitudes can resolve this inconsistency. To support this, a simple binary clustering algorithm is used. It shows that the clustering solution for each response is unique to itself and therefore helps to eliminate any ambiguity.

58.3.5 Binary Clustering

The peak of each spike in a spike train is considered as an object. The number of objects for each spike train is equal to the number of spikes.

$$Obj = [Firingtime, Amplitude] \tag{58.11}$$

$$d_{rs}^2 = (N_r - N_s)(N_r - N_s)' \tag{58.12}$$

The Euclidean distance between object-pairs is calculated using (Eq. (58.12)) where, N_r, N_S are the objects in the spike train. Once the distance between each pair of objects is determined, the objects are clustered based on the nearest neighbour approach using

$$\left.\begin{aligned} d(r,s) &= \min(dist(N_{ri} - N_{sj})) \\ i &\in (1,\ldots,n_r),\ j \in (1,\ldots,n_s) \end{aligned}\right\} \tag{58.13}$$

where n_r, n_s is the total number of objects in the respective clusters. The binary clusters are plotted to form a hierarchical tree whose vertical links indicate the distance between two objects linked to form a cluster. A number is assigned to each cluster as soon as it is formed. Numbering starts from (m + 1), where m = initial number of objects, till no more clusters can be formed. We investigated the case described in Section 58.3.3 for the response generated at $T_{ref} = 15$ ms and T = 16 ms (false coincidence). The coincidence-factor for these responses is 1 (Fig. 58.4) and indicates an exact match. The clustering tree shows that these responses are actually different from each other by a margin not captured by coincidence-factor (Fig. 58.6a, b). The clustered objects are shown on the X-axis and the distance between them is shown on the Y-axis. A comparison of the clustering solutions shows that the shape, form, height as well as linkages are different for the two spike trains. In Fig. 58.6a, objects 12 and 13 are clustered together at a height of 11.5 while in Fig. 58.6b, objects 11 and 12 are clustered at a height of 13.5 – shown in green circles. Also, objects 4 and 5 are clustered in Fig. 58.6a while objects 3 and 4 are clustered in Fig. 58.6b – shown in red circles. This means that the spike trains are inherently different from each other. The results hence prove that the two spike trains are not an exact match. We therefore believe that though determining coincidence-factor is important, a two-dimensional analysis is necessary for responses with fluctuating membrane voltages.

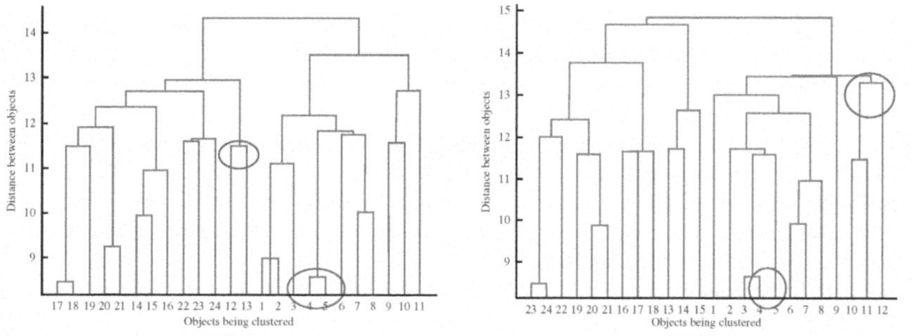

Fig. 58.6 (**a**) Clustering solution for T = 15 ms indicating objects being clustered. (**b**) Clustering solution for T = 16 ms indicating objects being clustered. The shape, form, height as well as linkages for each spike train are different

58.4 Conclusions

A synaptic stimulus known to induce fluctuations in the membrane voltage is used to stimulate an H-H neuron [3] to verify if firing time alone is enough to differentiate between these responses. The time constant of the pulse component of the external current is 2 ms and due to refractoriness of the neuron, coincidence-bound is also chosen as 2 ms. The coincidence-factors are calculated for time windows $t_1 = 14\text{–}16$ ms, $t_2 = 13\text{–}15$ ms and $t_3 = 15\text{–}17$ ms with reference spike trains at $T = 15$, 14 and 16 ms respectively. In all three sets of results, there is no consistent trend exhibited by the coincidence-factor. Also, the limits of faithful behaviour change and the percentage of acceptable results varies. The percentage of faithful behaviour for the three time windows is 30, 75 and 62.5 respectively. It is observed that: (a) the limits of faithful behaviour change with a change in the reference ISI. (b) The lower-bound limit of faithful behaviour increases with the increase of the reference ISI. (c) The difference between the reference ISI and the lower-bound limit of faithful behaviour decreases with the increase in the reference ISI. (d) In order to differentiate between these responses accurately, a two-dimensional analysis is required. A simple clustering algorithm easily differentiates two visually-distinct responses as opposed to the coincidence-factor. It effectively demonstrates the necessity of a two-dimensional approach to comparison of neural responses [33,34].

Clustering is primarily used to demonstrate the requirement of a two-dimensional analysis for comparing spike trains with fluctuating membrane voltages. We take this as a supporting claim for our future work.

References

1. Lundström I (1974). Mechanical wave propagation on nerve axons. *Journal of Theoretical Biology*, **45**: 487–499.
2. Abbott L F, Kepler T B (1990). Model neurons: From Hodgkin Huxley to Hopfield. *Statistical Mechanics of Neural Networks*, Edited by Garrido L, pp. 5–18.
3. Hasegawa H (2000). Responses of a Hodgkin-Huxley neuron to various types of spike-train inputs. *Physical Review E*, **61(1)**: 718.
4. Kepecs A, Lisman J (2003). Information encoding and computation with spikes and bursts. *Network: Computation in Neural Systems*, **14**: 103–118.
5. Fourcaud-Trocmé N, Hansel D, van Vreeswijk C, Brunel N (2003). How spike generation mechanisms determine the neuronal response to fluctuating inputs. *The Journal of Neuroscience*, **23(37)**: 11628–11640.
6. Bokil H S, Pesaran B, Andersen R A, Mitra P P (2006). A method for detection and classification of events in neural activity. *IEEE Transactions on Biomedical Engineering*, **53(8)**: 1678–1687.
7. Davies R M, Gerstein G L, Baker S N (2006). Measurement of time-dependent changes in the irregularity of neural spiking. *Journal of Neurophysiology*, **96**: 906–918.
8. Diba K, Koch C, Segev I (2006). Spike propagation in dendrites with stochastic ion channels. *Journal of Computational Neuroscience*, **20**: 77–84.
9. Dimitrov A G, Gedeon T (2006). Effects of stimulus transformations on estimates of sensory neuron selectivity. *Journal of Computational Neuroscience*, **20**: 265–283.

10. Rinzel J (1985). Excitation dynamics: Insights from simplified membrane models. *Theoretical Trends in Neuroscience Federal Proceedings*, **44(15)**: 2944–2946.
11. Gabbiani F, Metzner W (1999). Encoding and processing of sensory information in neuronal spike trains. *The Journal of Biology*, **202**: 1267–1279.
12. Panzeri S, Schultz S R, Treves A, Rolls E T (1999). Correlations and the encoding of information in the nervous system. *Proceedings of the Royal Society of London*, **B 266(1423)**: 1001–1012.
13. Agüera y Arcas B, Fairhall A L (2003). What causes a neuron to spike? *Neural Computation*, **15**: 1789–1807.
14. Agüera y Arcas B, Fairhall A L, Bialek W (2003). Computation in a single neuron: Hodgkin and Huxley revisited. *Neural Computation*, **15**: 1715–1749.
15. Izhikevich E M (2006). Polychronization: Computation with spikes. *Neural Computation*, **18**: 245–282.
16. Li X, Ascoli G A (2006). Computational simulation of the input-output relationship in hippocampal pyramidal cells. *Journal of Computational. Neuroscience*, **21**: 191–209.
17. Kepler T B, Abbott L F, Marder E (1992). Reduction of conductance-based neuron models. *Biological Cybernetics*, **66**: 381–387.
18. Joeken S, Schwegler H (1995). Predicting spike train responses of neuron models; in M.Verleysen (ed.), *Proceedings of the 3rd European Symposium on Artificial Neural Networks*, pp. 93–98.
19. Wang X J, Buzsáki G (1996). Gamma oscillation by synaptic inhibition in a hippocampal interneuronal network model. *The Journal of Neuroscience*, **16(20)**: 6402–6413.
20. Kistler W M, Gerstner W, Leo van Hemmen J (1997). Reduction of the Hodgkin-Huxley equations to a single-variable threshold model. *Neural Computation*, **9**: 1015–1045.
21. Izhikevich E M (2003). Simple model of spiking neurons. *IEEE Transactions on Neural Networks*, **14(6)**: 1569–1572.
22. Shriki O, Hansel D, Sompolinsky H (2003). Rate models for conductance-based cortical neuronal networks. *Neural Computation*, **15**: 1809–1841.
23. Jolivet R, Gerstner W (2004). Predicting spike times of a detailed conductance-based neuron model driven by stochastic spike arrival. *Journal of Physiology – Paris*, **98**: 442–451.
24. Jolivet R, Lewis T J, Gerstner W (2004). Generalized integrate-and-fire models of neuronal activity approximate spike trains of a detailed model to a high degree of accuracy. *Journal of Neurophysiology*, **92**: 959–976.
25. Jolivet R, Rauch A, Lüscher H-R, Gerstner W (2006). Integrate-and-fire models with adaptation are good enough: Predicting spike times under random current injection. *Advances in Neural Information Processing Systems*, **18**: 595–602.
26. Jolivet R, Rauch A, Lüscher H-R, Gerstner W (2006). Predicting spike timing of neocortical pyramidal neurons by simple threshold models. *Journal of Computational Neuroscience*, **21**: 35–49.
27. Clopath C, Jolivet R, Rauch A, Lüscher H-R, Gerstner W (2007). Predicting neuronal activity with simple models of the threshold type: Adaptive exponential integrate-and-fire model with two compartments. *Neurocomputing*, **70**: 1668–1673.
28. Djabella K, Sorine M (2007). Reduction of a cardiac pacemaker cell model using singular perturbation theory. *Proceedings of the European Control Conference 2007*, Kos, Greece, pp. 3740–3746.
29. Hodgkin A, Huxley A (1952). A quantitative description of membrane current and its application to conduction and excitation in nerve. *Journal of Physiology*, **117**:500–544.
30. Maršálek, P (2000). Coincidence detection in the Hodgkin–Huxley equations. *Biosystems*, **58(1–3)**.
31. Victor J D, Purpura K P (1997). Metric-space analysis of spike trains: Theory, algorithms and application. *Network: Computation in Neural Systems*, **8**: 127–164.
32. Park M H, Kim S (1996). Analysis of phase models for two coupled Hodgkin-Huxley neurons. *Journal of the Korean Physical Society*, **29(1)**: 9–16.

33. Sarangdhar M, Kambhampati C (2008). Spiking neurons: Is coincidence-factor enough to compare responses with fluctuating membrane voltage? In *World Congress on Engineering 2008: The 2008 International Conference of Systems Biology and Bioengineering*, London, UK, 2–4 July 2008, Vol. **2**, pp. 1640–1645.
34. Sarangdhar M, Kambhampati C (2008). Spiking neurons and synaptic stimuli: Determining the fidelity of coincidence-factor in neural response comparison. *Special Issue of IAENG Journal* (in print).

Chapter 59
Overcoming Neuro-Muscular Arm Impairment by Means of Passive Devices

Federico Casolo, Simone Cinquemani, and Matteo Cocetta

Abstract The present research was originated by the request of an organization of disables affected by muscular distrophy (UILDM) for the development of a passive system. Most of UILDM affiliated in fact, up to now, prefer to avoid any motor driven device except for the electrical wheelchair on which they are forced to live. Therefore, aim of the first part of the research is the development of a passive device as simple as possible, capable to enhance the upper limb mobility by acting against the gravity action. The device must be mounted on the frame of the subjects' wheelchair. Until now we designed and used only passive systems for preliminary analyses of patients response. These results are the basis for the development of our next passive device that will be optionally provided of an active module for the weaker subjects.

Keywords Neuro-Muscular Arm Impairment · Passive Device · muscular dystrophy

59.1 Introduction

Some degenerative neuromuscular diseases, such as dystrophy, affect muscles strength and force the subjects, few years after the appearance of the symptoms, to use electric wheelchairs. In fact, not only the legs muscles become too weak to sustain the upper body weight during walking, but also the upper limbs become soon inadequate to power an ordinary wheelchair. At a certain stage of the evolution of the pathology, due to the lack of muscular strength, the upper limbs are not able to counterbalance the gravity action, thus the subjects can only act, at the most, by

F. Casolo (✉)
Politecnico di Milano, Dipartimento di Meccanica – Campus Bovisa Sud, via La Masa 34, 20156 Milano, Italy
E-mail: federico.casolo@polimi.it

S.I. Ao and L. Gelman (Eds.), *Advances in Electrical Engineering and Computational Science*, Lecture Notes in Electrical Engineering.

moving the arm horizontally when it is supported by a plane. The forced immobility produces a faster degeneration of the muscular structure. The recovery of some active mobility of the arm is then important both to carry out some autonomous daily activity and to execute exercises as self physiotherapy. Assistive systems may help the subject in this task: they can be active, that means in general motor driven, or passive. Some active systems require to be driven by the controlateral limb – e.g. by replicating its movement or by means of a joystick – producing unnatural movements [1, 2]. The present research was originated by the request of an organization of disables affected by muscular distrophy (UILDM) for the development of a passive system. Most of UILDM affiliated in fact, up to now, prefer to avoid any motor driven device except for the electrical wheelchair on which they are forced to live. Therefore, aim of the first part of the research is the development of a passive device as simple as possible, capable to enhance the upper limb mobility by acting against the gravity action. The device must be mounted on the frame of the subjects' wheelchair. Until now we designed and used only passive systems for preliminary analyses of patients response. These results are the basis for the development of our next passive device that will be optionally provided of an active module for the weaker subjects.

59.2 Background

Presently some mechatronic devices are designed to increase subject physical abilities [3–7], but only few of them are suitable for subjects affected by muscular dystrophy.

In [8], the aim of the research is to increase the autonomy of people affected by muscular dystrophy and spinal muscular atrophy through a passive device. The system has four degrees of freedom: flexion extension of the shoulder, abduction adduction of the shoulder, flexion extension of the elbow and prono-supination of the forearm. The torque value at the joint is a non linear function of position, however gravity balance is achieved by means of linear springs (Fig. 59.1). To obtain

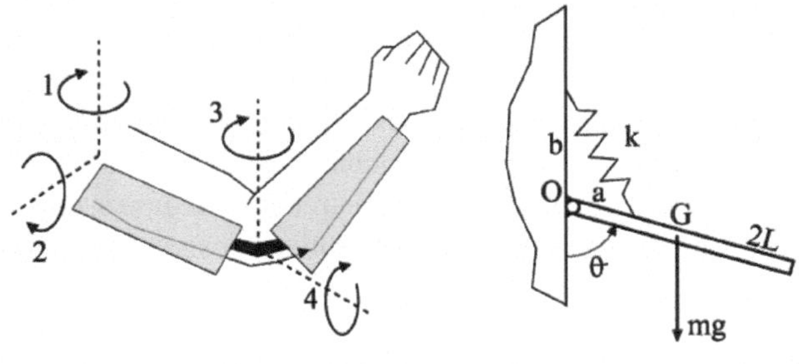

Fig. 59.1 (**a**) Rahman device – (**b**) schema of the static balance of a one d.o.f. structure with linear spring

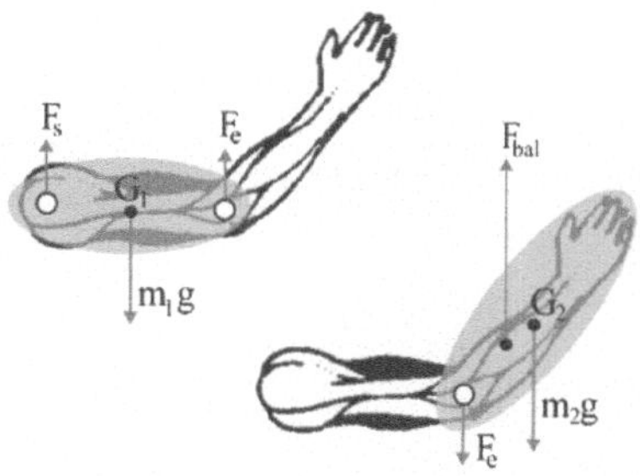

Fig. 59.2 Arm model

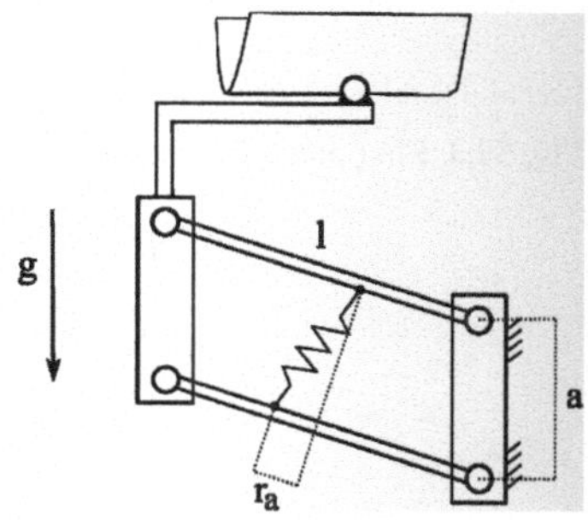

Fig. 59.3 DAS balancing system

the system balance regardless the value of θ, the spring stiffness value is set to: $k = mgl/(ab)$.

Armon orthosis [9, 10] is designed for people affected by spinal muscular atrophy. It assumes that arm weight can be sustained by natural shoulder and elbow (Fig. 59.2). Static balance is obtained by springs for about 75% of the arm weight, the interface between the device and the arm is located in the forearm, near the elbow. The system has two degrees of freedom in the vertical plane; the spring action is transmitted to the frame by a system of pulleys.

In [11], the Dynamic Arm Support (DAS) is designed to be mounted on the wheelchair of people affected by muscular dystrophy, muscular atrophy and amyotrophic lateral sclerosis. It's an hybrid device with an active operator-controlled electrical system to adjust weight compensation. Patient's forearm is balanced by a gravity compensation system, made by a planar linkage (Fig. 59.3) through a linear spring. The balance is guaranteed by equation: $r_a ak = mgl$.

59.3 Experimental Test Apparatus Set Up

Starting from a kinematic model of the human arm [12] with seven degrees of freedom – five d.o.f. for shoulder and two d.o.f. for elbow – a linkage system was designed to be coupled in parallel with the arm, avoiding to add constraints to the scapula-clavicle complex (Fig. 59.4).

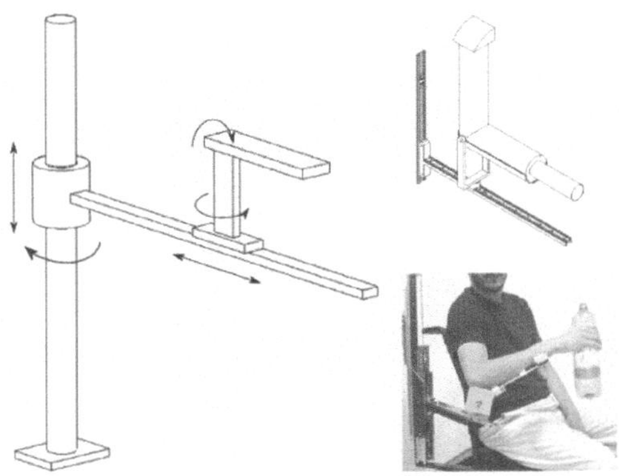

Fig. 59.4 First prototype mounted on a wheelchair

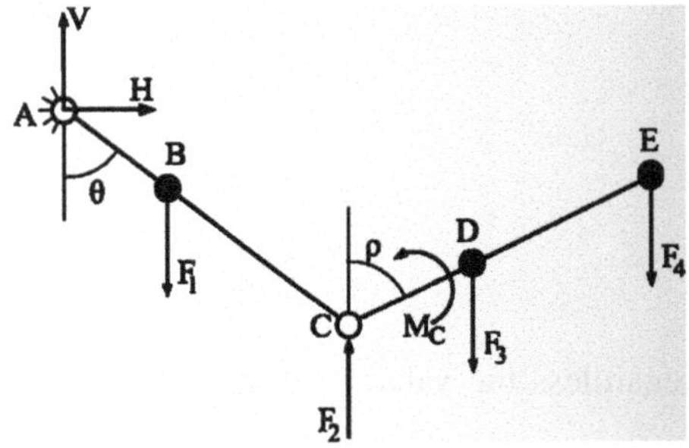

Fig. 59.5 Simplified scheme of loading actions on the arm

Aim of the system is to evaluate experimentally the gain of performance in arm motion due to the help of an anti gravity device. With this approach the device directly acts on the elbow (force F_2 on Fig. 59.5), providing the counterweight force and couple for balance, leaving subject's shoulder and trunk free to move. Thus the residual force becomes adequate to move the weight-compensated arm in its natural working volume, without any limitation of the motion due to the external device. The linkage includes two sliding pairs (for horizontal and vertical translations) and one revolute pair. Weight compensation at the elbow is achieved by a spiral spring at the revolute pair for the forearm flexo-extension and by a counterbalance moving in vertical to compensate the whole arm weight. In order to balance the system, forces and couples have been preliminary evaluated versus the arm position in space neglecting the inertia actions [13], which are small for the range of motion typical of this application.

The counterbalancing force F_2 and the counterbalancing torque M_C can be simply obtained as follow:

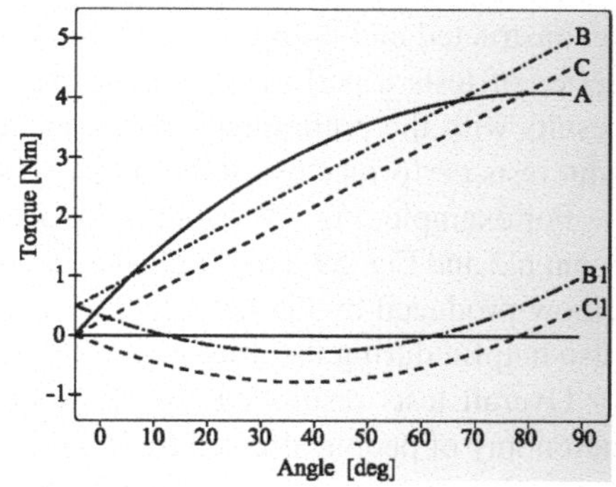

Fig. 59.6 Torque at the elbow

$$\sum F_V = 0 \rightarrow F_2 = \left(1 - \frac{AC - AB}{AC}\right) F_1 + F_3 + F_4 \tag{59.1}$$

$$\sum M_O = 0 \rightarrow M_C = F_3 CD \sin(\theta) + F_4 CF \sin(\theta) \tag{59.2}$$

The forearm counterbalancing torque (Fig. 59.6 – line A) is only function of the elbow angle $g\theta$, otherwise counterbalance weight is independent from the arm position and both of them are functions of the subject weight.

For the prototype a spiral spring has been designed to follow approximately the torque diagram and the vertical guide has been equipped with masses counterbalancing, through a cable and a pulley, the subject arm and the sliding structure weight. Line C in Fig. 59.6 shows the torque exerted by the spring and C1 represents the residual torques that the subject must exert to move the forearm. Lines B and B1 are related to a preloaded spring.

While counterweights can be easily adjusted on patients' weight and skills, spring action can only be tuned by varying the preload thanks to its adjustable support. Major variations of the torque can only be obtained by changing the spring. Preliminary tests allow to adjust the spring preload for each specific subject.

59.4 Experimental Tests

The prototype has been tested with some patients to evaluate its effectiveness for helping them to move the arm and to handle simple objects for daily tasks (e.g. drinking). The analyzed parameters are the changes produced by the external device to the joints excursions and to the trajectories followed by the hand to complete a task. The test protocol gives a description of the exercises performed by patients without any initial training. Every task is executed three times, starting from simple movements of a single joint, to more complex exercises (as to take a glass of water from a table and bring it to the mouth). Markers on upper body (arm and trunk) are acquired by an infrared motion capture system together with the EMG signals of the most important muscles. Therefore the system kinematics could be reconstructed and correlated with the timing of the muscular activity. For the first series of tests a performance gain index was also extrapolated by comparing the results with the preliminary tests obtained without using the assistive device. The four tests performed by all the subjects are listed in Table 59.1.

For example, Fig. 59.7 shows the data processed during the execution of movement n.2 and Fig. 59.8 highlights the increase of the range of motion of the subject's elbow produced by the helping device. Figure 59.9 shows how the same device is also helpful during the drinking task to reduce the lateral flexion of the trunk.

Overall tests results clearly show that an anti-gravity device can increase the autonomy of people affected by M.D. The subject's motion skills increase test after test and, moreover, in most cases the trunk posture improves.

Table 59.1 Test protocol

	Target movement	Involved muscles
1	Shoulder rotation to bring the forearm to the chest starting from a position where the elbow angle is 90°	Deltoid, pectorals et al.
2	Elbow flexo-extension	Biceps, triceps
3	Shoulder abdo-adduction	Deltoid, pectorals, triceps
4	Drinking: picking a glass it from a table in front of the subject	Deltoid, pectorals et al.

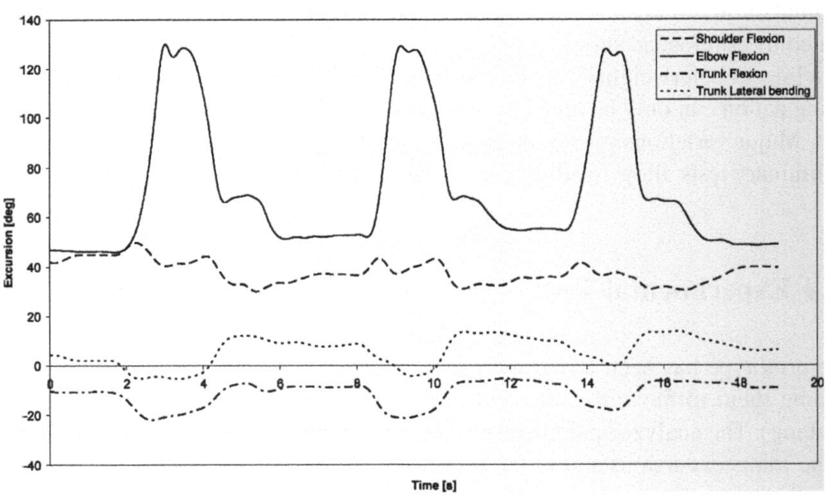

Fig. 59.7 Joints excursion with the helping device (movement 2)

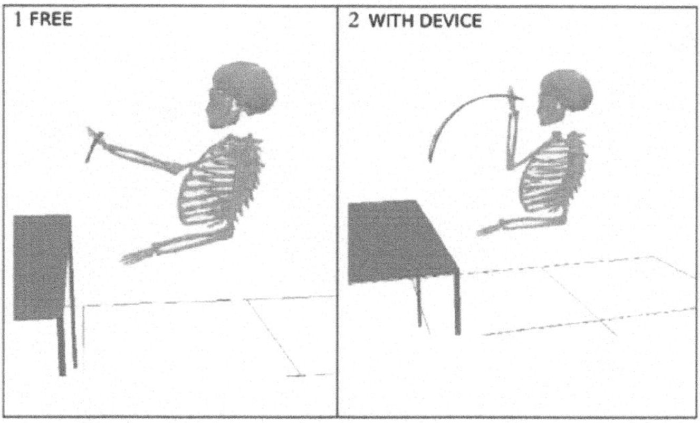

Fig. 59.8 Comparison of elbow max excursion

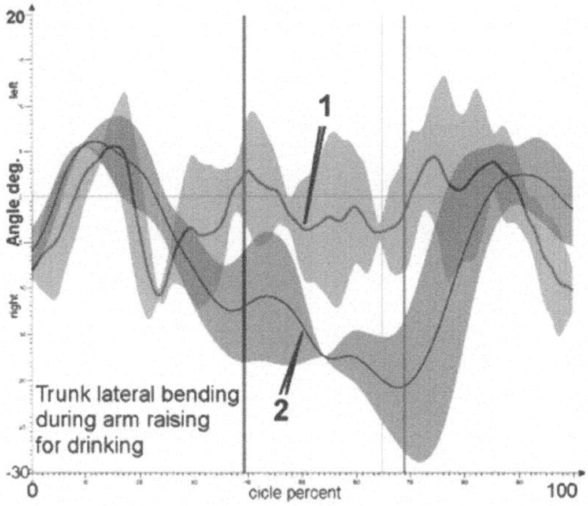

Fig. 59.9 Comparison of trunk bending (average values of one subject) for drinking task with the device (1) and without (2); areas show the st. deviat

59.5 Device Evolution

The prototype design is developed in order to obtain a better structure, statically balanced in every position, with a better interface with the operator and lower dimensions.

Two main solutions have been analyzed [14]: one with counterbalancing masses and the other with springs. Both structures support two degrees of freedom and provide static weight balance of the masses m_1 and m_2 of the arm and of the forearm in a vertical plane (Fig. 59.10).

The four-bars linkage provides weight compensation for every position of the structure (α and β angles). The second solution overcomes the problems due to counterweights, reducing the total weight of the device and its inertia. Relationships

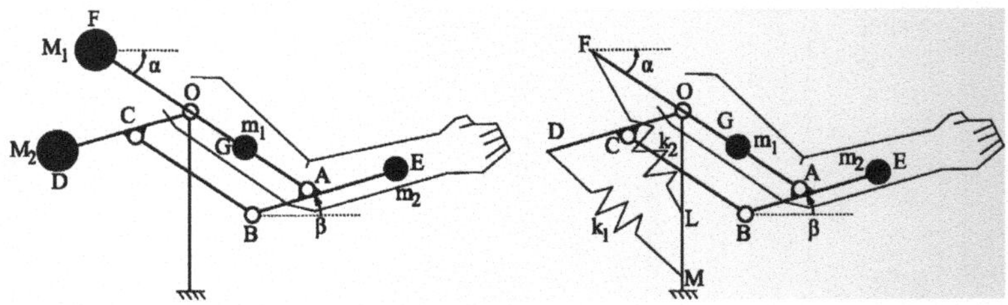

Fig. 59.10 Structures with counterweights and springs

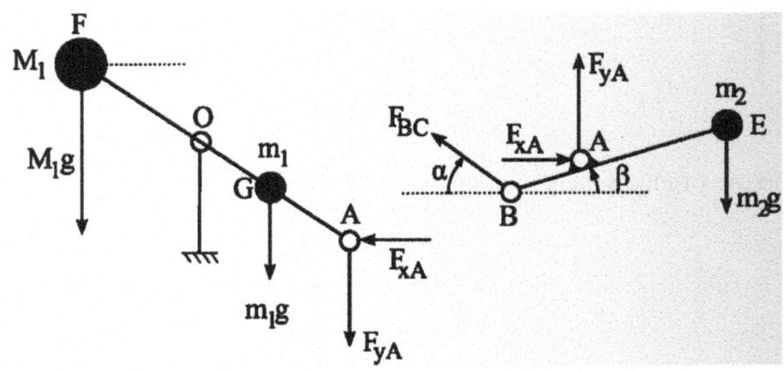

Fig. 59.11 Arm and forearm segments

Fig. 59.12 Counterbalance

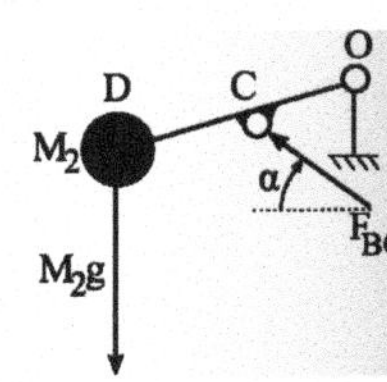

between masses, lengths and springs stiffness guarantee static equilibrium of the structure in every position.

Arm and forearm segments (respectively OA and AE in Fig. 59.11) are balanced for each value of α and β. For segment AE:

$$m_2 g AE \cos\beta + F_{BC} BA \sin\alpha \cos\beta + F_{BC} BA \sin\beta \cos\alpha = 0 \tag{59.3}$$

For the counterbalance segment (Fig. 59.12) equilibrium:

$$M_2 g OD \cos\beta + F_{BC} OC \sin\alpha \cos\beta + F_{BC} OC \sin\beta \cos\alpha = 0 \tag{59.4}$$

From Eqs. (59.3) and (59.9):

$$m_2 AE = M_2 OD \tag{59.5}$$

Analogously for balancing OA segment:

$$m_1 g OG \cos\alpha - M_1 g OF \cos\alpha + F_{y_A} OA \cos\alpha + F_{x_A} OA \sin\alpha = 0 \tag{59.6}$$

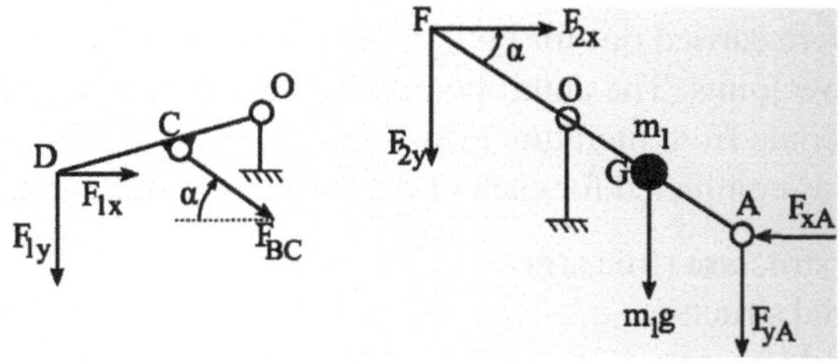

Fig. 59.13 Spring actions

From which:

$$m_1 OG + m_2 OA = M_1 OF \tag{59.7}$$

Equations (59.5) and (59.7) describe the static equilibrium of the structure; two values among M_1, M_2, OF and OD (directly related to structure dimensions and weight) can be arbitrarily chosen while the two others result from equations.

The equilibrium equations for the structure with springs can be achieved from analogous considerations.

Following notations on Fig. 59.13:

$$F_{1y} OD \cos\beta + F_{1x} OD \sin\beta + F_{BC} OC \sin\alpha \cos\beta + F_{BC} OC \cos\alpha \sin\beta = 0 \tag{59.8}$$

Where F_{1x} and F_{1y} are the components of elastic force: $F_{1x} = k_1 OD cos\beta$ and $F_{1y} = k_1 (OM - OD sin\beta)$. From Eq. (59.8), for the equilibrium:

$$m_2 g AE = k_1 OMOD \tag{59.9}$$

k_2 stiffness can be evaluated from $\sum M_O = 0$:

$$m_1 g OG \cos\alpha - F_{2y} OF \cos\alpha + F_{2x} OF \sin\alpha + F_{yA} OA \cos\alpha + F_{xA} OA \sin\alpha = 0 \tag{59.10}$$

With $F_{2x} = k_2 OF \cos\alpha$ and $F_{2y} = k_2 (OL + OF) \sin\alpha$.

Thus for OA segment:

$$(m_1 OG + m_2 OA)\, g = k_2 OFOL \tag{59.11}$$

Like the other configuration of the device, there are four unknown parameters (OM, OL, k_1, k_2), but only two of them are independent. Therefore, if the system geometry is constrained, it is enough to choose the spring stiffness and vice versa.

59.5.1 Simulations

To better evaluate and choose between the planned solutions, a multibody model of the human arm coupled with the device has been realized and torques in shoulder and elbow joints have been calculated. Inertia actions have been taken into account.

Simulations were carried out for both the new structures to calculate torques on shoulder and elbow joints. The anthropometric data (arm and forearm weights and centers of mass) come from literature [15].

Simulations was completed for each of the following configurations:

- Non balanced structure (free arm)
- Weight balanced structure
- Spring balanced structure

The torques supplied by the subjects have been evaluated for the following these different motion tasks:

1. Static initial position: $\alpha = 60°$, $\beta = 0°$, no motion
2. Motion 1: $\dot{\alpha} = 15°/\mathrm{s}$, $\beta = 0°$
3. Motion 2: $\dot{\alpha} = 0°/\mathrm{s}$, $\dot{\beta} = 10°/\mathrm{s}$
4. Motion 3: Arm flexion $\dot{\alpha} = 30°/\mathrm{s}$

Joint torque values with the balancing apparatus (either by weights or by springs) are lower than ones in unbalanced conditions, as confirmation of the device efficiency. Since in practice it is very hard to build a structure whose joints exactly remain overlapped to the arm articulations, the effect of the misalignment has also been investigated.

To evaluate how the torques exerted by shoulder and elbow can change, previous simulations were repeated with the new configurations where both human joints position have an offset with respect to the ideal one. Figure 59.14 shows the torques required during motion 2, starting for both joints with the same offset (30 mm horizontally and 70 mm vertically): although the new torques are higher than the ones evaluated in the ideal configuration (Figs. 59.14 and 59.15), they still remain acceptable.

59.6 Conclusions

Gravity counterbalancing systems can be profitable for increasing autonomous mobility of MD subjects. Also the built and tested simple preliminary device demonstrated to be able to recover the patient ability of performing some activities which are impossible without assistance. All the tested subjects demonstrated their satisfaction. The preliminary results are encouraging for the evolution of the passive counterbalancing systems optimized on subject anatomy and skills. The passive devices can help the subjects with enough residual muscular force, while for the late phases of MD an active system is required. The next step of the research requires a deeper analysis of the action requested to the muscles for the selected task, in order to exploit the muscles less weak because of MD, by means of specialized models. A new prototype is under development and in future it will include active modules. Active systems design will take advantage by the studies for the evolution of passive ones.

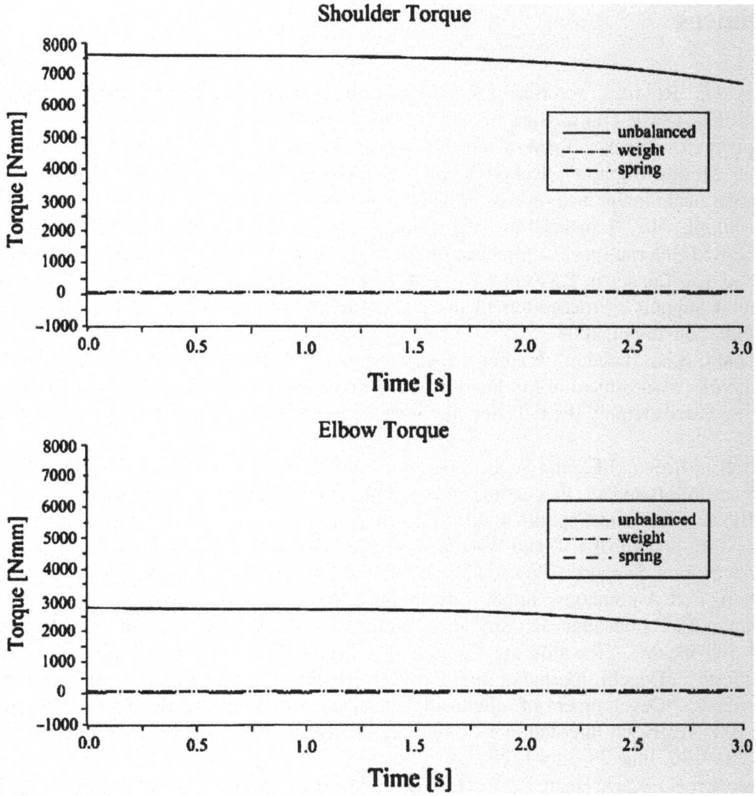

Fig. 59.14 Motion 2 – shoulder (up) and elbow (down) torques unbalanced and with weight and spring structure

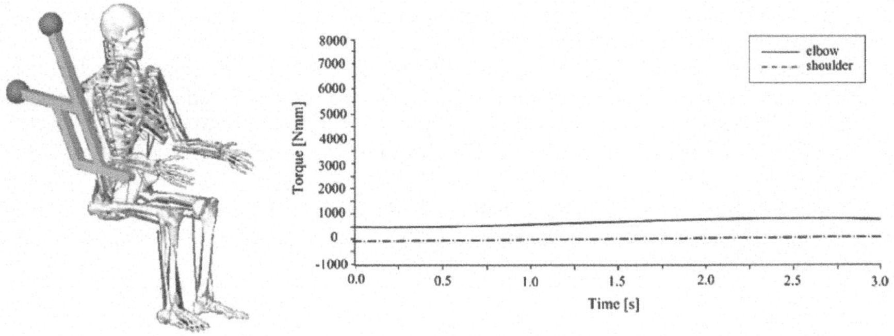

Fig. 59.15 Arm placed with joint off-set – influence on joints torques

References

1. Perry, J.C., Rosen, J., and Burns, S., "Upper-Limb Powered Exoskeleton Design", Mechatronics, IEEE/ASME Transactions on, vol. 12, no. 4, pp. 408–417, Aug. 2007.
2. Kiguchi, K., Tanaka, T., Watanabe, K., and Fukuda, T., "Exoskeleton for Human Upper-Limb Motion Support", Robotics and Automation, 2003, Proceedings. ICRA '03, IEEE International Conference on, vol. 2, pp. 2206–2211, 14–19 Sept. 2003.
3. Kazerooni, H., "Exoskeletons for Human Power Augmentation", Proceedings of the IEEE/RSJ. International Conference on Intelligent Robots and Systems, pp. 3120–3125.
4. Kazuo K., Takakazu T., Keigo W., and Toshio F., "Exoskeleton for Human Upper-Limb Motion Support", Proceedings of the 2003 IEEE International Conference on Robotics\& Automation Taipei, 2003.
5. Agrawal, S.K., Banala, S.K., Fattah, A., Sangwan, V., Krishnamoorthy, V., Scholz, J.P., and Hsu, W.L.,"Assessment of Motion of a Swing Leg and Gait Rehabilitation With a Gravity Balancing Exoskeleton", IEEE Transactions on Neural Systems and Rehabilitation Engineering, 2007.
6. Ball, S.J., Brown, I.E., and Scott, S.H., "A Planar 3DOF Robotic Exoskeleton for Rehabilitation and Assessment", Proceedings of the 29th Annual International Conference of the IEEE EMBS, Lyon, France, pp. 4024–4027, 23–26 Aug. 2007.
7. Kiguki, K. and Tanaka, T. and Watanabe, K. and Fukuda, T., "Exoskeleton for Human Upper-Limb Motion Support", Proceedings of the 2003 IEEE 10th International Conference on Robotics \& Automation, Taipei, Taiwan, pp. 2206–2211, 14–19 Sept.
8. Rahman T., Ramanathan R., Stroud S., Sample W., Seliktar R., Harwin W., Alexander M., and Scavina M., "Towards the Control of a Powered Orthosis for People with Muscular Dystrophy", IMechE Journal of Engineering in Medicine, vol. 215, Part H, pp. 267–274, 2001.
9. Herder, L., "Development of a Statically Balanced Arm Support: ARMON", Proceedings of the 2005 IEEE 9th International Conference on Rehabilitation Robotics, Chicago, IL, USA, pp. 281–286, June 28–July 1, 2005.
10. Mastenbroek, B., De Haan, E., Van Den Berg, M., and Herder, J.L., "Development of a Mobile Arm Support (Armon): Design Evolution and Preliminary User Experience", Proceedings of the 2007 IEEE 10th International Conference on Rehabilitation Robotics, Noordwijk, The Netherlands, pp. 1114–1120, 12–15 June.
11. Kramer, G., Romer, G., and Stuyt, H., "Design of a Dynamic Arm Support (DAS) for Gravity Compensation", Proceedings of the 2007 IEEE 10th International Conference on Rehabilitation Robotics, Noordwijk, The Netherlands, pp. 1042–1048, 12–15 June.
12. Keith K. and Barbara W., Anthropometry and Biomechanics, Man-Systems Integration Standards Revision B, NASA, July 1995.
13. Chandler, R.F., Clauser, C.E., McConville, J.T., Reynolds, H.M., and Young, J.W., Investigation of Inertial Properties of the Human Body. Technical Report DOT HS-801 430, Aerospace Medical Research Laboratory, Wright-Patterson Air Force Base, OH, March 1975.
14. Legnani, G., "Robotica Industriale", Milano, Casa Editrice Ambrosiana, 2003 ISBN 88-408-1262-8.
15. Chandler, R.F., Clauser, C.E., McConville, J.T., Reynolds, H.M., and Young, J.W. "Investigation of Inertial Properties of the Human Body". Wright Patterson Air Force Base, Ohio (AMRL-TR-75-137), 1975.

Chapter 60

EEG Classification of Mild and Severe Alzheimer's Disease Using Parallel Factor Analysis Method

PARAFAC Decomposition of Spectral-Spatial Characteristics of EEG Time Series

Charles-Francois Vincent Latchoumane, Francois-Benois Vialatte, Jaeseung Jeong, and Andrzej Cichocki

Abstract Electroencephalograms (EEG) recordings are now widely used more and more as a method to assess the susceptibility to Alzheimer's disease. In this study, we aimed at classifying control subjects from subjects with mild cognitive impairment (MCI) and from Alzheimer's disease (AD). For each subject, we computed the relative Fourier power of five frequency bands. Then for each frequency band, we estimated the mean power of five brain regions: frontal, left temporal, central, right temporal and posterior. There were an equivalent number of electrodes in each of the five regions. This grouping is very useful in normalizing the regional repartition of the information. We can form a three-way tensor, which is the Fourier power by frequency band and by brain region for each subject. From this tensor, we extracted characteristic filters for the classification of subjects using linear and nonlinear classifiers.

Keywords EEG Classification · Alzheimer's Disease · Parallel Factor Analysis Method · Spectral-Spatial Characteristics · Fourier power

60.1 Introduction

60.1.1 Diagnosis of Alzheimer's Disease and EEGs

Alzheimer's disease is the most prevalent neuropathology form leading to dementia; it affects approximately 25 million people worldwide and is expected to have fast recrudescence in near future [1]. The numerous clinical methods that are now

J. Jeong (✉)
Brain Dynamics Lab., Bio and Brain Engineering Dept., KAIST, Yuseong-gu, Guseong-dong, Daejeon, South Korea, 305-701
E-mail: jsjeong@kaist.ac.kr

S.I. Ao and L. Gelman (Eds.), *Advances in Electrical Engineering and Computational Science*, Lecture Notes in Electrical Engineering.

available to detect this disease including imaging [2, 3], genetic methods [4], and other physiological markers [5], however, do not allow a mass screening of the population. Whereas psychological tests such as Mini Mental State Evaluation (MMSE) in combination with an electrophysiological analysis (e.g. electroencephalograms or EEG) would be very efficient and inexpensive screening approach to detect the patients damaged by the disease.

EEG recordings are now widely used more and more as a method to assess the susceptibility to Alzheimer's disease, but are often obtained during steady states where temporal information does not easily reveal the relevant features for subject differentiation; interestingly, in those conditions, previous studies could obtain excellent classification results ranging from 84% to 100%, depending on the conditions (i.e. training or validation conditions for the classifier tests) and on the confronted groups (i.e. control subject, mild cognitive impairment (MCI), and different stages of Alzheimer's disease) demonstrating the promising use of resting EEGs for diagnosis of Alzheimer's disease [6–11].

The spectral spatial information estimated during resting states might contain valuable clues for the detection of demented subjects; however, the inter-subject variability, also influenced by the differences in the progression of the disease, might render the study difficult when undertaken subject-by-subject. In that case, a multi-way analysis would allow the extraction of information that is contained across subjects simultaneously considering the spectral spatial information. This methodology has been applied to epilepsy detection and has successfully characterized the epilepsy foci in a temporal-frequency-regional manner [12, 13]. Classification based on multi-way modeling has been performed on continuous EEGs [14] showing the power and versatility of multi-way analyses.

60.1.2 Multi-way Array Decomposition

Previous two-way analyses combining PCA-like techniques [6] (i.e. principal component analysis (PCA) and independent component analysis (ICA)) have shown very high performance in the classification of subjects and have assisted in early detection. These methods require data in the form of matrices, and then they limit the order (i.e. number of dimensions or modes) or mix several types of variables (e.g. using the unfolding method, also known as matricization). Either way, the naturally high-order of EEGs and interactions between the ways (or modes) is lost or destroyed, limiting further the understanding of underlying processes in the brain.

The multi-way array decomposition is a modeling tool that conserves the original high dimensional nature of the data. This method uses the common features and interactions between modes present in the data to create a model fit of the same data that can be decomposed into components. The decomposition into components provides the information of interactions between modes in the form of weight, which is relatively easier to interpret. The application of such methods to diagnosis of

Alzheimer's disease would allow characterization of the disease based on simple markers or weights.

Thus far, no application of multi-way array decomposition has been made in this type of database, dealing with the classification of Alzheimer's disease subjects based on EEG characteristics.

60.2 Subjects and EEG Recordings

The subjects analyzed in this study were obtained from a previously studied database [6, 9, 11] and consisted of eyes-open, steady state EEG recordings (20 s in duration), with over 21 leads disposed according to the 10–20 international system and digitalized at 200 Hz. The database contains 38 control (mean age: 71.7 ± 8.3) subjects, 22 mild cognitive impairment (MCI) subjects (mean age: 71.9 ± 10.2) who later contracted Alzheimer's disease, and 23 Alzheimer's disease patients (AD; mean age: 72.9 ± 7.5). The control subjects had no complaints or history of memory problems, and scored over 28 (mean score: 28.5 ± 1.6) on the mini mental state exam (MMSE). The MCI subjects had complaints about memory problems and scored over 24 at the MMSE (mean score: 26 ± 1.8). The inclusion criterion was set at 24 as suggested in [11], therefore encompassing MCI subjects with various cognitive deficits, but in the early stage of Alzheimer's disease. The AD patients scored below 20 on the MMSE and had had full clinical assessment. Thirty-three moderately or severely demented probable AD patients (mean MMSE score: 15.3 ± 6.4; range: 0–23) were recruited from the same clinic. After obtaining written informed consent from the patients and controls, all subjects underwent EEG and SPECT examination within 1 month of entering the study. All subjects were free of acute exacerbations of AD related co-morbidities and were not taking medication. The medical ethical committee of the Japanese National Center of Neurology and Psychiatry approved this study.

60.3 Method

60.3.1 Three-Way Tensor for Analysis

In this study, we aimed at classifying control subjects from subjects with MCI and from AD. For each subject, we computed the relative Fourier power of five frequency bands δ (1–4 Hz), θ (4–8 Hz), α_1 (8–10 Hz), α_2 (10–12 Hz) and β (12–25 Hz); then for each frequency band, we estimated the mean power of five brain regions: frontal, left temporal, central, right temporal and posterior. There were an equivalent number of electrodes in each of the five regions. This grouping is very useful in normalizing the regional repartition of the information (Fig. 60.1).

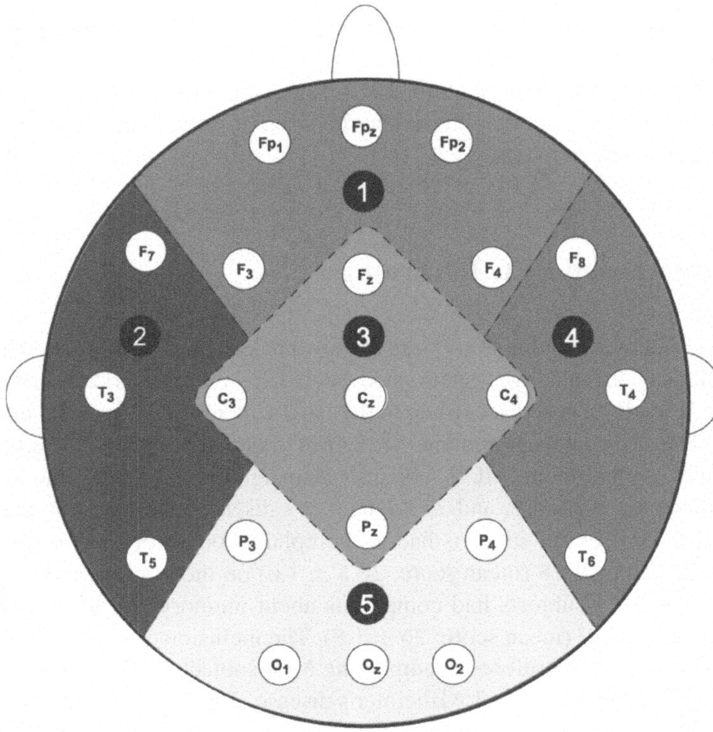

Fig. 60.1 Topological grouping of electrodes in five regions; the numbers 1, 2, 3, 4 and 5 denote the frontal, left temporal, central, right temporal and posterior regions, respectively

We can form a three-way tensor, *Subjects x Frequency Band Power x Brain Region*, which is the Fourier power by frequency band and by brain region for each subject. From this tensor, we extracted characteristic filters for the classification of subjects using linear and nonlinear classifiers. The detail of the method is described in the next sections.

60.3.2 Classification in Divide and Conquer Scheme

For simplicity and interpretation, we opted for two-step classification of the subjects (divide and conquer scheme): (1) we compared healthy subjects with the patient group (regrouping MCI and AD); (2) we compared MCI group with AD group. The schematic diagram of our approach is presented in Fig. 60.2.

For each step of classification, we extracted the features based on the Parallel Factor Analysis (PARAFAC) decomposition (unsupervised) and the reference filters (supervised). We estimated the accuracy in classification for subjects using a k-fold cross-validation with $k = 1$ (i.e. leave-one-out cross validation) on both

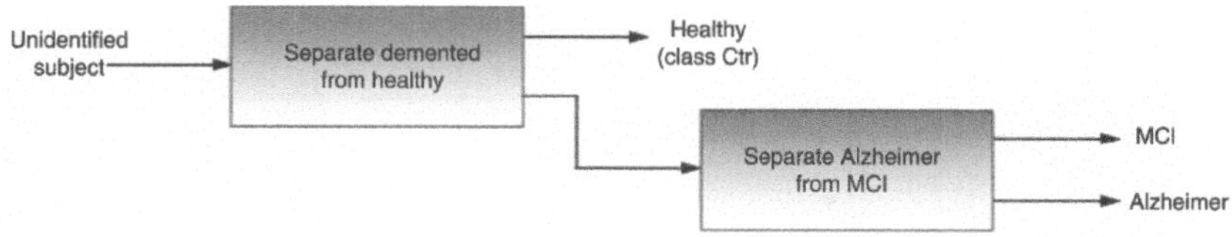

Fig. 60.2 Divide and conquer method for the classification of control subjects vs. MCI vs. AD subjects – cascade of classifiers

a quadratic naïve Bayes classifier (one of the simplest statistical classifier with quadratic boundary) and an artificial neural network (ANN; with $n = 1$ hidden neurons in one hidden layer, i.e. nonlinear classifier).

60.3.3 The PARAllel FACtor Analysis (PARAFAC)

The important idea underlying multi-way analysis is the extraction of multiple interactions within the structure of the data. Interactions and common patterns between modes of the data are often neglected cause studied separately (e.g. subject by subject, region by region) or folded on the same dimension (i.e. mixed, hence destroying interactions) as for method using two dimensional PCA or ICA. In this study, we constructed a three-way tensor in the form *Subjects x Frequency Band Power x Brain Region* to preserve relations between spectral and regional characteristics of each subject; we used a common modeling method for N-way analyses to extract common interaction in the tensor with relation to the subjects: the parallel factor analysis (PARAFAC) [15].

The PARAFAC is often referred to as a multilinear version of the bilinear factor models. From a given tensor, $X \in \mathbb{R}^{I \times J \times K}$, this model is able to extract linear decompositions of Rank-1 tensors.

$$\underline{\mathrm{X}} = \sum_{r=1}^{R} a_r \circ b_r \circ c_r + \underline{\mathrm{E}}, \tag{60.1}$$

where $\boldsymbol{a}_r$, $\boldsymbol{b}_r$, and $\boldsymbol{c}_r$ are the r-th column of the component matrices $A \in \mathbb{R}^{I \times \mathbb{R}}$, $B \in \mathbb{R}^{J \times \mathbb{R}}$, and $C \in \mathbb{R}^{K \times R}$, respectively, and $E \in \mathbb{R}^{I \times J \times K}$ is the residual error matrix (Fig. 60.3).

The operator $\circ$ designates the outer product of two vectors. The PARAFAC model under optimal fitting conditions (e.g. core consistency analysis) is capable of providing a model with the assumption of trilinearity relations between the dimensions (also called modes), thus providing a unique fit for the given data.

The fitting of the model depends on the number of components, R, chosen by the users, and for this approach we opted for validation of the model (i.e. number of components) based on the core consistency [16]. The optimal number R is chosen

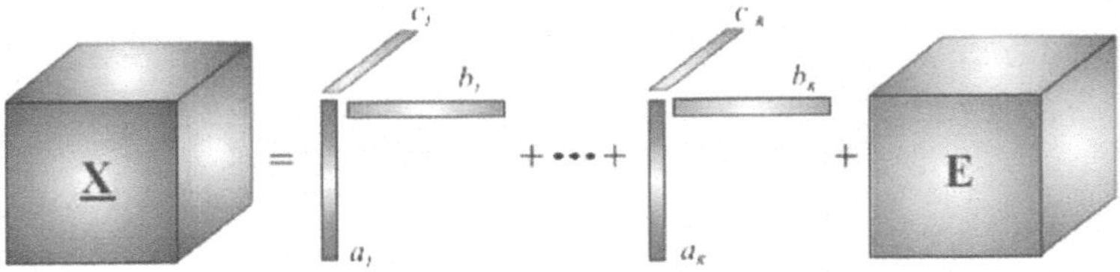

Fig. 60.3 PARAFAC modeling of a three-way tensor; each component (R = 2) is the outer product of Rank-1 tensor a, b, and c, and E is a residual tensor

for a core consistency value that is above 90% and with the core consistency value of $R + 1$ under 90%.

60.3.4 Filter Estimation and Feature Extraction

60.3.4.1 PARAFAC Filters

The PARAFAC model with a sufficient number of components and an appropriate fitting (i.e. core consistency [16] and component variability) is capable of fitting a model that decompose data in component of common characteristics with trilinear interaction between modes. We are interested in features that differentiate the subjects according to their clinical group. We assume that the common characteristic contain in the data could significantly discriminate the group based on the difference between the *Frequency Band Power x Brain Region* profile and normal control and demented subjects, i.e. shift in the frequency profile of subjects with AD or "slowing of frequencies" [17, 18].

The PARAFAC decomposition returns for each component and each mode a Rank-1 vector which can be considered as a weight at each component and each mode, respectively. The mode subject contains the discriminative features or weight of each subject to be compared with the combination of the mode Frequency Band Power and the mode Brain Region. More precisely, the component-wise combination (i.e. for each component) of "*Frequency Band Power*" and "*Brain Region*" could be interpreted as a characteristic filter. Similar to the Fourier Power Distribution which associates a power to each frequency, the decomposition obtained from PARAFAC can represent a distribution of filters (number of filters $R = 3$) associated with a weight (i.e. subject mode). We calculated the characteristic filters for each component, F_r, as described in Eq. (60.2):

$$F_r = b_r \times c_r^T, \tag{60.2}$$

where b_r and c_r are the Rank-1 vector of the frequency and region mode in the PARAFAC model, respectively. Following Eq. (60.2), we obtain a number of filters $R = 3$.

60.3.4.2 Reference Filters

The reference filter are supervised filters in contrast to the filters obtained using PARAFAC (unsupervised) because they integrate prior knowledge of the membership of the subjects. The reference filters of the control group and demented group were calculated as the average *Frequency Band Power x Brain Region* over the normal subjects and the demented subjects (regrouping MCI and AD), respectively. The reference filters of the MCI group and AD group were calculated as the average *Frequency Band Power x Brain Region* over MCI subjects and AD subjects, respectively.

60.3.5 From Filter to Feature

The reference filter and the PARAFAC filters contain common characteristics of the group under this study (i.e. control vs. demented or MCI vs. AD). We use a distance measure based on the matrix scalar product also known as normalized Frobenius inner product to compare the estimated filters (i.e. reference filters or PARAFAC filters) with the *Frequency Band Power x Brain Region* profile of each subject. The details on the distance measure are as follows:

$$Dis(F_1, F_2) = \frac{Trace(F_1^T F_2)}{||F_1||||F_2||}, \tag{60.3}$$

where F_1 and F_2 are two *Frequency Band Power x Brain Region* matrices, ^{T} denotes the conjugate transposition of a matrix and the Trace function returns the trace of a matrix. The function $||\cdot||$ indicates the Frobenius norm defined as $||F|| = \sqrt{Trace(F^T F)}$. For each subject, we obtained R features comparing the R PARAFAC filter to the subject's *Frequency Band Power x Brain Region* profile. Similarly, we obtained two features comparing the references filters to the subject's *Frequency Band Power x Brain Region* profile.

60.4 Results

The best classification results using the artificial neural network (ANN) and quadratic naïve Bayes classifier are displayed in Table 60.1. There was no noticeable difference between the performance of the two classifiers (both nonlinear) using either the reference filters or the PARAFAC filters; however, while comparing MCI with AD subjects, the ANN showed better performance using the PARAFAC filters (75.6%) than using the reference filters (68.9%).

Table 60.1 Classification results obtained using an ANN and quadratic naïve Bayes classifier with the leave-one-out cross-validation method; NC denotes normal controls and Patients denotes patient group (regrouping MCI group and AD group)

	NC vs. patients (PARAFAC) (%)	NC vs. patients (reference) (%)	AD vs. MCI (PARAFAC) (%)	AD vs. MCI (reference) (%)
ANN	74.7	61.4	75.6	64.4
Bayes	74.7	61.5	68.9	64.4

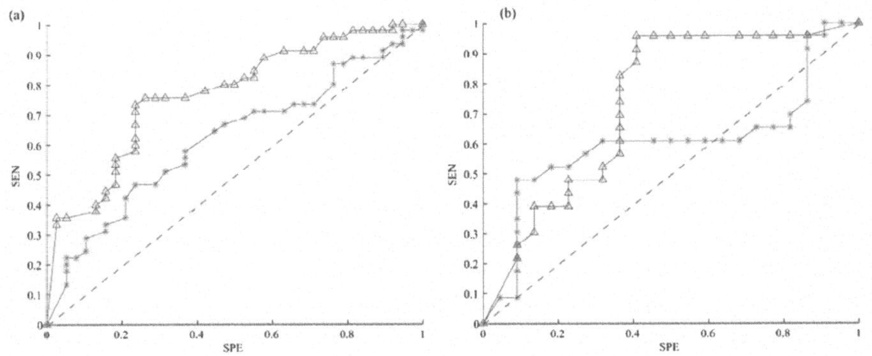

Fig. 60.4 ROC curve of classification accuracy of (**a**) control vs. demented subjects and (**b**) AD vs. MCI subjects using an ANN and leave-one-out cross-validation method; classification results obtained using the reference filters (stars) and using the filtered data extracted with a three-component PARAFAC ($R = 3$; triangle)

Generally, the PARAFAC filters showed an improvement in the accuracy of classification as compared with the reference filters and the two classification methods used.

As shown in Fig. 60.4, the performance in the classification based on the ANN exhibited higher accuracy using the information from the PARAFAC filters than when using the reference filters. The best performance was found to be 74.7% (75.6% sensitivity, 73.7% specificity) for the Ctr vs. demented classification (Fig. 60.4a) and 75.6% (95.6% sensitivity, 59.1% specificity) for MCI vs. AD classification (Fig. 60.4b).

Using PARAFAC on each separate group, a clustering of frequency bands (i.e. frequency band mode) can also be obtained [19] in order to study the interactions between the frequency ranges (Fig. 60.5). We observed the evolutions in the interrelations between the θ range and high frequency (α_2 and β) ranges: seemingly, for Control subjects and AD patients, the θ range activity is clustered with high frequencies (distances of 1.5 and 1, respectively); whereas for the MCI subjects, the θ range activity is much less (a distance of 2.5).

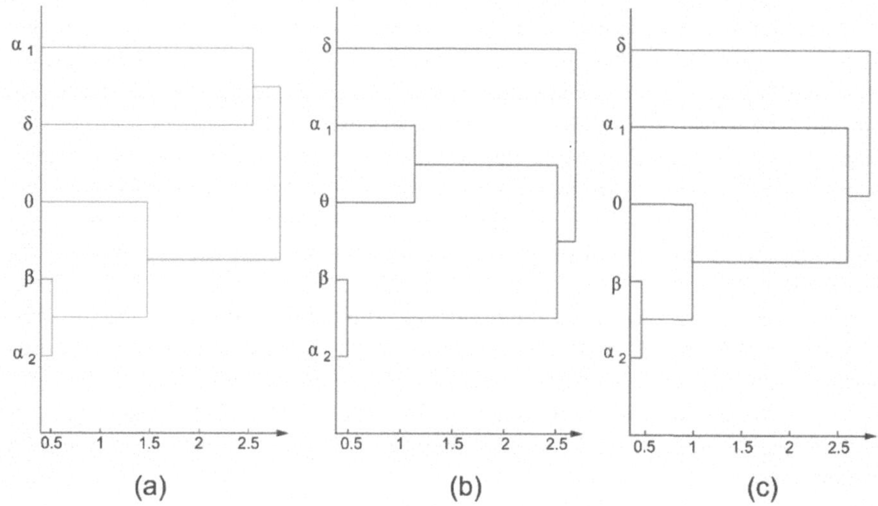

Fig. 60.5 Clustering of frequency bands; dendrograms extracted from the clustering of two-component PARAFAC models for (**a**) control subject, (**b**) MCI patients, and (**c**) Alzheimer patients

60.5 Discussion

In this study, we presented a method applied to the classification of subjects based on a PARAFAC decomposition of their EEG features. This type of application of multi-way analysis to EEG features has not yet been implemented for AD diagnosis. We also showed possible interpretation of the fitted model not only based on the spatial-frequency filters, but also based on the unimodal clustering for each group's model. The weight obtained using the PARAFAC decomposition could serve as a marker for differentiation between control and patient group but also between MCI and AD subjects.

Previous studies on Alzheimer's classification and compromising with different degrees of severity of the disease obtained classification accuracies ranging from 84% to 100% [6–11]. The classification results presented in this study (overall good classification of about 74%) are considerably good compared with the classification results using the same database in another study [6] (80%); however, this study proposes a three-class classification using a two step divide and conquer approach.

The capability of the method to make a stepwise differentiation of subjects illustrates the applicability of this study to the clinical domain. The method also demonstrates possibility of mass-screening of subjects for the diagnosis of AD as well as the differentiation of stages in the progression of the disease (MCI developed to turn out to be mild and severe).

In this study, we determined to use a three-way tensor confronting subjects, frequency band power and brain regions allowing simplicity of computation and interpretation. The method using multi-way array decomposition allows the integration of a large number of variables, organized on higher orders (dimensions of the tensor equal to D, typically with D higher than 3). The addition of more variables necessitate that those variables would commonly interact and characterize the data or, in our case, the subjects in the data, otherwise the extracted information might lose its property of separability. More importantly, increasing the number of variables or order would not seriously increase the number of features for classification (i.e. number of components of the PARAFAC model, R) which would avoid artificially increasing the separability of the groups and reducing the bias due to overfitting for the classifier.

Moreover, apart from its crucial uniqueness and the resulting easy interpretability [20], the PARAFEC model used might not be the best model to use in this analysis, as it imposes trilinearity conditions. Future investigation include the comparison of the accuracy of other models such as PARAFAC2 [21], Tucker3 [22], or the nonnegative tensor factorization method [20, 23] and its constrained model on sparsity [24].

Acknowledgements The first author would like to thank the Minister of Information and Technology of South Korea, Institute for Information and Technology Advancement (IITA) for his financial support. The authors would like to thank Dr. T. Musha for his contribution in subjects' analysis and the EEG recordings and Dr. M. Maurice and S. Choe for their help in the editing of the content.

References

1. Ferri, C.P. et al. Global prevalence of dementia: a Delphi consensus study. *The Lancet* **366**, 2112–2117 (2006).
2. Alexander, G.E. Longitudinal PET evaluation of cerebral metabolic decline in dementia: a potential outcome measure in Alzheimer's disease treatment studies. *American Journal of Psychiatry* **159**, 738–745 (2002).
3. Deweer, B. et al. Memory disorders in probable Alzheimer's disease: the role of hippocampal atrophy as shown with MRI. *British Medical Journal* **58**, 590 (1995).
4. Tanzi, R.E. & Bertram, L. New frontiers in Alzheimer's disease genetics. *Neuron* **32**, 181–184 (2001).
5. Andreasen, N. et al. Evaluation of CSF-tau and CSF-Aß42 as Diagnostic Markers for Alzheimer Disease in Clinical Practice, Archives of Neurology, **58**, pp. 373–379 (2001).
6. Cichocki, A. et al. EEG filtering based on blind source separation (BSS) for early detection of Alzheimer's disease. *Clinical Neurophysiology* **116**, 729–737 (2005).
7. Buscema, M., Rossini, P., Babiloni, C. & Grossi, E. The IFAST model, a novel parallel nonlinear EEG analysis technique, distinguishes mild cognitive impairment and Alzheimer's disease patients with high degree of accuracy. *Artificial Intelligence in Medicine* **40**, 127–141 (2007).
8. Huang, C. et al. Discrimination of Alzheimer's disease and mild cognitive impairment by equivalent EEG sources: a cross-sectional and longitudinal study. *Clinical Neurophysiology* **111**, 1961–1967 (2000).
9. Musha, T. et al. A new EEG method for estimating cortical neuronal impairment that is sensitive to early stage Alzheimer's disease. *Clinical Neurophysiology* **113**, 1052–1058 (2002).

10. Pritchard, W.S. et al. EEG-based, neural-net predictive classification of Alzheimer's disease versus control subjects is augmented by non-linear EEG measures. *Electroencephalography and Clinical Neurophysiology* **91**, 118–30 (1994).
11. Woon, W.L., Cichocki, A., Vialatte, F. & Musha, T. Techniques for early detection of Alzheimer's disease using spontaneous EEG recordings. *Physiological Measurement* **28**, 335–347 (2007).
12. Acar, E., Aykut-Bingol, C., Bingol, H., Bro, R. & Yener, B. Multiway analysis of epilepsy tensors. *Bioinformatics* **23**, i10–i18 (2007).
13. Acar, E., Bing, C.A., Bing, H. & Yener, B. in Proceedings of the 24th IASTED International Conference on Biomedical Engineering 317–322 (2006).
14. Lee, H., Kim, Y.D., Cichocki, A. & Choi, S. Nonnegative tensor factorization for continuous EEG classification. *International Journal of Neural Systems* **17**, 305 (2007).
15. Andersson, C.A. & Bro, R. The N-way toolbox for MATLAB. *Chemometrics and Intelligent Laboratory Systems* **52**, 1–4 (2000).
16. Bro, R. & Kiers, H.A.L. A new efficient method for determining the number of components in PARAFAC models. *Contract* **1999**, 10377 (1984).
17. Coben, L.A., Danziger, W.L. & Berg, L. Frequency analysis of the resting awake EEG in mild senile dementia of Alzheimer type. *Electroencephalography and Clinical Neurophysiology* **55**, 372–380 (1983).
18. Sloan, E.P., Fenton, G.W., Kennedy, N.S.J. & MacLennan, J.M. Electroencephalography and single photon emission computed tomography in dementia: a comparative study. *Psychological Medicine* **25**, 631 (1995).
19. Atkinson, A.C. & Riani, M. Exploratory tools for clustering multivariate data. *Computational Statistics and Data Analysis* **52**, 272–285 (2007).
20. Bro, R. PARAFAC. Tutorial and applications. *Chemometrics and Intelligent Laboratory Systems* **38**, 149–171 (1997).
21. Al Kiers, H., Ten Berge, J. & Bro, R. PARAFAC2: PART I. a direct fitting algorithm for the PARAFAC2 model. *Journal of Chemometrics* **13**, 275–294 (1999).
22. Kim, Y.D., Cichocki, A. & Choi, S. in IEEE International Conference on Acoustics, Speech, and Signal Processing (ICASSP-2008) (IEEE, Las Vegas, Nevada, 2008).
23. Cichocki, A., Zdunek, R. & Amari, S. Nonnegative matrix and tensor factorization. *Signal Processing Magazine, IEEE* **25**, 142–145 (2008).
24. Cichocki, A., Zdunek, R., Plemmons, R. & Amari, S. in ICANNGA-2007 (ed. Science, L.N.i.C.) 271–280 (Springer, Warsaw, Poland, 2007).

Chapter 61
Feature Selection of Gene Expression Data Based on Fuzzy Attribute Clustering

Elham Chitsaz, Mohammad Taheri, and Seraj D. Katebi

Abstract In this chapter, a novel approach which uses fuzzy version of k-mode has been proposed for grouping interdependent genes. The fuzzy approach considers uncertainty, achieves more stability and consequently improves the classification accuracy. In addition, other modifications have implemented to fuzzify the selection of the best features from each cluster. A new method to initialize cluster centers has also been applied in this work. Moreover, a novel discretization method based on Fisher criterion is also proposed.

Keywords Feature Selection · Gene Expression · Fuzzy Attribute Clustering · discretization method

61.1 Introduction

By now, many applications have been introduced in which, feature selection is utilized as a preprocessing stage for classification. This process speeds up both the training and reasoning stages, reduces memory space, and improves classification accuracy. Reducing the cost of gathering data is another advantage of feature selection. Biological data (e.g. micro-arrays) are usually wide and shallow.

Small number of samples narrows the acquirable knowledge. Hence it reduces the probability of correct reasoning whether a specified feature effects on the class label or not. Moreover, a classifier can generate the classification rules more easily with small number of features. But increasing the number of features may lead to ambiguity in training so that it would not even converge.

Finding proper features has been the subject of various approaches in the literature. (e.g. filter, wrapper, and embedded approaches [1], greedy [2], statistical

E. Chitsaz (✉)
Computer Science & Engineering Department of Shiraz University, Shiraz, Iran
E-mail: chitsaz@cse.shirazu.ac.ir

S.I. Ao and L. Gelman (Eds.), *Advances in Electrical Engineering and Computational Science*, Lecture Notes in Electrical Engineering.

approaches such as the typical principle component analysis (PCA) method [3] and the linear discriminant analysis (LDA) method [4], GA [5,6], Neural Network [7], Fuzzy Systems [8,9], mutual information-based feature selection [10,11]). Selection of the most influential genes, for classification of samples in micro-arrays, is one of the well known topics in which many investigations have been carried out. In the context of microarrays and LDA, wrapper approaches have been proposed [12]. Recently, biclustering algorithms [13,14] have been proposed to cluster both genes and samples simultaneously. Wai-Ho Au et al., in 2005 [15], proposed a feature selection technique based on clustering the correlated features by a semi-k-means method named k-modes.

In this paper a novel approach which uses fuzzy version of k-mode has been proposed for grouping interdependent genes. The fuzzy approach considers uncertainty, achieves more stability and consequently improves the classification accuracy. In addition, other modifications have implemented to fuzzify the selection of the best features from each cluster. A new method to initialize cluster centers has also been applied in this work. Moreover, a novel discretization method based on Fisher criterion [4] is also proposed.

C4.5 [16] classifier has been applied in the final stage to asses the feature selection process. Leukemia dataset [17], a micro-array data containing 73 samples and 7,129 genes, is used in all experimental results.

In next section, the previous work of feature selection by clustering on Microarrays [15] is explained following by its fuzzy approach in Section 61.3. Experimental results are presented and explained in Section 61.4. Finally we draw our conclusion in Section 61.5.

61.2 Related Work

The attribute clustering algorithm (ACA) has been applied by Wai-Ho Au et al., in 2005 [15], for grouping, selection, and classification of gene expression data which consists of a large number of genes (features) but a small number of samples.

This approach finds c disjoint clusters and assigns each feature to one of the clusters. The genes in each cluster should have high a correlation with each other while they are low correlated to genes in other clusters. This method uses the *interdependence redundancy measure* as the similarity measure.

To cluster features, the *k-modes* algorithm is utilized which is similar to the well known clustering method, *k-means* [18]. *Mode* of each cluster is defined as one of its features which has the largest *multiple interdependence redundancy* measure among other features in that cluster. The multiple interdependence redundancy measure is calculated for each feature by Eq.(61.1).

$$MR(A_i) = \sum_{\substack{A_j \in Cluster(i), \\ j \neq i}} R(A_i : A_j) \tag{61.1}$$

Where, *Cluster(i)* is the set of features which are in the same cluster with A_i and is the interdependence measure between the two features, A_i and A_j, which is defined by Eq. (61.2).

$$R\left(A_i : A_j\right) = \frac{I\left(A_i : A_j\right)}{H\left(A_i : A_j\right)} \tag{61.2}$$

Where, $I(A_i : A_j)$ is the mutual information between A_i and A_j as computed in Eq. (61.3).

$$\begin{aligned} I(A_i : A_j) = &\sum_{k=1}^{m_i}\sum_{l=1}^{m_j} \Pr\left(A_i = v_{ik}{}^\wedge A_j = v_{jl}\right) \\ &\log \frac{\Pr\left(A_i = v_{ik}{}^\wedge A_j = v_{jl}\right)}{\Pr\left(A_i = v_{ik}\right)\Pr\left(A_j = v_{jl}\right)} \end{aligned} \tag{61.3}$$

$H(A_i : A_j)$ is joint entropy of A_i and A_j which is given by Eq.(61.4).

$$\begin{aligned} H(A_i : A_j) = - &\sum_{k=1}^{m_i}\sum_{l=1}^{m_j} \Pr\left(A_i = v_{ik}{}^\wedge A_j = v_{jl}\right) \\ &\log \Pr\left(A_i = v_{ik}{}^\wedge A_j = v_{jl}\right) \end{aligned} \tag{61.4}$$

$H(A_i : A_j)$ is used to normalize $I(A_i : A_j)$ in Eq. (61.2). The larger value for $I(A_i : A_j)$, the higher interdependency of two features, A_i and A_j. Hence, there should be some pairs of values for these features which are simultaneously visited with high frequency and other pairs are less probable. Therefore, having one of the values, other one may be approximated considering value pairs with high probability.

K-modes is different from k-means in two points. First, mode of each cluster is selected as the cluster center instead of the mean. Second, use of Euclidean distance as the dissimilarity measure is substituted by the interdependency between attributes as a similarity measure.

In ACA method, genes are grouped into different clusters. A cluster is a set of features which are more correlated in comparison with features in other clusters. Hence, if there is a missed value for one of these features, it may be approximated considering other co-cluster features one by one. Therefore, a few features in each cluster may be sufficient to present properties of samples. But the selected features should be overall correlated with other features in the same cluster. This is the motivation of selecting features with highest *multiple interdependence redundancy* measure, as defined by Eq. (61.1), to represent the cluster.

Computing interdependency is defined on just discrete data types. Therefore, to determine the interdependence measure between two features, the range of all continuous features should be first discretized into a finite number of intervals. This is done by the Optimal Class-Dependent Discretization algorithm (OCDD).

61.3 The Proposed Fuzzy Approach

A novel method is proposed which combines the effectiveness of ACA with fuzzy k-means algorithm [19]. In this method each feature is assigned to different clusters with different degrees. This comes from the idea that each gene may not belong to just one cluster and it is much better to consider the correlation of each gene to features in entire clusters. Hence, during the selection of the best features, more accurate relations between genes are available. The main point here is that in selecting each feature, it is considered among the entire clusters not just one. Hence, in this case a feature, which is not correlated enough with members of one cluster but its correlation among entire clusters is high, gains more chance to be selected in comparison with crisp ACA. In this method better features might be selected for classification stage over crisp method. The stability of fuzzy clustering is also much higher in comparison with k-modes according to experimental results. In addition, Fuzzy Attribute Clustering Algorithm (FACA) converges smoothly whereas k-modes in ACA often oscillates between final states.

In the proposed method matrix $U_{k \times m}$represents the membership degree of each gene in each cluster where k is the number of clusters that is fixed and determined at first and m is the number of features. Matrix U is computed by Eq. (61.5).

$$u_{ri} = \frac{1}{\sum\limits_{c=1}^{k} \left(\frac{R(A_i, \eta_c)}{R(A_i, \eta_r)} \right)^{\frac{2}{m-1}}} \tag{61.5}$$

Where, k is number of clusters. u_{ri} is membership degree of *ith* feature in *rth* cluster and m is a weighting exponent. Afterwards, u_{ri} is normalized by Eq. (61.6).

$$u_{ri}^{new} = \frac{u_{ri}^{new}}{\sum\limits_{l=1}^{k} u_{li}^{old}} \tag{61.6}$$

According to this membership matrix, fuzzy multiple interdependence redundancy measure is defined by Eq. (61.7) which is a modified version of Eq. (61.1).

$$u_{ri}^{new} = \frac{u_{ri}^{new}}{\sum\limits_{l=1}^{k} u_{li}^{old}} \tag{61.7}$$

Where, p is the total number of features and r is the cluster number in which the multiple interdependence redundancy measure of feature A_i is calculated. Hence, in calculating $MR_r(A_i)$ the entire features are considered.

Indeed, fuzzy multiple interdependence redundancy measure should be computed for each cluster separately since each feature is not belonged to just one cluster. In this approach, mode of a cluster is updated to the feature with the highest fuzzy multiple interdependence redundancy in that cluster. Fuzzy multiple interdependence redundancy of each feature should be calculated regardless of its own membership degree in associated cluster. Considering this membership degree, mode of each cluster will never change, since it has high membership degree in that cluster.

The objective function in the proposed fuzzy method is computed as Eq. (61.8).

$$J = \sum_{r=1}^{k} \sum_{i=1}^{p} u_{ri}^{m} R\left(A_i : \eta_r\right) \tag{61.8}$$

Where, k and p are the number of clusters and features respectively and η_r is mode of *rth* cluster which represents center of that cluster.

Selection of the best features is based on the rank of each feature which is calculated as Eq. (61.9).

$$rank\left(A_i\right) = \sum_{r=1}^{k} u_{ri}^{m} MR_r\left(A_i\right) \tag{61.9}$$

The final stage is the classification of data points according to selected features. The following items encouraged us to use C4.5 as the final classifier for assessment of the proposed feature selection.

- Both C4.5 and proposed feature selection method can be applied just on discrete features.
- C4.5 is a well known decision tree technique with high performance which is used as a benchmark in many articles in the state of the art.
- To determine the priority of features, it uses Information gain and Entropy which are similar to mutual information (I) and joint entropy (H), respectively.

The flowchart of the proposed algorithm is depicted in Fig. 61.1. The stop condition is satisfied when the value of the cost function defined in Eq. (61.8) is the same as previous iteration with a predefined precision. A maximum number of iterations have also been defined.

Initialization of cluster centers (modes) is the only random event in ACA. But uncertainty concept in FACA and taking membership degree of each feature in each cluster into consideration leads to more stability in the final clustering result. Nevertheless, a simple initialization procedure is proposed here which seems to improve the final results as explained in the next section.

A method which is inspired by an initialization method proposed in [20] and [21], is utilized. All patterns are considered as just one cluster and the feature with the highest multiple interdependence redundancy measure is chosen as the first initial cluster center (η_1). Other centers are selected from features by Eq. (61.10).

$$\eta_r = \underset{A_i \notin S_{r-1}}{\arg\min} \left[\sum_{A_j \in S_{r-1}} R\left(A_i : A_j\right) \right] \quad 2 \leq r \leq k \tag{61.10}$$

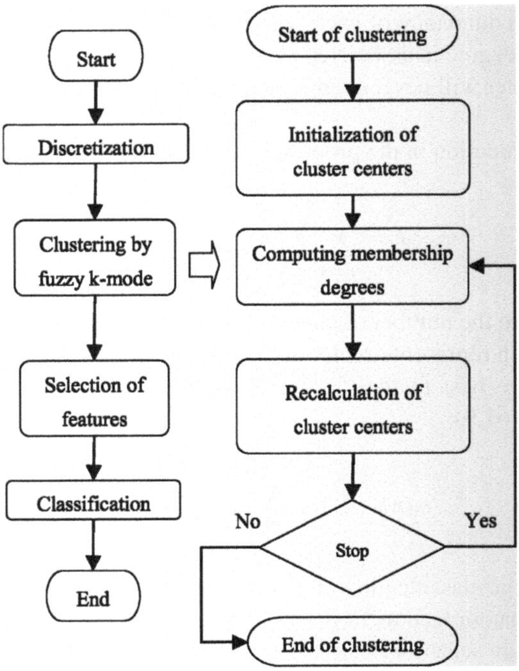

Fig. 61.1 Flowchart of the proposed method

Where, η_r is the center of *rth* cluster, A_i is the *ith* feature and k is the desired number of clusters. S_r represents the set $\{\eta_1, \eta_2, \ldots, \eta_r\}$. Indeed, after selecting the first cluster center, others are selected from unselected features one by one. In any step, the feature, which is the least interdependent feature with selected cluster centers by now, is selected as a new cluster center. It improves the final objective function as explained in the next section.

61.4 Experimental Results

In this paper, Leukemia dataset [17] is used as a well known biological dataset. It is a gene expression micro-array dataset with 73 samples and 7,129 genes (features). Due to memory constraints and dynamic implementations, only the first 1,000 features are considered.

61.4.1 Stability

As mentioned in last section, FACA seems to be more stable than ACA. Curve of objective function Eq. (61.8) (achieved in the final stage of clustering) vs. number

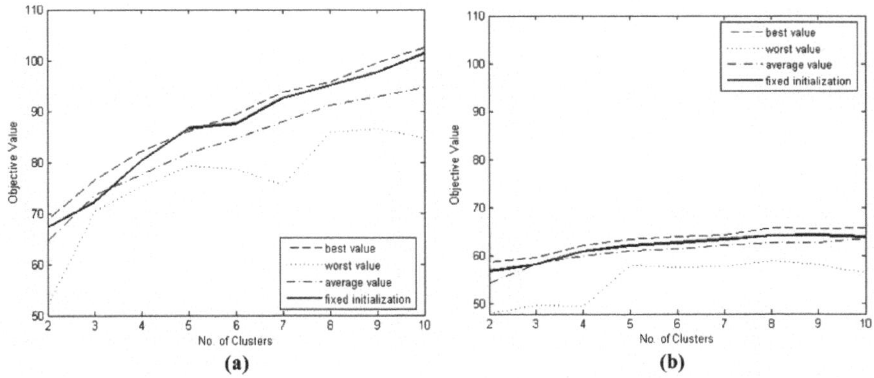

Fig. 61.2 Objective function vs. the number of clusters. Best, worst, average curves in 20 runs for random initialization and associated curve with proposed initialization technique. (**a**) Crisp. (**b**) Fuzzy

of clusters, is depicted Figs. 61.2a and b, for crisp and fuzzy approaches, respectively. In these figures, with a fixed number of clusters and specifying the initial cluster modes randomly, 20 separate clustering runs have been plotted. Best, worst and average values over these runs are depicted in separate curves. Less difference between best and worst values for fuzzy approach, in comparison with crisp version, may show less dependency of the clustering to initial values and consequently more stability is gained. Also considering the same curve, with the proposed initialization technique indicates completely better results than the average values for both crisp and fuzzy approaches, as shown in Fig. 61.2.

Also less slope of these curves for fuzzy approaches shows that the number of clusters is not influential on the objective function for fuzzy approach as much as crisp version. Although, objective function should be maximized but, higher objective value in crisp version is never an evidence of better performance in comparison with fuzzy version. Indeed, the objective function in the crisp version is summation of interdependence measure values between high correlated features; whereas, in fuzzy version, high interdependencies are reduced by membership degree which is less than one, and this reduction is tried to be counted with low interdependence measure value and low membership degrees of other feature pairs.

With a predefined number of clusters, objective function is expected to change after each iteration in fuzzy k-modes. As shown in Fig. 61.3, objective function oscillates in crisp version whereas, fuzzy objective function converges smoothly. This smoothness speeds up and guarantees convergence.

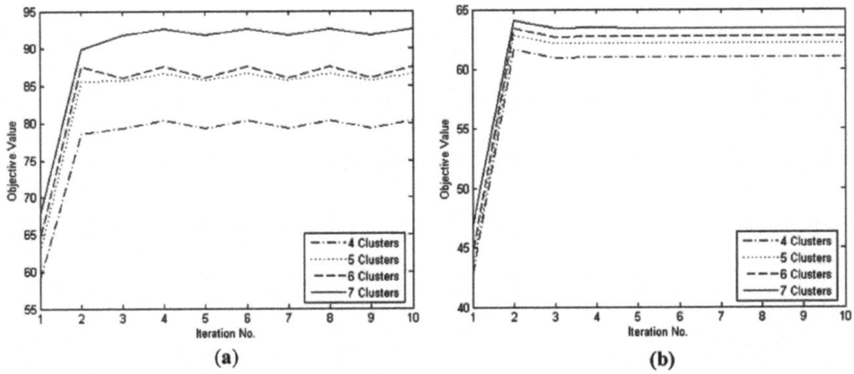

Fig. 61.3 Objective function vs. iterations, for 4, 5, 6 and 7 clusters. (**a**) Crisp. (**b**) Fuzzy

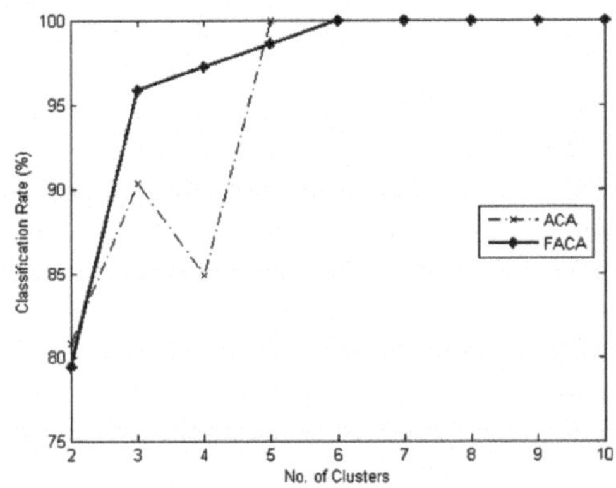

Fig. 61.4 Classification rate resulted by C4.5 on Leukemia dataset for two features per cluster

61.4.2 Classification Accuracy

As another experiment, C4.5 decision tree has been used to classify Leukemia dataset based on selected features. The resulted classification rate can be encountered as a measure of assessing the feature selection method. Hence, ACA and Fuzzy ACA are compared by the resulted classification rate on different number of clusters, as depicted in Figs. 61.4–61.6 for 2, 3 and 4 features per cluster, respectively. Wai-Ho Au et al. [15] proposed selection of a predefined number of features from each cluster. But in fuzzy approach, any feature is belonged to all clusters. Therefore, each feature is first belonged to the cluster with maximum membership degree before any feature selection based on their ranks. Leaved-One-Out method has been used here to assess the generalization of classification system. Increasing the number of features per cluster, improves the classification accuracy for both crisp and fuzzy approaches.

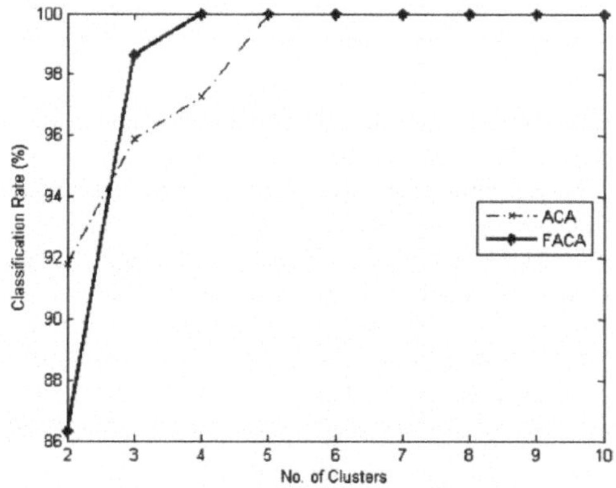

Fig. 61.5 Classification rate resulted by C4.5 on Leukemia dataset for three features per cluster

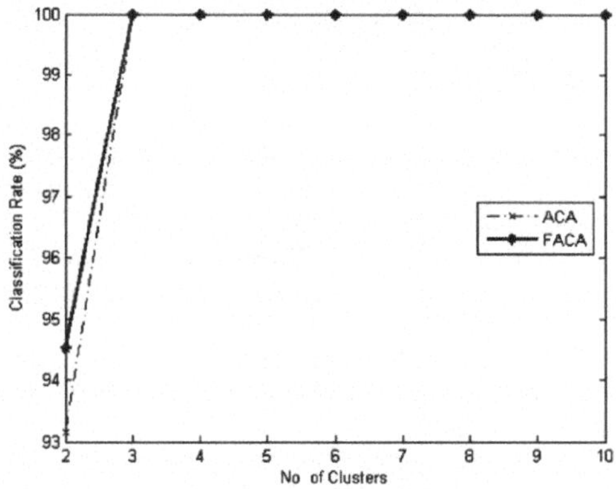

Fig. 61.6 Classification rate resulted by C4.5 on Leukemia dataset for four features per cluster

61.5 Conclusion

In this paper, a fuzzy approach is suggested to group features into clusters in order to select best ones for classification stage. It proposes a new clustering method which combines k-modes clustering used in ACA with fuzzy k-means clustering algorithm. Hence, it leads to more stability, faster convergence and greater classification rate, in comparison with the previous method.

Both ACA and FACA are based on the interdependence redundancy measure which is applicable only on discrete data types. Hence, a new method for discretization of continuous data has also been proposed. Moreover, the initialization of cluster centers is not random, but a new method has been applied for initialization in order to increase the probability of better clusters formation.

Further work will extend the suggested discretization method and improve the similarity measure. Discovering the influential genes on diseases regarding the selected genes is also worth working on.

References

1. I. Guyon and A. Elisseeff, 'An introduction to variable and feature selection', Journal of Machine Learning Research, vol. 3, pp. 1157–1182, 2003.
2. R. Caruana and D. Fratage, 'Greedy attribute selection', in Machine Learning: Proceedings of 11th International Conference, San Francisco, CA, pp. 283–288, 1994.
3. T. Joliffe, 'Principal Component Analysis', New York: Springer-Verlag, 1986.
4. K. Fukunaga, 'Introduction to Statistical Pattern Recognition', New York: Academic, 1972.
5. F. Z. Bril, D. E. Brown, and N. W. Worthy, 'Fast genetic selection of features for neural network classifiers', IEEE Transactions on Neural Networks, vol. 3, pp. 324–328, March 1992.
6. M. L. Raymer, W. F. Punch, E. D. Goodman, L. A. Kuhn, and L. C. Jain, 'Dimensionality reduction using genetic algorithms', IEEE Transactions on Evolutionary Computation, vol. 4, no. 2, pp. 164–171, 2000.
7. R. Setiono and H. Liu, 'Neural-network feature selector', IEEE Transactions on Neural Networks, vol. 8, pp. 654–662, May 1997.
8. E. C. C. Tsang, D. S. Yeung, and X. Z. Wang, 'OFFSS: Optimal Fuzzy-Valued Feature Subset Selection', IEEE Transactions on Fuzzy Systems, vol. 11, no. 2, pp. 202–213, 2003.
9. M. R. Rezaee, B. Goedhart, B. P. F. Lelieveldt, and J. H. C. Reiber, 'Fuzzy feature selection'. PR (32), no. 12, pp. 2011–2019, December 1999.
10. R. Battiti, 'Using mutual information for selecting features in supervised neural net learning', IEEE Transactions on Neural Networks, vol. 5, pp. 537–550, July 1994.
11. N. Kwak and C.-H. Choi, 'Input feature selection for classification problems', IEEE Transactions on Neural Networks, vol. 13, pp. 143–159, January 2002.
12. M. Xiong, W. Li, J. Zhao, L. Jin, and E. Boerwinkle, 'Feature (gene) selection in gene expression-based tumor classification', Molecular Genetics and Metabolism, vol. 73, no. 3, pp. 239–247, 2001.
13. S. C. Madeira and A. L. Oliveira, 'Biclustering algorithms for biological data analysis: A survey', IEEE/ACM Transactions on Computational Biology and Bioinformatics, vol. 1, no. 1, pp. 24–45, January-March 2004.
14. Y. Cheng and G. M. Church, 'Biclustering of expression data', Proceedings of Eighth International Conference Intelligent Systems for Molecular Biology, pp. 93–103, 2000.
15. W.-H. Au, K. C. C. Chan, A. K. C. Wong, Y. Wang, 'Attribute clustering for grouping, selection, and classification of gene expression data', IEEE/ACM Transactions on Computational Biology and Bioinformatics (TCBB), vol. 2, no. 2, pp. 83–101, 2005.
16. J. R. Quinlan, 'C4.5: Programs for Machine Learning', Morgan Kaufmann, San Francisco, CA, 1993.
17. T. R. Golub, D. K. Slonim, P. Tamayo, C. Huard, M. Gaasenbeek, J. P. Mesirov, H. Coller, M. L. Loh, J. R. Downing, M. A. Caligiuri, C. D. Bloomfield, and E. S. Lander, 'Molecular classification of cancer: class discovery and class prediction by gene expression monitoring', vol. 286, no. 5439, pp. 531–537, 1999.
18. T. Kanungo, D. Mount, N. Netanyahu, C. Piatko, R. Silverman, and A. Wu, 'A local search approximation algorithm for k-means clustering', Computational Geometry, vol. 28, pp. 89–112, 2004.
19. J. C. Bezdek, 'Fuzzy mathematics in pattern classification', Ph.D. thesis, Applied Mathematics Center, Cornell University, Ithaca, 1973.
20. M. L. Micó, J. Oncina, and E. Vidal, 'A new version of the nearest-neighbour approximating and eliminating search algorithm (AESA) with linear preprocessing time and memory requirements', Pattern Recognition, vol. 15, pp. 9–17, 1994.
21. M. Taheri and R. Boostani, 'Novel auxiliary techniques in clustering', International Conference on computer science and engineering, 2007.